T0235308

# Lecture Notes in Computer Science     11386

Commenced Publication in 1973
Founding and Former Series Editors:
Gerhard Goos, Juris Hartmanis, and Jan van Leeuwen

More information about this series at http://www.springer.com/series/7407

Ivan Dimov · István Faragó
Lubin Vulkov (Eds.)

# Finite Difference Methods

## Theory and Applications

7th International Conference, FDM 2018
Lozenetz, Bulgaria, June 11–16, 2018
Revised Selected Papers

 Springer

*Editors*
Ivan Dimov
IICT
Bulgarian Academy of Sciences
Sofia, Bulgaria

Lubin Vulkov
University of Rousse
Rousse, Bulgaria

István Faragó (ⓘ)
Eötvös Loránd University
Budapest, Hungary

ISSN 0302-9743                    ISSN 1611-3349   (electronic)
Lecture Notes in Computer Science
ISBN 978-3-030-11538-8            ISBN 978-3-030-11539-5   (eBook)
https://doi.org/10.1007/978-3-030-11539-5

Library of Congress Control Number: 2018967313

LNCS Sublibrary: SL1 – Theoretical Computer Science and General Issues

This Springer imprint is published by the registered company Springer Nature Switzerland AG
The registered company address is: Gewerbestrasse 11, 6330 Cham, Switzerland

# Preface

The papers published in this volume were presented at FDM: T&A 2018, the 7th International Conference on Finite Difference Methods: Theory and Applications, which was held in Lozenetz, Bulgaria, June 11–16, 2018. The conference was organized by the Division of Numerical Analysis and Statistics (DNAS), University of Rousse, in cooperation with the Department of Parallel Algorithms, Institute of Information and Communication Technologies, BAS, Sofia. This conference continued the tradition of four previous meetings: 1997 in Rousse (Bulgaria) organized by the Division of Numerical Analysis and Statistics; 1998 in Minsk (Belarus), organized by the Institute of Mathematics, Belarus Academy of Science; 2000 in Palanga, (Lithuania), organized by the Institute of Mathematics and Informatics (Vilnus); and 2006, 2010, and 2014 in Lozenetz, (Bulgaria) organized by DNAS.

The purpose of the first three conferences (held in Bulgaria, Belarus and Lithuania) was to bring together scientists from the East and West to exchange ideas and establish research cooperation for the next few years. Contact among scientists has become more regular and we can be proud of our contribution. The conference lectures were presented by the university researchers and practical industry engineers, including researchers working in applied mathematics, numerical analysis, and computer science. The key lectures reviewed recent results of theory and different applications of finite difference methods.

The general theme for FDM: T&A 2018 was finite difference and combined finite difference, finite element methods and various applications in physics, chemistry, engineering, biology, and finance. Many modern and new numerical techniques were discussed and presented at the conference such as splitting techniques, Green's function method, multigrid methods, immersed interface method, etc.

Nine Special Sessions were organized by leading specialists of the field.

1. Novel Methods in Computational Finance (by M. Ehrhardt, C. Mishra, and S. Wang)
2. Advanced Numerical and Applied Studies of Inverse Problems (by S. I. Kabanikhin and M. A. Shishlenin)
3. Numerical Methods for Fractional-Derivative Problems and Applications (by A. Alikhanov, R. Lazarov, and M. Stynes)
4. Meshfree Generalized Finite Difference Methods: From Theory to Applications (by I. Michel, J. Kuhnert, and P. Suchde)
5. Reliable Difference Methods for Singularly Perturbed Problems (by G. Shishkin and L. Shishkina)
6. Iterative Schemes for Solving Nonlinear Models (M. P. Vassileva, A. Cordero, and J. R. Torregrosa)
7. Finite Differences Methods in Life Sciences (by G. Pena and J. A. Ferreira)

VI      Preface

8. Asymptotic Analysis of Numerical Methods for Problems with Sharp Transition
   Layers (by N. Nefedov)
9. Numerical Methods for Propagation Processes (by P. Csomos and I. Farago)

We thank to our colleagues for their help in the organization of this conference. We
especially thank M. Koleva for preparation of this volume. We are also grateful to the
organizers of the Minisymposia.

The 8th International Conference FDM: T&A 2022 will be organized in June 2022.

October 2018                                                      Ivan Dimov
                                                                  István Faragó
                                                                  Lubin Vulkov

# Organization

FDM: T&A 2018 was organized by the Division of Numerical Analysis and Statistics, University of Ruse Angel Kanchev, Bulgaria, in cooperation with the Department of Parallel Algorithms, Institute of Information and Communication Technologies, Bulgarian Academy of Sciences, Sofia.

The conference was partly supported by the Bulgarian National Science Fund under Project DN 12/4 from 2017, "Advanced analytical and numerical methods for nonlinear differential equations with applications in finance and environmental pollution."

## Scientific Committee

| | |
|---|---|
| Anatoly Alikhanov | Russian Academy of Sciences, Russia |
| Claudio Cuevas | Universidade Federal Permabuco, Brazil |
| Ivan Dimov | Bulgarian Academy of Sciences, Bulgaria |
| Matthias Ehrhardt | Bergische Universität Wuppertal, Germany |
| István Faragó | Eötvös Loránd University, Hungary |
| Martin Gander | Université de Genève, Switzerland |
| Raytcho Lazarov | Texas A&M University, USA |
| Piotr Matus | Institute of Mathematics, NAS, Belarus |
| Nikolay Nefedof | Lomonosov Moscow State University, Russia |
| Vladimir Shaydurov | Russian Academy of Sciences, Russia |
| Grigorii Shishkin | Russian Academy of Sciences, Russia |
| Martin Stynes | Beijing Computational Science Research Center, China |
| Petr Vabishchevich | Russian Academy of Sciences, Russia |
| Song Wang | Curtin University, Australia |
| Ludmil Zikatanov | Pennsylvania State University, USA |

## Local Organizers

| | |
|---|---|
| Luben Vulkov (Chairperson) | Ruse University Angel Kanchev, Bulgaria |
| Ivanka Angelova | Ruse University Angel Kanchev, Bulgaria |
| Tatiana Chernogorova | Sofia University St. Kliment Ohridski, Bulgaria |
| Slavi Georgiev | Ruse University Angel Kanchev, Bulgaria |
| Miglena Koleva | Ruse University Angel Kanchev, Bulgaria |

# Contents

**Contributed Papers**

# Invited Papers

# From a Geometrical Interpretation of Bramble-Hilbert Lemma to a Probability Distribution for Finite Element Accuracy

Joel Chaskalovic[1]([✉])([iD]) and Franck Assous[2]

[1] ∂'alembert, Sorbonne University, Paris, France
`joel.chaskalovic@sorbonne-universite.fr`
[2] Department of Mathematics, Ariel University, Ariel, Israel
`franckassous55@gmail.com`
`http://www.dalembert.upmc.fr/home/chaskalovic/`

**Abstract.** The aim of this paper is to provide new perspectives on relative finite element accuracy which is usually based on the asymptotic speed of convergence comparison when the mesh size $h$ goes to zero. Starting from a geometrical reading of the error estimate due to Bramble-Hilbert lemma, we derive two probability distributions that estimate the relative accuracy, considered as a random variable, between two Lagrange finite elements $P_k$ and $P_m$, $(k < m)$. We establish mathematical properties of these probabilistic distributions and we get new insights which, among others, show that $P_k$ or $P_m$ is more likely accurate than the other, depending on the value of the mesh size $h$.

**Keywords:** Error estimates · Finite elements ·
Bramble-Hilbert lemma · Probability

## 1 Introduction

The past decades have seen the development of finite element error estimates due to their influence on improving both accuracy and reliability in scientific computing.

However, in these error estimates, an unknown constant is involved which depends, among others, on the basis functions of the considered finite element and on a given semi-norm of the exact solution one wants to approximate. Moreover, error estimates are only upper bounds of the approximation error yielding that the precise value of the approximation error is generally unknown.

Moreover, due to quantitative uncertainties which are generated in the process of the mesh generator and, as a consequence, in the corresponding approximation too, it gave us the idea of considering the approximation error as a random variable.

Therefore, we were able to evaluate the probability of the difference between two approximation errors corresponding to two different finite elements, and

© Springer Nature Switzerland AG 2019
I. Dimov et al. (Eds.): FDM 2018, LNCS 11386, pp. 3–14, 2019.
https://doi.org/10.1007/978-3-030-11539-5_1

then, we got a probabilistic way to compare the relative accuracy between these two finite elements.

The paper is organized as follows. We recall in Sect. 2 the mathematical problem we consider and a corollary of Bramble-Hilbert lemma to propose a geometrical interpretation of the error estimate which appears in this lemma. In Sect. 3 we derive two probability distributions to interpret and estimate the relative accuracy, considered as a random variable, between two Lagrange finite elements $P_k$ and $P_m, (k < m)$. Several mathematical properties of these probabilistic distributions are established in Sect. 4. Concluding remarks follow.

## 2    The Problem Model and a Geometrical Interpretation of an Error Estimate

Let $\Omega$ be an open bounded, and non empty subset of $\mathbb{R}^n$ and $\Gamma$ its boundary which we assumed to be $C^1$−piecewise, and let $u$ be the solution to the second order elliptic variational formulation:

$$(\textbf{VP}) \begin{cases} \text{Find } u \in V \text{ solution to:} \\ a(u,v) = l(v), \quad \forall v \in V, \end{cases} \tag{1}$$

where $V$ is a given Hilbert space endowed with a norm $\|.\|_V$, $a(\cdot,\cdot)$ is a bilinear, continuous and $V$−elliptic form defined on $V \times V$, and $l(\cdot)$ a linear continuous form defined on $V$. Classically, variational problem $(\textbf{VP})$ has one and only solution $u \in V$ (see for example [4]). In this paper and for simplicity, we will restrict ourselves to the case where $V$ is a usual Sobolev space of distributions.

Let us also consider an approximation $u_h$ of $u$, solution to the approximate variational formulation:

$$(\textbf{VP})_h \begin{cases} \text{Find } u_h \in V_h \text{ solution to:} \\ a(u_h, v_h) = l(v_h), \quad \forall v_h \in V_h, \end{cases} \tag{2}$$

where $V_h$ is a given finite-dimensional subset of $V$.

To state a corollary of Bramble-Hilbert's lemma and a corresponding error estimate, we follow [6] or [5], and we assume that $\Omega$ is exactly recovered by a mesh $\mathcal{T}_h$ composed by $N_K$ n-simplexes $K_\mu, (1 \leq \mu \leq N_K)$, which respect classical rules of regular discretization, (see for example [4] for the bidimensional case and [6] in $\mathbb{R}^n$). Moreover, we denote by $P_k(K_\mu)$ the space of polynomial functions defined on a given n-simplex $K_\mu$ of degree less than or equal to $k$, $(k \geq 1)$.

Then, we have the following result:

**Lemma 1.** *Suppose that there exists an integer $k \geq 1$ such that the approximation $u_h$ of $V_h$ is a continuous piecewise function composed by polynomials which belong to $P_k(K_\mu), (1 \leq \mu \leq N_K)$. Then, $u_h$ converges to $u$ in $H^1(\Omega)$:*

$$\lim_{h \to 0} \|u_h - u\|_{1,\Omega} = 0. \tag{3}$$

*Moreover, if the exact solution $u$ belongs to $H^{k+1}(\Omega)$, we have the following error estimate:*

$$\|u_h - u\|_{1,\Omega} \leq \mathscr{C}_k \, h^k \, |u|_{k+1,\Omega}, \tag{4}$$

*where $\mathscr{C}_k$ is a positive constant independent of $h$, $\|.\|_{1,\Omega}$ the classical norm in $H^1(\Omega)$ and $|.|_{k+1,\Omega}$ denotes the semi-norm in $H^{k+1}(\Omega)$.*

Let us now consider two families of Lagrange finite elements $P_k$ and $P_m$ corresponding to a set of values $(k, m) \in \mathbb{N}^2$ such that $0 < k < m$.

The two corresponding inequalities given by (4), assuming that the solution $u$ to **(VP)** belongs to $H^{m+1}(\Omega)$, are:

$$\|u_h^{(k)} - u\|_{1,\Omega} \leq \mathscr{C}_k h^k \, |u|_{k+1,\Omega}, \tag{5}$$

$$\|u_h^{(m)} - u\|_{1,\Omega} \leq \mathscr{C}_m h^m \, |u|_{m+1,\Omega}, \tag{6}$$

where $u_h^{(k)}$ and $u_h^{(m)}$ respectively denotes the $P_k$ and $P_m$ Lagrange finite element approximations of $u$.

Now, if one considers a given mesh for the finite element of $P_m$ which would contains whose of $P_k$ then, for the particular class of problems where **(VP)** is equivalent to a minimization formulation **(MP)** (see for example [4]), one can show that the approximation error of $P_m$ is always lower than those of $P_k$, and $P_m$ is more accurate than $P_m$ for all values of the mesh size $h$ corresponding to the largest diameter in the mesh $\mathcal{T}_h$.

Then, for a given mesh size value of $h$, we consider two independent meshes for $P_k$ and $P_m$ built be a mesh generator. So, usually, to compare the relative accuracy between these two finite elements, one asymptotically considers inequalities (5) and (6) to conclude that, when $h$ goes to zero, $P_m$ finite element is more accurate that $P_k$, as $h^m$ goes faster to zero than $h^k$.

However, for any application $h$ has a static fixed value and this way of comparison is not valid anymore. Therefore, our point of view will be to determine the relative accuracy between two finite elements $P_k$ and $P_m, (k < m)$, for any given value of $h$ for which two independent meshes have to be considered.

To this end, let us set:

$$C_k = \mathscr{C}_k |u|_{k+1,\Omega} \text{ and } C_m = \mathscr{C}_m |u|_{m+1,\Omega}. \tag{7}$$

Therefore, instead of (5) and (6), we consider in the sequel the two next inequalities:

$$\|u_h^{(k)} - u\|_{1,\Omega} \leq C_k h^k, \tag{8}$$

$$\|u_h^{(m)} - u\|_{1,\Omega} \leq C_m h^m. \tag{9}$$

Then, let us remark that inequalities (8) and (9) show that the two polynomial curves defined by $f_k(h) \equiv C_k h^k$ and $f_m(h) \equiv C_m h^m$ play a critical role regarding the values of the two norms $\|u_h^{(k)} - u\|_{1,\Omega}$ and $\|u_h^{(m)} - u\|_{1,\Omega}$.

More precisely, these inequalities indicate that the norm $\|u_h^{(k)} - u\|_{1,\Omega}$, (respectively the norm $\|u_h^{(m)} - u\|_{1,\Omega}$), is below the curve $f_k(h)$, (respectively below the curve $f_m(h)$), (see Fig. 1).

As we are interested in comparing the relative positions of these curves, we introduce their intersection point $h^*$ defined by:

$$h^* \equiv \left( \frac{C_k}{C_m} \right)^{\frac{1}{m-k}}. \tag{10}$$

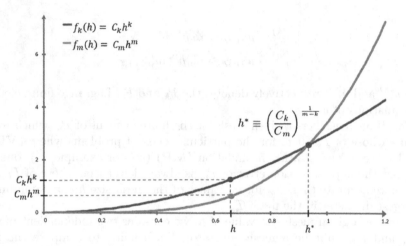

**Fig. 1.** Curves $f_k$ and $f_m$ and existence domain of $\|u_h^{(i)} - u\|_{1,\Omega}, i = k$ or $i = m$.

Now, as often in numerical analysis, there is no *a priori* information to surely or better specify the relative distance between $\|u_h^{(k)} - u\|_{1,\Omega}$, (respectively $\|u_h^{(m)} - u\|_{1,\Omega}$), and the curve $f_k$ or its precise value in the interval $[0, C_k h^k]$, (respectively the curve $f_m$ and the interval $[0, C_m h^m]$). Moreover, we have to deal with finite element methods that return quantitative uncertainties in their calculations. This mainly comes from the way the mesh grid generator will process the mesh to compute the approximation $u_h^{(k)}$, leading to a partial non control of the mesh, even for a given maximum mesh size. As a consequence, the corresponding grid is *a priori* random, and the corresponding approximation $u_h^{(k)}$ too.

For all of these reasons, we motivate that a probabilistic approach can provide a coherent framework for modeling quantitative uncertainties in finite element approximations.

This is the purpose of the following section where we will establish two probability distributions which will allowed us to estimate the relative accuracy between two Lagrange finite elements.

# 3 The Two Probabilistic Models for Relative Finite Elements Accuracy

In this section, we will introduce a convenient probabilistic framework to consider the possible values of the norm $\|u_h^{(k)} - u\|_{1,\Omega}$ as a random variable defined as follows:

- A *random trial* corresponds to the grid constitution and the associated approximation $u_h^{(k)}$.
- The probability space $\Omega$ contains therefore all the possible results for a given random trial, namely, all of the possible grids that the mesh generator may processed, or equivalently, all of the corresponding associated approximations $u_h^{(k)}$.

Then, for a fixed value of $k$, we define by $X^{(k)}$ the random variable as follows:

$$X^{(k)}: \quad \Omega \quad \rightarrow [0, C_k h^k] \tag{11}$$

$$\omega \equiv u_h^{(k)} \mapsto X^{(k)}(\omega) = X^{(k)}(u_h^{(k)}) = \|u_h^{(k)} - u\|_{1,\Omega}. \tag{12}$$

In the sequel, for simplicity, we will set: $X^{(k)}(u_h^{(k)}) \equiv X^{(k)}(h)$.

Now, regarding the absence of information concerning the more likely or less likely values of the norm $\|u_h^{(k)} - u\|_{1,\Omega}$ in the interval $[0, C_k h^k]$, we will assume that the random variable $X^{(k)}(h)$ has a uniform distribution on the interval $[0, C_k h^k]$.

So, our interest is to evaluate the probability of the event

$$\left\{ \|u_h^{(m)} - u\|_{1,\Omega} \leq \|u_h^{(k)} - u\|_{1,\Omega} \right\} \equiv \left\{ X^{(m)}(h) \leq X^{(k)}(h) \right\}, \tag{13}$$

which will allow us to estimate the relative accuracy between two finite elements of order $k$ and $m$, $(k < m)$.

To proceed it, let us now introduce the two random events $A$ and $B$ as follows:

$$A \equiv \left\{ \|u_h^{(m)} - u\|_{1,\Omega} \leq \|u_h^{(k)} - u\|_{1,\Omega} \right\}, \tag{14}$$

$$B \equiv \left\{ \|u_h^{(k)} - u\|_{1,\Omega} \in [C_m h^m, C_k h^k] \right\}. \tag{15}$$

Then, we have the following lemma:

**Lemma 2.** *Let $A$ and $B$ be the events defined by (14) and (15). Then, we have:*

$$\forall h < h^* : Prob\{A\} = \frac{Prob\{B\}}{Prob\{B/A\}}. \tag{16}$$

*Proof.* Let us use the following splitting:

$$Prob\{A\} = Prob\{A \cap B\} + Prob\{A \cap \bar{B}\}, \tag{17}$$

where $\bar{B}$ denotes the opposite event of $B$.

Now, by the definition of the conditional probability we have:

$$Prob\{A \cap B\} = Prob\{A/B\}.Prob\{B\} = Prob\{B\}, \tag{18}$$

since the probabilistic interpretation of Bramble-Hilbert lemma in the case $h < h^*$ corresponds to:

$$Prob\{A/B\} = 1. \tag{19}$$

Then, Eq. (17) can be written as:

$$Prob\{A\} = Prob\{B\} + Prob\{A \cap \bar{B}\}, \tag{20}$$

which can be transformed by the help of the conditional probability as follows:

$$Prob\{A\} = Prob\{B\} + Prob\{\bar{B}/A\}.Prob\{A\}, \tag{21}$$

or equivalently,

$$Prob\{A\} = \frac{Prob\{B\}}{1 - Prob\{\bar{B}/A\}} = \frac{Prob\{B\}}{Prob\{B/A\}}, \tag{22}$$

which corresponds to (16).

Then, we have two options regarding the nature of the dependency between the events $A$ and $B$ which will lead us to get two different distribution laws of probabilities of the event $\left\{X^{(m)}(h) \leq X^{(k)}(h)\right\}$.

The next two subsections are devoted to the dependency modeling between $A$ and $B$.

## 3.1  The Two Steps Model

The first case we will consider states that since, *a priori*, no information is available in numerical analysis to consider any kind of dependency between the events $A$ and $B$, we assume in this subsection that these events are independent.

**Corollary 1.** *Let $A$ and $B$ be the two events defined by (14) and (15) and let us assume they are independent. Then, the probability distribution of the event* $\left\{X^{(m)}(h) \leq X^{(k)}(h)\right\}$ *is given by:*

$$Prob\left\{X^{(m)}(h) \leq X^{(k)}(h)\right\} = \begin{vmatrix} 1 & \textit{if } 0 < h < h^*, \\ 0 & \textit{if } h > h^*. \end{vmatrix} \tag{23}$$

*Proof.* As the events $A$ and $B$ are supposed independent, we have:

$$Prob\{A/B\} = Prob\{A\}. \tag{24}$$

As a consequence, by Lemma 2 Eq. (22) gives after simplification:

$$Prob\{A\} = 1, \forall h < h^*. \tag{25}$$

With the same kind of arguments, when $h > h^*$ we get:

$$Prob\{A\} = 0, \forall h > h^*. \tag{26}$$

Let us now examine the main properties of probabilistic distribution (23):

- For any $h$ smaller than $h^*$, $P_m$ finite element is not only asymptotically better than $P_k$ finite element as $h$ becomes small, but they are *almost surely* more accurate for all these values of $h$ such that $h < h^*$.
- For any $h$ greater than $h^*$, $P_k$ finite element becomes *almost surely* more accurate than $P_m$ finite element, even if $k < m$.

This last feature upsets the widespread idea regarding the relative accuracy between $P_k$ and $P_m$, $(k < m)$, finite elements. It clearly indicates that there exist cases where $P_m$ finite elements *surely* must be overqualified and a significant reduction of implementation and execution cost can be obtained without a loss of accuracy.

Furthermore, one may expect to get a probabilistic distribution where more variations would appear, as it is in this two steps model, between the probability of the event $\{X^{(m)}(h) \leq X^{(k)}(h)\}$ and the mesh size $h$. It is certainly due to the assumption we considered regarding the independency between the events $A$ and $B$.

The purpose of the next subsection we will be devoted to relax this assumption by directly computing the probability of the event $\{X^{(m)}(h) \leq X^{(k)}(h)\}$.

## 3.2   The "Sigmoid" Model

To avoid the hypothesis of independency between the events $A$ and $B$ defined by (14) and (15), we will directly evaluate the probability of the event $A$ without considering anymore the splitting we wrote in formula (20).

However, we will assume that the two random variables $X^{(i)}(h), (i = k$ or $i = m)$, defined by (12) are independent and uniformly distributed on $[0, C_i h^i], (i = k$ or $i = m)$.

This is the aim of the following theorem.

**Theorem 1.** *Let $u$ be the solution to the second order variational elliptic problem (VP) defined in (1) and $u_h^{(i)}, (i = k$ or $i = m, k < m)$, the two corresponding Lagrange finite element $P_i$ approximations, solution to the approximated formulation (VP)$_h$ defined by (2).*

*We assume the two corresponding random variables $X^{(i)}(h), (i = k$ or $i = m)$, defined by (12) are independent and uniformly distributed on $[0, C_i h^i]$, where $C_i$ are defined by (8)–9).*

*Then, the probability of the event $\left\{ X^{(m)}(h) \leq X^{(k)}(h) \right\}$ is given by:*

$$
Prob\left\{ X^{(m)}(h) \leq X^{(k)}(h) \right\} = \left| \begin{array}{ll} 1 - \dfrac{1}{2}\left(\dfrac{h}{h^*}\right)^{m-k} & \text{if } 0 < h \leq h^*, \\[3mm] \dfrac{1}{2}\left(\dfrac{h^*}{h}\right)^{m-k} & \text{if } h \geq h^*. \end{array} \right. \tag{27}
$$

*Proof.*

▶ Let us first consider a fixed value of $h$ such that $h < h^*$. In this case, $f_m(h) < f_k(h)$, or in other words, $0 < C_m h^m < C_k h^k$ and due to Bramble-Hilbert lemma (see Fig. 1), one must deal with the following inequalities:

$$
X^{(k)}(h) \leq C_k h^k \text{ and } X^{(m)}(h) \leq C_m h^m < C_k h^k. \tag{28}
$$

Then, to compute the probability such that $X^{(m)}(h) \leq X^{(k)}(h)$, we consider inequalities (28) in the plane $(0; X^{(m)}(h), X^{(k)}(h))$, (see Fig. 2) in which the two random variables belong to the rectangle $R_t$ defined on $[0, C_m h^m] \times [0, C_k h^k]$.

Our purpose is to characterize the points in $R_t$ that satisfy $X^{(m)}(h) \leq X^{(k)}(h)$. Obviously, it only concerns the points which are above the bisector $X^{(k)}(h) = X^{(m)}(h)$, namely the points which belong to the trapezium $T_u$ (see Fig. 2) whose surface is given by:

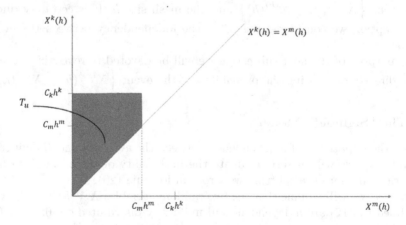

**Fig. 2.** Area corresponding to $X^{(m)}(h) \leq X^k(h)$.

$$
S(T_u) = C_m h^m (C_k h^k - C_m h^m) + \frac{C_m^2 h^{2m}}{2}, \tag{29}
$$

while the total surface of the rectangle $R_t$ is equal to $C_m C_k h^{m+k}$.

As we assume that the two random variables $X^{(k)}(h)$ and $X^{(m)}(h)$ are independent and uniformly distributed, the probability $Prob\left\{X^{(m)}(h) \leq X^{(k)}(h)\right\}$ corresponds to the ratio between the two surfaces of $T_u$ and $R_t$ and we have:

$$Prob\left\{X^{(m)}(h) \leq X^{(k)}(h)\right\} = \frac{S(T_u)}{S(R_t)} = \frac{C_m h^m (C_k h^k - C_m h^m) + C_m^2 h^{2m}/2}{C_m C_k h^{m+k}},$$

$$= 1 - \frac{1}{2}\frac{C_m}{C_k} h^{m-k}. \tag{30}$$

Using the definition (10) of $h^*$, we get:

$$\forall h < h^* : Prob\left\{X^{(m)}(h) \leq X^{(k)}(h)\right\} = 1 - \frac{1}{2}\frac{C_m}{C_k} h^{m-k} = 1 - \frac{1}{2}\left(\frac{h}{h^*}\right)^{m-k}. \tag{31}$$

▶ Let us consider now the second case where $h > h^*$.
The curve $f_m(h) = C_m h^m$ is above the curve $f_k(h) = C_k h^k$ and by the same arguments we used above, one must deal with the following inequalities:

$$X^{(m)}(h) \leq C_m h^m \text{ and } X^{(k)}(h) \leq C_k h^k < C_m h^m. \tag{32}$$

Then, if we change the role between $k$ and $m$, we can directly write :

$$Prob\left\{X^{(k)}(h) \leq X^{(m)}(h)\right\} = \frac{C_k h^k (C_m h^m - C_k h^k) + C_k^2 h^{2k}/2}{C_m C_k h^{m+k}}$$

$$= 1 - \frac{1}{2}\frac{C_k}{C_m} h^{k-m}. \tag{33}$$

Hence, the probability of the complementary event $X^{(m)}(h) \leq X^{(k)}(h)$ which interests us is given by:

$$Prob\left\{X^{(m)}(h) \leq X^{(k)}(h)\right\} = 1 - Prob\left\{X^k(h) \leq X^{(m)}(h)\right\}$$

$$= \frac{1}{2}\frac{C_k}{C_k} \cdot \frac{1}{h^{m-k}} = \frac{1}{2}\left(\frac{h^*}{h}\right)^{m-k}, \tag{34}$$

where we used the definition (10) of $h^*$.

The global shapes of the two probabilistic distributions (23) and (27) are plotted in Fig. 3 and particular features of (27) are described in the next section.

## 4  Properties of the Sigmoid Probability Distribution

We give now the main properties of the sigmoid probability distribution given by (27). To this end, we will denote by $\mathcal{P}(h)$ the probability defined by:

$$\mathcal{P}(h) \equiv Prob\left\{X^{(m)}(h) \leq X^{(k)}(h)\right\}. \tag{35}$$

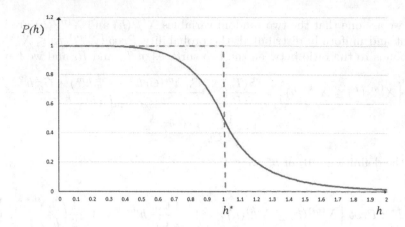

**Fig. 3.** Case $m - k \neq 1$: shape of the sigmoid distribution (27) and the two steps corresponding one (23), $(P(h) \equiv Prob\{X^{(m)}(h) \leq X^{(k)}(h)\})$.

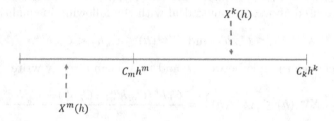

**Fig. 4.** Relative one dimensional position between $X^{(m)}$ and $X^{(k)}$ $(h < h^*)$.

– The first feature we observe concerns the global shape of P(h) together with (27), drawn in Fig. 3 for $m - k \neq 1$, which looks like a kind of sigmoid roughly approximated by a stepwise function given by (23) from Lemma 2 of Subsect. 3.1.

In this way, we achieve our objective to relax the dependency assumption between the events $A$ and $B$. As a consequence, non linearity appears in the relation described by (27) between the probability of the event "$P_m$ *finite element is more accurate than* $P_k$ *finite element*" and the mesh size $h$.

– Behavior of $\mathcal{P}(h)$ in the neighborhood of $0^+$: Directly, we get:

$$\lim_{h \to 0^+} \mathcal{P}(h) = \lim_{h \to 0^+} Prob\left\{X^{(m)}(h) \leq X^{(k)}(h)\right\} = 1, \tag{36}$$

which corresponds to the classical understanding of the error estimate (4) which derives from Bramble-Hilbert lemma, namely asymptotically when the maximum of the mesh size $h$ goes to zero.

Indeed, in these cases $h$ "*is sufficiently small*", and despite the unknown values of the constants $C_k$ and $C_m$ which appear in (8) and (9), one concludes as expected that the finite element $P_m$ is more accurate than the finite element $P_m$, if $k < m$.

But, the question is to determine what does it mean when $h$ "*is sufficiently small*". We will partially discuss about this in the next point regarding the behavior of $\mathcal{P}(h)$ at the neighborhood of $h^*$ given by (10).

From a probabilistic point of view the result (36) is also intuitive because, when $h$ goes to $0^+$, the quantity $C_m h^m$ goes to 0 faster than $C_k h^k, (k < m)$. Depicting the relative position of $X^{(m)}(h)$ and $X^{(k)}(h)$ in a one dimensional way, (see Fig. 4), it is clear that the probability of the event $\left\{X^{(m)} \leq X^{(k)}\right\}$ goes to 1 when $h$ goes to zero, as $X^{(m)} \leq C_m h^m$ due to Bramble-Hilbert lemma.

However, the interest of any probability distribution is to get additional information concerning the relative accuracy between two given finite elements, not only when $h$ goes to zero, as we will see further. Here, we just mentioned that we find again the well known conclusion to compare two finite elements when the mesh size is arbitrarily small.

Indeed, finite element $P_m$ is not only *asymptotically* more accurate than $P_k$ as $k < m$. Indeed, for all $h \leq h^*$, the probability for $P_m$ to be more accurate than $P_k$ is between 0.5 to 1. It means that $P_m$ is *more likely* accurate than $P_k$ for all of these values of $h$. We also notice that we have not anymore the event "$P_m$ is more accurate than $P_k$" as an almost sure event as we got in Subsect. 3.1 with the law (23). This is because we dropped the hypothesis of dependency between the events $A$ and $B$ which leads to a more general and realistic probabilistic distribution.

− Behavior of $\mathcal{P}(h)$ in the neighborhood of $h^*$:

The probabilistic stepwise law (23) did not described the case $h$ equals $h^*$. However, here, the sigmoid probability distribution (27) can be extended by continuity to $h = h^*$ as we simply have:

$$\lim_{h \to h^{*-}} \mathcal{P}(h) = \lim_{h \to h^{*+}} \mathcal{P}(h) = \frac{1}{2}, \tag{37}$$

and then, we extend $\mathcal{P}(h)$ by continuity at $h^*$ by setting:

$$\mathcal{P}(h^*) = Prob\left\{X^{(m)}(h^*) \leq X^k(h^*)\right\} \equiv \frac{1}{2}. \tag{38}$$

This feature illustrates that when $h = h^*$, $C_k h^{*k} = C_m h^{*m}$, and the two norms $\|u_h^{(k)} - u\|_{1,\Omega}$ and $\|u_h^{(m)} - u\|_{1,\Omega}$, which measures each approximation error of the two corresponding Lagrange finite elements, are somewhere below the two curves (see Fig. 1), or in other words, somewhere in the same interval as we here: $[0, C_k h^{*k}] = [0, C_m h^{*m}]$. Then, the probability to get $\left\{X^{(m)}(h^*) \leq X^k(h^*)\right\}$ is equal to 0.5.

This new behavior claims that when $h$ approaches the critical value $h^*$ the event "$P_m$ *finite element is more accurate than* $P_k$ *finite element*" is equally likely to occur or not to occur. As a consequence the accuracy between the two finite element $P_k$ and $P_m$ is equivalent.

It is clearly a new theoretical information because, as we mentioned above, the values of the two constants $C_k$ and $C_m$ are totally unknown. Indeed, we already suspected and pointed out by data mining techniques, (see for example [1–3]), that this situation would occur. Here, we complete this suspicion by a theoretical probabilistic framework.

- Despite the usual point of view which claims that $P_m$ finite element are more accurate than $P_k$ ones, we get here that $P_k$ finite element is *more likely* accurate than $P_m$ when $h > h^*$. This new point of view allows us to recommend that for specific situations, like for adaptive refinement meshes for example, $P_k$ finite element would be locally more appropriated as long as one will be able to detect the case $h > h^*$.

## 5 Conclusions

In this paper we present a new way to investigate the relative accuracy between two finite elements. Indeed, leaving the classical asymptotic point of view usually considered to compare the speed of convergence for different approximation errors, we got new insights for understanding error estimates. The way we thought the error estimates is not restricted to the finite element method but can be extended to other approximation methods. Indeed, the underlying idea is that, given a class of numerical schemes and their corresponding error estimates, one is able to rank them, not only in terms of asymptotic speed of convergence as usual, but also by evaluating the *almost surely* more accurate one.

For example, considering numerical schemes to approximate solution to ordinary differential equations, one would be able to argue, why (or why not!) RK4 scheme would be implemented rather than another simplest one.

**Homages:** The authors want to warmly dedicate this research to pay homage to the memory of Professors André Avez and Gérard Tronel who largely promote the passion of research and teaching in mathematics.

## References

1. Assous, F., Chaskalovic, J.: Data mining techniques for scientific computing: Application to asymptotic paraxial approximations to model ultra-relativistic particles. J. Comput. Phys. **230**, 4811–4827 (2011)
2. Assous, F., Chaskalovic, J.: Error estimate evaluation in numerical approximations of partial differential equations: a pilot study using data mining methods. C. R. Mec. **341**, 304–313 (2013)
3. Assous, F., Chaskalovic, J.: Indeterminate constants in numerical approximations of PDE's: a pilot study using data mining techniques. J. Comput. Appl. Math. **270**, 462–470 (2014)
4. Chaskalovic, J.: Mathematical and Numerical Methods for Partial Differential Equations. ME. Springer, Cham (2014). https://doi.org/10.1007/978-3-319-03563-5
5. Ciarlet, P.G.: Basic error estimates for elliptic problems. In: Ciarlet, P.G., Lions, J.L. (eds.) Handbook of Numerical Analysis, North Holland, vol. 2 (1991)
6. Raviart P.A., Thomas, J.M.: Introduction à l'analyse numérique des équations aux dérivées partielles, Masson (1982)

# A High Order Method on Graded Meshes for a Time-Fractional Diffusion Problem

Hu Chen and Martin Stynes$^{(\boxtimes)}$

Applied and Computational Mathematics Division,
Beijing Computational Science Research Center, Beijing 100193, China
{huchen90,m.stynes}@csrc.ac.cn

**Abstract.** In a recent paper we showed numerically and theoretically that a straightforward generalisation of Alikhanov's "L2-1$_\sigma$" scheme is $O(M^{-2})$ accurate on suitably chosen graded meshes (with $M$ time intervals) for initial-value problems (IVPs) and initial-boundary value problems (IBVPs) with a Caputo fractional time derivative of order $\alpha$, whose solutions typically exhibit a weak singularity at the initial time $t = 0$. The present paper constructs a better generalisation of Alikhanov's scheme that is demonstrated numerically to be $O(M^{-(3-\alpha)})$ accurate for these classes of IVPs and IBVPs, but its rigorous analysis remains an open problem.

**Keywords:** Alikhanov scheme · Caputo derivative

## 1 Introduction

Throughout this paper, let $\alpha \in (0, 1)$ be fixed. Fractional-derivative problems that involve a Caputo time derivative of order $\alpha$ are currently widely studied in the research literature. Recently, Alikhanov [1] devised and analysed a finite difference scheme—the "L2-1$_\sigma$" scheme—for discretising such derivatives, which has attracted much attention. In a recent work [3] we analysed the behaviour of this scheme for initial-value problems (IVP) and initial-boundary value problems (IBVP) with a Caputo fractional time derivative of order $\alpha$; here meshes that are graded near the initial time $t = 0$ were used to handle the weak singularity that is present in typical solutions of these classes of problems. In particular it was shown that for the IVP

$$D_t^\alpha w(t) = g(t) \text{ for } t \in (0, T], \quad w(0) = w_0,$$

where $g$ is a given smooth function, the computed solution $\{W^j\}_{j=0}^M$ on an optimal mesh $0 = t_0 < t_1 < \cdots < t_M = T$ satisfies

$$|w(t_k) - W^k| \lesssim M^{-(3-\alpha)} \text{ for } k = 0, \ldots, M. \tag{1}$$

But when this IVP is modified to

$$D_t^\alpha w(t) + b(t)w(t) = g(t) \text{ for } t \in (0, T], \quad w(0) = w_0, \tag{2}$$

© Springer Nature Switzerland AG 2019
I. Dimov et al. (Eds.): FDM 2018, LNCS 11386, pp. 15–27, 2019.
https://doi.org/10.1007/978-3-030-11539-5_2

where $b$ is a given smooth function—or if one considers the IBVP of [3]—then the "obvious" generalisation of Alikhanov's scheme on this mesh is shown in [3] to satisfy the estimate

$$|w(t_k) - W^k| \lesssim M^{-2} \text{ for } k = 0, \ldots, M, \tag{3}$$

which is less accurate than (1); furthermore, a numerical example shows that the bound (3) is sharp.

Our purpose in the present paper is to describe a better generalisaton of Alikhanov's scheme that retains the $O(M^{-(3-\alpha)})$ convergence of (1) for the more general problem (2) and for the IBVP of [3].

*Notation.* Throughout the paper, we use $A \lesssim B$ to mean that $A \leq cB$, where $c$ is a generic positive constant, and $A \simeq B$ to mean that $A \lesssim B$ and $B \lesssim A$. When we write, e.g., $\tau_{j+1} \lesssim t_j$ for all $j$, we mean that there is a single fixed constant $c$ such that $\tau_{j+1} \leq ct_j$ for all $j$.

## 2    The $L2\text{-}1_\sigma$ Scheme and Its Properties

In this section we describe Alikhanov's $L2\text{-}1_\sigma$ discretisation and present some basic results from [1,3] that will be used for its analysis.

Define the temporal Riemann-Liouville integral operator of order $\beta$, for any positive number $\beta$, by

$$J^\beta v(\cdot, t) := \frac{1}{\Gamma(\beta)} \int_0^t (t-s)^{\beta-1} v(\cdot, s) \, ds \text{ for } t > 0,$$

provided this integral is defined. Here $\Gamma(\cdot)$ is Euler's Gamma function. Then define the Caputo fractional derivative of order $\alpha$ with respect to $t$ by

$$D_t^\alpha v(\cdot, t) := J^{1-\alpha} \frac{\partial v(\cdot, t)}{\partial t} = \frac{1}{\Gamma(1-\alpha)} \int_0^t (t-s)^{-\alpha} \frac{\partial v(\cdot, s)}{\partial s} ds \text{ for } t > 0,$$

provided this integral is defined.

Let $M$ be a positive integer. Consider the graded mesh

$$\{t_j = T(j/M)^r\}_{j=0}^M \text{ with some user-chosen } r \geq 1. \tag{4}$$

(Our analysis can be extended to more general meshes such as those used in [3], but for simplicity and clarity in the present paper we shall consider only this family of graded meshes.) For the mesh (4), a calculation shows that

$$t_{j-1} \geq 2^{-r} t_j \text{ for } j \geq 2 \tag{5}$$

and

$$\tau_j := t_j - t_{j-1} \simeq M^{-1} t_j^{1-1/r} \quad \text{for } j = 1, \ldots, M. \tag{6}$$

In particular, $\tau_1 = t_1 \simeq M^{-r}$.

We discretise $D_t^\alpha v$ for $v \in C[0,T] \cap C^3(0,T]$ as in [1]. For $k = 0, \ldots, M-1$ and $0 \le \sigma \le 1$, set

$$t_{k+\sigma} = t_k + \sigma \tau_{k+1}.$$

Set

$$a_{k,k} = \frac{1}{\Gamma(1-\alpha)} \int_{t_k}^{t_{k+\sigma}} (t_{k+\sigma} - \eta)^{-\alpha} \, d\eta = \frac{\sigma^{1-\alpha}}{\Gamma(2-\alpha)} \tau_{k+1}^{1-\alpha} \quad \text{for } k \ge 0, \qquad (7)$$

and, for $k \ge 1$ and $0 \le j \le k-1$, set

$$a_{k,j} = \frac{1}{\Gamma(1-\alpha)} \int_{t_j}^{t_{j+1}} (t_{k+\sigma} - \eta)^{-\alpha} \, d\eta, \qquad (8)$$

$$b_{k,j} = \frac{1}{\Gamma(1-\alpha)} \frac{2}{t_{j+2} - t_j} \int_{t_j}^{t_{j+1}} (t_{k+\sigma} - \eta)^{-\alpha} (\eta - t_{j+1/2}) \, d\eta. \qquad (9)$$

Let

$$\delta_t^\alpha v(t_{k+\sigma}) := \sum_{j=0}^{k} g_{k,j} \big( v(t_{j+1}) - v(t_j) \big) \quad \text{for } k = 0, \ldots, M-1, \qquad (10)$$

where $g_{0,0} = \tau_1^{-1} a_{0,0}$, and for $k \ge 1$,

$$g_{k,j} = \begin{cases} \tau_{j+1}^{-1}(a_{k,0} - b_{k,0}) & \text{if } j = 0, \\ \tau_{j+1}^{-1}(a_{k,j} + b_{k,j-1} - b_{k,j}) & \text{if } 1 \le j \le k-1, \\ \tau_{j+1}^{-1}(a_{k,k} + b_{k,k-1}) & \text{if } j = k. \end{cases} \qquad (11)$$

For any grid function $\{V^j\}_{j=0}^M$ defined on our mesh (4) (so $V^j \equiv V(t_j)$), define

$$\delta_{t_{k+\sigma}}^\alpha V := g_{k,k} V^{k+1} - \sum_{j=0}^{k} (g_{k,j} - g_{k,j-1}) V^j \quad \text{for } k = 0, \ldots, M-1, \qquad (12)$$

where we set $g_{k,-1} := 0$. Then with $v^j := v(t_j)$ for $0 \le j \le M$, from (10) one sees that $\delta_t^\alpha v(t_{k+\sigma}) = \delta_{t_{k+\sigma}}^\alpha v$.

It can be proved [3, Lemma 4] that

$$g_{k,0} > \frac{t_{k+\sigma}^{-\alpha}}{\Gamma(1-\alpha)} \quad \text{for } k \ge 0, \qquad (13)$$

$$g_{k,k} > g_{k,k-1} > \cdots > g_{k,1} > g_{k,0} \quad \text{for } k \ge 1; \qquad (14)$$

and if $1 \ge \sigma \ge 1 - \alpha/2$, then

$$(2\sigma - 1)g_{k,k} - \sigma g_{k,k-1} > 0 \quad \text{for } k \ge 1. \qquad (15)$$

Using the properties of the coefficients $g_{k,j}$, we have the following lemmas:

**Lemma 1** ([3, Lemma 5]). *For any mesh function $\{V^j\}_{j=0}^M$, one has*

$$|V^{k+1}| \leq |V^0| + \Gamma(1-\alpha) \max_{j=0,\ldots,k} \{t_{j+\sigma}^\alpha \delta_{t_{j+\sigma}}^\alpha |V|\} \quad \text{for} \quad k = 0,\ldots,M-1.$$

**Lemma 2** ([1, Lemma 1] [3, Lemma 8]). *For any mesh function $\{V^j\}_{j=0}^M$, one has*

$$V^{k+1}\delta_{t_{k+\sigma}}^\alpha V \geq \frac{1}{2}\delta_{t_{k+\sigma}}^\alpha V^2 + \frac{1}{2g_{k,k}}|\delta_{t_{k+\sigma}}^\alpha V|^2$$

*and*

$$V^k \delta_{t_{k+\sigma}}^\alpha V \geq \frac{1}{2}\delta_{t_{k+\sigma}}^\alpha V^2 - \frac{1}{2(g_{k,k} - g_{k,k-1})}|\delta_{t_{k+\sigma}}^\alpha V|^2,$$

*for $k = 0,\ldots,M-1$.*

The convergence analysis in [3] of Alikhanov's scheme uses the following quantities. For any function $v \in C[0,T] \cap C^3(0,T]$, define

$$\psi_v^\sigma = \tau_1^\alpha \sup_{\eta \in (0,t_1)} \left(\eta^{1-\alpha}|\delta_t v(t_0) - v'(\eta)|\right), \tag{16a}$$

$$\psi_v^{j+\sigma} = \tau_{j+1}^{3-\alpha} t_{j+\sigma}^\alpha \sup_{\eta \in (t_j,t_{j+1})} |v'''(\eta)| \text{ for } j = 1,\ldots,M-1, \tag{16b}$$

$$\psi_v^{j,1} = \tau_1^\alpha \sup_{\eta \in (0,t_1)} \left(\eta^{1-\alpha}|(I_{2,1}v(\eta))' - v'(\eta)|\right) \text{ for } j = 1,\ldots,M-1, \tag{16c}$$

$$\psi_v^{j,s} = \tau_{j+1}^{-\alpha}\tau_s^2(\tau_s + \tau_{s+1})t_s^\alpha \sup_{\eta \in (t_{s-1},t_{s+1})} |v'''(\eta)| \text{ for } 2 \leq s \leq j \leq M-1. \tag{16d}$$

From now on we imitate [1] by making the particular choice

$$\boxed{\sigma = 1 - \alpha/2.}$$

We then have the following pair of results; the second one assumes typical behaviour of the derivatives of the unknown solutions to our problems.

**Lemma 3** ([3, Lemma 1]). *Assume that $\tau_{j+1} \lesssim t_j$ for $j \geq 2$ and $\tau_1/\tau_2 \leq \rho$, where $\rho$ is any fixed positive constant. Then for any function $v(t) \in C^3(0,T]$, one has*

$$|\delta_t^\alpha v(t_{j+\sigma}) - D_t^\alpha v(t_{j+\sigma})| \lesssim t_{j+\sigma}^{-\alpha}\left(\psi_v^{j+\sigma} + \max_{s=1,\ldots,j}\{\psi_v^{j,s}\}\right) \quad \text{for } j = 0,\ldots,M-1.$$

**Lemma 4** ([3, Lemma 7]). *Let $v \in C[0,T] \cap C^3(0,T]$. Assume that $|v^{(l)}(t)| \lesssim 1 + t^{\alpha-l}$ for $l = 0,1,2,3$ and $t \in (0,T]$. Then*

$$\psi_v^{j+\sigma} \lesssim M^{-\min\{r\alpha,3-\alpha\}} \quad \text{for } j = 0,\ldots,M-1,$$
$$\psi_v^{j,s} \lesssim M^{-\min\{r\alpha,3-\alpha\}} \quad \text{for } s = 1,\ldots,j, \text{ when } j \geq 1.$$

# 3 Analysis of Fractional IVPs

## 3.1 The Simplest IVP

Consider the IVP

$$D_t^\alpha w(t) = g(t) \text{ for } t \in (0, T], \quad w(0) = w_0. \tag{17}$$

and its discretisation on the graded mesh (4):

$$\delta_{t_{k+\sigma}}^\alpha W = g(t_{k+\sigma}) \text{ for } k = 0, \dots, M-1, \quad W^0 = w_0. \tag{18}$$

**Theorem 1** ([3, Theorem 1]). *Let $w$ and $\{W^j\}_{j=0}^M$ be defined by (17) and (18). Assume that $w \in C[0,T] \cap C^3(0,T]$ with $|w^{(l)}(t)| \lesssim 1 + t^{\alpha - l}$ for $l = 0, 1, 2, 3$ and $t \in (0, T]$. Then*

$$\max_{1 \le k \le M} |w(t_k) - W^k| \lesssim M^{-\min\{r\alpha, 3-\alpha\}}.$$

Thus our scheme (18) achieves better than second-order accuracy when (17) is solved numerically on graded meshes for which $r \ge (3 - \alpha)/\alpha$. Numerical results in [3] show that Theorem 1 is sharp.

## 3.2 A More General IVP

Consider the fractional-derivative initial-value problem

$$D_t^\alpha w(t) + b(t)w(t) = g(t) \text{ for } t \in (0, T], \quad w(0) = w_0, \tag{19}$$

where $b \ge 0$ on $[0, T]$.

For any mesh function $V$, set $V^{k,\sigma} = \sigma V^{k+1} + (1-\sigma)V^k$ for $k = 0, \dots, M-1$.

Let $W^j$ be the approximation of $w(t)$ at $t = t_j$ for $j = 0, \dots, M$. Then we define the following discretization of problem (19) on the graded mesh (4):

$$\delta_{t_{k+\sigma}}^\alpha W + b(t_{k+\sigma})W^{k,\sigma} = g(t_{k+\sigma}) \text{ for } k = 0, \dots, M-1, \quad W^0 = w_0. \tag{20}$$

This generalisation of (18) seems reasonable; it is analogous to the popular Crank-Nicolson method for parabolic PDEs.

Using Lemma 2 and inequality (15), it is straightforward to get

**Lemma 5** ([1, Corollary 1]). *For any function $V(\cdot, t)$ defined on the mesh $\{t_j\}_{j=0}^M$, one has*

$$(\sigma V^{k+1} + (1-\sigma)V^k)\delta_{t_{k+\sigma}}^\alpha V \ge \frac{1}{2}\delta_{t_{k+\sigma}}^\alpha V^2 \text{ for } k = 0, \dots, M-1.$$

**Lemma 6** ([3, Lemma 9]). *For any function $v(t) \in C^2(0,T]$, one has*

$$|\sigma v(t_{k+1}) + (1-\sigma)v(t_k) - v(t_{k+\sigma})| \le \frac{1}{8}\tau_{k+1}^2 \sup_{t \in (t_k, t_{k+1})} |v''(t)|$$

*for $k = 1, \dots, M-1$.*

Now we can prove a convergence result for our method (20), under realistic hypotheses on the behaviour of the derivatives of the unknown solution $w$ of (19).

**Theorem 2.** *Let $w$ and $\{W^j\}_{j=0}^M$ be defined by (19) and (20). Assume that $w \in C[0,T] \cap C^3(0,T]$ with $|w^{(l)}(t)| \lesssim 1 + t^{\alpha-l}$ for $l = 0,1,2,3$ and $t \in (0,T]$. Then*

$$\max_{1 \leq k \leq M} |w(t_k) - W^k| \lesssim M^{-\min\{r\alpha,2\}}.$$

*Proof.* Set $e^k = w(t_k) - W^k$ for $k = 0, \ldots, M$. Then (19) and (20) yield easily

$$\delta_{t_{m+\sigma}}^\alpha e + b(t_{k+\sigma})e^{m,\sigma} = \underbrace{\delta_{t_{m+\sigma}}^\alpha w - D_t^\alpha w(t_{m+\sigma})}_{=:r^{m+\sigma}} + \underbrace{b(t_{k+\sigma})[w^{m,\sigma} - w(t_{m+\sigma})]}_{=:R^{m+\sigma}}$$

for $m = 0, \ldots, M-1$, with $e^0 = 0$. Multiply both sides of this equation by $e^{m,\sigma}$; then using Lemma 5 we get

$$\frac{1}{2}\delta_{t_{m+\sigma}}^\alpha e^2 + b(t_{k+\sigma})|e^{m,\sigma}|^2 \leq |r^{m+\sigma}| \cdot |e^{m,\sigma}| + |R^{m+\sigma}| \cdot |e^{m,\sigma}|.$$

Hence

$$\delta_{t_{m+\sigma}}^\alpha e^2 \leq 2\big(|r^{m+\sigma}| + |R^{m+\sigma}|\big)|e^{m,\sigma}| \leq 2\big(|r^{m+\sigma}| + |R^{m+\sigma}|\big)\max_{1 \leq j \leq M}|e^j|.$$

Now Lemma 1 gives

$$|e^{k+1}|^2 \lesssim \max_{0 \leq m \leq k}\{t_{m+\sigma}^\alpha \delta_{t_{m+\sigma}}^\alpha |e|^2\}$$

$$\lesssim \max_{0 \leq m \leq M-1}\{t_{m+\sigma}^\alpha\big(|r^{m+\sigma}| + |R^{m+\sigma}|\big)\}\max_{1 \leq j \leq M}|e^j| \text{ for } k = 0, \ldots, M-1.$$

It follows that

$$\max_{1 \leq k \leq M}|e^k|^2 \lesssim \max_{0 \leq m \leq M-1}\{t_{m+\sigma}^\alpha\big(|r^{m+\sigma}| + |R^{m+\sigma}|\big)\}\max_{1 \leq j \leq M}|e^j|.$$

Dividing both sides by $\max_{1 \leq k \leq M}|e^k|$ gives

$$\max_{1 \leq k \leq M}|e^k| \lesssim \max_{0 \leq m \leq M-1}\{t_{m+\sigma}^\alpha\big(|r^{m+\sigma}| + |R^{m+\sigma}|\big)\}. \tag{21}$$

Recalling Lemmas 3 and 4, we have

$$\max_{0 \leq m \leq M-1}\{t_{m+\sigma}^\alpha |r^{m+\sigma}|\} \lesssim \max_{0 \leq m \leq M-1}\{|\psi_w^{m+\sigma}| + \max_{s=1,\ldots,m}\{|\psi_w^{m,s}|\}\}$$

$$\lesssim M^{-\min\{r\alpha, 3-\alpha\}}. \tag{22}$$

It remains to estimate the term $\max_{0 \leq m \leq M-1}\{t_{m+\sigma}^\alpha |R^{m+\sigma}|\}$ in (21). In the case $m = 0$, because $|w(t)| \lesssim 1$, we have $t_\sigma^\alpha |R^\sigma| \lesssim t_1^\alpha \simeq M^{-r\alpha}$. When $m \geq 1$, by using (5) and (6), Lemma 6 yields

$$\max_{1\leq m\leq M-1}\{t_{m+\sigma}^{\alpha}|R^{m+\sigma}|\} \lesssim M^{-2} \max_{1\leq m\leq \hat{M}-1}\{t_{m+1}^{2\alpha-2/r}\}$$

$$\lesssim \begin{cases} M^{-2} & \text{if } r \geq 1/\alpha, \\ M^{-2}t_1^{2\alpha-2/r} \simeq M^{-2r\alpha} & \text{if } 1 \leq r < 1/\alpha. \end{cases}$$

Combining all these cases, we get

$$\max_{0\leq m\leq M-1}\{t_{m+\sigma}^{\alpha}|R^{m+\sigma}|\} \lesssim M^{-\min\{r\alpha,2\}}. \qquad (23)$$

The theorem now follows from (21)–(23).

*Example 1.* Let $w(t) = 1 + t^{\alpha} + t^{2\alpha} + t^{1+\alpha}$ be the exact solution, and take $b(t) \equiv 1/4$ and $T = 1$. The function $w$ exhibits the weak singularity at $t = 0$ that is typical of solutions to (19).

The numerical experiments test several values of $\alpha, r$ and $M$. The maximum error in each computed solution is $\max_{1\leq n\leq M} |W^n - w(t_n)|$ and the rates of convergence are then computed in the usual way. Tables 1, 2, 3 and 4 demonstrate that the convergence rate predicted by Theorem 2 is sharp.

**Table 1.** Maximum error and convergence rate with $r = 1$.

| $M$ | $\alpha = 0.2$ | | $\alpha = 0.4$ | | $\alpha = 0.6$ | | $\alpha = 0.8$ | |
|---|---|---|---|---|---|---|---|---|
| | *Error* | *Rate* | *Error* | *Rate* | *Error* | *Rate* | *Error* | *Rate* |
| 64 | 3.01e−02 | 0.19 | 1.66e−02 | 0.38 | 6.63e−03 | 0.57 | 1.81e−03 | 0.77 |
| 128 | 2.63e−02 | 0.19 | 1.28e−02 | 0.38 | 4.47e−03 | 0.58 | 1.06e−03 | 0.78 |
| 256 | 2.30e−02 | 0.20 | 9.83e−03 | 0.39 | 2.99e−03 | 0.59 | 6.16e−04 | 0.79 |
| 512 | 2.01e−02 | 0.20 | 7.52e−03 | 0.39 | 1.99e−03 | 0.59 | 3.56e−04 | 0.79 |
| 1024 | 1.75e−02 | 0.20 | 5.74e−03 | 0.39 | 1.32e−03 | 0.59 | 2.05e−04 | 0.80 |
| 2048 | 1.52e−02 | * | 4.37e−03 | * | 8.76e−04 | * | 1.18e−04 | * |

## 4   An Improved High-Order Scheme

Theorems 1 and 2 show that the performance of our scheme (20) for the problem (19) is slightly disappointing, since it is at best only $O(M^{-2})$ convergent although its parent scheme (18) for the simpler problem (17) achieves $O(M^{-(3-\alpha)})$ convergence when the mesh is suitably graded. Thus we now present an *improved scheme* for the problem (19).

**Table 2.** Maximum error and convergence rate with $r = 1/\alpha$.

| $M$ | $\alpha = 0.2$ | | $\alpha = 0.4$ | | $\alpha = 0.6$ | | $\alpha = 0.8$ | |
|---|---|---|---|---|---|---|---|---|
| | Error | Rate | Error | Rate | Error | Rate | Error | Rate |
| 64 | 1.09e−03 | 1.00 | 1.46e−03 | 1.00 | 1.32e−03 | 0.99 | 8.09e−04 | 0.98 |
| 128 | 5.44e−04 | 1.00 | 7.30e−04 | 1.00 | 6.65e−04 | 1.00 | 4.09e−04 | 0.99 |
| 256 | 2.72e−04 | 1.00 | 3.65e−04 | 1.00 | 3.34e−04 | 1.00 | 2.05e−04 | 1.00 |
| 512 | 1.36e−04 | 1.00 | 1.83e−04 | 1.00 | 1.67e−04 | 1.00 | 1.03e−04 | 1.00 |
| 1024 | 6.80e−05 | 1.00 | 9.14e−05 | 1.00 | 8.36e−05 | 1.00 | 5.15e−05 | 1.00 |
| 2048 | 3.40e−05 | * | 4.57e−05 | * | 4.18e−05 | * | 2.58e−05 | * |

**Table 3.** Maximum error and convergence rate with $r = 2/\alpha$.

| $M$ | $\alpha = 0.2$ | | $\alpha = 0.4$ | | $\alpha = 0.6$ | | $\alpha = 0.8$ | |
|---|---|---|---|---|---|---|---|---|
| | Error | Rate | Error | Rate | Error | Rate | Error | Rate |
| 64 | 6.07e−05 | 2.00 | 5.59e−05 | 2.00 | 4.74e−05 | 1.99 | 6.00e−05 | 1.98 |
| 128 | 1.51e−05 | 2.00 | 1.40e−05 | 2.00 | 1.20e−05 | 2.00 | 1.52e−05 | 1.99 |
| 256 | 3.78e−06 | 2.00 | 3.50e−06 | 2.00 | 2.99e−06 | 2.00 | 3.81e−06 | 2.00 |
| 512 | 9.46e−07 | 2.00 | 8.75e−07 | 2.00 | 7.49e−07 | 2.00 | 9.51e−07 | 2.01 |
| 1024 | 2.36e−07 | 2.00 | 2.19e−07 | 2.00 | 1.87e−07 | 2.00 | 2.36e−07 | 2.01 |
| 2048 | 5.91e−08 | * | 5.47e−08 | * | 4.68e−08 | * | 5.87e−08 | * |

**Table 4.** Maximum error and convergence rate with $r = (3 - \alpha)/\alpha$.

| $M$ | $\alpha = 0.2$ | | $\alpha = 0.4$ | | $\alpha = 0.6$ | | $\alpha = 0.8$ | |
|---|---|---|---|---|---|---|---|---|
| | Error | Rate | Error | Rate | Error | Rate | Error | Rate |
| 64 | 5.30e−05 | 2.48 | 3.86e−05 | 2.35 | 5.89e−05 | 2.03 | 6.50e−05 | 1.99 |
| 128 | 9.48e−06 | 1.58 | 7.57e−06 | 2.35 | 1.44e−05 | 2.03 | 1.64e−05 | 2.00 |
| 256 | 3.17e−06 | 1.56 | 1.49e−06 | 2.32 | 3.54e−06 | 2.02 | 4.11e−06 | 2.00 |
| 512 | 1.07e−06 | 1.77 | 2.98e−07 | 2.28 | 8.72e−07 | 2.02 | 1.03e−06 | 2.00 |
| 1024 | 3.16e−07 | 1.87 | 6.13e−08 | 2.09 | 2.16e−07 | 2.01 | 2.57e−07 | 2.00 |
| 2048 | 8.67e−08 | * | 1.44e−08 | * | 5.34e−08 | * | 6.40e−08 | * |

For any $v \in C[0,T] \cap C^3(0,T]$, set $\delta_t v(t_j) := (v(t_{j+1}) - v(t_j))/\tau_{j+1}$ for $j \geq 0$. For $s \geq 1$, let $I_{2,s}v(t)$ be the quadratic interpolant to $v(t)$ at the three points $(t_{s-1}, v(t_{s-1}))$, $(t_s, v(t_s))$, $(t_{s+1}, v(t_{s+1}))$. Then one can write

$$I_{2,s}v(t) = v(t_{s-1}) + (t - t_{s-1})\delta_t v(t_{s-1}) + \frac{\delta_t v(t_s) - \delta_t v(t_{s-1})}{t_{s+1} - t_{s-1}}(t - t_{s-1})(t - t_s).$$

It is well known (see, e.g., [2, p. 122]) that for $t \in [t_{s-1}, t_{s+1}]$ one has

$$v(t) - I_{2,s}v(t) = \frac{v'''(\bar{\xi}_s)}{6}(t - t_{s-1})(t - t_s)(t - t_{s+1}) \tag{24}$$

for some $\bar{\xi}_s(t) \in (t_{s-1}, t_{s+1})$. Consequently using $I_{2,k}v(t)$ to approximate $v(t)$ at $t = t_{k+\sigma}$ should give a high-order scheme.

Now

$$
\begin{aligned}
I_{2,k}v(t_{k+\sigma}) &= v(t_{k-1}) + (\tau_k + \sigma\tau_{k+1})\delta_t v(t_{k-1}) \\
&\quad + \frac{\delta_t v(t_k) - \delta_t v(t_{k-1})}{\tau_k + \tau_{k+1}}(\tau_k + \sigma\tau_{k+1})(\sigma\tau_{k+1}) \\
&= v(t_{k+1})\frac{\sigma(\tau_k + \sigma\tau_{k+1})}{\tau_k + \tau_{k+1}} + v(t_k)\frac{(1 - \sigma)(\tau_k + \sigma\tau_{k+1})}{\tau_k} \\
&\quad + v(t_{k-1})\frac{\sigma(\sigma - 1)\tau_{k+1}^2}{\tau_k(\tau_k + \tau_{k+1})}.
\end{aligned}
$$

Thus, for an arbitrary grid function $V(t)$ on our mesh, for $k \geq 1$ define

$$V^{\bar{k},\sigma} := \frac{\sigma(\tau_k + \sigma\tau_{k+1})}{\tau_k + \tau_{k+1}}V^{k+1} + \frac{(1 - \sigma)(\tau_k + \sigma\tau_{k+1})}{\tau_k}V^k - \frac{\sigma(1 - \sigma)\tau_{k+1}^2}{\tau_k(\tau_k + \tau_{k+1})}V^{k-1}.$$

Then define an *improved scheme* on the graded mesh (4) by

$$\delta_{t_{k+\sigma}}^{\alpha} W + b(t_{k+\sigma})W^{\bar{k},\sigma} = g(t_{k+\sigma}) \text{ for } k = 1, \ldots, M - 1, \tag{25a}$$

$$\delta_{t_\sigma}^{\alpha} W + b(t_{k+\sigma})W^{0,\sigma} = g(t_\sigma) \text{ for } k = 0, \tag{25b}$$

$$W^0 = w_0. \tag{25c}$$

*Suppose we can prove* the following analogue of Lemma 5: for all mesh functions $V$, one has

$$V^{\bar{k},\sigma}\delta_{t_{k+\sigma}}^{\alpha} V \geq c_0 \, \delta_{t_{k+\sigma}}^{\alpha} V^2 \text{ for some fixed positive constant } c_0. \tag{26}$$

Then repeating the proof of Theorem 2, with Lemmas 5 and 6 replaced by (26) and (24) respectively, shows that the convergence order of the improved scheme (25) is $O(M^{-\min\{r\alpha, 3-\alpha\}})$.

We now test numerically how our improved scheme (25) performs on the problem that was used in Example 1.

*Example 2.* Let $w(t) = 1 + t^\alpha + t^{2\alpha} + t^{1+\alpha}$ be the exact solution, and take $b(t) = 1/4$ and $T = 1$. The results in Tables 5, 6 and 7 show that the improved scheme (25) attains the rate of convergence $O(M^{-\min\{r\alpha, 3-\alpha\}})$. Thus it is superior numerically to our earlier scheme (20) for typical solutions of (19); but for (20), unlike (25), we have a complete error analysis.

**Table 5.** Maximum error and convergence rate with $r = 2/\alpha$.

| M | $\alpha = 0.2$ | | $\alpha = 0.4$ | | $\alpha = 0.6$ | | $\alpha = 0.8$ | |
|---|---|---|---|---|---|---|---|---|
| | Error | Rate | Error | Rate | Error | Rate | Error | Rate |
| 64 | 6.48e−05 | 2.08 | 5.64e−05 | 2.01 | 4.78e−05 | 2.00 | 3.37e−05 | 1.95 |
| 128 | 1.53e−05 | 2.01 | 1.40e−05 | 2.00 | 1.20e−05 | 2.00 | 8.73e−06 | 1.98 |
| 256 | 3.79e−06 | 2.00 | 3.50e−06 | 2.00 | 3.00e−06 | 2.00 | 2.21e−06 | 1.99 |
| 512 | 9.46e−07 | 2.00 | 8.75e−07 | 2.00 | 7.49e−07 | 2.00 | 5.55e−07 | 2.00 |
| 1024 | 2.36e−07 | 2.00 | 2.19e−07 | 2.00 | 1.87e−07 | 2.00 | 1.39e−07 | 2.00 |
| 2048 | 5.91e−08 | * | 5.47e−08 | * | 4.68e−08 | * | 3.48e−08 | * |

**Table 6.** Maximum error and convergence rate with $r = (3 - \alpha)/\alpha$.

| M | $\alpha = 0.2$ | | $\alpha = 0.4$ | | $\alpha = 0.6$ | | $\alpha = 0.8$ | |
|---|---|---|---|---|---|---|---|---|
| | Error | Rate | Error | Rate | Error | Rate | Error | Rate |
| 64 | 1.36e−04 | 2.61 | 4.49e−05 | 2.47 | 3.08e−05 | 2.27 | 2.55e−05 | 2.05 |
| 128 | 2.24e−05 | 2.64 | 8.11e−06 | 2.50 | 6.37e−06 | 2.31 | 6.15e−06 | 2.10 |
| 256 | 3.58e−06 | 2.67 | 1.43e−06 | 2.53 | 1.28e−06 | 2.34 | 1.44e−06 | 2.12 |
| 512 | 5.63e−07 | 2.70 | 2.48e−07 | 2.55 | 2.53e−07 | 2.36 | 3.30e−07 | 2.15 |
| 1024 | 8.68e−08 | 2.71 | 4.24e−08 | 2.56 | 4.93e−08 | 2.37 | 7.45e−08 | 2.16 |
| 2048 | 1.32e−08 | * | 7.18e−09 | * | 9.53e−09 | * | 1.67e−08 | * |

**Table 7.** Maximum error and convergence rate with $r = 3/\alpha$.

| M | $\alpha = 0.2$ | | $\alpha = 0.4$ | | $\alpha = 0.6$ | | $\alpha = 0.8$ | |
|---|---|---|---|---|---|---|---|---|
| | Error | Rate | Error | Rate | Error | Rate | Error | Rate |
| 64 | 1.61e−04 | 2.61 | 5.60e−05 | 2.48 | 2.99e−05 | 2.32 | 1.67e−05 | 2.15 |
| 128 | 2.64e−05 | 2.64 | 1.00e−05 | 2.51 | 6.01e−06 | 2.35 | 3.77e−06 | 2.17 |
| 256 | 4.23e−06 | 2.67 | 1.76e−06 | 2.54 | 1.18e−06 | 2.37 | 8.36e−07 | 2.19 |
| 512 | 6.65e−07 | 2.69 | 3.03e−07 | 2.55 | 2.29e−07 | 2.38 | 1.84e−07 | 2.19 |
| 1024 | 1.03e−07 | 2.71 | 5.17e−08 | 2.57 | 4.40e−08 | 2.39 | 4.02e−08 | 2.20 |
| 2048 | 1.57e−08 | * | 8.74e−09 | * | 8.42e−09 | * | 8.78e−09 | * |

## 5  Time Fractional Diffusion Problem

Consider a time fractional diffusion equation in two dimensional spatial domain as follows:

$$D_t^\alpha u + \mathcal{L}u = f(x,t), \quad x := (x_1, x_2) \in \Omega, \, 0 < t \le T, \tag{27a}$$

subject to the following initial and boundary conditions:

$$u(x,0) = u_0(x), \quad x \in \Omega, \tag{27b}$$

$$u|_{\partial\Omega} = 0, \quad 0 \le t \le T, \tag{27c}$$

where $\Omega = (-1,1) \times (-1,1) \in \mathbb{R}^2$, $D_t^\alpha u$ $(0 < \alpha < 1)$ is the Caputo fractional partial derivative of order $\alpha$ with respect to $t$, and $f$ is a given function. The spatial operator $\mathcal{L}$ is a linear second-order elliptic operator: $\mathcal{L}u := -\nu\Delta u + c(x)u$, where $\nu > 0$ is the diffusion coefficient and $c \ge 0$ is a smooth function.

To discretise $D_t^\alpha u$ we use the graded mesh (4) and our previous discretisation $\delta_{t_{k+\sigma}}^\alpha$. For the discretisation in space we use a spectral method as in [3]. Set $\Lambda := (-1,1)$. Let $N$ be a positive integer. Denote by $\mathbb{P}_N(\Lambda)$ the space of all polynomials of degree at most $N$. Set $\mathbb{P}_N^0 := \{\phi \in \mathbb{P}_N(\Lambda) : \phi(\pm 1) = 0\}$ and $H_0^1(\Omega) := \{v \in H^1(\Omega) : v|_{\partial\Omega} = 0\}$, where $H^1(\Omega)$ is the standard Sobolev space.

Let $\pi_N^{1,0}$ be the $H_0^1$-orthogonal projector from $H_0^1(\Omega)$ into $(\mathbb{P}_N^0)^2$, viz., for each $v \in H_0^1(\Lambda)$ one has

$$\left(\nabla\pi_N^{1,0}v, \nabla v_N\right) = (\nabla v, \nabla v_N) \quad \forall v_N \in (\mathbb{P}_N^0)^2, \tag{28}$$

with $(\cdot, \cdot)$ denoting the $L^2(\Omega)$ inner product.

At each discrete time level $t_k$, the computed solution will lie in $(\mathbb{P}_N^0)^2$ and is denoted by $u_N^k$. Imitating Sect. 4, set

$$u_N^{\overline{k},\sigma} := \frac{\sigma(\tau_k + \sigma\tau_{k+1})}{\tau_k + \tau_{k+1}}u_N^{k+1} + \frac{(1-\sigma)(\tau_k + \sigma\tau_{k+1})}{\tau_k}u_N^k - \frac{\sigma(1-\sigma)\tau_{k+1}^2}{\tau_k(\tau_k + \tau_{k+1})}u_N^{k-1}$$

for $k = 1, \ldots, M - 1$, and $f^{k+\sigma}(x) = f(x, t_{k+\sigma})$ for $k = 0, \ldots, M - 1$.

The *improved scheme* for problem (27) in its weak formulation is as follows: find $u_N^{k+1} \in (\mathbb{P}_N^0)^2$, with $u_N^0 = \pi_N^{1,0}u_0$, such that for $k = 1, \ldots, M - 1$ one has

$$(\delta_{t_{k+\sigma}}^\alpha u_N, v_N) + \nu(\nabla u_N^{\overline{k},\sigma}, \nabla v_N) + (cu_N^{\overline{k},\sigma}, v_N) = (f^{k+\sigma}, v_N) \quad \forall v_N \in (\mathbb{P}_N^0)^2, \tag{29a}$$

and for $k = 0$,

$$(\delta_{t_\sigma}^\alpha u_N, v_N) + \nu(\nabla u_N^{0,\sigma}, \nabla v_N) + (cu_N^{0,\sigma}, v_N) = (f^\sigma, v_N) \quad \forall v_N \in (\mathbb{P}_N^0)^2, \tag{29b}$$

where $u_N^{0,\sigma} := \sigma u_N^1 + (1-\sigma)u_N^0$.

*Example 3.* We consider the problem (27) with $\nu = 0.1$, $c(x) = x_1^2 + x_2^2$, equipped with an exact solution that displays a typical weak singularity at $t = 0$:

$$u(x,t) = (1 + t^\alpha + t^{2\alpha} + t^{1+\alpha})(1 - x_1^2)(1 - x_2^2)\exp(x_1 + x_2).$$

Take $T = 1$.

Set $E^{M,N} := \max_{1 \le k \le M} \|u_N^k - u(t_k)\|$, where $\|\cdot\|$ is the $L^2(\Omega)$ norm. In our numerical experiments we use Gaussian quadrature (see [4, Eq. (3.180)]) to compute each term $\|u_N^k - u(t_k)\|$ in $E^{M,N}$.

To investigate the temporal accuracy of our improved scheme, we choose $N = 15$ which is big enough to render negligible the error caused by the spectral spatial discretization as the solution is very smooth in the spatial direction. From Tables 8, 9 and 10, it is clear that the temporal accuracy in $L^2(\Omega)$ is of order $M^{-\min\{r\alpha, 3-\alpha\}}$.

**Table 8.** Maximum $L^2(\Omega)$ error and convergence rate with $r = 2/\alpha$.

| $M$ | $\alpha = 0.2$ | | $\alpha = 0.4$ | | $\alpha = 0.6$ | | $\alpha = 0.8$ | |
|---|---|---|---|---|---|---|---|---|
| | *Error* | *Rate* | *Error* | *Rate* | *Error* | *Rate* | *Error* | *Rate* |
| 64 | 9.62e−05 | 2.13 | 8.06e−05 | 2.02 | 6.72e−05 | 2.00 | 4.65e−05 | 1.94 |
| 128 | 2.19e−05 | 2.03 | 1.98e−05 | 2.01 | 1.69e−05 | 2.00 | 1.21e−05 | 1.97 |
| 256 | 5.36e−06 | 2.01 | 4.93e−06 | 2.00 | 4.22e−06 | 2.00 | 3.10e−06 | 1.99 |
| 512 | 1.33e−06 | 2.00 | 1.23e−06 | 2.00 | 1.05e−06 | 2.00 | 7.81e−07 | 2.00 |
| 1024 | 3.33e−07 | 2.00 | 3.08e−07 | 2.00 | 2.64e−07 | 2.00 | 1.96e−07 | 2.00 |
| 2048 | 8.32e−08 | * | 7.70e−08 | * | 6.59e−08 | * | 4.89e−08 | * |

**Table 9.** Maximum $L^2(\Omega)$ error and convergence rate with $r = (3 - \alpha)/\alpha$.

| $M$ | $\alpha = 0.2$ | | $\alpha = 0.4$ | | $\alpha = 0.6$ | | $\alpha = 0.8$ | |
|---|---|---|---|---|---|---|---|---|
| | *Error* | *Rate* | *Error* | *Rate* | *Error* | *Rate* | *Error* | *Rate* |
| 64 | 2.09e−04 | 2.77 | 5.76e−05 | 2.50 | 4.16e−05 | 2.26 | 3.45e−05 | 2.04 |
| 128 | 3.07e−05 | 2.75 | 1.02e−05 | 2.50 | 8.67e−06 | 2.30 | 8.37e−06 | 2.08 |
| 256 | 4.57e−06 | 2.75 | 1.81e−06 | 2.51 | 1.76e−06 | 2.33 | 1.97e−06 | 2.12 |
| 512 | 6.79e−07 | 2.75 | 3.17e−07 | 2.53 | 3.50e−07 | 2.35 | 4.55e−07 | 2.14 |
| 1024 | 1.01e−07 | 2.76 | 5.50e−08 | 2.54 | 6.85e−08 | 2.37 | 1.03e−07 | 2.16 |
| 2048 | 1.49e−08 | * | 9.43e−09 | * | 1.33e−08 | * | 2.32e−08 | * |

**Table 10.** Maximum $L^2(\Omega)$ error and convergence rate with $r = 3/\alpha$.

| $M$ | $\alpha = 0.2$ | | $\alpha = 0.4$ | | $\alpha = 0.6$ | | $\alpha = 0.8$ | |
|---|---|---|---|---|---|---|---|---|
| | *Error* | *Rate* | *Error* | *Rate* | *Error* | *Rate* | *Error* | *Rate* |
| 64 | 2.50e−04 | 2.77 | 7.04e−05 | 2.55 | 3.79e−05 | 2.33 | 2.15e−05 | 2.16 |
| 128 | 3.66e−05 | 2.75 | 1.20e−05 | 2.55 | 7.54e−06 | 2.35 | 4.82e−06 | 2.17 |
| 256 | 5.42e−06 | 2.75 | 2.05e−06 | 2.55 | 1.48e−06 | 2.37 | 1.07e−06 | 2.19 |
| 512 | 8.05e−07 | 2.75 | 3.50e−07 | 2.56 | 2.86e−07 | 2.38 | 2.35e−07 | 2.19 |
| 1024 | 1.19e−07 | 2.76 | 5.92e−08 | 2.57 | 5.49e−08 | 2.39 | 5.14e−08 | 2.20 |
| 2048 | 1.76e−08 | * | 9.96e−09 | * | 1.05e−08 | * | 1.12e−08 | * |

# 6    Conclusion

We have constructed a variant of Alikhanov's L2-1$_\sigma$ scheme for Caputo IVPs and IBVPs that numerically exhibits order $3 - \alpha$ convergence on suitably chosen graded meshes for typical solutions that have a weak singularity at the initial time $t = 0$, but the analysis of our scheme is incomplete because inequality (26) remains unproved.

**Acknowledgements.** The work of the first author was funded by the Chinese Postdoc Foundation Grant 2018M631316 and the National Natural Science Foundation of China young scientists fund Grant 11801026. The work of the second author was funded by the National Natural Science Foundation of China under grants 91430216 and NSAF-U1530401.

# References

1. Alikhanov, A.A.: A new difference scheme for the time fractional diffusion equation. J. Comput. Phys. **280**, 424–438 (2015). https://doi.org/10.1016/j.jcp.2014.09.031
2. Atkinson, K., Han, W.: Theoretical Numerical Analysis. A Functional Analysis Framework. Texts in Applied Mathematics, 3rd edn. Springer, New York (2009). https://doi.org/10.1007/978-1-4419-0458-4
3. Chen, H., Stynes, M.: Error analysis of a second-order method on fitted meshes for a time-fractional diffusion problem. J. Sci. Comput. (2018). https://doi.org/10.1007/s10915-018-0863-y
4. Shen, J., Tang, T., Wang, L.L.: Spectral Methods, Algorithms, Analysis and Applications. Springer Series in Computational Mathematics. Springer, Heidelberg (2011). https://doi.org/10.1007/978-3-540-71041-7

# Numerical Solving a Boundary Value Problem for the Eikonal Equation

Alexander G. Churbanov[1] and Petr N. Vabishchevich[1,2(✉)]

[1] Nuclear Safety Institute, Russian Academy of Sciences,
52, B. Tulskaya, Moscow, Russia
[2] North–Eastern Federal University, 58, Belinskogo, Yakutsk, Russia
vabishchevich@gmail.com

**Abstract.** In the present work, a Dirichlet problem is studied for the eikonal equation. A nonlinear boundary value problem formulated here can be treated as the limit of the diffusion–reaction problem with a diffusion parameter tending to zero. For numerical solving the singularly perturbed diffusion–reaction problem, monotone approximations are used. Predictions for a 3D model problem are presented to demonstrate possibilities of the developed numerical algorithm. The standard piecewise-linear finite-element approximation is employed to constructed discretization in space.

**Keywords:** Eikonal equation · Finite-element method ·
Diffusion–reaction equation ·
Singularly perturbed boundary value problem ·
Monotone approximation

## 1  Introduction

Many problems of practical interest result in boundary value problems for the eikonal equation. This nonlinear partial differential equation is applied to describe wave propagation in the approximation of geometric optics [4]. In computational fluid dynamics, image processing and computer graphics (see, e.g., [10]), the solution of BVPs for the eikonal equation is used for calculating the normal distance to boundaries of a computational domain.

The eikonal equation is a typical example of steady-state Hamilton–Jacobi equations. The issues of the existence and uniqueness of the solution for BVPs for such equations are considered, e.g., in [14]. To solve numerically the eikonal equation, standard approaches are employed using finite-difference methods or finite-element approximations. In this approach based on using partial differential equations, the main attention is paid to problems of nonlinearity.

Marching methods (the first class of algorithms) are in common use and employ the hyperbolic nature of the eikonal equation. In this case, the desired solution is obtained by successive moving into the interior of the domain from its boundary, using, e.g., first-order upwind finite differences [20]. Another popular

© Springer Nature Switzerland AG 2019
I. Dimov et al. (Eds.): FDM 2018, LNCS 11386, pp. 28–34, 2019.
https://doi.org/10.1007/978-3-030-11539-5_3

method is the fast sweeping method [19] based on a Gauss–Seidel-style update strategy to progress across the domain. Recently (see, e.g., [13]), a fast iterative method for eikonal equations is actively developed using triangular [8] and tetrahedral [9] grids. Other modern variants of the fast marching method, which are adapted to modern computing systems of parallel architecture, have been studied and compared, e.g., in [11].

Next, the second class of algorithms is based on a transition from the initial BVP for the nonlinear equation of first order to a linear or nonlinear BVP for an elliptic equation [1]. The solution of the BVP for the eikonal equation can be related to the solution of the homogeneous Dirichlet problem for $p$-Laplacian [2].

In the present work, we focus on solving auxiliary BVPs for linear equations. This approach (see [12]) employs a connection between the nonlinear Hamilton–Jacobi equation and the linear Schrodinger equation. A similar approach is used in [7], where the auxiliary functions $v_\alpha(x)$ are associated with the solution of the unsteady heat equation.

## 2   Singularly Perturbed Diffusion–Reaction Problem

In a bounded polygon $\Omega \subset R^m$, $m = 1, 2, 3$ with the Lipschitz continuous boundary $\partial\Omega$, we solve the BVP for the eikonal equation

$$|\nabla u|^2 = 1, \quad x \in \Omega. \tag{1}$$

Equation (1) is supplemented with the homogeneous Dirichlet boundary condition

$$u(x) = 0, \quad x \in \partial\Omega. \tag{2}$$

The main issue of numerical solving the problem (1, 2) is connected with the nonlinearity of the equation.

As it was done in [12], let us apply the transformation

$$v_\alpha(x) = \exp\left(-\frac{u_\alpha(x)}{\alpha}\right) \tag{3}$$

with a numerical parameter $\alpha > 0$. Such a transformation is widely used in studying elliptical equations with quadratic nonlinearity (see, e.g., [3]).

For (3), we obtain

$$\alpha^2 \triangle v_\alpha - v_\alpha = \exp\left(-\frac{u_\alpha(x)}{\alpha}\right)(|u_\alpha|^2 - 1 - \alpha\triangle u_\alpha).$$

Let $u_\alpha(x)$ satisfies the equation

$$\alpha\triangle u_\alpha - |u_\alpha|^2 = -1, \quad x \in \Omega, \tag{4}$$

and the boundary conditions

$$u_\alpha(x) = 0, \quad x \in \partial\Omega. \tag{5}$$

Under these conditions, for $v_\alpha(\boldsymbol{x})$, we get the equation

$$\alpha^2 \triangle v_\alpha - v_\alpha = 0, \quad \boldsymbol{x} \in \Omega. \tag{6}$$

In view of (3), from (5), we obtain the following boundary condition:

$$v_\alpha(\boldsymbol{x}) = 1, \quad \boldsymbol{x} \in \partial\Omega. \tag{7}$$

Equation (4) can be treated as a regularization of the Hamilton–Jacobi equation via the method of vanishing viscosity. The problem (4, 5) provides an approximate solution of the problem (1, 2) for small values of $\alpha$:

$$u_\alpha(\boldsymbol{x}) \to u(\boldsymbol{x}) \text{ as } \alpha \to 0, \quad \boldsymbol{x} \in \Omega.$$

In this case, $u_\alpha(\boldsymbol{x})$ is evaluated according to (3) from the solution of the linear BVP (6, 7).

## 3   Numerical Algorithm

For an approximate solution of the boundary value problem (1, 2), we introduce (see (3)) the representation

$$u_\alpha(\boldsymbol{x}) = -\alpha \ln(v_\alpha(\boldsymbol{x})), \tag{8}$$

with a sufficiently low value of $\alpha$. Here $v_\alpha(\boldsymbol{x})$ is the solution of the BVP (6, 7). In the present study, the numerical implementation is constructed using standard finite-element discretizations in space. The main features of the computational algorithm result from the fact that the boundary value problem of diffusion–reaction (6, 7) at small $\alpha$ is singularly perturbed, i.e., we have a small parameter at higher derivatives.

Introduce a standard quasi-uniform triangulation of the domain $\Omega$ into tetrahedra for the 3D case. Next, introduce

$$V_0 = \{ v \in H^1(\Omega) \mid v(\boldsymbol{x}) = 0, \ \boldsymbol{x} \in \partial\Omega \},$$

$$V_1 = \{ v \in H^1(\Omega) \mid v(\boldsymbol{x}) = 1, \ \boldsymbol{x} \in \partial\Omega \}.$$

Denote by $V_0^h \subset V_0$ and $V_1^h \subset V_1$ the linear finite-element spaces.

For the BVP (6, 7), we put into the correspondence the variational problem of finding the numerical solution $y \in V_1^h$ from the conditions

$$a(y, v) = 0, \quad \forall v \in V_0^h. \tag{9}$$

For the bilinear form, we have

$$a(y, v) = \int_\Omega \left( \alpha^2 \nabla y \nabla v + yv \right) d\boldsymbol{x}.$$

For the differential problem (6, 7), the maximum principle holds that guarantees the positiveness of the solution. This imply (see, e.g., [16]) that at the internal points of the domain $\Omega$, we have

$$0 < v_\alpha(\boldsymbol{x}) < 1, \quad \boldsymbol{x} \in \Omega.$$

This essential property must be also fulfilled for the solution of the discrete problem (9):

$$0 < y(\boldsymbol{x}) < 1, \quad \boldsymbol{x} \in \Omega. \tag{10}$$

If (10) holds, we obtain monotone approximations for the solution of the diffusion–reaction problem.

If the parameter $\alpha$ in (6) is not small (the problem is regular), it is also possible to construct monotone approximations using linear finite elements with restrictions on the computational grid (Delaunay-type mesh, see, e.g., [15]). Additional restrictions appear (see, e.g., [6]) on the value of the reaction coefficient. For the problem (6, 7), restrictions on the grid size seem like $h \leq \mathcal{O}(\alpha)$.

These restrictions on the grid size can be removed. The standard approach is based on a correction of approximations for the reaction coefficient using the lumping procedure (see, e.g., [18]).

The standard approach for solving singularly perturbed diffusion–reaction problems (see [17]) employs computational grids with refinements in the vicinity of boundaries. A refinement of the grid is directly related to the value of the small parameter $\alpha$.

Next, it is possible to monotonize the solution of the problem (6, 7) at small values of $\alpha$ is the following way. Following the work [5], for singularly perturbed problems to the diffusion–convection equation, finite-element approximations of higher order not only increase the accuracy of the numerical solution, but also improve the monotonicity property. It seems reasonable to study the approximation order impact in solving singularly perturbed problems for the diffusion–reaction equations.

## 4   Predictions

Previously, the above numerical algorithm for solving the singularly perturbed BVP (4, 5) has been verified on solving model 2D problems. In particular, the key issue is a practical choice of the small parameter $\alpha$. Here we will illustrate the possibilities of the developed approach on solving a model 3D problem.

As a model 3D test, let us consider the problem of calculating the distance to solid walls in a long channel of rectangular cross-section, where a row of spheres is located along the channel axis. This problem is formulated as the BVP (1, 2) taking into account the symmetry of the problem. Namely, we introduce symmetry planes, where the homogeneous Neumann condition is imposed. The computational domain is shown in Fig. 1.

**Fig. 1.** Computational domain

The computational grid contains 505,507 nodes and 2,547,255 tetrahedra with piecewise linear finite elements. The solution of the auxiliary BVP (4, 5) is given in Fig. 2. The corresponding approximate solution of the problem for the eikonal equation (distance to solid walls) is depicted in Fig. 3. These calculations are performed for $\alpha = 0.02$. Decreasing $\alpha$ in half gives a nonmonotonic solution for the singularly perturbed problem (4, 5).

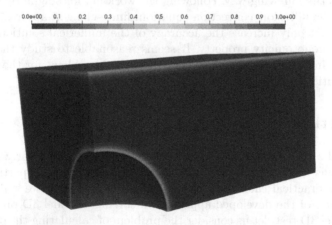

**Fig. 2.** Solution $v_\alpha(\boldsymbol{x})$ of the diffusion–reaction problem for $\alpha = 0.02$

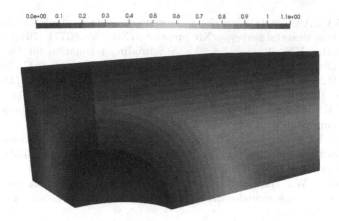

**Fig. 3.** Solution $u_\alpha(x)$ at $\alpha = 0.02$

**Acknowledgements.** Petr Vabishchevich gratefully acknowledges support from the the Russian Federation Government (# 14.Y26.31.0013).

# References

1. Belyaev, A.G., Fayolle, P.A.: On variational and PDE-based distance function approximations. In: Computer Graphics Forum, vol. 34, pp. 104–118. Wiley Online Library (2015)
2. Bhattacharya, T., DiBenedetto, E., Manfredi, J.: Limits as $p \to \infty$ of $\triangle_p u_p = f$ and related extremal problems. Rend. Sem. Mat. Univ. Polytec. Torino **47**, 15–68 (1989)
3. Bitzadze, A.V.: Some Classes of Partial Differential Equations. Nauka, Moscow (1981). (in Russian)
4. Born, M., Wolf, E.: Principles of Optics: Electromagnetic Theory of Propagation, Interference and Diffraction of Light. Cambridge University Press, Cambridge (2005)
5. Cai, Q., Kollmannsberger, S., Sala-Lardies, E., Huerta, A., Rank, E.: On the natural stabilization of convection dominated problems using high order Bubnov-Galerkin finite elements. Comput. Math. Appl. **66**(12), 2545–2558 (2014)
6. Ciarlet, P.G., Raviart, P.A.: Maximum principle and uniform convergence for the finite element method. Comput. Methods Appl. Mech. Eng. **2**(1), 17–31 (1973)
7. Crane, K., Weischedel, C., Wardetzky, M.: Geodesics in heat: a new approach to computing distance based on heat flow. ACM Trans. Graph. (TOG) **32**(5), 152 (2013)
8. Fu, Z., Jeong, W.K., Pan, Y., Kirby, R.M., Whitaker, R.T.: A fast iterative method for solving the eikonal equation on triangulated surfaces. SIAM J. Sci. Comput. **33**(5), 2468–2488 (2011)
9. Fu, Z., Kirby, R.M., Whitaker, R.T.: A fast iterative method for solving the eikonal equation on tetrahedral domains. SIAM J. Sci. Comput. **35**(5), C473–C494 (2013)
10. Gilles, A., Pierre, K.: Mathematical Problems in Image Processing: Partial Differential Equations and the Calculus of Variations. Applied Mathematical Sciences, 2nd edn. Springer, New York (2006). https://doi.org/10.1007/978-0-387-44588-5

11. Gómez, J.V., Alvarez, D., Garrido, S., Moreno, L.: Fast methods for eikonal equations: an experimental survey. arXiv preprint arXiv:1506.03771 (2015)
12. Gurumoorthy, K.S., Rangarajan, A.: A Schrödinger equation for the fast computation of approximate euclidean distance functions. In: Tai, X.-C., Mørken, K., Lysaker, M., Lie, K.-A. (eds.) SSVM 2009. LNCS, vol. 5567, pp. 100–111. Springer, Heidelberg (2009). https://doi.org/10.1007/978-3-642-02256-2_9
13. Jeong, W.K., Whitaker, R.T.: A fast iterative method for eikonal equations. SIAM J. Sci. Comput. **30**(5), 2512–2534 (2008)
14. Kružkov, S.N.: Generalized solutions of the Hamilton-Jacobi equations of eikonal type. I. Formulation of the problems; existence, uniqueness and stability theorems; some properties of the solutions. Sb.: Math. **27**(3), 406–446 (1975)
15. Letniowski, F.W.: Three-dimensional Delaunay triangulations for finite element approximations to a second-order diffusion operator. SIAM J. Sci. Stat. Comput. **13**(3), 765–770 (1992)
16. Protter, M.H., Weinberger, H.F.: Maximum Principles in Differential Equations. Springer, New York (2012). https://doi.org/10.1007/978-1-4612-5282-5
17. Roos, H., Stynes, M., Tobiska, L.: Robust Numerical Methods for Singularly Perturbed Differential Equations: Convection-Diffusion-Reaction and Flow Problems. Springer Series in Computational Mathematics, vol. 24. Springer, Heidelberg (2008). https://doi.org/10.1007/978-3-540-34467-4
18. Thomée, V.: Galerkin Finite Element Methods for Parabolic Problems. Springer Series in Computational Mathematics. Springer, Berlin (2006). https://doi.org/10.1007/978-3-662-03359-3
19. Tsai, Y.H.R., Cheng, L.T., Osher, S., Zhao, H.K.: Fast sweeping algorithms for a class of Hamilton-Jacobi equations. SIAM J. Numer. Anal. **41**(2), 673–694 (2003)
20. Tsitsiklis, J.: Fast marching methods. IEEE Trans. Autom. Control **40**, 1528–1538 (1995)

# Iterative Semi-implicit Splitting Methods for Stochastic Chemical Kinetics

Jürgen Geiser[✉]

Department of Electrical Engineering and Information Technology,
Ruhr University of Bochum, Universitätsstraße 150, 44801 Bochum, Germany
juergen.geiser@ruhr-uni-bochum.de

**Abstract.** In this paper, we present splitting methods that are based on iterative schemes and applied to stochastic models for chemical kinetics. The motivation arose of solving chemical kinetics with respect to stochastic influences in their models. The parameters and variables that describe the concentrations are based on extending the deterministic models to stochastic models. Such an extension is important to simulate the uncertainties of the concentrations. For the modelling equations, we deal with stochastic differential equations and it is important to extend the deterministic methods to stochastic methods. Here, we consider iterative splitting methods, based on Picard's successive approximations, to solve the underlying stochastic differential equations. The benefit of relaxation behaviour of the iterative solvers is also obtained in the stochastic method and based on the stochastic process we obtain $1/2$ of the accuracy as for the deterministic method, which is also given for non-iterative methods. We present the numerical analysis of the schemes and verified the results in numerical experiments of different chemical reaction systems.

**Keywords:** Splitting methods · Stochastic differential equations · Iterative splitting schemes · Convergence analysis · Chemical reaction systems

## 1 Introduction

We are motivated to develop fast algorithms to solve stochastic chemical reaction kinetics, see [2,11] and [7]. Such stochastic chemical reaction kinetics are approximated chemical Langevin equations with the thermodynamical limit of the deterministic chemical reaction kinetics, see [7].

We consider the nonlinear stochastic differential equation, which is based on the nonlinearities and multiscale problems delicate to solve. Here, we have to apply higher order stochastic solvers, e.g., Milstein scheme, see [10] and [14]. Such solvers are direct or non-iterative and have the drawback in missing relaxations of such nonlinear parts, see [4] and [9].

Therefore, we propose new iterative splitting schemes, see [6], which allow to obtain higher order accuracy with a nonlinear solver effect which is related

© Springer Nature Switzerland AG 2019
I. Dimov et al. (Eds.): FDM 2018, LNCS 11386, pp. 35–47, 2019.
https://doi.org/10.1007/978-3-030-11539-5_4

to the fixpoint scheme, see [8]. The increase in the formal order with additional iterative steps of the iterative splitting scheme is an important and valuable property of the class of iterative scheme, see [4]. Here, we can extend such a benefit to the stochastic differential equations, see also [6].

In the paper, we discuss the nonlinear stochastic differential equations of the Langevin-type, see [7], with the following solver methods:

– Direct methods: Euler-Maruyama and Milstein schemes, see [10],
– Indirect methods: Iterative splitting schemes, see [4].

The paper is outlined as following. In the Sect. 2, we derive the underlying modelling equations. In the Sect. 3, we discuss the iterative splitting method for the stochastic differential equations and the convergence analysis. The numerical algorithms of the direct and indirect methods are presented in Sect. 4. The numerical results are discussed in Sect. 5 and we conclude our results in Sect. 6

## 2   Modelling Equations

We assume to have a well-stirred mixture of $N$ molecular species $\{S_1, \ldots, S_N\}$, which chemically interact through $M \geq 0$ reaction channels $\{R_1, \ldots, R_M\}$. Further, we specify the dynamical state of the system by $X(t) = (X_1(t), \ldots, X_N(t))^t$, where $X_k(t)$ is the number of $S_k$ molecules in the system at time $t$.

The molecular populations $X_k(t)$ are random variables and we describe the evolution of $X(t)$ from an initial state $X_0(t)$.

The time-evolution equation for the probability $P(x, t|x_0, t_0)$ that $X(t)$ will equal $x$ given that $X_0(t) = x_0$ is modelled by the chemical master equation:

$$\frac{\partial}{\partial t} P(x, t|x_0, t_0) = \sum_{j=1}^{M} \left( a_j(x - \nu_j) \, P(x - \nu_j), t|x_0, t_0) - a_j(x) \, P(x, t|x_0, t_0) \right), \quad (1)$$

where $a_j(x)dt$ is the probability density, given $X(t) = x$, that $R_j$ reaction occurs inside a fixed volume $\Omega$ in the next infinitesimal time interval $[t, t+dt)$, $with\ j = 1, \ldots, M$.

$\nu_j$ is the state-change vector, which is given with the components: $\nu_{jk}$ is the change in the number of $S_k$ molecules produced by one $R_j$ reaction with $j = 1, \ldots, M,\ k = 1, \ldots, N$.

For solving such chemical reaction system, the underlying stochastic simulation algorithms (SSA) used the Direct method, First Reaction method or Next Reaction method, see the discussion of the methods in [2] and [7]. By the way, all the algorithms are very time-consuming, while we solve a microscopically scale, see [7].

Therefore, we assume to deal with macroscopically infinitesimal time scale, such that during any time increment $dt$ the reaction channel can fire many more times, such that the propensity functions did not change appreciable. Therefore we obtain, that the jump Markov process $X(t)$ is approcimated by a continuous

Markov process. Such a continuous process can be defined by the standard form of a chemical Langevin equation (CLE):

$$X_k(t + \Delta t) = X_k(t) + \sum_{j=1}^{M} \nu_{jk} a_j(X(t)) \, dt + \sum_{j=1}^{M} \nu_{jk} a_j^{1/2}(X(t)) \, dW_{j,t}, \quad (2)$$

where $k = 1, \ldots, N$ and $W_{j,t}$ are Wiener processes with $\Delta W_{j,t} = \sqrt{\Delta t} \xi_j(t)$ and $\xi_j$ are temporally uncorrelated, statistically independent normal random variables with mean 0 and variance 1.

We obtain the thermodynamical limit in the deterministic chemical reaction equation:

$$\frac{dX_k(t)}{dt} = \sum_{j=1}^{M} \nu_{jk} a_j(X(t)), \ k = 1, \ldots, N, \ \frac{dX(t)}{dt} = A(X(t)), \quad (3)$$

where $A(X)$ is nonlinear vector of rang $N$.

In the following, we dicuss the numerical methods to solve such stochastic differential equations.

## 3    Iterative Splitting Method for Stochastic Ordinary Differential Equations

We consider the following stochastic differential equations (SDEs):

$$dX = A(X)dt + B(X)dW, \ X(0) = X_0, \quad (4)$$

where we assume to have the following decomposition of the vectors $A(X) = A_1 X + A_2(X)$ and $B(X) = B_1 X + B_2(X)$.

We have the following definition, see also [12]:

**Definition 1.** *We have the probability space* $(\Omega, \mathcal{F}, P$ *given with* $\Omega$ *as fixed volume,* $\mathcal{F}_{\sqcup t \geq 0}$ *as filtration and* $P$ *as probability. Further* $W$ *is a* $M$-*dimensional Brownian motion. We have* $B(X) \in \mathbb{R}^M \times \mathbb{R}^N$ *and* $A(X) \in \mathbb{R}^N$. *Further, we assume that we have smooth functions on* $\mathbb{R}^N$.

We have the following Assumptions 1, see also [12]:

**Assumption 1.** *We assume that the first order derivatives of* $B$ *and* $A$ *are bounded. Then, we can assume that* $B$ *and* $A$ *are Lipschitz:*

$$|B(x) - B(y)| + |A(x) - A(y)| \leq K(|x - y|), \quad (5)$$

*where* $K = \max\{\|\frac{\partial \sigma}{\partial x}, \frac{\partial b}{\partial x}\}$, *and linear growth, i.e.,*

$$\|\sigma(x)\| + \|b(x)\| \leq L(1 + \|x\|), \quad (6)$$

*where* $L$ *is a positive constant and* $\frac{\partial B}{\partial x}$ *and* $\frac{\partial A}{\partial x}$ *are the derivatives with respect to* $x$. *Further* $\| \cdot \|$ *is the usual Euclidean norm.*

The following algorithm is based on the iteration with a fixed-splitting discretization step-size $\tau$. For the time-interval $[t^n, t^{n+1}]$, we solve the following sub-problems consecutively for $i = 1, 2, \ldots m$, (cf. [4]).

We apply the following two ideas:

– Picard's successive approximation

$$dc_i(t) = A_1 c_i(t)dt + B_1 c_i dW_t(t) + A_2(c_{i-1}(t))dt + B_2(c_{i-1})dW_t(t), \quad (7)$$
$$\text{with } c_i(t^n) = c^n, \text{ and } c_0(t) = 0, \quad (8)$$

where $c^n$ is the known split approximation at the time-level $t = t^n$. The split approximation at the time-level $t = t^{n+1}$ is defined as $c^{n+1} = c_{2m+2}(t^{n+1})$. Furthermore, $W$ is a Wiener process, see [10].

We can rewrite this into the form of the following linear scheme as:

$$c_i(t^{n+1}) = c(t^n) + (A_1 c_{i-1}(t^{n+1}) + A_2(c_{i-1}(t^{n+1})))\Delta t \quad (9)$$
$$+ (B_1 c_{i-1}(t^{n+1}) + B_2(c_{i-1}(t^{n+1})))\Delta W_t,$$
$$\text{with } c_i(t^n) = c^n \text{ and } c_0(t) = 0, \quad (10)$$

where $\Delta W_t = \sqrt{\Delta t}\xi$ is a Gaussian normal distributed variable with $\langle \xi \rangle = 0$ and $\langle \xi^2 \rangle = 1$.

– Iterative exponential splitting approximation

$$dc_i(t) = A_1 c_i(t)dt + B_1 c_{i-1}dW_t(t) + A_2(c_{i-1}(t))dt + B_2(c_{i-1})dW_t(t), \quad (11)$$
$$\text{with } c_i(t^n) = c^n, \text{ and } c_0(t) = 0, \quad (12)$$

where $c^n$ is the known split approximation at the time-level $t = t^n$. We assume that we can compute the linear part with $\phi(t)c^n = \exp(A_1 t)c^n$.

The split approximation at the time-level $t = t^{n+1}$ is defined as $c^{n+1} = c_{2m+2}(t^{n+1})$. Furthermore, $W$ is a Wiener process, see [10].

We can rewrite this into the form of the following linear scheme as:

$$c_i(t^{n+1}) = \phi(\Delta t)c(t^n) + \int_{t^n}^{t^{n+1}} \phi(t^{n+1} - s)A_2(c_{i-1}(s))\,ds \quad (13)$$

$$+ \int_{t^n}^{t^{n+1}} \phi(t^{n+1} - s)(B_1 c_{i-1}(s) + B(c_{i-1}(s)))\,dW_s,$$
$$\text{with } c(t^n) = c^n, \text{ and } c_0(t) = 0, \quad (14)$$

where $\Delta W_t = \sqrt{\Delta t}\xi$ is a Gaussian normal distributed variable with $\langle \xi \rangle = 0$ and $\langle \xi^2 \rangle = 1$.

For simplicity, we assume the system of operators are generators of a $C_0$-semigroup based on their underlying operator norms.

## 3.1   Convergence for the Picard's Approximation

In the following Theorem 2, we discuss the convergence of the Picard's approximation.

**Theorem 2.** *Let us consider the following stochastic differential equation of Ito type:*

$$\partial_t X = A_1 X dt + A_2(X) dt + B(X) dW_t, \; X(0) = X_0, \tag{15}$$

*where $A_1 \in \mathbb{R}^{d \times d}, A_2 : \mathbb{R}^d \to \mathbb{R}^d, B : \mathbb{R}^d \to \mathbb{R}^d \times \mathbb{R}^l$ are Borel measurable functions, $W_t$ is an l-dimensional standard Brownian motion defined Wiener space $(\Omega, \mathcal{F}, P)$ with $\Omega = C_0(\mathbb{R}^+, \mathbb{R}^l)$ with uniform convergence topology, $P$ is the Wiener measure and $\mathcal{F}$ is the P-completion of Borel field $\mathcal{B}(\Omega)$.*

*Then the iterative operator splitting method (9) has the following splitting error:*

$$||\Gamma(X^i) - \Gamma(X^{i-1})||_{p,T}^p \le (C(p))^i K^{p\,i} (T^{p/2} - T^p)^i ||X_0||_{p,T}^p, \tag{16}$$

*where $i$ is the number of iterative steps, $\Gamma$ is the solution operator, $C^p$ is a constant depends only on $p$.*

*Proof.* We have given:

$$||\Gamma(X^i) - \Gamma(X^{i-1})||_{p,T}^p \le \tag{17}$$

$$\le C(p) \left( E \left( \sup_{t \in [0,T]} | \int_0^t (\sigma(X^i(s,x)) - \sigma(X^{i-1}(s,x))) \, dw_s|^p \right) \right.$$

$$\left. + E \left( \sup_{t \in [0,T]} | \int_0^t (b(X^i(s,x)) - b(X^{i-1}(s,x))) \, ds|^p \right) \right)$$

$$\le C(p) K^p (T^{p/2} - T^p) ||X^i - X^{i-1}||_{p,T}^p. \tag{18}$$

Then, we apply the recursion with $i \to i - 1 \to \ldots \to 1$:

$$||\Gamma(X^i) - \Gamma(X^{i-1})||_{p,T}^p \le ||\Gamma^i(X^0) - \Gamma^i(X^{-1})||_{p,T}^p$$

$$\le C(p) K^p (T^{p/2} - T^p) ||\Gamma^{i-1}(X^i) - \Gamma^{i-1}(X^{i-1})||_{p,T}^p$$

$$\le (C(p))^i K^{p\,i} (T^{p/2} - T^p)^i ||X_0||_{p,T}^p, \tag{19}$$

where $X^{-1} = 0$, we also assumed $\lim_{i \to \infty} \Gamma^i(X^0) \to X^*$, where $X^*$ is the fixpoint of the iteration, see [3] and [12]. $\quad\square$

*Remark 1.* For the stochastic equations, we obtain a convergence order of $\mathcal{O}(T^{p \frac{i}{2}})$ for the Picard's approximation method, while for the deterministic equations, we obtain a convergence order of $\mathcal{O}(T^{p\,i})$ for the Picard's approximation method, see [4] and [13].

## 3.2  Convergence for the Iterative Exponential Splitting Approximation

In the following Theorem 3, we discuss the convergence of the iterative exponential splitting approximation.

**Theorem 3.** *Let us consider the following stochastic differential equation of Ito type:*

$$\partial_t X = A_1 X dt + A_2(X)dt + B(X)dW_t, \ X(0) = X_0, \qquad (20)$$

*where* $A_1 \in \mathbb{R}^{d \times d}$, $A_2 : \mathbb{R}^d \to \mathbb{R}^d$, $B : \mathbb{R}^d \to \mathbb{R}^d \times \mathbb{R}^l$ *are Borel measurable functions,* $W_t$ *is an l-dimensional standard Brownian motion defined Wiener space* $(\Omega, \mathcal{F}, P)$ *with* $\Omega = C_0(\mathbb{R}^+, \mathbb{R}^l)$ *with uniform convergence topology,* $P$ *is the Wiener measure and* $\mathcal{F}$ *is the P-completion of Borel field* $\mathcal{B}(\Omega)$.

*Then the iterative operator splitting method (13) has the following splitting error:*

$$||\Gamma(X^i) - \Gamma(X^{i-1})||_{p,T}^p \leq (\tilde{C}(p))^i \tilde{K}^{p\,i}(T^{p/2} - T^p)^i ||X_0||_{p,T}^p, \qquad (21)$$

*where* $i$ *is the number of iterative steps,* $\Gamma$ *is the solution operator,* $C^p$ *is a constant depends only on* $p$.

*Proof.* We define the operator $\Gamma$ as following:

$$\Gamma X = \phi(t)\xi(x) + \int_0^t \phi(t-s)\sigma(X(s,x))\,dw_s + \int_0^t \phi(t-s)\tilde{b}(X(s,x))\,ds. \ (22)$$

We estimate the operator $\Gamma$ as following:

$$||\Gamma X||_{p,T}^p \leq \tilde{C}(p) \left( E|\phi(t)\xi(x)|^p + E \left( \sup_{t \in [0,T]} |\int_0^t \phi(t-s)\sigma(X(s,x))\,dw_s|^p \right) \right.$$

$$\left. + E \left( \sup_{t \in [0,T]} |\int_0^t \phi(t-s)\tilde{b}(X(s,x))\,ds|^p \right) \right). \qquad (23)$$

We estimate the different terms and obtain:

$$E \left( \sup_{t \in [0,T]} |\int_0^t \phi(t-s)\sigma(X(s,x))\,dw_s|^p \right) \leq \tilde{C}E \left( \sup_{t \in [0,T]} |\int_0^t \sigma(X(s,x))\,dw_s|^p \right),$$

where $\tilde{C} = E(\sup_{t \in [0,T]} |\int_0^t \phi(t-s)|ds)$.

We apply the Burkholder-Davis-Gundy Inequality, see [1] and the Hölder's inequality, see [12]. Further we apply the Assumption 1, then we have:

$$E \left( \sup_{t \in [0,T]} |\int_0^t \phi(t-s)\sigma(X(s,x))\,dw_s|^p \right) \leq C(p)\tilde{C}E \left( |\int_0^T \sigma^2(X(s,x))\,dw_s|^{p/2} \right)$$

$$\leq C(p)\tilde{C}T^{p/2-1} \int_0^T E(|\sigma^2(X(s,x))|^p)ds \leq C(p)\tilde{C}T^{p/2-1}(1 + ||X(s,x)||_{p,T}^p), \quad (24)$$

where $\tilde{C}(p) = C(p)\tilde{C}$.

Further, we have:

$$E\left(\sup_{t\in[0,T]}|\int_0^t \phi(t-s)b(X(s,x))\,ds|^p\right) \le C(p)\tilde{C}T^{p-1}\int_0^T E(|b(X(s,x))|^p)\,ds$$

$$\le C(p)\tilde{C}T^{p-1}\int_0^T (1+E(|X(s,x)|^p))\,ds \le C(p)\tilde{C}T^p(1+||X(s,x)||_{p,T}^p), \quad (25)$$

then we obtain:

$$||\Gamma X||_{p,T}^p \le \tilde{C}(p)\left(E|\phi(t)\xi(x)|^p + (T^{p/2}+T^p)(1+||X(s,x)||_{p,T}^p)\right). \quad (26)$$

We have given:

$$||\Gamma(X^i)-\Gamma(X^{i-1})||_{p,T}^p$$

$$\le \tilde{C}(p)\left(E\left(\sup_{t\in[0,T]}|\int_0^t (\sigma(X^i(s,x))-\sigma(X^{i-1}(s,x)))\,dw_s|^p\right)\right.$$

$$\left.+E\left(\sup_{t\in[0,T]}|\int_0^t (b(X^i(s,x))-b(X^{i-1}(s,x)))\,ds|^p\right)\right)$$

$$\le \tilde{C}(p)K^p(T^{p/2}-T^p)||X^i-X^{i-1}||_{p,T}^p. \quad (27)$$

Then, we apply the recursion with $i \to i-1 \to \ldots \to 1$:

$$||\Gamma(X^i)-\Gamma(X^{i-1})||_{p,T}^p \le ||\Gamma^i(X^0)-\Gamma^i(X^{-1})||_{p,T}^p$$

$$\le \tilde{C}(p)K^p(T^{p/2}-T^p)||\Gamma^{i-1}(X^i)-\Gamma^{i-1}(X^{i-1})||_{p,T}^p$$

$$\le (\tilde{C}(p))^i K^{p\,i}(T^{p/2}-T^p)^i||X_0||_{p,T}^p, \quad (28)$$

where $X^{-1}=0$, we also assumed $\lim_{i\to\infty}\Gamma^i(X^0) \to X^*$, where $X^*$ is the fixpoint of the iteration, see [3] and [12].

*Remark 2.* For the exponential version of the Picard's approximation, we have also for the stochastic equations a convergence order of $\mathcal{O}(T^{p\,\frac{i}{2}})$, while for the deterministic equations, we obtain a convergence order of $\mathcal{O}(T^{p\,i})$, see [4] and [13]. Such that we need 2-times more iterative steps to obtain the same convergence results as in the deterministic case.

## 4  Numerical Algorithms for the Nonlinear Stochastic Ordinary Differential Equations

In the following, we deal with the different numerical algorithms to solve the nonlinear stochastic differential equations.

We deal with the underlying nonlinear stochastic differential equation, which is given as:

$$dX = A(X)dt + B(X)dW, \quad (29)$$

where $A, B$ are matrices in $I\!R^{m \times m}$ with $m$ is the number of unknown. We have $A(X) = A_1 + A_2(X)$ and $B(X) = B_1 + B_2(X)$.

Further, the components of the matrices are dependent of the solution $X$. Further, the initial values are given as $X_{t_0} = X_0$ and $W$ is Wiener process, see [10].

In the following, we deal with the direct and indirect algorithms, which are implemented in the numerical experiments. The direct methods are numerical standard methods, which are used in the numerical approximation of stochastic differential equations. They are simply to implement and obtain direct the numerical solutions (one-step methods), while they have their drawback in the resolution of the nonlinear solutions, while the linearization is given by the time-step. Instead the indirect methods are iterative solvers and obtain higher order resolutions with additional iterative cycles (multi-step methods), such that they allow to resolve the nonlinear solution in the time-step approach, see [4].

### 4.1   Algorithms

In the following, we deal with the following algorithms:

1. Direct Algorithms are given as:
   - Euler-Maruyama,
   - Milstein scheme.

   *Remark 3.* The direct methods are fast to implement and obtain lower order results. The numerical scheme have the following accuracy: $\mathcal{O}(t^{\frac{1}{2}})$ for the Euler-Maruyama scheme, $\mathcal{O}(t)$ for the Milstein scheme and $\mathcal{O}(t^{\frac{1}{2}})$ for the AB-splitting scheme for large n $n \to \infty$. Here, the approach to higher order schemes are delicate, see [10].

2. Indirect Algorithms are given as:
   - Iterative Semi-EM scheme,
   - Iterative exponential splitting scheme.

   *Remark 4.* The indirect methods are based on the iterative approaches related to fixpoint-schemes and obtained higher order accuracy: $\mathcal{O}(t^{i+\frac{1}{2}})$, where $i$ is the number of iterative steps. Based on their recursive behavior numerical approaches in previous iterative steps can be used. Such a clever combination of the previous computed iterative cycles allows to obtain fast iterative methods, see [5].

### 4.2   Numerical Error Analysis (Strong and Weak Errors)

For the verification of the theoretical results for the iterative splitting scheme in Sect. 3, we deal with the following numerical error analysis.

We present the convergence rates of the following strong and weak errors:

- Strong error: $err_{strong,\Delta t} = E(X_{\Delta t,Scheme} - X_{\Delta t_{fine},Mil})$,
  where $N$ are the number of the time-steps.
- Weak errors: $err_{weak,\Delta t,f} = |E(f(X_{\Delta t,Scheme})) - E(f(X_{\Delta t_{fine},Mil})))|$,
  where $N$ are the number of the time-steps and $f$ is a polynomial function,
  e.g., $f(x) = \frac{1}{N}x$ or $f(x) = \frac{1}{N^2}x$.
- Convergence order (weak error): $\rho_{Scheme,\Delta t} = \log_2\left(\frac{err_{weak,\Delta t}}{err_{weak,\Delta t/2}}\right)$,
  where $Scheme = \{EM, Mil, Iter - Semi, Iter - Exp\}$.

*Remark 5.* In the numerical examples, we obtain the highest sensitivity of the convergence rates with the strong error. Therefore, we deal with different weak errors, which damped such high sensitivity. The best results, we obtain with the weighting function of $f(x) = \frac{1}{N}x$ and $f(x) = \frac{1}{N^2}x$, where $N$ is the number of the time-grid points. Such weighting function relax the oscillations of the results, see [10].

# 5  Numerical Examples

In the following numerical examples, we verify the theoretical results and the benefits of the novel iterative solvers for the stochastic differential equations.

We deal with the following examples and discuss the methodological sense of the different schemes:

- Scalar benchmark problem (scalar multiplicative noise): The stochastic differential equations are based on $m \times m$ operator matrices, while we have a scalar multiplicative noises term. For such a benchmark examples, we can detailed analyze the benefit of the iterative schemes.
- Real-life problem (Lotka system), see [15]: We apply a stochastic chemical system with multiplicative noise. We have nonlinear terms and such examples are delicate to solve. We present approximation of the solutions with the Picard's schemes and obtain accurate results.

## 5.1  Scalar Multiplicative Noise

In the following, we deal with a simple chemical reaction model, while the reaction part is influenced via stochastic noise.

We deal first with an ordinary differential equation and separate the complex operator into two simpler operators: the $m \times m$ ordinary differential equation system given as:

$$dy(t) = Ay(t) + Py(t)\,dW(t), \tag{30}$$

$$A = \mu \begin{pmatrix} -\lambda_{1,1} & \lambda_{2,1} & \cdots & \lambda_{1,10} \\ \lambda_{2,1} & -\lambda_{2,2} & \cdots & \lambda_{2,10} \\ \vdots & & & \\ \lambda_{10,1} & \lambda_{10,2} & \cdots & -\lambda_{10,10} \end{pmatrix}, P = \sigma \begin{pmatrix} \sigma_{1,1} & \sigma_{1,2} & \cdots & \sigma_{1,10} \\ \sigma_{2,1} & \sigma_{2,2} & \cdots & \sigma_{2,10} \\ \vdots & & & \\ \sigma_{10,1} & \sigma_{10,2} & \cdots & \sigma_{10,10} \end{pmatrix}, \tag{31}$$

$$\lambda_{i,i} = 1.0, \ \sigma_{i,i} = 0.01, \ i = 1\ldots,10, \tag{32}$$

$$\lambda_{i,j} = 0.1, \ \sigma_{i,j} = 0.005, \ i = 2\ldots,10, \ j = 1,\ldots,i-1, \tag{33}$$

$$\lambda_{i,j} = 0.0, \ \sigma_{i,j} = 0.0, \ i = 1\ldots,9, \ j = i+1,\ldots,10, \tag{34}$$

where the initial conditions are $\mathbf{y}(0) = (1,\ldots,1)^t$, the scalar components are $\mu = 0.1$, $\sigma = 0.01$ and the matrix components are $\lambda_{11}\ldots\lambda_{10,10} \in I\!\!R^+$ as decay factors and $\sigma_{11},\ldots,\sigma_{10,10} \in I\!\!R^+$ as parameters of the perturbations. Further $W$ is a white stochastic noise force with $\Delta W = W_{t_{n+1}} - W_{t_n} = \sqrt{\Delta t}N(0,1)$, where $N(0,1) = rand$ is a normally distributed random variable. We deal with non-commutation matrices $[A, P] \neq AP - PA$ as given with the tridiagonal matrices in the experiment.

We have the time interval $t \in [0, T]$ and $m \in I\!\!N$.

In the Table 1, we have the convergence orders with the following values:

**Table 1.** Convergence results for the different schemes.

| Scheme | $\rho_{Scheme,\Delta t}$ | $\rho_{Scheme,\Delta t/2}$ | $\rho_{Scheme,\Delta t/4}$ |
|---|---|---|---|
| EM | 0.1621 | 0.3419 | 0.6717 |
| Mil | 0.1446 | 0.3401 | 0.6638 |
| Iter semi-implicit | 0.1712 | 0.3125 | 0.7115 |
| Iter exponential | 0.1712 | 0.3125 | 0.7115 |

The strong and weak errors are presented in Fig. 1. Here, we obtain large difference of the numerical errors between the implicit schemes (iterative splitting methods with $i = 2$ and $i = 3$) and the explicit schemes (Euler-Maruyama and Milstein scheme). While the implicit methods can relax the oscillations and results to small numerical errors, the explicit methods fail to relax such oscillations and obtain larger numerical errors.

*Remark 6.* In the multiplicative noise example, we present the benefits of the iterative splitting schemes, which resolves the stochastic behavior more accurate as the standard schemes.

## 5.2 Lotka System: Vectorial Linear Noise

In the following, we deal with a simple chemical reaction model, while the reaction part is influenced via additive noise, see also [15].

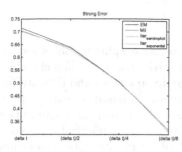 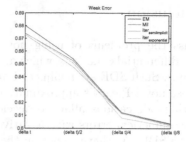

**Fig. 1.** The left figure presents the convergence of the strong errors for the different schemes and the right figure presents the convergence of the weak errors of the different schemes.

We deal first with an simple Lotka system, one species $A$ with a constant concentration and $X, Y$ as chemical intermediaries.

The SODE's are given as:

$$dX = q_2\, A\, X - q_1\, X\, Y + \epsilon_1\, X\, dW_1(t),\ X(0) = 1.0,\ A = 4, \qquad (35)$$
$$dY = q_1\, X\, Y - q_3 Y + \epsilon_2\, Y\, dW_2(t),\ Y(0) = 1.5, \qquad (36)$$

where the scalar components are $q_1 = 0.5$, $q_2 = 0.6$, $q_3 = 0.7$ and $\epsilon_1 = \epsilon_2 = 0.01$. Further $W_1$, $W_2$ are two white noise stochastic forces with $\Delta W = W_{t_{n+1}} - W_{t_n} = \sqrt{\Delta t} N(0,1)$, where $N(0,1) = rand$ is a normally distributed random variable.

We have the time interval $t \in [0, T]$ with $T = 40$ and we assume to have $N = 10000$ time points.

The strong errors and the phase-space diagram are presented in Fig. 2.

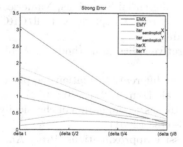 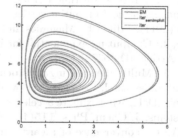

**Fig. 2.** The left figure presents the convergence of the strong errors for the different schemes the right figure presents the phase-space diagram $(X, Y)$ of the limit cycle for the Lotka system.

*Remark 7.* In the Lotka system, we present the benefits of the iterative splitting schemes. While the system is more delicate to solve with respect of the nonlinear terms, we need more iterative steps. At least, we have a benefit to the non-iterative schemes, while we could more relax the nonlinear behaviour.

## 6    Conclusion

We discuss the problems of using novel iterative splitting schemes to solve stochastic differential equations, which are motivated of modelling chemical reaction systems. Such SDE are nonlinear Langevin equations and delicate to solve. We propose novel Picard's approximation methods based on exponential functions, such modifications allow to accelerate the numerical computation and reduce the numerical errors with iterative steps. The numerical analysis shows, that for the SDEs, we have to obtain the doubled iterative steps to achieve the same results as for deterministic ODEs. The numerical examples present the advantages of the iterative schemes and their computational costs with respect to their relaxation effects. Here, we obtain an increase of the convergence order of the iterative splitting method with the number of the iteration steps. Further, we have verified the numerical results with different numerical experiments, which show the a benefit of the iterative schemes. In future, we see an area to optimize such novel Picard's approximation schemes with their benefit of relaxing the nonlinear solutions and to apply larger time steps.

## References

1. Burkholder, D.L., Davis, B., Gundy, R.F.: Integral inequalities for convex functions of operators on martingales. In: Proceedings of 6th Berkeley Symposium on Mathematical Statistics and Probability, vol. 2, pp. 223–240 (1972)
2. Cao, Y., Li, H., Petzold, L.: Efficient formulation of the stochastic simulation algorithm for chemically reacting systems. J. Chem. Phys. **121**(9), 4059–4067 (2004)
3. Fan, Z.: SOR waveform relaxation methods for stochastic differential equations. Appl. Math. Comput. **219**, 4992–5003 (2013)
4. Geiser, J.: Iterative Splitting Methods for Differential Equations, Numerical Analysis and Scientific Computing Series. CRC Press, Chapman & Hall/CRC, Boca Raton (2011). Edited by Magoules and Lai
5. Geiser, J.: Computing exponential for iterative splitting methods. J. Appl. Math. **2011**, Article ID 193781 (2011)
6. Geiser, J.: Multiscale splitting for stochastic differential equations: applications in particle collisions. J. Coupled Syst. Multiscale Dyn. **1**, 241–250 (2013)
7. Gillespie, D.T.: Approximate accelerated stochastic simulation of chemically reacting systems. J. Chem. Phys. **115**(4), 1716–1733 (2001)
8. Izzo, A.: $C^r$ convergence of Picard's sucessive approximations. Proc. Am. Math. Soc. **127**(7), 2059–2063 (1999)
9. Kelley, C.T.: Iterative Methods for Linear and Nonlinear Equations. SIAM Frontiers in Applied Mathematics, vol. 16. SIAM, Philadelphia (1995)
10. Kloeden, P.E., Platen, E.: The Numerical Solution of Stochastic Differential Equations. Springer, Heidelberg (1992). https://doi.org/10.1007/978-3-662-12616-5
11. Kafash, B., Lalehzari, R., Delavarkhalafi, A., Mahmoudi, E.: Application of stochastic differential system in chemical reactions via simulation. MATCH Commun. Math. Comput. Chem. **71**, 265–277 (2014)
12. Li, Z., Liu, J.: $C^\infty$-convergence of Picard's successive approximations to solutions of stochastic differential equations. Stat. Probab. Lett. **129**, 203–209 (2017)

13. Miekkala, U., Nevanlinna, O.: Iterative solution of systems of linear differential equations. Acta Numerica **5**, 259–307 (1996)
14. Oksendal, B.: Stochastic Differential Equations. Springer, Berlin (2003). https://doi.org/10.1007/978-3-642-14394-6
15. Silva-Dias, L., Lopez-Castillo, A.: Practical stochastic model for chemical kintetics. Quim. Nova **38**(9), 1232–1236 (2015)

# The Application of a Special Hermite Finite Element Coupled with Collocation to the Diffusion Equation

Lidiya Gileva[1] , Evgeniya Karepova[1,2]($\boxtimes$) , and Vladimir Shaydurov[1]

[1] Institute of Computational Modelling of SB RAS, 660036 Akademgorodok,
Krasnoyarsk, Russia
{gileva,e.d.karepova}@icm.krasn.ru, shaidurov04@mail.ru
[2] Siberian Federal University, 79 Svobodny pr., 660041 Krasnoyarsk, Russia
http://icm.krasn.ru

**Abstract.** In the paper, we propose an efficient method based on the use of a bicubic Hermite finite element coupled with collocation for the diffusion equation. This enables one to reduce the dimension of the system of equations in comparison with the standard finite element scheme. Numerical experiments confirm a theoretical convergence estimate and demonstrate the advantage of the proposed method.

**Keywords:** Diffusion equation · Finite element method · Hermite finite element · Collocation method

## 1 Introduction

The application of Hermite finite elements to the solution of a partial differential equation results in a system of linear algebraic equations whose dimension is several times less than in the case of full or incomplete (serendipity) Lagrange elements of the same degree which provide the same order of convergence [8]. Besides, a finite element with the values of second-order derivatives at the nodes in the set of degrees of freedom (DoF) provides a way to use the collocation technique for some equations. Requiring the residual to vanish at the nodes, we express one DoF in terms of other ones from the original equation and eliminate the related equations from the system. This enables one to reduce still further the dimension of the system of equations. In [4] we propose this approach for the Poisson equation and confirm its efficiency with numerical experiments.

In the paper, we extend this approach to the diffusion equation with a constant coefficient on a rectangular domain. To construct a discrete problem, we use the bicubic Hermite element on a rectangular cell proposed in [4] and eliminate the unknowns being the values of the second-order derivative with respect to one of variables. We prove an error estimate for the modified problem. Numerical experiments confirm that the elimination of 1/3 of the unknowns and the equations retains the order of convergence of an approximate solution.

© Springer Nature Switzerland AG 2019
I. Dimov et al. (Eds.): FDM 2018, LNCS 11386, pp. 48–59, 2019.
https://doi.org/10.1007/978-3-030-11539-5_5

A direct application of Hermite elements on a rectangular cell is restricted to the case of domains composed of rectangles. However, to extend the area of the application of the bicubic Hermite element considered in the paper to the case of a polygonal domain or a domain with the curved boundary, we can complement it with the appropriate triangular elements proposed and studied in [5].

## 2    Notations

Let $\Omega = (0, H_1) \times (0, H_2) \subset R^2$ be a rectangle with the boundary $\Gamma$. To construct a uniform triangulation $\mathcal{T}_h$, we subdivide $\bar{\Omega}$ into $N_1 \times N_2$ closed rectangles by the lines

$$x = ih_1, \quad y = jh_2, \qquad i = 0, ..., N_1, \quad j = 0, ..., N_2,$$

where $h_k = H_k/N_k$, $k = 1, 2$. Denote

$$h = \max\{h_1, h_2\}.$$

For functions $u$, $v$ defined and measurable on $\Omega$, introduce the inner product and the induced norm:

$$(u, v) = \int_\Omega uv \, d\Omega, \qquad |u|_{0,\Omega} = (u, u)^{1/2}.$$

Denote by $L_2(\Omega)$ the space of all functions $u$ defined and measurable on $\Omega$ with the finite norm $|u|_{0,\Omega}$. Introduce also the semi-norms

$$|u|_{k,\Omega} = \left( \sum_{i_1+i_2=k} \left| \frac{\partial^k u}{\partial x^{i_1} \partial y^{i_2}} \right|^2_{0,\Omega} \right)^{1/2}.$$

Denote by $H^k(\Omega)$ the space of all functions $u$ defined and measurable on $\Omega$ together with their partial derivatives up to order $k$ with the finite norm [1]

$$\| u \|_{k,\Omega} = (|u|^2_{0,\Omega} + |u|^2_{1,\Omega} + ... + |u|^2_{k,\Omega})^{1/2}.$$

Denote by $P_k$ the space of all polynomials in two variables up to degree $k$:

$$\sum_{0 \le i_1+i_2 \le k} a_{i_1,i_2} x^{i_1} y^{i_2}.$$

Define a "reference" finite elements as a triple $(\hat{e}, \hat{P}, \hat{\Sigma})$ where $\hat{e}$ is a "reference" cell, $\hat{P}$ is a space of polynomials defined on $\hat{e}$, and $\hat{\Sigma}$ is a set of functionals called degrees of freedom.

## 3   A Bicubic Hermite Element

In [4] we propose a new Hermite element defined as follows (Fig. 1):

$$\hat{e} = [0,1]^2,$$
$$\hat{P} = \text{span}\{1, \hat{x}, \hat{y}, \hat{x}\hat{y}, \hat{x}^2, \hat{y}^2, \hat{x}^2\hat{y}, \hat{x}\hat{y}^2, \hat{x}^3, \hat{y}^3, \hat{x}^3\hat{y}, \hat{x}\hat{y}^3\},$$
$$\hat{\Sigma} = \{\hat{\psi}_{s,i}(s=0,1,2): \ \hat{\psi}_{0,i}(p) = p(\hat{a}_i), \ \hat{\psi}_{1,i}(p) = \partial^2 p/\partial \hat{x}^2(\hat{a}_i),$$
$$\hat{\psi}_{2,i}(p) = \partial^2 p/\partial \hat{y}^2(\hat{a}_i), \ i = 1,...,4, \ \forall p \in \hat{P}\}. \tag{1}$$

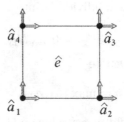

**Fig. 1.** A double arrow shows the DoF being the value of the second-order derivative it the corresponding direction at the node.

The Lagrange basis for (1) is given by

$$\begin{aligned}
\hat{\varphi}_{0,1} &= (1-\hat{x})(1-\hat{y}), & \hat{\varphi}_{0,3} &= \hat{x}\hat{y}, \\
\hat{\varphi}_{1,1} &= \hat{x}(1-\hat{x})(\hat{x}-2)(1-\hat{y})/6, & \hat{\varphi}_{1,3} &= \hat{x}(\hat{x}^2-1)\hat{y}/6, \\
\hat{\varphi}_{2,1} &= (1-\hat{x})\hat{y}(1-\hat{y})(\hat{y}-2)/6, & \hat{\varphi}_{2,3} &= \hat{x}\hat{y}(\hat{y}^2-1)/6, \\
\hat{\varphi}_{0,2} &= \hat{x}(1-\hat{y}), & \hat{\varphi}_{0,4} &= (1-\hat{x})\hat{y}, \\
\hat{\varphi}_{1,2} &= \hat{x}(\hat{x}^2-1)(1-\hat{y})/6, & \hat{\varphi}_{1,4} &= \hat{x}(1-\hat{x})(\hat{x}-2)\hat{y}/6, \\
\hat{\varphi}_{2,2} &= \hat{x}\hat{y}(1-\hat{y})(\hat{y}-2)/6, & \hat{\varphi}_{2,4} &= (1-\hat{x})\hat{y}(\hat{y}^2-1)/6.
\end{aligned} \tag{2}$$

For the partition $\mathcal{T}_h$, the usual affine mapping of the "reference" element onto an elementary cell $[x_i, x_i + h_1] \times [y_j, y_j + h_2]$ is of the form

$$x = x_i + h_1\hat{x}, \quad y = y_j + h_2\hat{y}. \tag{3}$$

Since $\hat{P} \supset P_3$, for the interpolant $\tilde{u}_I$ of a function $u \in H^4(\Omega)$ on the triangulation $\mathcal{T}_h$ we have the usual accuracy estimates [2,3]

$$|u - \tilde{u}_I|_{0,\Omega} \le ch^4 \parallel u \parallel_{4,\Omega},$$
$$\parallel u - \tilde{u}_I \parallel_{k,\Omega} \le ch^{4-k} \parallel u \parallel_{4,\Omega}, \quad k = 1,2. \tag{4}$$

From here on, $c$ denotes a positive constant independent of $u$ and $h$.

A feature of element (1) is that for some partial differential equations we can use the collocation technique requiring the residual to vanish at the nodes of the elements. This results in considerably reducing the number of unknowns in the system of linear algebraic equations of the finite element method. In [4] this approach is applied to the Poisson equation with numerical results confirming its efficiency.

# 4    A Discrete Problem for the Diffusion Equation

Without loss of generality, for simplicity sake we assume that $\Omega = (0,1) \times (0,1)$. Consider the problem

$$-\Delta u + ku = f \text{ in } \Omega, \tag{5}$$

$$u = 0 \text{ on } \Gamma, \tag{6}$$

where $\Delta = \partial^2/\partial x^2 + \partial^2/\partial y^2$ is the Laplacian, $f \in H^2(\Omega)$, $k = \text{const} \geq c > 0$.

For (5)–(6) we construct the standard scheme of the finite element method with Hermite element (1). Notice that when constructing the partition of $\Omega$, we have $N_1 = N_2 = N$, $h_1 = h_2 = h$. We look for an approximate solution in the form

$$\tilde{u}^h = \sum_{i,j=1}^{N-1} \tilde{u}^h(z_{i,j})\varphi_{0,i,j}(x,y)$$

$$+ h^2 \sum_{i,j=0}^{N} \left( \tilde{u}^h_{xx}(z_{i,j})\varphi_{1,i,j}(x,y) + \tilde{u}^h_{yy}(z_{i,j})\varphi_{2,i,j}(x,y) \right)$$

where $z_{i,j} = (x_i, y_j)$, $x_i = ih$, $y_j = jh$; the unknowns $\tilde{u}^h(z_{i,j})$, $\tilde{u}^h_{xx}(z_{i,j})$, and $\tilde{u}^h_{yy}(z_{i,j})$ are the values of the approximate solution and the approximate values of the second-order derivatives of the solution of (5)–(6) at the grid nodes, respectively. The functions $\varphi_{s,i,j}$ form a basis of the finite element space. They are obtained from basis functions (2) of the "reference" element with the affine mapping of the form (3).

As a result, we get a system of linear algebraic equations. The number of equations (and the number of unknowns) equals $3N_{\text{int}}$ where $N_{\text{int}}$ is the number of interior nodes. The values of $\tilde{u}^h$, $\tilde{u}^h_{xx}$, and $\tilde{u}^h_{yy}$ at the boundary nodes are determined from (5)–(6).

For convenience sake, we introduce some notations. Consider the grid expression

$$\sum_{l=j-1}^{j+1} \sum_{k=i-1}^{i+1} a_{k,l} g^h(z_{k,l})$$

where $a_{k,l}$ are coefficients and $g^h(z_{k,l})$ is a grid function. Introduce the nine-point stencil

$$[A] = \begin{bmatrix} a_{i-1,j+1} & a_{i,j+1} & a_{i+1,j+1} \\ a_{i-1,j} & a_{i,j} & a_{i+1,j} \\ a_{i-1,j-1} & a_{i,j-1} & a_{i+1,j-1} \end{bmatrix} \tag{7}$$

where the coefficients $a_{k,l}$ are arranged according to the position of the related nodes (Fig. 2).

Put

$$[A]g^h = \sum_{l=j-1}^{j+1} \sum_{k=i-1}^{i+1} a_{k,l} g^h(z_{k,l}).$$

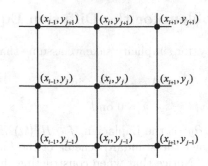

**Fig. 2.** The nodes of nine-point stencil (7).

Besides, we use the notation $[A^T]$ for the stencil with the coefficients $a_{k,l}^T = a_{l,k}$.

Introduce the grid functions $u^h$, $u_{xx}^h$, and $u_{yy}^h$ defined by $u^h(z_{i,j}) = \tilde{u}^h(z_{i,j})$, $u_{xx}^h(z_{i,j}) = \tilde{u}_{xx}^h(z_{i,j})$, and $u_{yy}^h(z_{i,j}) = \tilde{u}_{yy}^h(z_{i,j})$. Calculating directly the entries of the stiffness matrix, we arrive at the grid equations of three types

$$([A_0] + kh^2[A_1])u^h + ([A_2] + kh^2[A_4])\, h^2 u_{xx}^h \tag{8}$$

$$+ ([A_2^T] + kh^2[A_4^T])\, h^2 u_{yy}^h = \int_\Omega f\varphi_{0,i,j}\, d\Omega,$$

$$([A_2] + kh^2[A_4])u^h + ([A_3] - [A_4] + kh^2[A_5])\, h^2 u_{xx}^h \tag{9}$$

$$+ kh^2[A_6]h^2 u_{yy}^h = \int_\Omega f\varphi_{1,i,j}\, d\Omega,$$

$$([A_2^T] + kh^2[A_4^T])u^h + kh^2[A_6^T]h^2 u_{xx}^h \tag{10}$$

$$+ ([A_3^T] - [A_4^T] + kh^2[A_5^T])\, h^2 u_{yy}^h = \int_\Omega f\varphi_{2,i,j}\, d\Omega,$$

for all $i,j = 1, ..., N-1$ with the stencils

$$[A_0] = \begin{bmatrix} -1/3 & -1/3 & -1/3 \\ -1/3 & 8/3 & -1/3 \\ -1/3 & -1/3 & -1/3 \end{bmatrix}, \quad [A_1] = \begin{bmatrix} 1/36 & 1/9 & 1/36 \\ 1/9 & 4/9 & 1/9 \\ 1/36 & 1/9 & 1/36 \end{bmatrix},$$

$$[A_2] = \begin{bmatrix} 7/360 & 2/45 & 7/360 \\ -7/180 & -4/45 & -7/180 \\ 7/360 & 2/45 & 7/360 \end{bmatrix}, \quad [A_3] = \begin{bmatrix} -31/15120 & -4/945 & -31/15120 \\ 31/7560 & 8/945 & 31/7560 \\ -31/15120 & -4/945 & 31/15120 \end{bmatrix},$$

$$[A_4] = \begin{bmatrix} -7/2160 & -1/135 & -7/2160 \\ -7/540 & -4/135 & -7/540 \\ -7/2160 & -1/135 & -7/2160 \end{bmatrix}, \; [A_5] = \begin{bmatrix} 31/90720 & 2/2835 & 31/90720 \\ 31/22680 & 8/2835 & 31/22680 \\ 31/90720 & 2/2835 & 31/90720 \end{bmatrix},$$

$$[A_6] = \begin{bmatrix} 49/129600 & 7/8100 & 49/129600 \\ 7/8100 & 4/2025 & 7/8100 \\ 49/129600 & 7/8100 & 49/129600 \end{bmatrix}.$$

At the nodes near the boundary Eqs. (8)–(10) are modified in an obvious way.

Notice that the values of $u_{xx}^h$ and $u_{yy}^h$ are equipped with the factor $h^2$. We use the DoF $\psi_{0,i,j}(\tilde{u}^h) = \tilde{u}^h(z_{i,j})$, $\psi_{1,i,j}(\tilde{u}^h) = h^2 \tilde{u}_{xx}^h(z_{i,j})$, $\psi_{2,i,j}(\tilde{u}^h) = h^2 \tilde{u}_{yy}^h(z_{i,j})$ as the unknowns of system (8)–(10). This is convenient for several reasons, in particular, this improves properties of the stiffness matrix.

## 5    The Elimination of Unknowns by Collocation

The grid functions $u^h$, $u_{xx}^h$, and $u_{yy}^h$ satisfy the algebraic identity

$$- u_{xx}^h(z_{i,j}) - u_{yy}^h(z_{i,j}) + ku^h(z_{i,j}) = f^h(z_{i,j}) - r^h(z_{i,j}), \qquad (11)$$

where $r^h$ is the residual, $f^h$ is a grid function defined by $f^h(z_{i,j}) = f(z_{i,j})$.

Assuming the residual is known, neglect one group of equations, for instance, (10); express $u_{yy}^h(z_{i,j})$ from (11) and substitute it into (8)–(9). As a result, we obtain the following system

$$
\begin{aligned}
([A_0] &+ kh^2 \left([A_1] + [A_2^T]\right) + k^2 h^4 [A_4^T])u^h \\
&+ \left([A_2] - [A_2^T] + kh^2 \left([A_4] - [A_4^T]\right)\right) h^2 u_{xx}^h \\
&= \int_\Omega f\varphi_{0,i,j} \, d\Omega + \left([A_2^T] + kh^2 [A_4^T]\right) h^2 (f^h - r^h),
\end{aligned}
\qquad (12)
$$

$$
\begin{aligned}
([A_2] &+ kh^2 [A_4] + k^2 h^4 [A_6])u^h + \left([A_3] - [A_4] + kh^2 \left([A_5] - [A_6]\right)\right) h^2 u_{xx}^h \\
&= \int_\Omega f\varphi_{1,i,j} \, d\Omega + kh^4 [A_6](f^h - r^h),
\end{aligned}
\qquad (13)
$$

for all $i, j = 1, ..., N - 1$, reducing the number of equations by a factor of $1/3$.

Now introduce grid functions $v^h$, $v_{xx}^h$, and $v_{yy}^h$ being the values of a new approximate solution and new approximate values of the second-order derivatives at the nodes, respectively, and require that the residual vanish at the nodes:

$$- v_{xx}^h(z_{i,j}) - v_{yy}^h(z_{i,j}) + kv^h(z_{i,j}) = f^h(z_{i,j}). \qquad (14)$$

In (12)–(13), replace $u^h$ and $u_{xx}^h$ by $v^h$ and $v_{xx}^h$, respectively, and put $r^h = 0$. This gives a system of equations for the new unknowns. Once the system has been solved, $v_{yy}^h(z_{i,j})$ can readily be determined from (14).

We are interested in accuracy of the new approximate solution. In [4] we use finite element (1) coupled with collocation for the Poisson equation. Numerical experiments show that a solution on the reduced system is of the same order of accuracy as that of the original system of the finite element method. Below we prove this result.

# 6  An Error Estimate

Consider an arbitrary function

$$
\tilde{w} = \sum_{i,j=1}^{N-1} w_{i,j}\varphi_{0,i,j}(x,y) + h^2 \sum_{i,j=0}^{N} \left( (w_{xx})_{i,j}\,\varphi_{1,i,j}(x,y) + (w_{yy})_{i,j}\,\varphi_{2,i,j}(x,y) \right)
$$

of the finite element space. To this function, there corresponds the vector of the coefficients

$$
W = \left( \left\{ w_{i,j} \right\}_{i,j=1}^{N-1},\ \left\{ h^2\,(w_{xx})_{i,j} \right\}_{i,j=0}^{N},\ \left\{ h^2\,(w_{yy})_{i,j} \right\}_{i,j=0}^{N} \right)
$$

of a linear vector space $M$. The function $\tilde{w}$ is called a prolongation of the vector $W$.

In the vector space $M$ we introduce the weighted Euclidean norm

$$
\| W \| = \left( \sum_{i,j=1}^{N-1} w_{i,j}^2 + h^4 \left( \sum_{i,j=0}^{N} \left( (w_{xx})_{i,j}^2 + (w_{yy})_{i,j}^2 \right) \right) \right)^{1/2}.
$$

It is equivalent to the norm $|\tilde{w}|_{0,\Omega}$ of the prolongation up to the factor $h$ [7]:

$$
c^* h \| W \| \le |\tilde{w}|_{0,\Omega} \le c^{**} h \| W \|. \tag{15}
$$

Denote the prolongation of the vector

$$
V^h = \left( \left\{ v^h(z_{i,j}) \right\}_{i,j=1}^{N-1},\ \left\{ h^2 v_{xx}^h(z_{i,j}) \right\}_{i,j=0}^{N},\ \left\{ h^2 v_{yy}^h(z_{i,j}) \right\}_{i,j=0}^{N} \right)
$$

by $\tilde{v}^h$.

**Theorem 1.** *An approximate solution $\tilde{v}^h$ of problem (5)–(6) satisfies the estimate*

$$
|u - \tilde{v}^h|_{0,\Omega} \le ch^4 \| u \|_{4,\Omega}. \tag{16}
$$

*Proof.* We have system (12)–(13) for the unknowns $u^h$ and $h^2 u_{xx}^h$ and the system with the same matrix for the unknowns $v^h$ and $h^2 v_{xx}^h$. The only difference between these systems is in the terms

$$
\delta^{0,h} = \left( [A_2^T] + kh^2 [A_4^T] \right) h^2 r^h, \tag{17}
$$

$$
\delta^{1,h} = kh^4 [A_6] r^h. \tag{18}
$$

in the right-hand side.

Introduce the vectors

$$
U^h = \left( \left\{ u^h(z_{i,j}) \right\}_{i,j=1}^{N-1},\ \left\{ h^2 u_{xx}^h(z_{i,j}) \right\}_{i,j=0}^{N},\ \left\{ h^2 u_{yy}^h(z_{i,j}) \right\}_{i,j=0}^{N} \right),
$$

$$\bar{U}^h = \left( \{u^h(z_{i,j})\}_{i,j=1}^{N-1}, \ \{h^2 u_{xx}^h(z_{i,j})\}_{i,j=0}^{N} \right)$$

$$\bar{V}^h = \left( \{v^h(z_{i,j})\}_{i,j=1}^{N-1}, \ \{h^2 v_{xx}^h(z_{i,j})\}_{i,j=0}^{N} \right).$$

The matrix of system (12)–(13) is equivalent to the matrix of finite element system with less number of DoF which are reduced by equality (11). Therefore it has the same property of stability. Thus, taking into account (17)–(18), we have the estimate [6]

$$\| \bar{U}^h - \bar{V}^h \| \le ch^2 \| r^h \| . \tag{19}$$

All vector norms are the Euclidean norms in the vector spaces of corresponding dimension.

Now we estimate $\| U^h - V^h \|$. Using (11), (14), and the algebraic inequality $(a + b + c)^2 \le 4(a^2 + b^2 + c^2)$, we get

$$\| U^h - V^h \|^2 = \sum_{i,j=1}^{N-1} \left( u^h(z_{i,j}) - v^h(z_{i,j}) \right)^2$$

$$+ \sum_{i,j=0}^{N} \left( h^4 \left( u_{xx}^h(z_{i,j}) - v_{xx}^h(z_{i,j}) \right)^2 \right.$$

$$+ h^4 \left( k \left( u^h(z_{i,j}) - v^h(z_{i,j}) \right) + v_{xx}^h(z_{i,j}) - u_{xx}^h(z_{i,j}) + r^h(z_{i,j}) \right)^2 \right)$$

$$\le c \left( \sum_{i,j=1}^{N-1} \left( u^h(z_{i,j}) - v^h(z_{i,j}) \right)^2 + h^4 \sum_{i,j=0}^{N} \left( u_{xx}^h(z_{i,j}) - v_{xx}^h(z_{i,j}) \right)^2 \right)$$

$$+ ch^4 \sum_{i,j=0}^{N} \left( r^h(z_{i,j}) \right)^2 = c \| \bar{U}^h - \bar{V}^h \|^2 + ch^4 \| r^h \|^2 .$$

Combining this inequality with (19) gives

$$\| U^h - V^h \| \le ch^2 \| r^h \| . \tag{20}$$

From (5) and (11) we have

$$r^h(z_{i,j}) = u_{xx}^h(z_{i,j}) - \frac{\partial^2 u}{\partial x^2}(z_{i,j}) + u_{yy}^h(z_{i,j}) - \frac{\partial^2 u}{\partial y^2}(z_{i,j}) + k \left( u(z_{i,j}) - u^h(z_{i,j}) \right) .$$

Hence,

$$\| r^h \|^2 \le c \left( \sum_{i,j=1}^{N-1} \left( u^h(z_{i,j}) - u(z_{i,j}) \right)^2 + \sum_{i,j=0}^{N} \left( \left( u_{xx}^h(z_{i,j}) - \frac{\partial^2 u}{\partial x^2}(z_{i,j}) \right)^2 \right. \right.$$

$$\left. \left. + \left( u_{yy}^h(z_{i,j}) - \frac{\partial^2 u}{\partial y^2}(z_{i,j}) \right)^2 \right) \right)$$

$$\leq ch^{-4} \left( \sum_{i,j=1}^{N-1} \left(u^h(z_{i,j}) - u(z_{i,j})\right)^2 + \sum_{i,j=0}^{N} \left(h^4 \left(u_{xx}^h(z_{i,j}) - \frac{\partial^2 u}{\partial x^2}(z_{i,j})\right)^2 \right.\right.$$

$$\left.\left. + h^4 \left(u_{yy}^h(z_{i,j}) - \frac{\partial^2 u}{\partial y^2}(z_{i,j})\right)^2 \right) \right) = ch^{-4} \parallel U^h - U_I \parallel^2$$

where the vector $U_I$ consists of the coefficients of the interpolant $\tilde{u}_I$. Taking into account norm equivalence (15), we obtain

$$\parallel r^h \parallel \leq ch^{-3} |\tilde{u}^h - \tilde{u}_I|_{0,\Omega}. \tag{21}$$

For the interpolant $\tilde{u}_I$ we have estimate (4). Besides, the approximate solution $\tilde{u}^h$ satisfies the estimate [4]

$$|u - \tilde{u}^h|_{0,\Omega} \leq ch^4 \parallel u \parallel_{4,\Omega}. \tag{22}$$

Hence, with the help of the triangle inequality from (21) we get

$$\parallel r^h \parallel \leq ch^{-3} \left(|u - \tilde{u}^h|_{0,\Omega} + |u - \tilde{u}_I|_{0,\Omega}\right) \leq ch \parallel u \parallel_{4,\Omega}. \tag{23}$$

Finally, from (20), (15), and (23) we conclude that

$$|\tilde{u}^h - \tilde{v}^h|_{0,\Omega} \leq ch \parallel U^h - V^h \parallel \leq ch^4 \parallel u \parallel_{4,\Omega},$$

which together with (22) proves the theorem. $\qquad\square$

## 7  Numerical Results

Let $\Omega = (0,1) \times (0,1)$ and $\Gamma$ be the boundary of $\Omega$. Consider the boundary value problem

$$-\Delta u + 0.5u = f \quad \text{in } \Omega, \tag{24}$$

$$u = \begin{cases} 0 & \text{for } x = 0, \\ 0 & \text{for } y = 0, \\ -10000y \sin y & \text{for } x = 1, \\ -10000y \sin x & \text{for } y = 1 \end{cases} \tag{25}$$

with the right-hand side

$$f(x,y) = 5000 \left(4(x+y)\cos(1-x-y) + 5xy \sin(1-x-y)\right). \tag{26}$$

The exact solution of (24)–(26) has the form

$$u(x,y) = 10000xy \sin(1-x-y). \tag{27}$$

To construct a discrete problem, we subdivide $\Omega$ into elementary squares for mesh size $h = 1/N$. In numerical experiments we use $N = 5, 10, 20, 40, 80$. On the obtained partition, along with the standard system of the finite element method for Hermite element (1), we construct and solve the modified system of smaller dimension obtained with the help of collocation.

Since exact solution (27) is known, the error $u(z_{i,j}) - u^h(z_{i,j})$ or $u(z_{i,j}) - v^h(z_{i,j})$ of an approximate solution at the node $z_{i,j}$ can be calculated explicitly. Besides, we can calculate $\partial^2 u/\partial x^2(z_{i,j}) - u^h_{xx}(z_{i,j})$ and $\partial^2 u/\partial y^2(z_{i,j}) - u^h_{yy}(z_{i,j})$ or $\partial^2 u/\partial x^2(z_{i,j}) - v^h_{xx}(z_{i,j})$ and $\partial^2 u/\partial y^2(z_{i,j}) - v^h_{yy}(z_{i,j})$ to estimate convergence of the second-order derivatives. We use the discrete norms

$$\| u^h \|_h = \left( \sum_{i,j=1}^{N-1} \left( u^h(z_{i,j}) \right)^2 h^2 \right)^{1/2},$$

$$\| u - u^h \|_h = \left( \sum_{i,j=1}^{N-1} \left( u(z_{i,j}) - u^h(z_{i,j}) \right)^2 h^2 \right)^{1/2}.$$

In addition, we also use the following discrete norm for a relative error:

$$\| u - u^h \|_h^{rel} = \frac{\| u - u^h \|_h}{\| u^h \|_h}.$$

In Tables 1, 2 and 3 we present results of calculations for the standard system of the finite element method. We use the following notations:

$$\varepsilon^h(u^h) = \| u - u^h \|_h, \qquad \bar{\varepsilon}^h(u^h) = \| u - u^h \|_h^{rel},$$

$$\sigma_x^h(u^h) = \left\| \frac{\partial^2 u}{\partial x^2} - u^h_{xx} \right\|_h, \qquad \bar{\sigma}_x^h(u^h) = \left\| \frac{\partial^2 u}{\partial x^2} - v^h_{xx} \right\|_h^{rel},$$

**Table 1.** Accuracy of the approximate solution for the standard system.

| $h$ | $\varepsilon^h(u^h)$ | $\sigma_x^h(u^h)$ | $\sigma_y^h(u^h)$ | $\sigma_\Delta^h(u^h)$ |
|---|---|---|---|---|
| 0.2 | $4.22E-1$ | $1.46E+2$ | $1.46E+2$ | $2.86E+2$ |
| 0.1 | $2.97E-2$ | $3.90E+1$ | $3.90E+1$ | $7.60E+1$ |
| 0.05 | $1.97E-3$ | $9.93E+0$ | $9.93E+0$ | $1.94E+1$ |
| 0.025 | $1.27E-4$ | $2.51E+0$ | $2.51E+0$ | $4.91E+0$ |
| 0.0125 | $7.17E-6$ | $6.30E-1$ | $6.30E-1$ | $1.24E+0$ |

**Table 2.** Relative error for the standard system.

| $h$ | $\bar{\varepsilon}^h(u^h)$ | $\bar{\sigma}_x^h(u^h)$ | $\bar{\sigma}_y^h(u^h)$ | $\bar{\sigma}_\Delta^h(u^h)$ |
|---|---|---|---|---|
| 0.2 | 1.86E−4 | 1.50E−2 | 1.50E−2 | 1.68E−2 |
| 0.1 | 1.47E−5 | 4.02E−3 | 4.02E−3 | 4.38E−3 |
| 0.05 | 1.05E−6 | 1.02E−3 | 1.02E−3 | 1.11E−3 |
| 0.025 | 6.99E−8 | 2.58E−4 | 2.58E−4 | 2.78E−4 |
| 0.0125 | 4.03E−9 | 6.49E−5 | 6.49E−5 | 6.98E−5 |

**Table 3.** A numerical convergence estimate for the standard system.

| $h$ | $\log\left(\dfrac{\varepsilon^h(u^{2h})}{\varepsilon^h(u^h)}\right)$ | $\log\left(\dfrac{\sigma_x^h(u^{2h})}{\sigma_x^h(u^h)}\right)$ | $\log\left(\dfrac{\sigma_y^h(u^{2h})}{\sigma_y^h(u^h)}\right)$ | $\log\left(\dfrac{\sigma_\Delta^h(u^{2h})}{\sigma_\Delta^h(u^h)}\right)$ |
|---|---|---|---|---|
| 0.1 | 3.8 | 1.9 | 1.9 | 1.9 |
| 0.05 | 3.9 | 2.0 | 2.0 | 2.0 |
| 0.025 | 4.0 | 2.0 | 2.0 | 2.0 |
| 0.0125 | 4.1 | 2.0 | 2.0 | 2.0 |

$$\sigma_y^h(u^h) = \left\|\frac{\partial^2 u}{\partial y^2} - u_{yy}^h\right\|_h, \qquad \bar\sigma_y^h(u^h) = \left\|\frac{\partial^2 u}{\partial y^2} - v_{yy}^h\right\|_h^{rel},$$

$$\sigma_\Delta^h(u^h) = \left\|\Delta u - \Delta u^h\right\|_h, \qquad \bar\sigma_\Delta^h(u^h) = \left\|\Delta u - \Delta u^h\right\|_h^{rel},$$

where $\Delta u^h(z_{i,j}) = u_{xx}^h(z_{i,j}) + u_{yy}^h(z_{i,j})$ is an approximate value of $\Delta u$ at the node.

Observe that the error $\varepsilon^h(u^h)$ tends to be of order 4 as $h$ decreases while $\sigma_x^h(u^h)$ and $\sigma_y^h(u^h)$ tend to be of order 2. This is in good agreement with theoretical estimate (22).

Another series of calculations is related to the solution of the reduced system for the unknowns $v^h(z_{i,j})$ and $v_{xx}^h(z_{i,j})$. In this case the values of $v_{yy}^h$ are determined from (14). Besides, from (14) we also directly determine the approximate values of $\Delta u$ at the nodes:

$$\Delta v^h(z_{i,j}) = k v^h(z_{i,j}) - f^h(z_{i,j}). \tag{28}$$

The results of calculations are presented in Tables 4, 5 and 6.

**Table 4.** Accuracy of the approximate solution for the reduced system.

| $h$ | $\varepsilon^h(v^h)$ | $\sigma_x^h(v^h)$ | $\sigma_y^h(v^h)$ | $\sigma_\Delta^h(v^h)$ |
|---|---|---|---|---|
| 0.2 | 2.45E−1 | 1.39E+2 | 1.39E+2 | 1.22E−1 |
| 0.1 | 1.87E−2 | 3.84E+1 | 3.84E+1 | 9.34E−3 |
| 0.05 | 1.21E−3 | 9.88E+0 | 9.88E+0 | 6.04E−4 |
| 0.025 | 7.59E−5 | 2.50E+0 | 2.50E+0 | 3.79E−5 |
| 0.0125 | 5.18E−6 | 6.30E−1 | 6.30E−1 | 2.59E−6 |

**Table 5.** Relative error for the reduced system.

| $h$ | $\bar\varepsilon^h(v^h)$ | $\bar\sigma_x^h(v^h)$ | $\bar\sigma_y^h(v^h)$ | $\bar\sigma_\Delta^h(v^h)$ |
|---|---|---|---|---|
| 0.2 | 1.07E−4 | 1.44E−2 | 1.44E−2 | 7.20E−6 |
| 0.1 | 9.28E−6 | 3.96E−3 | 3.96E−3 | 5.38E−7 |
| 0.05 | 6.42E−7 | 1.02E−3 | 1.02E−3 | 3.44E−8 |
| 0.025 | 4.19E−8 | 2.58E−4 | 2.58E−4 | 2.15E−9 |
| 0.0125 | 2.92E−9 | 6.49E−5 | 6.49E−5 | 1.46E−10 |

**Table 6.** A numerical convergence estimate for the reduced system.

| $h$ | $\log\left(\dfrac{\varepsilon^h(v^{2h})}{\varepsilon^h(v^h)}\right)$ | $\log\left(\dfrac{\sigma_x^h(v^{2h})}{\sigma_x^h(v^h)}\right)$ | $\log\left(\dfrac{\sigma_y^h(v^{2h})}{\sigma_y^h(v^h)}\right)$ | $\log\left(\dfrac{\sigma_\Delta^h(v^{2h})}{\sigma_\Delta^h(v^h)}\right)$ |
|---|---|---|---|---|
| 0.1 | 3.7 | 1.9 | 1.9 | 3.7 |
| 0.05 | 4.0 | 2.0 | 2.0 | 4.0 |
| 0.025 | 4.0 | 2.0 | 2.0 | 4.0 |
| 0.0125 | 3.9 | 2.0 | 2.0 | 3.9 |

Notice that the errors $\varepsilon^h(v^h)$, $\sigma_x^h(v^h)$, and $\sigma_y^h(v^h)$ are close to those for $u^h$ which confirms estimate (16). Moreover, in the case of the reduced system the accuracy is even slightly better. In addition, from (28) we can calculate approximate values of $\Delta u$ within $O(h^4)$.

Thus, numerical experiments confirm the theoretical estimate and demonstrate the advantage of the proposed approach.

**Acknowledgements.** Supported by Project 17-01-00270 of Russian Foundation for Basic Research.

# References

1. Adams, R., Fournier, I.: Sobolev Spaces. Academic Press, New York (2003)
2. Brenner, S., Scott, L.: The Mathematical Theory of Finite Element Method. TAM. Springer, Berlin (1994). https://doi.org/10.1007/978-1-4757-4338-8
3. Ciarlet, P.: The Finite Element Method for Elliptic Problems. North Holland, Amsterdam (1978)
4. Gileva, L., Karepova, E., Shaidurov, V.: New Hermite finite elements on rectangles. In: Todorov, M.D. (ed.) AIP Conference Proceedings Application of Mathematics in Technical and Natural Sciences: 8th International Conference for Promoting the Application of Mathematics in Technical and Natural Sciences - AMiTaNS 2016, vol. 1773, pp. 100005-1–100005-7. AIP Publishing, NY, USA (2016)
5. Gileva, L., Shaydurov, V.: Bicubic Hermite elements in a domain with the curved boundary. Lobachevskii J. Math. **39**(7), 893–903 (2018)
6. Samarskii, A.: The Theory of Difference Schemes. CRC Press, Boca Raton (2001)
7. Shaidurov, V.: Multigrid Method for Finite Elements. Kluwer Academic Publishers, Dordrecht (1995)
8. Shaidurov, V., Shut, S., Gileva, L.: Some properties of Hermite finite elements on rectangles. In: Todorov, M.D. (ed.) AIP Conference Proceedings Application of Mathematics in Technical and Natural Sciences: 6th International Conference for Promoting the Application of Mathematics in Technical and Natural Sciences - AMiTaNS 2014, vol. 1629, pp. 32–43. AIP Publishing, NY, USA (2014)

# A 2nd-Order Numerical Scheme for Fractional Ordinary Differential Equation Systems

W. Li and S. Wang$^{(\boxtimes)}$

Department of Mathematics and Statistics, Curtin University, GPO Box U1987,
Perth 6845, Australia
{Wen.Li,Song.Wang}@curtin.edu.au

**Abstract.** We propose a new numerical method for fractional ordinary differential equation systems based on a judiciously chosen quadrature point. The proposed method is efficient and easy to implement. We show that the convergence order of the method is 2. Numerical results are presented to demonstrate that the computed rates of convergence confirm our theoretical findings.

## 1 Introduction

We consider the following system of fractional ordinary differential equations:

$$_0D_t^\alpha x(t) = f(t, x(t)), \ t \in (0, T], \text{ satisfying } x(0) = x^0, \tag{1}$$

where $x(t) = (x_1(t), x_2(t), \ldots, x_n(t))^\top \in \mathbb{R}^n$ for a positive integer $n$, $T > 0$ is a fixed constant, $f : \mathbb{R}^{n+1} \mapsto \mathbb{R}^n$ a given mapping, $x^0 \in \mathbb{R}^n$ a given initial condition, and $_0D_t^\alpha x(t) = (_0D_t^{\alpha_1} x_1(t), _0D_t^{\alpha_2} x_2(t), \ldots, _0D_t^{\alpha_n} x_n(t))^\top$ for $\alpha_i \in (0, 1)$ with $_0D_t^{\alpha_i} x_i(t)$ denoting the following Caputo's $\alpha_i$-th derivative

$$_0D_t^{\alpha_i} x_i(t) = \frac{1}{\Gamma(1 - \alpha_i)} \int_0^t \frac{x_i'(\tau)}{(t - \tau)^{\alpha_i}} \, d\tau$$

for $t > 0$ and $i = 1, 2, \ldots, n$, where $\Gamma(\cdot)$ denotes the Gamma function.

In the open literature, there are a number of methods for solving (1). Adomian decomposition method [3,5,9], variational iteration method [11,12], differential transform method [2] and homotopy analysis method [10,14] have been used for the problem. Recently, we proposed a new one-step numerical integration scheme for (1) [8]. This method is easy to implement and computationally inexpensive. In this paper, we will show that the global error of the method is of order $\mathcal{O}(h^2)$, where $h$ denotes the maximal mesh size to be defined.

The rest of the paper is organized as the follows. In Sect. 2, we first transform Eq. (1) into an equivalent Volterra integral equation, we then propose an approximation of Volterra integral equation based on a Taylor expansion. An error analysis of the approximation is also presented. In Sect. 3, we propose an algorithm for implementing the approximate equation and analyse its convergence. In Sect. 4, numerical examples are presented. Section 5 concludes the paper.

© Springer Nature Switzerland AG 2019
I. Dimov et al. (Eds.): FDM 2018, LNCS 11386, pp. 60–71, 2019.
https://doi.org/10.1007/978-3-030-11539-5_6

## 2    Approximation

We first rewrite (1) as the following Volterra integral equation:

$$x_i(t) = x_i^0 + \frac{1}{\Gamma(\alpha_i)} \int_0^t (t-\tau)^{\alpha_i - 1} f_i(\tau, x(\tau))\, d\tau, \tag{2}$$

for $t \in (0, T]$ and $i = 1, 2, ..., n$. It has been proven [1,4,6] that solving Eq. (1) is equivalent to solving (2). In this section, we will develop a numerical method based on a Taylor expansion to approximate (2) and estimate the approximation error.

Let $N$ be a given positive integer. We divide $(0, T)$ into $N$ sub-intervals with mesh points $t_i = ih$ for $i = 0, 1, ..., N$, where $h = T/N$. Thus, we have

$$x_i(t_j) = x_i^0 + \frac{1}{\Gamma(\alpha_i)} \int_0^{jh} (jh - \tau)^{\alpha_i - 1} f_i(\tau, x(\tau))\, d\tau$$

$$= x_i^0 + \frac{1}{\Gamma(\alpha_i)} \sum_{k=1}^{j} \int_{(k-1)h}^{kh} (jh - \tau)^{\alpha_i - 1} f_i(\tau, x(\tau))\, d\tau. \tag{3}$$

To approximate the integral on the RHS of (3), we assume that $f_i(t, x(t))$ is twice continuously differentiable with respect to both $t$ and $x$. For any $k$, we expend $f_i(\tau, x(\tau))$ at any point in $((k-1)h, kh)$, denoted as $\tau_{jk}^i$, into

$$f_i(\tau, x(\tau)) = f_i(\tau_{jk}^i, x(\tau_{jk}^i)) + K_{jk}^i(\tau - \tau_{jk}^i) + c_{jk}^i(\tau - \tau_{jk}^i)^2, \tag{4}$$

where $c_{jk}^i$ is the coefficient of the reminder of the expansion and

$$K_{jk}^i = \frac{\partial f_i}{\partial \tau}\Big|_{(\tau_{jk}^i, x(\tau_{jk}^i))} + \sum_{l=1}^{n} \frac{\partial f_i}{\partial x_l}\Big|_{(\tau_{jk}^i, x(\tau_{jk}^i))} \frac{\partial x_l}{\partial \tau}\Big|_{(\tau_{jk}^i)}.$$

Therefore, replacing $f_i(\tau, x(\tau))$ in (3) with the RHS of (4) and by direct integration we have

$$\frac{1}{\Gamma(\alpha_i)} \int_{(k-1)h}^{kh} (jh - \tau)^{\alpha_i - 1} f_i(\tau, x(\tau))\, d\tau$$

$$= \frac{1}{\Gamma(\alpha_i)} \int_{(k-1)h}^{kh} (jh - \tau)^{\alpha_i - 1} [f_i(\tau_{jk}^i, x(\tau_{jk}^i)) + K_{jk}^i(\tau - \tau_{jk}^i)]\, d\tau + R_{jk}^i$$

$$= \frac{h^{\alpha_i}}{\Gamma(\alpha_i + 1)} f_i(\tau_{jk}^i, x(\tau_{jk}^i))[(j - k + 1)^{\alpha_i} - (j - k)^{\alpha_i}]$$

$$+ \frac{K_{jk}^i}{\Gamma(\alpha_i)} \int_{(k-1)h}^{kh} (jh - \tau)^{\alpha_i - 1}(\tau - \tau_{jk}^i)\, d\tau + R_{jk}^i, \tag{5}$$

where $R_{jk}^i = \frac{1}{\Gamma(\alpha_i)} \int_{(k-1)h}^{kh} (jh - \tau)^{\alpha_i - 1} c_{jk}^i (\tau - \tau_{jk}^i)^2\, d\tau$.

From (5) it is clear that $\tau_{jk}^i$ should be chosen such that the integral term in (5) becomes zero so that the truncation error is $R_{jk}^i$. The choice of $\tau_{jk}^i$ is given in the following theorem.

**Theorem 1.** *For any feasible $j$ and $k$, the unique solution to*

$$\int_{(k-1)h}^{kh} (jh - \tau)^{\alpha_i - 1}(\tau - \tau_{jk}^i)d\tau = 0$$

*is*

$$\tau_{jk}^i = h\frac{[(j-k+1)^{\alpha_i+1} - (j-k)^{\alpha_i+1}] + (\alpha_i + 1)[(j-k+1)^{\alpha_i}(k-1) - (j-k)^{\alpha_i}k]}{(\alpha_i + 1)[(j-k+1)^{\alpha_i} - (j-k)^{\alpha_i}]}.$$

$$(6)$$

*Furthermore, $(k-1)h < \tau_{jk}^i < kh$.*

*Proof.* See the proof of Theorem 2.1 in [7].

Substituting the expression for $\tau_{jk}^i$ in (6) into (5) and combining the resulting expression with (3), we have the following representation for $x_i(t_j)$.

$$x_i(t_j) = x_i^0 + \frac{h^{\alpha_i}}{\Gamma(\alpha_i + 1)} \sum_{k=1}^{j} f_i(\tau_{jk}^i, x(\tau_{jk}^i))[(j-k+1)^{\alpha_i} - (j-k)^{\alpha_i}] + R_j^i, \quad (7)$$

for $j = 1, 2, \ldots, N$, where $\tau_{jk}^i$ is given in (6) for $k = 1, 2, \ldots, j$ and $R_j^i = \sum_{k=1}^{j} R_{jk}^i$. Omitting $R_j^i$ in (7), we have an approximation to (3) with the truncation error $R_j^i$. An upper bound for $R_j^i$ is given in the following theorem.

**Theorem 2.** *If $f(t, x)$ is twice continuously differentiable in $t$ and $x$, then we have $|R_j^i| \leq Ch^2$, where $C$ denotes a positive constant independent of $h$.*

*Proof.* See the proof of Theorem 2.2 in [7].

From (7) it is clear that to compute $x_i(t_j)$, we need to calculate $f_i(\tau_{jk}^i, x(\tau_{jk}^i))$. However, $x(\tau_{jk}^i)$ is not available directly from the scheme. Thus, approximations to $x(\tau_{jk}^i)$ need to be determined. In the next section, we propose a single step numerical scheme for implementing (7) when the remainder $R_j^i$ is omitted.

## 3    Algorithm and Its Convergence

For any $j$ and $k$ satisfying $1 \leq k \leq j \leq N$, since $\tau_{jk}^i \in (t_{k-1}, t_k)$ by Theorem 1, we use the following linear interpolation to approximate $x_i(\tau_{jk}^i)$:

$$x(\tau_{jk}^i) = x(t_{k-1}) + \rho_{jk}^i(x(t_k) - x(t_{k-1})) + \mathcal{O}(h^2)E_n, \quad (8)$$

where $\rho_{jk}^i := \frac{\tau_{jk}^i - t_{k-1}}{h} \in (0, 1)$ and $E_n = (1, 1, \ldots, 1)^\top \in \mathbb{R}^n$. Using (8), we approximate $f_i(\tau_{jk}^i, x(\tau_{jk}^i))$ as follows.

$$f_i(\tau_{jk}^i, x(\tau_{jk}^i)) = f_i\left(\tau_{jk}^i, x(t_{k-1}) + \rho_{jk}^i(x(t_k) - x(t_{k-1}))\right) + \mathcal{O}(h^2). \quad (9)$$

Replacing $f_i(\tau_{jk}^i, x(\tau_{jk}^i))$ in (7) with the RHS of (9), we have

$$x_i(t_j) = x_i^0 + h_{\alpha_i} \sum_{k=1}^{j} \left[ f_i\left(\tau_{jk}^i, x(t_{k-1}) + \rho_{jk}^i(x(t_k) - x(t_{k-1}))\right) \right.$$

$$\left. \cdot \left((j-k+1)^{\alpha_i} - (j-k)^{\alpha_i}\right) \right] + \mathcal{O}(h^2) \tag{10}$$

for $j = 1, 2, \ldots, N$, where $h_{\alpha_i} = \frac{h^{\alpha_i}}{\Gamma(\alpha_i+1)}$ and $\tau_{jk}^i$ is defined in (6). Clearly, (10) defines a time-stepping scheme for (2) if we omit the term $\mathcal{O}(h^2)$.

The above scheme is implicit as it is a nonlinear system in $x(t_j)$. We now define an explicit single step scheme by further approximating the $j$th term in the sum in (10) by the following Taylor expansion:

$$f_i(\tau_{jj}^i, x(t_{j-1}) + \rho_{jj}^i(x(t_j) - x(t_{j-1})))$$

$$= f_i(\tau_{jj}^i, x(t_{j-1})) + \sum_{l=1}^{n} \left.\frac{\partial f_i}{\partial x_l}\right|_{(\tau_{jj}^i, x(t_{j-1}))} (\rho_{jj}^i(x_l(t_j) - x_l(t_{j-1}))) + \mathcal{O}(h^2). \tag{11}$$

Thus, combining (11) and (10) yields

$$x_i(t_j) = x_i^0 + h_{\alpha_i} \sum_{k=1}^{j-1} \left[ f_i\left(\tau_{jk}^i, x(t_{k-1}) + \rho_{jk}^i(x(t_k) - x(t_{k-1}))\right) \right.$$

$$\left.\left((j-k+1)^{\alpha_i} - (j-k)^{\alpha_i}\right)\right] + h_{\alpha_i} f_i\left(\tau_{jj}^i, x(t_{j-1})\right))$$

$$+ h_{\alpha_i} \sum_{l=1}^{n} \left[\left.\frac{\partial f_i}{\partial x_l}\right|_{(\tau_{jj}^i, x(t_{j-1}))} (\rho_{jj}^i(x_l(t_j) - x_l(t_{j-1})))\right] + \mathcal{O}(h^2). \tag{12}$$

Let $x^j := (x_1^j, x_2^j, \ldots, x_n^j)^\top$ for $j = 0, 1, \ldots, N$ and omitting the truncation error terms of order $\mathcal{O}(h^2)$ in (12), we define the following single step time-stepping scheme for approximating (2):

$$x_i^j = x_i^0 + h_{\alpha_i} \sum_{k=1}^{j-1} \left[ f_i\left(\tau_{jk}^i, x^{k-1} + \rho_{jk}^i(x^k - x^{k-1})\right)\left((j-k+1)^{\alpha_i} - (j-k)^{\alpha_i}\right)\right]$$

$$+ h_{\alpha_i} f_i\left(\tau_{jj}^i, x^{j-1}\right) + h_{\alpha_i} \sum_{l=1}^{n} \left[\left.\frac{\partial f_i}{\partial x_l}\right|_{(\tau_{jj}^i, x^{j-1})} (\rho_{jj}^i(x_l^j - x_l^{j-1}))\right] \tag{13}$$

Re-organising (13), we have the following linear system for $x^j$:

$$B^j x^j = C^j, \quad j = 1, 2, \ldots, N. \tag{14}$$

where $B^j$ is the $n \times n$ matrix given by

$$B^j = \begin{pmatrix} 1 - b_{11}^j & -b_{12}^j & \cdots & -b_{1n}^j \\ -b_{21}^j & 1 - b_{22}^j & \cdots & -b_{2n}^j \\ \vdots & \vdots & \ddots & \vdots \\ -b_{n1}^j & -b_{n2}^j & \cdots & 1 - b_{nn}^j \end{pmatrix} \tag{15}$$

with

$$b_{il}^j = \rho_{jj}^i h_{\alpha_i} \left. \frac{\partial f_i}{\partial x_l} \right|_{(\tau_{jj}^i, x^{j-1})} \tag{16}$$

for $i = 1, 2, \ldots, n$, $l = 1, 2, \ldots, n$ and $C^j = (c_1^j, c_2^j, \ldots, c_n^j)^\top$ with

$$c_i^j = x_i^0 + h_{\alpha_i} \sum_{k=1}^{j-1} \left[ f_i \left( \tau_{jk}^i, x^{k-1} + \rho_{jk}^i (x^k - x^{k-1}) \right) \left( (j-k+1)^{\alpha_i} - (j-k)^{\alpha_i} \right) \right]$$

$$+ h_{\alpha_i} f_i \left( \tau_{jj}^i, x^{j-1} \right) - \sum_{l=1}^{n} x_l^{j-1} b_{il}^j. \tag{17}$$

It is clear that to calculate $x^j$, we need to solve the system of equations (14)–(17). It has been shown in [8] that (14)–(17) is uniquely solvable when $h$ is sufficiently small.

For a given initial condition $x^0$, using the above results, we propose the following algorithm for solving (3) numerically.

## Algorithm A

1. For a given positive integer $N$, let $t_j = jh$ for $j = 0, 1, \ldots, N$, where $h = T/N$.
2. Calculate $x^j$ for $j = 1, \ldots, N$ using (14)–(17).

Using a linear interpolation and Taylor's theorem, we are able to prove in the following theorem that, for any $j = 1, 2, \ldots, N$, $x^j$ generated by the above algorithm converges to the solution of (2) at the rate $\mathcal{O}(h^2)$ when $h \to 0^+$.

**Theorem 3.** *Let $x(t_j)$ and $x^j$ be respectively the solution to (3) and the sequence generated by Algorithm A. If $f(t, x)$ is twice continuously differentiable in $t$ and $x$, then there exists an $\bar{h} > 0$ such that when $h < \bar{h}$*

$$\|x(t_j) - x^j\|_\infty \le Ch^2, \quad j = 1, 2, \ldots, N. \tag{18}$$

*Proof.* In what follows, we let $C$ denote a generic positive constant, independent of $h$. We now prove this theorem by mathematical induction.

When $j = 1$, from (12) we have

$$x_i(t_1) = x_i^0 + h_{\alpha_i} \left[ f_i(\tau_{11}^i, x^0) + \sum_{l=1}^{n} \left. \frac{\partial f_i}{\partial x_l} \right|_{(\tau_{11}^i, x^0)} (\rho_{11}^i (x_l(t_1) - x_l^0)) \right] + \mathcal{O}(h^2) \tag{19}$$

Re-organising (19) and using the definitions for $B^j$ and $C^j$, we have

$$B^1 x(t_1) = C^1 + \mathcal{O}(h^2) E_n. \tag{20}$$

Solving (20) yields

$$x(t_1) = (B^1)^{-1} C^1 + \mathcal{O}(h^2)(B^1)^{-1} E_n.$$

From (14)–(17), we see that

$$x^1 = (B^1)^{-1}C^1.$$

Therefore,

$$\|x(t_1) - x^1\|_\infty = \mathcal{O}(h^2)\|(B^1)^{-1}E_n\|_\infty \leq Ch^2\|(B^1)^{-1}\|_\infty.$$

It has been proven [8] that $B^j, j = 1, 2, \ldots, N$ satisfies, when $h < \bar{h}$,

$$\sigma^j := \min_{1 \leq i \leq n} \left\{ |b_{ii}^j| - \sum_{j=1, j \neq i}^{n} |b_{ij}^j| \right\} \geq \beta > 0 \tag{21}$$

for a constant $\beta$, independent of $h$, where $\bar{h} = \min_{1 \leq i \leq n} \left( \frac{\Gamma(\alpha_i + 1)}{nM} \right)^{\frac{1}{\alpha_i}}$ and $M = \max_{\substack{1 \leq i \leq n \\ 1 \leq l \leq n}} \left| \frac{\partial f_i}{\partial x_l} \right|$. Thus, using [13] and (21), we have

$$\|(B^1)^{-1}\|_\infty \leq \frac{1}{\sigma^1} \leq \frac{1}{\beta}.$$

Therefore, we have

$$\|x(t_1) - x^1\|_\infty \leq Ch^2.$$

When $i \geq 2$ and $h \leq \bar{h}$, we assume that

$$\|x(t_j) - x^j\|_\infty \leq Ch^2, \quad 1 \leq j \leq i - 1. \tag{22}$$

We now show that $\|x(t_j) - x^j\|_\infty \leq Ch^2$ for $1 \leq j \leq i$.

Note that (12) can be re-written in the following form:

$$x_i(t_j) = x_i^0 + A_i^j + D_i^j + \mathcal{O}(h^2), \tag{23}$$

where

$$A_i^j = h\alpha_i \sum_{k=1}^{j-1} \left[ f_i \left( \tau_{jk}^i, x(t_{k-1}) + \rho_{jk}^i(x(t_k) - x(t_{k-1})) \right) \left( (j-k+1)^{\alpha_i} - (j-k)^{\alpha_i} \right) \right], \tag{24}$$

$$D_i^j = h\alpha_i f_i \left( \tau_{jj}^i, x(t_{j-1}) \right) + h\alpha_i \sum_{l=1}^{n} \left[ \frac{\partial f_i}{\partial x_l} |_{(\tau_{jj}^i, x(t_{j-1}))} (\rho_{jj}^i (x_l(t_j) - x_l(t_{j-1}))) \right]. \tag{25}$$

Similarly, (13) can be re-written as follows.

$$x_i^j = x_i^0 + \tilde{A}_i^j + \tilde{D}_i^j, \tag{26}$$

where

$$\tilde{A}_i^j = h\alpha_i \sum_{k=1}^{j-1} \left[ f_i \left( \tau_{jk}^i, x^{k-1} + \rho_{jk}^i(x^k - x^{k-1}) \right) \left( (j-k+1)^{\alpha_i} - (j-k)^{\alpha_i} \right) \right], \tag{27}$$

$$\tilde{D}_i^j = h_{\alpha_i} f_i\left(\tau_{jj}^i, x^{j-1}\right) + h_{\alpha_i} \sum_{l=1}^n \left[\frac{\partial f_i}{\partial x_l}\big|_{(\tau_{jj}^i, x^{j-1})}(\rho_{jj}^i(x_l^j - x_l^{j-1}))\right]. \tag{28}$$

Subtracting (26) from (23) gives

$$x_i(t_j) - x_i^j = (A_i^j - \tilde{A}_i^j) + (D_i^j - \tilde{D}_i^j) + \mathcal{O}(h^2). \tag{29}$$

Let us first estimate $D_i^j - \tilde{D}_i^j$. From (25) and (28), we have

$$
\begin{aligned}
D_i^j - \tilde{D}_i^j &= \left[h_{\alpha_i} f_i\left(\tau_{jj}^i, x(t_{j-1})\right) + h_{\alpha_i} \sum_{l=1}^n \left[\frac{\partial f_i}{\partial x_l}\big|_{(\tau_{jj}^i, x(t_{j-1}))}(\rho_{jj}^i(x_l(t_j) - x_l(t_{j-1})))\right]\right] \\
&\quad - \left[h_{\alpha_i} f_i\left(\tau_{jj}^i, x^{j-1}\right) + h_{\alpha_i} \sum_{l=1}^n \left[\frac{\partial f_i}{\partial x_l}\big|_{(\tau_{jj}^i, x^{j-1})}(\rho_{jj}^i(x_l^j - x_l^{j-1}))\right]\right] \\
&= h_{\alpha_i}[f_i(\tau_{jj}^i, x(t_{j-1})) - f_i(\tau_{jj}^i, x^{j-1})] \\
&\quad + h_{\alpha_i}\rho_{jj}^i\left[\sum_{l=1}^n \left[\frac{\partial f_i}{\partial x_l}\big|_{(\tau_{jj}^i, x(t_{j-1}))}(x_l(t_j)) - \frac{\partial f_i}{\partial x_l}\big|_{(\tau_{jj}^i, x^{j-1})}(x_l^j)\right]\right] \\
&\quad - h_{\alpha_i}\rho_{jj}^i\left[\sum_{l=1}^n \left[\frac{\partial f_i}{\partial x_l}\big|_{(\tau_{jj}^i, x(t_{j-1}))}(x_l(t_{j-1})) - \frac{\partial f_i}{\partial x_l}\big|_{(\tau_{jj}^i, x^{j-1})}(x_l^{j-1})\right]\right]. \tag{30}
\end{aligned}
$$

Since $f_i$ is twice continuously differentiable, using a Taylor expansion we get

$$\frac{\partial f_i}{\partial x_l}\big|_{(\tau_{jj}^i, x(t_{j-1}))} = \frac{\partial f_i}{\partial x_l}\big|_{(\tau_{jj}^i, x^{j-1})} + r_j^i, \tag{31}$$

where

$$r_j^i = \sum_{p=1}^n \frac{\partial^2 f_i}{\partial x_l \partial x_p}\big|_{(\tau_{jj}^i, \xi)}(x_p(t_{j-1}) - x_p^{t-1}),$$

where $\xi = x(t_{j-1}) + \theta(x(t_{j-1}) - x^{j-1})$ with $\theta \in (0,1)$. From the assumption (22) we have $r_j^i = \mathcal{O}(h^2)$. Similarly, since $f$ is twice differentiable, using (22) it is easy to show $f_i(\tau_{jj}^i, x_i(t_{i-1})) - f_i(\tau_{jj}^i, x_i^{j-1}) = \mathcal{O}(h^2)$.

Using (31) and the above estimates we have, from (30),

$$
\begin{aligned}
D_i^j - \tilde{D}_i^j &= h_{\alpha_i}\mathcal{O}(h^2) + h_{\alpha_i}\rho_{jj}^i\left[\sum_{l=1}^n \left[\frac{\partial f_i}{\partial x_l}\big|_{(\tau_{jj}^i, x^{j-1})}(x_l(t_j) - x_l^j + x_l(t_j)r_j^i)\right]\right] \\
&\quad - h_{\alpha_i}\rho_{jj}^i\left[\sum_{l=1}^n \left[\frac{\partial f_i}{\partial x_l}\big|_{(\tau_{jj}^i, x^{j-1})}(x_l(t_{j-1}) - x_l^{j-1} + x_l(t_{j-1})r_j^i)\right]\right] \\
&= h_{\alpha_i}\rho_{jj}^i\left[\sum_{l=1}^n \left[\frac{\partial f_i}{\partial x_l}\big|_{(\tau_{jj}^i, x^{j-1})}(x_l(t_j) - x_l^j)\right]\right] \\
&\quad + h_{\alpha_i}\rho_{jj}^i\left[\sum_{l=1}^n \left[\frac{\partial f_i}{\partial x_l}\big|_{(\tau_{jj}^i, x^{j-1})}(x_l^{j-1} - x_l(t_{j-1}))\right]\right] + \mathcal{O}(h^{2+\alpha_i}) \\
&= h_{\alpha_i}\rho_{jj}^i\left[\sum_{l=1}^n \left[\frac{\partial f_i}{\partial x_l}\big|_{(\tau_{jj}^i, x^{j-1})}(x_l(t_j) - x_l^j)\right]\right] + \mathcal{O}(h^{2+\alpha_i}),
\end{aligned}
$$

since $h_{\alpha_i} = h^{\alpha_i}/\Gamma(1 + \alpha_i)$ and (22). Thus, from the above expression and (29), we get

$$x_i(t_j) - x_i^j = (A_i^j - \tilde{A}_i^j) + h_{\alpha_i}\rho_{jj}^i \left[ \sum_{l=1}^n \left[ \frac{\partial f_i}{\partial x_l}|_{(\tau_{jj}^i, x^{j-1})}(x_l(t_j) - x_l^j) \right] \right] + \mathcal{O}(h^2).$$

Re-organising the above equation gives

$$B^j\big(x(t_j) - x^j\big) = A^j + \mathcal{O}(h^2)E_n, \quad j = 1, 2, \ldots, N,$$

where $B^j$ is defined in (15)–(16) and $A^j = (A_1^j - \tilde{A}_1^j, A_2^j - \tilde{A}_2^j, \ldots, A_n^j - \tilde{A}_n^j)^\top$. From this equation we have

$$x(t_j) - x^j = (B^j)^{-1}\left(A^j + \mathcal{O}(h^2)E_n\right), \quad j = 1, 2, \ldots, N.$$

Thus, we have

$$\|x(t_j) - x^j\|_\infty = \|(B^j)^{-1}(A^j + \mathcal{O}(h^2)E_n)\|_\infty \leq \|(B^j)^{-1}\|_\infty \left(\|A^j\|_\infty + \mathcal{O}(h^2)\right)$$

for $j = 1, 2, \ldots, N$. Using [13] and (21), we have

$$\|(B^j)^{-1}\|_\infty \leq \frac{1}{\sigma^j} \leq \frac{1}{\beta}.$$

Therefore, we obtain

$$\|x(t_j) - x^j\|_\infty \leq \frac{1}{\beta}\|A^j\|_\infty + Ch^2. \tag{32}$$

We now examine $\|A^j\|_\infty = \|A_i^j - \tilde{A}_i^j\|_\infty$. For notational simplicity, we let $x^{jk} = x^{k-1} + \rho_{jk}^i(x^k - x^{k-1})$ and $x(t_{jk}) = x(t_{k-1}) + \rho_{jk}^i(x(t_k) - x(t_{k-1}))$. From (24) and (27), we have

$$|A_i^j - \tilde{A}_i^j| \leq h_{\alpha_i} \sum_{k=1}^{j-1} \left| [f_i(\tau_{jk}^i, x(t_{jk})) - f_i(\tau_{jk}^i, x^{jk})][(j - k + 1)^{\alpha_i} - (j - k)^{\alpha_i}] \right|$$

$$= h_{\alpha_i} \sum_{k=1}^{j-1} \left| f_i(\tau_{jk}^i, x(t_{jk})) - f_i(\tau_{jk}^i, x^{jk}) \right| [(j - k + 1)^{\alpha_i} - (j - k)^{\alpha_i}],$$

$$\tag{33}$$

since $z^{\alpha_i}$ is an increasing function of $z$ for $\alpha_i \in (0, 1)$. Because $f$ is twice continuously differentiable, we have, using a Taylor expansion,

$$|f_i(\tau_{jk}^i, x(t_{jk})) - f_i(\tau_{jk}^i, x^{jk})| \leq C\|x(t_{jk}) - x^{jk}\|_\infty$$

$$= C\|[x(t_{k-1}) + \rho_{jk}^i(x(t_k) - x(t_{k-1}))] - [x^{k-1} + \rho_{jk}^i(x^k - x^{k-1})]\|_\infty$$

$$= C\|[x(t_{k-1}) - x^{k-1}] + \rho_{jk}^i[x(t_k) - x^k] + \rho_{jk}^i[x^{k-1} - x(t_{k-1})]\|_\infty$$

$$\leq C\left(\|x(t_{k-1}) - x^{k-1}\|_\infty + \|x(t_k) - x_k\|_\infty + \|x(t_{k-1}) - x^{k-1}\|_\infty\right),$$

since $\rho_{jk}^i \in (0,1)$. Thus, from Assumption (22), we have

$$|f_i(\tau_{jk}, x(t_{jk})) - f_i(\tau_{jk}, x^{jk})| \le Ch^2.$$

Replacing $|f_i(\tau_{ij}, x(t_{jk})) - f_i(\tau_{ij}, x^{jk})|$ in (33) with the above upper bound yields

$$|A_i^j - \tilde{A}_i^j| \le h_{\alpha_i} Ch^2 \sum_{k=1}^{j-1} [(j - k + 1)^{\alpha_i} - (j - k)^{\alpha_i}] = \frac{h^{\alpha_i}}{\Gamma(\alpha_i + 1)} Ch^2(j^{\alpha_i} - 1)$$

$$\le \frac{h^{\alpha_i}}{\Gamma(\alpha_i + 1)} Ch^2 N^{\alpha_i} = \frac{C}{\Gamma(\alpha_i + 1)} h^2 (hN)^{\alpha_i}$$

$$= \frac{C}{\Gamma(\alpha_i + 1)} h^2 T^{\alpha_i} \le Ch^2$$

for $i = 1, 2, ..., n$. Thus, we have

$$\|A^j - \tilde{A}^j\|_\infty \le Ch^2.$$

Combining the above error bound with (32), we have the estimate (18). Thus, the theorem is proved.

## 4   Numerical Results

In this section, we will use Algorithm A to solve two non-trivial examples. All the computations have been performed in double precision under Matlab environment on a PC with Intel Xeon 3.3 GHz CPU and 16 GB RAM.

**Example 1.** Consider the following system of fractional differential equations:

$$\begin{cases} {}_0D_t^{\alpha_1} x_1(t) = x_2(t), \\ {}_0D_t^{\alpha_2} x_2(t) = x_3(t), \\ {}_0D_t^{\alpha_3} x_3(t) = \frac{\Gamma(5)}{\Gamma(5-\alpha_1-\alpha_2-\alpha_3)} t^{4-(\alpha_1+\alpha_2+\alpha_3)}, \ t \in (0,1], \end{cases}$$

$$x_1(0) = x_2(0) = x_3(0) = 0,$$

where $\alpha_i \in (0,1)$, $i = 1, 2, 3$. The exact solution is

$$x_1(t) = t^4, \ x_2(t) = \frac{\Gamma(5)}{\Gamma(5 - \alpha_1)} t^{4-\alpha_1}, \ x_3(t) = \frac{\Gamma(5)}{\Gamma(5 - \alpha_1 - \alpha_2)} t^{4-(\alpha_1+\alpha_2)}.$$

We solve the problem using Algorithm A for various values of $\alpha_i, i = 1, 2, 3$ and $h_k = 1/(2^k \times 10), k = 1, ..., 6$. The computed errors $E_{h_k}^i = \max_{1 \le j \le 1/h_k} |x_i^j - x_i(t_j)|$ for the chosen values of $\alpha_i$'s are listed in Table 1. To estimate the rates of convergence, we calculate $\log_2(E_{h_k}/E_{h_{k+1}})$ for $k = 1, ..., 5$ and the computed rates of convergence, as well as CPU times, are also listed in Table 1. From the results in Table 1 we see that our method has a 2nd-order convergence rate for all the chosen values of $\alpha$, as predicted by Theorem 3, indicating that our method is very robust in $\alpha$. The CPU time in Table 1 shows that our method is very efficient.

**Table 1.** Computed errors, convergence order and CPU time for Example 1.

| $h$ | $x_1$ | Order | $x_2$ | Order | $x_3$ | Order | CPU (seconds) |
|---|---|---|---|---|---|---|---|
| $\alpha_1 = \alpha_2 = \alpha_3 = 0.9$ | | | | | | | |
| 1/20 | 0.0041 | - | 0.00296 | - | 0.0022 | | 0.2496 |
| 1/40 | 0.0010 | 2.00 | 7.1406e−04 | 2.05 | 5.7828e−04 | 1.93 | 0.2652 |
| 1/80 | 2.4856e−04 | 2.01 | 1.7308e−04 | 2.04 | 1.4936e−04 | 1.95 | 0.3588 |
| 1/160 | 6.1488e−05 | 2.02 | 4.2162e−05 | 2.04 | 3.8314e−05 | 1.96 | 0.4212 |
| 1/320 | 1.5238e−05 | 2.01 | 1.0315e−05 | 2.03 | 9.7768e−06 | 1.97 | 0.5928 |
| 1/640 | 3.7822e−06 | 2.01 | 2.5328e−06 | 2.03 | 2.4845e−06 | 1.98 | 1.6848 |
| $\alpha_1 = 0.8, \alpha_2 = 0.6, \alpha_3 = 0.4$ | | | | | | | |
| 1/20 | 0.0036 | - | 0.0021 | - | 0.0028 | - | 0.2184 |
| 1/40 | 9.1509e−04 | 1.98 | 5.4008e−04 | 1.96 | 7.0318e−04 | 1.99 | 0.2808 |
| 1/80 | 2.3067e−04 | 1.99 | 1.3723e−04 | 1.98 | 1.7557e−04 | 2.00 | 0.3588 |
| 1/160 | 5.7981e−05 | 1.99 | 3.4680e−05 | 1.98 | 4.3850e−05 | 2.00 | 0.3900 |
| 1/320 | 1.4547e−05 | 1.99 | 8.7325e−06 | 1.99 | 1.0954e−05 | 2.00 | 0.6864 |
| 1/640 | 3.6451e−06 | 2.00 | 2.1936e−06 | 1.99 | 2.7371e−06 | 2.00 | 1.6692 |
| $\alpha_1 = 0.1, \alpha_2 = 0.3, \alpha_3 = 0.2$ | | | | | | | |
| 1/20 | 0.0018 | - | 9.6495e−04 | - | 0.00145 | - | 0.2652 |
| 1/40 | 4.9549e−04 | 1.86 | 2.6383e−04 | 1.87 | 3.7287e−04 | 1.96 | 0.3120 |
| 1/80 | 1.3425e−04 | 1.88 | 7.0393e−05 | 1.90 | 9.5368e−05 | 1.97 | 0.3432 |
| 1/160 | 3.5823e−05 | 1.91 | 1.8472e−05 | 1.93 | 2.4300e−05 | 1.97 | 0.4524 |
| 1/320 | 9.4522e−06 | 1.92 | 4.7897e−06 | 1.95 | 6.1738e−06 | 1.98 | 0.6552 |
| 1/640 | 2.4728e−06 | 1.93 | 1.2311e−06 | 1.96 | 1.5648e−-06 | 1.98 | 1.5912 |

**Example 2.** Consider the following fractional differential equation used in [14].

$$\begin{cases} {}_0D_t^{\alpha_1}x_1(t) = x_1(t), \\ {}_0D_t^{\alpha_2}x_2(t) = 2x_1^2(t), \\ {}_0D_t^{\alpha_3}x_3(t) = 3x_1(t)x_2(t), \, t \in (0,1] \end{cases}$$
$$x_1(0) = 1, x_2(0) = 1, x_3(0) = 0.$$

The exact solution when $\alpha_1 = \alpha_2 = \alpha_3 = 1$ is

$$x_1(t) = e^t, \quad x_2(t) = e^{2t}, \quad x_3(t) = e^{3t} - 1.$$

It is solved using Algorithm A for various values of $h$ and $\alpha_i, i = 1,2,3$. The computed errors and rates of convergence when $\alpha_1 = \alpha_2 = \alpha_3 = 1$ are listed in Table 2 from which we see that the computed rates of convergence is $\mathcal{O}(h^2)$, confirming our theoretical result.

Since the exact solution to this Example 2 is unknown when $\alpha_i < 1$ for any $i$, we are unable to compute the rates of convergence. Instead, we solve the problem

**Table 2.** Computed errors, convergence order and CPU time for Example 2.

| $h$ | $\alpha_1 = \alpha_2 = \alpha_3 = 1$ | | | | | | |
|---|---|---|---|---|---|---|---|
| | $x_1$ | Order | $x_2$ | Order | $x_3$ | Order | CPU (seconds) |
| 1/20 | 5.6658e−04 | - | 0.0026 | - | 0.0159 | - | 0.2496 |
| 1/40 | 1.4159e−04 | 2.00 | 7.1349e−04 | 1.87 | 0.0045 | 1.82 | 0.2808 |
| 1/80 | 3.5395e−05 | 2.00 | 1.8530e−04 | 1.94 | 0.0012 | 1.91 | 0.3120 |
| 1/160 | 8.8486e−06 | 2.00 | 4.7210e−05 | 1.97 | 3.1080e−04 | 1.95 | 0.3588 |
| 1/320 | 2.2121e−06 | 2.00 | 1.1914e−05 | 1.99 | 7.8895e−05 | 1.98 | 0.5616 |
| 1/640 | 5.5304e−07 | 2.00 | 2.9925e−06 | 1.99 | 1.9874e−05 | 1.99 | 0.8112 |

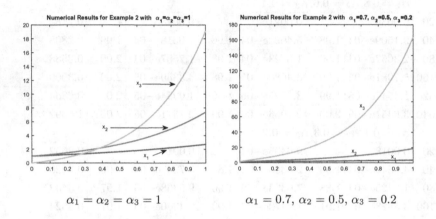

$$\alpha_1 = \alpha_2 = \alpha_3 = 1 \qquad\qquad \alpha_1 = 0.7,\ \alpha_2 = 0.5,\ \alpha_3 = 0.2$$

**Fig. 1.** Computed solutions of Example 2

for $\alpha_1 = 0.7, \alpha_2 = 0.5, \alpha_3 = 0.2$ using $h = 1/640$ and plot the computed solution, along with the solution for $\alpha_i = 1$ for $i = 1, 2, 3$, in Fig. 1. From Fig. 1 we see that $x_1, x_2$ and $x_3$ from the fractional system grow much faster than those from the integer system.

## 5    Conclusion

We have proposed and analysed a 1-step numerical integration method for a system of fractional differential equations, based a superconvergent quadrature point we have derived recently. The proposed method is unconditionally stable and easy to implement. We have shown the method has a 2nd-order accuracy. Non-trivial examples have been solved by our method and the numerical results show that our method is 2nd-order accurate for all the chosen values of the fractional orders, demonstrating our method is very robust.

**Acknowledgment.** This work is partially supported by US Air Force Office of Scientific Research Project FA2386-15-1-4095.

# References

1. Diethelm, K., Ford, N.J.: Analysis of fractional differential equations. J. Math. Anal. Appl. **265**, 229–248 (2002)
2. Erturk, V.S., Momani, S.: Solving systems of fractional differential equations using differential transform method. J. Comput. Appl. Math. **215**, 142–151 (2008)
3. Gejji, V.D., Jafari, H.: Adomian decomposition: a tool for solving a system of fractional differential equations. J. Math. Anal. Appl. **301**(2), 508–518 (2005)
4. Guo, B., Pu, X., Huang, F.: Fractional Partial Differential Equations and Their Numerical Solutions. World Scientific, Singapore (2015)
5. Jafari, H., Gejji, V.D.: Solving a system of nonlinear fractional differential equation using adomain decomposition. Appl. Math. Comput. **196**, 644–651 (2006)
6. Kilbas, A.A., Marzan, S.A.: Cauchy problem for differential equation with caputo derivative. Fract. Calc. Appl. Anal. **7**(3), 297–321 (2004)
7. Li, W., Wang, S., Rehbock, V.: A 2nd-order one-point numerical integration scheme for fractional ordinary differential Equation. Numer. Algebra Control Optim. **7**(3), 273–287 (2017)
8. Li, W., Wang, S., Rehbock, V.: Numerical solution of fractional optimal control. J. Optim. Theory Appl. (2018). https://doi.org/10.1007/s10957-018-1418-y
9. Momani, S., Al-Khaled, K.: Numerical solutions for systems of fractional differential equations by the decomposition method. Appl. Math. Comput. **162**(3), 1351–1365 (2005)
10. Momani, S., Odibat, Z.: Homotopy perturbation method for nonlinear partial differential equations of fractional order. Phys. Lett. A **365**(5–6), 345–350 (2007)
11. Momani, S., Odibat, Z.: Numberical approach to differential equations of fractional order. J. Comput. Appl. Math. **207**, 96–110 (2007)
12. Odibat, Z., Momani, S.: Application of variational iteration method to nonlinear differential equations of fractional order. Int. J. Nonlinear Sci. Numer. Simulat. **1**(7), 15–27 (2006)
13. Varga, R.S.: On Diagonal dominance arguments for bounding $\|A^{-1}\|_\infty$. Linear Algebra Appl. **14**(3), 211–217 (1976)
14. Zurigat, M., Momani, S., Odibat, Z., Alawneh, A.: The homotopy analysis method for handling systems of fractional differential equations. Appl. Math. Model. **34**, 24–35 (2010)

# Blow-Up of Fronts in Burgers Equation with Nonlinear Amplification: Asymptotics and Numerical Diagnostics

Dmitry Lukyanenko and Nikolay Nefedov[✉]

Department of Mathematics, Faculty of Physics,
Lomonosov Moscow State University, Moscow 119991, Russia
nefedov@phys.msu.ru

**Abstract.** This work develops a theory of asymptotic-numerical investigations of moving fronts in reaction-diffusion-advection models. We present the result of consideration of singularly perturbed parabolic Burgers-type equations with nonlinear forcing. Conditions of solution blow-up are formulated. Numerical algorithm which allows to recognise and describe the solutions blow-up is presented. In particular, in order to demonstrate the proposed method, we apply our approach to the problem with cubic forcing.

**Keywords:** Blow-up phenomena · Burgers equation ·
Singularly perturbed · Interior layer · Richardson extrapolation

## 1 Introduction

Regimes with sharpening, or "blow-up" have been actively studied in recent years in connection with a large number of practical applications. Physically such solutions describe explosions, the evolution of instability, and so on. The main approaches for analytical study of the blow-up phenomena and their extensions one can find in [1–3] (see also [4]). Numerical methods, which was designed for applications, have extensive considerations in literature. In particular, adaptive approaches, connected with the reduction of grid step (in time and possibly in space) [5,6] and with *a posteriori* estimates [6] are often used. Some of them are based on the transformation from a time variable to a stretched variable [7], for example, to the length of the arc of the graph. In some cases, the solution is scaled according its grows [8]. However, it turned out that each of these methods has a limited range of applicability.

At this paper we consider the new important for applications class of problems (singularly perturbed parabolic Burgers-type equations), where the blow-up phenomena was stated in the papers in [9] and [10]. Our goal is to use the asymptotic analysis for the numerical investigations of the solutions blow-up. The asymptotic analysis gives *a priori* information for the blow-up time $T_{bl}$. This *a priori* information can be used for numerical description of the process of

© Springer Nature Switzerland AG 2019
I. Dimov et al. (Eds.): FDM 2018, LNCS 11386, pp. 72–79, 2019.
https://doi.org/10.1007/978-3-030-11539-5_7

the solutions blow-up. Our numerical method is based on the ideas of *a posteriori* estimates of order of accuracy. The main ideas of this method are presented in the works [11–14]. It was shown, that sharp change of the effective order of accuracy indicates the solutions blow-up. Thus, this method is universal (it is applicable, perhaps, to an arbitrary ordinary differential equation or partial differential equation) and needs minimal requirements on *a priori* information. It suffices to know the upper estimate of $T_{bl}$. We note, that this method is closed to the method in [19], which indicate the solutions blow-up by means of the growth of the numerical solution with the grid parameter tending to zero (for a smooth solution the discrepancy should decrease).

The structure of the paper is as follows. In Sect. 2 we give the sufficient conditions for the blow-up phenomena and the asymptotic estimate for the blow-up time in the considered problems. In Sect. 3 we discuss and explain in details the numerical methods to determine the process of the solutions blow-up more accurately in comparison with the analytical approach, which can give us asymptotic estimate of the blow-up solution behavior.

## 2    Blow-Up Phenomena

We consider the following class of problems:

$$\begin{cases} \varepsilon \frac{\partial^2 u}{\partial x^2} - \frac{\partial u}{\partial t} = A(u,x)\frac{\partial u}{\partial x} + B(u,x), & x \in (0,1), \quad t \in (0,T], \\ u(0,t) = u_{left}, \quad u(1,t) = u_{right}, & t \in (0,T], \\ u(x,0) = u_{init}(x), & x \in [0,1], \end{cases} \tag{1}$$

where functions $A(u,x)$ and $B(u,x)$ are sufficiently smooth for $(x,u) \in [0;1] \times (-\infty; +\infty)$.

The rigorous treatment of problem (1) was obtained in [9], where the existence and asymptotics of the front type solution were established.

We assume

$(H_1)$ The equation

$$A(u,x)\frac{du}{dx} + B(u,x) = 0, \tag{2}$$

with the initial condition $u(0) = u_{left}$ has the solution $u = \varphi^l(x)$, and with the initial condition $u(1) = u_{right}$ has the solution $u = \varphi^r(x)$, moreover, $\varphi^l(x) < \varphi^r(x)$, $x \in [0,1]$ and $A(\varphi^l(x),x) > 0$, $A(\varphi^r(x),x) < 0$, $x \in [0,1]$.

$(H_2)$ $I(x) := \int_{\varphi^l(x)}^{\varphi^r(x)} A(u,x)\,du > 0.$

$(H_3)$ The initial value problem

$$\frac{dx^0}{dt} = \frac{I(x^0)}{\varphi^r(x^0) - \varphi^l(x^0)} \equiv V(x^0), \quad x^0(0) = x_0, \tag{3}$$

where $x_0$ is the initial location of the front, has the solution $x^0(t)$: $[0;T] \to [0,1]$ such that $x^0(t) \in (0,1)$ for $t \in [0,T]$.

The main result for this problem is the Theorem of existence and asymptotic approximation of the moving front with the principal term of the front location $x^0(t)$. The main result of [9] is that asymptotically stable stationary points of the equation of motion

$$\frac{dx^0}{dt} = V(x^0) \tag{4}$$

produce the stable stationary solutions with fronts of the original parabolic problem, located near this points.

If the assumptions $(H_1)$ is changed by the following way:

$(H_1')$ The equation

$$A(u,x)\frac{du}{dx} + B(u,x) = 0, \tag{5}$$

with the initial condition $u(0) = u_{left}$ has the solution $u = \varphi^l(x)$ with blow-up near some point $x_c$: $\varphi^l(x) \to -\infty$ for $x \to x_c$, and with the initial condition $u(1) = u_{right}$ has the solution $u = \varphi^r(x)$, moreover, $\varphi^l(x) < \varphi^r(x)$, $x \in [0, x_c)$ and $A(\varphi^l(x), x) > 0$, $x \in [0, x_c)$, $A(\varphi^r(x), x) < 0$, $x \in [0, 1]$

we have shown (see [10]) that the Burgers type equation can exhibit the blow-up of the front type solution with the jump at the front tending to infinity, where $T_{bl} = \int_{x_0}^{x_c} \frac{dx}{V(x)}$.

We illustrate this phenomena by Burgers-type equation with the cubic forcing at the following Section.

## 3  Numerical Localization of Solution's Blow-Up

In this Section we discuss and explain in details the methods which can help us to localize numerically the time and/or spatial points of the solution's blow-up.

At first, we rewrite our main Eq. (1):

$$\begin{cases} \varepsilon\frac{\partial^2 u}{\partial x^2} - \frac{\partial u}{\partial t} = A(u,x)\frac{\partial u}{\partial x} + B(u,x), & x \in (0,1), \quad t \in (0, T_{bl}], \\ u(0,t) = u_{left}, \quad u(1,t) = u_{right}, \\ u(x,0) = u_{init}(x). \end{cases} \tag{6}$$

Note, that we write this problem up to the moment $T_{bl}$ inclusive, though we know that the solution may not exist at this moment and may blow up even earlier. But the essence of this part of the paper is numerical localization of the blow-up, so we should obtain the *numerical* solution up to this moment [13,14].

*Stiff Method of Lines and Rosenbrock Scheme with Complex Coefficient.* For numerical solving of the system (6) we apply the stiff method of lines (SMOL) [20] in order to reduce the initial PDE to the system of ODEs that can be solved by Rosenbrock scheme with complex coefficient [21], which is the most suitable and very efficient scheme for stiff systems of ODEs [22].

At first, we introduce uniform mesh $X_N$ only on $x$–dimension with step $h = (1-0)/N$ that has number of nodes $N+1$ (i.e. $N$ intervals): $X_N = \{x_n, \ 0 \le$

$n \leq N: \ x_n = 0 + nh\}$. So, after finite-difference approximations of the spatial derivatives with the second order of accuracy in (6) we obtain the following system of ODEs from which we should determine $N - 1$ unknown functions $u_n \equiv u_n(t) \equiv u(x_n, t)$, $n = \overline{1, N - 1}$, ($u_0$ and $u_N$ we know from the boundary conditions):

$$
\begin{cases}
\frac{du_n}{dt} = \varepsilon \frac{u_{n+1} - 2u_n + u_{n-1}}{h^2} - A(u_n, x_n) \frac{u_{n+1} - u_{n-1}}{2h} - B(u_n, x_n), & n = \overline{1, N - 1}, \\
u_0 = u_{left}, \quad u_N = u_{right}, \\
u_n(0) \equiv u(x_n, 0) = u_{init}(x_n), \quad n = \overline{0, N}.
\end{cases}
$$

(7)

This system can be rewritten as

$$
\begin{cases}
\frac{du}{dt} = f(u), \\
u(0) = u_{init},
\end{cases}
$$

(8)

where

$$
u = \begin{pmatrix} u_1 & u_2 & \cdots & u_{N-1} \end{pmatrix}^T,
$$

$$
f = \begin{pmatrix} f_1 & f_2 & \cdots & f_{N-1} \end{pmatrix}^T,
$$

(9)

$$
u_{init} = \begin{pmatrix} u_{init}(x_1) & u_{init}(x_2) & \cdots & u_{init}(x_{N-1}) \end{pmatrix}^T.
$$

(10)

The vector-function $f$ has the following structure.
For $n = 1$:

$$
f_1 = \varepsilon \frac{u_2 - 2u_1 + u_{left}}{h^2} - A(u_1, x_1) \frac{u_2 - u_{left}}{2h} - B(u_1, x_1);
$$

(11)

for $n = \overline{2, N - 2}$:

$$
f_n = \varepsilon \frac{u_{n+1} - 2u_n + u_{n-1}}{h^2} - A(u_n, x_n) \frac{u_{n+1} - u_{n-1}}{2h} - B(u_n, x_n);
$$

(12)

and for $n = N - 1$:

$$
f_{N-1} = \varepsilon \frac{u_{right} - 2u_{N-1} + u_{N-2}}{h^2} - A(u_{N-1}, x_{N-1}) \frac{u_{right} - u_{N-2}}{2h} - B(u_{N-1}, x_{N-1}).
$$

(13)

For numerical solving of this system of ODEs (8) we use Rosenbrock scheme with complex coefficient (CROS1) [21] that is the good choice for solving such kind of problems because of its order of accuracy $(O(\tau^2))$, monotonicity and stability [22].

In order to apply this scheme we introduce uniform mesh (it is also possible to introduce a quasi-uniform mesh without any change in further algorithm) $T_M$ on $t$–dimension that has number of nodes $M + 1$ (i.e. $M$ intervals): $T_M = \{t_m, \ 0 \leq m \leq M: \ 0 = t_0 < t_1 < t_2 < \ldots < t_{M-1} < t_M = T_{bl}\}$.

Then we are able to apply the CROS1 scheme for solving the system (8):

$$
u(t_{m+1}) = u(t_m) + (t_{m+1} - t_m) \, Re \, w,
$$
where $w$ is a solution of the SLAE

$$
\left[ E - \frac{1 + i}{2} (t_{m+1} - t_m) f_u \big( u(t_m) \big) \right] w = f \big( u(t_m) \big).
$$

(14)

Here $E$ is the identity matrix, $f_u$ is the Jacobian matrix.

*Computations on Condensing Grids.* In numerical computations, it is important not only to obtain the result of calculations but to estimate its accuracy with guarantee. The computation method on condensing grids [11] allows one to do this. The main formulas of this paragraph one can find in the works [11, 12], now we use it for our purpose [13–18].

In compliance with the fact that we approximate all spatial variables in (6) by finite difference with accuracy $O(h^2)$ and we use the CROS1 scheme with accuracy $O(\tau^2)$ for numerical integration of (8), the constructed method of solving the equation (6) is of accuracy $O(\tau^2 + h^2)$.

Let us perform the computation of the solution $u(x,t)$ on the start grid $\{x_n, t_m\}$, $0 \le n \le N$, $0 \le m \le M$. Since the theoretical accuracy order in the time is equal to 2 and spatial variables are equal to 2, we perform a sequential condensation of the time grid by integer number of times $r_t$ and spatial grid by integer number of times $r_x$ (the most convenient is $r_t = 2$ and $r_x = 2$ as $r_t^2 = r_x^2$). In this case, the nodes of the start grid are those of all subsequent grids. At these points, we estimate the error

$$\Delta^{(r_x^s N, r_t^s M)}(x,t) = \frac{u^{(r_x^s N, r_t^s M)}(x,t) - u^{(r_x^{s-1} N, r_t^{s-1} M)}(x,t)}{r_t^2 - 1} + o(\tau^2 + h^2) \quad (15)$$

and the effective accuracy order

$$p_s^{eff}(x,t) = \log_{r_t} \frac{u^{(r_x^{s-1} N, r_t^{s-1} M)}(x,t) - u^{(r_x^{s-2} N, r_t^{s-2} M)}(x,t)}{u^{(r_x^s N, r_t^s M)}(x,t) - u^{(r_x^{s-1} N, r_t^{s-1} M)}(x,t)}. \quad (16)$$

At points $(x,t)$ at which

$$p_s^{eff}(x,t) \xrightarrow{s \to \infty} p^{theor} = 2 \quad (17)$$

the solution of the problem has bounded second derivative in time and second derivative in the spatial variables, and the error estimate is asymptotically sharp as $s \to \infty$ (that is same to $N, M \to \infty$). The violation of convergence (17) indicates the loss of smoothness of the exact solution. In particular, for a power singularity $u(x,t) \sim (t^* - t)^{-\beta}$ and for any $t > t^*$, the effective accuracy order $p_s^{eff}(x,t) \longrightarrow -\beta$ allows us to uniquely find the degree $\beta$. As $p_s^{eff}(x,t) \longrightarrow -\infty$, for $t > t^*$, we can assert that the exact solution exponentially grows, $u(x,t) = \infty$; as $p_s^{eff}(x,t) \longrightarrow 0$, for $t > t^*$, the growth of the solution is logarithmic: $u(x,t) \sim \ln(t^* - t)$. The singularity moment $t^*$ can be found with accuracy up to the step of the control grid.

If the violation of the smoothness of the solution occurs on the whole range of the spatial variable simultaneously, then a considerable deviation of $p_s^{eff}(x,t)$ from 2 is detected at all nodes of the grid $\{x_n\}$ at the first temporal layer $t \ge t^*$. If the singularity of the solution arises first at a single point $x^*$, then the method described also allows us to trace its propagation [13]. Such a diagnosis is possible owing to the fact that the use of the CROS scheme does not lead to an overflow even if the exact solution of the problems tends to infinity [11].

*Numerical Example.* Let us consider numerical example for

$$A(u,x) = -u, \quad B(u,x) = -u^3, \quad \varepsilon = 10^{-1.5},$$
$$u_{left} = -\tfrac{8}{7}, \quad u_{right} = \tfrac{8}{7},$$
$$u_{init}(x) \equiv \tfrac{8}{7}\tanh\tfrac{x-x_0}{\varepsilon}, \quad x_0 = 0.5 \pm 0.1.$$

In this case the exact solution of the Eq. (6) can not be calculated analytically, but all hypotheses above are satisfied, and we have the asymptotic approximation of the solution with blow-up (see [10]). Let's implement our numerical algorithm for recognizing of time and spatial points of solution's blow-up.

For numerical computations in problem (6) we took the initial data as follows: $T_{bl} = 0.7$, $N = 300$, $M = 200$, $r_x = 2$, $r_t = 2$, $S = 5$ (number of subsequent grids including the initial grid). The result of calculations are represented on Fig. 1.

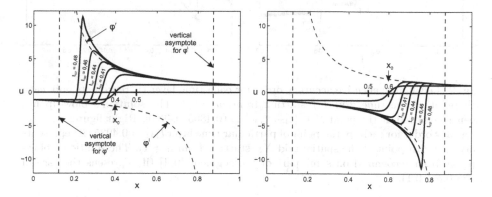

**Fig. 1.** Left figure: solution for $x_0 = 0.5 - 0.1$. Right figure: solution for $x_0 = 0.5 + 0.1$.

We are able to control the effective accuracy order for each time layer by the formula

$$p_s^{eff}(t_m) = \log_{r_t} \frac{\sqrt{\sum_{n=1}^{N}\left(u^{(r_x^{s-1}N, r_t^{s-1}M)}(x_n, t_m) - u^{(r_x^{s-2}N, r_t^{s-2}M)}(x_n, t_m)\right)^2}}{\sqrt{\sum_{n=1}^{N}\left(u^{(r_x^{s}N, r_t^{s}M)}(x_n, t_m) - u^{(r_x^{s-1}N, r_t^{s-1}M)}(x_n, t_m)\right)^2}}, \tag{18}$$

where $s$ is a number of the corresponding subsequent grid. After calculations on $S$ subsequent grids the effective accuracy orders $p^{eff}$ for each time layer $t_m$ converge to $p^{theor} = 2$ except $m \geq 140$, for which $p^{eff}(t_m) \longrightarrow -1$ (see Fig. 2). That means that blow-up of the solution appears from the time $t^* \in (t_{139}, t_{140}] \equiv (0.4865, 0.4900]$ and the solution has at this time point $t^*$ a power singularity $u(x,t) \sim (t^* - t)^{-1}$ (what is confirmed by the theory).

We are also able to estimate the effective accuracy order for each spatial point for particular time layer $t_m$ by the formula

$$p_s^{eff}(x_n, t_m) = \log_{r_t} \frac{u^{(r_x^{s-1}N, r_t^{s-1}M)}(x_n, t_m) - u^{(r_x^{s-2}N, r_t^{s-2}M)}(x_n, t_m)}{u^{(r_x^{s}N, r_t^{s}M)}(x_n, t_m) - u^{(r_x^{s-1}N, r_t^{s-1}M)}(x_n, t_m)}. \tag{19}$$

We can also use this formula for the different time layers in order to examine whether blow-up appears in each spatial point or not. We can see on Fig. 2 that solution's blow-up is recognized in the left side of the segment $[0, 1]$ for the case of $x_0 = 0.5 - 0.1$.

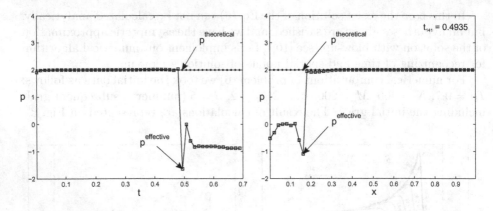

**Fig. 2.** Left figure: effective accuracy order for each time layer (it represents only each 4th point of the time grid $T_M$ starting from $m = 1$). The solution's blow-up has been recognized at the point $t^* \in (t_{139}, t_{140}] \equiv (0.4865, 0.4900]$. Right figure: effective accuracy order for each spatial point of particular time layer $t_{141} = 0.4935$ (it represents only each 6th point of the spatial grid $X_N$ starting from $n = 1$). The solution's blow-up has been recognized on some part of the segment $[0, 1]$ (it represens the case of $x_0 = 0.5 - 0.1$).

We can claim that our numerical treatment have a good correspondence with the asymptotic results and also give more detailed description of the solution near the blow-up points.

**Acknowledgements.** The work was supported by the Russian Science Foundation [grant number 18-11-00042].

# References

1. Mitidieri, E., Pokhozhaev, S.I.: A priori estimates and the absence of solutions of nonlinear partial differential equations and inequalities. Tr. Mat. Inst. Steklova **234**(3), 1–384 (2001)
2. Levine, H.A.: Some nonexistence and instability theorems for solutions of formally parabolic equations of the form $pu_t = -au + \mathcal{F}(u)$. Arch. Rat. Mech. Anal. **51**(5), 371–386 (1973)
3. Samarskii, A.A., Galaktionov, V.A., Kurdyumov, S.P., Mikhailov, A.P.: Blow-Up in Quasilinear Parabolic Qquations. Gruyter, Berlin (1995)
4. Korpusov, M.O.: Blow-up of ion acoustic waves in a plasma. Mat. Sb. **202**(1), 37–64 (2011)

5. Pelinovsky, D.E., Xu, C.: On numerical modelling and the blow-up behavoir of contact lines with a 180° contact angle. J. Eng. Math. **92**, 31–44 (2015)
6. Cangiani, A., Georgoulis, E.H., Kyza, I., Metcalfe, S.: Adaptivity and blow-up detection for nonlinear evolution problems. SIAM J. Sci. Comput. **38**(6), 3833–3856 (2016)
7. Haynes, R., Turner, C.: A numerical and theoretical study of blow-up for a system of ordinary differential equations using the sundman transformation. Atlantic Electron. J. Math. **2**(1), 1–13 (2007)
8. Berger, M., Kohn, R.V.: A rescaling algorithm for the numerical calcultaion of blowing-up solutions. Commun. Pure Appl. Math. **41**(6), 841–863 (1988)
9. Nefedov, N.N.: Asymptotic analysis of reaction-diffusion-advection problems: fronts with periodic motion and blow-up. IOP Conf. Ser.: J. Phys.: Conf. Ser. **811**, 012008 (2017)
10. Nefedov, N.N.: Multiple scale reaction-diffusion-advection problems with moving fronts. J. Phys.: Conf. Ser. **727**(1), 012011 (2016)
11. Al'shin, A.B., Kalitkin, N.N., Koryakin, P.V.: Diagnostics of singularities of exact solutions in computations with error control. Comput. Math. Math. Phys. **45**(19), 1769–1779 (2005)
12. Al'shin, A.B., Al'shina, E.A.: Numerical diagnosis of blow-up of solutions of pseudoparabolic equations. J. Math. Sci. **148**(1), 143–162 (2008)
13. Korpusov, M.O., Lukyanenko, D.V., Panin, A.A., Yushkov, E.V.: Blow-up for one sobolev problem: theoretical approach and numerical analysis. J. Math. Anal. Appl. **442**(2), 451–468 (2016)
14. Korpusov, M.O., Lukyanenko, D.V., Panin, A.A., Yushkov, E.V.: Blow-up phenomena in the model of a space charge stratification in semiconductors: analytical and numerical analysis. Math. Methods Appl. Sci. **40**(7), 2336–2346 (2017)
15. Korpusov, M.O., Lukyanenko, D.V., Ovsyannikov, E.A., Panin, A.A.: Local solvability and decay of the solution of an equation with quadratic noncoercive nonlineatity. Bull. South Ural State University Ser.-Math. Modell. Program. Comput. Softw. **10**(2), 107–123 (2017)
16. Korpusov, M.O., Lukyanenko, D.V.: Instantaneous blow-up versus local solvability for one problem of propagation of nonlinear waves in semiconductors. J. Math. Anal. Appl. **459**(1), 159–181 (2018)
17. Korpusov, M.O., Lukyanenko, D.V., Panin, A.A., Shlyapugin, G.I.: On the blow-up phenomena for a one-dimensional equation of ion-sound waves in a plasma: analytical and numerical investigation. Math. Methods Appl. Sci. **41**, 2906–2929 (2018). https://doi.org/10.1002/mma.4791
18. Korpusov, M.O., Lukyanenko, D.V., Panin, A.A., Yushkov, E.V.: On the blow-up of solutions of a full non-linear equation that describes ion-sound waves in plasma with non-coercive non-linearities, Izvestiya. Mathematics **82**, 283 (2018). https://doi.org/10.1070/IM8579
19. Hoffman, J., Johnson, C.: Blow up of incompressible euler solutions. BIT Numer. Math. **48**(2), 285–307 (2008)
20. Hairer, E., Wanner, G.: Solving of Ordinary Differential Equations. Stiff and Differential-Algebraic Problems. Springer, Heidelberg (2002)
21. Rosenbrock, H.H.: Some general implicit processes for the numerical solution of differential equations. Comput. J. **5**(4), 329–330 (1963)
22. Al'shin, A.B., Al'shina, E.A., Kalitkin, N.N., Koryagina, A.B.: Rosenbrock schemes with complex coefficients for stiff and differential algebraic systems. Comput. Math. Math. Phys. **46**(8), 1320–1340 (2006)

# Improved Computer Scheme
# for a Singularly Perturbed Parabolic
# Convection–Diffusion Equation

Grigorii Shishkin[✉] and Lidia Shishkina

Krasovskii Institute of Mathematics and Mechanics, UB RAS,
S. Kovalevskaya Str., 16, 620990 Yekaterinburg, Russia
shishkin@imm.uran.ru, Lida@convex.ru

*Dedicated to the academician A. A. Samarskii.*

**Abstract.** For a singularly perturbed parabolic convection–diffusion equation with a perturbation parameter $\varepsilon$, $\varepsilon \in (0,1]$, multiplying the highest-order derivative in the equation, we construct an improved computer difference scheme (with approximation of the first-order spatial derivative in the convective term by the central difference operator) on uniform meshes and study the behavior of discrete solutions in the presence of perturbations in the problem data. When solving such a problem numerically, errors in the grid solution depend on the parameter $\varepsilon$, on the parameters of the difference scheme, and also on the value of perturbations introduced in the process of computations (computer perturbations). For small values of the parameter $\varepsilon$, such errors, in general, significantly exceed the solution itself. For the computer perturbations, the conditions imposed on these admissible perturbations are obtained, under which accuracy of the computer solution in order is the same as for the solution of the unperturbed improved difference scheme, namely, $\mathcal{O}(\varepsilon^{-2} N^{-2} + N_0^{-1})$. As a result, we have been constructed the improved computer difference scheme suitable for practical use.

**Keywords:** Perturbation parameter ·
Singularly perturbed initial-boundary value problem ·
Parabolic convection–diffusion equation · Boundary layer ·
Difference scheme on uniform meshes ·
Perturbations in data of the grid problem · Computer perturbations ·
Stability of schemes to perturbations · Conditioning of schemes ·
Computer difference scheme · Maximum norm

## 1  Introduction

An initial–boundary value problem is considered for a singularly perturbed parabolic convection–diffusion equation with a perturbation parameter $\varepsilon$, $\varepsilon \in (0,1]$, multiplying the highest-order derivative (see problem (2), (1) in Sect. 2).

© Springer Nature Switzerland AG 2019
I. Dimov et al. (Eds.): FDM 2018, LNCS 11386, pp. 80–91, 2019.
https://doi.org/10.1007/978-3-030-11539-5_8

For small values of the parameter $\varepsilon$, a boundary layer appears in a neighborhood of the set $S_1$ [1]. Standard difference schemes on uniform grids developed for regular problems are very effective in their numerical solving (see, for example, [2,3]).

These schemes are rather frequently applied to solve singularly perturbed problems of the type (2), (1) by choosing the step-size with respect to the spatial variable $x$ sufficiently small, much less than $\varepsilon$ (see, for example, [1,4,5]). It is known that for convergence of the solution of a standard difference scheme in the absence of perturbations, it is required to use the meshes in $x$ with the number of nodes $N$, unboundedly increasing when $\varepsilon \to 0$, as $N \gg \varepsilon^{-1}$. However, when the step-size of a uniform grid with respect to $x$ becomes much less than the perturbation parameter $\varepsilon$, such schemes converge in the maximum norm, but provided that all the computations are executed precisely. The presence of computer perturbations in the calculations leads to serious difficulties due to loss of stability of standard schemes.

In [6], for a boundary value problem for a singularly perturbed ordinary differential convection–diffusion equation, a standard difference scheme was considered in the presence of computer perturbations. There a difference scheme was constructed with controlled computer perturbations, which is convergent with the first-order accuracy; results of numerical experiments were given consistent with theoretical results.

In [7,8], for the problem (2), (1), a standard difference scheme on uniform meshes was investigated in the presence of computer disturbances. The conditions imposed on the "parameters" of the difference scheme and computer (determined by the value of the parameter $\varepsilon$, by the quantities $N$, $N_0$, where $N + 1$ and $N_0 + 1$ are the numbers of nodes in the meshes in $x$ and $t$, respectively, as well as admissible perturbations for the data of the grid problem and admissible computer perturbations) were derived, which ensure convergence of the perturbed solutions. Conditions were also obtained under which the convergence rate of the perturbed numerical solution is the same in order as for the solution of the standard difference scheme in the absence of perturbations. The convergence rate of these schemes using the first-order (forward and backward) difference operators when approximating the first-order spatial derivatives [2], is of the first-order accuracy in the informative variables $\varepsilon^{-1} h$ and $\tau$, where $h$ and $\tau$ are the step-sizes in $x$ and $t$, respectively.

In the present paper, for the problem (2), (1), an improved computer difference scheme is constructed which converges for fixed values of the parameter $\varepsilon$ at the second-order accuracy with respect to the spatial variable and at the first order with respect to the temporal variable. Unlike [7,8], when approximating the first-order spatial derivative in the convective member, the central difference operator is used, that allows us to improve the order of the convergence rate of the scheme with respect to the spatial variable.

The influence of the perturbation in the data of the constructed difference scheme on the perturbation of its solution is investigated. The value $\triangle$ is defined as an upper boundary for perturbations in the data of the perturbed problem, introduced at computer calculations; $\triangle$ is called the threshold of admissible computer disturbances. The threshold value $\triangle$ as a function $\triangle = \triangle(\varepsilon, N, N_0)$, is chosen sufficiently small so that the computer solution was close to the solution of the initial–boundary value problem, thereby admissible computer perturbations become controlled. For the computer perturbations, the conditions are obtained which are determined by the value of the parameter $\varepsilon$, by the scheme parameters $N$, $N_0$ and the controlled threshold for admissible computer perturbations $\triangle = \triangle(\varepsilon, N, N_0)$, under which accuracy of the computer solution is the same as for the solution of the unperturbed improved difference scheme, namely, $\mathcal{O}(\varepsilon^{-2} N^{-2} + N_0^{-1})$. As a result, an improved computer difference scheme is constructed, suitable for practical use.

**Content of the Paper.** In Sect. 3, an improved difference scheme for problem (2), (1) is constructed and its convergence is studied. In Sect. 4, the conditioning of the matrix of the improved scheme is considered. Estimates for the error in the solution of the perturbed difference scheme are derived in Sect. 5. Errors of the grid solution in the presence of computer perturbations are studied in Sect. 6. Here also, the improved computer difference scheme is constructed and estimates of its convergence rate are obtained. In Appendix, some *a priori* estimates for the solution and derivatives in initial–boundary value problem (2), (1) are given that are used to justify convergence of the scheme.

## 2   Problem Formulation. Aim of the Research

On the set
$$\overline{G} = G \bigcup S, \; G = D \times (0, T], \; D = (0, d), \tag{1}$$
we consider the initial–boundary value problem for the parabolic convection–diffusion equation[1]
$$L_{(2)} u(x, t) = f(x, t), \; (x, t) \in G; \; u(x, t) = \varphi(x, t), \; (x, t) \in S. \tag{2}$$
Here
$$L_{(2)} = \varepsilon \, a(x, t) \frac{\partial^2}{\partial x^2} + b(x, t) \frac{\partial}{\partial x} - c(x, t) - p(x, t) \frac{\partial}{\partial t}, \quad (x, t) \in G,$$

$S = S^L \cup S_0$, $S_0$ and $S^L$ are the lower lateral parts of the boundary; $S^L = S_1 \cup S_2$, $S_1$ and $S_2$ are the left and right sides of the lateral boundary $S^L$, $S^L = \overline{S}^L$; the functions $a(x, t)$, $b(x, t)$, $c(x, t)$, $p(x, t)$, $f(x, t)$ and $\varphi(x, t)$ are assumed to be sufficiently smooth on the sets $\overline{G}$ and $S$, respectively, and satisfy[2]

$$m \le a(x, t), \; b(x, t), \; p(x, t) \le M, \quad 0 \le c(x, t) \le M,$$

---

[1] The notation $L_{(k)}$ ($m_{(k)}$, $M_{(k)}$, $G_{h(k)}$) means that these sets (operators, constants, grids) were introduced in formula $(k)$.

[2] By $M$ ($m$) we denote sufficiently large (small) positive constants independent of $\varepsilon$. In the case of grid problems, these constants also do not depend on the stencils of the difference schemes.

$$|f(x,t)| \le M, \ (x,t) \in \overline{G}; \quad |\varphi(x,t)| \le M, \ (x,t) \in S;$$

the parameter $\varepsilon$ takes arbitrary values from the open-closed interval $(0,1]$. Assume also that on the set of corner points $S^* = S^L \cap \overline{S}_0$, compatibility conditions are fulfilled, ensuring the required smoothness for the solution of problem (2), (1) on $\overline{G}$. For small values of the parameter $\varepsilon$, a boundary layer appears in a neighborhood of the set $S_1$ [1].

Our aim in this paper is for the singularly perturbed initial–boundary value problem (2), (1),

- to construct the improved difference scheme convergent for fixed values of the parameter $\varepsilon$ at the second-order accuracy in $x$ and at the first order in $t$, in the absence of perturbations, on the basis of the approximation of the first-order spatial derivative in the convective term by the central difference operator;
- to study the influence of perturbations in the scheme data and computer perturbations on the perturbations of grid solutions;
- to find conditions imposed on admissible computer perturbations, under which the computer solution converges in the maximum norm with the same accuracy order as the solution of the grid problem in the absence of perturbations that allows us to construct the improved computer scheme.

## 3   Improved Difference Scheme

To construct an improved difference scheme for problem (2), (1), we use the uniform grid

$$\overline{G}_h = \overline{G}_h^u = \overline{G}_{h(3)}^u \equiv \overline{\omega} \times \overline{\omega}_0 \tag{3}$$

with the step-sizes $h = d/N$ and $\tau = T/N_0$ in $x$ and $t$, where $N+1$ and $N_0+1$ are the numbers of nodes $x = x^i$ and $t = t^k$ in the meshes $\overline{\omega}$ in $x$ and $\overline{\omega}_0$ in $t$, respectively, in the grid $\overline{G}_h^u$ for $i = 0, 1, \ldots, N$ and $k = 0, 1, \ldots, N_0$.

We approximate problem (2), (1) by the following difference scheme:

$$\Lambda z(x,t) = f(x,t), \ (x,t) \in G_h, \quad z(x,t) = \varphi(x,t), \ (x,t) \in S_h, \tag{4a}$$

$$\Lambda \equiv \varepsilon\, a(x,t)\, \delta_{\overline{x}\widehat{x}} + b(x,t)\, \delta_{\widetilde{x}} - c(x,t) - p(x,t)\, \delta_{\overline{t}},$$

$$G_h = G \cap \overline{G}_h, \quad S_h = S \cap \overline{G}_h,$$

where $\delta_{\overline{x}\widehat{x}}\, z(x,t)$ is the second order central difference derivative in $x$ on the uniform mesh:

$$\delta_{\overline{x}\widehat{x}}\, z(x,t) = 2\,(h^i + h^{i-1})^{-1}[\delta_x\, z(x,t) - \delta_{\overline{x}}\, z(x,t)], \quad (x,t) = (x^i, t) \in G_h;$$

$\delta_x\, z(x,t)$ and $\delta_{\overline{x}}\, z(x,t)$, $\delta_{\overline{t}}\, z(x,t)$ are the first-order (forward and backward) difference derivatives.

Unlike [7,8], when approximating the first-order spatial derivative in the convective member, the central difference operator $\delta_{\widetilde{x}}\, z(x,t)$ is used.

Difference scheme (4a), (3) satisfies the monotonicity condition under the additional condition

$$h \leq m\varepsilon, \ m = \min_{\overline{G}} \{a(x,t) \, b^{-1}(x,t)\}, \tag{4b}$$

that is, scheme (4), (3) refers to conditionally convergent schemes.

For the solution $z(x,t)$ of difference scheme (4), (3), similar to the constructs in [1,8], using *a priori* estimates (23), (24) for the solution $u(x,t)$ of problem (2), (1), we obtain an improved estimate

$$\|u - z\|_{\overline{G}_h} \leq M \left( \min[\varepsilon^{-2} \, N^{-2}, \ 1] + N_0^{-1} \right). \tag{5}$$

The scheme (4), (3), approximating initial–boundary value problem (2), (1) with the second-order accuracy in $x$, is called an improved difference scheme.

The following estimate is also valid

$$\|u - z\|_{\overline{G}_h} \leq M \, [\delta_1 + \delta_0], \tag{6}$$

where $\delta_1 = \delta_1(\varepsilon, N) = \min[\varepsilon^{-2} \, N^{-2}, 1]$, $\delta_0 = N_0^{-1}$.

Thus, the scheme (4), (3) converges conditionally in $x$, namely, under the condition $N \gg \varepsilon^{-1}$, and unconditionally in $t$, for $N_0 \gg 1$.

We call the variables $N$, $N_0$ and $\delta_1$, $\delta_0$ primitive and informative variables, respectively.

**Theorem 1.** *Let for the solution $u(x,t)$ of problem (2), (1), a priori estimate (23) is fulfilled. Then for the solution of the difference scheme (4), (3), the estimates (5) and (6) are valid in primitive and informative variables, respectively.*

For fixed values of the parameter $\varepsilon$, the improved difference scheme (4), (3), according to estimate (5), converges with the second-order accuracy with respect to the quantity $\varepsilon^{-1} \, N$. The scheme converges with the first-order accuracy with respect to $N_0$ for any values $\varepsilon$.

The scheme (4), (3) does not converge $\varepsilon$–uniformly as $N, N_0 \to \infty$; for its convergence it is required to use grids in $x$ satisfying the condition $N \gg \varepsilon^{-1}$, $\varepsilon \in (0,1]$.

# 4 Conditioning of the Matrix of Improved Difference Scheme (4), (3)

In this section, we consider the conditioning of the matrix of the improved scheme (4), (3), using a matrix notation.

## 4.1 Matrix Notation of Difference Scheme (4), (3)

To estimate the condition number for the matrix of difference scheme (4), (3), we write it as a system of algebraic equations.

Let $N+1$ components of the function $z(x,t)$ at the nodes $x \in \overline{\omega}$ for $t = t^{k-1}$, $(x,t) \in \overline{G}_h$ correspond to an $(N+1)$-dimensional vector $Y = Y_k$, $k = 1, \ldots, N_0 + 1$. Ordering the elements $z(x,t)$, $(x,t) \in \overline{G}_h$, in the matrix notation of difference scheme (4), (3), we obtain the system

$$\mathbf{A}\mathbf{Y} = \mathbf{F}. \tag{7a}$$

Here $\mathbf{A}$ is an $\{(N+1)(N_0+1)\} \times \{(N+1)(N_0+1)\}$-matrix, $\mathbf{Y}$ and $\mathbf{F}$ are $\{(N+1)(N_0+1)\}$-dimensional vectors from the normed space $R^{(N+1)(N_0+1)}$ with the maximum norm $\|\cdot\|$.

The matrix $\mathbf{A}$ is a block one with blocks $(A_{kl})$ of size $(N+1) \times (N+1)$, where $k$ and $l$ are the indices of the block row and column containing the element $A_{kl}$, $k, l = 1, \ldots, N_0 + 1$; the blocks $A_{kk}$ and $A_{k,k-1}$ are the only nonzero, matrices $A_{kk}$, $k \geq 2, \ldots, N_0+1$ are tridiagonal matrices, and $A_{11}$ and $A_{k,k-1}$ are diagonal matrices. The vectors $\mathbf{Y}$ and $\mathbf{F}$ consist of the block vectors $Y = Y_k$ and $F = F_k$, $k = 1, \ldots, N_0 + 1$.

In block matrix notation, difference scheme (4), (3) can be written as

$$A_{kk}Y_k + A_{k,k-1}Y_{k-1} = F_k, \ k = 2, \ldots, N_0 + 1; \ Y_k = F_k, \ k = 1. \tag{7b}$$

The components of the matrices $A_{kk}$ and $A_{k,k-1}$ and of the vectors $Y_k$ and $F_k$, where $Y_k = (y_1^k, \ldots, y_{N+1}^k)'$ and $F_k = (f_1^k, \ldots, f_{N+1}^k)'$ are defined by the following relations (similar to [7,8]):

$$a_{ij}^{kk} = 1, \ y_i^k = z(x_i, t_1), \ f_i^k = \varphi(x_i, t_1), \ i,j = 1, \ldots, N+1 \ for \ k = 1;$$

$$a_{11}^{kk} = 1, \ a_{i,i-1}^{kk} = -\varepsilon h^{-2} a(x_i, t_k) + 2^{-1} h^{-1} b(x_i, t_k),$$

$$a_{i,i+1}^{kk} = -\varepsilon h^{-2} a(x_i, t_k) - 2^{-1} h^{-1} b(x_i, t_k),$$

$$a_{ii}^{kk} = 2\varepsilon h^{-2} a(x_i, t_k) + c(x_i, t_k) + \tau^{-1} p(x_i, t_k), \tag{7c}$$

$$a_{N,N}^{kk} = 1, \ a_{ii}^{k,k-1} = -\tau^{-1} p(x_i, t_k), \ i = 2, \ldots, N;$$

$$y_i^k = z(x_i, t_k), \ i = 1, \ldots, N+1; \ f_1^k = \varphi(x_1, t_k),$$

$$f_i^k = -f(x_i, t_k), \ i = 2, \ldots, N, \ f_{N+1}^k = \varphi(x_{N+1}, t_k) \ for \ k = 2, \ldots, N_0 + 1.$$

Here $x_{i\,(7)} = x^{i-1}$, $t_{k\,(7)} = t^{k-1}$, $(x^{i-1}, t^{k-1}) \in \overline{G}_h$, $i = 1, \ldots, N+1$, $k = 1, \ldots, N_0 + 1$.

## 4.2 Estimates of the Norm and the Condition Number for the Matrix A on the Grid $\overline{G}_h = \overline{G}_{h(3)}^u$

On the uniform grid $\overline{G}_{h(3)}^u$, the norm $\|\mathbf{A}(\overline{G}_h^u)\|$ of the matrix $\mathbf{A}_{(7)}$ and its condition number $æ(\mathbf{A})$, where

$$æ(\mathbf{A}) = æ_M(\mathbf{A}) = \|\mathbf{A}\| \, \|\mathbf{A}^{-1}\|,$$

satisfy the following estimate in the primitive variables $N$, $N_0$ (see [[1], Chap. 12]):

$$\|\mathbf{A}(\overline{G}_h^u)\|, \; æ_M(\mathbf{A}; \overline{G}_h^u) \leq M \left[ N \left( 1 + \varepsilon N \right) + N_0 \right],$$

which is unimprovable with respect to $\varepsilon$, $N$ and $N_0$.

In the informative variables $\delta_1$ and $\delta_0$, for the norm of the matrix $\mathbf{A}$ and its condition number $æ_M$ we obtain the estimate

$$\| \mathbf{A}(\overline{G}_h^u) \|, \; æ_M(\mathbf{A}; \overline{G}_h^u) \leq M \left[ \varepsilon^{-1} \delta_1^{-1} + \delta_0^{-1} \right], \tag{8}$$

that is, the matrix of the improved difference scheme (4), (3) is not $\varepsilon$-uniformly well conditioned with respect to the informative variable $\delta_1$.

**Theorem 2.** *The condition number* $æ_M(\mathbf{A}; \overline{G}_h^u)$ *of the matrix* $\mathbf{A}$ *for the scheme* (4), (3) *satisfies the estimate* (8).

# 5   Estimation of the Error in the Solution of the Perturbed Difference Scheme

In this section, we discuss the influence of perturbations in the data of difference scheme (4), (3) on the perturbation of its solution.

## 5.1   Matrix Notation of the Perturbed Difference Scheme

Consider the perturbed matrix problem corresponding to matrix problem (7):

$$\mathbf{A}^* \mathbf{Y}^* = \mathbf{F}^*. \tag{9}$$

Here $\mathbf{A}^* = \mathbf{A} + \delta \mathbf{A}$ is a perturbed matrix, while $\mathbf{Y}^* = \mathbf{Y} + \delta \mathbf{Y}$ and $\mathbf{F}^* = \mathbf{F} + \delta \mathbf{F}$ are perturbed vectors. The perturbations of the coefficient $a(x_i, t_k)$ involved in the components $a_{ij}^{kk}$, $j = i - 1, i, i + 1$, $i = 2, \ldots, N$ of the matrix $A_{kk\,(7)}$, in general, differ; these perturbations of $a(x_i, t_k)$ in the perturbed components $a_{ij}^{kk\,*}$ are denoted by $\delta a_i^{kk;j}$. In a similar way, the perturbations of the coefficient $b(x_i, t_k)$ in the components $a_{ij}^{kk}$, $j = i, i + 1$ are denoted by $\delta b_i^{kk;j}$, while the perturbations of the coefficient $c(x_i, t_k)$ in the components $a_{ii}^{kk}$ are denoted by $\delta c_i^{kk;i}$; $i = 2, \ldots, N$. The perturbations of the coefficient $p(x_i, t_k)$ involved in the components $a_{ii}^{kk}$ and $a_{ii}^{k,k-1}$ of the matrices $A_{kk\,(7)}$ and $A_{k,k-1\,(7)}$ are denoted by $\delta p_i^{kk;i}$ and $\delta p_i^{k,k-1;i}$, respectively. Thus, with the block matrices $\delta A_{kk}$, $\delta A_{k,k-1}$ and vectors $\delta F_k$ and $\delta Y_k$ written componentwise, we have

$$\delta a_{ij}^{kk} = 0, \; \delta f_i^k = \delta \varphi_{i,1}, \; \delta y_i^k = \delta z(x_i, t_1), \; i, j = 1, \ldots, N + 1 \; \text{for } k = 1;$$

$$\delta a_{11}^{kk} = 0, \; \delta a_{i,i-1}^{kk} = -\varepsilon \, h^{-2} \, \delta a_i^{kk;i-1} + 2^{-1} \, h^{-1} \, \delta b_i^{kk;i-1},$$

$$\delta a_{ii}^{kk} = 2 \, \varepsilon \, h^{-2} \, \delta a_i^{kk;i} + \delta c_i^{kk;i} + \tau^{-1} \, \delta p_i^{kk;i}, \tag{10}$$

$$\delta a_{i,i+1}^{kk} = -\varepsilon \, h^{-2} \, \delta a_i^{kk;i+1} - 2^{-1} \, h^{-1} \, \delta b_i^{kk;i+1}, \; \delta a_{N+1,N+1}^{kk} = 0,$$

$$\delta a_{ii}^{k,k-1} = -\tau^{-1} \, \delta p_i^{k,k-1;i}, \qquad\qquad i = 2, \ldots, N;$$

$$\delta f_1^k = \delta \varphi_{1,k}, \ \delta f_i^k = -\delta f_{i,k}, \ i = 2, \ldots, N, \ \delta f_{N+1}^k = \delta \varphi_{N+1,k};$$

$$\delta y_i^k = \delta z(x_i, t_k), \qquad i = 1, \ldots, N+1 \ for \ k = 2, \ldots, N+1,$$

where $\varphi_{ik} = \varphi(x_i, t_k)$, $f_{ik} = f(x_i, t_k)$.

## 5.2   Estimates for Perturbations in the Grid Solution

Consider the following perturbed difference scheme corresponding to perturbed matrix problem (9):

$$\Lambda^* z^*(x,t) \equiv \left[ \varepsilon \, a^*(x,t) \, \delta_{\overline{x}\widehat{x}} + b^*(x,t) \, \delta_{\widehat{x}} - c^*(x,t) \right. \tag{11}$$

$$\left. - p^*(x,t) \, \delta_{\overline{t}} \right] z^*(x,t) = f^*(x,t), \quad (x,t) \in G_h,$$

$$z^*(x,t) = \varphi^*(x,t), \quad (x,t) \in S_h.$$

Taking into account relations (10), for the perturbation of the grid solution $z^*(x,t) - z(x,t)$, where

$$z^*(x^i, t^k) - z(x^i, t^k) = \delta Y_{i+1}^{k+1}, \ (x^i, t^k) \in \overline{G}_h, \ i = 1, \ldots, N+1,$$

$$k = 1, \ldots, N_0 + 1, \qquad z^*(x,t) = z_{(11)}^*(x,t), \quad (x,t) \in \overline{G}_h,$$

we obtain the following estimate in primitive variables:

$$\| z^* - z \|_{\overline{G}_h} \le M \left[ \varepsilon \, N^2 \max_{i,j,k; i \neq 1, N+1, k \neq 1} |\delta a_i^{kk;j}| + \right. \tag{12}$$

$$\left. N \max_{i,j,k} |\delta b_i^{kk;j}| + N_0 \max_{i,k,l} |\delta p_i^{kl;i}| + \max_{i,k} \widehat{\psi}_{ik} \right].$$

Here

$$\max_{i,k} \widehat{\psi}_{ik} = \max \left[ \max_{i,k} |\delta c_i^{kk;i}|, \max_{i,k} |\delta f_{ik}|, \max_{i,k} |\delta \varphi_{ik}| \right],$$

$$j = i - 1, \ i, \ i + 1; \quad i = 1, \ldots, N+1, \ k = 1, \ldots, N_0 + 1; \ l = k - 1, \ k.$$

Taking into account the estimate (12) for the perturbation $z^*(x,t) - z(x,t)$, we have the following estimate in informative variables, unimprovable with respect to the quantities $\varepsilon$, $\delta_1$, $\delta_0$:

$$\| z^* - z \|_{\overline{G}_h} \le M \left[ \varepsilon^{-1} \delta_1^{-1} \max_{i,j,k; i \neq 1, N+1, k \neq 1} |\delta a_i^{kk;j}| + \right. \tag{13}$$

$$\left. \varepsilon^{-1} \delta_1^{-1/2} \max_{i,j,k} |\delta b_i^{kk;j}| + \delta_0^{-1} \max_{i,k,l} |\delta p_i^{kl;i}| + \max_{ik} \widehat{\psi}_{ik} \right].$$

Here

$$\max_{ik} \widehat{\psi}_{ik} = \max_{ik} \widehat{\psi}_{ik \, (12)}, \quad \delta_1 = \delta_{1 \, (6)}(\varepsilon, N), \ \delta_0 = \delta_{0 \, (6)}(N_0).$$

For the condition number of the difference scheme (4), (3), we obtain the estimate:

$$æ_P(\mathbf{A}; \overline{G}_h) \leq M\left(\varepsilon^{-1}\delta_1^{-1} + \delta_0^{-1}\right), \tag{14}$$

that is, the scheme (4), (3) is not $\varepsilon$-uniformly well-conditioned.

Thus, we have the following theorem:

**Theorem 3.** *When the data in the difference scheme* (4), (3) *are perturbed, the perturbation of the grid solution satisfies the estimate* (12) *in primitive variables and the estimate* (13) *in informative variables. For the condition number of the difference scheme* (4), (3), *the estimate* (14) *in informative variables is valid.*

Taking into account the estimates (6), (13), we obtain the following estimate for the error in the solution of the perturbed difference scheme in informative variables:

$$\|u - z^*\|_{\overline{G}_h} \leq M\left[\delta_1 + \delta_0 + \varepsilon^{-1}\delta_1^{-1} \max_{i,j,k;\,i\neq 1,N+1,k\neq 1} |\delta a_i^{kk;j}|\right.$$

$$\left. + \varepsilon^{-1}\delta_1^{-1/2} \max_{i,j,k}|\delta b_i^{kk;j}| + \delta_0^{-1}\max_{i,k,l}|\delta p_i^{kl;i}| + \max_{ik}\widehat{\psi}_{ik}\right], \tag{15}$$

**Theorem 4.** *Let the solution* $u(x,t)$ *of the problem* (2), (1) *satisfy the estimate* (23). *Then for the error* $\|u - z^*\|_{\overline{G}_h}$ *in the solution of the perturbed difference scheme, estimate* (15) *holds in informative variables.*

## 6   Errors of the Grid Solution in the Presence of Computer Perturbations

Consider the difference scheme (4), (3) in the presence of computer perturbations.

We denote by $\triangle$ the upper bound of the perturbations in the data of perturbed problem (9), taking into account perturbations (10), introduced in computer calculations, that is, computer perturbations,

$$|\delta a_i^{kk;j}|, \ |\delta b_i^{kk;j}|, \ |\delta c_i^{kk;i}|, \ |\delta p_i^{kl;i}|, \ |\delta f_{ik}|, \ |\delta \varphi_{ik}| \leq \triangle. \tag{16}$$

The quantity $\triangle$ is called the threshold of computer perturbations; $\triangle$ determines admissible computer perturbations.

Let $z_\triangle^*(x,t)$ be the solution of the perturbed difference scheme in the matrix notation (9), (10) under the condition (16); we call it a computer solution. Then for the function $z_\triangle^*(x,t)$, we obtain the estimate for the perturbation of the grid solution in informative variables

$$\| z_\triangle^* - z \|_{\overline{G}_h} \leq M\left(\varepsilon^{-1}\delta_1^{-1} + \delta_0^{-1}\right)\triangle, \tag{17a}$$

equivalent to the following estimate in primitive variables

$$\| z_\triangle^* - z \|_{\overline{G}_h} \leq M\left(\varepsilon N^2 + N_0\right)\triangle. \tag{17b}$$

By virtue of estimates (15), (16) and (17a), we obtain the following estimate of the error of the computer solution $z_\triangle^*$ in the variables $\varepsilon$, $\delta_1$, $\delta_0$, $\triangle$:

$$\| u - z_\triangle^* \|_{\overline{G}_h} \leq M \left[ \delta_1 + \delta_0 + ( \varepsilon^{-1} \delta_1^{-1} + \delta_0^{-1}) \triangle \right]. \tag{18}$$

It follows from estimate (18) that the error $\| u - z_\triangle^* \|_{\overline{G}_h}$ of the computer solution is not $\varepsilon$-uniformly bounded.

**Theorem 5.** *Let the condition (16) be satisfied for computer perturbations. Then for the error of the computer solution $z_\triangle^*(x,t)$, the estimate (18) is valid.*

## 7  Improved Computer Difference Scheme

In this section, we construct an improved computer difference scheme in the presence of controlled computer perturbations.

According to estimate (18), the value $\triangle$ of the threshold for the computer perturbations can be chosen sufficiently small so that the computer solution was close to the solution of the initial–boundary value problem (2), (1); in this case the computer perturbations become controlled.

We choose $\triangle$ satisfying the following condition:

$$\triangle \leq M \, \min[\varepsilon \, \delta_1^2, \, \delta_0^2], \quad \triangle = \triangle(\varepsilon \, \delta_1^2, \, \delta_0^2), \tag{19a}$$

equivalent to the condition

$$\triangle \leq M \, \min[\varepsilon^{-1} N^{-2}, N_0^{-2}], \tag{19b}$$

Under condition (19), we have the estimate for the error $\| u - z_\triangle^* \|_{\overline{G}_h}$ of the computer solution $z_\triangle^*(x,t)$, similar to estimate (6), in informative variables

$$\| u - z_\triangle^* \|_{\overline{G}_h} \leq M \, (\delta_1 + \delta_0); \tag{20}$$

in primitive variables, we have the improved estimate

$$\| u - z_\triangle^* \|_{\overline{G}_h} \leq M \, (\varepsilon^{-2} N^{-2} + N_0^{-1}). \tag{21}$$

If $\triangle \gg \varepsilon \, \delta_1^2$, or $\triangle \gg \delta_0^2$, then $\|u - z_\triangle^*\|_{\overline{G}_h} \gg \delta_1$, or $\|u - z_\triangle^*\|_{\overline{G}_h} \gg \delta_0$, respectively, that is, the estimate (20) is violated.

Thus, the estimates (20) and (21) are unimprovable.

Under condition (19), providing estimates (20) and (21), admissible computer perturbations become controlled.

**Definition 1.**  *A computing system consisting of the improved difference scheme (4), (3) with data perturbations (determined by $N$, $N_0$ and the parameter $\varepsilon$), and a "consistent" computer with computer perturbations (with computer parameters determined by the values $N$, $N_0$, and the parameter $\triangle$, $\triangle = \triangle(\varepsilon, \delta_1, \delta_0)$ satisfying (19) is the controlled threshold for admissible computer perturbations), under which accuracy of the computer solution in order is the same as for the solution of the unperturbed improved difference scheme (4), (3), is called an improved computer difference scheme.*

Thus, the difference scheme {{(4), (3)}; (16), (19)} (that is, the difference scheme (4), (3) under the condition (16) for the admissible computer perturbations, as well as an additional condition (19) for the threshold of the admissible computer perturbations, providing controlled computer disturbances) is the improved computer difference scheme. Such a scheme becomes reliable in computer calculations, suitable for practical use.

The following theorem holds.

**Theorem 6.** *The solution $z^*_\triangle(x,t)$ of the improved computer difference scheme {{(4), (3)}; (16), (19)} converges with the estimate (20) in informative variables and with improved estimate (21) in primitive variables.*

Note that in [7,8], where in the problem (2), (1), when approximating the first-order spatial derivatives, standard difference derivatives were used, and for the computer solution, the following unimprovable estimate was obtained in primitive variables:

$$\| u - z^*_\triangle \|_{\overline{G}_h} \le M \left( \varepsilon^{-1} N^{-1} + N_0^{-1} \right), \tag{22}$$

that is weaker in the order than the estimates (21) with respect to the primitive variable $\varepsilon^{-1} N^{-1}$ in this paper.

**Acknowledgements.** This research was partially supported by the Russian Foundation for Basic Research under grant No. 16-01-00727.

## Appendix

Here we give some *a priori* estimates for solutions and derivatives in initial–boundary value problem (2), (1) that were used to justify convergence of the scheme under construction. These estimates are derived in a similar way as it was done, for example, in [1]; their complete derivation can be found, for example, in [9].

The solution of problem (2), (1) is represented as a sum of functions:

$$u(x,t) = U(x,t) + V(x,t), \quad (x,t) \in \overline{G}, \tag{23}$$

where $U(x,t)$ and $V(x,t)$ are the regular and singular components of the solution.

When the data of this problem are sufficiently smooth and the compatibility conditions at the corner points are satisfied, for the solution of the problem and its component from (23), the following estimates are valid (see, for example, [1]):

$$\left| \frac{\partial^{k+k_0}}{\partial x^k \partial t^{k_0}} u(x,t) \right| \le M \varepsilon^{-k}, \quad (x,t) \in \overline{G}, \ k + 2k_0 \le 4; \tag{24}$$

$$\left| \frac{\partial^{k+k_0}}{\partial x^k \partial t^{k_0}} U(x,t) \right| \le M, \tag{25}$$

$$\left| \frac{\partial^{k+k_0}}{\partial x^k \partial t^{k_0}} V(x,t) \right| \le M \varepsilon^{-k} \exp^{-m\varepsilon^{-1} x}, \quad (x,t) \in \overline{G}, \ k + 2\,k_0 \le 4,$$

where $m \le \min_{\overline{G}}(b(x,t)/c(x,t))$.

# References

1. Shishkin, G., Shishkina, L.: Difference Methods for Singular Perturbation Problems. Chapman and Hall/CRC Monographs and Surveys in Pure and Applied Mathematics, vol. 140. CRC Press, Boca Raton (2009)
2. Samarskii, A.: The Theory of Difference Schemes. Nauka, Moskva (1989). (in Russian)
3. Samarskii, A.: On monotonic difference schemes for elliptic and parabolic equations in the case of a non-selfadjoint elliptic operator. In: Selected Works, pp. 233–238. MAKS Press, Moskva (2003). (in Russian)
4. Miller, J., O'Riordan, E., Shishkin, G.: Fitted Numerical Methods for Singular Perturbation Problems. Error Estimates in Maximum Norm for Linear Problems in One and Two Dimensions, Revised edn. World Scientific Publishing Co., Inc., Singapore (2012)
5. Farrel, P., Hegarty, A., Miller, J., O'Riordan, E., Shishkin, G.: Robust Computational Techniques for Boundary Layers. Chapman and Hall/CRC, Boca Raton (2000)
6. Shishkin, G.: Use of standard difference scheme on uniform grids for solving singularly perturbed problems under computer perturbations. In: Knobloch, P. (ed.) Boundary and Interior Layers, Computational and Asymptotic Methods - BAIL 2014. LNCSE, vol. 108, pp. 269–279. Springer, Cham (2015). https://doi.org/10.1007/978-3-319-25727-3_21
7. Shishkin, G.: Standard scheme for a singularly perturbed parabolic convection-diffusion equation with computer perturbations. Dokl. Math. **91**(3), 273–276 (2015). https://link.springer.com/article/10.1134%2FS1064562415030059
8. Shishkin, G.: Difference scheme for a singularly perturbed parabolic convection–diffusion equation in the presence of perturbations. Comput. Math. Math. Phys. **55**(11), 1842–1856 (2015). https://link.springer.com/article/10.1134%2FS0965542515110159
9. Shishkin, G.: Conditioning and stability of finite difference schemes on uniform meshes for a singularly perturbed parabolic convection-diffusion equation. Comput. Math. Math. Phys. **53**(4), 431–454 (2013). https://link.springer.com/article/10.1134%2FS096554251304009X

# On Ignorance Regions and Spatial Aspects for South American sea lion (Otaria byronia) Operational Interaction with the Artisan Gillnet Fishery in Chile

Milan Stehlík[1,2,3(✉)] [iD], Jean Paul Maidana[3,4] [iD], Claudia Navarro Villarroel[3],
Maritza Sepúlveda[5,6] [iD], and Doris Oliva[5] [iD]

[1] Linz Institute of Technology (LIT) and Department of Applied Statistics,
Johannes Kepler University in Linz, Linz, Austria
mlnstehlik@gmail.com
[2] School of Mathematical and Statistical Sciences, Arizona State University,
Tempe, AZ, USA
[3] Instituto de Estadística, Universidad de Valparaíso, Valparaíso, Chile
[4] Facultad de Ingeniería, Universidad Andrés Bello, Quillota 980, Viña del Mar, Chile
[5] Instituto de Biología, Centro de Investigación y Gestión de Recursos Naturales
(CIGREN), Facultad de Ciencias, Universidad de Valparaíso, Gran Bretaña 1111,
Playa Ancha, Valparaíso, Chile
[6] Núcleo Milenio INVASAL, Concepción, Chile

**Abstract.** The South Pacific Hake (*Merluccius gayi*) is the main artisan gillnet fishery in central Chile and the South American sea lion (*SASL, Otaria byronia*) has a strong operational interaction with this fishery. This was analyzed in paper [2] and several issues, like ignorance regions and spatial variation of such conflicts have not been studied in detail. In this paper we provide such analyses in order to complement the study. In particular, we show that construction of ignorance regions and its boundaries in the parameter space for *SASL, dist, season4* and *depth* variables of interest could play an important role in order to characterize and possibly to eliminate the inaccuracies in the results and decisions that could be made.

We also checked Ripley's K and Moran's I in order to conduct spatial analysis using GIS tools. This may help to determine areas of conflict and how these vary in time and space. The interactions with sea lions are not a determining factor in the variation of artisan fishery catches and such observations are important for managing fisheries interactions and protection of marine species.

**Keywords:** South Pacific Hake (*Merluccius gayi*) ·
Southern sea lion (*Otaria byronia*) · Ignorance regions · Spatial analysis

© Springer Nature Switzerland AG 2019
I. Dimov et al. (Eds.): FDM 2018, LNCS 11386, pp. 92–101, 2019.
https://doi.org/10.1007/978-3-030-11539-5_9

# 1  Introduction

The South Pacific Hake (*Merluccius gayi*) is the main artisan gillnet fishery in central Chile and the South American sea lion (*Otaria byronia*) has a strong operational interaction with this fishery. To evaluate the factors affecting the operational interaction between the South American sea lion (SASL) and the artisan gillnet fishery in central Chile, 122 fishing trips were sampled onboard fishing coves: Higuerillas (32°55′S–71°32′W), El Membrillo (33°1′S–71°37′W) and San Antonio (33°35′S–71°36′W). Onboard surveys were carried out between May 2014 and January 2016 and in each haul the following information was collected: geographic position, number and surface area of monofilament and multifilament nets, immersion time of the fishing net, number and biomass of hakes, number and biomass of the damage fish, number of SASL observed near the vessel, SASL mortality and use of artisanal deterrent systems. In this manuscript we provide both ignorance regions and spatial analysis in order to complement results from [2]. Another open problems can be related to application of random elements on lattices, see [4].

# 2  Methods

## 2.1  Ignorance Region

Incomplete data models typically involve strong intestable assumptions about the missing data distributions. The setup of the software R can go beyond standard tuning for Generalized Linear Models (GLM), however, interested reader can found more in [3]. We show that construction of ignorance regions and its boundaries in the parameter space for *SASL*, *dist*, *season4* and *depth* variables of interest could play an important role in order to characterize and possibly to eliminate the inaccuracies in the results.

## 2.2  Homogeneity and Correlation in a Spatial Model

In compared 3 fishing villages, there is not spatial analysis included using for instance GIS tools. This could be added to determine areas of conflict and how these vary in time and space. We provide the analysis of spatial variation for locations of SASLs. We use Ripley's K. Moran's I and Correlogram analysis. We can conclude the logically (and biologically) expected spatial variation because of geometry of Chilean coast and SASL habitual behavior.

# 3  Results

## 3.1  Ignorance Region

We estimate boundaries for the region of ignorance in the parameter space for *SASL*, *dist*, *season4* and *depth* variables of interest, in order to show that if we extract some of the observations which possess missing or not observed values for the *SASL* variable, the adjusted parameters involved in the model will change producing inaccuracies in the results and decisions. Observations of these changes in the estimation shows a better perspective of what is ignored if the not observed variable is considered or even in the process of estimation.

In order to reveal the subspace to which the region of ignorance belongs, we used a heuristic sampling approach, which consist in performing $10^4$ repetitions of the GLM's parameters estimation by randomly selecting data points which has not observed variable *SASL*, and then removing them from the data set. The procedure, for each repetition, consists in two steps, namely

(1) drawing a uniform random variable between zero and seven which will represent the number of observations that will be removed from the data set,
(2) this number of observations is extracted randomly from the total of not observed data points, finally this process is repeated $10^4$ times, saving the estimated parameters of the GLM model. For the GLM estimation we used the library `statsmodels` from Python [8] and for Figs. 1 and 2 the library `matplotlib` and `numpy` from Python [5].

In Fig. 1 it can be seen the $10^4$ estimations for a GLM with Poisson family, each of the points in the figure represent the estimated value of the parameter of interest and the region formed by this points is the region of ignorance. A similar result can be seen in Fig. 2 for a GLM's estimations for the Gamma family, for which again the region of ignorance is the ellipsoid formed by the points. Also, it can be observed that variables *depth* and *SASL* markedly change its magnitude between families, noting that for example in the Poisson family, the variable *depth* is between 0.055 and 0.075 but for the Gamma family this variable is between 0.25 and 0.55.

## 3.2   Ripley's K

Given the location of each SASL in GIS coordinates the question is to check for spatial homogeneity with the Ripley's K function.

Using the `spatstat` package from the Comprehensive R Archive Network [6] we estimate for exploratory purposes the Ripley's K function in order to check graphically the null hypothesis that data follows a homogeneous Poisson process (Complete Spatial Randomness). In Fig. 3 part **A** is the map of the region where the SASL were seen and in **B** it can be seen the lower and upper K function envelopes for 99 independent simulations from completely random point process. Graphically this shows that data do not lie between this theoretical envelope, thus showing a deviation from randomness.

This apparently non randomness of data could be due to clustering of observations in the two main locations Higuerillas-Membrillo and San Antonio, in Fig. 4 we separate data in these two groups and then we apply the same analysis as before. It is showing the map with SASL locations in **A** and **C** and the Ripley's K function for these two groups in **B** and **D**.

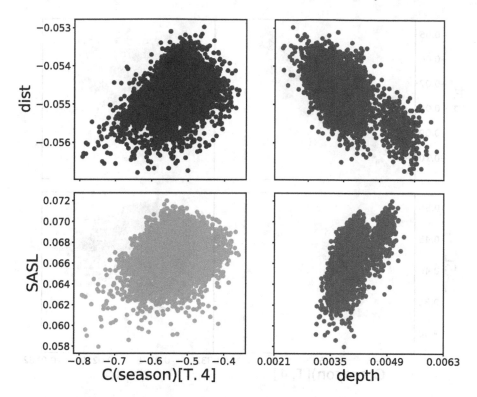

**Fig. 1.** GLM fitting values of the parameters *SASL*, *dist*, *season4* and *depth* for the $10^4$ repetitions with the Poisson family. Each of the four figures represent a combination of the parameter space of interest showing the broad range of values that the GLM fitting can take only by removing from 1 to 7 missing values.

Despite the fact that the two main locations were analyzed separately, results show a huge deviation from spatial randomness in the SASL locations.

### 3.3 Moran's I

Given the location of each SASL in GIS coordinates the task is to check spatial correlations using the Moran's I function.

Using the `ape`, `lctools` and `ncf` packages from the Comprehensive R Archive Network [6] we calculate the Moran's I measure of spatial autocorrelation for the complete data observations and also for the two groups Higuerillas-Membrillo and San Antonio. Tables 1, 2 and 3 show the results for the package `ape`, in this package a matrix of weights must be given, in this tables we put three different distances based on Euclidean, inverse of Euclidean and a Binary like matrix with 1 if the two points are close (Euclidean distance between 0 and 0.0001) and 0 if not.

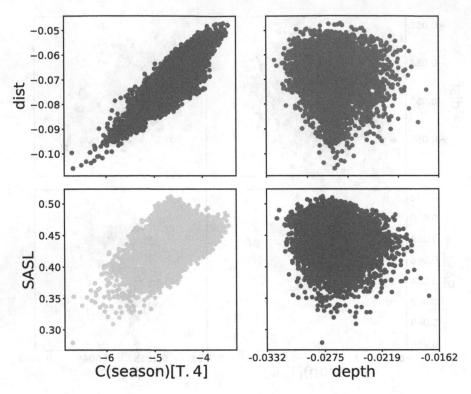

**Fig. 2.** GLM fitting values of the parameters *SASL*, *dist*, *season4* and *depth* for the $10^4$ repetitions with the Gamma family. Each of the four figures represent a combination of the parameter space of interest showing the broad range of values that the GLM fitting can take only by removing from 1 to 7 missing values.

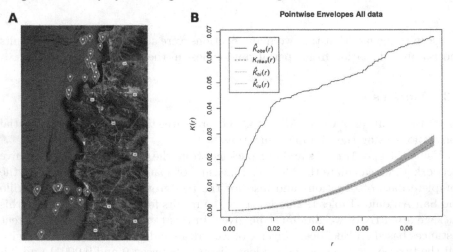

**Fig. 3.** In **A** is the Google Map [7] with marked SASL locations and in **B** is the Ripley's K function for homogeneity in the complete data set.

**Table 1.** Results obtained by the `ape` package from the Comprehensive R Archive Network [6], for the complete data sets

| Weights | Observed | Expected | sd | p-value |
|---|---|---|---|---|
| Binary | −0.0005554022 | −0.004081633 | 0.00294868 | 0.2317483 |
| Euclidean | −0.003766666 | −0.004081633 | 0.0006536919 | 0.6299284 |
| Inverse Euclidean | −0.00122596 | −0.004081633 | 0.001884331 | 0.1296501 |

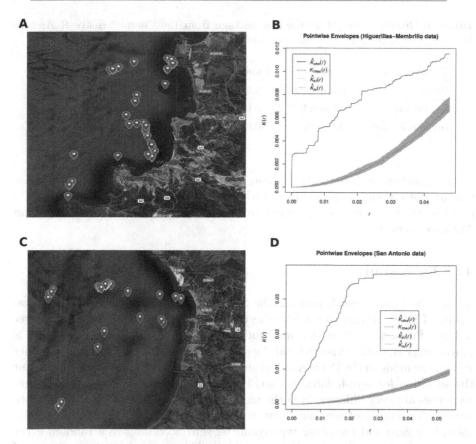

**Fig. 4.** In **A** and **C** is the Google Map [7] with marked SASL locations from Higuerillas-Membrillo and San Antonio respectively. **B** and **D** show the Ripley's K function for homogeneity in the Higuerillas-Membrillo and San Antonio, respectively.

In Table 4 is the result for the `lctools` package from the Comprehensive R Archive Network [6], for which we put the Moran's I, the Expected I and also p-values for 500 resampling.

98    M. Stehlík et al.

**Table 2.** Results obtained by the `ape` package from the Comprehensive R Archive Network [6], for the Higuerillas-Membrillo data sets

| Weights | Observed | Expected | sd | p-value |
|---|---|---|---|---|
| Binary | −0.005474335 | −0.007518797 | 0.004575645 | 0.6550094 |
| Euclidean | −0.009113253 | −0.007518797 | 0.001683345 | 0.3435395 |
| Inverse Euclidean | −0.006186662 | −0.007518797 | 0.002872761 | 0.6428539 |

**Table 3.** Results obtained by the `ape` package from the Comprehensive R Archive Network [6], for the San Antonio data sets

| Weights | Observed | Expected | sd | p-value |
|---|---|---|---|---|
| Binary | 0.9356902 | −0.009009009 | 0.08043459 | 0 |
| Euclidean | −0.09242782 | −0.009009009 | 0.007618862 | 6.719644e−28 |
| Inverse Euclidean | 0.8172632 | −0.009009009 | 0.06535924 | 0 |

The package `ncf` from the Comprehensive R Archive Network [6] gives as output the correlogram as a function of discrete distance classe. Figures 5, 6 and 7 show the results for the complete data set, Higuerillas-Membrillo and San Antonio, respectively.

## 4  Discussion

As our results confirmed, very specific techniques should be developed for measuring of SASL interactions with fisheries. One of the potential models is Zero-inflated Poisson (ZIP) model, used for count data when the zero outcomes is much larger than the expected and the data also present overdispersion. In other words, the mode in the Poisson model should be the integer part of the mean, but the situation for zero-inflated is that, the modal response is 0 and many other responses are very different from the mode, and overall mean will be not near to zero [1]. Two types of zero can occur in the data, one type, as mentioned before, should be zero and the other type could be zero according to a random event (are generated according to a Poisson model). These cases leads to a mixture of

**Table 4.** Results obtained by the `lctools` package from the Comprehensive R Archive Network [6]

| Data | Moran's I | Expected I | p-value resampling | p-value randomization |
|---|---|---|---|---|
| Complete | −0.0001996478 | −0.004081633 | 0.9070651 | 0.07897985 |
| Higuerillas-Membrillo | −0.005207385 | −0.007518797 | 0.9584588 | 0.5467232 |
| San Antonio | 0.8139802 | −0.009009009 | 4.829738e−65 | 4.303355e−65 |

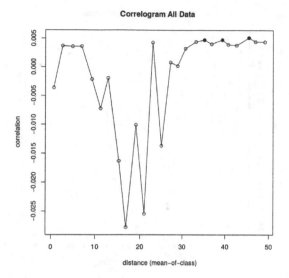

**Fig. 5.** Results obtained with the complete data set by the `ncf` package from the Comprehensive R Archive Network [6]. Black dots indicate that the permutation two-sided p-value for each distance-class is less than 0.05, the opposite is represented by white dots.

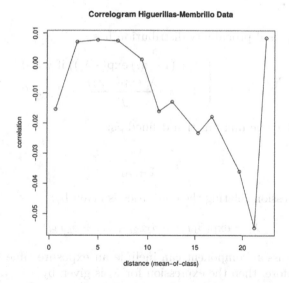

**Fig. 6.** Results obtained with the Higuerillas-Membrillo by the `ncf` package from the Comprehensive R Archive Network [6]. Black dots indicate that the permutation two-sided p-value for each distance-class is less than 0.05, the opposite is represented by white dots.

the ordinary Poisson model with one that indicates the presence of zero:

$$y_i = \begin{cases} 0 & \text{with probability } \pi_i; \\ \text{Poisson}(\lambda_i) & \text{with probability } (1 - \pi_i). \end{cases} \tag{1}$$

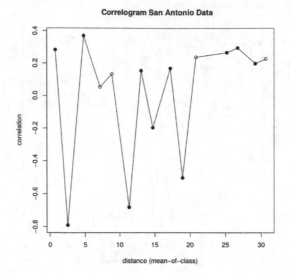

**Fig. 7.** Results obtained with the San Antonio data set by the **ncf** package from the Comprehensive R Archive Network [6]. Black dots indicate that the permutation two-sided p-value for each distance-class is less than 0.05, the opposite is represented by white dots.

The unconditional probability distribution has

$$P(y_i = j) = \begin{cases} \pi_i + (1 - \pi_i)\exp(-\lambda_i)) & \text{if } j = 0; \\ (1 - \pi_i)\dfrac{\lambda_i^j \exp(-\lambda_i)}{j!} & \text{if } j > 0. \end{cases} \tag{2}$$

where $\pi_i$ is the logistic link function defined as:

$$\pi_i = \frac{\lambda_i}{1 + \lambda_i} \tag{3}$$

and the expression relating the quantities is given by:

$$\lambda_i = \exp(\beta_1 x_{1i} + \beta_2 x_{2i} + \ldots + \beta_k x_{ki}) \tag{4}$$

Note: The Poisson component can include an exposure time $t$ and a set of regressors, as before, then the expression for $\lambda_i$ is given by:

$$\lambda_i = \exp(\ln(t_i) + \beta_1 x_{1i} + \beta_2 x_{2i} + \ldots + \beta_k x_{ki}) \tag{5}$$

When a ZIP model is used but there is overdispersion, standard error estimates can be badly biased downward. A zero-inflated negative binomial model (ZINB) model is then more appropriate. In this situation, with probability $\pi_i$, $y_i = 0$, and with probability $1 - \pi_i$, $y_i$ has a negative binomial distribution with mean $\lambda_i$ and dispersion parameter $\gamma$.

For the data in study, a ZIP model was used in order to obtain a first approximation to the estimates. Following the structure of the data, a ZINB was replicated for the data. The results were similar by controlling and considering the overdispersion, in both models. One observation that should be made, is the values were considered as zero since this correspond to the no observation of the SASL on fishing operation, but it was part of their habitat.

**Acknowledgements.** This study was supported by the Undersecretariat for Fisheries and Aquaculture, Chilean Government [Grant number 2013-115-DAP-35, "Characterization of the effects of principal artisan fisheries on marine ecosystems"], and Fondo de Investigación Pesquera [Grant number FIP 2014-29 "Population estimates for sea lions in Regions V, VI, VII and VIII"]. The authors acknowledge Tamara Martínez and Pablo Couve for collecting the SASL data in the field. Milan Stehlík acknowledges the support of project Fondecyt Regular No. 1151441 and LIT-2016-1-SEE-023 mODEC. Jean Paul Maidana acknowledges the support of the PhD. grant FIB-UV from the Universidad de Valparaíso. Maritza Sepúlveda acknowledges the support of the Iniciativa Científica Milenio from Chile's Ministerio de Economía, Fomento y Turismo. The authors were supported by the bilateral projects Bulgaria - Austria, 2016 2019, Feasible statistical modelling for extremes in ecology and finance, Contract number 01/8, 23/08/2017 and WTZ Project No. BG 09/2017.

# References

1. Agresti, A.: Foundations of Linear and Generalized Linear Models. Wiley Series in Probability and Statistics. Wiley, Hoboken (2015)
2. Sepúlveda, M., et al.: Factors affecting the operational interaction between the South American sea lions and the artisan gillnet fishery in Chile. Fish. Res. **201**, 147–152 (2018)
3. Vansteelandt, S., Goetghebeur, E.: Analyzing the sensitivity of generalized linear models to incomplete outcomes via the IDE algorithm. J. Comput. Graph. Stat. **10**(4), 656–672 (2001)
4. Potocký, R., Villarroel, C.N., Sepúlveda, M., Luna, G., Stehlík, M.: Random elements on lattices: review and statistical applications. In: AIP Conference Proceedings, vol. 1863, p. 100007 (2017). https://doi.org/10.1063/1.4992283
5. Python Software Foundation. Python Language Reference, version 2.7. http://www.python.org
6. The Comprehensive R Archive Network (CRAN). http://CRAN.R-project.org
7. Google Maps (2018). https://www.google.cl/maps
8. Seabold, S., Perktold, J.: Statsmodels: econometric and statistical modeling with python. In: 9th Python in Science Conference (2010)

# Finite Difference Scheme for Stochastic Differential Games with Several Singular Control Variables and Its Environmental Application

Hidekazu Yoshioka[1]([✉]) and Yuta Yaegashi[2,3]

[1] Faculty of Life and Environmental Science, Shimane University, Nishikawatsu-cho 1060, Matsue 690-8504, Japan
yoshih@life.shimane-u.ac.jp
[2] Graduate School of Agriculture, Kyoto University, Kitashirakawa-oiwake-cho, Sakyo-ku, Kyoto 606-8502, Japan
yaegashi.yuta.54s@st.kyoto-u.ac.jp
[3] Tokyo, Japan

**Abstract.** Stochastic differential games have recently been key mathematical tools for resolution of environmental and ecological optimization problems. A finite difference scheme is proposed for solving a variational inequality arising in a stochastic differential game having several singular control variables: optimization of algae population management under model ambiguity. The present scheme employs fitted-exponential and upwind discretization methods to generate stable numerical solutions. Accuracy of the scheme is verified against an exact solution to a simplified problem and an asymptotic solution to a more complicated problem. The scheme is finally applied to numerical computation of the optimal algae population management policy against a range of an incurred cost. The computational results suggest that qualitatively different optimal policies are obtained depending on the magnitude of the incurred cost.

## 1 Introduction

Usually, we have several methods for solving an environmental or an ecological problem. For example, population of some waterfowl, a predator of inland fishery resources, can be controlled through direct (kill individuals) or indirect (not kill individuals) countermeasures (Yaegashi et al. 2017). Another example is management of the harmful benthic algae population in dam-downstream (Yoshioka and Yaegashi 2018a). One can decrease the population through cleaning up the riverbed by residents and/or local governments, or through controlling river water quality so that the population growth is suppressed.

Stochastic fluctuations are inherent in environmental and ecological dynamics. Stochastic optimal control and stochastic differential game have recently

Y. Yaegashi—Research Fellow of Japan Society for the Promotion of Science.

© Springer Nature Switzerland AG 2019
I. Dimov et al. (Eds.): FDM 2018, LNCS 11386, pp. 102–113, 2019.
https://doi.org/10.1007/978-3-030-11539-5_10

been useful candidates for approaching such problems. The concept of the multiplier robust control (Hansen and Sargent 2001) has been effectively utilized in particular (Roseta-Palma and Xepapadeas 2004; Yoshioka and Yaegashi 2018b). These problems are not exactly solvable except for simplified cases. This fact motivates us to employing a numerical method for their resolution.

The objective of this paper is to present a finite difference scheme for solving a variational inequality (VI) arising in a stochastic differential game of an algae population management problem under model ambiguity. The algae population dynamics is described with a jump-diffusion stochastic differential equation (SDE) subject to the direct and indirect population decreases, which are represented by singular control variables. The population growth rate is ambiguous in the model. A performance index to be maximized by the most cost-effective algae population management policy is then formulated. The dynamic programming principle leads to the VI to be solved. The present scheme is based on fitted-exponential and upwind methods to generate stable numerical solutions. Accuracy of the scheme is verified against exact and asymptotic solutions. The scheme is finally applied to numerical computation of the optimal algae population management policy.

## 2    Mathematical Model

### 2.1    Stochastic Differential Game

The problem considered here is a simplified counterpart of that in Yoshioka and Yaegashi (2018b). The basic mathematical setting in this paper follows those in the standard literatures (Øksendal and Sulem-Bialobroda 2005; Hansen and Sargent 2001). Growth of the benthic algae population in dam-downstream is subject to continuous fluctuations by base river flows and discontinuous fluctuations by floods. Let $t$ be the time and $X_t$ be the algae population at $t$. Let $B_t$ be the standard Brownian motion and $N_t$ be the Poisson process with the intensity of $T^{-1}$ where $T$ is the mean time interval between successive jumps. $B_t$ and $N_t$ are independent with each other. The control variables $\eta_t^{(1)}$ and $\eta_t^{(2)}$ are adapted, right continuous processes, which represent the direct and indirect countermeasures, respectively. We set $\eta_{-0}^{(1)} = \eta_{-0}^{(2)} = 0$. The growth rate of the population is denoted as $\mu > 0$ and its fluctuation, the volatility, as $\sigma > 0$ with $2\mu > \sigma^2$.

The SDE without model ambiguity is set as

$$\mathrm{d}X_t = X_{t-0}\left(\mu\mathrm{d}t + \sigma\mathrm{d}B_t - \gamma\mathrm{d}N_t\right) - \mathrm{d}\eta_t^{(1)} - X_{t-0}\mathrm{d}\eta_t^{(2)}, \quad X_{-0} = x \geq 0, \quad (1)$$

where $0 < \gamma < 1$ is the jump strength and the last term is understood in the sense of Al Morairi and Zervos (2017). The first through last terms of the SDE (1) represent the population growth, continuous population fluctuation, discontinuous population decrease by floods, population decrease by the direct countermeasure, and population decrease by the indirect countermeasure.

Assume that the decision-maker, the observer of the population dynamics, has only a distorted model (Hansen and Sargent 2001) where the growth rate $\mu$ is ambiguous. Based on the concept of multiplier-robust control, the growth rate $\mu$ is replaced by $\mu + z_t$ where $z_t$ is the variable that represents the ambiguity. The range of $z_t$ is $[-z_{\max}, z_{\max}]$ with a constant $z_{\max} > 0$. Now, the SDE (1) is extended to an ambiguous counterpart as

$$\mathrm{d}X_t = X_{t-0}\left((\mu + z_t)\,\mathrm{d}t + \sigma\mathrm{d}B_t - \gamma\mathrm{d}N_t\right) - \mathrm{d}\eta_t^{(1)} - X_{t-0}\mathrm{d}\eta_t^{(2)}, \quad X_{-0} = x \geq 0. \quad (2)$$

The control variables in this model are $(\eta_t^{(1)}, \eta_t^{(2)}, z_t)$, where $(\eta_t^{(1)}, \eta_t^{(2)})$ are chosen by the observer and $z_t$ by nature. These variables are chosen so that the SDE (2) has a unique strong solution. The observer and nature are the maximizing and minimizing agents, respectively, of the differential game presented below.

## 2.2   Variational Inequality

The performance index to be optimized by choosing the controls is set as

$$J\left(x; \eta^{(1)}, \eta^{(2)}, z\right) = \mathrm{E}^x\left[-a\int_0^\tau e^{-\delta t}X_t^m\mathrm{d}t - \int_0^\tau e^{-\delta t}\mathrm{d}(b\eta_t^{(1)} + c\eta_t^{(2)})\right] + J_p, \quad (3)$$

$$J_p = \mathrm{E}^x\left[d\int_0^\tau e^{-\delta t}\frac{1}{2}X_t^M z_t^2\mathrm{d}t\right], \quad (4)$$

where $\mathrm{E}^x$ represents the expectation conditioned on $X_{-0} = x$. Here, $\delta > 0$ is the discount rate that is assumed to be sufficiently large, $a, b, c, d > 0$ are weight constants, $m, M > 1$ are constant parameters, and $\tau$ is the first hitting time of $X_t = 0$ for $t \geq 0$. The first and second terms in the right-hand side of (3) evaluate the disutility caused by the existence of the algae (Yoshioka and Yaegashi 2018b) and the costs of the countermeasures, respectively. The term $J_p$ represents the penalty due to the model ambiguity; large (small) $d$ implies less (more) ambiguity-aversion. The limit $d \to +\infty$ formally leads to a problem without model ambiguity. Hereafter, we assume $M = m$ so that the problem becomes more tractable.

The value function $u = u(x)$ is the saddle-point of $J$:

$$u(x) = \sup_{\eta^{(1)}, \eta^{(2)}} \inf_z J\left(x; \eta^{(1)}, \eta^{(2)}, z\right). \quad (5)$$

Here, $\delta$ is assumed to be sufficiently large so that $u(x)$ is locally bounded with respect to $x \geq 0$. The control variables chosen to optimize $J$ is indicated by the super-script $*$ like $z_t^*$. The dynamic programming principle leads to the VI of $u$:

$$\min\left(\max_{-z_{\max} \leq z \leq z_{\max}} L_z u + ax^m, \frac{\mathrm{d}u}{\mathrm{d}x} + b, x\frac{\mathrm{d}u}{\mathrm{d}x} + c\right) = 0, \quad x > 0, \quad u(0) = 0 \quad (6)$$

with

$$L_z u = \delta u - (\mu + z)x\frac{\mathrm{d}u}{\mathrm{d}x} - \frac{\sigma^2}{2}x^2\frac{\mathrm{d}^2u}{\mathrm{d}x^2} + \frac{1}{T}(u - u((1-\gamma)x)) - \frac{d}{2}x^m z^2. \quad (7)$$

The present VI is of a non-local degenerate elliptic type. Actually, the optimal controls $\eta_t^{(1)}$ and $\eta_2^{(1)}$ are of the threshold type exerted at some free boundaries of the VI (6) as shown later. Therefore, both $u$ and its free boundaries have to be accurately computed. Threshold-type policies are common in environmental management (McAllister et al. 2016). We assume that the value function $u$ is a continuous viscosity solution (Øksendal and Sulem-Bialobroda 2005) to the VI (6).

## 2.3  On the Value Function

Here, several results on the value function $u$ are presented. At least formally, most of them can be derived with the techniques in Yoshioka and Yaegashi (2018b). These results are utilized for verifying the numerical scheme presented later. The characteristic function for the set $S$ is denoted as $\chi(S)$. The optimizer that achieves $\max_z L_z u$ is denoted as $z = z^*(x)$ with an abuse of notations.

Firstly, parameter dependence of $u$ is found as follows.

**Proposition 1.** *$u(x)$ for each $x > 0$ is increasing with respect to $\sigma$, $\gamma$, $\delta$, $d$. In addition, it is decreasing with respect to $\mu$, $T$, $a$, $b$, $c$, $z_{\max}$.*

Secondly, $u(x)$ is non-increasing and Lipschitz continuous with respect to $x \geq 0$.

**Proposition 2.** *$u$ satisfies $u(0) = u(+0) = 0$ and*

$$0 \leq u(x) - u(y) \leq \min\left(b\,(y - x), c\,(\ln y - \ln x)\right), \quad y \geq x > 0. \tag{8}$$

Thirdly, for the no ambiguity case ($d \to +\infty$), the VI (6) is exactly-solvable in a viscosity sense under certain conditions.

**Proposition 3.** *For sufficiently large $b$, the viscosity solution (classical solution) $u \in C[0, +\infty) \cap C^2(0, +\infty)$ is found as*

$$u(x) = \chi(0 \leq x < \bar{x}_1)\left(C_1 x^k + C_2 x^m\right) + \chi(x \geq \bar{x}_1)\left(C_3 - c\ln x\right). \tag{9}$$

*Here, $k$ is the unique positive solution to*

$$\Omega(k) = \delta - \mu k - k(k-1)\sigma^2/2 + T^{-1}\left(1 - (1-\gamma)^k\right) = 0, \tag{10}$$

*where $C_1 x^m$ is the particular solution to $L_0 u + ax^m = 0$ with $x > 0$, and $C_2$, $C_3$, and $\bar{x}_1 > 0$ are determined from the twice continuous differentiability of $u$ at $x = \bar{x}_1$. Similarly, for sufficiently small $b$, the viscosity solution $u \in C[0, +\infty) \cap C^2(0, +\infty)$ is found as*

$$u(x) = \chi(0 \leq x < \bar{x}_2)\left(C_1 x^k + C_4 x^m\right)$$
$$+ \chi(\bar{x}_2 \leq x < \bar{x}_3)\left(C_5 - bx\right) + \chi(x \geq \bar{x}_3)\left(C_6 - c\ln x\right), \tag{11}$$

*where $C_4$, $C_5$, and $\bar{x}_2 > 0$ are determined from the twice continuous differentiability of $u$ at $x = \bar{x}_2$, and $C_6$ from the continuity of $u$ at $\bar{x}_3 = c/b$.*

Finally, for $d < +\infty$, an asymptotic estimate of $u$ is derived.

**Proposition 4.** *Assume $d$ and $z_{\max}$ are sufficiently large and $u$ is twice contin-uously differentiable near $x = 0$. Then, for small $x > 0$, $u$ and $z^*$ are asymptot-ically expressed as*

$$u(x) = C_1 x^m, \quad z^*(x) = m^{-1}\Omega(m)\left(1 - \sqrt{1 - d^{-1}d_c}\right), \tag{12}$$

$$d_c = 2am^2\left[\Omega(m)\right]^{-2}. \tag{13}$$

Propositions 1 and 2 characterize the relationship among the solution profile and the parameters and variables. The exact solution with $d \to +\infty$ for the moderate range of $b$ has not been analytically obtained but numerical computa-tions presented later indicate that there exist three free boundaries in this range. Proposition 3 can be utilized for accuracy verification of numerical schemes. The free boundary $x = \bar{x}_1$ is the threshold at which the threshold-type singular control is carried out so that $X_t$ is confined in $[0, \bar{x}_1]$. Proposition 3 gives the asymptotic solution to the VI (6), which can also be used for the accuracy verifi-cation. In addition, it shows that the asymptotic solution breaks down for $z < z_c$: namely, for small risk-aversion. Similar results have been observed in some dif-ferential game problems in finance (Athanassoglou and Xepapadeas 2012; Miao and Rivera 2012). This asymptotic estimate is sharp as demonstrated later.

# 3    Finite Difference Scheme

## 3.1    Discretization

The domain for computation of the VI (6) is set as $(0, A)$ with a sufficiently large constant $A > 0$. Hence, $x = A$ is an artificial boundary. The boundary condition

$$\min\left(\frac{\mathrm{d}u}{\mathrm{d}x} + b, x\frac{\mathrm{d}u}{\mathrm{d}x} + c\right) = 0 \tag{14}$$

is specified at $x = A$. This boundary condition is based on the conjecture that $\max_z L_z u + ax^m > 0$ for sufficiently large $x$, which can be proven at least for the no ambiguity case ($d \to +\infty$) with the help of a simple proof by contradiction.

The finite difference scheme here is based on the Conforming Petrov-Galerkin Finite Element (CPGFE) formalism (Yoshioka et al. 2014), which has been applied to Hamilton-Jacobi-Bellman equations (Yoshioka and Yaegashi 2018a,b). The trial and test functions of the CPGFE scheme are constructed based on the exact solutions to local two-point boundary value problems as in De Falco and O'Riordan (2011). The domain is discretized into $I \geq 1$ cells $I_i = (x_i, x_{i+1})$ ($i = 0, 1, 2, ..., I - 1$) with $0 = x_0 < x_1 <, ..., < x_I = A$. The quantity discretized at the vertex $x_i$ is represented with the sub-script $i$ like $u_i$. The length of $I_i$ is denoted as $h_i$. The coefficients are attributed to the cells.

Consider the discretization at an inner vertex $x_i$ ($1 \leq i \leq I - 1$). The scheme employs a exponentially-fitted discretization for the first element in min of the

VI (6) and the conventional upwind discretization for the others. For the latter, we simply obtain the one-sided discretization

$$\frac{du}{dx} + b \rightarrow \frac{u_i - u_{i-1}}{h_{i-1}} + b, \quad x\frac{du}{dx} + c \rightarrow x_i\frac{u_i - u_{i-1}}{h_{i-1}} + c. \tag{15}$$

For the former, the discretization is obtained with the CPGFE formalism:

$$\delta u - \mu x\frac{du}{dx} - \frac{\sigma^2}{2}x^2\frac{d^2u}{dx^2} \rightarrow -p_i u_{i-1} - r_i u_{i+1} + (p_i + q_i + r_i)u_i, \tag{16}$$

$$\frac{1}{T}\left(u - u\left((1 - \gamma)x\right)\right) + ax^m \rightarrow \frac{s_i}{T}\left(u_i - u_{l(i)}\right) + s_i a x_i^m, \tag{17}$$

$$\max_z\left(-z\frac{du}{dx} - \frac{d}{2}x^m z^2\right) \rightarrow -z_i^*\frac{u_{i+1} - u_i}{h_i} - \frac{d}{2}x_i^m(z_i^*)^2 \tag{18}$$

with

$$z_i^* = \max\left(-z_{\max}, \min\left(-\frac{1}{dx_i^m}\frac{u_{i+1} - u_i}{h_i}, z_{\max}\right)\right). \tag{19}$$

The coefficients $p_i$, $q_i$, $r_i$, $s_i$ are given as follows:

$$p_i = \frac{W_{\kappa(i,j)}}{h^2_{\kappa(i,j)}}\frac{\zeta^+_{\kappa(i,j)} - \zeta^-_{\kappa(i,j)}}{e^{\zeta^+_{\kappa(i,j)}} - e^{\zeta^-_{\kappa(i,j)}}}\Bigg|_{j=0}, \quad r_i = \frac{W_{\kappa(i,j)}}{h^2_{\kappa(i,j)}}\frac{\zeta^+_{\kappa(i,j)} - \zeta^-_{\kappa(i,j)}}{e^{\zeta^+_{\kappa(i,j)}} - e^{\zeta^-_{\kappa(i,j)}}}\Bigg|_{j=1}, \tag{20}$$

$$q_i = -\sum_{j=0}^{1}\left(\frac{W_{\kappa(i,j)}}{h^2_{\kappa(i,j)}}\frac{\zeta^+_{\kappa(i,j)}e^{\zeta^-_{\kappa(i,j)}} - \zeta^-_{\kappa(i,j)}e^{\zeta^+_{\kappa(i,j)}}}{e^{\zeta^+_{\kappa(i,j)}} - e^{\zeta^-_{\kappa(i,j)}}}\right) - p_i - r_i, \tag{21}$$

$$s_i = \sum_{j=0}^{1}\frac{1}{e^{\zeta^+_{\kappa(i,j)}} - e^{\zeta^-_{\kappa(i,j)}}}\left(\frac{e^{\zeta^+_{\kappa(i,j)}} - 1}{\zeta^+_{\kappa(i,j)}} - \frac{e^{\zeta^-_{\kappa(i,j)}} - 1}{\zeta^-_{\kappa(i,j)}}\right), \tag{22}$$

respectively. Here, $\kappa(i,0) = i - 1$, $\kappa(i,1) = i$, $\sigma_{i,0} = -1$, $\sigma_{i,1} = 1$,

$$\zeta^{\pm}_{\kappa(i,j)} = \frac{-P_{\kappa(i,j)} \pm \sqrt{P^2_{\kappa(i,j)} + 4D_{\kappa(i,j)}}}{2}, \quad P_{\kappa(i,j)} = \frac{\sigma_{i,j}V_{\kappa(i,j)}h_{\kappa(i,j)}}{W_{\kappa(i,j)}}, \quad D_{\kappa(i,j)} = \frac{\delta h^2_{\kappa(i,j)}}{W_{\kappa(i,j)}}, \tag{23}$$

$V = \mu x$, $W = 0.5\sigma^2 x^2$. The quantity $u_{l(i)}$ in the non-local term is given as

$$u_{l(i)} = w_i u_{i_0} + (1 - w_i)u_{i_1}, w_i = \frac{x_{i_1} - (1 - \gamma)x_i}{x_{i_1} - x_{i_0}}, \tag{24}$$

where the non-negative integers $i_0$ and $i_1$ satisfy $i_1 = i_0 + 1$ and $x_{i_0} \leq (1 - \gamma)x_i \leq x_{i_1}$. Such $i_0$ and $i_1$ are uniquely determined for each $i$. What is important in the present scheme is the positive coefficient condition

$$p_i, q_i, r_i, s_i > 0, \tag{25}$$

and the fact that the present scheme can generate exact solutions to linear advection-diffusion-decay equations with constant coefficients. At the boundary $x = 0$, the boundary condition $u_0 = 0$ is directly specified. On the other hand, the boundary condition at $x = A$ in (14) is discretized following (15).

Assembling the discretized equations at all the vertices with the help of a scaled direct control formulation (Huang et al. 2012) yields the nonlinear system

$$\mathbf{Mu} = \mathbf{f}, \tag{26}$$

where $\mathbf{u} = [u_i]$ is the $(I + 1)$-dimensional solution vector, $f$ is the vector that does not explicitly involve $\mathbf{u}$, and $\mathbf{M} = [m_{i,j}]$ is the $(I+1) \times (I+1)$-dimensional coefficient matrix that depends on $\mathbf{u}$. Each $z_i^*$, which depends on $u_i$ and $u_{i+1}$, is involved in the matrix $\mathbf{M}$ and the non-local term is involved in $\mathbf{f}$. The condition (25) guarantees that the present discretization for $\max_z L_z u$ is of the positive coefficient type, and that the matrix $\mathbf{M}$ is an M-matrix. Notice that Oberman (2006) discussed Dirichlet problems while ours is a Dirichlet-Neumann problem. However, the mathematical analysis results of Oberman (2006) apply to our scheme by seeing $u_N - u_{N-1}$ but not $u_N$ as a new nodal unknown. Then, the present finite difference scheme is Lipschitz, proper, and degenerate elliptic in the sense of Oberman (2006), indicating its unique solvability.

**Proposition 5.** *The present scheme admits a unique numerical solution.*

The system (26) is nonlinear, and its solution has to be found with an iterative algorithm. A policy iteration algorithm (Forsyth and Vetzal 2012) is employed to efficiently solve the system (26). Let $n \geq 0$ be the step number in the policy iteration algorithm. Starting from an initial guess $\mathbf{u}^{(0)}$, the sequence $\mathbf{u}^{(n)}$ ($n = 1, 2, 3, ...$) is iteratively computed as

$$\mathbf{M}^{(n-1)} \mathbf{u}^{(n)} = \mathbf{f}^{(n-1)} \quad (n = 1, 2, 3, ...), \tag{27}$$

which is expected to give the unique solution to the system (26), the desired numerical solution. The linear system (27) is solved with the TriDiagonal Matrix Algorithm method (Thomas 1949). The iteration step in the policy iteration algorithm is terminated at the $n$th step ($n \geq 1$) such that

$$\max_{1 \leq i \leq I} |u_i^{(n)} - u_i^{(n-1)}| < \epsilon \tag{28}$$

with a sufficiently small error threshold $\epsilon$, which is set as $10^{-13}$ in this paper. Performance of the present finite difference scheme is examined in what follows.

## 4    Numerical Computation

Hereafter, the computational domain is uniformly discretized into $I$ cells with the length of $h = AI^{-1}$ with $A = 4$. The model parameters are set as follows unless otherwise specified: $\mu = 0.1$, $\sigma = 0.2$, $\gamma = 0.33$, $T = 1.5$, $a = 0.4$, $b = 1.0$, $c = 2.8$, $d = 10$, $m = 2$, and $z_{\max} = 10$. In each computation, numerical free boundaries are placed at the cell-centers.

**Table 1.** Errors $E_s$ and $E_f$ and convergence rate $R_s$ for the large $b$ case.

| $I$ | $h$ | $E_s$ | $R_s$ | $E_f$ |
|---|---|---|---|---|
| 100 | 0.04 | 1.32E$-$2 | 9.9E$-$1 | 2.62E$-$3 |
| 200 | 0.02 | 6.64E$-$3 | 1.0E+0 | 7.38E$-$3 |
| 400 | 0.01 | 3.32E$-$3 | 1.0E+0 | 2.38E$-$3 |
| 800 | 0.005 | 1.66E$-$3 | 1.0E+0 | 1.24E$-$4 |
| 1,600 | 0.0025 | 8.29E$-$4 | 1.0E+0 | 1.13E$-$3 |
| 3,200 | 0.00125 | 4.15E$-$4 | | 5.01E$-$4 |

## 4.1 Comparison with an Exact Solution

The proposed finite difference scheme is firstly examined against the exact solutions for sufficiently large $b(= 1.5)$ and sufficiently small $b(= 1.0)$ presented in Proposition 3. Table 1 shows the computed $l^\infty$ error $(E_s)$ for the solution $u$ and its convergence rate $(R_s)$, and the absolute error $(E_f)$ for the free boundary $x = \bar{x}_1$ for the large $b$ case. Similarly, Table 2 shows the computed error $E_s$, its convergence rate $C_s$, and the absolute errors $E_{fL}$ and $E_{fR}$ for the free boundaries $\bar{x}_2$ and $\bar{x}_3$, respectively, for the small $b$ case. Figure 1 compares the numerical and exact solutions for the small $b$ case with $I = 100$, indicating that the scheme generates non-oscillatory numerical solutions with reasonably capturing the free boundaries. Qualitatively similar results have been obtained for the large $b$ case. The present scheme is first-order accurate for $u$ and computes the free boundaries with the error smaller than the cell length $h$. The computational results suggest satisfactory ability of the scheme to handle the VI. In this sense, accuracy of the scheme on the free boundaries is first-order.

**Table 2.** Errors $E_s$, $E_{fL}$, $E_{fR}$, and convergence rate $R_s$ for the small $b$ case.

| $I$ | $h$ | $E_s$ | $R_s$ | $E_{fL}$ | $E_{fR}$ |
|---|---|---|---|---|---|
| 100 | 0.04 | 5.97E$-$3 | 1.0E+0 | 1.79E$-$2 | 2.00E$-$2 |
| 200 | 0.02 | 3.00E$-$3 | 1.0E+0 | 7.92E$-$3 | 1.00E$-$2 |
| 400 | 0.01 | 1.50E$-$3 | 1.0E+0 | 2.92E$-$3 | 5.00E$-$3 |
| 800 | 0.005 | 7.51E$-$4 | 1.0E+0 | 4.24E$-$4 | 2.50E$-$3 |
| 1,600 | 0.0025 | 3.74E$-$4 | 1.0E+0 | 8.26E$-$4 | 1.25E$-$3 |
| 3,200 | 0.00125 | 1.88E$-$4 | | 2.01E$-$4 | 6.25E$-$4 |

## 4.2 Comparison with an Asymptotic Solution

Accuracy of the finite difference scheme is verified against the asymptotic results of Proposition 4. The focus here is on the ability of the scheme to capture the

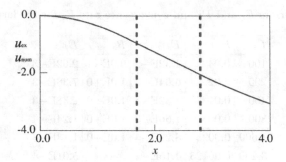

**Fig. 1.** Comparison of the exact (Red) and numerical solutions (Blue) for the small $b$ case with $I = 100$. The solid and dotted lines show the solutions and free boundaries, respectively. (Color figure online)

asymptotic monomial profile of $u$ and the constant profile of $z^*$. For the sake of brevity, the scaled optimal control $Z^*$ is introduced as

$$Z^* = z^* \left(1 - \sqrt{1 - d^{-1}d_c}\right)^{-1}, \tag{29}$$

which does not depend on the weight constant $d$. The domain is uniformly discretized into $I = 400$ cells. Figure 2 compares the asymptotic and computed $u$ for the weight constants $d$ of 10, 100, and 500. Similarly, Fig. 3 compares the asymptotic and computed $Z^*$ for the same values of $d$. The threshold value of $d$ for the given model parameter values is $d_c = 2.504$, and hence the computational condition here complies with $d > d_c$. The computational results suggest reasonable agreement between the asymptotic and numerical solutions. The computed $Z^*$ has slight oscillations near $x = 0$, which disappear as $x$ increases. Their agreement becomes better for larger $d$: namely, for less risk-aversion. It is remarkable that numerical solutions approximately satisfy $\max_z L_z u > 0$ when $d$ is small, implying that Proposition 4 indeed does not hold true for such cases.

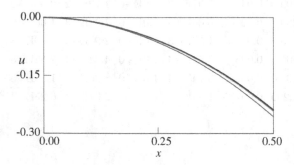

**Fig. 2.** Comparison of the asymptotic (Black) and numerical solutions for $a(= 0.10)$ and $I = 400$ with $d = 10$ (Blue), $d = 50$ (Pink), and $d = 100$ (Red). (Color figure online)

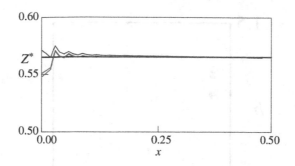

**Fig. 3.** Comparison of the asymptotic (Black) scaled optimal control $Z^*$ for $b = 1.0$ and $I = 400$ with $d = 10$ (Blue), $d = 50$ (Pink), and $d = 100$ (Red). (Color figure online)

### 4.3   Optimal Management Policy

The optimal management policy for the range of $1.0 \leq b \leq 1.5$ is computed using the present scheme with $I = 1,000$, which has been found to be a sufficiently fine resolution by preliminary computations. Figure 4 presents the free boundaries that govern the optimal management policy. The results suggest that the following policy is optimal. On the free boundaries $\bar{x}_1$ through $\bar{x}_4$, see Fig. 4.

- (For large $b$)
  (a) $0 \leq x \leq \bar{x}_1$: Let $\eta^{(2),*}$ be the local time measure at $x = \bar{x}_1$ and confine $X_t$ in $[0, \bar{x}_1]$.
  (b) $\bar{x}_1 < x$: Immediately decrease $X_t$ to $\bar{x}_1$ by $\eta^{(2),*}$. Then, follow (a).

- (For small $b$)
  (a) $0 \leq x \leq \bar{x}_2$: Let $\eta^{(1),*}$ be the local time measure at $x = \bar{x}_2$ and confine $X_t$ in $[0, \bar{x}_2]$
  (b) $\bar{x}_2 < x \leq \bar{x}_3$: Immediately decrease $X_t$ to $\bar{x}_2$ by $\eta^{(1),*}$. Then, follow (a).
  (c) $\bar{x}_3 < x$: Immediately decrease $X_t$ to $\bar{x}_3$ by $\eta^{(2),*}$. Then, follow (b).

- (For moderate $b$)
  (a) $0 \leq x \leq \bar{x}_2$: Let $\eta^{(1),*}$ be the local time measure at $x = \bar{x}_2$ and confine $X_t$ in $[0, \bar{x}_2]$.
  (b) $\bar{x}_2 < x \leq \bar{x}_4$: Immediately decrease $X_t$ to $\bar{x}_2$ by $\eta^{(1),*}$. Then, follow (a).
  (c) $\bar{x}_4 < x \leq \bar{x}_1$: Let $\eta^{(2),*}$ be the local time measure at $x = \bar{x}_1$ and confine $X_t$ in $(\bar{x}_4, \bar{x}_1]$. When $X_t$ hits $x = \bar{x}_4$, then follow (b).
  (d) $\bar{x}_1 < x$: Immediately decrease $X_t$ to $\bar{x}_1$ by $\eta^{(2),*}$. Then, follow (c).

The results thus suggest that the algae population, when it is large, should be firstly indirectly decreased, and then they should be directly controlled through the direct countermeasure. In addition, the direct countermeasure should not be performed for large $b$. Similar computational results have been obtained for smaller and moderately larger $d$, implying robustness of the optimal policy.

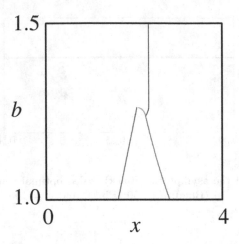

**Fig. 4.** Free boundaries for a range of $(x, b)$. Blue: $x = \bar{x}_1$, Red: $x = \bar{x}_2$, Pink: $x = \bar{x}_3$, and Green: $x = \bar{x}_4$. (Color figure online)

## 5    Conclusions

A finite difference scheme was developed for solving a variational inequality arising in a stochastic differential game of the algae population management under model ambiguity. Accuracy of the present scheme was verified numerically. The obtained analysis results suggest satisfactory performance of the scheme. Optimal algae population management policy against a range of the incurred cost was finally computed with the scheme. The scheme can be easily applied to problems with more than two singular control variables.

In reality, river environment seasonally changes. The present problem should then be extended to a time-dependent stochastic differential game. The proposed finite difference scheme can be applied to time-dependent variational inequalities without significant difficulties. We recently found that the present scheme is applicable to numerical computation of quasi-variational inequalities arising in impulse control problems (Cadenillas et al. 2006) as well. The topics above are currently in progress from both theoretical and practical standpoints.

**Acknowledgments.** This work was supported by The River Foundation under grant The River Fund No. 285311020, The Japan Society for the Promotion Science under grant KAKENHI No. 17K15345 and No. 17J09125, and Water Resources Environment Center under grant The WEC Applied Ecology Research Grant No. 2016-02.

# References

Yaegashi, Y., Yoshioka, H., Unami, K., Fujihara, M.: Optimal policy of predator suppression for sustainable inland fishery management. In: Proceedings of 12th SDEWES Conference, pp. 309-1–309-11 (2017)

Yoshioka, H., Yaegashi, Y.: Stochastic control model of dam discharge for algae growth management. J. Biol. Dyn. **12**, 242–270 (2018a)

Hansen, L., Sargent, T.J.: Robust control and model uncertainty. Am. Econ. Rev. **91**, 60–66 (2001)

Yoshioka, H., Yaegashi, Y.: Robust stochastic control modeling of dam discharge to suppress overgrowth of downstream harmful algae. Appl. Stoch. Model. Bus. (2018b, to appear)

Roseta-Palma, C., Xepapadeas, A.: Robust control in water management. J. Risk Uncertain. **29**, 21–34 (2004)

Øksendal, B., Sulem-Bialobroda, A.: Applied Stochastic Control of Jump Diffusions. Springer, Heidelberg (2005). https://doi.org/10.1007/b137590

Al Morairi, H., Zervos, M.: Irreversible capital accumulation with economic impact. Appl. Math. Optim. **75**, 525–551 (2017)

McAllister, T.G., Wood, S.A., Hawes, I.: The rise of toxic benthic Phormidium proliferations: a review of their taxonomy, distribution, toxin content and factors regulating prevalence and increased severity. Harmful Algae **55**, 282–294 (2016)

Athanassoglou, S., Xepapadeas, A.: Pollution control with uncertain stock dynamics: when, and how, to be precautious. J. Environ. Econ. Manag. **63**, 304–320 (2012)

Miao, J., Rivera, A.: Robust contracts in continuous time. Econometrica **84**, 1405–1440 (2016)

Yoshioka, H., Unami, K., Fujihara, M.: Mathematical analysis on a conforming finite element scheme for advection-dispersion-decay equations on connected graphs. J. JSCE. Ser. A2 **70**, I265–I274 (2014)

De Falco, C., O'Riordan, E.: A parameter robust Petrov-Galerkin scheme for advection-diffusion-reaction equations. Numer. Algorithms **56**, 107–127 (2011)

Oberman, A.M.: Convergent difference schemes for degenerate elliptic and parabolic equations: Hamilton-Jacobi equations and free boundary problems. SIAM J. Numer. Anal. **44**, 879–895 (2006)

Huang, Y., Forsyth, P.A., Labahn, G.: Iterative methods for the solution of a singular control formulation of a GMWB pricing problem. Numer. Math. **122**, 133–167 (2012)

Forsyth, P.A., Vetzal, K.R.: Numerical methods for nonlinear PDEs in finance. In: Duan, J.C., Härdle, W., Gentle, J. (eds.) Handbook of Computational Finance, pp. 503–528. Springer, Heidelberg (2012). https://doi.org/10.1007/978-3-642-17254-0_18

Thomas, L.H.: Elliptic problems in linear difference equations over a network. Watson Sci. Comp. Lab. Rep. Columbia University, New York (1949)

Cadenillas, A., Choulli, T., Taksar, M., Zhang, L.: Classical and impulse stochastic control for the optimization of the dividend and risk policies of an insurance firm. Math. Financ. **16**, 181–202 (2006)

# Stability Properties of Repeated Richardson Extrapolation Applied Together with Some Implicit Runge-Kutta Methods

Zahari Zlatev[1] , Ivan Dimov[2], István Faragó[3,4] , Krassimir Georgiev[2] ,
and Ágnes Havasi[3,4(✉)]

[1] Department of Environmental Science, Aarhus University, Roskilde, Denmark
zz@envs.au.dk
[2] Institute of Information and Communication Technologies,
Bulgarian Academy of Sciences, Sofia, Bulgaria
ivdimov@bas.bg, georgiev@parallel.bas.bg
[3] MTA-ELTE Numerical Analysis and Large Networks Research Group,
Budapest, Hungary
[4] Department of Applied Analysis and Computational Mathematics,
Eötvös Loránd University, Budapest, Hungary
{faragois,havasia}@cs.elte.hu

**Abstract.** Repeated Richardson Extrapolation can successfully be used in the efforts to improve the efficiency of the numerical treatment of systems of ordinary differential equations (ODEs) mainly by increasing the accuracy of the computed results. It is assumed in this paper that Implicit Runge-Kutta Methods (IRKMs) are used in the numerical solution of systems of ODEs. If the order of accuracy of the selected IRKM is $p$, then the order of accuracy of its combination with the Repeated Richardson Extrapolation is at least $p + 2$ (assuming here that the right-hand-side of the system of ODEs is sufficiently many times continuously differentiable). However, it is additionally necessary to establish that the absolute stability properties of the new numerical methods (that are combinations of the Repeated Richardson Extrapolation and the selected IRKMs) are preserved, and this is an extremely difficult problem. Results related to the stability of the computations are derived and numerical tests with a two-parameter system of three ODEs and an atmospheric chemical scheme with 56 compounds, which is defined mathematically by a very stiff and ill-conditioned system of non-linear ODEs, are presented. The research results described in this paper can be considered as a continuation of the study carried out in Zlatev et al.: Richardson Extrapolation: Practical Aspects and Applications. De Gruyter, Berlin (2017).

**Keywords:** Systems of ordinary differential equations (ODEs) ·
Implicit Runge-Kutta Methods · Repeated Richardson Extrapolation ·
Absolute stability properties · Atmospheric chemical schemes

I. Dimov et al. (Eds.): FDM 2018, LNCS 11386, pp. 114–125, 2019.
https://doi.org/10.1007/978-3-030-11539-5_11

# 1   Implicit Runge-Kutta Methods (IRKMs)

Assume that systems of first-order ordinary differential equations (ODEs):

$$\frac{dy}{dt} = f(t,y), \ t \in [a,b], \ y \in \mathbb{R}^s, \ s \geq 1, \ f \in D \subset \mathbb{R} \times \mathbb{R}^s, \ y(a) = \eta \qquad (1)$$

are to be solved on the following set of equidistant grid-points:

$$t_0 = a, \ t_n = t_{n-1} + h \ (n = 1, 2, \dots, N), \ t_N = b, \ h = (b-a)/N \qquad (2)$$

by applying an arbitrary one-step numerical method of order $p$. The fact that one-step numerical methods are used means that only the approximation $y_{n-1} \approx y(t_{n-1})$ is used in the calculation of the next approximation $y_n \approx y(t_n)$ for any step $n \in \{1, 2, \dots, N\}$.

The $m$-stage Fully Implicit Runge-Kutta Methods (FIRKMs) are one-step numerical methods for solving systems of ODEs (1) that are based on the following formula:

$$y_n = y_{n-1} + h \sum_{i=1}^{m} c_i k_i^n. \qquad (3)$$

The coefficients $c_i$ are constants, while the stage vectors $k_i^n$ are defined by

$$k_i^n = f(t_{n-1} + ha_i, y_{n-1} + h \sum_{j=1}^{m} b_{ij} k_j^n), \ a_i = \sum_{j=1}^{m} b_{ij}, \qquad (4)$$

where $i = 2, 3, \dots, m$ and $b_{ij}$ are constants depending on the particular numerical method.

By using a fully implicit Runge-Kutta method one can achieve good stability and high accuracy, but the computational cost is high. One possibility to reduce the costs is to apply a high-order Diagonally Implicit Runge-Kutta Method (DIRKM). DIRKMs form a sub-class of IRKMs where

$$k_i^n = f(t_{n-1} + ha_i, y_{n-1} + h \sum_{j=1}^{i-1} b_{ij} k_j^n + \gamma k_i^n), \ a_i = \sum_{j=1}^{i-1} b_{ij} + \gamma. \qquad (5)$$

Their major advantage is that $m$ systems of order $s$ are to be solved instead of one large system the coefficient matrix of which is of order $ms$. This leads, roughly speaking, to a reduction of the computational time by a factor of $\mathcal{O}(m^2)$. It is expected that the factorization computed during the solution of the first system in (5) can also be applied in the solution of the remaining systems because the same $\gamma$ is used in all stages. In this paper we investigate another possibility to enhance the efficiency: to apply a low-order IRKM, and combine it with Repeated Richardson Extrapolation.

## 2    Definition of the Repeated Richardson Extrapolation

If the Repeated Richardson Extrapolation (RRE) is to be used for the solution of problem (1), then the approximation $y_n$ can be computed by using the previous approximation $y_{n-1}$ and by performing successively the following four calculation processes:

**Calculation process 1:** Compute an approximation $z_n^{[1]}$ of the solution of (1) at the point $t = t_n$ by using the selected one-step numerical method with a stepsize $h$ (i.e. one step is to be performed during this process).

**Calculation process 2:** Compute an approximation $z_n^{[2]}$ of the solution of (1) at the point $t = t_n$ by using the selected one-step numerical method with a stepsize $h/2$ (i.e. two steps are to be performed during this process).

**Calculation process 3:** Compute an approximation $z_n^{[3]}$ of the solution of (1) at the point $t = t_n$ by using the selected one-step numerical method with a stepsize $h/4$ (i.e. four steps are to be performed during this process).

**Calculation process 4:** Compute an approximation $y_n$ of the solution of (1) at the point $t = t_n$ by calculating a linear combination of the approximations $z_n^{[1]}$, $z_n^{[2]}$ and $z_n^{[3]}$ obtained in the previous three calculation processes.

The weights in the linear combination used in step 4 are to be specified so as to get a method of order $p + 2$. If the right-hand side function $f$ in (1) is twice continuously differentiable, then the following relations hold:

$$y(t_n) - z_n^{[1]} = h^p K_1 + h^{p+1} K_2 + \mathcal{O}(h^{p+2}) \tag{6}$$

$$y(t_n) - z_n^{[2]} = (0.5h)^p K_1 + (0.5h)^{p+1} K_2 + \mathcal{O}(h^{p+2}) \tag{7}$$

$$y(t_n) - z_n^{[3]} = (0.25h)^p K_1 + (0.25h)^{p+1} K_2 + \mathcal{O}(h^{p+2}) \tag{8}$$

where $y(t_n)$ is the exact value of the solution of (1) at the point $t = t_n$, while $K_1$ and $K_2$ are constants, which do not depend on the stepsize $h$. Now one has to eliminate the terms in the right-hand-sides of (6)–(8) that contain these two constants. Straight-forward transformations lead to the following relationship:

$$y(t_n) = \frac{2^{2p+1} z_n^{[3]} - 3 \cdot 2^p z_n^{[2]} + z_n^{[1]}}{2^{2p+1} - 3 \cdot 2^p + 1} + \mathcal{O}(h^{p+2}), \tag{9}$$

which show that

$$y(t_n) = y_n + \mathcal{O}(h^{p+2}), \tag{10}$$

where

$$y_n := \frac{2^{2p+1} z_n^{[3]} - 3 \cdot 2^p z_n^{[2]} + z_n^{[1]}}{2^{2p+1} - 3 \cdot 2^p + 1} \tag{11}$$

According to (10), the order of accuracy of the approximation $y_n$ is greater than or equal to $p + 2$, while (11) shows how this approximation can be calculated. It should be mentioned here that the classical RE (based essentially on the first two calculation processes) was first introduced in [4], while the RRE was studied in [1,8].

We shall demonstrate by using several numerical examples that the high accuracy of the RRE is sometimes allowing us to increase the stepsize and, by solving the problem (1) with a sufficiently large stepsize, to achieve both the required accuracy and a very good compensation for the need to use much more computations in the performance of the four calculation processes. However, it is necessary first to study the absolute stability properties of IRKMs when used in combination with the RRE.

## 2.1   Absolute Stability of IRKMs Combined with the RRE

The stability of a one-step numerical method on a fixed mesh can be investigated by using Dahlquist's test problem [2]

$$\frac{dy}{dt} = \lambda y, \ t \in [0, \infty), \ y \in \mathbb{C}, \ \lambda = \alpha + \beta i \in \mathbb{C}^-, \ \alpha \le 0, \ y(0) = \eta \in \mathbb{C} \quad (12)$$

on the grid

$$t_0 = 0, \ t_n = t_{n-1} + h = t_0 + nh \ (n = 1, 2, \ldots) \quad (13)$$

The exact solution of (12) is $y(t) = e^{\lambda t}$, and it is bounded when the condition $\alpha \le 0$ is satisfied. Therefore, it is necessary to require that the selected numerical method produces arbitrarily long sequences of approximations $\{y_1, y_2, \ldots, y_N\}$ which are also bounded as $N \to \infty$ when $\alpha \le 0$. The application of any one-step numerical method in the treatment of (12) leads to the following recursive relation [3]:

$$y_n = R(\nu)y_{n-1} = [R(\nu)]^n y_0, \ \nu = \lambda h, \ n = 1, 2, \ldots \quad (14)$$

The function $R(\nu)$ from (14) is called stability function. Clearly, if $|R(\nu)| \le 1$ for some value of $\nu = h\lambda \in \mathbb{C}^-$, then the selected one-step numerical method will produce a bounded sequence $\{y_1, y_2, \ldots, y_N\}$ of approximations to the solution of (12) for the applied value $h$ of the time-stepsize. The set of all values $\nu \in \mathbb{C}^-$ for which the relationship $|R(\nu)| \le 1$ holds is forming the absolute stability region of the chosen one-step numerical method [3].

Now the following theorem can be proved:

**Theorem 1.** *The stability function $\tilde{R}(\nu)$ of the RRE, when it is applied to solve (12) in combination with an arbitrary one-step numerical method with a stability function $R(\nu)$, is given by the following expression:*

$$\tilde{R}(\nu) = \frac{2^{2p+1}[R(\frac{\nu}{4})]^4 - 3 \cdot 2^p[R(\frac{\nu}{2})]^2 + R(\nu)}{2^{2p+1} - 3 \cdot 2^p + 1} \quad (15)$$

*Proof.* It can be concluded from (14) that the following three relationships hold:

$$z_n^{[1]} = R(\nu)y_{n-1}, \ z_n^{[2]} = [R(\tfrac{\nu}{2})]^2 y_{n-1}, \ z_n^{[3]} = [R(\tfrac{\nu}{4})]^4 y_{n-1}.$$

The following relationship can be obtained by using the above equalities and (11):

$$y_n = \frac{2^{2p+1}[R(\frac{\nu}{4})]^4 - 3 \cdot 2^p[R(\frac{\nu}{2})]^2 + R(\nu)}{2^{2p+1} - 3 \cdot 2^p + 1}y_{n-1}. \tag{16}$$

It is clear that the coefficient of $y_{n-1}$ in (16) is the stability polynomial $\tilde{R}(\nu)$ of the RRE.

If the problems solved is stiff, then one often requires that the condition $|\tilde{R}(\nu)| \leq 1$ is satisfied for all points $\nu = h\lambda \in \mathbb{C}^-$, which leads to the concept of A-stability. This is, of course, a very useful property of the numerical methods, since it provides a sufficient condition which will normally ensures stable computations. However, the absolute stability regions of IRKMs combined with RRE would be very difficult to determine analitically. Moreover, it is not a necessary condition. Instead of imposing the demand that the condition $|\tilde{R}(\nu)| \leq 1$ is satisfied for all points $\nu = h\lambda \in \mathbb{C}^-$, one can require that this condition is satisfied in an extremely large domain $P$, for example within the square with corners: $(0,0), (0,10^5), (-10^5,0), (-10^5,10^5)$. This is normally enough to ensure stable computations in most cases.

By the procedure described in [7] and [10] it is possible to examine whether $|\tilde{R}(\nu)| \leq 1$ holds in very densely chosen points $\nu$ of the above defined large domain $P$. This algorithm can be explained in the following way. Assume that an arbitrary Runge-Kutta Method is selected. Let $\nu = \overline{\alpha} + \overline{\beta}i$ with $\overline{\alpha} \leq 0$ and suppose that $\varepsilon = 0.01$ is an increment (other small values of this parameter can also be used). Set $\overline{\alpha} = 0$ and calculate the values of $\tilde{R}(\nu)$ of the chosen IRKM for $\overline{\alpha} = 0$ and $\overline{\beta} = 0, \varepsilon, 2\varepsilon, 3\varepsilon, \ldots$. The computations are carried on as long as both $|\tilde{R}(\nu)| \leq 1$ and $\overline{\beta} \leq 10^5$. Continue the process by setting $\overline{\alpha} = -\varepsilon$ and by repeating the computations for the new value of $\overline{\alpha}$ and for $\overline{\beta} = 0, \varepsilon, 2\varepsilon, 3\varepsilon, \ldots$. Let $\overline{\beta}_1$ be the largest value of $\overline{\beta}$ for which the requirements $|\tilde{R}(\nu)| \leq 1$ and $\overline{\beta} \leq 10^5$ are satisfied. Set $\overline{\alpha}$ equal to $-2\varepsilon, -3\varepsilon, \ldots$ and for each $\overline{\alpha}$ repeat the computations described above. Continue this process until both $|\tilde{R}(\nu)| \leq 1$ and $\overline{\alpha} \leq -10^5$ hold. A very large set of points $\{(0, \overline{\beta}_0), (-\varepsilon, \overline{\beta}_1), (-2\varepsilon, \overline{\beta}_2), \ldots\}$ located in the negative part of the complex plane and over the real axis will be calculated in this manner. It is clear that if the computations are always stopped when $\overline{\beta} \leq 10^5$ for each values of $\overline{\alpha}$ and continue until the point $\overline{\alpha} = -10^5$ is reached, then the considered version of the Richardson Extrapolation will be stable in the square with vertices: $(0,0), (0,10^5), (-10^5,0)$ and $(-10^5,10^5)$, which was required. The procedure is very straight-forward, but also very time-consuming.

Figure 1(a) shows that the absolute stability region of any 2-stage 3rd order DIRKM combined with RRE, obtained as described above, includes the domain $P$. For comparison, the most popular four-stage fourth-order Explicit Runge-Kutta Method (ERKM) has much smaller absolute stability regions even when combined with the classical RE, the RRE, and even the Two-Times, Three-Times and Four-Times Repeated RE, see the boundaries of the corresponding absolute stability regions in Fig. 1(b). These latter versions of the RRE are not described and studied in detail here, but they can be defined following the description of the RRE, combining numerical solutions obtained by more than three different stepsizes.

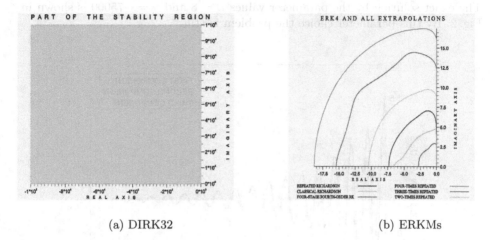

(a) DIRK32                                                    (b) ERKMs

**Fig. 1.** Absolute stability regions for (a) 2-stage third-order DIRKMs combined with RRE and (b) four-stage fourth-order ERKMs and their combinations with the classical RE, the RRE and the Two-Times, Three-Times and Four-Times Repeated RE.

# 3   Numerical Examples

The performance of IRKMs used in combination with RRE was investigated on two examples: (1) a two-parameter set of three ODEs and (2) a chemical scheme of 56 species, used in the Unified Danish Eulerian Model (UNI-DEM).

## 3.1   Example 1

Consider the two-parameter family of systems of ODEs defined in the following way:

$$\frac{dy}{dt} = Ay, \ t \in [0, 13.1072], \ y(0) = (1, 0, 2)^T, \tag{17}$$

where

$$A = \begin{bmatrix} -\gamma - \beta - 0.6 & -\gamma - 0.3 & \gamma + \beta + 0.3 \\ \gamma - 2\beta + 0.3 & \gamma - \beta & -\gamma + \beta - 0.3 \\ -\gamma - 3\beta - 0.3 & -\gamma - \beta - 0.3 & \gamma + 2\beta \end{bmatrix}. \tag{18}$$

The eigenvalues of $A$ are $\mu_1 = \gamma$, $\mu_2 = -0.3 + \beta i$, $\mu_3 = -0.3 - \beta i$. The exact solution of (17) can be given as

$$y_1(t) = e^{-0.3t} \sin \beta t + e^{\gamma t}, \tag{19}$$

$$y_2(t) = e^{-0.3t}\cos \beta t + e^{\gamma t}, \tag{20}$$

$$y_3(t) = e^{-0.3t}(\sin \beta t + \cos \beta t) + e^{\gamma t}. \tag{21}$$

The exact solution by the parameter values $\beta = 8$ and $\gamma = -75000$ is shown in Fig. 2. By this parameter choice the problem is stiff.

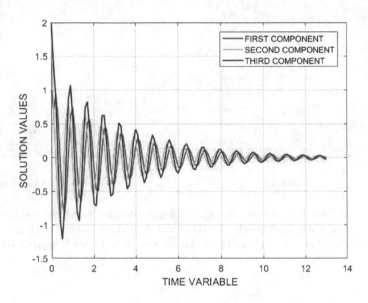

**Fig. 2.** The exact solution of Example 1 for $\gamma = -75000$ and $\beta = 8$.

Table 1 shows the maximum errors obtained when the fifth-order three-stage Fully Implicit Runge-Kutta Method (FIRKM35) and its combinations with (a) the classical RE and (b) the RRE are run with $\beta = 8$ and $\gamma = -75000$ by using twelve different stepsizes.

One can draw the following major conclusions from Table 1:

– The expected orders of accuracy (five, six and seven) are easily achieved: the convergent rates being 32, 64 and 128.
– The rounding errors are starting to be dominating when the Repeated Richardson Extrapolation is used with small stepsizes.
– The achieved accuracy is very high.

We ran the experiments by $\beta = 512$ and $\gamma = 75000$ as well, where the accuracy requirements are higher. The results are shown in Table 2. The first rows if this table show that the RRE can sometimes produce stable results in cases where the underlying IRKM or even when its combination with the classical RE is unstable.

**Table 1.** Maximum errors obtained in the treatment of the first example when the fifth-order three-stage implicit Runge-Kutta method and its combinations with (a) the classical RE and (b) the RRE are run with $\beta = 8$ and $\gamma = -75000$ and twelve different stepsizes. In parentheses the ratios between two successive errors are given where applicable (the perfect convergent rates are 32, 64 and 128).

| Job | No. of steps | FIRKM35 | RE | RRE |
|-----|-----|-----|-----|-----|
| 1 | 640 | 2.77E−07 | 1.26E−10 | 8.62E−14 |
| 2 | 1280 | 8.65E−09 (31.99) | 1.97E−12 (64.22) | 6.74E−16 (127.96) |
| 3 | 2560 | 2.70E−10 (32.01) | 3.07E−14 (64.07) | 5.26E−18 (127.99) |
| 4 | 5120 | 8.44E−12 (32.00) | 4.79E−16 (64.00) | 4.11E−20 (128.03) |
| 5 | 10240 | 2.64E−13 (32.01) | 7.49E−18 (64.00) | 3.21E−22 (128.00) |
| 6 | 20480 | 8.24E−15 (32.00) | 1.17E−19 (63.99) | 2.51E−24 (127.98) |
| 7 | 40960 | 2.58E−16 (32.00) | 1.83E−21 (64.00) | 1.96E−26 (128.08) |
| 8 | 81920 | 8.05E−18 (32.00) | 2.86E−23 (64.01) | 1.48E−28 (132.28) |
| 9 | 163240 | 2.52E−19 (32.00) | 4.46E−25 (63.99) | 4.21E−30 (35.19) |
| 10 | 327680 | 7.86E−21 (32.01) | 6.98E−27 (63.99) | 5.29E−30 (0.80) |
| 11 | 655360 | 2.46E−22 (31.99) | 1.11E−28 (63.01) | 5.20E−30 (1.02) |
| 12 | 1310720 | 7.68E−24 (32.00) | 6.08E−30 (18.21) | 5.31E−30 (0.98) |

**Table 2.** Global errors obtained in the treatment of the first example when the fifth-order three-stage Fully Implicit Runge-Kutta Method and its combinations with (a) the classical RE and (b) the RRE are run with $\beta = 512$ and $\gamma = -75000$ and twelve different stepsizes. In parentheses the ratios between two successive errors are given where applicable (the perfect convergent rates are 32, 64 and 128). "V.L" means "very large".

| Job | No. of steps | FIRKM35 | RE | RRE |
|-----|-----|-----|-----|-----|
| 1 | 640 | V.L | V.L | V.L |
| 2 | 1280 | V.L | V.L | V.L |
| 3 | 2560 | V.L | V.L | 1.41E−03 |
| 4 | 5120 | V.L | 2.28E−03 (64.00) | 1.12E−15 (126.16) |
| 5 | 10240 | V.L | 3.38E−05 (67.41) | 8.81E−08 (126.63) |
| 6 | 20480 | 5.57E−04 | 5.19E−07 (65.20) | 6.90E−10 (127.69) |
| 7 | 40960 | 1.75E−05 (31.79) | 8.09E−09 (64.15) | 5.42E−12 (127.37) |
| 8 | 81920 | 5.49E−07 (31.93) | 1.26E−10 (64.10) | 4.24E−14 (127.78) |
| 9 | 163240 | 1.72E−08 (31.98) | 1.97E−12 (64.06) | 3.32E−16 (127.90) |
| 10 | 327680 | 5.36E−10 (31.99) | 3.08E−14 (64.02) | 2.59E−18 (127.98) |
| 11 | 655360 | 1.68E−11 (32.00) | 4.81E−16 (64.02) | 2.03E−20 (127.95) |
| 12 | 1310720 | 5.24E−13 (31.99) | 7.51E−18 (64.01) | 1.58E−22 (128.00) |

## 3.2   Example 2: An Atmospheric Chemical Scheme

The performance of IRKMs used in combination with RRE was investigated on one of the three chemical schemes used in UNI-DEM, the Unified Danish Eulerian Model, developed for studying the long-range transport of air pollutants over different parts of Europe and the impact of climate changes on some high pollution levels, see, e.g., [5,6]. This scheme contains 56 chemical species. It is a non-linear, stiff, badly-scaled and very ill-conditioned system of ODEs. Some species vary very rapidly in relatively short periods (during the changes from day-time to night-time and from night-time to day-time), as illustrated by Table 3 and Fig. 3. The condition numbers of the Jacobian matrix of the system vary in the interval $[4.66 \cdot 10^8, 9.27 \cdot 10^{12}]$.

**Table 3.** Orders of magnitude and variations of some chemical species during a period of 24 h (from the noon of a given day to the noon of the next day). The units are molecules/cubic centimetre.

| Chemical species | Max. conc. | Min. conc. | Mean conc. |
|---|---|---|---|
| $O_3$ | $1.8 \times 10^{12}$ | $1.4 \times 10^{12}$ | $1.5 \times 10^{12}$ |
| PAN | $1.3 \times 10^{10}$ | $9.4 \times 10^5$ | $2.3 \times 10^9$ |
| ISOPRENE | $3.7 \times 10^9$ | $1.1 \times 10^6$ | $1.5 \times 10^9$ |
| OP | $1.6 \times 10^4$ | $1.7 \times 10^{-35}$ | $5.9 \times 10^3$ |

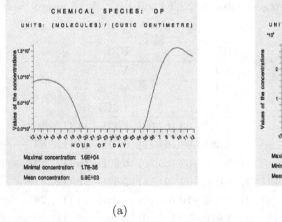

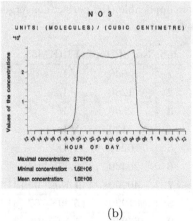

(a)                                              (b)

**Fig. 3.** Typical diurnal variation of some chemical species: (a) OP and (b) $NO_3$.

In the experiments the length of the time interval was 24 h (86400 s): from 12 o'clock on a given day to 12 o'clock on the next day. IRKM1 (the Backward Euler Formula, BEF), the classical RE and the RRE were used in the computations. The Newton Iterative Procedure was used to solve the non-linear systems of algebraic equations at every step during the outer loop. A special sparse matrix technique was used in the solution of the systems of linear algebraic equations in the inner iterative loop (during each iteration with the Newton Procedure).

Nineteen runs with different stepsizes were successively completed with each of the above numerical methods, both directly and in combination with the Classical RE and the RRE. The global errors were evaluated as follows. The integration interval $[0, 13.1072]$ was divided into 168 equal sub-intervals, and the accuracy of the results obtained by any of the selected numerical methods was evaluated at the end of each sub-interval. Let $\bar{t}_j$, $j = 1, 2, \ldots, 168$, be the end of any of the 168 sub-intervals, and $\bar{y}_{ij} \approx y_i(\bar{t}_j)$. Then the following formula is used to evaluate the accuracy achieved by the selected numerical method at this point:

$$ERROR_j = \frac{\sqrt{\sum_{i=1}^{3}(y_i(\bar{t}_j) - \bar{y}_{ij})^2}}{\max[\sqrt{\sum_{i=1}^{3}(y_i(\bar{t}_j))^2}, 1.0]} \tag{22}$$

The global error is computed by using the formula:

$$ERROR = \max_{1,2,\ldots,168}(ERROR_j) \tag{23}$$

The errors for the Backward Euler Method are given in Table 4.

The following important conclusions can be drawn from the results given in Table 4:

(a) The rates of convergence are nearly perfect (2 and 4) when the Backward Euler Formula (BEF) is used directly or in combination with the classical Richardson Extrapolation. The correct convergence rate (8) is achieved for sufficiently small stepsizes when the Repeated Richardson Extrapolation (RRE) is used.

(b) If a requirement for getting two correct digits is imposed, then the BEF, the RE and the RRE must be run with 43008, 672 and 336 steps respectively. These numbers are increased to 44040192, 43008 and 2688 when the requirement is to get five correct digits.

(c) The use of Two-times, Three-times or Four-times Repeated Richardson Extrapolation will be useful if high accuracy is required: the results being 2.35E−03, 1.61E−04 and 1.01E−05, respectively, when only 168 steps are performed.

**Table 4.** Global errors obtained in the UNI-DEM when the first-order one-stage explicit Runge-Kutta method (the Backward Euler Formula) and its combinations with (a) the classical RE and (b) the RRE. In parentheses the ratios between two successive errors are given where applicable (the perfect convergent rates are 2, 4 and 8).

| Job | No. of steps | IRKM1 | RE | RRE |
|---|---|---|---|---|
| 1 | 168 | 2.92E+00 | 1.31E+00 | 2.76E−02 |
| 2 | 336 | 1.32E+00 (2.20) | 3.45E−01 (3.80) | 6.44E−03 (4.29) |
| 3 | 672 | 6.23E−01 (2.13) | 4.39E−02 (7.85) | 1.41E−03 (4.58) |
| 4 | 1344 | 3.06E−01 (2.03) | 9.66E−03 (4.55) | 3.10E−04 (4.54) |
| 5 | 2688 | 1.52E−01 (2.02) | 3.05E−03 (3.16) | 5.46E−05 (5.68) |
| 6 | 5376 | 7.54E−02 (2.01) | 5.47E−04 (5.58) | 1.71E−05 (3.20) |
| 7 | 10752 | 3.76E−02 (2.01) | 1.46E−04 (3.76) | 6.93E−06 (2.46) |
| 8 | 21504 | 1.88E−02 (2.00) | 3.77E−05 (3.87) | 2.03E−06 (3.42) |
| 9 | 43008 | 9.37E−03 (2.00) | 9.58E−06 (3.93) | 4.41E−07 (4.70) |
| 10 | 86016 | 4.68E−03 (2.00) | 2.42E−06 (3.96) | 7.68E−08 (5.75) |
| 11 | 172032 | 2.34E−03 (2.00) | 6.07E−07 (3.98) | 1.15E−08 (6.66) |
| 12 | 344064 | 1.17E−03 (2.00) | 1.52E−07 (3.99) | 1.59E−09 (7.26) |
| 13 | 688128 | 5.84E−04 (2.00) | 3.81E−08 (4.00) | 2.09E−10 (7.61) |
| 14 | 1376256 | 2.93E−04 (2.00) | 9.53E−09 (4.00) | 2.68E−11 (7.80) |
| 15 | 2752512 | 1.46E−04 (2.00) | 2.38E−09 (4.00) | 3.39E−12 (7.90) |
| 16 | 5505024 | 7.32E−05 (2.00) | 5.96E−10 (4.00) | 4.26E−13 (7.95) |
| 17 | 11010048 | 3.66E−05 (2.00) | 1.49E−10 (4.00) | 5.35E−14 (7.97) |
| 18 | 22020096 | 1.83E−05 (2.00) | 3.72E−11 (4.00) | 6.69E−05 (7.99) |
| 19 | 44040192 | 9.14E−06 (2.00) | 9.31E−12 (4.00) | 8.37E−16 (7.99) |

## 4    General Conclusions

The Repeated Richardson Extrapolation (RRE) proved to be an efficient tool to enhance the accuracy of Implicit Runge-Kutta Methods. The order of accuracy that can be achieved by the this device is at least $p+2$ when the order of accuracy of the underlying method is only $p$.

It was found that IRKMs combined with RRE have better stability properties than those of the Explicit Runge-Kutta Methods (ERKMs) used directly or in combination with RRE. Although the A-stability of the combined IRKM + RRE schemes was not proved, it was numerically established that the absolute stability regions of several IRKMs combined with RRE are very large.

A natural extension of this work will be the investigation of several times Repeated Richardson Extrapolations. Some results have been included in this work.

**Acknowledgement.** The work is partially supported by the Bulgarian National Science Fund under grant DFNI 12/5 and by the Hungarian Research Fund OTKA under grant no. K-112157 and SNN-125119.

# References

1. Christiansen, E., Petersen, H.G.: Estimation of convergence orders in repeated Richardson extrapolation. BIT **29**, 48–59 (1989)
2. Dahlquist, G.: A special stability problem for linear multistep methods. BIT **3**, 27–43 (1963)
3. Lambert, J.D.: Numerical Methods for Ordinary Differential Equations: The Initial Values Problem. Wiley, New York (1991)
4. Richardson, L.F.: The deferred approach to the limit, I-Single Lattice. Philos. Trans. R. Soc. Lond. Ser. A **226**, 299–349 (1927)
5. Zlatev, Z.: Computer Treatment of Large Air Pollution Models. Kluwer Academic Publishers, Dordrecht, Boston, London (1995). (Now Distributed by Springer-Verlag, Berlin)
6. Zlatev, Z., Dimov, I.: Computational and Numerical Challenges in Environmental Modelling. Elsevier, Amsterdam, Boston, Heidelberg, London, New York, Oxford, Paris, San Diego, San Francisco, Singapore, Sidney, Tokyo (2006)
7. Zlatev, Z., Georgiev, K., Dimov, I.: Studying absolute stability properties of the Richardson Extrapolation combined with Explicit Runge-Kutta Methods. Comput. Math. Appl. **67**(12), 2294–2307 (2014)
8. Zlatev, Z., Dimov, I., Faragó, I., Georgiev, K., Havasi, Á.: Stability properties of the Repeated Richardson Extrapolation combined with some Explicit Runge-Kutta methods. Talk Presented at the SIAM Conference in Sofia, 22 December 2017
9. Zlatev, Z., Dimov, I., Faragó, I., Georgiev, K., Havasi, Á.: Stability of the Richardson extrapolation combined with some implicit Runge-Kutta methods. J. Comput. Appl. Math. **310**, 224–240 (2017)
10. Zlatev, Z., Dimov, I., Faragó, I., Havasi, Á.: Richardson Extrapolation: Practical Aspects and Applications. De Gruyter, Berlin (2017)

# Contributed Papers

# Computer Modeling of Control the Thermal Imaging System Surveillance

Tatiana Akimenko$^{(\boxtimes)}$ and Ekaterina Filippova

Department of Robotics and Automation,
Tula State University, Tula 300012, Russia
tantan72@mail.ru, kisskin@bk.ru

**Abstract.** The diagnostics of the thermal imaging device, consisting in obtaining a thermal image of test objects and comparing it with the thermal image of a reference thermal imaging device, is presented. Relative coordinates, radiation temperature of elements and other characteristics are determined on the thermal image of the test objects of the thermovision device to be diagnosed, then the thermal imaging of the test objects of the reference thermal imager is selected.

**Keywords:** Thermal imaging system surveillance · Test object ·
Thermal image model · Principle of control of thermal imaging systems

## 1 Principle of Monitoring the Thermal Imaging System Surveillance

The thermal imaging system surveillance (TISS) is a hardware-software complex that performs a two-dimensional transformation of thermal radiation in the wavelength range 3–5 or 8–14 microns from the object and terrain or background into a visible image with the representation of the corresponding two-dimensional image of the temperature distribution of the observable objects on the screen video control device. TISS has a number of advantages and features inherent only to it: the detection of remote heat-emitting objects (or targets) regardless of the level of natural illumination, and also to a certain extent - thermal or other interference (backlight, smoke, rain, fog, snow, dust, P.) [1–3].

### 1.1 Principle of Operation the Thermal Imaging System Surveillance

The task of the thermal imaging surveillance system is to create an thermal image model that accurately represents the thermal image of the observed picture. The work of TISS can be described as follows. Invisible to the human eye infrared radiation emanating from the object, is focused by the lens on the infrared detector. This detector sends a signal to the electronic unit for image processing. The electronic unit converts the signals coming from the detector

© Springer Nature Switzerland AG 2019
I. Dimov et al. (Eds.): FDM 2018, LNCS 11386, pp. 129–136, 2019.
https://doi.org/10.1007/978-3-030-11539-5_12

into an image model that is converted into a visible image of the thermal field of the object for its visual and quantitative evaluation on a standard video monitor, LCD display or on the screen of any computer connected to the network. The information visualization system has a color palette in which each signal value is assigned a certain color. After that, a dot appears on the monitor, the color of which corresponds to the numerical value of the infrared radiation that hit the photodetector. The scanning system (mirrors or semiconductor matrix) sequentially traverses all points within the field of view of the device, and as a result, an apparent picture of the infrared radiation of the object is obtained. Thus, on the TISS monitor, we see the infrared radiation power at each point of the TISS field of view, displayed according to the specified color palette (black and white or color).

## 1.2   Principle of Monitoring the Thermal Imaging System Surveillance

The effectiveness of TISS is an important indicator when using this information-measuring device in any of the spheres of human activity. There is a significant amount of TISS having, nevertheless, similar constructive elements and inter-relations between them, which allows to form generalized functional schemes of thermal imaging devices and to apply to them the same principle of constructing a control system for their operation. The thermal image depends both on the distribution of temperature over the surface of the observed object, and on the radiation coefficient and orientation of the elements of its surface that are being sighted-its shape. In addition, the quality of the thermal image depends on the transmission characteristics of the optical system and all links of the thermal imaging system. The common principle of controlling thermal imaging systems is shown in Fig. 1.

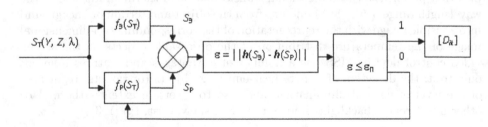

**Fig. 1.** Principle of monitoring the thermal imaging system surveillance

In accordance with the principle of control [6,7]:

(1) the reference test signal $S_T(Y, Z, \lambda)$ is formed, the test signal can be a thermal test object representing a heat-radiating object with a given spatial frequency or shape, a temperature contrast on a uniformly radiating background with a known value of its temperature and emissivity of the background and object;

(2) the reference test signal $S_T(Y, Z, \lambda)$ is fed to the inputs of the real and reference systems, which, as a result of the functional transformations $f_p(S_T)$ and $f_\ni(S_T)$, form the signals $S_p$ and $S_\ni$, respectively;

(3) an error is determined in the form of the norm of the difference in the similar characteristics of the signals at the outputs of the real and reference systems

$$\varepsilon = ||h(S_\ni) - h(S_p)||; \tag{1}$$

where $h(S_\ni)$ and $h(S_p)$ is the $K$-dimensional vector of signal characteristics, calculated as its vector function;

(4) the error value is compared with the threshold $\varepsilon_\Pi$, and if $\varepsilon \leq \varepsilon_\Pi$, then the picture of the reference test signal is recognized as suitable for further use, and the thermal imaging system is correctly recognized;

(5) if $\varepsilon \leq \varepsilon_\Pi$, then the picture of the reference test signal is recognized as unfit for further use, and the thermal imaging system is to be adjusted.

The following requirements apply to the norm (1):

(1) $||h(S_\ni) - h(S_p)|| \geq 0$;
(2) the function $||h(S_\ni) - h(S_p)|| = 0$, if and only if $S_\ni = S_p$;
(3) the function is symmetric, i.e. $||h(S_\ni) - h(S_p)|| = ||h(S_p) - \chi(S_\ni)||$;
(4) $||h(S_\ni) - h(S_p)|| \leq ||h(S_\ni) - h(S_0)|| + ||h(S_0) - h(S_p)||$, where $S_0$ is some signal.

To calculate the error, we can use the sum of the squares of distances

$$\varepsilon = \sum_{k=1}^{K} [h_k(S_\ni) - h_k(S_p)]^2, \tag{2}$$

where $h_k(S_\ni)$ - $k$ component of the reference signal characteristic vector, $1 \leq k \leq K$; $h_k(S_p)$ - $k$ component of the characteristics vector of the real signal, $1 \leq k \leq K$.

In the event that the characteristics have different weight in assessing the proximity of the functioning of the real and reference information systems, the norm can be calculated from the dependence:

$$\varepsilon = \sum_{k=1}^{K} c_k \cdot [h_k(S_\ni) - h_k(S_p)]^2, \tag{3}$$

where $0 \leq c_k \leq 1$ - "the weight" $k - i$ signal characteristics in the overall evaluation

$$\sum_{k=1}^{K} c_k = 1. \tag{4}$$

Dependencies (2) and (3) are chosen as more acceptable, since they are smooth differentiable functions and do not contain a square root calculation.

The system that conducts the conversion $f_{\ni}(S_T)$ can be a real physical system, for example, a reference thermal imager from a given class of thermal imagers, or it can be implemented as a virtual system on a computer into which a virtual test object is entered. When monitoring TISS, methods and test objects should be used, focused on automation and the shortest time for performing control procedures. All parameters should be evaluated automatically as a result of monitoring a single test object using a thermal imaging system, measuring signal parameters and comparing parameters with reference values.

## 2 Test Object for Testing and Measuring the Parameters of the TISS

The test object is a control image with known parameters, refers to means for assessing the quality and (or) calibration of image forming devices. The technical solution offered by the authors relates to the field of information and measuring technology and, in particular, to test objects that are intended for testing and measuring the main parameters and characteristics of instruments sensitive to infrared radiation of objects - thermal imagers. Figure 2 shows the test object for measuring the resolution of thermal imagers [4].

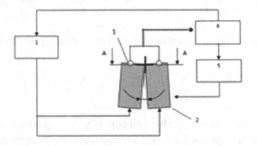

**Fig. 2.** Test object for measuring the resolution of thermal imagers.

The thermal test object for measuring the resolution of the thermal imager includes a power source 1 whose output is connected to a heat radiator 2 made in the form of fuel elements - plates arranged at an angle to each other by a micrometer screw mounted on a panel of a material with a low thermal conductivity, temperature sensors 3 mounted on the fuel elements - plates of the heat radiator 2, supplying the temperatures of the plates to the input of the control device 4 on the basis of the microprocessor, communications is a power source 1 implemented using digital signal from the temperature sensor 3 and the cooling unit 5, the input of which is connected to the control unit 4. The device works as follows. The plates of the heat radiator 2 are installed at a certain angle using a micrometer screw. From the power source 1, a current is supplied to the plates of the heat radiator 2, which heats the plates of the heat radiator 2. The digital temperature sensors 3 mounted on the heat-emitting elements-plates of the heat radiator 2, monitor the heating temperatures of each of the plates of the heat radiator 2. The temperatures are applied to the input of the device control 4 based on a microprocessor, where it is compared with predetermined values. By feedback of the control device 4 to the power supply 1, the necessary heating of the plates of the heat radiator 2 is created. The temperature of each of the plates of the heat radiator 2 is maintained in accordance with the values set by the control device 4. The temperature of the plates of the heat radiator 2 is reduced by blowing its surface by the cooling unit 5. Thus, the plates of the heat radiator 2 installed at a certain angle with the aid of a micrometer screw are heated by the power supply 1 up to the temperature set by the control device 4. Correction and maintenance of the set temperature are carried out by means of digital temperature sensors 3, control device 4 and cooling unit 5. The proposed technical solution makes it possible to increase the accuracy of measuring the resolving power of the thermal imager, and also to simplify the design of the test object.

# 3  Control of TISS Operation Modes

When monitoring the operating modes (checking the operation of the thermal imager under various conditions, determining the angle of the field of view, determining the spatial (angular) resolution, checking the range and determining the basic error in measuring the radiation temperature, determining the threshold for temperature sensitivity, determining the resolving power of the thermal imager, determining the unevenness of the thermal imager sensitivity field, the determination of the convergence of the thermal imager) TISS great laboriousness is the processing of the result in and calculations for control and diagnostic tests [5–7].

## 3.1  Determination of the Field of View Angle

The angle of the field of view can be determined in two ways. The first way. With the help of a thermal test object and TISS installed on a turntable providing the ability to rotate and record the angle of rotation of the table relative to the fixed base in two planes. The rotation axis must coincide with the vertical plane passing through the front surface of the TISS input lens. Rotating the TISS using a turntable in a horizontal (vertical) plane, combine the vertical (horizontal) axis of the location of the marks on the thermal test object with the edges of the thermogram and record the corresponding angles on the table scale $\vartheta_{X1}$ and $\vartheta_{X2}$ degrees ($\vartheta_{Y1}$ and $\vartheta_{Y2}$, deg.) Angles of the field of view horizontally $\varphi_X$ and vertically $\varphi_Y$, deg., are calculated accordingly according to the formulas:

$$\varphi_X = |\vartheta_{X1} - \vartheta_{X2}|. \tag{5}$$

The second way. Measure the distance between the extreme marks of the thermal test object in millimeters and the distance between the extreme marks of the thermal test object on the thermogram in the elements of the thermogram decomposition (hereinafter - e.). The instantaneous angle of view $\gamma$, rad., is calculated from the formula

$$\gamma = \frac{2}{a} \cdot \arctan \frac{A}{2R}, \tag{6}$$

where $a$ is the distance between the extreme marks of the thermal test object on the thermogram, el.; $A$ is the distance between the extreme marks of the heat test object, mm; $R$ - working distance, determined earlier, mm. Angles of the rhenium field horizontally $\varphi_X$ and vertically $\varphi_Y$, deg., are calculated according to the formulas:

$$\varphi_X = \gamma X \frac{180}{\pi}, \quad \varphi_Y = \gamma Y \frac{180}{\pi}, \tag{7}$$

where $\gamma$ is the instantaneous field of view, rad; $X$ - the number of elements of the decomposition of the thermogram horizontally; $Y$ is the number of elements of the decomposition of the thermogram vertically.

## 3.2   Determination of Spatial (Angular) Resolution

The maximum temperature of the gap $t'_{max}$, °C and the average temperature of the shutters $t'_0$, °C, are determined from the thermogram. Record the measured results. Based on the measurement results, the contrast of the slit on the thermograms $K_S$ is calculated from the formula

$$K_S = \frac{t'_{max} - t'_0}{t_s - t'_0}, \tag{8}$$

where $t'_{max}$ is the maximum value of the gap temperature, determined from the thermogram, °C; $t'_0$ - average value of the temperature of the shutters, determined from the thermogram, °C; $t_s$ - the maximum value of the gap temperature, with the maximum opening of the gap, determined from the thermogram, °C. From the recorded values of the width of the slit $A$, mm, the slit width $a$, el, calculated in the image plane (thermogram) is calculated from

$$a = Ak, \tag{9}$$

where $A$ is the width of the slit, mm; $k$ - is the scale factor (the ratio of the distance between the corresponding holes on the thermogram, el., and the thermal test object, mm.), el/mm. By decreasing and registering the width of the slit $A$, mm, measurements and calculations are made by (9) at least five times. Spatial resolution horizontally $F_X$ (vertically $F_Y$) is defined as the slit width at which the contrast on the thermogram $K_S$ is 99% of the maximum value.

## 3.3   Checking the Range and Determining the Basic Error in Measuring the Radiation Temperature

The measurements are carried out at a distance between a reference (model) radiator and a thermal imager, which ensures that the aperture of the emitter overlaps not less than 20% of the field of view of the TISS. The basic error of TISS is determined at five points in the operating temperature range (bottom, top and three points within the range). The average value of the radiation temperature of the reference radiator is determined from the thermogram, $t'_{cp}$, °C, taking into account its emissivity and the radiation background temperature. The basic error, $\Delta t$ °C, for each temperature of TISS, is calculated by the formula

$$\Delta t = t'_{cp} - t_{cp}, \tag{10}$$

where $t'_{cp}$ - is the average value of the temperature over the area limiting the image of the emitter aperture on the thermogram, °C; $t_{cp}$ - average value of the temperature of the reference (model) radiator, °C.

### 3.4   Determination of the Threshold of Temperature Sensitivity

Write two thermograms in the memory of TISS in a short period of time. Determine the temperature difference, $\Delta t_j$ °C, for each element of the decomposition of the registered thermograms using the software attached to the TISS, or calculated by the formula

$$\Delta t_j = t_j^{(1)} - t_j^{(2)}, \tag{11}$$

where $t_j^{(1)}$ - is the temperature of the decomposition element of the first thermogram with coordinates $(i; j)$, °C; $t_j^{(2)}$ - temperature of the decomposition element of the second thermogram with coordinates $(i; j)$, °C the matrix of temperature differences $\Delta t_j$ is represented as a numerical series Dti. The threshold of temperature sensitivity, $\Delta t_{nop}$ °C, in this case is calculated from the formula

$$\Delta t_{nop} = 0.707 \sqrt{\sum_{i=1}^{n} \frac{(\Delta t_i - \Delta \bar{t})^2}{n}}, \tag{12}$$

where $\Delta t_i$ - is the temperature difference of the i-th element of the decomposition of thermograms, °C; $\Delta \bar{t}$ - average temperature difference, °C; $n$ - is the number of decomposition elements in the thermogram. The value $\Delta t_{nop}$ is determined for each TISS band.

## 4   TISS Control Algorithm

The formation of control and diagnostic tests for the control of TISS is performed cyclically for each mode of operation. In this case, the operations are performed in the following order [6,7]:

(1) reference digital thermal images of TISS are entered in the computer database of the control unit;
(2) for each mode diagnosed TISS form thermal test objects and produce a measurement corresponding to obtain thermal images;

(3) the reference thermal images of various states are identified by appropriate codes and stored in the computer's memory of the diagnostic and control system;

(4) the thermal image on diagnostiruemooy TISS determine relative coordinates values radiation temperature of elements, and other characteristics, and then choose the thermal image of the test object reference TISS, S calculated similarity function between the radiation temperature and other characteristics of the respective elements to be diagnosed and a reference to the thermal image.

(5) continue stage 4 before evaluating all control states of the diagnosed TISS.

Reference CH thermal images are formed and recorded in the data base once and used many times for all corresponding types of TISS. Proceeding from the foregoing, it is necessary to propose and use methods and test objects that are oriented towards automation and the shortest time of execution of control procedures. All parameters should be evaluated automatically as a result of monitoring a single test object using a thermal imaging system, measuring signal parameters and comparing parameters with reference values.

During the formation of diagnostic tests, digital infrared (thermal) images of a reference (control) sample of this type of TISS are used as criteria for the healthy state of TISS. The formed thermal portrait of the reference TISS is stored in the memory of the computer of the diagnostic and control unit and is used in the subsequent diagnosis of samples of this type of TISS as a criterion of health.

**Acknowledgements.** This article was written within the framework of project 2.3121/$\Gamma$3 "Parallel semi-Markov processes in mobile robot control systems".

# References

1. Akimenko, T.A., Dunaev, V.A., Larkin, E.V.: Computer simulation of the surface heating process by the movable laser. In: V International Workshop on Mathematical Models and their Applications 2016, IOP Conference Series: Materials Science and Engineering, vol. 173, pp. 012002. IOP Publishing (2017). https://doi.org/10.1088/1757-899X/173/1/012002
2. Numerical Analysis And Its Applications. Springer (2017)
3. Tarasov, V.V., Yakushenkov, Y.G.: Infrared Systems of the "Looking" Type. Logos, Moscow (2004)
4. The patent of the Russian Federation for utility model -176316 The Russian Federation. "Test object for measuring the resolution of thermal imagers" - / Akimenko TA, Larkin EV, Filippova EV, applicant and patent holder FGBOU VO "Tula State University", Tula. Accessed 25 July 2017
5. GOST of the Russian Federation 8.619-2006
6. Filippova, E.V., Akimenko, T.A.: Evaluation of the thermal imaging system. In: Proceedings of Tula State University. Series of Engineering, pp. 229–236, no. 9. Publishing House of Tula State University, Tula (2016)
7. Akimenko, T.A., Filippova, E.V.: The image formation on the receiver of thermal radiation. In: Proceedings of Tula State University. Series of Engineering, pp. 13–20, no. 9–1. Publishing House of Tula State University, Tula (2017)

# Generalized Multiscale Finite Element Method for Elasticity Problem in Fractured Media

V. Alekseev[1]([✉]) [ID], A. Tyrylgin[1] [ID], and M. Vasilyeva[1,2] [ID]

[1] Multiscale Model Reduction Laboratory,
North-Eastern Federal University, Yakutsk, Russia
alekseev.valen@mail.ru, koc9tk@mail.ru
[2] Institute for Scientific Computation,
Texas A&M University, College Station, TX, USA
vasilyevadotmdotv@gmail.com

**Abstract.** In this work, we consider the elasticity problem in fractured media. For the efficient numerical solution, we present a Generalized Multiscale Finite Element Method (GMsFEM). GMsFEM is used for the construction of a coarse grid approximation of the problem by solution of the local spectral problems. We consider two types of the multiscale basis functions: (1) CG-GMsFEM with continuous multiscale basis functions and (2) DG-GMsFEM with discontinuous multiscale basis functions. The result of the numerical solution for the two-dimensional model problem is presented to show the performance of the presented multiscale method for fractured media. We compute error between the multiscale solution with the fine-scale solutions by choosing different numbers of multiscale basis functions.

**Keywords:** Fractured media · Elasticity problem ·
Discontinuous Galerkin · Multiscale method · GMsFEM

## 1 Introduction

One important class of multiscale problem consists of problems in fractured domains. Due to the variable sizes and geometries of these fractures, solutions to these problems have multiscale features [1–4]. Direct numerical methods in multiscale media are too expensive and some type of model reduction is needed. The main aim of this work is to use Generalized Multiscale Finite Element Method for elasticity equations in fractured media [5–9].

To approximate the model problem with respect to spatial variables, we use the Galerkin discontinuous method for fine grid. In GMsFEM, we solve problem on a coarse grid where each coarse grid consists of fine-grid blocks. The multiscale basis function construction is local and uses both local snapshot solutions and local spectral problems. To couple multiscale basis functions constructed, we consider two methods Continuous Galerkin (CG) approach and Discontinuous

© Springer Nature Switzerland AG 2019
I. Dimov et al. (Eds.): FDM 2018, LNCS 11386, pp. 137–144, 2019.
https://doi.org/10.1007/978-3-030-11539-5_13

Galerkin (DG) approach based on symmetric interior penalty method for coarse grid approximation [10–13].

The work is organized as follows. In Sect. 2, we present a model problem and fine grid approximation. In Sect. 3, we consider both the CG-GMsFEM and DG-GMsFEM for coarse grid approximation. Finally, numerical results are presented in Sect. 4.

## 2   Problem Formulation and Fine Grid Approximation

We consider a mathematical model describing the elastic state of a body with allowance for the presence of defects (fracture). We present the formulation of the problem for the stationary elasticity equation in the computational domain $\Omega$ [14]

$$\text{div } \sigma(u) = f, \quad x \in \Omega, \tag{1}$$

where $f = f(x, t)$ is a given source term.

The Eq. (1) is supplemented by the relation between the stress tensor $\sigma$ and the strain tensor $\epsilon$

$$\epsilon(u) = \frac{1}{2}(\nabla u + (\nabla u)^T), \tag{2}$$

$$\sigma(u) = 2\mu\epsilon(u) + \lambda \text{ div } uE \tag{3}$$

where $E$ is the identity tensor, $\lambda$ and $\mu$ are the Lame parameters.

Substituting the relations (2) and (3) in the Eq. (1), we obtain the following elliptic equation

$$\mu\Delta u + (\lambda + \mu)\text{grad div } u = f. \tag{4}$$

To approximate the problem with respect to spatial variables, we use the discontinuous Galerkin method (IPDG). Let $E$ be an edge (face) between the elements $K_1$ and $K_2$, then the average and the jump of the vector $u$ on the edge $E$ is given by the following formula.

$$\{u\} = \frac{u\big|_{K_1} + u\big|_{K_2}}{2}, \quad [u] = u\big|_{K_1} - u\big|_{K_2}.$$

Let $\mathcal{T}_h$ be a triangulation of the computational domain $\Omega$ and $\Gamma_h$ be the set of all interior faces between the elements $\mathcal{T}_h$. Let $\Gamma_c \subset \Gamma_h$ be the subset of all faces, where the displacement field is a continuous function. On the fracture surface, we suppose discontinuous displacements.

The variational formulation of the elasticity equation using the discontinuous Galerkin method (IPDG, Interior Penalty Discontinuous Galerkin) in a fractured medium is defined as follows: find $u \in V_h$ such that

$$\sum_{K \in \mathcal{T}^h} \int_K (\sigma(u), \varepsilon(v))dx - \sum_{E \in \Gamma_c} \int_E \{\tau(u)\} [v] \, ds$$

$$- \sum_{E \in \Gamma_c} \int_E \{\tau(v)\} [u] \, ds + \gamma_f \sum_{E \in \Gamma_c} \int_E (\lambda + 2\mu) [u] [v] \, ds = \sum_{K \in \mathcal{T}^c} \int_K fv dx,$$

where $\gamma_f$ is the penalty parameter and $\tau(u) = \sigma(u)n$.

# 3 Coarse Grid Approximation Using GMsFEM

For construction of the coarse grid solver, we use a Generalized Multicale Finite Element Method(GMsFEM). In this section we will consider the CG-GMsFEM and DG-GMsFEM.

## 3.1 Multiscale Basis Functions for CG-GMsFEM

We generate a coarse grid approximation using construction of the multiscale space. Let $\mathcal{T}_H$ is the coarse grid and $\omega_i$ is the local domain, where $i = 1, ..., N_v$ and $N_v$ is the number of coarse grid nodes. A local domain $\omega_i$ is obtained by the combining all the coarse cells around one vertex of the coarse grid. We begin by the construction of local snapshot space $V_{snap}^{\omega_i}$. The snapshot space is constructed by solution of the following local problems

$$-\nabla \cdot \sigma(u_{k1}) = 0, \ in \ \omega_i$$
$$u_{k_1} = (\delta_j, 0)^T, \ on \ \partial\omega_i \tag{5}$$

and

$$-\nabla \cdot \sigma(u_{k2}) = 0, \ in \ \omega_i$$
$$u_{k_2} = (0, \delta_j)^T, \ on \ \partial\omega_i \tag{6}$$

where $\delta_j$ is the fine-grid delta function by

$$\delta_j(x_l) = \begin{cases} 1, & l = j \\ 0, & l \neq j, \end{cases}$$

We write

$$V_{snap}^{\omega_i} = span\{\psi_j^{\omega_i}, j = 1, 2, ..., M^{\omega_i}\},$$

where $M^{\omega_i}$ is the number of basis functions in $V_{snap}^{\omega_i}$.

We will perform a dimension reduction on the above snapshot spaces by use of a spectral problem. First, we will need a partition of unity function $\chi_i$ for the local domain $\omega_i$. Partition of unity is the piecewise bi-linear function on the coarse grid that equals to 1 at the coarse vertex $x_i$, equals to 0 at all other coarse vertices and linear between them. Next, we define the local spectral problem on the snapshot space as

$$A\phi^i = \lambda S\phi^i, \tag{7}$$

where the elements the matrices $A = \{a_{i,j} = a(\phi_i, \phi_j)\}$ and $S = \{s_{i,j} = s(\phi_i, \phi_j)\}$ are defined as follow

$$a(u, v) = \sum_{K \in \mathcal{T}_h^{\omega_i}} \int_K (\sigma(u), \varepsilon(v)) dx - \sum_{E \in \Gamma_c^{\omega_i}} \int_E \{\tau(u)\} [v] ds$$

$$- \sum_{E \in \Gamma_c^{\omega_i}} \int_E \{\tau(v)\} [u] ds + \gamma \sum_{E \in \Gamma_c^{\omega_i}} \int_E (\lambda + 2\mu) [u] [v] ds$$

$$s(u, v) = \sum_{K \in \mathcal{T}_h^{\omega_i}} \int_K (\lambda + 2\mu) uv dx$$

where $\gamma = R/h$ is the penalty parameter, $R > 0$. To construct a reduced space $V_{snap}^{\omega_i}$, we select the first $L_i$ eigenvectors $\phi_1, \phi_2, ..., \phi_{L_i}$, corresponding to the first $L_i$ smallest eigenvalues $\lambda_1 \leq \lambda_2 \leq ... \leq \lambda_{L_i}$. We define

$$\psi_l^{\omega_i} = \sum_{j=1}^{M^{\omega_i}} (\phi_{lj} \psi_j^{\omega_i}), \quad l = 1, 2, ... L_i, \tag{8}$$

where $\phi_{lj}$ is the $j$-th component of $\phi_l$. The local offline space is then defined as

$$V_{off}^{\omega_i} = span\{\chi_i \psi_l^{\omega_i}, l = 1, 2..., L_i\}.$$

Next, we define the global continuous Galerkin offline space as

$$V_{off} = span\{V_{off}^{\omega_i}, i = 1, 2..., N_v\}.$$

The coarse-scale system can be calculated by projecting the fine-scale matrices onto coarse grid with global projection matrix assembled from the calculated multiscale basis functions

$$R = (R_1, R_2, ..., R_{N_v})^T, R_i = [\phi_1, \phi_2, ..., \phi_{L_i}], \tag{9}$$

where $R_i$ is the local projection matrix and $\phi_l = \chi_i \psi_l^{\omega_i}$. Using the global projection matrix $R$, we can define the coarse-scale system. After calculation of the coarse-scale solution $u_H$, we can recover the fine-scale solution.

$$u_{ms} = R^T u_H$$

## 3.2    Multiscale Basis Functions for DG-GMsFEM

We construct the local basis functions required for the DG coarse grid coupling. Let $K_i$ is the local domain, where $i = 1, ..., N_c$ and $N_c$ is the number of coarse grid cells. Let $V_H$ be a finite dimensional function space, which consists of functions on each coarse grid cells. For the local basis functions, a snapshot space $V_{snap}^{K_i}$ is first constructed for each coarse grid cells $K_i \in T_H$. We have two types of local snapshot spaces, $V_{snap}^{K_i} = V_{i,snap}^{K_i} + V_{b,snap}^{K_i}$.

We perform a dimension reduction on the above snapshot spaces by use of a spectral problem. As we mentioned, we construct two types of the multiscale space $V_H$, $V_H = V_{H,b}^K + V_{H,i}^K$.

**Boundary Basis Functions.** The first space $V_{H,b}^K$ is constructed in the local snapshot space $V_{b,snap}^{K_i}$. The snapshot space $V_{b,snap}^{K_i}$ is constructed by solution of the following local problems for each-grid edge $e_l$ on the boundary of $K_i$

$$\begin{aligned} -\nabla \cdot \sigma(\psi_j^{K_i}) &= 0, \ in \ K_i \\ \psi_j^{K_i} &= (\delta_j, 0)^T \ or \ (0, \delta_j)^T, \ on \ \partial K_i \end{aligned} \tag{10}$$

where $\delta_j$ is the fine-grid delta function.

We write

$$V_{b,snap}^{K_i} = span\{\psi_j^{K_i}, j = 1, 2, ..., M^{K_i}\},$$

where $M^{K_i}$ is the number of basis functions in $V_{b,snap}^{K_i}$. Then, we define the local spectral problem on the snapshot space

$$A\phi^b = \lambda^b S\phi^b, \tag{11}$$

where $\phi^b \in V_{b,snap}^{K_i}$, $A = \{a_{i,j}\}$ and $S = \{s_{i,j}\}$.

To construct a reduced space $V_{b,H}(K_i)$, we select the first $M_b$ eigenvectors $\phi_1, \phi_2, ..., \phi_{M_b}$, corresponding to the first $M_b$ smallest eigenvalues $\lambda_1^b \leq \lambda_2^b \leq ... \leq \lambda_{M_b}^b$. We define the space $V_{b,H}(K_i)$ by

$$V_{b,H}(K_i) = span\{\phi_{l,K_i}^b, l = 1, 2..., M_b\}$$

These multiscale basis functions from $V_{b,H}(K_i)$ are called boundary basis functions.

**Interior Basis Functions.** The second space $V_{H,i}^{K_i}$ is defined to capture interior eigenmodes for $K_i$. The local snapshot space $V_{i,snap}^{K_i}$ for the coarse grid cells K is defined as $V_{i,snap}^{K_i} = V_h^0(K_i)$. We use following spectral problem with homogeneous Dirichlet boundary conditions to identify the important modes

$$a(\phi^{K_i,i}, v) = \lambda^i s(\phi^{K_i,i}, v), \ in \ \in K_i,$$
$$\phi^{K_i,i} = (0, 0), \ on \ \partial K_i. \tag{12}$$

To construct a reduced space $V_{i,H}(K_i)$, we select the first $M_i$ eigenvectors $\phi_1, \phi_2, ..., \phi_{M_i}$, corresponding to the first $M_i$ smallest eigenvalues $\lambda_1^i \leq \lambda_2^i \leq ... \leq \lambda_{M_i}^i$. We define the space $V_{i,H}(K_i)$ by

$$V_{i,H}(K_i) = span\{\phi_{l,K_i}^i, l = 1, 2..., M_i\}$$

These multiscale basis functions from $V_{i,H}(K_i)$ are called interior basis functions.

The coarse-scale system can be calculated by projecting the fine-scale matrices onto the coarse grid with the global projection matrix assembled from the calculated multiscale basis functions

$$R = (R_1, R_2, ..., R_N)^T, \quad R_i = \left[\phi_1^{K,i}, \phi_2^{K,i}, ..., \phi_{M_i}^{K,i}, \phi_1^{K,b}, \phi_2^{K,b}, ..., \phi_{M_b}^{K,b}\right], \tag{13}$$

where $R_i$ is the local projection matrix in a coarse grid element $K_i$ and $N$ is the number of coarse grid elements. Using the global projection matrix $R$, we can define the coarse-scale system After calculation of the coarse-scale solution $u_H$, we can recover the fine-scale solution.

$$u_{ms} = R^T u_H. \tag{14}$$

# 4   Numerical Results

We consider numerical simulation using the proposed models. Geometric regions with a computational mesh are shown in Fig. 1. The computational domain $\Omega = [0, L_x] \times [0, L_y]$ c $L_x = L_y = 1$. The computational mesh contains about 13825 vertices and 27248 triangular cells.

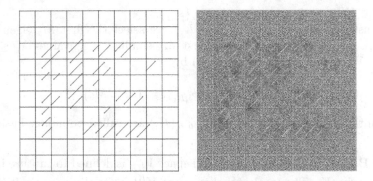

**Fig. 1.** Computational domain with a computational mesh.

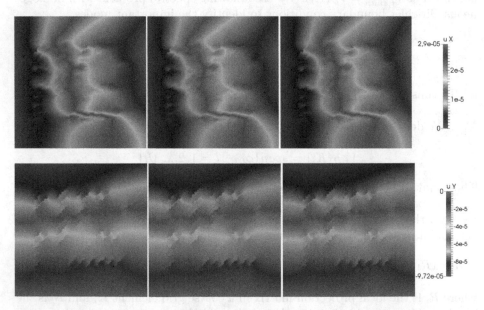

**Fig. 2.** The distribution of the X and Y components of the displacements ($u_x$ - first row and $u_y$ - second row). First column: fine grid solution. Second column: multiscale solution using CG-GMsFEM. Third column: multiscale solution using DG-GMsFEM

**Table 1.** Relative errors for solutions with different number of multiscale basis functions using CG-GMsFEM. Left: $5 \times 5$ coarse grid. Right: $10 \times 10$ coarse grid

| Basis | $L_2$ norm (%) | $H_1$ norm (%) | Basis | $L_2$ norm (%) | $H_1$ norm (%) |
|---|---|---|---|---|---|
| 1 | 95.567 | 93.632 | 1 | 95.747 | 93.455 |
| 2 | 82.387 | 79.005 | 2 | 19.896 | 30.377 |
| 4 | 66.210 | 62.535 | 4 | 11.938 | 21.760 |
| 6 | 55.996 | 55.301 | 6 | 6.028 | 13.158 |
| 8 | 6.017 | 17.178 | 8 | 1.591 | 8.597 |
| 12 | 2.767 | 8.487 | 12 | 0.400 | 4.447 |
| 16 | 2.497 | 4.629 | 16 | 0.170 | 2.625 |

**Table 2.** Relative errors for displacement for different number of multiscale basis functions for DG-GMsFEM

| $M_i = 0, M_b$ | $L_2$ norm (%) | $H_1$ norm (%) | $M_i + M_b$ | $L_2$ norm (%) | $H_1$ norm (%) |
|---|---|---|---|---|---|
| 5 | 44.614 | 51.015 | 5 | 42.525 | 48.648 |
| 10 | 7.946 | 17.249 | 10 | 5.422 | 12.915 |
| 15 | 4.183 | 12.409 | 15 | 2.395 | 8.685 |
| 20 | 2.195 | 8.275 | 20 | 1.440 | 6.673 |
| 25 | 1.662 | 7.099 | 25 | 1.092 | 5.825 |

For the calculations, we used the following parameters: $\mu = 1$, $\lambda = 1$ and source term $f = (0,0)$. We set the following boundary conditions

$$u_x = 0, x \in \Gamma_1$$
$$u_y = 0, x \in \Gamma_4 \tag{15}$$
$$u \cdot n = g, x \in \Gamma_2$$

where $g = (0, -0.0002)$. Here $\Gamma_1$, $\Gamma_4$ are the upper and lower boundaries, $\Gamma_2$ and $\Gamma_3$ are the right and left boundaries.

The results of calculations in the region with fractures using the mathematical model of (4), (15) are shown in Fig. 2. In Fig. 2, we present the X and Y components of displacement distributions for the numerical solution using CG-GMsFEM, DG-GMsFEM and using fine grid approximation. Calculations was performed using 25 multiscale basis functions for DG-GMsFEM and 16 multiscale functions for CG-GMsFEM.

Relative $L_2$ errors for CG-GMsFEM are shown in Table 1 for $5 \times 5$ and $10 \times 10$ coarse grids. Relative $L_2$ errors for DG-GMsFEM are shown in Table 2, where $M_i$ and $M_b$ are the numbers of the multiscale interior and boundary basis functions. From the numerical results, we observe that the interior multiscale basis functions can reduce errors. We see that a solution using CG-GMsFEM bases is better than solution using DG-GMsFEM.

We present numerical results for fractured domain to demonstrate a robustness of proposed method with two types of the multiscale basis functions. We construct reduced order model using Generalized Multiscale Finite Element Method using continuous and discontinuous multiscale basis functions. Our results show that the presented method demonstrate high accuracy for elasticity problem.

**Acknowledgments.** This work is supported by the grant of the Russian Scientific Found (N 17-71-20055).

# References

1. Grechka, V., Kachanov, M.: Effective elasticity of fractured rocks: a snapshot of the work in progress. Geophysics **71**(6), W45–W58 (2006)
2. Nikolaevskij, V.N.: Mechanics of Porous and Fractured Media. World Scientific, Singapore (2014)
3. Coates, R.T., Schoenberg, M.: Finite-difference modeling of faults and fractures. Geophysics **60**, 1514–1526 (1995)
4. Schoenberg, M., Sayers, C.M.: Seismic anisotropy of fractured rock. Geophysics **60**, 204–211 (1995)
5. Chung, E.T., Efendiev, Y., Fu, S.: Generalized multiscale finite element method for elasticity equations. GEM-Int. J. Geomath. **5**, 225–254 (2014)
6. Chung, E.T., Efendiev, Y., Gibson, R.L., Vasilyeva, M.: A generalized multiscale finite element method for elastic wave propagation in fractured media. GEM-Int. J. Geomath. **7**, 163–182 (2016)
7. Efendiev, Y., Galvis, J., Hou, T.Y.: Generalized multiscale finite element methods (GMsFEM). J. Comput. Phy. **251**, 116–135 (2013)
8. Hou, T., Efendiev, Y.: Multiscale Finite Element Methods. Theory and Applications. Springer, New York (2009). https://doi.org/10.1007/978-0-387-09496-0
9. Efendiev, Y., Galvis, J., Lazarov, R., Moon, M., Sarkis, M.: Generalized multiscale finite element method Symmetric interior penalty coupling. J. Comput. Phys. **255**, 1–15 (2013)
10. Chung, E.T., Efendiev, Y., Leung, W.T.: An online generalized multiscale discontinuous Galerkin method (GMsDGM) for flows in heterogeneous media. Commun. Comput. Phys. **21**, 401–422 (2017)
11. Chung, E.T., Efendiev, Y., Vasilyeva, M., Wang, Y.: A multiscale discontinuous Galerkin method in perforated domains. In: Proceedings of the Institute of Mathematics and Mechanics (2016)
12. Riviere, B., et al.: Discontinuous Galerkin finite element methods for linear elasticity and quasistatic linear viscoelasticity. Numer. Math. **95**, 347–376 (2003)
13. Chung, E.T., et al.: Online adaptive local multiscale model reduction for heterogeneous problems in perforated domains. Appl. Anal. **96**, 2002–2031 (2017)
14. Oleinik, O.A., Shamaev, A.S., Yosifian, G.A.: Mathematical Problems in Elasticity and Homogenization. Elsevier, New York City (2008)

# Automatic Time Step Selection for Numerical Solution of Neutron Diffusion Problems

A. V. Avvakumov[1], V. F. Strizhov[2], P. N. Vabishchevich[2,3], and A. O. Vasilev[3(✉)]

[1] National Research Center Kurchatov Institute, Moscow, Russia
[2] Nuclear Safety Institute of RAS, Moscow, Russia
[3] North-Eastern Federal University, Yakutsk, Russia
haska87@gmail.com

**Abstract.** An automatic algorithm of time step control for solving the boundary value problems for nonstationary parabolic equations is presented. The solution is obtained using complete stable implicit schemes, and the time step is evaluated using of the explicit scheme solution. The time step evaluation formulas are derived using the estimation of the approximation error at next time step. Calculation results obtained for several neutron diffusion problems demonstrate reliability of the proposed algorithm for time step control.

**Keywords:** Time step selection · Parabolic equation ·
Approximation error · Neutron diffusion

## 1 Introduction

For the second order parabolic equations unconditionally stable schemes are constructed on the basis of implicit approximations [1]. In computational practice two-layer schemes are mostly used, compared with three-layered and multilayered schemes which are not so often used. The problem of the time step control is relatively well developed for the Cauchy problem solution of differential equations systems [2]. The basic approach is to use additional calculations at a new time step to estimate the approximate solution. The time step is estimated using the theoretical asymptotic dependence of the accuracy on time step [3]. The algorithm takes into account the features of neutron diffusion problems [7], for instance, fast changes in the solution or instability with respect to the initial data. The proposed algorithm allows a gain in CPU time with respect to the fine mesh calculation at the same calculation accuracy.

## 2 Problem Description

Consider the Cauchy problem for the linear equation

$$\frac{du}{dt} + A(t)u = f(t), \quad 0 < t \le T, \tag{1}$$

© Springer Nature Switzerland AG 2019
I. Dimov et al. (Eds.): FDM 2018, LNCS 11386, pp. 145–152, 2019.
https://doi.org/10.1007/978-3-030-11539-5_14

with initial condition $u(0) = u^0$.

The problem is considered in a finite-dimensional Hilbert space. We assume that $A(t) \geq 0$. Let's use an irregular time grid

$$t^0 = 0, \quad t^{n+1} = t^n + \tau^{n+1}, \quad n = 0, 1, ..., N-1, \quad t^N = T.$$

For approximate solution the implicit scheme are used

$$\frac{y^{n+1} - y^n}{\tau^{n+1}} + A^{n+1} y^{n+1} = f^{n+1}, \quad n = 0, 1, ..., N-1, \tag{2}$$

and initial condition $y^0 = u^0$. For the approximate solution we can use the following estimate

$$\|y^{n+1}\| \leq \|y^n\| + \tau^{n+1} \|f^{n+1}\|.$$

Then we can obtain a difference analogue of the estimate

$$\|y^{n+1}\| \leq \|u^0\| + \sum_{k=0}^{n} \tau^{k+1} \|f^{k+1}\| \tag{3}$$

for problem (2). For the error of the approximate solution $z^n = y^n - u^n$ we have

$$\frac{z^{n+1} - z^n}{\tau^{n+1}} + A^{n+1} z^{n+1} = \psi^{n+1}, \quad n = 0, 1, ..., N-1, \quad z^0 = 0.$$

Here $\psi^{n+1}$ is approximation error

$$\psi^{n+1} = f^{n+1} - \frac{u^{n+1} - u^n}{\tau^{n+1}} - A^{n+1} u^{n+1}. \tag{4}$$

Similarly (3) we have an estimate for the error

$$\|z^{n+1}\| \leq \sum_{k=0}^{n} \tau^{k+1} \|\psi^{k+1}\|. \tag{5}$$

Checking the error we can focus on the total error $\delta \tau^{n+1}$ in interval $t^n < t < t^{n+1}$. Then from (5) we obtain $\|z^{n+1}\| \leq \delta t^{n+1}$. The error accumulates and increases linearly.

The solution is obtained using the unconditionally stable implicit scheme. The major part of computing costs is related with this scheme. The step control is performed using the explicit scheme solution. The algorithm stability is not violated and is determined by the implicit scheme properties.

The error accumulation from the time layer $t^n$ to the new layer $t^{n+1}$ is defined as

$$\|z^{n+1}\| \leq \|z^n\| + \tau^{n+1} \|\psi^{n+1}\|. \tag{6}$$

Therefore, we have to control the local error $\psi^{n+1}$. Comparing $\psi^{n+1}$ with a given level of error $\delta$ we can control the time step choice. If $\psi^{n+1}$ is significantly larger (or smaller) then the $\delta$ - it means that the time step is too large (or small). Thereby

$$\tau^{n+1} : \ \|\psi^{n+1}\| \approx \delta. \tag{7}$$

Consider the basic algorithm to control the time step. We select the time step based on the analysis of the previous step solutions. The predicted time step is determined as following

$$\widetilde{\tau}^{n+1} = \gamma \tau^n, \tag{8}$$

where $\gamma$ is numerical parameter. The default value of $\gamma$ is 1.25 or 1.5. Using the explicit scheme we can obtain a solution $\widetilde{y}^{n+1}$ at time $\widetilde{t}^{n+1} = t^n + \widetilde{\tau}^{n+1}$. The calculation is performed only at single time step; therefore the possible computational instability does not appear. We estimate the approximation error using the calculated $\widetilde{y}^{n+1}$ by the implicit scheme. The $\tau^{n+1}$ is estimated by the proximity of the error norm to $\delta$. The solution at a new time step $t^{n+1}$ is calculated with a $\tau^{n+1}$ by the implicit scheme.

## 3  Calculated Formulas

We present the calculated formulas for the time step control for neutron diffusion equation.

**Neutron Diffusion Equation.** Let's consider neutron diffusion equation in one-group approximation with one-group delayed neutron sources. Neutron flux dynamics is considered within a bounded 2D domain $\Omega$ ($\boldsymbol{x} = \{x_1, x_2\} \in \Omega$) with a convex boundary $\partial\Omega$.

$$\begin{aligned} \frac{1}{v}\frac{\partial\phi}{\partial t} - \nabla \cdot D\nabla\phi + \Sigma_a\phi &= (1-\beta)\nu\Sigma_f\phi + \lambda c, \\ \frac{\partial c}{\partial t} + \lambda c &= \beta\nu\Sigma_f\phi. \end{aligned} \tag{9}$$

Here $\phi(\boldsymbol{x}, t)$ is the neutron flux at point $\boldsymbol{x}$ and time $t$, $v$ is the neutron velocity, $D(\boldsymbol{x})$ is the diffusion coefficient, $\Sigma_a(\boldsymbol{x}, t)$ is the absorption cross-section, $\beta$ is the effective fraction of delayed neutrons, $\nu\Sigma_{fg}(\boldsymbol{x}, t)$ is the generation cross-section, $c$ is density of source of delayed neutrons, $\lambda$ is decay constant of delayed neutrons source.

The conditions so-called albedo-type are set at the boundary $\partial\Omega$:

$$D\frac{\partial\phi}{\partial n} + \gamma\phi = 0. \tag{10}$$

where $n$ is the outer normal to the boundary $\partial\Omega$. Let's consider problem (9) with boundary conditions (10) and initial conditions: $\phi(0) = \phi^0, c(0) = c^0$. Space discretization is performed using the standard Lagrange finite elements (for example, see [4]).

**Time Step Estimate.** In our case, the approximation error $\psi^{n+1} = \{\psi_1^{n+1}, \psi_2^{n+1}\}$ is

$$\psi_1^{n+1} = \frac{1}{v}\frac{\phi^{n+1} - \phi^n}{\tau^{n+1}} - \nabla \cdot D^{n+1}\nabla\phi^{n+1} + \Sigma_a^{n+1}\phi^{n+1}$$
$$- (1-\beta)\nu\Sigma_f^{n+1}\phi^{n+1} - \lambda c^{n+1}, \tag{11}$$

$$\psi_2^{n+1} = \frac{c^{n+1} - c^n}{\tau^{n+1}} + \lambda c^{n+1} - \beta\nu\Sigma_f^{n+1}\phi^{n+1},$$

where $\phi^{n+1} = \{\phi^{n+1}, c^{n+1}\}$ is exact solution. The predictive solution is

$$\frac{1}{v}\frac{\widetilde{\varphi}^{n+1} - \varphi^n}{\widetilde{\tau}^{n+1}} - \nabla \cdot D^n\nabla\varphi^n + \Sigma_a^n\varphi^n = (1-\beta)\nu\Sigma_f^n\varphi^n + \lambda s^n,$$
$$\frac{\widetilde{s}^{n+1} - s^n}{\widetilde{\tau}^{n+1}} + \lambda s^n = \beta\nu\Sigma_f^n\varphi^n. \tag{12}$$

This solution is used to estimate the approximation error of the implicit scheme in the transition from time $t^n$ to time $\widetilde{t}^{n+1}$.

In accordance with (11), the approximation error is calculated from the exact solution for two time moments: in our case, for $t^n$ and $\widetilde{t}^{n+1}$. To estimate the error we take $\varphi^n = \{\varphi^n, s^n\}$ instead of $\phi(t^n)$. An exact solution at a new time step $\phi(\widetilde{t}^{n+1})$, is matched by an approximate solution $\widetilde{\varphi}^{n+1} = \{\widetilde{\varphi}^{n+1}, \widetilde{s}^{n+1}\}$, which is obtained by the explicit scheme. By virtue of this, we set

$$\widetilde{\psi}_1^{n+1} = \frac{1}{v}\frac{\widetilde{\varphi}^{n+1} - \varphi^n}{\widetilde{\tau}^{n+1}} - \nabla \cdot D^{n+1}\nabla\widetilde{\varphi}^{n+1} + \Sigma_a^{n+1}\widetilde{\varphi}^{n+1}$$
$$- (1-\beta)\nu\Sigma_f^{n+1}\widetilde{\varphi}^{n+1} - \lambda\widetilde{s}^{n+1}, \tag{13}$$

$$\widetilde{\psi}_2^{n+1} = \frac{\widetilde{s}^{n+1} - s^n}{\widetilde{\tau}^{n+1}} + \lambda\widetilde{s}^{n+1} - \beta\nu\Sigma_f^{n+1}\widetilde{\varphi}^{n+1}.$$

We match the approximation error $\widetilde{\psi}^{n+1} = \{\widetilde{\psi}_1^{n+1}, \widetilde{\psi}_2^{n+1}\}$ at the time step $\widetilde{\tau}^{n+1}$ and $\psi^{n+1}$ at the time step $\tau^{n+1}$. Taking into account (7), we set

$$\widetilde{\tau}^{n+1} = \gamma_{n+1}\tau^n, \quad \gamma_{n+1} = \frac{\delta}{\|\widetilde{\psi}^{n+1}\|}\gamma. \tag{14}$$

The needed time step can not exceed the predicted time step, therefore

$$\tau^{n+1} \leq \widetilde{\tau}_{n+1}, \quad \tau^{n+1} \leq \widetilde{\tau}_{n+1}.$$

We limit the allowable time step by the minimum step $\tau^0$:

$$\tau^{n+1} = \max\{\tau^0, \min\{\gamma_{n+1}, \gamma\}\tau^n\}. \tag{15}$$

Let's determine the calculation formulas for the step selection algorithm in accordance with (12)–(15):

$$\widetilde{\psi}_1^{n+1} = (-\nabla \cdot D^{n+1}\nabla + \Sigma_a^{n+1} - (1-\beta)\nu\Sigma_f^{n+1} + \nabla \cdot D^n\nabla - \Sigma_a^n + (1-\beta)\nu\Sigma_f^n)\varphi^n$$
$$+ (-\nabla \cdot D^{n+1}\nabla + \Sigma_a^{n+1} - (1-\beta)\nu\Sigma_f^{n+1})(\widetilde{\varphi}^{n+1} - \varphi^n) - \lambda(\widetilde{s}^{n+1} - s^n).$$

$$\widetilde{\psi}_2^{n+1} = (-\beta\nu\Sigma_f^{n+1} + \beta\nu\Sigma_f^n)\varphi^n + \lambda(\widetilde{s}^{n+1} - s^n) - \beta\nu\Sigma_f^{n+1}(\widetilde{\varphi}^{n+1} - \varphi^n).$$

For convenience, rewrite it in the operator notation

$$\widetilde{\psi}^{n+1} = (A^{n+1} - A^n)\varphi^n + A^{n+1}(\widetilde{\varphi}^{n+1} - \varphi^n)$$

$$= \widetilde{\tau}^{n+1}\left(\frac{A^{n+1} - A^n}{\widetilde{\tau}^{n+1}}\varphi^n + A^{n+1}\frac{\widetilde{\varphi}^{n+1} - \varphi^n}{\widetilde{\tau}^{n+1}}\right),$$

where

$$A = \begin{pmatrix} -\nabla \cdot D\nabla + \Sigma_a - (1-\beta)\nu\Sigma_f - \lambda & 0 \\ 0 & \lambda - \beta\nu\Sigma_f \end{pmatrix}.$$

Thus, the approximation error has the first order for time variable: $\widetilde{\psi}^{n+1} = \mathcal{O}(\widetilde{\tau}_{n+1})$. By virtue of this, we set

$$\|\widetilde{\psi}^{n+1}\| \leq \|(A^{n+1} - A^n)\varphi^n + A^{n+1}(\widetilde{\varphi}^{n+1} - \varphi^n)\|. \tag{16}$$

Taking into account (16) from (14), we obtain the calculated formula for time step (15), in which

$$\gamma_{n+1} = \frac{\delta}{\|(A^{n+1} - A^n)\varphi^n + A^{n+1}(\widetilde{\varphi}^{n+1} - \varphi^n)\|}\gamma. \tag{17}$$

This formula (the denominator of the expression) clearly shows the corrective actions, which are associated with the change of the problem operator (the first part), with the dynamics of the solution (second part).

## 4   Test Problem

The test problem for reactor IAEA-2D without a reflector in a one-dimensional approximation ($\Omega$—is the reactor core area) is considered [5]. The geometrical model of the IAEA-2D reactor core is presented in Fig. 1. The fuel assembly pitch equals 20 cm. Diffusion neutronics constants in the common notations are given in Table 1. The boundary conditions (2) are used at $\gamma_g = 0.5$. The following delayed neutrons parameters are used: $\beta = 6.5 \cdot 10^3$, $\lambda = 0.08$ s$^{-1}$ and $v = 1.0 \cdot 10^6$ cm/s. Modeling effect of insersion or withdrawal of control rods (depending on the sign of the perturbation). Define the scenario of the process:

1. The spectral problem is solved [6], the solution is taken as the initial condition;
2. Calculation for the non-stationary model at the time range 0 to 0.1 s;
3. At a moment of 0.1 s the value $\Sigma_a$ for the zone 3 changes to $\pm 0.000625$;
4. The dynamic regime is calculated starting from 0.1 s to 0.5 s.

At each time the integrated power is calculated

$$P(t) = a \int_\Omega \Sigma_f\varphi dx,$$

where $a$ is the normalization coefficient by a given value of the integrated power.

**Table 1.** Diffusion neutronics constants for IAEA-2D test problem

| Zone | 1 | 2 | 3 |
|---|---|---|---|
| $D$, cm | 1.03 | 1.03 | 1.03 |
| $\Sigma_a$, cm$^{-1}$ | 0.02 | 0.02125 | 0.02625 |
| $\nu\Sigma_f$, cm$^{-1}$ | 0.0225 | 0.0225 | 0.0225 |

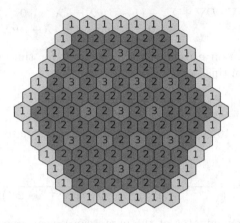

**Fig. 1.** Geometrcial model of the IAEA-2D reactor core

**Computational Results.** The accuracy of the solution was evaluated by a reference solution, which uses a numerical solution on a sufficiently detailed grid in time ($\tau_{ref} = 1 \cdot 10^{-5}$) by the implicit scheme with a fixed time step. The initial value of $k_{eff}$ was 1.0005063. Figure 2 show the integral powers for insersion or withdrawal of the control rods respectively.

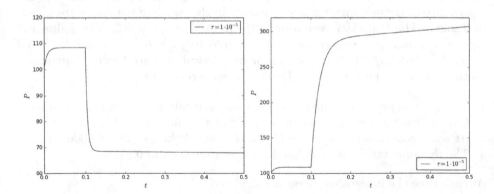

**Fig. 2.** The integral powers.

The error is estimated as $\varepsilon_P(t) = |P_{ref} - P|$, where $P_{ref}$ is the reference solution, $P$ is the solution obtained by using the time stepping algorithm. We took a minimum time step $\tau_0 = \tau_{ref}$.

Figure 3 show the error $\varepsilon_P$ when the control rods are taken for immersion and extraction, respectively, for different values of the parameter $\delta$. Here we see that the error converges as the parameter $\delta$ decreases.

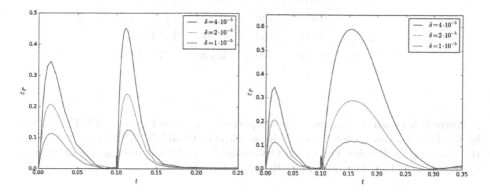

**Fig. 3.** Error by time.

Figure 4 shows the time steps for insertion or withdrawal of rods, respectively. It is seen that first there is a rapid growth of the time step with a specified accuracy of $\delta$. Then the *catch* of a sudden change in power occurs with a strong decrease in the time step. Further, the time step grows to a certain point and remains at the same level, which is controlled by the error $\delta$.

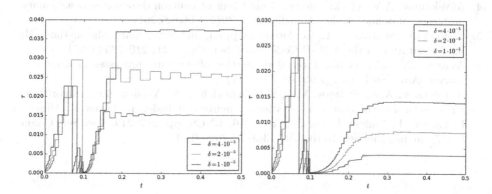

**Fig. 4.** Steps by time.

Table 2 shows the various output data for different values of the parameter $\delta$, where $\max(\epsilon_P)$ is the maximum power error, $n$ is the number of steps in time, and $t$ is the calculation time in seconds. The reference solution: the number of steps in time is 50000, the counting time is 2130 s.

**Table 2.** Counting time and number of steps.

| $\delta$ | Insersion | | | Withdrawal | | |
|---|---|---|---|---|---|---|
| | $\max(\epsilon_P)$ | $n$ | $t$ | $\max(\epsilon_P)$ | $n$ | $t$ |
| $4 \cdot 10^{-5}$ | 0.450 | 136 | 16 | 0.590 | 241 | 35 |
| $2 \cdot 10^{-5}$ | 0.241 | 159 | 20 | 0.290 | 373 | 62 |
| $1 \cdot 10^{-5}$ | 0.125 | 270 | 37 | 0.120 | 773 | 145 |

**Acknowledgements.** This work are supported A.V. Avvakumov and V.F. Strizhov by the Russian Foundation for Basic Research #16-08-01215, P.N. Vabishchevich by the grant of the Russian Federation Government #14.Y26.31.0013 and A.O. Vasilev by the Russian Foundation for Basic Research #18-31-00315.

# References

1. Samarskii, A.A., Matus, P.P., Vabishchevich, P.N.: Difference Schemes with Operator Factors. Kluwer, Dordrecht (2002)
2. Ascher, U.M.: Computer Methods for Ordinary Differential Equations and Differential-Algebraic Equations. Society for Industrial Mathematics, Philadelphia (1998)
3. Vabishchevich, P.N.: A priori estimation of a time step for numerically solving parabolic problems. Math. Model. Anal. **20**(1), 94–111 (2015)
4. Avvakumov, A.V., et al.: Numerical modeling of neutron diffusion non-stationary problems. Matematicheskoe Modelirovanie **29**(7), 44–62 (2017)
5. Chao, Y.A., Shatilla, Y.A.: Conformal mapping and hexagonal nodal methods-II: Implementation in the ANC-H Code. Nucl. Sci. Eng. **121**, 210–225 (1995)
6. Avvakumov, A.V., et al.: Spectral properties of dynamic processes in a nuclear reactor. Ann. Nucl. Energy **99**, 68–79 (2017)
7. Avvakumov, A.V., Strizhov, V.F., Vabishchevich, P.N., Vasilev, A.O.: Algorithms for numerical simulation of non-stationary neutron diffusion problems. In: Dimov, I., Faragó, I., Vulkov, L. (eds.) NAA 2016. LNCS, pp. 212–219. Springer, Cham (2017). https://doi.org/10.1007/978-3-319-57099-0_21

# Local Convergence Results for an Optimal Iterative Method for Multiple Roots

Ramandeep Behl[1], Eulalia Martínez[2($\boxtimes$)], Fabricio Cevallos[3],
and Ali Saleh Alshomrani[1]

[1] Department of Mathematics, King Abdulaziz University,
Jeddah 21577, Saudi Arabia
ramanbehl87@yahoo.in
[2] Instituto Universitario de Matemática Multidisciplinar,
Universitat Politècnica de València, Valencia, Spain
eumarti@mat.upv.es
[3] Facultad de Ciencias Económicas, Universidad laica "Eloy Alfaro" de Manabí,
Manta, Ecuador
alfa2205@gmail.com

**Abstract.** In this paper our aim is to perform a local convergence study of a fourth order iterative method in the case of multiple roots. As far as we know, these kind of studies have only been performed for iterative methods of second and third order of convergence in the case of multiple roots. So it is our purpose to analyze the radius of local convergence for higher-order methods. Usually the local convergence radius decreases when the order of the method increases, so it is necessary to study its behavior when we propose a new iterative method. In this sense, we introduce in this paper a new idea for establishing local convergence results of iterative methods for locating multiple zeros, under the assumption of a bounding condition for the $(m+1)-th$ derivative of the function $f(x)$ in its existence domain. We apply this technique to the modification of the Maheshwari fourth order method for the case of multiple roots. Finally, we perform some numerical examples that confirm the theoretical results established in this paper.

## 1 Introduction

The study of the local convergence for optimal multi-point iterative methods for approximating zeros with multiplicity $m \geq 1$ of the involved function $f$ (where $f : D \subset \mathbb{R} \to \mathbb{R}$ is analytic in the enclosed region enclosing the required zero) is the aim of this work. We can find in the literature high order methods for multiple roots, see [2–10] where one can find from fourth to eighth order of convergence algorithms for approximating multiple roots.

But, none of of these schemes have been studied from their local convergence treatment in Banach spaces. Studies of these type have a special interest from a

© Springer Nature Switzerland AG 2019
I. Dimov et al. (Eds.): FDM 2018, LNCS 11386, pp. 153–158, 2019.
https://doi.org/10.1007/978-3-030-11539-5_15

mathematical point of view, since these studies allow us to get the local convergence balls, which are balls contained in the domain of the function with center at the solution and where any point of this ball can be taken as a starting point for getting the sequence of iterates that converges to the root.

As far as we know, these kind of studies can be found in literature for method of order two and three, but not for high order methods. We can find a local convergence study for obtaining the convergence radius of the well-known modified Newton's method for multiple zeros, when the involved function satisfies center-Holder continuity condition, see [11]. This result is improved in [12]. Similar results for the third order method due to Halley are obtained in [13,14].

We are now interested in this kind of local convergence studies for a high order method for multiple roots. We focus on the modification of the Maheshwari fourth order method for the case of multiple roots proposing a local convergence study by using a characterization of the function that describes the problem with a root $\alpha$ multiplicity $m$ in terms of its $m-th$ derivative.

Summarizing, our aim is to perform a local convergence study of a fourth order iterative method in the case of multiple roots. Since the radius of local convergence for higher-order methods decreases when the order increases, it is necessary to study its behavior when we introduce a new iterative method. In this sense, we introduce in this paper a new idea for establishing local convergence results of iterative methods for locating multiple zeros, under the assumption of a bounding condition for the $(m+1)$th derivative of the function $f(x)$ in its existence domain.

Finally, we present several numerical tests for applying the theoretical results established in this paper and we give some concluding remarks.

## 2    Modified Maheshwari Iterative Method for Multiple Roots

Firstly, we extend in the following way the fourth order iterative method due to Maheshwari to the case of multiple roots, where the iterative expression is a two step methods given by:

$$y_n = x_n - mu_n,$$
$$x_{n+1} = x_n - mu_n \left[ v_n^2 - \frac{1}{v_n - 1} \right], \qquad (1)$$

where $u_n = \frac{f(x_n)}{f'(x_n)}$, $v_n = \left( \frac{f(y_n)}{f(x_n)} \right)^{\frac{1}{m}}$.

The error equation for this method is the following:

$$e_{n+1} = -\frac{c_1 \left( (m+7)c_1^2 - 2mc_2 \right)}{2m^3} e_n^4 + \sum_{i=1}^{4} L_i e_n^{i+4} + O(e_n^9), \qquad (2)$$

where $L_i = L_i(m, c_1, c_2, \ldots, c_8)$.

Notice that it is an optimal method in the sense of Kung-Traub conjecture, [1].

## 2.1   Local Convergence

Our aim is to obtain a local convergence study for the optimal fourth-order method defined above, in order to find a root of the problem $\alpha$, with multiplicity $m$ for the equation $f(x) = 0$.

So, we look for the radius of the local convergence ball, that is, a real positive number $r$ such that the sequence $\{x_n\}$ generated by this iterative method, starting from any point in the open ball $B(\alpha, r)$, remains in this ball and converges to $\alpha$. For this kind of study the best value of $r$ is the largest; however, this will obviously depend on the conditions that the nonlinear function satisfies.

This kind of study has been performed for methods of second and third order. See [11] where the authors obtain an estimate of the convergence radius of the well-known modified Newton's method for multiple zeros when the involved function satisfies a Hölder and center-Hölder continuity condition. This result is improved in [12]. Third-order methods have been considered in [13,14]. In these papers the authors establish the local convergence study by using different properties of divided differences. We do not follow this line of inquiry. We instead utilize Taylor's developments to introduce a new set of bounds to guarantee convergence. Notice that the authors in [14] also use Taylor's developments, but work in a different way from our technique. We characterized the function that define the problem with multiple roots as follows:

**Lemma 1.** *Let $f : I \subseteq \mathbb{R} \to \mathbb{R}$ be a sufficiently differentiable function in the open interval $I = ]\alpha - r, \alpha + r[$, with $r > 0$ and $\alpha$ a zero with multiplicity $m$ with $m > 1$, of the nonlinear equation $f(x) = 0$. Then, for all $x \in I$, the function $f(x)$ can be expressed as*

$$f(x) = (x - \alpha)^m g(x), \quad g(\alpha) \neq 0, \tag{3}$$

*where*

$$g(x) = \frac{f^{(m)}(\alpha)}{m!} + \frac{1}{(m-1)!} \int_0^1 [f^{(m)}(\alpha + \theta(x - \alpha)) - f^{(m)}(\alpha)](1 - \theta)^{m-1} d\theta. \tag{4}$$

**Lemma 2.** *Under the same conditions of Lemma 1, the function $g(x)$ satisfies*

$$g(\alpha) = \frac{f^{(m)}(\alpha)}{m!}$$

$$g'(x) = \frac{1}{(m-1)!} \int_0^1 f^{(m+1)}(\alpha + \theta(x - \alpha))\theta(1 - \theta)^{m-1} d\theta.$$

This expression suggests that we consider the following bound property for obtaining the local convergence study:

$$\left| f^{(m)}(\alpha)^{-1} f^{(m+1)}(x) \right| \leq k_1, \quad \forall x \in I \tag{5}$$

where, $k_1$ is positive real number. We first obtain certain bounds that we will use for our local convergence study. We define $r_0 = \frac{m+1}{k_1}$, by taking a starting point $x_0 \in ]\alpha - r_0, \alpha + r_0[ = I_0$ and denoting the local error by $|e_0| = |x_0 - \alpha| < r_0$, we can proof the following result.

**Lemma 3.** *Under the same conditions of Lemma 1 for all $x$, $x_0 \in I_0$ we obtain the following bounds:*

$$(B_1) \qquad \left| g(\alpha)^{-1} g(x_0) \right| \leq \frac{m + 1 + k_1 |e_0|}{m + 1}$$

$$(B_2) \qquad \left| g(\alpha)^{-1} g'(x_0) \right| \leq \frac{k_1}{m + 1}$$

$$(B_3) \qquad \left| g(x_0)^{-1} g(\alpha) \right| \leq= \frac{m + 1}{m + 1 - k_1 |e_0|}$$

$$(B_4) \qquad \left| g(x_0)^{-1} g'(x) \right| \leq \frac{k_1}{m + 1 - k_1 |e_0|}$$

These previous Lemmas allow us to obtain the main result, that gives us the local convergence radius.

**Theorem 1.** *Let $I \subset \mathbb{R}$ be an open, convex, and non-empty set and $f : I \longrightarrow \mathbb{R}$ be in $\mathcal{R}^m(I)$ with $\alpha$ a root of multiplicity $m$ for the equation $f(x) = 0$. If boundary condition (5) is satisfied, let $r_0 = \frac{m+1}{K_1}$. Then there exists $r \leq r_0$ such that, for any initial point $x_0 \in B(\alpha, r)$, the sequence $\{x_n\}$, $n \geq 0$ generated by (1) is well-defined, remains in $B(\alpha, r)$, and converges to solution $\alpha$, that is the unique solution in $B(\alpha, r_0)$.*

## 3  Numerical Experience

Now, in order to perform the local convergence study we take the following examples from the literature, see [13], where the local convergence radius have been calculated for a third order iterative methods, so we compare the local convergence radius $r_H$ of modified Halley's iterative method for multiple roots, with the one obtained in this paper for a fourth order method, denoted by $r$.

$$f_a(x) = \cos(x) - 1,$$
$$f_b(x) = x^2(x^2 - 1)$$

and

$$f_c(x) = x + \cos(x) - \frac{\pi}{2}.$$

See the results in Table 1, where we have indicated by columns the root, its multiplicity, the interval of definition of the function, the constant $k_1$ defined in (5) the sequence of values $r_1, r_2, r_3$ defined in this study and $r$ the local convergence radius.

*Remark 1.* In Table 1 we show the values that restrict the local convergence radius. Notice that the value $r_3$ in all the examples coincides with $r_2$. This gave us an open interval where one can choose the initial guess for our iterative method, but more importantly, it proves that despite it being a higher-order method, the interval of local convergence still remains considerably good.

**Table 1.** Radius of the convergence ball

| Example | $\alpha$ | $m$ | $[a,b]$ | $k_1$ | $r_H$ | $r_0$ | $r_1$ | $r_2 = r_3 = r$ |
|---------|----------|-----|---------|-------|-------|-------|-------|-----------------|
| $f_a$ | 0 | 2 | $[-\pi/2, \pi/2]$ | 1 | 1.2679 | 3 | 2 | 0.374785 |
| $f_b$ | 0 | 2 | $\mathbb{R}$ | 12 | 0.1152 | 0.25 | 0.166667 | 0.031221 |
| $f_c$ | $\pi/2$ | 3 | $\mathbb{R}$ | 1 | 1.972 | 4 | 3 | 0.540692 |

**Acknowledgements.** Supported by the project of Generalitat Valenciana Prometeo/2016/089 and MTM2014-52016-C2-2-P of the Spanish Ministry of Science and Innovation.

# References

1. Traub, J.F.: Iterative Methods for the Solution of Equations. Prentice-Hall, Englewood Cliffs (1964)
2. Li, S., Liao, X., Cheng, L.: A new fourth-order iterative method for finding multiple roots of nonlinear equations. Appl. Math. Comput. **215**, 1288–1292 (2009)
3. Neta, B.: Extension of Murakami's high-order non-linear solver to multiple roots. Int. J. Comput. Math. **87**(5), 1023–1031 (2010)
4. Li, S.G., Cheng, L.Z., Neta, B.: Some fourth-order nonlinear solvers with closed formulae for multiple roots. Comput. Math. Appl. **59**, 126–135 (2010)
5. Zhou, X., Chen, X., Song, Y.: Constructing higher-order methods for obtaining the muliplte roots of nonlinear equations. J. Comput. Math. Appl. **235**, 4199–4206 (2011)
6. Sharifi, M., Babajee, D.K.R., Soleymani, F.: Finding the solution of nonlinear equations by a class of optimal methods. Comput. Math. Appl. **63**, 764–774 (2012)
7. Soleymani, F., Babajee, D.K.R.: Computing multiple zeros using a class of quartically convergent methods. Alex. Eng. J. **52**, 531–541 (2013)
8. Soleymani, F., Babajee, D.K.R., Lofti, T.: On a numerical technique for finding multiple zeros and its dynamic. J. Egypt. Math. Soc. **21**, 346–353 (2013)
9. Zhou, X., Chen, X., Song, Y.: Families of third and fourth order methods for multiple roots of nonlinear equations. Appl. Math. Comput. **219**, 6030–6038 (2013)
10. Hueso, J.L., Martínez, E., Teruel, C.: Determination of multiple roots of nonlinear equations and applications. J. Math. Chem. **53**, 880–892 (2015)
11. Argyros, I.: On the convergence and application of Newton's method under weak Hölder continuity assumptions. Int. J. Comput. Math. **80**, 767–780 (2003)
12. Zhou, X., Chen, X., Song, Y.: On the convergence radius of the modified Newton method for multiple roots under the center-Hölder condition. Numer. Algorithms **65**, 221–232 (2014)
13. Bi, W., Ren, H., Wu, Q.: Convergence of the modified Halley's method for multiple zeros under Hölder continuous derivative. Numer. Algorithms **58**, 497–512 (2011)
14. Zhou, X., Son, Y.: Convergence radius of Osada's method under center-Hölder continuous condition. Appl. Math. Comput. **243**, 809–816 (2014)
15. Maheshwari, A.K.: A fourth order iterative method for solving nonlinear equations. Appl. Math. Comput. **211**(2), 383–391 (2009)
16. Cordero, A., Torregrosa, J.R.: Variants of Newton's method using fifth-order quadrature formulas. Appl. Math. Comput. **190**(1), 686–698 (2007)

158    R. Behl et al.

17. Balaji, G.V., Seader, J.D.: Application of interval Newton's method to chemical engineering problems. Rel. Comput. **1**(3), 215–223 (1995)
18. Shacham, M.: An improved memory method for the solution of a nonlinear equation. Chem. Eng. Sci. **44**(7), 1495–1501 (1989)

# Spline Iterative Method for Pantograph Type Functional Differential Equations

Alexandru Mihai Bica[1]($\boxtimes$) [ID], Mircea Curila[2] [ID], and Sorin Curila[3] [ID]

[1] Department of Mathematics and Computer Science, University of Oradea,
Universitatii Str. no. 1, 410087 Oradea, Romania
abica@uoradea.ro
[2] Department of Environmental Protection, University of Oradea, Oradea, Romania
mirceacurila@yahoo.com
[3] Department of Electronics and Telecommunications, University of Oradea, Oradea,
Romania
scurila@uoradea.ro

**Abstract.** Initial value problems and two-point boundary value problems for nonlinear pantograph type differential equations are investigated by presenting a new iterative numerical method based on constructing a sequence of splines that converges to the solution. The convergence of the method was proved by providing an error estimate and is tested on some numerical experiments.

**Keywords:** Pantograph type equations · Iterated spline method · Convergence analysis

## 1 Introduction

Functional differential equations with proportional delays are usually referred to as pantograph equations and arise in many applications such as electrodynamics, engineering, astrophysics, nonlinear dynamical systems, biology, control problems, quantum mechanics, number theory (see [8] and [10]). Such equations were firstly considered by Ockendon and Tayler for the study of the curent collection by the pantograph head of an electric locomotive (see [8]) being intensively studied (see [4]). The existing numerical methods for pantograph type equations include: Runge-Kutta-Nystrom techniques, piecewise polynomial collocation, spline functions, shooting techniques, and finite differences approach (see [1,4,5,7]). Recently, two-point boundary value problems for second order pantograph equations were investigated based on their applications in elasticity problems (see [10]). Since the fourth order two-point boundary value problems govern the deflection of a uniform elastic clamped beam supporting a distributed load (see [6]) such problems received special attention. Recently, their pantograph variant was approached for constructing the numerical solution (see [2]).

In this paper we present a new iterative numerical method for pantograph type initial and two-point boundary value problems based on constructing a

© Springer Nature Switzerland AG 2019
I. Dimov et al. (Eds.): FDM 2018, LNCS 11386, pp. 159–166, 2019.
https://doi.org/10.1007/978-3-030-11539-5_16

sequence of splines that uniformly converges to the solution. The convergence of the method is proved by providing an error estimate that indicates the order of convergence and the numerical stability regarding the choice of the first iteration is investigated. Since the aim of this paper is to show that the Picard's method of successive approximations becomes an effective tool when it is combined with suitable quadrature rules and spline interpolation procedures, the theoretical results concerning the convergence of the method were tested by presenting some numerical examples.

## 2   The Iterative Algorithm

The pantograph type initial value problem

$$\begin{cases} x^{(p)}\left(t\right) = f\left(t, x\left(t\right), x\left(qt\right)\right), & t \in [0,a], \ q \in (0,1) \\ x\left(0\right) = x_0 = x_0^{(0)}, ..., x^{(p-1)}\left(0\right) = x_0^{(p-1)} \end{cases} \tag{1}$$

and the two-point boundary value problems of the second and fourth order

$$\begin{cases} x^{(2p)}\left(t\right) = f\left(t, x\left(t\right), x\left(qt\right)\right), & t \in [0,a], \ p = \overline{1,2} \\ x^{(i)}\left(0\right) = a_i, \quad x^{(i)}\left(a\right) = b_i, \quad i = \overline{0, p-1}, \ q \in (0,1) \end{cases} \tag{2}$$

are written in the equivalent integral equation form

$$x\left(t\right) = g\left(t\right) + \int\limits_0^{b(t)} H\left(t,s\right) \cdot f\left(s, x\left(s\right), x\left(qs\right)\right) ds, \quad t \in [0,a] \tag{3}$$

with $b\left(t\right) = t$, $g\left(t\right) = \sum\limits_{i=0}^{p-1} x_0^{(p-1)} \cdot \frac{t^{p-1}}{(p-1)!}$ and $H\left(t,s\right) = \frac{(t-s)^{p-1}}{(p-1)!}$ for initial value problems. For two-point boundary value problems $b\left(t\right) = a$ and $H\left(t,s\right)$ is the corresponding Green function (see [2] and [6]). For the numerical solution consider a uniform mesh of $[0,a]$ with the knots $t_i = i \cdot h$, $i = \overline{0,n}$ and stepsize $h = \frac{a}{n}$. The iterative method consist of the construction of a recurrent sequence of splines $(S_k)_{k \in \mathbb{N}^*}$ that converges to the exact solution on the grid points and uniformly on $[0,a]$. Define the set of functions $f_{0,i} : [0,a] \to \mathbb{R}$, $i = \overline{0,n}$, by $f_{0,i}\left(s\right) = H\left(t_i, s\right) \cdot f\left(s, g\left(s\right), g\left(qs\right)\right)$, and the starting values are

$$\overline{x_1\left(t_0\right)} = g\left(t_0\right), \quad \overline{x_1\left(t_i\right)} = g\left(t_i\right) + Q_n\left(f_{0,i}\right), \ i = \overline{1,n}, \tag{4}$$

with $Q_n$ be a composite quadrature rule operator. At each iterative step, as interpolation procedure $(S_k)_{k \geq 1}$ we can use the natural cubic spline from [2], page 130, or the quadratic spline from [3] and the iterative algorithm has the following structure presented here for initial value problems:

   *Step 1*: Construct the corresponding spline procedure that interpolates the starting values (4): $\overline{x_1\left(t_i\right)} = S_1\left(t_i\right)$, $i = \overline{0,n}$.

*Step 2* (the discrete numerical solution): For $k \in \mathbb{N}^*$ and $i = \overline{0, n}$, define the functions $F_{k,i} : [0, a] \to \mathbb{R}$, by $F_{k,i}(s) = H(t_i, s) \cdot f(s, S_k(s), S_k(qs))$ where the sequence of cubic splines $(S_k)_{k \in \mathbb{N}^*}$ is given as in [2], page 130. For $k \geq 2$ applying the composite quadrature rule $Q_n$ to the set of functions $F_{k-1,i}$ we compute

$$S_k(t_0) = g(t_0), \quad S_k(t_i) = g(t_i) + Q_n(F_{k-1,i}), \quad i = \overline{1, n}. \tag{5}$$

*Step 3* (the iterated splines): For $k \geq 2$ construct the spline $S_k$, specified at Step 2, interpolating the values $x_k(t_i) = S_k(t_i)$, $i = \overline{0, n}$.

*Stopping criterion* ("do-while" cycle): for given $n \in \mathbb{N}^*$ and $\varepsilon' > 0$, find the first natural number $k \in \mathbb{N}^*$ for which we have $|S_k(t_i) - S_{k-1}(t_i)| < \varepsilon'$, $\forall i = \overline{1, n}$, and stop at this step $k$.

*Step 4* (the continuous numerical solution): By using the above considered spline procedure, construct the spline $S_k$ interpolating the values $S_k(t_i)$, $i = \overline{0, n}$ obtained at the last iterative step.

This iterative algorithm was related in [2,3] to the name of iterated splines method.

## 3   The Convergence Analysis

### 3.1   The Error Estimate

Consider the following conditions:

(i) exist $L_1, L_2 > 0$ such that for all $s \in [0, a], u, u', v, v' \in \mathbb{R}$ we have
$|f(s, u, v) - f(s, u', v')| \leq L_1 |u - u'| + L_2 |v - v'|$

(ii) $f \in C([0, a] \times \mathbb{R} \times \mathbb{R})$ and $a M_H (L_1 + L_2) < 1$, where $|H(t, s)| \leq M_H$, $\forall t, s \in [0, a]$

(iii) $f \in C^m([0, a] \times \mathbb{R} \times \mathbb{R})$, with $m \geq 2$.

In the case $f \in C^2([0, a] \times \mathbb{R} \times \mathbb{R})$, considering natural cubic splines and the composite trapezoidal quadrature rule, for second order boundary value problems (2) the order of convergence is $O(h^2)$ (see [2] page 138, Theorem 7, part (b)). Considering $f \in C^4([0, a] \times \mathbb{R} \times \mathbb{R})$ in the case of fourth order boundary value problems the order of convergence is $O(h^4)$ according to [2], page 139, Theorem 8, part (b). For initial value problems (1), involving quadratic splines and the composite trapezoidal quadrature rule, the order of convergence is $O(h^2)$ (see [3]). In the following we approach the first order initial value problem (1) by using natural cubic splines and considering the composite Simpson quadrature rule.

**Theorem 1.** *Under the conditions (i)–(ii) the pantograph type first order initial value problem has unique bounded solution $x^* \in C^1[0, a]$ and if $f \in C^3([0, a] \times \mathbb{R} \times \mathbb{R})$ and $a\left(\frac{5}{3}L_1 + L_2\right) < 1$, the error estimate in the discrete and continuous approximation of the solution is*

$$|x^*(t_i) - S_k(t_i)| \leq \frac{a^k (L_1 + L_2)^k a M_0}{1 - a(L_1 + L_2)}$$
$$+ \frac{a M''' h^3}{192\left[1 - a\left(\frac{5}{3}L_1 + L_2\right)\right]} + \frac{13a\left(\frac{5}{3}L_1 + L_2\right) \cdot M''' h^4}{4\left[1 - a\left(\frac{5}{3}L_1 + L_2\right)\right]}, \quad k \in \mathbb{N}^*, \, i = \overline{1, n} \tag{6}$$

*and*

$$|x^*(t) - S_k(t)| \leq \frac{a^k(L_1 + L_2)^k a M_0}{1 - a(L_1 + L_2)} + \frac{5a M''' h^3}{384[1 - a(\frac{5}{3}L_1 + L_2)]}$$

$$+ \frac{75a(\frac{5}{3}L_1 + L_2) \cdot M''' h^4}{8[1 - a(\frac{5}{3}L_1 + L_2)]} + \frac{13 M''' h^4}{4}, \quad k \in \mathbb{N}^*, \ t \in [0, a], \tag{7}$$

*respectively.*

*Proof.* Applying the Banach fixed point principle under the conditions (i)–(ii) we obtain the existence and uniqueness of the solution $x^* \in C^1[0, a]$ and the uniform convergence of the sequence of successive approximations

$$x_k(t) = x_0 + \int_0^t f(s, x_k(s), x_{k-1}(qs))\, ds, \quad t \in [0, a] \tag{8}$$

to this solution with the apriori error estimate

$$|x^*(t) - x_k(t)| \leq \frac{a^k(L_1 + L_2)^k a M_0}{1 - a(L_1 + L_2)}, \quad t \in [0, a], \ k \in \mathbb{N}^* \tag{9}$$

where $M_0 \geq 0$ is such that $|f(t, x_0, x_0)| \leq M_0, \ \forall t \in [0, a]$. Similarly as in [3], page 24, we obtain the uniform boundedness of the sequence of successive approximations $|x_k(t)| \leq |x_0| + \frac{a M_0}{1 - a(L_1 + L_2)} = R$, for all $t \in [0, a], \ k \in \mathbb{N}^*$. Since $x'_k(t) = f(t, x_{k-1}(t), x_{k-1}(qt))$, by $f \in C^3([0, a] \times \mathbb{R} \times \mathbb{R})$ we infer that $x_k \in C^4[0, a]$ and therefore, on the compact $[0, a] \times [-R, R] \times [-R, R]$, the function $f$ and their partial derivatives are bounded. In order to obtain the error estimate in the discrete approximation of the solution (6), by (9) we see that it remains to estimate $|x_k(t_i) - S_k(t_i)| = \overline{R_{k,i}}, \ i = \overline{1, n}, \ k \in \mathbb{N}^*$. In this purpose we see that

$$|\overline{R_{k,i}}| \leq \left| \int_0^{t_i} f_{k-1}(s)\, ds - Q_n(f_{k-1}) \right| + |Q_n(f_{k-1}) - Q_n(F_{k-1})|$$

$$\leq |R_{k,i}| + \frac{a}{6n} \cdot \sum_{j=1}^{i} [L_1 \sum_{p=1}^{3} |\alpha_p(x_{k-1} - S_{k-1})(t_{j-1} + \beta_p h)| \tag{10}$$

$$+ L_2 \sum_{p=1}^{3} \alpha_p |(x_{k-1} - S_{k-1})(q(t_{j-1} + \beta_p h))|]$$

for all $i = \overline{1, n}, \ k \in \mathbb{N}^*$, where $\alpha_1 = \alpha_3 = 1, \ \alpha_2 = 4, \ \beta_1 = 0, \ \beta_2 = \frac{1}{2}, \ \beta_3 = 1, \ f_k(t) = f(t, x_k(t), x_k(qt)), \ F_k(t) = f(t, S_k(t), S_k(qt))$ and $|R_{k,i}| \leq \frac{a^4}{192 n^3} \cdot \|f'''_{k-1}\|_\infty$. So, we need to obtain the estimate for $\max_{t \in [0, a]} |x_{k-1}(t) - S_{k-1}(t)|$. Therefore we introduce the sequence of auxiliary functions $(V_k)_{k \in \mathbb{N}^*}, \ V_k : [0, a] \to \mathbb{R}$ given by their restrictions to the subintervals $[t_{i-1}, t_i], \ i = \overline{1, n}$, as follows:

$$V_k(t) = x_k(t) + \frac{(t_i - t)(t - t_{i-1}) \cdot [(t_i - t) m_{k,i-1} - (t - t_{i-1}) m_{k,i}]}{h^2}$$

$$+ \frac{(t - t_{i-1})^2 [2(t_i - t) + h] \cdot y_i + (t_i - t)^2 [2(t - t_{i-1}) + h] \cdot y_{i-1}}{h^3} \tag{11}$$

where $m_{k,i}$, $i = \overline{0,n}$, are obtained by the smoothness condition $V_k \in C^2[0,a]$ with $V_k''(t_0) = x_k''(t_0)$, $V_k''(t_n) = x_k''(t_n)$. Therefore the values $m_{k,i}$, $i = \overline{0,n}$, are the solutions of the system

$$\begin{cases} 2m_{k,0} + m_{k,1} = \frac{3}{h}(y_1 - y_0) \\ \frac{1}{2}m_{k,i-1} + 2m_{k,i} + \frac{1}{2}m_{k,i+1} = \frac{3(y_{i+1}-y_i)}{2h} + \frac{3(y_i-y_{i-1})}{2h}, \ i = \overline{1,n-1} \\ m_{k,n-1} + 2m_{k,n} = \frac{3}{h}(y_n - y_{n-1}). \end{cases}$$

where $y_i = \overline{x_k(t_i)} - x_k(t_i)$, $i = \overline{0,n}$. Analogous to the manner presented in [9], page 104, it obtains $\max\limits_{i=\overline{1,n-1}} |S_k''(t_i) - V_k''(t_i)| \leq \frac{3}{4} \cdot \|x_k^{IV}\|_\infty h^2$. Now, by using the Taylor's formula we get $\left|S_k'''(t) - V_k'''(t)\right| \leq \frac{13}{4} \cdot \|x_k^{IV}\|_\infty h^2$. Since $S_k(t_i) - V_k(t_i) = 0$, $\forall i = \overline{0,n}$, by the Rolle's mean value theorem will be $\xi_i \in (t_{i-1}, t_i)$ such that $S_k'(\xi_i) - V_k'(\xi_i) = 0$. Therefore $|V_k(t) - S_k(t)| \leq \frac{13h^4}{4} \cdot \|x_k^{IV}\|_\infty$, for all $t \in [t_{i-1}, t_i]$, $i = \overline{1,n}$. Now, with $|x_{k-1}(t) - S_{k-1}(t)| \leq |x_{k-1}(t) - V_{k-1}(t)| + |V_{k-1}(t) - S_{k-1}(t)|$ and since $\max\limits_{i=\overline{0,n}} |m_{k-1,i}| \leq \frac{6}{h} \cdot \overline{R_{k-1}}$, by the relations (11) we obtain $\max\limits_{t\in[0,a]} |x_{k-1}(t) - S_{k-1}(t)| \leq \frac{5}{2}\overline{R_{k-1}} + \frac{13h^4}{4} \cdot \|x_k^{IV}\|_\infty$, where $\overline{R_{k-1}} = \max\limits_{i=\overline{1,n}} |\overline{R_{k-1,i}}|$. By (10) in inductive manner for $k \in \mathbb{N}^*$, $k \geq 2$, we get $|\overline{R_{k,i}}| \leq |R_{k,i}| + \frac{aL_1}{3} \cdot \overline{R_{k-1}} + \left(\frac{2aL_1}{3} + aL_2\right) \cdot \|x_{k-1} - S_{k-1}\|_\infty$, and since $x_k^{IV}(t) = f_{k-1}'''(t)$, $\forall t \in [0,a]$, we obtain

$$|\overline{R_{k,i}}| \leq \frac{aM'''h^3}{192\left[1 - a\left(\frac{5}{3}L_1 + L_2\right)\right]} + \frac{13a\left(\frac{5}{3}L_1 + L_2\right) \cdot M'''h^4}{4\left[1 - a\left(\frac{5}{3}L_1 + L_2\right)\right]}$$

for all $i = \overline{1,n}$, $k \in \mathbb{N}^*$, where $M''' = \max\limits_{t\in[0,a]} |f_{k-1}'''(t)|$ is obtained similarly as in [2], pages 142–143.

*Remark 1.* By the estimates (6) and (7), denoting $e_{k,i} = |x^*(t_i) - S_k(t_i)|$, $e_k(t) = |x^*(t) - S_k(t)|$, $k \in \mathbb{N}^*$, $i = \overline{1,n}$, we obtain

$$\lim\limits_{k\to\infty, h\to 0} e_{k,i} = 0, \ \forall i = \overline{1,n}, \quad \text{and} \quad \lim\limits_{k\to\infty, h\to 0} e_k(t) = 0, \forall t \in [0,a]$$

that is the uniform convergence of $(S_k)_{k\in\mathbb{N}^*}$ to the solution on $[0,a]$. Moreover, we see that the order of convergence is $O(h^3)$ when $f \in C^3([0,a] \times \mathbb{R} \times \mathbb{R})$.

## 3.2 Numerical Stability

Taking into account that the present method is an iterative one, in what concerns the suitable numerical stability concept it is appropriate to investigate how the numerical solution varies depending on the choice of the first iteration. Since the first iteration is described by the function $g$, we consider another first iteration done by a continuous function $h$ with $|g(t) - h(t)| < \varepsilon$, $\forall t \in [0,a]$ and the computed values on the knots in this case, $\overline{S_k}(t_i)$, $i = \overline{0,n}$, are obtained with the same algorithm (4–5) by starting with $x_0(t) = h(t)$. This kind of stability involves the estimation of $\left|S_k(t_i) - \overline{S_k}(t_i)\right|$, $i = \overline{1,n}$.

**Definition 1.** *The iterative method (4–5) applied to the problem (1) of first order is numerically stable regarding the choice of the first iteration if exist the constants $K_1, K_2, K_3 \geq 0$ and the natural numbers $m, m' \in \mathbb{N}^*$ such that*

$$\left| S_k\left(t_i\right) - \overline{S_k\left(t_i\right)} \right| < K_1 \cdot \varepsilon + K_2 h^m + K_3 h^{m'}, \quad \text{for all } k \in \mathbb{N}^*, \ i = \overline{1, n}.$$

Under the conditions of Theorem 1, the estimates (6) and (7) provide the numerical stability property for the first order initial value problem (1) similarly as in Theorem 7 from [3]. For two-point boundary value problems (2) this kind of numerical stability can be proved in analogous way to that presented in Theorem 6 from [2]. In the case of first order initial value problems the numerical stability is related to the choice of the initial value, while for second order two-point boundary value problems this type of numerical stability describes the sensitivity to the choice of the boundary values.

## 4   Numerical Experiments

In order to illustrate the performances of the method and to test the theoretical results concerning the order of convergence and the numerical stability we present some numerical examples. The convergence will be tested by choosing $n = 10$, $n = 100$, and $n = 1000$, the numerical stability regarding the choice of the first iteration being well-tried for $\varepsilon = 10^{-2}$ and $\varepsilon = 10^{-1}$.

*Example 1.* The nonlinear pantograph initial value problem

$$\begin{cases} x'\left(t\right) = -q\left[x\left(t\right) + \frac{(1-q)x^2(t)}{q}\right] \cdot x\left(qt\right), \quad t \in [0, \tfrac{1}{2}] \\ x\left(0\right) = 1 \end{cases}$$

has the exact solution $x^*\left(t\right) = \frac{1}{t+1}$ and applying the algorithm (4–5) with $q = 0.5$, $q = 0.25$, $q = 0.75$, for $\varepsilon' = 10^{-16}$ we get $k = 15$ iterations. The numerical stability with respect to the choice of the first iteration is illustrated for $q = 0.5$ by $d_i = \left| S_k\left(t_i\right) - \overline{S_k\left(t_i\right)} \right|$, $i = \overline{0, n}$, while the errors on the knots $e_i = |x^*\left(t_i\right) - S_k\left(t_i\right)|$, $i = \overline{0, n}$, for $n = 10$ and $n = 100$ are presented for $q = 0.5$, $q = 0.25$, $q = 0.75$ on the columns of Table 1. We observe the accuracy $O\left(10^{-5}\right)$ for $n = 10$ and $O\left(10^{-8}\right)$ for $n = 100$. In the case $n = 1000$ the obtained accuracy is $O\left(10^{-11}\right)$, revealing the order of convergence $O\left(h^3\right)$ as was asserted in Remark 1.

*Example 2.* The solution of the second order pantograph type two-point boundary value problem

$$\begin{cases} x''\left(t\right) = \tfrac{2}{3}e^{-t} + \tfrac{1}{3}\left[x\left(qt\right)\right]^{\frac{1}{q}}, \quad t \in [0, \tfrac{1}{2}] \\ x\left(0\right) = 1, \quad x\left(0.5\right) = \frac{1}{\sqrt{e}} \end{cases}$$

**Table 1.** Testing the convergence for Example 1

| $t_i$ | $e_i, 10$ | $e_i, 10^2$ | $e_i, 10$ | $e_i, 10^2$ | $e_i, 10$ | $e_i, 10^2$ | $d_i$ |
|---|---|---|---|---|---|---|---|
| 0 | $q = 0.5$ | 0 | $q = 0.25$ | 0 | $q = 0.75$ | 0 | 1.00e−02 |
| 0.15 | 5.91e−05 | 5.31e−08 | 6.94e−05 | 9.91e−08 | 3.92e−05 | 3.64e−08 | 7.06e−03 |
| 0.30 | 4.32e−05 | 3.85e−08 | 8.82e−05 | 6.92e−08 | 2.96e−05 | 2.76e−08 | 5.11e−03 |
| 0.40 | 3.54e−05 | 3.16e−08 | 7.57e−05 | 5.44e−08 | 2.49e−05 | 2.33e−08 | 4.19e−03 |
| 0.50 | 2.77e−05 | 2.45e−08 | 6.32e−05 | 4.10e−08 | 1.99e−05 | 1.85e−08 | 3.46e−03 |

is $x^*(t) = e^{-t}$ and by using the trapezoidal quadrature rule, for $\varepsilon' = 10^{-16}$ the last iteration was $k = 8$. The algorithm is tested taking $q = 0.5$, $q = 0.35$, $q = 0.65$ and here the numerical stability was considered for $\varepsilon = 10^{-1}$ being illustrated for $q = 0.65$. The results are presented in Table 2 and the accuracy is $O\left(10^{-6}\right)$ for $n = 10$, $O\left(10^{-8}\right)$ for $n = 100$, and $O\left(10^{-10}\right)$ for $n = 1000$, which means an order of convergence $O\left(h^2\right)$.

**Table 2.** Numerical results for Example 2

| $t_i$ | $e_i, 10$ | $e_i, 10^2$ | $e_i, 10$ | $e_i, 10^2$ | $e_i, 10$ | $e_i, 10^2$ | $d_i$ |
|---|---|---|---|---|---|---|---|
| 0 | $q = 0.5$ | 0 | $q = 0.35$ | 0 | $q = 0.65$ | 0 | 1.00e−01 |
| 0.1 | 3.31e−06 | 3.39e−08 | 3.03e−06 | 3.38e−08 | 3.37e−06 | 3.40e−08 | 9.90e−02 |
| 0.2 | 4.87e−06 | 4.91e−08 | 4.78e−06 | 4.90e−08 | 4.89e−06 | 4.92e−08 | 9.85e−02 |
| 0.3 | 4.71e−06 | 4.75e−08 | 4.65e−06 | 4.73e−08 | 4.74e−06 | 4.75e−08 | 9.86e−02 |
| 0.4 | 3.04e−06 | 3.06e−08 | 2.99e−06 | 3.05e−08 | 3.06e−06 | 3.07e−08 | 9.91e−02 |
| 0.5 | 0 | 0 | 0 | 0 | 0 | 0 | 1.00e−01 |

*Example 3.* For the pantograph variant of the beam equation with clamped end-conditions

$$\begin{cases} x^{IV}(t) = e^{-t}\left[x(t)\right]^{2-q} \cdot x(qt), & t \in [0,1] \\ x(0) = 1, \ x(1) = e, \ x'(0) = 1, \ x'(1) = e \end{cases}$$

and the exact solution $x^*(t) = e^t$ we have applied the Euler-Mac Laurin composite quadrature rule and for $\varepsilon' = 10^{-16}$, $q = \frac{1}{4}$, $q = \frac{1}{2}$, $q = \frac{3}{4}$, we get $k = 8$ iterations. The numerical stability was tested for $\varepsilon = 10^{-1}$ considering $q = 0.25$ and the results presented in Table 3 confirm an order of convergence $O\left(h^4\right)$.

Inspecting the above presented examples we see that the convergence of the method proved in Theorem 1 and the numerical stability regarding the choice of the first iteration are confirmed. The order of convergence depends by the choice of the quadrature rule and by the involved interpolation procedure.

**Table 3.** Numerical results for the pantograph clamped beam model

| $t_i$ | $e_i$, 10 | $e_i$, $10^2$ | $e_i$, $10^3$ | $e_i$, 10 | $e_i$, $10^2$ | $e_i$, 10 | $e_i$, $10^2$ | $d_i$ |
|---|---|---|---|---|---|---|---|---|
| 0 | $q = \frac{1}{2}$ | 0 | 0 | $q = \frac{1}{4}$ | 0 | $q = \frac{3}{4}$ | 0 | 1.0e−1 |
| 0.2 | 2.6e−8 | 2.2e−12 | $O\left(10^{-16}\right)$ | 2.1e−8 | 1.7e−12 | 2.3e−8 | 2.3e−12 | 6.7e−1 |
| 0.4 | 7.1e−8 | 6.6e−12 | $O\left(10^{-16}\right)$ | 7.9e−8 | 7.2e−12 | 6.6e−8 | 6.5e−12 | 3.9e−1 |
| 0.6 | 1.7e−7 | 1.6e−11 | $O\left(10^{-15}\right)$ | 1.8e−7 | 1.7e−11 | 1.6e−7 | 1.6e−11 | 6.2e−2 |
| 0.8 | 1.8e−7 | 1.7e−11 | $O\left(10^{-15}\right)$ | 1.8e−7 | 1.8e−11 | 1.7e−7 | 1.7e−11 | 3.4e−1 |
| 1 | 0 | 0 | 0 | 0 | 0 | 0 | 0 | 1.0e−1 |

# References

1. Agarwal, R.P., Chow, Y.M.: Finite-difference methods for boundary-value problems of differential equations with deviating arguments. Comput. Math. Appl. **12A**, 1143–1153 (1986)
2. Bica, A.M., Curila, M., Curila, S.: Two-point boundary value problems associated to functional differential equations of even order solved by iterated splines. Appl. Numer. Math. **110**, 128–147 (2016)
3. Bica, A.M.: Initial value problems with retarded argument solved by iterated quadratic splines. Appl. Numer. Math. **101**, 18–35 (2016)
4. Brunner, H.: Collocation Methods for Volterra Integral and Related Functional Differential Equations. Cambridge University Press, Cambridge (2004)
5. Brunner, H.: Recent advances in the numerical analysis of Volterra functional differential equations with variable delays. J. Comput. Appl. Math. **228**, 524–537 (2009)
6. Korman, P.: Computation of displacements for nonlinear elastic beam models using monotone iterations. Internat. J. Math. Math. Sci. **11**, 121–128 (1988)
7. Micula, G., Micula, S.: Handbook of Splines. Mathematics and its Applications, vol. 462. Kluwer Academic Publishers, Dordrecht (1999). https://doi.org/10.1007/978-94-011-5338-6
8. Ockendon, J.R., Tayler, A.B.: The dynamics of a current collection system for an electric locomotive. Proc. Roy. Soc. Lond. A **322**, 447–468 (1971)
9. Stoer, J., Bulirsch, R.: Introduction to Numerical Analysis. Texts in Applied Mathematics, 2nd edn. Springer, New York (1993). https://doi.org/10.1007/978-1-4757-2272-7
10. Wazwaz, A.-M., Raja, M.A.Z., Syam, M.I.: Reliable treatment for solving boundary value problems of pantograph delay differential equation. Rom. Rep. Phys. **69**, 102 (2017)

# On Gas Dynamic Hierarchy

S. V. Bogomolov$^{(\boxtimes)}$, N. B. Esikova, A. E. Kuvshinnikov, and P. N. Smirnov

M.V. Lomonosov Moscow State University, Moscow, Russia
bogomo@cs.msu.su,{esikova.nata,kuvsh90}@yandex.ru,
pavel.smirnov91@mail.ru

**Abstract.** On the example of a simple and clear, but far from being trivial, model of hard sphere gas, we will try to show the main stages in constructing the mathematical formalization of a complex physical system.

We are considering a set of about $10^{25}$ solid balls that just fly and collide. A mathematical description of the evolution of such a system inevitably leads to the necessity of using the apparatus of the theory of random processes. To identify the mathematical and computational features of the problem under study it is important to write it in a dimensionless form. This procedure leads to the appearance of the Knudsen number, the physical meaning of which is the ratio of the mean free path of molecules to the characteristic size of the problem. The hierarchy of micro-macro models is constructed in accordance with the change in this parameter from values of the order of unity (micro) to magnitudes of the order of 0.1 (meso) and further to 0.01 (macro). Accurate movement along this path leads to more accurate mathematical models, in comparison with traditional ones, which affects their greater computational fitness - nature pays for a careful attitude towards it. In particular, obtained macroscopic equations are softer for simulations than the classical Navier-Stokes equations.

This hierarchy of mathematical statements generates a corresponding chain of computational methods. Microscopic problems are most often solved using Monte Carlo methods, although there are research groups that are committed to nonrandom methods for solving the Boltzmann equation. Recently, much attention has been paid to mesomodels based on modeling the Brownian motion or solving the deterministic Fokker - Planck - Kolmogorov equations. To solve the problems of a continuous medium, difference methods, finite element methods, and particle methods are used. The latter ones, in our opinion, are particularly promising for the entire hierarchy, uniting different statements with a single computational ideology. A discontinuous particle method is particularly effective.

**Keywords:** Boltzmann equation · Kolmogorov – Fokker –
Planck equation · Navier – Stokes equation · Random processes ·
Stochastic differential equations with respect to Poisson and Wiener
measures · Discontinuous particle method

© Springer Nature Switzerland AG 2019
I. Dimov et al. (Eds.): FDM 2018, LNCS 11386, pp. 167–174, 2019.
https://doi.org/10.1007/978-3-030-11539-5_17

# 1  Introduction

A description of gas dynamic phenomena, base on hierarchies of micro – macro models has become a classical part of theoretical physics a long time ago [1] as well as a foundation for high performance industrial calculations recently. In the last years more attention was drawn to "meso" models in the phase space. That models are often called Kolmogorov – Fokker – Planck equations. They are used by theoreticians and applied mathematicians [2,3]. The last ones mostly implement the models of so-called Maxwell molecules, not hard sphere ones, that lead to considerably different results [4]. We'll try to briefly show our approach to that matters.

# 2  Micro - Macro Transition Scheme

For simple transport, that scheme is quite clear without comments.
**Micro for Transport** (Lagrange approach):

$$dx_i(t)/dt = v_i, x_i(0) = x_i^0, i = 1, ..., N.$$

**Micro - Macro Connection** (determination of $u(x,t)$ by $x_i(t)$):

$$\forall\varphi(x) : \int \varphi(x)u(x,t) = \frac{1}{N}\sum_{i=1}^{N} \varphi(x_i(t)).$$

**Macro:** $\frac{d}{dt} \to \int \varphi\frac{\partial u}{\partial t}dx = \frac{1}{N}\sum_{i=1}^{N}\frac{\partial\varphi}{\partial x}\frac{dx_i}{dt} = \int \frac{\partial\varphi}{\partial x}(vu)dx;$

generalized equation: $\int \varphi\frac{\partial u}{\partial t}dx - \int \frac{\partial\varphi}{\partial x}(vu)dx = 0, \int \varphi(\frac{\partial u}{\partial t} + \frac{\partial vu}{\partial x})dx = 0, \forall\varphi \to$
partial differential equation: $\frac{\partial u}{\partial t} + \frac{\partial vu}{\partial x} = 0.$

# 3  Micro

Our starting point is **Collisions.** Jump function for hard sphere collisions is $\mathbf{f}(v_i, v_j, \omega) = \omega(\omega, v_i - v_j)$ (Fig. 1).

**Stochastic Molecular Dynamics**

$$dx_i(t) = v_i dt, \quad dv_i(t) = \sum_{j=1}^{N}\int_{\Omega} f(v_i, v_j, \omega)p_{ij}(d\omega \times dt),$$

$$\lambda_{ij} = N_{ij}^{collisions}/N, \quad N_{ij}^{collisions} = N_{ij}^{reached},$$

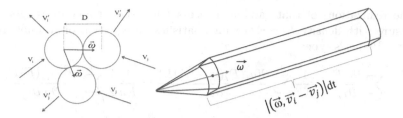

**Fig. 1.** Collisions with different $\omega$ and (to the right) a cylinder for $N_{ij}^{reached}$ counting

**Micro Jump Model** (if we calculate $N_{ij}^{reached}$ introducing $F(\cdot)$)

$$\lambda_{ij} = \frac{1}{Kn}\frac{1}{2}F(x_i, v_j, t)\Delta x_i \Delta v_j,$$

$$\forall \varphi(x, v) : \int \varphi(x, v) F(x, v, t) dt = \frac{1}{N}\sum_{i=1}^{N} \varphi(x_i(t), v_i(t)),$$

where $x_i(t)$, positions, and $v_i(t)$, velocities of molecules, are 3D random processes, $f(\cdot)$ - jump function or increment of velocity $v_i$ caused by a collision with a molecule having velocity $v_j$, $p_{ij}$ - Poisson measures (with intensities $\lambda_{ij}$) which indicate a collision, $Kn(x,t) = 1/D^2 n_* x_*$ - Knudsen number, $D$ - diameter of a molecule, $F(\cdot)$ - "distribution" function or density of measure, generated by $(x(t), v(t))$, $\Delta x_i = |(\omega, v_i - v_j)| dt D^2 d\omega$ - a volume of the cylinder where molecules that can reach $i$-molecule during $dt$ (it means "collide") are situated.

Inside domains with large $Kn$ our model can be written as a Boltzmann type of equation with fluctuations for $F$ which is random as well as $\lambda_{ij}$. It is the main difference in comparison to traditional kinetic theory.

## 4   MESO

If $Kn$ is small then $1/Kn$ is large. That's why instead of the jump process we implement a diffusion process in phase space. We consider a model valid, in our opinion, at moderate Knudsen numbers, transient between a molecular description and an imagination of a gas as continuous medium. $Kn$ is a parameter of nondimensionalization depending on a space subdomain. Its physical meaning is a ratio of an average mean free pass to a character dimension of the subdomain. Our model [5–8,13,14] is a system of stochastic differential equations (SDE) with respect to Wiener measure $dw(t)$ describing a movement of a particle in the phase space at moderately small $Kn$:

$$dx(t) = v(t)dt, \quad dv(t) = -\frac{1}{Kn}a(c)(v(t) - V)dt + \frac{1}{\sqrt{Kn}}\sigma(c)dw(t), \quad (1)$$

where $c$ is an absolute value of the dimensionless heat velocity $\mathbf{c} \equiv v(t) - V$, $V(x,t)$ is a macroscopic velocity, the coefficients in the second equation (vector $\mathbf{a}(c) = a(c)\mathbf{c}$) and matrix $\sigma(c)$ are determined in [7,9].

The realizations of that random process (the set of trajectories) generate a measure with density $F(x, v, t)$ which satisfies an equation of Kolmogorov – Fokker – Planck's type:

$$\frac{\partial F}{\partial t} + \sum_{i=1}^{3} \frac{\partial v_i F}{\partial x_i} - \frac{1}{Kn} \sum_{i=1}^{3} \frac{\partial (a_i(F)(v_i - V_i)F)}{\partial v_i} = \frac{1}{Kn} \frac{1}{2} \sum_{i,j=1}^{3} \frac{\partial^2 (\sigma_{ij}^2(F)F)}{\partial v_i \partial v_j}. \quad (2)$$

We study that, computationally more efficient than Boltzmann equation, diffusive in velocity space, gas model which is a link in a chain of multi scale algorithms based on micro – macro models, depending on different subdomains (distinguished by their Knudsen numbers) of a whole problem under consideration. That model, on the one hand, is connected to microscopic model and, on the other hand, leads to more accurate macroscopic equations.

That equation is well known as an heuristic model Boltzmann equation with Fokker – Planck collision integral. But before now its coefficients $a$, $\sigma^2$ and the limits of its applicability have not been specified. In [7] it was shown that at moderate Knudsen numbers Boltzmann equation can be approximated by Kolmogorov – Fokker – Planck one (2). Its coefficients are integrals in the phase space representing moments of a jump random process describing molecules collisions formulized by means of stochastic integration with respect to Poisson measures [8]. At small $Kn$ for hard sphere gas (at an assumption of distribution function $F$ local maxwellity and isotropy by thermal velocity $c$ inside eight – foled integrals calculations of "drift" vector $\mathbf{a}$ and "diffusion" matrix $\sigma^2$ in velocity space), these coefficients in Eqs. (1), (2) are obtained in [9], the computations [10] have shown that they are quite adequate.

We've got more accurate, than in [7] and than Navier – Stokes, system of macroscopic gas dynamics equations in case of hard sphere gas. The accuracy, higher than in [7], is based on analytical solving system (1), more precisely, its version, simplified to a level when appropriate calculations can be done by hand.

Note also that our approach with the help of SDE technique differs from other approaches for obtaining gas dynamics equations connected to application of deterministic equations for distribution function in the phase space [11,12,15] as well as other hierarchical models [16]. The models like ours with the coefficients $\mathbf{a}$ and $\sigma$ depending on velocity are in use by the physics theoreticians to study phenomena in turbulent flows (f. e. [17]).

## 5    Macro

**Macro as a Process in X - Space.** We'll rearrange the system (1) in a way that makes it possible to get macroscopic equations keeping maximum of microscopic information. The system (1) is a system of equations for unknown functions $x(t)$ and $v(t)$. Express $v(t)$ through $x(t)$ from the second equation and substitute it in the first one having got an equation only for $x(t)$.

That equation will give us macroscopic equations. To do it we need to make some simplifications. Analytical solving SDE is more difficult than solving ordinary differential equations. Only some successful examples are known.

The coefficients in (1) at large $c$ behave themselves as $\mathbf{a}(\mathbf{c}) \sim a_1 \mathbf{c}\mathbf{c}, a_1 \equiv \pi/2$ and $\sigma \sim \sigma_1 c^{3/2}$ (we denote $\sigma_1 \equiv (2\sqrt{\pi}/(3\sqrt{3}))T^{1/4}$). To derive them to a form enabling to get an exact solution let us put $\sigma = k\sigma_1 c$, introducing a parameter $k$ which can be taken so, for example, that one of the terms in our macro – model coincides with the thermo dynamical equation of state. We underline that the introduction of the parameter $k$ was done to get an analytical solution, we do not need the thermo dynamical equation of state – all the coefficients in our equations are obtained from the model of hard spheres.

We come to the system:

$$dx(t) = V dt + \sqrt{Kn}\,\widetilde{\sigma}\,(dw + d\widetilde{w}), \quad dv(t) = -\frac{1}{Kn}a(v(t) - V)dt + \frac{1}{\sqrt{Kn}}\sigma dw,$$

where $\widetilde{\sigma}_{ij} \equiv \sigma_{0ij}/a_{0i} = 0,43\ kT^{1/4}$. In the second equation instead of $a(c)$ and $\sigma(c)$ we take their computed values, averaged in velocity space with respect to the local maxwellian: $a \approx 2,979\ T^{1/2}$, $\sigma \approx 1,73\ T^{3/4}$.

The choice of coefficients depending only on $x$ and $t$ is frequently used, for instance, in the context of model collision integral in Fokker – Planck form. Note, that with our simplifications we get Einstein's fluctuation – dissipation relation (in dimensional form for explicity): $\sigma^2/a = 2RT$.

The presence of the increment of stochastic term in the first equation in the form of two independent processes $(dw + d\widetilde{w})$, which has appeared at the calculation of $\int_0^t w_s/s ds$, is not trivial, in our view.

We derive the equations of stochastic gas dynamics for that set of coefficients. Our goal is an obtaining macroscopic equations for non – random macro – parameters. It means that we need to construct the equations for measures in 3D space which are generated by the random processes $x(t)$ and $v(x(t))$ belonging to the phase space. A physical meaning of that measures is the evolution of mass, momentum and energy distributions.

**Continuity Equation with Self Diffusion.** An amount of gas in a domain $D$ is, from one side, a whole mass of molecules and, from the other side, an integral with respect to a measure: $\sum_{l:x_l \in D} m_l = \int_D \mu_t(dx)$, or, if all the particles possess the equal masses $1/N$: $\frac{1}{N}\sum_{l=1}^{N} \chi(x_l(t) \in D) = \int_D \mu_t(dx)$, where $\chi$ is a characteristic function. $N$ can be considered as a number of realizations of the random process $x(t)$ which is a solution of the system (4). We define a stochastic empirical measure $\mu_t(dx)$ by an expression: for any function $\psi \in C_b^{(2)}(\mathbf{R}^3)$ (a space of continuously differentiable finite functions) $\int \psi(x)\mu_t(dx) = \frac{1}{N}\sum_{l=1}^{N} \psi(x_l(t))$. That expression, connecting the measure distribution to realizations of particle positions at time moment $t$, is a Chebyshev quadrature formula (the weights are known and the nodes are parameters) if to read if from left to right.

For obtaining an equation for measure $\mu_t(dx)$, let us take a stochastic differentials from both of two sides of measure definition. Using Ito's formula for complex function differentiation, assuming existence of a density $\rho(x,t)$ of stochastic empirical measure $\mu_t(dx)$ (taking the usual steps while deriving from a generalized equation an equation in partial derivatives: having integrated by parts one or two times in appropriate places) we get a stochastic continuity equation in the form:

$$
d\rho = \left[ -\sum_{i=1}^{3} \frac{\partial}{\partial x_i} (V_i \rho) + \sum_{i,j=1}^{3} \frac{\partial^2}{\partial x_i \partial x_j} \left( Kn\, \tilde{\sigma}_{ij}^2 \rho \right) \right] dt
$$

$$
- \sum_{i,j=1}^{3} \frac{\partial}{\partial x_i} \left( \sqrt{Kn}\, \tilde{\sigma}_{ij} \rho \right) (dw_j + d\tilde{w}_j),
$$

and having averaged over the time we get a deterministic continuity equation for time averaged deterministic mass density $\overline{\rho}(x,t)$. The right hand side reflects the "trace" left by the thermal motion of molecules, or self – diffusion. It does not spoil conservativeness property because it describes diffusion and has a divergent form [20].

If we assume that the time averaging leads to the values using by traditional gas dynamics, then we get a continuity equation taking into account the self – diffusion (further on we omit the lines denoting time averaging above the macro parameters). If one performs our derivations for a gas of Maxwell molecules [4] (which do not exist in nature), the self – diffusion corrector in the right hand side will be equal to zero.

**Stochastic Gas Dynamics System.** Proceeding in a similar way with momentum and energy we get a new system of gas dynamic equations [13] in a simplest form (the matrix $\widetilde{\sigma}_{ij}^2$ in general is not diagonal). Denote $A \equiv \widetilde{\sigma}_{ij}^2 = 0.085\, T^{1/2}$, $B \equiv \overline{\sigma_{ij}\widetilde{\sigma}_{ij}} = 0.5\, T$. The latter means our choice of the parameter $k = 0.675$ for denoting a combination $B\rho$ as $p$ calling it "pressure". Then the equality $p = B\rho$ or $p = \rho RT$ (in dimensional form) can be called an equation of state.

$$
\frac{\partial \rho}{\partial t} + \sum_{j=1}^{3} \frac{\partial}{\partial x_j} (\rho V_j) = \sum_{j=1}^{3} \frac{\partial^2}{\partial x_j^2} (KnA\rho)
$$

$$
\frac{\partial(\rho V_i)}{\partial t} + \sum_{j=1}^{3} \frac{\partial}{\partial x_j} (V_j(\rho V_i)) = -\frac{\partial}{\partial x_i} (B\rho) + \sum_{j=1}^{3} \frac{\partial^2}{\partial x_j^2} (KnA(\rho V_i)),
$$

$$
\frac{\partial(\rho E)}{\partial t} + \sum_{j=1}^{3} \frac{\partial}{\partial x_j} (V_j(\rho E)) = -\sum_{j=1}^{3} \frac{\partial}{\partial x_j} (B(\rho V_j)) + \sum_{j=1}^{3} \frac{\partial^2}{\partial x_j^2} (KnA(\rho E)).
$$

Introducing the notions $p \equiv B\rho$, $\nu \equiv KnA$ and calling $p$ by pressure and $\nu$ by coefficient of kinematic viscosity, we'll have the macroscopic system in more traditional form.

**A Numerical Example Using Discontinuous Particle Method.** We take a well - known problem of shock wave structure as a test. The original model (1) is of interest not only because of the possibility to construct new stochastic and deterministic macro – models but also is a base for direct modelling by the help of stochastic particle method, attractable for high performance simulation. In the present paper we use a deterministic particle method, alternative to stochastic one, free from parasitic fluctuations. The applied explicit particle method [21] has a minimal dissipation that makes it possible to get a solution with high accuracy for correct comparison of different models. We do not use limiters because of sufficient natural viscosity. Our system turned out to be less demanding to the values of time steps than Navier – Stokes one. Note that our system gives a result [13] close to the experiment (Fig. 2).

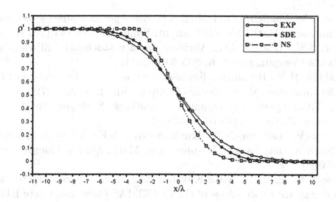

**Fig. 2.** The profiles of the normed density in a shock wave for the stochastic gas dynamics (SDE) and Navier – Stokes (NS) systems, compared to EXPeriment.

## 6   Conclusions

Regardless to our quite severe simplifications we obtained a gas dynamics system having clear microscopic origin and gives more adequate than usual results at the well - known test. Moreover, we've got the hierarchy of micro – macro stochastic and deterministic models each of which has its own place in a row of unified solvers.

## References

1. Boltzmann, L.: Weitere Studien über das Wärme gleichgenicht unfer Gasmoläkuler. Sitzungsberichte der Akademie der Wissenschaften **66**, 275–370 (1872)
2. Jun, E., Hossein Gorji, M., Grabe, M., Hannemann, K.: Assessment of the cubic Fokker-Planck-DSMC hybrid method for hypersonic rarefied flows past a cylinder. Comput. Fluids **168**, 1–13 (2018)

3. Zhang, J., Zeng, D., Fan, J.: Analysis of transport properties determined by Langevin dynamics using Green-Kubo formulae. Physica A: Stat. Mech. Appl. **411**, 104–112 (2014)
4. Gupta, V.K., Torrilhon, M.: Comparison of relaxation phenomena in binary gas-mixtures of Maxwell molecules and hard spheres. Comput. Math. Appl. **70**, 73–88 (2015)
5. Bogomolov, S.V.: An approach to deriving stochastic gas dynamics models. Doklady Math. **78**, 929–931 (2008)
6. Arsen'yev, A.A.: On the approximation of the solution of the Boltzmann equation by solutions of the ito stochastic differential equations. USSR Comput. Math. Math. Phys. **27**, 51–59 (1987)
7. Bogomolov, S.V., Dorodnitsyn, L.V.: Equations of stochastic quasi-gas dynamics: viscous gas case. Math. Models Comput. Simul. **3**, 457–467 (2011)
8. Skorokhod, A.V.: Stochastic Equations for Complex Systems. Kluwer Academic, Dordrecht (1987)
9. Bogomolov, S.V., Gudich, I.G.: Diffusion model of gas in a phase space for moderate Knudsen numbers. Math. Models Comput. Simul. **5**, 130–144 (2013)
10. Bogomolov, S.V., Gudich, I.G.: Verification of a stochastic diffusion gas model. Math. Models Comput. Simul. **6**, 305–316 (2014)
11. Chetverushkin, B.N.: Resolution limits of continuous media mode and their mathematical formulations. Math. Models Comput. Simul. **5**, 266–279 (2013)
12. Elizarova, T.G.: Quasi-Gas Dynamic Equations. Springer, Heidelberg (2009). https://doi.org/10.1007/978-3-642-00292-2
13. Bogomolov, S.V., Esikova, N.B., Kuvshinnikov, A.E.: Micro-macro Kolmogorov-Fokker-Planck models for a rigid-sphere gas. Math. Models Comput. Simul. **8**(5), 533–547 (2016)
14. Bogomolov, S.V., Esikova, N.B., Kuvshinnikov, A.E.: Meso - Macro models for a hard sphere gas. In: Proceedings of the ECCOMAS Congress, Crete Island, Greece (2016)
15. Mathiaud, J., Mieussens, L.: A Fokker-Planck model of the Boltzmann equation with correct Prandtl number. J. Stat. Phys. **162**, 397–414 (2016)
16. Morinishi, K.: A continuum/kinetic hybrid approach for multi-scale flow. In: Proceedings of the ECCOMAS CFD, Egmond aan Zee, Netherlands (2006)
17. Aringazin, A.K., Mazhintov, M.I.: Stochastic models of Lagrangian acceleration of fluid particle in developed turbulence. Int. J. Mod. Phys. B **18**, 3095–3168 (2004)
18. Oksendal, B.: Stochastic Differential Equations, 6th edn. Springer, Heidelberg (2000). https://doi.org/10.1007/978-3-662-13050-6
19. Stepanov, S.S.: Stochastic World. Springer, Switzerland (2013). https://doi.org/10.1007/978-3-319-00071-8
20. Dadzie, S.K., Reese, J.M.: Spatial stochasticity and non-continuum effects in gas flows. Phys. Lett. A **376**, 967–972 (2012)
21. Bayev, A.Z., Bogomolov, S.V.: On the stability of the discontinuous particle method for the transfer equation. Math. Models Comput. Simul. **10**(2), 186–197 (2018)

# The Error Analysis of Finite Difference Approximation for a System of Singularly Perturbed Semilinear Reaction-Diffusion Equations with Discontinuous Source Term

S. Chandra Sekhara Rao$^{(\boxtimes)}$ and Sheetal Chawla

Department of Mathematics, Indian Institute of Technology Delhi,
Hauz Khas, New Delhi 110 016, India
scsr@maths.iitd.ac.in, chawlaasheetal@gmail.com

**Abstract.** We consider a coupled system of two singularly perturbed semilinear reaction-diffusion equations with a discontinuous source term. The leading term in each equation is multiplied by a small positive parameter, but these parameters have different order of magnitude. The solution of these system of equations have overlapping and interacting boundary and interior layers. Based on the discrete Green's function theory, the properties of the discretized operator are established. The error estimates are derived in the maximum norm for a central difference scheme on layer-adapted meshes, and the method is proved to be almost second order uniformly convergent independently of both the perturbation parameters. Numerical results validate the theoretical results.

**Keywords:** Singular perturbation · Semilinear · Coupled system ·
Discontinuous source term · Uniformly convergent · Shishkin mesh ·
Interior layers · Discrete Green's function

## 1 Introduction

Consider a coupled system of two singularly perturbed semilinear reaction-diffusion equations with a discontinuous source term on an interval of unit length $\Omega = (0,1)$, and assume that $d \in \Omega$ be the point of discontinuity in the source term. Let $\Omega_1 = (0,d)$ and $\Omega_2 = (d,1)$ and jump at $d$ in any function $\eta$ be given by $[\eta](d) = \eta(d+) - \eta(d-)$. The corresponding boundary value problem is to find $\boldsymbol{u} \in C(\overline{\Omega})^2 \cap C^1(\Omega)^2 \cap C^4(\Omega_1 \cup \Omega_2)^2$ such that

$$\boldsymbol{T}\boldsymbol{u} := -\boldsymbol{E}\boldsymbol{u}'' + \boldsymbol{a}(x,\boldsymbol{u}) = \boldsymbol{f} \quad \text{in} \quad \Omega_1 \cup \Omega_2, \quad \boldsymbol{u}(0) = \boldsymbol{p}, \quad \boldsymbol{u}(1) = \boldsymbol{q}, \quad (1)$$

where $\boldsymbol{E} = \text{diag}(\varepsilon_1, \varepsilon_2)$ with small parameters $\varepsilon_1, \varepsilon_2$ are such that $0 < \varepsilon_1 \leq \varepsilon_2 \leq 1$, $\boldsymbol{a}(x,\boldsymbol{u}) = (a_1(x,\boldsymbol{u}), a_2(x,\boldsymbol{u}))^T$, $\boldsymbol{f} = (f_1, f_2)^T$. The solution $\boldsymbol{u} = (u_1, u_2)^T$ satisfies the following interface conditions:

$$[u_i] = 0, \qquad -\varepsilon_i[u_i'] = 0, \qquad \text{at} \quad x = d, \quad \text{for} \quad i = 1, 2. \qquad (2)$$

© Springer Nature Switzerland AG 2019
I. Dimov et al. (Eds.): FDM 2018, LNCS 11386, pp. 175–184, 2019.
https://doi.org/10.1007/978-3-030-11539-5_18

The reduced problem correspond to (1) is $a(x, u) = f$. Define the Jacobian matrix $J$ and the linear operator $L$ in terms of Jacobian matrix using mean value theorem: $J(x, u) := \left(\frac{\partial a_i}{\partial u_j}\right)(x, u),$ $\quad Lu \equiv -Eu'' + \int_{s=0}^{1} J(x, su)ds \ u = -a(x, 0) + f(x),$
which is natural extension of the linear case [5]. The Jacobian matrix $J$ is assumed to satisfy the following conditions for all $(x, u) \in \overline{\Omega} \times \mathbb{R}^2$:

$$\frac{\partial a_i}{\partial u_j} \leq 0, \ i \neq j, \quad \frac{\partial a_i}{\partial u_i} > 0, \quad \sum_{j=1}^{2} \frac{\partial a_i}{\partial u_j}(x, u) > \alpha > 0, \quad i = 1, 2.$$

Gracia et al. [4] considered an arbitrary coupled system of singularly perturbed semilinear reaction-diffusion equations having continuous source term. The numerical approximations were proved to be almost first order uniformly convergent. Further, in the special case of two equations and in the case of system of equations with equal diffusion parameters, an almost second order convergence was established. A singularly perturbed linear system of reaction-diffusion equations with continuous source term was considered in [5]. Properties of the discretized operator were established using Green's function, and the central difference scheme on layer-adapted meshes was proved to be almost second order uniformly convergent. In [2,3], a scalar nonlinear problem with discontinuous source term was considered. An almost first order of uniform convergence for the locally exact schemes on piecewise uniform and log-meshes was proved in [2], while based on the classical central difference and piecewise-uniform Shishkin mesh almost first order of uniform convergence was derived in [3]. In [6–9], a singularly perturbed linear system of the steady state and unsteady state reaction-diffusion equations and initial value problems with a discontinuous source term were considered.

This paper is arranged as follows. Section 2 presents the properties of the exact solution of the singular perturbation problem. Based on the discrete Green's function theory, the properties of the discretized operator are established in Sect. 3. A central difference scheme on layer-adapted meshes is proved to be second order uniformly convergent in Sect. 4. In Sect. 5, supporting numerical examples confirm the theoretical results.

**Notations:** We shall use $C$ to denote a generic positive constant and $C = (C, C)^T$ a generic positive constant vector such that both are independent of perturbation parameters $\varepsilon_1, \varepsilon_2$ and also of the discretization parameter $N$, but may not be same at each occurrence. Define $v \leq w$ if $v_1(x) \leq w_1(x)$ and $v_2(x) \leq w_2(x)$ $x \in [0, 1]$. We consider the maximum norm and denote it by $\|.\|_S$, where S is a closed subset in $[0, 1]$. For a real valued function $v \in C(S)$ and for a vector valued function $v = (v_1, v_2)^T \in C(S)^2$, we define $\|v\|_S := \max_{x \in S} |v(x)|$ and $\|v\|_S := \max\{\|v_1\|_S, \|v_2\|_S\}$.

## 2    Properties of the Exact Solution

**Theorem 1.** *The problem* (1)–(2) *has a solution* $\boldsymbol{u} = (u_1, u_2)^T$ *with* $u_1, u_2 \in C(\overline{\Omega}) \cap C^1(\Omega) \cap C^4(\Omega_1 \cup \Omega_2)$.

**Theorem 2.** *Suppose* $\boldsymbol{u} = (u_1, u_2)^T \in C(\overline{\Omega})^2 \cap C^2(\Omega_1 \cup \Omega_2)^2$. *Further suppose that* $\boldsymbol{u}$ *satisfies* $\boldsymbol{u}(0) \geq \boldsymbol{0}, \boldsymbol{u}(1) \geq \boldsymbol{0}, \boldsymbol{L}\boldsymbol{u}(x) \geq \boldsymbol{0}$, *for all* $x \in \Omega_1 \cup \Omega_2$ *and* $[\boldsymbol{u}'](d) \leq \boldsymbol{0}$. *Then* $\boldsymbol{u}(x) \geq \boldsymbol{0}$, *for all* $x \in \overline{\Omega}$.

**Lemma 1.** *Let* $\boldsymbol{u}$ *be the solution of* (1)–(2), *then*

$$\|\boldsymbol{u}\|_{\overline{\Omega}} \leq \max(\|\boldsymbol{u}(0)\|, \|\boldsymbol{u}(1)\|) + \frac{1}{\alpha}(\|\boldsymbol{a}(\cdot, 0)\|_{\Omega_1 \cup \Omega_2} + \|\boldsymbol{f}\|_{\Omega_1 \cup \Omega_2}).$$

**Lemma 2.** *Let* $\boldsymbol{u}$ *be the solution of* (1)–(2), *then for each* $i = 1, 2$, *and* $x \in \Omega_1 \cup \Omega_2$ *and* $k = 0, 1, 2$,

$$|u_i^{(k)}(x)|_{\Omega_1 \cup \Omega_2} \leq C \, \varepsilon_i^{\frac{-k}{2}} (\max(|u_i(0)|, |u_i(1)|) + \| \, a_i(\cdot, 0)\|_{\Omega_1 \cup \Omega_2} + \|f_i\|_{\Omega_1 \cup \Omega_2}),$$

$$|u_i^{(3)}(x)|_{\Omega_1 \cup \Omega_2} \leq C\varepsilon_1^{\frac{-1}{2}} \varepsilon_i^{-1}(\max(|u_i(0)|, |u_i(1)|) + \|a_i(\cdot, 0)\|_{\Omega_1 \cup \Omega_2} + \|f_i\|_{\Omega_1 \cup \Omega_2} + \sqrt{\varepsilon_1}\|f_i'\|_{\Omega_1 \cup \Omega_2}),$$

*and*

$$|u_i^{(4)}(x)|_{\Omega_1 \cup \Omega_2} \leq C \, \varepsilon_1^{-1}\varepsilon_i^{-1}(\max(|u_i(0)|, |u_i(1)|) + \|a_i(\cdot, 0)\|_{\Omega_1 \cup \Omega_2} + \|f_i\|_{\Omega_1 \cup \Omega_2} + \varepsilon_1\|f_i''\|_{\Omega_1 \cup \Omega_2}).$$

To derive the parameter-uniform convergence of the numerical method, a decomposition of the exact solution $\boldsymbol{u}$, into regular component $\boldsymbol{v}$ and a singular component $\boldsymbol{w}$ is required. That is, $\boldsymbol{u} = \boldsymbol{v} + \boldsymbol{w}$. The corresponding reduced problem is $\boldsymbol{a}(x, \boldsymbol{v_0}) = \boldsymbol{f}$. The regular component $\boldsymbol{v}$, is defined as the solution of the following problem:

$$\boldsymbol{T}\boldsymbol{v} = \boldsymbol{f}, \qquad x \in \Omega_1 \cup \Omega_2, \tag{3}$$

$$\boldsymbol{v}(0) = \boldsymbol{v_0}(0), \quad \boldsymbol{v}(d-) = \boldsymbol{v_0}(d-), \quad \boldsymbol{v}(d+) = \boldsymbol{v_0}(d+), \quad \boldsymbol{v}(1) = \boldsymbol{v_0}(1), \tag{4}$$

and the singular component $\boldsymbol{w}$ is defined as the solution of the following problem:

$$- \boldsymbol{E}\boldsymbol{w}'' + \boldsymbol{a}(x, \boldsymbol{v} + \boldsymbol{w}) - \boldsymbol{a}(x, \boldsymbol{v}) = \boldsymbol{0}, \qquad x \in \Omega_1 \cup \Omega_2, \tag{5}$$

$$\boldsymbol{w}(x) = \boldsymbol{u}(x) - \boldsymbol{v}(x), \quad x \in \{0, 1\}, [\boldsymbol{w}](d) = -[\boldsymbol{v}](d), \ [\boldsymbol{w}'](d) = -[\boldsymbol{v}'](d). \tag{6}$$

Define the following layer functions for deriving the bounds on derivatives of regular and singular components for $i = 1, 2$:

$$B^1_{\varepsilon_{l_i}}(x) := e^{-x\sqrt{\alpha/\varepsilon_i}}, B^2_{\varepsilon_{l_i}}(x) := e^{-(d-x)\sqrt{\alpha/\varepsilon_i}}, B_{\varepsilon_{l_i}}(x) := B^1_{\varepsilon_{l_i}}(x) + B^2_{\varepsilon_{l_i}}(x),$$

$$B^1_{\varepsilon_{r_i}}(x) := e^{(d-x)\sqrt{\alpha/\varepsilon_i}}, B^2_{\varepsilon_{r_i}}(x) := e^{-(1-x)\sqrt{\alpha/\varepsilon_i}}, B_{\varepsilon_{r_i}}(x) := B^1_{\varepsilon_{r_i}}(x) + B^2_{\varepsilon_{r_i}}(x).$$

**Theorem 3.** *The regular component* $\boldsymbol{v}$ *and the singular component* $\boldsymbol{w}$ *and its derivatives satisfy the bounds for* $i = 1, 2$,

$$|v_i^{(k)}(x)|_{\Omega_1 \cup \Omega_2} \leq C(1 + \varepsilon_i^{(1-\frac{k}{2})}), \quad for \ k = 0, \dots, 4,$$

$$|w_i^{(k)}(x)| = C \begin{cases} \sum\limits_{m=i}^{2} \varepsilon_i^{\frac{-k}{2}} B_{\varepsilon l_i}(x) & x \in \Omega_1, \\ \sum\limits_{m=i}^{2} \varepsilon_i^{\frac{-k}{2}} B_{\varepsilon r_i}(x) & x \in \Omega_2, \end{cases} \quad for \ k = 0, 1, 2,$$

$$|w_i^{(k)}(x)| = C \begin{cases} \varepsilon_i^{1-\frac{k}{2}}(\varepsilon_1^{-1} B_{\varepsilon l_1}(x) + \varepsilon_2^{-1} B_{\varepsilon l_2}(x)) & x \in \Omega_1, \\ \varepsilon_i^{1-\frac{k}{2}}(\varepsilon_1^{-1} B_{\varepsilon r_1}(x) + \varepsilon_2^{-1} B_{\varepsilon r_2}(x)) & x \in \Omega_2, \end{cases} \quad for \ k = 3, 4.$$

*Proof.* The result can be proved by following similar arguments given in [5] for $x \in \Omega_1 \cup \Omega_2$.

## 3  Discretization of the Problem

A piecewise-uniform variant of Shishkin mesh with $N$ mesh-intervals is constructed. The interior points of the mesh are

$$\Omega^N = \{x_i : 1 \le i \le \frac{N}{2} - 1\} \cup \{x_i : \frac{N}{2} + 1 \le i \le N - 1\} = \Omega_1^N \cup \Omega_2^N.$$

Let $h_i = x_i - x_{i-1}$ be the $i^{th}$ mesh step and $\hbar_i = \frac{h_i + h_{i+1}}{2}$, clearly $x_{\frac{N}{2}} = d$ and $\overline{\Omega}^N = \{x_i : i = 0, 1, 2, \ldots .N\}$. Let $N = 2^l$, $l \ge 5$ be any positive integer. The transition parameters are defined like this:

$$\sigma_{\varepsilon l_2} := \min \left\{ \frac{d}{4}, 2\sqrt{\frac{\varepsilon_2}{\alpha}} \ln N \right\}, \quad \sigma_{\varepsilon r_2} := \min \left\{ \frac{1-d}{4}, 2\sqrt{\frac{\varepsilon_2}{\alpha}} \ln N \right\},$$

$$\sigma_{\varepsilon l_1} := \min \left\{ \frac{\sigma_{\varepsilon l_2}}{4}, 2\sqrt{\frac{\varepsilon_1}{\alpha}} \ln N \right\}, \quad \sigma_{\varepsilon r_1} := \min \left\{ \frac{\sigma_{\varepsilon r_2}}{4}, 2\sqrt{\frac{\varepsilon_1}{\alpha}} \ln N \right\}.$$

The interval $[0, d]$ is subdivided into five subintervals as follows:

$$[0, \sigma_{\varepsilon l_1}] \cup (\sigma_{\varepsilon l_1}, \sigma_{\varepsilon l_2}] \cup (\sigma_{\varepsilon l_2}, d - \sigma_{\varepsilon l_2}] \cup (d - \sigma_{\varepsilon l_2}, d - \sigma_{\varepsilon l_1}] \cup (d - \sigma_{\varepsilon l_1}, d].$$

On each subinterval $[0, \sigma_{\varepsilon l_1}]$, $(\sigma_{\varepsilon l_1}, \sigma_{\varepsilon l_2}]$, $(d - \sigma_{\varepsilon l_2}, d - \sigma_{\varepsilon l_1}]$, $(d - \sigma_{\varepsilon l_1}, d]$ a uniform mesh of $N/16$ mesh intervals is placed. On the subinterval $(\sigma_{\varepsilon l_2}, d - \sigma_{\varepsilon l_2}]$ a uniform mesh of $N/4$ mesh intervals, is placed.

The interval $[d, 1]$ is divided in the same manner. The location of the layers is shown in the Fig. 1.

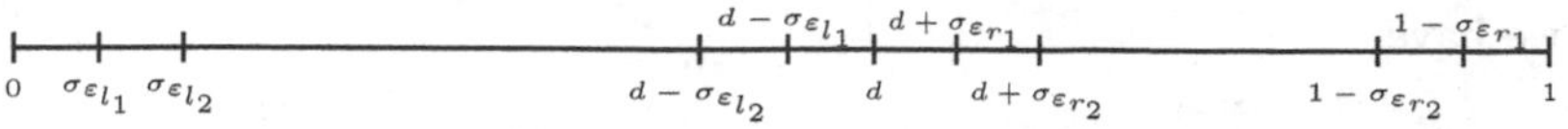

**Fig. 1.** Location of the overlapping and interacting boundary and interior layers at $x = 0, 1$ and to the left and right hand sides of the point of discontinuity at $x = d$.

The discretization of the problem (1) satisfies the following classical difference scheme

$$T^N U(x_i) \equiv -E\mathfrak{D}^2 U(x_i) + a(x_i, U(x_i)) = \mathfrak{D}^0 f(x_i); \quad x_i \in \overline{\Omega}^N \qquad (7)$$

$$U(x_0) = p, \quad U(x_N) = q \qquad (8)$$

where

$$\mathfrak{D}^2 Z(x_i) = \left[ \frac{\mathfrak{D}^+ Z(x_i) - \mathfrak{D}^- Z(x_i)}{\hbar_i} \right]; \quad \mathfrak{D}^+ Z(x_i) = \left[ \frac{z(x_{i+1}) - z(x_i)}{h_{i+1}} \right];$$

$$\mathfrak{D}^- Z(x_i) = \left[ \frac{z(x_i) - z(x_{i-1})}{h_i} \right]; \quad \mathfrak{D}^0 f(x_i) = \frac{1}{\hbar_i} \left[ \frac{h_{i+1} f(x_i+0) + h_i f(x_i-0)}{2} \right].$$

Consider two arbitrary mesh functions $V^N$ and $W^N$, then it holds true that

$$T^N V^N(x_i) - T^N W^N(x_i) = L^N \big( V^N(x_i) - W^N(x_i) \big)$$

where $L^N$ is a linear operator defined by,

$$L^N U^N(x_i) := -E\mathfrak{D}^2 U^N(x_i) + \int_0^1 J\big(x_i, s U^N(x_i)\big) ds \, U^N(x_i). \qquad (9)$$

**Lemma 3.** *Let $V^N$ and $W^N$ be any two mesh functions such that $V^N(0) = W^N(0)$ and $V^N(1) = W^N(1)$. Then*

$$\| V^N - W^N \| \le C \| L^N V^N - L^N W^N \|.$$

### 3.1   Discrete Green's Function and Its Properties

The discrete Green's function $G(x_i, \xi_j)$, associated with the operator $L^N$ as a function of $x_i \in \Omega^N$ for a fixed $\xi_j \in \Omega^N$ is defined by the following relations

$$L^N G(x_i, \xi_j) = \frac{\delta(x_i, \xi_j)}{\hbar_i}, \quad G(0, \xi_j) = G(1, \xi_j) = 0, \quad i = 1, \ldots, N-1.$$

where the $\delta(x_i, \xi_j)$ is a Kronecker's delta. The discrete Green's function $G(x_i, \xi_j)$ is a function of the variable $\xi_j$ for a fixed $x_i$, is the solution of the adjoint problem:

$$L^{N*} G(x_i, \xi_j) = \frac{\delta(x_i, \xi_j)}{\hbar_j}, \quad G(x_i, 0) = G(x_i, 1) = 0, \quad j = 1, \ldots, N-1.$$

Then we have,

$$(\boldsymbol{V}^N - \boldsymbol{W}^N)(x_i) = \sum_{j=1}^{N-1} \hbar j \, \boldsymbol{G}(x_i, \xi_j)((\boldsymbol{T}^N \boldsymbol{V}^N - \boldsymbol{T}^N \boldsymbol{W}^N)(\xi_j)), \quad i = 1, \ldots, N-1.$$

Using the discrete Green's function, the solution of the problem (9) is given by the following formula

$$\boldsymbol{U}(x_i) = \sum_{j=1}^{N-1} \boldsymbol{G}(x_i, \xi_j)[\boldsymbol{L}^N \boldsymbol{U}](\xi_j) \hbar_j.$$

**Lemma 4.** *For the discrete Green's function* $\boldsymbol{G}(x_i, \xi_j)$ *on the mesh* $\Omega^N$, *we have*

$$\int_0^1 \boldsymbol{G}(x_i, \xi) d\xi = \sum_{j=1}^{N-1} \hbar_j |\boldsymbol{G}(x_i, \xi_j)| \leq \frac{1}{\alpha},$$

$$\int_0^1 |(\boldsymbol{G}_1'(x_i, \xi))| d\xi \leq \frac{1}{\sqrt{\alpha \varepsilon_1}}, \qquad \int_0^1 |(\boldsymbol{G}_2'(x_i, \xi))| d\xi \leq \frac{1}{\sqrt{\alpha \varepsilon_2}}.$$

*Proof.* Following the similar arguments given in [1], this result can be proved.

## 4    Error Analysis

**At the mesh points, for** $i \neq \frac{N}{2}$,

$$-\boldsymbol{E}\mathfrak{D}^2 \boldsymbol{U}(x_i) + \boldsymbol{a}(x_i, \boldsymbol{U}) = -\boldsymbol{E}\boldsymbol{u}''(x_i) + \boldsymbol{a}(x_i, \boldsymbol{u}),$$

$$-\boldsymbol{E}\mathfrak{D}^2(\boldsymbol{U} - \boldsymbol{u})(x_i) + \left( \int_{s=0}^1 \boldsymbol{J}(x_i, \boldsymbol{u} + s(\boldsymbol{U} - \boldsymbol{u}) ds)(\boldsymbol{U} - \boldsymbol{u}) \right) = -\boldsymbol{E}\boldsymbol{u}''(x_i) +$$

$$\boldsymbol{E}\mathfrak{D}^2 \boldsymbol{u}(x_i).$$

**Similarly, for the mesh point** $i = \frac{N}{2}$,

$$-\boldsymbol{E}\mathfrak{D}^2(\boldsymbol{U} - \boldsymbol{u})(d) + \left( \int_{s=0}^1 \boldsymbol{J}(d, \boldsymbol{u} + s(\boldsymbol{U} - \boldsymbol{u}) ds)(\boldsymbol{U} - \boldsymbol{u}) \right) = \boldsymbol{E}\mathfrak{D}^2 \boldsymbol{u}(d) + \mathfrak{D}^0 \boldsymbol{f}(d) -$$

$$\boldsymbol{a}(d, \boldsymbol{u}).$$

Define the discrete linear operator for $i \neq \frac{N}{2}$

$$\boldsymbol{L}^N \boldsymbol{V}(x_i) := -\boldsymbol{E}\mathfrak{D}^2 \boldsymbol{V}(x_i) + \sum_{j=1}^m \boldsymbol{J}(x_i, \boldsymbol{u} + s_j(\boldsymbol{V})) \boldsymbol{V}(x_i).$$

We consider the error at the mesh node $x_i$ by the means of discrete Green's functions as a basis for $[(\boldsymbol{L}^N \gamma)_i]_k = -\varepsilon_i [\mathfrak{D}^2 u_i - u_i'']_k$ where $k = 1, \ldots, \frac{N}{2} - 1$ and $i = 1, 2$, we have

$$(\gamma_i)_k = -\sum_{j=1}^{\frac{N}{2}-1} \hbar_j G_{i,j} \varepsilon_i [\mathfrak{D}^2 u_i - u_i'']_j = -\varepsilon_i \int_0^d u_i''(x)(G_i)(x) dx + \varepsilon_i \sum_{j=0}^{\frac{N}{2}-1} \hbar_j G_{i,j} u_{i,j}''.$$

Then, it follows that

$$\gamma_{i;k} = \int_0^d \{(G_i\varphi_i)(x) - (G_i\varphi_i)^I(x)\}dx \quad \text{with} \quad \varphi_i = -\varepsilon_i u_i''. \tag{10}$$

The interpolation error to $G_i\varphi_i$ for $x \in [x_{k-1}, x_k]$ can be written as

$$(G_i\varphi_i)(x) - (G_i\varphi_i)^I(x) = 2G_i'|_{[x_{k-1},x_k]}\frac{1}{h_k}\int_{x_{k-1}}^{x_k}\int_{x_{k-1}}^x\int_\xi^s \varphi_i'(t)dtd\xi ds$$

$$+ \frac{1}{h_k}\int_{x_{k-1}}^{x_k}\int_{x_{k-1}}^x\int_\xi^s (G_i\varphi_i'')(t)dtd\xi ds. \tag{11}$$

Split $\varphi_i'$ into two functions such that one decreasing and other increasing to bound the first integral on the right hand side of (11). Let $\varphi_i' = \varphi_{i,D}(x) + \varphi_{i,I}(x)$, where

$$\varphi_{i,D}(x) = \begin{cases} \varphi_i'(x) & \text{for } x \le \frac{d}{2} \\ 0 & \text{for } x \ge \frac{d}{2} \end{cases} \quad \text{and} \quad \varphi_{i,I}(x) = \begin{cases} 0 & \text{for } x \le \frac{d}{2} \\ \varphi_i'(x) & \text{for } x \ge \frac{d}{2}. \end{cases}$$

Using the bounds on the regular and singular components defined in Theorem (3) for $k = 3$, we get

$$\frac{|\varphi_{i,D}(x)|}{\sqrt{\varepsilon_i}} \le C(1 + \varepsilon_1^{-1}B_{\varepsilon l_1}^1(x) + \varepsilon_2^{-1}B_{\varepsilon l_2}^1(x)) =: \bar{\varphi}_{i,D}(x).$$

For $x \in [x_{k-1}, x_k]$, we have

$$\left|\frac{1}{h_k}\int_{x_{k-1}}^{x_k}\int_{x_{k-1}}^x\int_\xi^s \varphi_{i,D}(t)dtd\xi ds\right| \le \sqrt{\varepsilon_i}\int_{x_{k-1}}^{x_k}\int_\xi^{x_k}\bar{\varphi}_{i,D}(t)dtd\xi$$

$$\le \frac{\sqrt{\varepsilon_i}}{2}\left\{\int_{x_{k-1}}^{x_k}\bar{\varphi}_{i,D}(t)^{1/2}dt\right\}^2.$$

A similar expression can be obtained for the increasing function $\varphi_{i,I}(x)$. The second integral on the right hand side of (11) can be bounded in a similar manner by defining the monotone functions as in [5].
Finally, for $x \in [x_{k-1}, x_k]$, we obtain

$$|(G_i\varphi_i)(x) - (G_i\varphi_i)^I(x)| \le C\vartheta(\Omega_1^N)^2(\sqrt{\varepsilon_i}G_i'|_{[x_{k-1},x_k]} + G_{i,k-1} + G_{i,k}), \tag{12}$$

where $\quad \vartheta(\Omega_1^N) := \max_{k\in\Omega_1^N}\int_{x_{k-1}}^{x_k}\left(1 + \varepsilon_1^{\frac{-1}{2}}B_{\varepsilon l_1}(t) + \varepsilon_2^{\frac{-1}{2}}B_{\varepsilon l_2}(t)\right)dt.$

Now to find a bound on $(\gamma_i)_k$ integrate (12) over (0,d), and then use the Lemma (4), to obtain

$$\|\gamma_i\|_{\Omega_1^N} \le C\vartheta(\Omega_1^N)^2. \tag{13}$$

Proceeding in the same way for $x \in \Omega_2^N$, we obtain

$$\|\gamma_i\|_{\Omega_2^N} \leq C\vartheta(\Omega_2^N)^2, \tag{14}$$

where    $\vartheta(\Omega_2^N) := \max_{k \in \Omega_2^N} \int_{x_{k-1}}^{x_k} \left(1 + \varepsilon_1^{\frac{-1}{2}} B_{\varepsilon_{r_1}}(t) + \varepsilon_2^{\frac{-1}{2}} B_{\varepsilon_{r_2}}(t)\right) dt.$

Also, it is shown that $\vartheta(\Omega_i^N) \leq C(N^{-1}\ln N)$, $i = 1, 2$.

Now at the point $x_{N/2} = d$,

$$\varepsilon_i \mathfrak{D}^2 u_i(d) - a_i(d, \boldsymbol{u}) + \mathfrak{D}^0 f_i(d)$$

$$= \left(-\frac{1}{h^2} \int_{t=d}^{d+h} \int_{s=d}^{t} \int_{p=d}^{s} + \frac{1}{h^2} \int_{t=d-h}^{d} \int_{s=t}^{d} \int_{p=s}^{d}\right) \frac{d}{dp}(f_i(p) - a_i(p, u)) dp \, ds \, dt +$$

$O(h),$

from which it follows that

$$|(\boldsymbol{L}^N(U-u))_i(d)| \leq C(N^{-1}\ln N) \text{ for } i = 1, 2.$$

**Theorem 4.** *Let $\boldsymbol{u}$ be the solution of the problem (1)–(2) and $\boldsymbol{U}$ be the solution of discrete problem on a variant of Shishkin mesh defined in Sect. 2, then*

$$\|\boldsymbol{U} - \boldsymbol{u}\|_{\overline{\Omega}^N} \leq C(N^{-1}\ln N)^2.$$

*Proof.* Define the mesh functions $\eta_3, \eta_4$ to be

$$\eta_3(x_i) = \Pi_{j=1}^i \left(1 + \sqrt{\frac{\alpha}{2\varepsilon_1}} h_j\right), \qquad \eta_4(x_i) = \Pi_{j=1}^i \left(1 + \sqrt{\frac{\alpha}{2\varepsilon_1}} h_j\right)^{-1}.$$

Define the barrier function $\theta_d$ as follows

$$\theta_d(x_i) = \begin{cases} \frac{\eta_3(x_i)}{\eta_3(d)}, & 0 \leq x_i \leq d, \\ \frac{\eta_4(x_i)}{\eta_4(d)}, & d \leq x_i \leq 1. \end{cases}$$

Now define the mesh function

$$\Theta^{\pm}(x_i) = C(N^{-1}\ln N)^2(1 + \theta_d(x_i))(1, 1)^T \pm (\boldsymbol{U} - \boldsymbol{u})(x_i).$$

Using the discrete maximum principle, we conclude that

$$\|\boldsymbol{U} - \boldsymbol{u}\|_{\overline{\Omega}^N} \leq C(N^{-1}\ln N)^2.$$

## 5    Numerical Results

*Example 1.* Consider the singularly perturbed semilinear reaction-diffusion equations with discontinuous source term

$$\begin{aligned} -\varepsilon_1 u_1'' + u_1 - (1 - u_1)^3 + \exp(u_1 - u_2) &= f_1(x), \\ -\varepsilon_2 u_2'' + u_2 - (0.5 - u_2)^5 + \exp(u_2 - u_1) &= f_2(x), \\ \boldsymbol{u}(0) = \boldsymbol{0}, \qquad \boldsymbol{u}(1) &= \boldsymbol{0}, \end{aligned}$$

where

$$f_1(x) = \begin{cases} 2 & \text{for } 0 \le x \le 0.5 \\ 3 & \text{for } 0.5 < x \le 1, \end{cases} \quad f_2(x) = \begin{cases} 1.5 & \text{for } 0 \le x \le 0.5 \\ 2.5 & \text{for } 0.5 < x \le 1. \end{cases}$$

To solve the corresponding semilinear system, the Newton's method is used with zero as an initial guess. The stopping criterion is $\| U^{(k)} - U^{(k-1)} \| < N^{-2}$. Here $U^{(k)}$, for $k = 1, 2, \ldots$ represents the successive approximates to $U$ computed iteratively. For the construction of a variant of piecewise-uniform Shishkin mesh $\overline{\Omega}^N$, we take $\alpha = 0.99$. Since the Exact solution of the Example is not known, to obtain the maximum point-wise errors and rates of convergence, we use the double mesh principle [8]. For different values of $N$ and $\varepsilon_1, \varepsilon_2$, we compute $D^N_{\varepsilon_1,\varepsilon_2} := \|(U - \widetilde{U})(x_j)\|_{\overline{\Omega}^N}$. If $\varepsilon_1 = 10^{-j}$ for some non-negative integer $j$, set $D^N_{\varepsilon_1} := \max\{D^N_{\varepsilon_1,1}, D^N_{\varepsilon_1,10^{-1}}, D^N_{\varepsilon_1,10^{-2}}, \ldots, D^N_{\varepsilon_1,10^{-j}}\}$. Then the parameter-uniform error is computed as $D^N := \max\{D^N_1, D^N_{10^{-1}}, \ldots, D^N_{10^{-16}}\}$, and the order of convergence is calculated using the formula $p^N = \dfrac{\ln(D^N) - \ln(D^{2N})}{\ln(2\ln N) - \ln(\ln(2N))}$. For different values of $N, \epsilon_1, \epsilon_2$ Table 1 represent the maximum point-wise errors $D^N_{\epsilon_1}$, parmeter uniform errors $D^N$ and $\epsilon_1, \epsilon_2$-uniform rate of convergence $P^N$ for Example 1.

**Table 1.** Maximum point-wise errors $D^N_{\varepsilon_1}$, $D^N$ and $\varepsilon_1, \varepsilon_2$–uniform rates of convergence $p^N$ for Example 1.

| $\varepsilon_1 = 10^{-j}$ | $N = 64$ | $N = 128$ | $N = 256$ | $N = 512$ | $N = 1024$ | $N = 2048$ |
|---|---|---|---|---|---|---|
| $j = 0$ | 5.63E−06 | 1.41E−06 | 3.52E−07 | 8.79E−08 | 2.20E−08 | 5.50E−09 |
| 1 | 7.97E−05 | 1.99E−05 | 4.99E−06 | 1.25E−06 | 3.12E−07 | 7.79E−08 |
| 2 | 7.79E−04 | 1.99E−04 | 4.99E−05 | 1.25E−05 | 3.13E−06 | 7.82E−07 |
| 3 | 6.53E−03 | 1.87E−03 | 4.89E−04 | 1.24E−04 | 3.10E−05 | 7.75E−06 |
| 4 | 1.90E−02 | 1.16E−02 | 3.91E−03 | 1.19E−03 | 3.07E−04 | 7.73E−05 |
| 5 | 1.99E−02 | 1.70E−02 | 1.01E−02 | 3.93E−03 | 1.34E−03 | 4.17E−04 |
| 6 | 5.98E−02 | 2.48E−02 | 1.01E−02 | 3.93E−03 | 1.34E−03 | 4.17E−04 |
| 7 | 7.81E−02 | 3.60E−02 | 1.38E−02 | 4.66E−03 | 1.34E−03 | 4.17E−04 |
| 8 | 7.81E−02 | 3.60E−02 | 1.38E−02 | 4.69E−03 | 1.49E−03 | 4.53E−04 |
| . | . | . | . | . | . | . |
| . | . | . | . | . | . | . |
| . | . | . | . | . | . | . |
| 16 | 7.81E−02 | 3.60E−02 | 1.38E−02 | 4.69E−03 | 1.49E−03 | 4.53E−04 |
| $D^N$ | 7.81E−02 | 3.60E−02 | 1.38E−02 | 4.69E−03 | 1.49E−03 | 4.53E−04 |
| $p^N$ | 1.44E+00 | 1.71E+00 | 1.87E+00 | 1.95E+00 | 1.99E+00 | . |

# References

1. Andreev, V.B.: On the uniform convergence of a classical difference scheme on a nonuniform mesh for the one-dimensional singularly perturbed reaction-diffusion equation. Comput. Math. Phys. **44**, 449–464 (2001)
2. Boglaev, I., Pack, S.: A uniformly convergent method for a singularly perturbed semilinear reaction-diffusion problem with discontinuous data. Appl. Math. Comput. **182**, 244–257 (2006)
3. Farrell, P.A., O'Riordan, E., Shishkin, G.I.: A class of singularly perturbed semilinear differential equations with interior layers. Math. Comput. **74**, 1759–1776 (2005)
4. Gracia, J.L., Lisbona, F., Madaune-Tort, M., O'Riordan, E.: A system of singularly perturbed semilinear equations. In: Hegarty, A., Kopteva, N., O'Riordan, E., Stynes, M. (eds.) BAIL 2008 - Boundary and Interior Layers. LNCSE, vol. 69, pp. 163–172. Springer, Heidelberg (2009). https://doi.org/10.1007/978-3-642-00605-0_12
5. Linß, T., Madden, N.: Layer-adapted meshes for a system of coupled singularly perturbed reaction-diffusion problems. IMA J. Numer. Anal. **29**, 109–125 (2009)
6. Rao, S.C.S., Chawla, S.: Interior layers in coupled system of two singularly perturbed reaction-diffusion equations with discontinuous source term. In: Dimov, I., Faragó, I., Vulkov, L. (eds.) NAA 2012. LNCS, vol. 8236, pp. 445–453. Springer, Heidelberg (2013). https://doi.org/10.1007/978-3-642-41515-9_50
7. Rao, S.C.S., Chawla, S.: Second order uniformly convergent numerical method for a coupled system of singularly perturbed reaction-diffusion problems with discontinuous source term. In: Knobloch, P. (ed.) BAIL 2014. LNCSE, vol. 108, pp. 233–244. Springer, Cham (2015). https://doi.org/10.1007/978-3-319-25727-3_18
8. Chandra Sekhara Rao, S., Chawla, S.: Numerical solution for a coupled system of singularly perturbed initial value problems with discontinuous source term. In: Agrawal, P., Mohapatra, R., Singh, U., Srivastava, H. (eds.) Mathematical Analysis and its Applications. Springer Proceedings in Mathematics & Statistics, vol. 143, pp. 753–764. Springer, New Delhi (2015). https://doi.org/10.1007/978-81-322-2485-3_60
9. Rao, S.C.S., Chawla, S.: Numerical solution of singularly perturbed linear parabolic system with discontinuous source term. Appl. Numer. Math. **127**, 249–265 (2018)

# Efficiency and Stability of a Family of Iterative Schemes for Solving Nonlinear Equations

Alicia Cordero[1] , Ivan Giménez[2], and Juan R. Torregrosa[1](✉)

[1] Instituto Matemática Multidisciplinar, Universitat Politècnica de València,
Camino de Vera s/n, 46022 València, Spain
{acordero,jrtorre}@mat.upv.es
[2] Facultat de Matemàtiques, Universitat de València, València, Spain
igipa@alumni.uv.es

**Abstract.** In this paper, we construct a family of iterative methods with memory from one without memory, analyzing their convergence and stability. The main aim of this manuscript yields in the advantage that the use of real multidimensional dynamics gives us to decide among the different classes designed and, afterwards, to select its most stable members. Some numerical tests confirm the theoretical results.

## 1 Introduction

Usually real problems are modeled by nonlinear equations, scalar or vectorial, that must be solved numerically. Because of their nonlinear nature, these problems are frequently solved by means of iterative methods. Newton's scheme is the best known one, being a one-step method. Based on it, extensive research has been developed over the years, not only of one-step procedures, but also of multi-step methods. In addition to this, so-called memory methods have emerged in recent years. That is, iterative methods in which two or more previous iterations are used to calculate the new one.

The problem of solving a nonlinear equation $f(x) = 0$, being $f : D \subset \mathbb{R} \to \mathbb{R}$, that is, of finding a simple zero $\alpha$ of $f$, usually requires the use of an approximating method. An iterative method consists of obtaining a new better approximation from the last calculated, by means of a fixed-point or iteration function, $x_{k+1} = g(x_{k-p}, x_{k-p+1}, \ldots, x_{k-1}, x_k), k > 0, p \geq 0$. If $p = 0$, the scheme is classified as an iterative method without memory and therefore, it is considered as a method with memory if $p > 0$. The advantage of using one of the later methods is a faster convergence with the same functional evaluations as a method without memory, although if the increasing of order is very high, it usually is followed by a high complexity of the iterative expression. Some good texts about both kind of procedures can be found in [1,7].

Many methods have been designed in the last decades, improving those existing previously in different aspects as order, efficiency or applicability, among

© Springer Nature Switzerland AG 2019
I. Dimov et al. (Eds.): FDM 2018, LNCS 11386, pp. 185–192, 2019.
https://doi.org/10.1007/978-3-030-11539-5_19

other. In the last years, the analysis of the stability by means of dynamical techniques is a new trend that, not only enriches the area of research, but it is revealing as necessary to select those classes or methods with better qualitative behavior. The most of these analysis have been made with the tools of complex dynamics, based on the knowledge in the area developed in the last century. Also real dynamics has shown to be useful for this kind of analysis, as can be seen in diverse works as [2,6].

Nevertheless, none of these techniques can be applied to analyze the behavior of iterative methods with memory, beyond the mere plot of the basins of attraction on specific functions. It has been very recently that the authors in [4,5] have developed a technique to study their stability on low-degree polynomials based on the real multidimensional dynamics.

Our goal in this paper is, after designing different classes of iterative methods with memory by using the known technique of introducing accelerating parameters in a without memory class, to study its dynamical behavior and detect how slight differences in the design to get several classes of iterative schemes with the same order of convergence involve very different qualitative behavior, varying from the chaos in strange attractors to a "boring" but desirable stable behavior.

Moreover, to arrange the dynamical analysis, some elements must be defined, as the technique consists of describing a multidimensional discrete dynamical system from the real fixed-point operator of the method under analysis and studying it by using multidimensional real dynamics tools.

## 1.1  Discrete Dynamical Systems

The expression of an iterative method with memory, which uses two previous iterations to calculate the following estimation, is $x_{k+1} = g(x_{k-1}, x_k)$, $k \geq 1$, where $x_0$ and $x_1$ are the initial estimations. In order to obtain the fixed points of this method, we define the *fixed point function* $G : \mathbb{R}^2 \to \mathbb{R}^2$ by means of:

$$G(x_{k-1}, x_k) = (x_k, x_{k+1}) = (x_k, g(x_{k-1}, x_k)), \quad k = 1, 2, \ldots$$

This definition can be extended in a natural way to adapt it to iterative schemes with memory using more than two previous iterations per step. Let us remark that, as $(x_{k-1}, x_k)$ is a fixed point of $G$ if $G(x_{k-1}, x_k) = (x_{k-1}, x_k)$, then $x_{k+1} = x_k$ and $x_{k-1} = x_k$.

We have defined a discrete dynamical system in the real plane from function $G : \mathbb{R}^2 \to \mathbb{R}^2$ given by $G(z, x) = (x, g(z, x))$, where $g$ is the operator of the iterative method with memory. Fixed points $(z, x)$ of $G$ satisfy $z = x$ and $x = g(z, x)$.

In the following, we recall some basic real dynamics concepts. If a fixed point $(z, x)$ of operator $G$ is different from $(r, r)$, where $r$ is a zero of $f(x)$, it is called *strange fixed point*. On the other hand, the orbit of a point $\overline{x} \in \mathbb{R}^2$ is defined as the set of successive images of $\overline{x}$ by the vector function, $\{\overline{x}, G(\overline{x}), \ldots, G^m(\overline{x}), \ldots\}$.

The dynamical performance of a point of $\mathbb{R}^2$ is classified depending on its asymptotical behavior. So, a point $x^* \in \mathbb{R}^2$ is a *k-periodic point* if $G^k(x^*) = x^*$

and $G^p(x^*) \neq x^*$, for $p = 1, 2, \ldots, k-1$. The stability of fixed points for multivariable nonlinear operators, see for example [8], satisfies the following statements.

**Theorem 1.** *Let $G : \mathbb{R}^n \to \mathbb{R}^n$ be $C^2$. Let us assume $x^*$ is a k-periodic point. Let $\lambda_1, \lambda_2, \ldots, \lambda_n$ be the eigenvalues of $G'(x^*)$.*

1. *If all the eigenvalues $\lambda_j$ have $|\lambda_j| < 1$, then $x^*$ is attracting.*
2. *If one eigenvalue $\lambda_{j_0}$ has $|\lambda_{j_0}| > 1$, then $x^*$ is unstable.*
3. *If all eigenvalues $\lambda_j$ have $|\lambda_j| > 1$, then $x^*$ is repelling.*

Furthermore, a fixed point is called *hyperbolic* if all the eigenvalues $\lambda_j$ of $G'(x^*)$ have $|\lambda_j| \neq 1$. In particular, if there exist an eigenvalue $\lambda_i$ such that $|\lambda_i| < 1$ and an eigenvalue $\lambda_j$ such that $|\lambda_j| > 1$, the hyperbolic point is called *saddle* point.

Moreover, a point $x$ is a *critical* point of $G$ if the associate Jacobian matrix $G'(x)$ satisfies $det(G'(x)) = 0$. One particular case of critical points, for iterative methods of convergence order higher than two, are those fixed points whose associated eigenvalues $\lambda_j = 0$, $\forall j$ are null. These points are called *superattracting*.

Then, if $x^*$ is an attracting fixed point of function $G$, its *basin of attraction* $\mathcal{A}(x^*)$ is defined as the set of pre-images of any order such that

$$\mathcal{A}(x^*) = \left\{ x^{(0)} \in \mathbb{R}^n : G^m \left( x^{(0)} \right) \to x^*, m \to \infty \right\}.$$

In the following, we design three different families of methods with memory and study their order of convergence (Sect. 2). This order is not as high as it could be as the main aim of this paper is to compare the stability of the resulting schemes, and it is more feasible when their iterative expressions are simpler. Then we make a multidimensional real dynamical analysis of the classes (Sect. 3) applied on second-degree polynomial, showing some bifurcation diagrams and dynamical planes because of the dependence on a parameter in those families. In this way, different intervals have been identified where the involved parameter defines methods with stable (convergence only to the roots) or unstable methods, with convergence to strange fixed points, strange attractors or chaotic behavior.

## 2    Modified Parametric Family with Memory

From the method presented in [3] we add some accelerating parameters obtaining the following iterative methods:

$$y_k = x_k - \frac{f(x_k)}{f'(x_k) + \gamma_1 f(x_k)}, \quad t_k = y_k - \beta \frac{f(y_k)}{f'(x_k) + \gamma_2 f(x_k)},$$

$$x_{k+1} = t_k - \frac{1}{\beta} \frac{f(t_k) - (\beta - 1)^2 f(y_k)}{f'(x_k) + \gamma_3 f(x_k)}, \quad k = 1, 2, \ldots \tag{1}$$

**Proposition 1.** *Let $\alpha$ be a simple zero of a sufficiently differentiable $f : D \subset \mathbb{R} \to \mathbb{R}$ in an open interval $D$. If $x_0$ is sufficiently close to $\alpha$, and fixing $\gamma_2 = \gamma_3$, $\gamma_1, \gamma_2 \in \mathbb{R}$, the fourth-order of convergence of (1) is held $\forall \beta \in \mathbb{R} \setminus \{0\}$ and the error equation is (being $c_2 = \frac{f''(\alpha)}{2f'(\alpha)}$)*

$$e_{k+1} = -(c_2 + \gamma_1)\left(c_2\left((\beta - 1)\gamma_1 - 4\gamma_2\right) + (\beta - 5)c_2^2 - \gamma_2^2\right)e_k^4 + \mathcal{O}(e_k^5).$$

Now, we want to transform these methods including memory to increase this fourth-order of convergence, without adding new functional evaluations. It is clear that for $\gamma_1 = -c_2$, we have at least order five, but it is useless because we ignore the root $\alpha$ in practice. If we show now the error equation with Taylor series expansions with higher order and fixing $\gamma_2 = 2\gamma_1$, we have

$$\begin{aligned}
e_{k+1} = &-(c_2 + \gamma_1)^2\left((\beta - 5)c_2 - 4\gamma_1\right)e_k^4 \\
&+2(c_2 + \gamma_1)\left(2(4\beta - 19)\gamma_1 c_2^2 + c_2\left((4\beta - 31)\gamma_1^2 - 2(\beta - 6)c_3\right)\right. \\
&+\left.(5\beta - 18)c_2^3 + 10\gamma_1(c_3 - \gamma_1^2)\right)e_k^5 + \mathcal{O}(e_k^6).
\end{aligned}$$

By using divided differences or derivatives in the quotient, we can estimate $\gamma_1 \approx -c_2$. In the first case, that is, replacing $f''(\alpha)$ by $f'[x_k, x_{k-1}]$ and $f'(\alpha)$ by $f[x_k, x_{k-1}]$, we have

$$\gamma_{1k} = -\frac{1}{2}\frac{f'(x_k) - f'(x_{k-1})}{f(x_k) - f(x_{k-1})}. \tag{2}$$

Therefore, we get the class of iterative methods

$$\begin{aligned}
y_k &= x_k - \frac{f(x_k)}{f'(x_k) + \gamma_{1k}f(x_k)}, \quad t_k = y_k - \beta\frac{f(y_k)}{f'(x_k) + 2\gamma_{1k}f(x_k)}, \\
x_{k+1} &= t_k - \frac{1}{\beta}\frac{f(t_k) - (\beta - 1)^2 f(y_k)}{f'(x_k) + 2\gamma_{1k}f(x_k)}, \quad k = 1, 2, \ldots
\end{aligned} \tag{3}$$

denoted by M41.

**Theorem 2.** *Let $\alpha$ be a simple zero of a sufficiently differentiable $f : D \subset \mathbb{R} \to \mathbb{R}$ in an open interval $D$. If $x_0$ and $x_1$ are sufficiently close to $\alpha$, then the order of convergence of methods with memory (3) is at least $2 + \sqrt{6}$. The error equation is in this case $e_{k+1} = -\frac{1}{4}(\beta - 1)c_2(2c_2^2 - 3c_3)^2 e_{k-1}^2 e_k^4 + \mathcal{O}_6(e_{k-1}e_k)$, where $c_j = \frac{1}{j!}\frac{f^{(j)}(\alpha)}{f'(\alpha)}$, $j = 2, 3, \ldots$ and $\mathcal{O}_6(e_{k-1}e_k)$ indicates that the sum of exponents of $e_{k-1}$ and $e_k$ in the rejected terms of the development is at least 6. Also, if $\beta = 1$, the error equation is $e_{k+1} = \frac{1}{2}(2c_2^2 - 3c_3)^3 e_{k-1}^3 e_k^4 + \mathcal{O}_7(e_{k-1}e_k)$, being the local order $2 + \sqrt{7}$.*

In the following sections, we analyze the dynamics of the rational operator associated to family M41 on polynomials $p_1(x) = x^2 - 1$, $p_2(x) = x^2 + 1$ and $p_3(x) = x^2$. We choose these polynomials since it is known that any quadratic polynomial, by an affine change of variables, can be reduced to one of them. Moreover, the dynamics of operators associated to affine conjugate functions are equivalent. All the rational operators depend on the last iteration, $x_k$ (denoted by $x$), the previous one $x_{k-1}$ (denoted by $z$) and parameter $\beta$.

# 3    Multidimensional Dynamical Analysis of Family M41

The operator associated to M41 on $p_1(x) = x^2 - 1$ is,

$$M_1^1(z, x, \beta) = \left( x, C - \frac{(x+z)(C^2 - D - 1)}{2\beta(1 + xz)} \right),$$

where $C = -\frac{\beta(x^2-1)^2(x+z)(z^2-1)}{2(1+xz)(1+x^2+2xz)^2} + \frac{z+x(2+xz)}{1+x^2+2xz}$ and $D = \frac{(\beta-1)^2(x^2-1)^2(z^2-1)}{(1+x^2+2xz)^2}$.

**Proposition 2.** *The fixed points (and their stability) of the operator $M_1^1(z, x, \beta)$ are:*

1. *Points $(1,1)$ and $(-1,-1)$ associated to the roots of $p_1(x)$, are superattracting.*
2. *The origin $(z, x) = (0,0)$, which is a saddle fixed point.*
3. *Those points $(r, r)$ being $r$ one of the real roots of polynomial $m(t) = 4 + (30 - \beta)t^2 + (114 + 5\beta)t^4 + (268 - 10\beta)t^6 + (328 + 10\beta)t^8 + (214 - 5\beta)t^{10} + (66 + \beta)t^{12}$, whose number varies depending of $\beta$:*
   *(i) There are two saddle points if $\beta < -66$.*
   *(ii) None if $-66 \leq \beta < 138.032649$.*
   *(iii) Two non-hyperbolic points if $\beta = 138.032649$.*
   *(iv) Four real roots if $\beta > 138.032649$: two are repelling and the other two are attracting for $138.032649 < \beta < 141.2412$, non-hyperbolic for $\beta = 141.2412$ and repulsive if $\beta > 141.2412$.*

The rational operator of M41 on $p_2(x) = x^2 + 1$ is expressed as

$$M_2^1(z, x, \beta) = \left( x, C - \frac{(x+z)(C^2 - D + 1)}{2(\beta xz - \beta)} \right),$$

being $C = \frac{-2x+(-1+x^2)z}{-1+x^2+2xz} - \frac{\beta(1+x^2)^2(x+z)(1+z^2)}{2(-1+xz)(-1+x^2+2xz)^2}$ and $D = \frac{(\beta-1)^2(1+x^2)^2(1+z^2)}{(-1+x^2+2xz)^2}$.

**Proposition 3.** *The fixed point (and their stability) of operator $M_2^1(z, x, \beta)$ are:*

1. *The origin $(z, x) = (0,0)$, which is a saddle fixed point.*
2. *The real roots of polynomial $r(t) = 4 + (-30 + \beta)t^2 + (114 + 5\beta)t^4 + (-268 + 10\beta)t^6 + (328 + 10\beta)t^8 + (-214 + 5\beta)t^{10} + (66 + \beta)t^{12}$, whose number varies depending on the range of parameter $\beta$:*
   *(i) There are two real repelling points if $\beta \leq -66$.*
   *(ii) Four strange fixed points, if $-66 < \beta < 0$: two real repelling points and other two real roots which are: attracting for $-66 < \beta < -24.5740$, non-hyperbolic points if $\beta \approx -24.5740$, repulsive if $-24.5740 < \beta \leq 0$.*
   *(iii) None if $\beta = 0$, two saddle and two repelling points for $0 < \beta < 0.311664$.*
   *(iv) One saddle and one repelling point if $\beta = 0.311664$ and none for $\beta > 0.311664$.*

Finally, the associate fixed point operator associated to M41 on $p_3(x) = x^2$ is

$$M_3^1(z, x, \beta) = \left( x, -\frac{xz((4 - \beta)x^3 + 3(\beta - 8)x^2z + (3\beta - 52)xz^2 + (\beta - 40)z^3)}{8(x + 2z)^4} \right).$$

**Proposition 4.** *The fixed points (and their stability) of operator are:*

1. *The origin $(z,x) = (0,0)$, whose stability depends on parameter $\beta$: saddle for $\beta < -66$, non-hyperbolic if $\beta = -66$, attracting for $-66 < \beta < 63$, non-hyperbolic if $\beta = 63$, and saddle for $\beta > 63$.*
2. *All $(x,z) \in \mathbb{R}^2$ when $\beta = -66$, which are non-hyperbolic fixed points.*

Let us remark that some pathological aspects have been already found, as $M_1^1(z,x,\beta)$ and $M_2^1(z,x,\beta)$ have strange fixed points that can be attracting and the root in case $M_3^1(z,x,\beta)$ can be saddle. We deep in this analysis by using other tools, as bifurcation diagrams.

## 3.1   Bifurcation Diagrams and Dynamical Planes

To study the bifurcation phenomena, we use Feigenbaum diagrams of the map associated to the family M41 on polynomials $p_i(x)$, $i = 1,2,3$ by using as a starting point each one of the strange fixed points (slightly perturbed) of the map and observing the ranges of the parameter $\beta$ where changes of stability or other behavior happen. When we use $p_1(x)$, all the real strange fixed points are roots of $m(t)$, so we have at most four real roots depending on the interval where $\beta$ is defined. All the bifurcation diagrams have a similar form, so we analyze only the first root of $m(t)$. In Fig. 1, we have the bifurcation diagram on an interval where all the stability changes are observed. It shows convergence

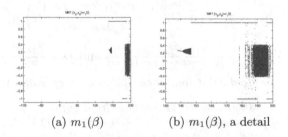

(a) $m_1(\beta)$                    (b) $m_1(\beta)$, a detail

**Fig. 1.** Bifurcation diagrams of M41 on $p_1(x)$

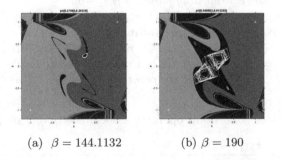

(a)  $\beta = 144.1132$                    (b) $\beta = 190$

**Fig. 2.** Dynamical planes of M41 on $p_1(x)$ for different $\beta$

to the roots of $p_1(x)$ and, in some ranges as the interval $[140, 200]$, we can see period-doubling bifurcations and regions of chaotic behavior. In Figs. 2a and b we show the dynamical planes associated to some values of $\beta$ in two different regions of Fig. 1. These attractors have been isolated and the convergence to these periodic orbits and strange attractors can be detected. Let us see also Fig. 3 for the dynamical plane corresponding to a value of $\beta$ where two strange fixed points are attracting, although their basin of attraction is so small that it must be zoomed to be observed. But, in spite of this chaotic behavior, for the values of $\beta$ out of the interval $[138, 200]$, is mainly stable. To visualize this stable behavior, we observe in Fig. 4 some dynamical planes have been obtained by using the rational function $M_1^1(z, x, \beta)$, for different values of parameter $\beta$. In them, there is no other basin of attraction than those of the roots, being simpler when $\beta < 0$. Now, when we use $p_2(x)$, we have at the most four real strange fixed points, depending on $\beta$. All the bifurcation diagrams are similar, as can be seen in Fig. 5, where two of them are shown; the other strange fixed points are the opposite of these ($r_{1,2}(\beta) = \pm\sqrt{s_1}(\beta)$, $r_{3,4}(\beta) = \pm\sqrt{s_2}(\beta)$), being $s_i(\beta)$, $i = 1, 2$ the only real roots of polynomial $r(\sqrt{t})(\beta)$) and their behavior is symmetric, so they are omitted. We observe chaotic behavior in Fig. 5 with period-doubling bifurcations in $\beta \in [-20, 0]$.

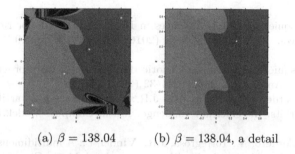

(a) $\beta = 138.04$          (b) $\beta = 138.04$, a detail

**Fig. 3.** Dynamical planes of M41 on $p_1(x)$ for $\beta = 138.04$

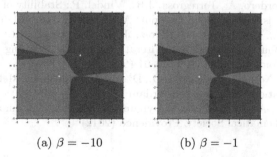

(a) $\beta = -10$          (b) $\beta = -1$

**Fig. 4.** Dynamical planes of family M41 (stable) on $p_1(x)$

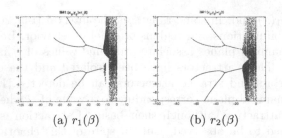

(a) $r_1(\beta)$            (b) $r_2(\beta)$

**Fig. 5.** Bifurcation diagrams of family M41 on $p_2(x)$

Finally, for $p_3(x)$, the dynamical planes show convergence to $(0,0)$ in the interval where it is attractor, as it has been previously stated in the analysis of stability.

**Acknowledgements.** This research was partially supported by Ministerio de Economía y Competitividad under grants MTM2014-52016-C2-2-P and Generalitat Valenciana PROMETEO/2016/089.

# References

1. Amat, S., Busquier, S. (eds.): Advances in Iterative Methods for Nonlinear Equations. SSSS, vol. 10. Springer, Cham (2016). https://doi.org/10.1007/978-3-319-39228-8
2. Amat, S., Busquier, S., Plaza, S.: Chaotic dynamics of a third-order Newton-type method. Math. Anal. Appl. **366**(1), 24–32 (2010)
3. Arroyo, V., Cordero, A., Torregrosa, J.R.: Approximation of artificial satellites preliminary orbits: the efficiency challenge. Math. Comput. Model. **54**, 1802–1807 (2011)
4. Campos, B., Cordero, A., Torregrosa, J.R., Vindel, P.: A multidimensional dynamical approach to iterative methods with memory. Appl. Math. Comput. **271**, 701–715 (2015)
5. Campos, B., Cordero, A., Torregrosa, J.R., Vindel, P.: Stability of King's family of iterative methods with memory. Comput. Appl. Math. **318**, 504–514 (2017)
6. Magreñán, Á.A., Cordero, A., Gutiérrez, J.M., Torregrosa, J.R.: Real qualitative behavior of a fourth-order family of iterative methods by using the convergence plane. Math. Comput. Simul. **105**, 49–61 (2014)
7. Petković, M.S., Neta, B., Petković, L.D., Džunić, J.: Multipoint Methods for Solving Nonlinear Equations. Elsevier, Amsterdam (2013)
8. Robinson, R.C.: An Introduction to Dynamical Systems: Continuous and Discrete. American Mathematical Society, Providence (2012)

# Stability of a Family of Iterative Methods of Fourth-Order

Alicia Cordero[ID], Lucía Guasp$^{(\boxtimes)}$, and Juan R. Torregrosa[ID]

Instituto Universitario de Matemática Multidisciplinar, Universitat Politècnica de València, Camino de Vera s/n, 46022 València, Spain
{acordero,jrtorre}@mat.upv.es, luguaal@ade.upv.es

**Abstract.** In this paper, we analyze the dynamical anomalies of a family of iterative methods, for solving nonlinear equations, designed by using weight function procedure. All the elements of the family are optimal schemes (in the sense of Kung-Traub conjecture) of fourth-order, but not all have the same stability properties. So, we describe the dynamical behavior of this family on quadratic polynomials. The study of fixed points and their stability, joint with the critical points and their associated parameter planes, show the richness of the presented class and allow us to select the members of the family with good stability properties.

**Keywords:** Nonlinear equation · Iterative method ·
Dynamical behavior · Fatou and Julia sets · Basin of attraction

## 1    Introduction

Nonlinear equations $f(x) = 0$, where $f : I \subseteq \mathbb{R} \to \mathbb{R}$ is a real function defined in an open interval $I$, are often used for modeling real problems arising in science and engineering. For obtaining the solutions $\bar{x}$ of these equations, iterative schemes must be used. The best known iterative approach is Newton's method. In last decades, many researchers have proposed different iterative methods to improve Newton's scheme (see, for example, the reviews [1,6] and the references therein). These variants of Newton's method have been designed by means of different techniques providing, in the most of cases, multistep schemes.

Recently, the weight-function procedure has been used to increase the order of convergence of known methods [6], allowing to get optimal methods, under the point of view of Kung-Traub's conjecture [5]. These authors conjectured that an iterative method, without memory, which uses $d$ functional evaluations per iteration can reach, at most, order of convergence $2^{d-1}$. When this bound is reached, the scheme is called optimal. Although the aim of many researches in this area is to design optimal high-order methods, it is also known that the higher the order is, the more sensitive the scheme to initial estimations will be.

© Springer Nature Switzerland AG 2019
I. Dimov et al. (Eds.): FDM 2018, LNCS 11386, pp. 193–200, 2019.
https://doi.org/10.1007/978-3-030-11539-5_20

In this paper, by using Newton's scheme and the weight function procedure, we design a new family of iterative methods, whose iterative expression is

$$
\begin{aligned}
y_n &= x_n - \alpha \frac{f(x_n)}{f'(x_n)}, \\
x_{n+1} &= x_n - H(t_n)\frac{f(x_n)}{f'(x_n)}, \quad n = 0, 1, \dots,
\end{aligned}
\tag{1}
$$

where the variable of the weight function $H(t)$ is $t = \frac{f'(y)}{f'(x)}$.

The following result establishes the order of convergence of this family.

**Theorem 1.** *Let $f : I \subseteq \mathbb{R} \to \mathbb{R}$ be a sufficiently derivable function in an open interval $I$ and let $\bar{x} \in I$ be a simple solution of the nonlinear equation $f(x) = 0$. We consider that $x_0$ is an initial approximation close enough to $\bar{x}$. If $\alpha = 2/3$ and function $H$ satisfies $H(1) = 1$, $H'(1) = -3/4$ and $H''(1) = 9/4$, then sequence $\{x_n\}_{n\geq 0}$ obtained by using (1) converges to $\bar{x}$ with order of convergence four, being the error equation*

$$
e_{n+1} = (5c_2^3 - c_2c_3 + c_4/9)e_n^4 + O(e_n^5).
$$

*where $c_j = \frac{f^{(j)}(\bar{x})}{j!f'(\bar{x})}$, $j = 2, 3, \dots$ and $e_n = x_n - \bar{x}$.*

If we replace $f'(x_n)$ by the divided difference $f[x_n, w_n]$, where $w_n = x_n + \gamma f(x_n)$, a similar family of derivative-free methods can be constructed

$$
\begin{aligned}
y_n &= x_n - \alpha \frac{f(x_n)}{f[x_n, w_n]}, \\
x_{n+1} &= x_n - H(t_n)\frac{f(x_n)}{f[x_n, w_n]}, \quad n = 0, 1, \dots,
\end{aligned}
\tag{2}
$$

where, in this case, the variable of the weight function is $t = \frac{f[y, w]}{f[x, w]}$.

On the other hand, family (1) can be extended in a natural way to the multidimensional case, obtaining a family of iterative methods for solving a nonlinear system $F(x) = 0$, where $F : D \subseteq \mathbb{R}^n \to \mathbb{R}^n$, whose iterative expression is:

$$
\begin{aligned}
y^{(n)} &= x^{(n)} - \alpha[F'(x^{(n)})]^{-1}F(x^{(n)}), \\
x^{(n+1)} &= x^{(n)} - H(t^{(n)})[F'(x^{(n)})]^{-1}F(x^{(n)}), \quad n = 0, 1, \dots,
\end{aligned}
\tag{3}
$$

where $F'(x^{(n)})$ denotes the Jacobian matrix of $F$ evaluated in the iteration $x^{(n)}$ and $t = [F'(x)]^{-1}F'(y)$.

For both families (2) and (3) we can establish a similar result to Theorem 1 which guarantees, under certain conditions, the order 4 of both classes.

In the next sections, we are going to study the dynamical behavior of a sub-class of family (1) obtained by using $\alpha = 2/3$ and

$$
H(t) = 1 - \frac{3}{4}(t - 1) + \frac{9}{8}(t - 1)^2 + \frac{1}{6}h(t - 1)^3,
$$

where $h$ is a free parameter. This class is denoted by CGT.

## 2    Basic Concepts

Under the point of view of complex dynamics, we will study the general convergence of family CGT on quadratic polynomials. It is known that the roots of a polynomial can be transformed by an affine map with no qualitative changes on the dynamics of the family. So, we can use the generic quadratic polynomial $p(z) = (z-a)(z-b)$. For $p(x)$, the operator of the family is the rational function $T_{p,h,a,b}(z)$, depending on parameters $h$, $a$ and $b$.

Blanchard in [2] considered the conjugacy map $\phi(z) = \dfrac{z-a}{z-b}$, (a Möbius transformation) with the following properties:

$$\text{(i)} \quad \phi(\infty) = 1, \quad \text{(ii)} \quad \phi(a) = 0, \quad \text{(iii)} \quad \phi(b) = \infty,$$

and proved that, for quadratic polynomials, Newton's operator is conjugate to the rational map $z^2$. In an analogous way, operator $T_{p,h,a,b}$ on quadratic polynomials is conjugated to operator $O_h(z)$,

$$O_h(z) = \left(\phi \circ T_{p,h,a,b} \circ \phi^{-1}\right)(x)$$

$$= z^4 \frac{405 + 32h + 1134z + 1134z^2 + 486z^3 + 81z^4}{81 + 486z + 1134z^2 + 1134z^3 + 405z^4 + 32hz^4}. \tag{4}$$

We observe that parameters $a$ and $b$ have been obviated in $O_h$.

Now, we are going to recall some dynamical concepts of complex dynamics (see [3]) that we use in this work. Given a rational function $R : \hat{\mathbb{C}} \to \hat{\mathbb{C}}$, where $\hat{\mathbb{C}}$ is the Riemann sphere, the *orbit of a point* $z_0 \in \hat{\mathbb{C}}$ is defined as:

$$\{z_0, R(z_0), R^2(z_0), ..., R^n(z_0), ...\}.$$

We analyze the phase plane of the map $R$ by classifying the starting points from the asymptotic behavior of their orbits. A $z_0 \in \hat{\mathbb{C}}$ is called a *fixed point of R* if $R(z_0) = z_0$. A *periodic point of R* $z_0$ of period $p > 1$ is a point such that $R^p(z_0) = z_0$ and $R^k(z_0) \neq z_0$, for $k < p$. A *pre-periodic point of R* is a point $z_0$ that is not periodic but there exists a $k > 0$ such that $R^k(z_0)$ is periodic. A *critical point* $z_0$ is a point where the derivative of the rational function $R$ vanishes, $R'(z_0) = 0$. Moreover, a fixed point of $R$ $z_0$ is called *attractor* if $|R'(z_0)| < 1$, *superattractor* if $|R'(z_0)| = 0$, *repulsor* if $|R'(z_0)| > 1$ and *parabolic* if $|R'(z_0)| = 1$.

*The basin of attraction* of an attractor $\alpha$ is defined as:

$$\mathcal{A}(\alpha) = \{z_0 \in \hat{\mathbb{C}} \ : \ R^n(z_0) \to \alpha, \ n \to \infty\}.$$

The Fatou set of the rational function $R$, $\mathcal{F}(R)$, is the set of points $z \in \hat{\mathbb{C}}$ whose orbits tend to an attractor (fixed point, periodic orbit or infinity). Its complement in $\hat{\mathbb{C}}$ is the *Julia set*, $\mathcal{J}(R)$. That means that the basin of attraction of any fixed point belongs to the Fatou set and the boundaries of these basins of attraction belong to the Julia set.

The following theorem establishes a classical result of Fatou and Julia that we use in the study of parameter space associated to the family.

**Theorem 2.** *Let $R$ be a rational function. The immediate basin of attraction of an attracting fixed or periodic point holds, at least, a critical point.*

By using this result, one can be sure to find all the stable behavior associated to a rational function $R$, by analyzing the performance of $R$ on the set of critical points.

# 3     Analysis of the Fixed and Critical Points

In the next sections we are going to analyze, under the dynamical point of view, the stability and reliability of the members of the proposed family CGT. Firstly, we will study the fixed points of the rational function $O_h(z)$ that are not related with the original roots of the polynomial $p(z)$ (called *strange fixed points*), and the *free critical points*, that is, the critical points of $O_h(z)$ different from 0 and $\infty$.

Some relations between the strange fixed points are described in the following result.

**Theorem 3.** *Fixed points of $O_h(x)$ are the roots of the equation $O_h(z) = z$, that is, $z = 0$, $z = \infty$ and the strange fixed points:*

- *$ex_1(h) = 1$,*
- *the roots of symmetric sixth-degree polynomial $r(t) = 81 + 567t + 1701t^2 + (2430 - 32h)t^3 + 1701t^4 + 567t^5 + 81t^6$, or equivalently,*

$$ex_2(h) = \frac{s_1(h) \pm \sqrt{s_1(h)^2 - 4}}{2} = \frac{1}{ex_3(h)},$$

$$ex_4(h) = \frac{s_2(h) \pm \sqrt{s_2(h)^2 - 4}}{2} = \frac{1}{ex_5(h)},$$

$$ex_6(h) = \frac{s_3(h) \pm \sqrt{s_3(h)^2 - 4}}{2} = \frac{1}{ex_7(h)},$$

*where $s_1(h)$, $s_2(h)$ and $s_3(h)$ are the roots of the third-degree polynomial $s(t) = 81t^3 + 567t^2 + 1458t + 1296 - 32h$.*

*So, the number of strange fixed points of operator $O_h(x)$ is seven, except in some cases:*

(i) *If $h = 0$ or $h = \frac{891}{4}$, there are only five strange fixed points.*
(ii) *If $h = \frac{3}{16}i\left(8i + 5\sqrt{5}\right)$ or $h = -\frac{3}{16}i\left(-8i + 5\sqrt{5}\right)$ the strange fixed points are $ex_1 = 1$ and the roots of a sixth-degree polynomial, which are two simple and two double roots.*

In order to determine the critical points, we calculate the first derivative of $O_h(x)$,

$$O_h'(z) = (1 + z)^6 \frac{324z^3 405(1 + z)^2 + 16h\left(2 - 3z + 2z^2\right)}{(81 + 486z + 1134z^2 + 1134z^3 + (405 + 32h)z^4)^2}.$$

It is clear that $z = 0$ and $z = \infty$ (related to the roots of the polynomial by means of Möbius map) are critical points and give rise to their respective Fatou components, but there exist in the family some free critical points, some of them depending on the value of the parameter $h$.

**Proposition 1.** *Rational function $O_h(z)$ has a different number of free critical points, depending on the value of parameter $h$:*

*(a) If $h = 0$, then $z = -1$ is the only free critical point, being a pre-image of the fixed point $z = 1$.*

*(b) In any other case, the free critical points are $cr_1(h) = -1$,*

$$cr_2(h) = \frac{-405 + 24h - 4\sqrt{7}\sqrt{-h(405 + 4h)}}{405 + 32h} = \frac{1}{cr_3},$$

*that is, there exists only one independent free critical point.*

As we will see in the following section, not only the number but also the stability of the fixed points depend on the parameter of the family. The relevance of this study yields in the fact that the existence of attracting strange fixed points can make the iterative scheme converge to a "false" solution.

## 4    Stability of the Fixed Points

It is clear that $z = 0$ and $z = \infty$ are always superattracting fixed points, but the stability of the other fixed points gives us interesting numerical information. In the following results we show the stability of the strange fixed points.

**Theorem 4.** *The character of the strange fixed point $ex_1(h) = 1$, for $h \neq \frac{405}{4}$, is as follows:*

*(i) If $\left| h + \frac{405}{4} \right| < 324$, then $ex_1(h) = 1$ is a repulsor.*
*(ii) When $\left| h + \frac{405}{4} \right| = 324$, $ex_1(h) = 1$ is a parabolic point.*
*(iii) If $\left| h + \frac{405}{4} \right| > 324$, then $ex_1(h) = 1$ is an attractor.*

In Fig. 1, we represent the stability regions of all strange fixed points $ex_i(h)$, $i = 1, 2, \ldots, 7$. The analysis of the stability of strange fixed points $ex_i(h), i = 4, 5, 6, 7$ shows that they are repulsive for any value of parameter $h$.

## 5    The Parameter Space

As we have seen, the dynamical behavior of operator $O_h(z)$ depends on the values of the parameter $h$. The parameter space associated with a free critical point of operator (4) is obtained by associating each point of the parameter plane with a complex value of $h$, i.e., with an element of family (1). Every value of $h$ belonging to the same connected component of the parameter space gives rise to subsets of schemes of family (1) with similar dynamical behavior. So, it

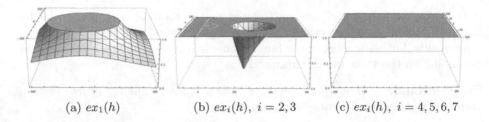

(a) $ex_1(h)$          (b) $ex_i(h)$, $i = 2, 3$          (c) $ex_i(h)$, $i = 4, 5, 6, 7$

**Fig. 1.** 3D-view of stability functions of strange fixed points

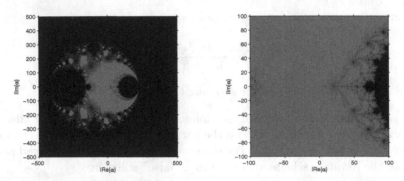

**Fig. 2.** Parameter plane associated to $cr_i(h)$, $i = 2, 3$, and a detail

is interesting to find regions of the parameter plane as much stable as possible, because these values of $h$ will give us the best members of the family in terms of numerical stability.

As $cr_2(h) = \dfrac{1}{cr_3(h)}$, we have at most one free independent critical point, so there exist only one parameter plane. When we consider the independent free critical points as a starting guess of all the iterative schemes of the family associated to each complex value of $h$, we paint this value of the complex plane in red if the method converges to any of the roots (zero and infinity) and they are black in other cases. The color used is brighter when the number of iterations is lower. Then, the parameter plane presented in Fig. 2 is obtained. We also show a zoom of this parameter plane. The parameter plane has been generated by using the routines described in [4]. A mesh of $2000 \times 2000$ points has been used, 500 has been the maximum number of iterations involved and $10^{-3}$ the tolerance used as a stopping criterium.

## 5.1   Dynamical Planes

In this section we will show, by means of dynamical planes, the qualitative behavior of the different elements of family. We will select this elements by using the conclusions obtained by analyzing the parameter planes of the family.

As in case of parameter planes, these dynamical planes has been generated by using the routines appearing in [4]. The dynamical plane associated to a value of

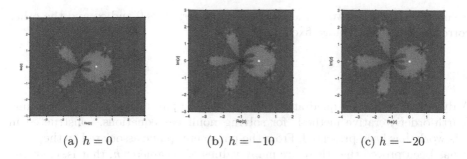

(a) $h = 0$                (b) $h = -10$                (c) $h = -20$

**Fig. 3.** Some dynamical planes with stable behavior (Color figure online)

parameter $h$, that is, obtained by iterating an element of family (1), is generated by using each point of the complex plane as initial estimation (we have used a mesh of $500 \times 500$ points). We paint in blue the points whose orbit converges to infinity, in orange those points converging to zero (with a tolerance of $10^{-3}$), in green those points whose orbit converges to one of the strange fixed points (all the attracting fixed points appear marked as a white star in the figures) and in black if it reaches the maximum number of 40 iterations without converging to any of the fixed points.

In Fig. 3 we show some dynamical planes corresponding to values of parameter $h$ which, from parameter plane, give us elements of the family with stable behavior. So, we can see only two basins of attraction, that correspond to zero (orange basin) and infinity (blue basin).

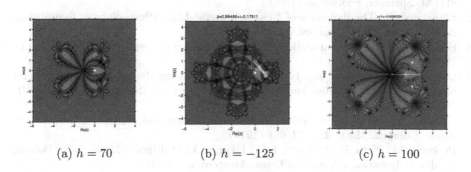

(a) $h = 70$                (b) $h = -125$                (c) $h = 100$

**Fig. 4.** Some dynamical planes with unstable behavior

On the other hand, unstable behavior is found when we choose values of $h$ in the black region of parameter plane (Fig. 2). In Fig. 4(a), the dynamical plane of the iterative method corresponding to $h = 70$ is presented, with regions of slow convergence. In Fig. 4(b), corresponding to $h = -125$, we can observe an orbit of period two. Finally, Fig. 4(c) and (d), show the existence of three different basins of attraction, two of them of the superattractors 0 and $\infty$ and the other corresponding to a strange fixed point.

# 6    Conclusions

A dynamical study on quadratic polynomials of a parametric family of optimal fourth-order iterative methods for solving nonlinear equations, constructed in this work, has been presented. From the parameter plane associated to the class, it has been proved that there are many values of parameter $h$, that is, elements of the family, with good stability properties and other ones with no convergence to the roots of the polynomial, and the existence of periodic orbits of period two has been showed and its analytical expression has been obtained in terms of parameter $h$. As we have mentioned in the Introduction, the presented family can be extended in two directions: a class of derivative-free methods for solving nonlinear equations and a multidimensional family for nonlinear systems, preserved in both cases the order of convergence. The dynamical analysis in these areas will be object of study in future works.

**Acknowledgement.** This research was partially supported by Ministerio de Economía y Competitividad under grants MTM2014-52016-C2-2-P and Generalitat Valenciana PROMETEO/2016/089.

# References

1. Amat, S., Busquier, S.: Advances in Iterative Methods foir Nonlinear Equations. SIMAI. Springer, Switzerland (2016)
2. Blanchard, P.: The dynamics of Newton's method. In: Proceedings of Symposium Applied Mathematics, vol. 49, pp. 139–154 (1994)
3. Blanchard, P.: Complex analytic dynamics on the Riemann sphere. Bull. AMS **11**(1), 85–141 (1984)
4. Chicharro, F., Cordero, A., Torregrosa, J.R.: Drawing dynamical and parameter planes of iterative families and methods. Sci. World J. **2013** (2013). Article ID 780153
5. Kung, H.T., Traub, J.F.: Optimal order of one-point and multipoint iteration. J. Assoc. Comput. Math. **21**, 634–651 (1974)
6. Petković, M., Neta, B., Petković, L.D., Džunić, J.: Multipoint Methods for Solving Nonlinear Equations. Academic Press, Amsterdam (2013)

# Multidimensional Real Dynamics
# for High-Order Processes

Alicia Cordero$^{1(\boxtimes)}$ ⓘ, Javier G. Maimó$^2$ ⓘ, Juan R. Torregrosa$^1$ ⓘ,
and Maria P. Vassileva$^2$ ⓘ

$^1$ Instituto Universitario de Matemática Multidisciplinar,
Universitat Politècnica de València, Camino de Vera s/n, 46022 València, Spain
{acordero,jrtorre}@mat.upv.es
$^2$ Instituto Tecnológico de Santo Domingo,
Avda. Los Próceres 49, Santo Domingo, Dominican Republic
javiermaimo@hotmail.com, maria.vassilev@gmail.com

**Abstract.** In this manuscript, we design a parametric family of iterative
methods for solving nonlinear problems, that does not need to evaluate
Jacobian matrices and needs to solve three linear systems per iteration
with the same divided difference operator as coefficient matrix. The stability performance of the class is analyzed on a quadratic polynomial
system and it is shown that for a wide set of values (including positive
ones), there exist only convergence to the roots of the problem.

**Keywords:** Nonlinear systems · Real multidimensional dynamics ·
Stability

## 1 Introduction

The most of nonlinear models appearing in Science and Engineering can be
approached, by means of a discretization process, to a system of nonlinear equations. The solutions of these systems are usually estimated by means of iterative methods, mainly Newton's scheme, due to its simplicity and efficiency. In
the last years, many researchers have designed new iterative schemes trying to
improve the original Newton's procedure, in terms of order of convergence, efficiency or stability. Many tools have been used to get this aim: composition or
weight-function procedures, non-direct extension of scalar schemes, pseudocom-
position (that is, using different schemes as stating point of a predictor-corrector
method), etc.

In 2016, the authors proposed in [1] a fourth-order family of parametric
methods specifically designed for solving nonlinear systems that included a fifth-
order scheme with the best computational efficiency index, up to our knowledge.
Once the convergence and efficiency analysis is made, it is useful to study the
dependence of the elements of the family of iterative methods to the initial
estimations used.

© Springer Nature Switzerland AG 2019
I. Dimov et al. (Eds.): FDM 2018, LNCS 11386, pp. 201–207, 2019.
https://doi.org/10.1007/978-3-030-11539-5_21

Then, the real dynamical tools allow us to study the stability of the members of the class and decide which ones are the best, under this point of view. Specifically, we analyze the stability of the fixed points, the existence of free critical points depending on the parameter and the chaotic behavior found (period-doubling bifurcations, strange attractors, ...) in some specific methods by means of Feigenbaum diagrams. Finally, some dynamical planes are presented to show in practice the theoretical performance of the schemes.

## 1.1    Introductory Concepts

Let us consider the problem of finding a real solution $\bar{x} \in D$ of the nonlinear system $F(x) = 0$, of $n$ equations with $n$ variables with coordinate functions $f_i$, $i = 1, 2, \ldots, n$. This solution can be obtained as a fixed point of some function $\bar{G} : \mathbb{R}^n \longrightarrow \mathbb{R}^n$ by means of the fixed-point iteration method

$$x^{(k+1)} = \bar{G}(x^{(k)}), \quad k = 0, 1, \ldots, \tag{1}$$

where $x^{(0)}$ is the initial estimation.

In order to analyze the dynamical behavior of a fixed-point iterative method for nonlinear systems when is applied to n-variable polynomial $p(x)$, $p : \mathbb{R}^n \to \mathbb{R}^n$, $x \in \mathbb{R}^n$, it is necessary to recall some basic dynamical concepts.

Let us denote by $G(x)$ the vectorial fixed-point function associated to the iterative method on polynomial $p(x)$. Let us note that the next concepts and results are also valid when the iterative method is applied on a general function $F(x)$.

**Definition 1.** *Let $G : \mathbb{R}^n \to \mathbb{R}^n$ be a vectorial rational function. The orbit of a point $x^{(0)} \in \mathbb{R}^n$ is defined as the set of successive images of $x^{(0)}$ by the vectorial rational function, $\{x^{(0)}, G(x^{(0)}), \ldots, G^m(x^{(0)}), \ldots\}$.*

The dynamical behavior of the orbit of a point of $\mathbb{R}^n$ can be classified depending on its asymptotic behavior. In this way, a point $x^* \in \mathbb{R}^n$ is a fixed point of $G$ if $G(x^*) = x^*$.

We recall a known result in Discrete Dynamics that gives the stability of fixed points for nonlinear operators.

**Theorem 1** ([3], p. 558). *Let $G$ from $\mathbb{R}^n$ to $\mathbb{R}^n$ be $C^2$. Assume $x^*$ is a period-k point. Let $\lambda_1, \lambda_1, \ldots, \lambda_n$ be the eigenvalues of $G'(x^*)$.*

(a) *If all the eigenvalues $\lambda_j$ have $|\lambda_j| < 1$, then $x^*$ is attracting.*
(b) *If one eigenvalue $\lambda_{j_0}$ has $|\lambda_{j_0}| > 1$, then $x^*$ is unstable, that is, repelling or saddle.*
(c) *If all the eigenvalues $\lambda_j$ have $|\lambda_j| > 1$, then $x^*$ is repelling.*

In addition, a fixed point is called hyperbolic if all the eigenvalues $\lambda_j$ of $G'(x^*)$ have $|\lambda_j| \neq 1$. In particular, if there exist an eigenvalue $\lambda_i$ such that $|\lambda_i| < 1$ and an eigenvalue $\lambda_j$ such that $|\lambda_j| > 1$, the hyperbolic point is called saddle point.

Let us note that the entries of $G'(x^*)$ are the partial derivatives of each coordinate function of the vectorial rational operator that defines the iterative scheme. To avoid the calculation of spectrum of $G'(x^*)$ the authors in [2] proposed the following result that, being consistent with the previous theorem, gives us a practical tool for classifying the stability of fixed points in many cases.

**Proposition 1.** *Let $x^*$ be a fixed point of $G$. Then,*

*(a)* *If* $\left| \frac{\partial g_i(x^*)}{\partial x_j} \right| < \frac{1}{n}$ *for all* $i, j \in \{1, \ldots, n\}$, *then* $x^* \in \mathbb{R}^n$ *is attracting.*

*(b)* *If* $\left| \frac{\partial g_i(x^*)}{\partial x_j} \right| = 0$, *for all* $i, j \in \{1, \ldots, n\}$, *then* $x^* \in \mathbb{R}^n$ *is superattracting.*

*(c)* *If* $\left| \frac{\partial g_i(x^*)}{\partial x_j} \right| > \frac{1}{n}$, *for all* $i, j \in \{1, \ldots, n\}$, *then* $x^* \in \mathbb{R}^n$ *is unstable and lies at the Julia set.*

*being $g_i(x)$, $i = 1, 2, \ldots, n$, the coordinate functions of the fixed point multivariate function $G$.*

The proof of this result is based in Theorem 1 and on the fact that $\rho(G'(x^*)) \leq \|G'(x^*)\|$, where $\rho(A)$ denotes the spectral radius of matrix $A$ and the unstable points (repelling and saddle) are always on Julia set.

It is obvious that, if the order of the iterative method is at least two, then the roots of the nonlinear function are superattracting fixed points of the vectorial rational function associated to the iterative method. If a fixed point is not a root of the nonlinear function, it is called strange fixed point and its character can be analyzed in the same manner.

Then, if $x^*$ is an attracting fixed point of the rational function $G$, its basin of attraction $\mathcal{A}(x^*)$ is defined as the set of pre-images of any order such that

$$\mathcal{A}(x^*) = \left\{ x^{(0)} \in \mathbb{R}^n : G^m(x^{(0)}) \to x^*, m \to \infty \right\}.$$

In the same way as in the scalar case, the set of points whose orbits tend to an attracting fixed point $x^*$ is defined as the Fatou set, $\mathcal{F}(G)$. The complementary set, the Julia set $\mathcal{J}(G)$, is the closure of the set consisting of its repelling fixed points, and establishes the borders between the basins of attraction.

The concept of critical point can be defined following the idea of multivariate convergence of iterative methods.

**Definition 2.** *A fixed point $x \in \mathbb{R}^n$ is a critical point of $G$ if the eigenvalues of the associate Jacobian matrix are null. So, if coordinate functions $g_i$ satisfy $\frac{\partial g_i(x)}{\partial x_j} = 0$ for all $i, j \in \{1, \ldots, n\}$, then $x$ is a critical point.*

In this terms, a superattracting fixed point will be also a critical point and, from the numerical point of view, the iterative method involved will be, at least, of second order of convergence. A critical point that is not root of the polynomial $p(x)$ will be called free critical point.

In this paper we apply these dynamical concepts extended to nonlinear systems, to a Jacobian-free family M4 of iterative schemes. Specifically, we analyze the stability of the fixed points, the existence of free critical points depending on the parameter and the chaotic behavior found (period-doubling bifurcations, strange attractors, ...) in some specific methods, that is, for several values of the parameter of the fourth-order class, by means of bifurcation diagrams. Finally, some dynamical planes are presented to show in practice the performance of the members of the class.

## 2    M4 Jacobian-Free Parametric Family

In this section we will apply the extended dynamical concepts to the family of iterative schemes called M4, whose iterative expression is

$$y^{(k)} = x^{(k)} - \left[x^{(k)}, x^{(k)} + F(x^{(k)}); F\right]^{-1} F(x^{(k)}), \quad k \geq 0$$

$$z^{(k)} = y^{(k)} - \beta \left[x^{(k)}, x^{(k)} + F(x^{(k)}); F\right]^{-1} F(y^{(k)}), \tag{2}$$

$$x^{(k+1)} = z^{(k)} + \frac{1}{\beta} \left[x^{(k)}, x^{(k)} + F(x^{(k)}); F\right]^{-1} \left((\beta - 1)^2 F(y^{(k)}) - F(z^{(k)})\right)$$

Because of the technique used in the proof, This method has fourth-order of convergence in a neighborhood of the solution where the Jacobian matrix $F'(x)$ is non-singular. In spite of this, it is a Jacobian-free version of a fourth-order iterative class designed by the authors in [1]. The following result shows its local convergence.

**Theorem 2.** *Let $\bar{x} \in D$ be a simple zero of a sufficiently differentiable Frechét function $F : D \subset \mathbb{R}^n \to \mathbb{R}^n$ in an open convex set $D$ and $x^{(0)}$ an initial guess close enough to $\bar{x}$. The method defined by (2) has fourth-order of convergence for any $\beta \in \mathbb{R}$. The error equation of the method is*

$$e_{k+1} = (1 + F'(\bar{x})) \left(F'(\bar{x})^2 - (F'(\bar{x}) + 1)(-5 + \beta)\right) C_2^3 e_k^4 + O(e_k^5),$$

*where $C_k = (1/k!)[F'(\bar{x})]^{-1} F^{(k)}(\bar{x})$, $k = 1, 2, \ldots$ and $e_k = x^{(k)} - \bar{x}$.*

In the following, we will denote by $M(x, \beta) = (M_1(x, \beta), M_2(x, \beta), \ldots, M_n(x, \beta))$ the fixed point function associated to M4 class applied on $n$-dimensional quadratic polynomial $p(x) = 0$, where

$$p_i(x) = x_i^2 - 1, \quad i = 1, 2, \ldots, n. \tag{3}$$

As the polynomial system has separated variables, all the components of $M(x, \beta)$ has the same expression, with the only difference of the sub-index corresponding to the component of $x$:

$$M_j(x,\beta) = x_j + \frac{1 - x_j^2}{-1 + 2x_j + x_j^2} - \frac{x_j(2 + x_j)\left(-1 + x_j^2\right)^2 \beta}{\left(-1 + 2x_j + x_j^2\right)^3}$$

$$+ \frac{1}{\left(-1 + 2x_j + x_j^2\right)^7} x_j(2 + x_j)\left(-1 + x_j^2\right)^2$$

$$\left(x_j(6 - 10\beta) + x_j^3(14 - 4\beta) + x_j^8(-2 + \beta) + 6x_j^5(-1 + \beta) + \beta\right.$$

$$\left. - 6x_j^4(-7 + 4\beta) + 2x_j^7(-7 + 4\beta) + x_j^6(-30 + 19\beta) + x_j^2(-26 + 19\beta)\right),$$

for $j = 1, 2, \ldots, n$.

By solving the equation $M(x,\beta) = x$, the fixed points of the multidimensional rational function are obtained, that is, the roots of $M(x) = x$, are $x_j = \pm 1$ and also the roots of the polynomial $r(t) = 1 - 12t + (66 - 4\beta)t^2 + (-146 - 4\beta)t^3 + (5 + 11\beta)t^4 + (312 + 12\beta)t^5 + (-48 - 9\beta)t^6 + (-332 - 12\beta)t^7 + (-73 + \beta)t^8 + (148 + 4\beta)t^9 + (110 + \beta)t^{10} + 30t^{11} + 3t^{12}$. All this information, including the stability analysis of these fixed points is summarized in the following result.

**Theorem 3.** *The n-dimensional rational function associated to family M4 on the polynomial $p(x)$ has $2^n$ superattracting fixed points with components $x_j = \pm 1$ (the real solutions of the nonlinear system) and also different number of real strange fixed points depending on parameter $\beta$:*

- *If $\beta < -118.1782$, six roots of $r(t)$, denoted by $r_i(\beta)$, $i = 1, 2, \ldots, 6$ are real, being their respective eigenvalues of the associate Jacobian matrix $\lambda_i(\beta) > 1$, $i = 1, 2, \ldots, 5$ and $\lambda_6(\beta) < 1$.*
- *If $\beta = -118.1782$, five roots of $r(t)$, are real, being their respective eigenvalues of the associate Jacobian matrix $\lambda_i(-118.1782) > 1$, $i = 1, 2, 3, 4$ and $\lambda_5(-118.1782) = 1$.*
- *For $-118.1782 < \beta < -0.4991$, there are four real roots of $r(t)$, being their respective eigenvalues of the associate Jacobian matrix $\lambda_i(\beta) > 1$, $i = 1, 2, 4$ and $\lambda_3(\beta) < 1$ if $-0.5242 < \beta < -0.4991$, $\lambda_3(\beta) > 1$ for $118.1782 < \beta < -0.5242$ and $\lambda_3(\beta) = 1$ if $\beta = -0.5242$.*
- *If $\beta = -0.4991$, three roots of $r(t)$ are real, being their respective eigenvalues of the associate Jacobian matrix $\lambda_i(-0.4991) > 1$, $i = 1, 2$ and $\lambda_3(-0.4991) = 1$.*
- *For $-0.4991 < \beta < 0$ and $0 < \beta < 62.0613$, there are two real roots of $r(t)$ satisfying $\lambda_i(\beta) > 1$, $i = 1, 2$.*
- *If $\beta = 62.0613$, four roots of $r(t)$ are real, being their respective eigenvalues of the associate Jacobian matrix $\lambda_i(62.0613) = 1$, $i = 1, 2$ and $\lambda_j(62.0613) > 1$, $j = 3, 4$.*
- *When $\beta > 62.0613$, there are four real roots of $r(t)$ satisfying $\lambda_i(\beta) > 1$, $i = 1, 2, 3, 4$.*

The existence of fixed points different to the roots is not the only dangerous behavior that can be found in an iterative method, as they can bifurcate into periodic orbits that can be eventually attracting. This kind of problems is focused in the next section.

## 3   Dynamical Planes for n = 2

In Figs. 1 and 2, the dynamical planes associated with M4 family on $p(x)$ for different values of parameter $\beta$ are showed. These planes has been generated by slightly modifying the routines described in [4].

In them, a mesh of $800 \times 800$ points has been used, 200 has been the maximum number of iterations involved and $10^{-3}$ the tolerance used as a stopping criterium. Then, if an starting point of this mesh converges to one of the roots of the polynomial, it is painted in the color assigned to the root which has converged to. The color used is brighter when the number of iterations is lower. If it reaches the maximum number of iterations without converging to any of the roots, it is painted in black.   It can be observed in Fig. 1, corresponding

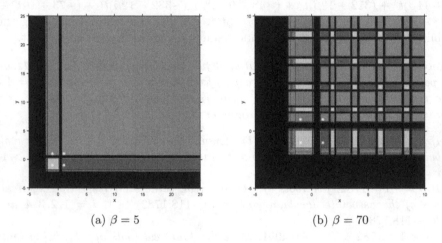

(a) $\beta = 5$                    (b) $\beta = 70$

**Fig. 1.** Dynamical plane for stable values of $\beta$ (Color figure online)

to $\beta = 5$, that the four roots of the vectorial polynomial have their respective wide basins of attraction (colored in orange, cyan, blue and purple) with several connected components for each root, separated by the Julia set. There are also four repulsive fixed points and 16 saddle points. In case $\beta = 70$, there exist four superattracting fixed points, sixteen repulsive and other sixteen saddle fixed points. It can be observed in Fig. 1b that the number of connected components in the plotted area is much higher and the immediate basin of attraction (the component holding the fixed point) is smaller in general.

On the other hand, unstable behavior appears in Fig. 2, corresponding to the values $\beta = -130$ and $\beta = -0.5$, where nine strange attracting fixed points had been found, respectively. They are located in the colored areas of no convergence to the roots, presented in yellow, brown, gray, pink, green, among others. There are also 25 repulsive and 30 saddle fixed points for $\beta = -130$ and, in case of $\beta = -0.5$, there are 9 repulsive and 18 saddle fixed points. Regarding the symbols, fixed points are marked with a white circle, meanwhile those attracting

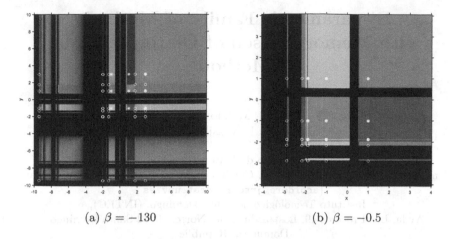

(a) $\beta = -130$                    (b) $\beta = -0.5$

**Fig. 2.** Dynamical plane for unstable values of $\beta$

are shown with a white star. Let us remark that repulsive and non-hyperbolic points are always in the Julia set and attracting ones lay in their respective basin of attraction.

By summarizing, the stability of the elements of Jacobian-free class M4 of iterative methods has been studied for a quadratic polynomial system. Unstable behavior, in terms of attracting strange fixed points, have been located for $\beta \leq -0.4991$, being the rest of elements stable, with wide sets of convergent initial estimations.

**Acknowledgement.** This research was partially supported by Ministerio de Economia y Competitividad under grants MTM2014-52016-C2-2-P, Generalitat Valenciana PROMETEO/2016/089 and FONDOCYT, Dominican Republic.

# References

1. Cordero, A., Gutiérrez, J.M., Magrenán, Á.A., Torregrosa, J.R.: Stability analysis of a parametric family of iterative methods for solving nonlinear models. Appl. Math. Comput. **285**, 26–40 (2016)
2. Cordero, A., Soleymani, F., Torregrosa, J.R.: Dynamical analysis of iterative methods for nonlinear systems or how to deal with the dimension? Appl. Math. Comput. **244**, 398–412 (2014)
3. Robinson, R.C.: An Introduction to Dynamical Systems, Continous and Discrete. Americal Mathematical Society, Providence (2012)
4. Chicharro, F.I., Cordero, A., Torregrosa, J.R.: Drawing dynamical and parameters planes of iterative families and methods. Sci. World J. **2013**, 11 (2013). Article ID 780153

# Bi-parametric Family of Methods with Memory Based of Ostrowski-Chun Method

Alicia Cordero[1], Javier G. Maimó[2], Juan R. Torregrosa[1], and Maria P. Vassileva[2(✉)]

[1] Instituto Universitario de Matemática Multidisciplinar,
Universitat Politècnica de València, Camino de Vera s/n, 46022 València, Spain
{acordero,jrtorre}@mat.upv.es
[2] Instituto Tecnológico de Santo Domingo, (INTEC),
Avda. Los Próceres 49, Los Jardines del Norte, 10602 Santo Domingo,
Dominican Republic
{javier.garcia,maria.penkova}@intec.edu.do

**Abstract.** In this work, we design a family of new iterative methods with memory, using some known schemes without memory keeping or increasing its order of convergence. As starting point we use the Ostrowski-Chun bi-parametric family of methods without memory, to design a new bi-parametric family of methods with memory, increasing the original order of convergence without adding new functional evaluations.

**Keywords:** Nonlinear equation · Multipoint iterative method ·
Divided differences · Method with memory · Stability

## 1 Introduction

The problem of solving nonlinear equations and systems of nonlinear equations appears in a wide variety of problems in different fields of Science and Technology. The most well-known iterative method used to solve this kind of problems is Newton's method. It is an efficient, optimal, second-order method, whose iterative expression has a single step without memory. For its implementation it is necessary to know the derivative of the function. For solving many problems, methods that have no derivatives in their iterative formula are required, like Steffensen's method [1].

These methods described are methods without memory. The most known variant of Newton's method with memory is the secant method, replacing the derivative in the iterative formula by its approximation using Newton's first-degree interpolation polynomial. In this process, the order of convergence decreases, becoming superlinear.

In recent years, the interest in designing new iterative methods to solve these problems has increased. To obtain methods with high order of convergence, multipoint methods have been developed using various techniques: interpolation

I. Dimov et al. (Eds.): FDM 2018, LNCS 11386, pp. 208–215, 2019.
https://doi.org/10.1007/978-3-030-11539-5_22

quadrature, Adomian decomposition and weight functions, among others [2–7]. Also, several techniques have been developed to transfer methods that can not be transferred directly to systems, [8]. For the study of the efficiency and stability of the methods, the tools of complex and real dynamics are used [9–11].

In this work, we describe a new family of iterative methods with memory to find a simple root $\xi$ of a nonlinear equation $f(x) = 0$, where $f : I \subset \mathbb{R} \to \mathbb{R}$ is a scalar function on an open interval $I$. The starting point of this work is to design a new family of iterative methods with memory for nonlinear equations using some of the known methods preserving or increasing its order of convergence. For this purpose we use the Ostrowski-Chum bi-parametric family of methods without memory, as well in its with-derivatives version as in its derivative-free form, presented in [8] and designed using Ostrowski' [12] and Chun's [13] schemes.

The initial proposal of the combination of these two methods given by the authors for an iterative method with derivatives is

$$y_k = x_k - \alpha \frac{f(x_k)}{f'(x_k)}, \tag{1}$$

$$x_{k+1} = y_k - M \frac{f(y_k)}{f'(x_k)}, \tag{2}$$

where $M = \left[ \dfrac{f(x_k)}{a_1 f(x_k) + a_2 f(y_k)} + \dfrac{b_1 f(x_k) + b_2 f(y_k)}{f(x_k)} \right]$ and real parameters $\alpha$, $a_1$, $a_2$, $b_1$ and $b_2$. The derivative-free class of iterative methods is

$$y_k = x_k - \alpha \frac{f(x_k)}{f[w_k, x_k]}, \tag{3}$$

$$x_{k+1} = y_k - M \frac{f(y_k)}{f[w_k, x_k]}, \tag{4}$$

where $w_k = x_k + f^2(x_k)$. If $\alpha = 1$, $a_2 = a_1^2(b_2 - 2)$, $b_1 = 1 - \frac{1}{a_1}$ and for all $a_1, b_2 \in \mathbb{R}$ with $a_1 \neq 0$ it is shown in [8] that the local order of convergence of families (1) and (3) is at least four.

The paper is organized as follows. In Sect. 2, we design two families of iterative methods and we analyze your order of convergence. The first family of iterative methods contains derivatives, the second is derivative-free and both use memory. In Sect. 3, we present the numerical tests that confirm the theoretical results and the excellent convergence properties of the presented methods in comparison with some optimal iterative methods, without memory, of order eight. Also, we confirmed the better stability properties with the dynamical planes of all the compared methods applied on the test functions.

## 2    Design and Convergence Analysis of the Families with Memory

Let us consider the problem of finding a zero of the nonlinear function $f : I \subset \mathbb{R} \to \mathbb{R}$, that is, a solution $\xi \in I$ of the nonlinear equation $f(x) = 0$, using the bi-parametric family with iterative scheme is the Eq. (1). Let us now modify this method introducing some accelerating parameters, one per step. We use two parameters as accelerators: each one is a factor of the functional evaluation at the previous iterate. The first product is added to the denominator in the first step (Newton's step) and the second in the denominator of the Newton's factor of the second step of (1). The modified iterative expression is

$$
\begin{aligned}
y_k &= x_k - \alpha \frac{f(x_k)}{f'(x_k) + \gamma f(x_k)} \\
x_{k+1} &= y_k - M \frac{f(y_k)}{f'(x_k) + \delta f(x_k)}
\end{aligned}
\tag{5}
$$

where $\gamma$ and $\delta$ are real parameters. It can be easily checked that this family preserves the fourth-order of convergence under conditions established in the following result.

**Theorem 1.** *Let $\xi$ be a simple zero of a sufficiently differentiable function $f : I \subset \mathbb{R} \to \mathbb{R}$ in an open interval $I$. If $x_0$ is sufficiently close to $\xi$, then the order of convergence of family (5) is at least four if: $\alpha = 1$, $b_1 = (a_1 - 1)/a_1$, $a_2 = a_1^2(b_2 - 2)$ and $\delta = 2\gamma$ and the error equation obtained have the follow form:*

$$
e_{k+1} = (\gamma + c_2)Q_k e_k^4 + O[e_k^5]
\tag{6}
$$

*where $Q_k = (2 - a(b-2)^2)\gamma^2 + (7 - 2a(b-2)^2)\gamma c_2 + (5 - a(b-2)^2)c_2^2 + c_3$, $\gamma$ a real parameter, $e_k = x_k - \xi$ and $c_k = \dfrac{f^{(k)}(\xi)}{k! f'(\xi)}$, for $k \geq 0$.*

If we modify method (3) introducing accelerating parameters, then its iterative scheme takes the form

$$
\begin{aligned}
y_k &= x_k - \alpha \frac{f(x_k)}{f[w_k, x_k] + \delta f(w_k)}, \\
x_{k+1} &= y_k - M \frac{f(y_k)}{f[y_k, w_k] + \eta f(w_k)},
\end{aligned}
\tag{7}
$$

where $w_k = x_k + \gamma f(x_k)$. Its local order of convergence is stated in the following result.

**Theorem 2.** *Let $\xi$ be a simple zero of a sufficiently differentiable function $f : I \subset \mathbb{R} \to \mathbb{R}$ in an open interval $I$. If $x_0$ is sufficiently close to $\xi$, then the order of convergence of family (7) is at least four if $\alpha = 1$, $b_1 = (a_1 - 1)/a_1$, $a_2 = a_1^2(b_2 - 1)$ and $\delta = \gamma$. The error equation is*

$$
e_{k+1} = -[1 + \gamma f'(\xi)]^2(\eta + c_2)Q_k e_k^4 + O[e_k^5],
\tag{8}
$$

*where $Q_k = (-1 + a_1(b_2 - 1)^2)(1 + \gamma f'(\xi))\eta^2 - 2(2 + \gamma f'(\xi) - a_1(b_2 - 1)^2(1 + \gamma f'(\xi)))\eta c_2 - (3 + \gamma f'(\xi)) - a_1(b_2 - 1)^2(1 + \gamma f'(\xi))c_2^2 + c_3$.*

Hereinafter we denote $a = a_1$ and $b = b_2$. If we approximate $\gamma = -\frac{1}{f'(\xi)}$ the order of convergence could reach 6 with error equation $e_{k+1} = c_2^2(\eta + c_2)(2\eta c_2 + 2c_2^2 - c_3)e_k^6 + O[e_k^7]$ and if we might take $\eta = -c_2 = -\frac{f''(\xi)}{2f'(\xi)}$, we could get an order of convergence up to seven. The error equation take the form $e_{k+1} = -c_2^2 c_3^2 e_k^7 + O[e_k^8]$. As $\xi$ is not known in practice, we consider the approximation of parameters $\gamma_k = -\frac{1}{N_3'(x_k)}$ and $\eta_k = -\frac{1}{2}\frac{N_4''(x_k)}{N_4(x_k)}$, where $N_3(t) = N_3(t; x_k, y_{k-1}, x_{k-1}, w_{k-1})$ and $N_4(t) = N_4(t; w_k, x_k, y_{k-1}, w_{k-1}, x_{k-1})$ are Newton's interpolating polynomials of third and fourth degree. The set of given points $x_k, y_{k-1}, x_{k-1}, w_{k-1}$ and $w_k, x_k, y_{k-1}, w_{k-1}, x_{k-1}$ are four and five best available nodes for generating Newton's interpolating polynomials of third and fourth degree, respectively. In this case, it can be proven that we get seven R-order of convergence and the iterative scheme for this bi-parametric family results

$$
\begin{cases}
x_0, \gamma_0, \eta_0 \text{ are given, then } w_0 = x_0 + y_0 f(x_0), \\
\gamma_k = -\frac{1}{N_3'(x_k)}, \quad w_k = x_k + \gamma_k f(x_k), \quad \eta_k = -\frac{1}{2}\frac{N_4''(x_k)}{N_4(x_k)}, \quad k = 1, 2, \ldots \\
y_k = x_k - \frac{f(x_k)}{f[w_k, x_k] + \eta_k f(x_k)}, \\
M_k = \frac{f(x_k)}{f(x_k) + a(b-2)f(y_k)} + \frac{(a-1)f(x_k) + abf(y_k)}{f(x_k)}, \\
x_{k+1} = y_k - \frac{f(y_k)}{f[y_k, w_k] + \eta_k f(w_k)}.
\end{cases}
\tag{9}
$$

## 3   Numerical Results

In this section we show the convergence behavior and the efficiency of two elements of the bi-parametric family that we have chosen from family (9): that satisfying $a = 1$ and $b = 0$, namely OMM7 and the iterative method with $a = 1$ and $b = 1$, called CMM7. In the numerical test made, variable precision arithmetics has been used, with 2000 digits of mantissa in Matlab R2014a. In our numerical experiments, for comparison purposes, we use the following derivative-free methods without memory with optimal order of convergence eight:

1. Soleymani et al. designed in [14] the scheme, denoted by SK8, whose iterative expression is

$$
y(x_k) = x_k - \frac{f(x_k)}{f[w_k, x_k]}, \quad u_k = y_k - H(x_k)\frac{f(y_k)}{f[w_k, x_k]}, \tag{10}
$$

$$
x_{k+1} = u_k - G(x_k)\frac{f(u_k)}{f[w_k, x_k]},
$$

where $H(x_k) = 1 + w_1(x_k) + w_2(x_k)$, $G(x_k) = g_1(x_k) + g_2(x_k) + g_3(x_k)$, $w_k = x_k - f(x_k)$, $w_1(x_k) = \frac{f(y_k)}{f(x_k)}$, $w_2(x_k) = \frac{f(y_k)}{f(w_k)}$, $w_3(x_k) = \frac{f(u_k)}{f(y_k)}$ and $w_4(x_k) = \frac{f(u_k)}{f(w_k)}$.

2. On the other hand, Thukral in [15] constructed the method, that we denote
   by T8, with iterative expression

$$y(x_k) = x_k - \frac{f^2(x_k)}{f(w_k) - f(x_k)}, \quad u_k = y_k - \frac{f(y_k)f[w_k, x_k]}{f[y_k, x_k]f[y_k, w_k]}, \tag{11}$$

$$x_{k+1} = u_k - G(x_k)H(x_k)\frac{f(u_k)}{f[u_k, y_k] - f[y_k, x_k] + f[u_k, x_k]},$$

where $w_k = x_k + f(x_k)$, $G(x_k) = [1 - f(u_k)/f(w_k)]^{-1}$ and $H(x_k) = [1 + 2f^3(y_k)/(f(x_k)f^3(x_k))]^{-1}$.

In Table 1, the results of the numerical tests on the different functions and initial points are given. The test functions used are: $f_1(x) = \sin(x) - x^2 + 1$, $f_2(x) = (x - 1)(x^{10} + x^3 + 1)\sin x$, $f_3(x) = e^x \sin 5x - 2$, $f_4(x) = \frac{1}{3}x^4 - x^2 - \frac{1}{3}x + 1$. The results obtained by using the different methods are compared by showing the difference in norm between the two last iterations, the number of iterations needed to satisfy the stopping criterium, and the computational order of convergence, the test function, the root and the initial estimation. The computational order of convergence $\rho$ is calculated by (see [16]):

$$\rho = \frac{\log(|f(x_k)/f(x_{k-1})|)}{\log(|f(x_{k-1}/f(x_{k-2})|)}$$

The results obtained in Table 1 are calculated with initial values of the accelerating parameters $\gamma_0 = \eta_0 = 0.01$ for the methods with memory OMM7 and CMM7.

In general, the numerical results show that iterative methods OMM7 and CMM7 have good stability and convergence properties. One of the advantages shown in these numerical tests is that the proximity of the initial approach for the OMM7 and CMM7 methods is not as important as it is in the eight-order optimal derivative-free iterative scheme, without memory.

## 3.1    Dynamical Behavior

The order of convergence is not the only criterion to take into account when evaluating an iterative method. The suitability of a method depends also on how it behaves depending on the initial estimates that are taken, for that reason it is necessary to introduce tools that allow for a more in-depth study.

One way to visualize the stability of an iterative method is to use a tool called dynamical plane. A dynamical plane is constructed from a mesh of points, each of which is taken as the initial estimation of the iterative method. The abscissa axis corresponds to the real part of the initial point and the real part to the imaginary one. If starting from the initial estimation the iterative method converges to some zero of the function, it will be assigned to a certain color, whereas if it converges to some point that is not a zero of the function, or if it simply does not converge after a maximum number of iterations established, it will be assigned to the black color.

**Table 1.** Results of numerical tests

| | OMM7 | CMM7 | SS8 | T8 |
|---|---|---|---|---|
| $f_1(x) = 0, \ \xi_1 \approx -0.63673\ldots, \ \xi_2 \approx 1.40962\ldots, \ x_0 = 0.1$ | | | | |
| $\xi$ | $\xi_1$ | $\xi_1$ | $\xi_1$ | n.c. |
| $|x_{k+1} - x_k|$ | $4.9583 \times 10^{-48}$ | $6.3172 \times 10^{-199}$ | $2.9289 \times 10^{-60}$ | |
| $\rho$ | 6.8751 | 6.9994 | 7.8847 | |
| iter | 4 | 5 | 4 | |
| $f_1(x) = 0, \ \xi_1 \approx -0.63673\ldots, \ \xi_2 \approx 1.40962\ldots, \ x_0 = 0.2$ | | | | |
| $\xi$ | $\xi_1$ | $\xi_1$ | $\xi_1$ | $\xi_2$ |
| $|x_{k+1} - x_k|$ | $4.5031 \times 10^{-221}$ | $5.0151 \times 10^{-127}$ | $1.3080 \times 10^{-156}$ | $4.9815 \times 10^{-106}$ |
| $\rho$ | 6.9986 | 7.0485 | 7.9989 | 8.0041 |
| iter | 5 | 5 | 5 | 4 |
| $f_2(x) = 0, \ \xi_1 = 0, \ \xi_2 = 1, \ x_0 = 1.6$ | | | | |
| $\xi$ | $\xi_2$ | $\xi_2$ | $\xi_2$ | $\xi_2$ |
| $|x_{k+1} - x_k|$ | $4.0486 \times 10^{-69}$ | $3.7269 \times 10^{-83}$ | $6.7496 \times 10^{-96}$ | $6.4205 \times 10^{-90}$ |
| $\rho$ | 7.0554 | 7.0061 | 7.9889 | 7.9851 |
| iter | 6 | 6 | 6 | 6 |
| $f_2(x) = 0, \ \xi_1 = 0, \ \xi_2 = 1, \ x_0 = 1.55$ | | | | |
| $\xi$ | $\xi_2$ | $\xi_2$ | $\xi_2$ | $\xi_2$ |
| $|x_{k+1} - x_k|$ | $7.7965 \times 10^{-151}$ | $2.3538 \times 10^{-116}$ | $1.9292 \times 10^{-149}$ | $1.3100 \times 10^{-132}$ |
| $\rho$ | 7.0182 | 7.0160 | 7.9990 | 7.9979 |
| iter | 6 | 6 | 6 | 6 |
| $f_3(x) = 0, \ \xi_1 \approx 1.81888\ldots, \ \xi_2 = 6.91111, \ x_0 = 2$ | | | | |
| $\xi$ | $\xi_1$ | $\xi_1$ | $\xi_1$ | n.c. |
| $|x_{k+1} - x_k|$ | $1.9531 \times 10^{-133}$ | $9.9717 \times 10^{-133}$ | $2.4914 \times 10^{-280}$ | |
| $\rho$ | 7.1269 | 7.1284 | 8.0000 | |
| iter | 4 | 4 | 4 | |
| $f_3(x) = 0, \ \xi_1 \approx 1.81888\ldots, \ \xi_2 = 6.91111, \ x_0 = 1.9$ | | | | |
| $\xi$ | $\xi_1$ | $\xi_1$ | $\xi_1$ | n.c. |
| $|x_{k+1} - x_k|$ | $3.3346 \times 10^{-174}$ | $1.9286 \times 10^{-171}$ | $2.8176 \times 10^{-50}$ | |
| $\rho$ | 7.1218 | 7.1264 | – | |
| iter | 4 | 4 | 3 | |
| $f_4(x) = 0, \ \xi_1 = 1, \ \xi_2 = 1.54682\ldots, \ x_0 = 0.5$ | | | | |
| $\xi$ | $\xi_1$ | $\xi_1$ | $\xi_1$ | $\xi_1$ |
| $|x_{k+1} - x_k|$ | $2.3909 \times 10^{-110}$ | $2.8328 \times 10^{-110}$ | $3.0226 \times 10^{-235}$ | $1.1742 \times 10^{-51}$ |
| $\rho$ | 7.0053 | 7.0052 | 8.0001 | – |
| iter | 4 | 4 | 4 | 3 |
| $f_4(x) = 0, \ \xi_1 = 1, \ \xi_2 = 1.54682\ldots, \ x_0 = 0.45$ | | | | |
| $\xi$ | $\xi_1$ | $\xi_1$ | $\xi_1$ | $\xi_1$ |
| $|x_{k+1} - x_k|$ | $6.7079 \times 10^{-88}$ | $1.2423 \times 10^{-87}$ | $6.3675 \times 10^{-199}$ | $9.4785 \times 10^{-50}$ |
| $\rho$ | 7.0096 | 7.0093 | 8.0002 | – |
| iter | 4 | 4 | 4 | 3 |

In Figs. 1 and 2, we show the dynamical planes associated to the method applied on functions $f_1(z)$ and $f_4(z)$. We use $800 \times 800$ initial estimation in the rectangle $[-1.5, 2.5] \times [-2, 2]$ of the complex plane. We use tolerance of $10^{-4}$ and one maximum number of iterations 40. The roots of each function are marked

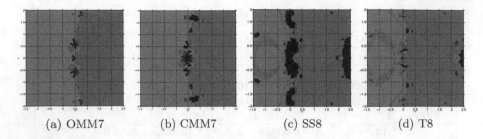

(a) OMM7            (b) CMM7            (c) SS8            (d) T8

**Fig. 1.** Dynamical planes of different schemes on $f_1(x) = \sin x - x^2 + 1$ (Color figure online)

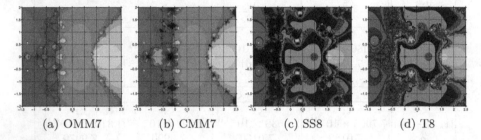

(a) OMM7            (b) CMM7            (c) SS8            (d) T8

**Fig. 2.** Dynamical planes of different schemes on $f_4(x) = \frac{1}{3}x^4 - x^2 - \frac{1}{3}x + 1$ (Color figure online)

with white stars. We have selected in advance the color that should be given to each point depending on which root it converges to. The basins of attraction of the iterative schemes drawn in these two figures show their stability and the complexity of the associated Fatou (the union of the basins of attraction) and Julia (the boundaries between basins of attraction) sets. We note that the attraction regions, in general, are larger for OMM7 and CMM7 methods, which justifies their better convergence when the initial estimate is not very close to the solution. The behavior of the two new schemes is much more stable than the others. This can be seen in Figs. 1a, b; 2a and b. The reason for this improvement in the stability of the proposed methods with memory may be the introduction of memory factors that increase the order of convergence without new functional evaluations.

## 4    Concluding Remarks

From the optimal fourth-order Ostrowski-Chun family, we have designed a biparametric family of seventh-order of convergence which is derivative-free and memory-based. Numerical and dynamic tests have been made comparing two of the methods of the new family with memory and four methods without memory with order of convergence 8, optimal. The results obtained show that OMM7 and CMM7 methods converge to the roots even when the initial estimate is far from the solution. This behavior has been justified in the dynamic planes, since their attraction basins are wider than those of the optimal eighth order schemes used to compare.

**Acknowledgments.** This research was partially supported by Ministerio de Economia y Competitividad under grants MTM2014-52016-C2-2-P, Generalitat Valenciana PROMETEO/2016/089 and FONDOCYT, Dominican Republic.

# References

1. Steffensen, J.F.: Remarks on iteration. Skand. Aktuar Tidskr. **16**, 64–72 (1933)
2. Adomian, G.: Solving Frontier Problem of Physics: The Decomposition Method. Kluwer Academic Publishers, Dordrecht (1994)
3. Babolian, E., Biazar, J., Vahidi, A.R.: Solution of a system of nonlinear equations by Adomian decomposition method. Appl. Math. Comput. **150**, 847–854 (2004)
4. Abbasbandy, S.: Extended Newton's method for a system of nonlinear equations by modified Adomian decomposition method. Appl. Math. Comput. **170**, 648–656 (2005)
5. Cordero, A., Torregrosa, J.R.: Extended Newton's method using fifth-order quadrature formulas. Appl. Math. Comput. **190**, 686–698 (2007)
6. Artidiello, S., Cordero, A., Torregrosa, J.R., Vassileva, M.P.: Two weighted eight-order classes of iterative root-finding methods. Int. J. Comput. Math. **92**(9), 1790–1805 (2015)
7. Petković, M.S., Neta, B., Petković, L.D., Džunić, J.: Multipoint Methods for Solving Nonlinear Equations. Academic Press, Cambridge (2013)
8. Cordero, A., Maimó, J.G., Torregrosa, J.R., Vassileva, M.P.: Solving nonlinear problems by Ostrowski-Chun type parametric families. Math. Chem. **53**, 430–449 (2015)
9. Blanchard, P.: The dynamics of Newton's method. In: Proceedings of Symposia Applied Mathematics, vol. 49, pp. 139–152 (1994)
10. Chicharro, F., Cordero, A., Gutiérrez, J.M., Torregrosa, J.R.: Complex dynamics of derivative-free methods for nonlinear equations. Appl. Math. Comput. **219**, 7023–7035 (2013)
11. Cordero, A., Gutiérrez, J.M., Magreñán, Á.A., Torregrosa, J.R.: Stability analysis of a parametric family of iterative methods for solving nonlinear models. Appl. Math. Comput. **285**, 26–40 (2016)
12. Ostrowski, A.M.: Solution of Equations and System of Equations. Academic Press, Cambridge (1966)
13. Chun, C.: Construction of Newton-like iterative methods for solving nonlinear equations. Numer. Math. **104**, 297–315 (2006)
14. Soleymani, F., Khattri, S.K.: Finding simple roots by seventh- and eighth-order derivative-free methods. Int. J. Math. Models Appl. Sci. **1**, 45–52 (2012)
15. Thukral, R.: Eighth-order iterative methods without derivatives for solving nonlinear equations. ISRN Appl. Math. **2011**, 12 (2011). Article ID 693787
16. Jay, I.O.: A note on Q-order of convergence. BIT Numer. Math. **41**, 422–429 (2001)

# Asymptotically Lyapunov-Stable Solutions with Boundary and Internal Layers in the Stationary Reaction-Diffusion-Advection Problems with a Small Transfer

M. A. Davydova[✉][iD], N. N. Nefedov[iD], and S. A. Zakharova[iD]

Moscow State University, Moscow, Russia
{m.davydova,sa.zakharova}@physics.msu.ru, nefedov@phys.msu.ru

**Abstract.** The stationary reaction-diffusion-advection problems, modeling the processes of the transport and chemical transformation of active and passive impurities in the surface layer of the atmosphere, to which the asymptotic methods are applicable (to the problems), are considered. We study the multidimensional asymptotically Lyapunov-stable solutions of the boundary layer type and the contrast structures by constructing the formal asymptotic approximations of an arbitrary-order accuracy based on the boundary-function method. To justify the constructed asymptotics, we use an asymptotic method of differential inequalities. The results of the study are illustrated by the example of the two-dimensional boundary value problem with a cubic nonlinearity. They can be used to create a numerical algorithm that uses asymptotic analysis to construct spatially inhomogeneous mashes when describing the internal layer of contrast structure, and also for the purposes of constructing the test examples.

**Keywords:** Stationary reaction-diffusion-advection problems ·
Solutions with boundary and internal layers ·
Transport and chemical transformation (or decay) of the antropogenic impurities in the atmosphere

## 1 Introduction

In the last years, the progress in the numerical modeling of the atmospheric composition is evident (see, for example, [1]). The chemical transport models, describing the processes of transport and chemical transformation of the gas impurities and aerosols emitted into the atmosphere by urban sources, are developed. The main problem of using models to predict the sudden changes in the air quality is the lack of the information of space-time structure of the emission distribution and the vertical stratification of the impurities over the city. In the some cases, the necessary information can be obtained by using the asymptotic

© Springer Nature Switzerland AG 2019
I. Dimov et al. (Eds.): FDM 2018, LNCS 11386, pp. 216–224, 2019.
https://doi.org/10.1007/978-3-030-11539-5_23

solutions of the singularly perturbed problems, describing the processes of transport, decay and chemical transformation of the anthropogenic impurities in the atmosphere.

The stationary distribution of the gas impurity concentration $u(x)$ in the dimensionless variables is described by the problem for the stationary equation of the reaction-diffusion-advection type:

$$\nabla(k(x)\nabla u) - \nabla(u\mathbf{A}) + f(u,x) = 0, \quad x \in D \subset R^3,$$
$$-k(x)(\nabla u, \mathbf{n}) = G(x), \quad x \in S,$$

where $\mathbf{A}(x)$ is a velocity of transfer, $f(u,x)$ is a volume density of sources (or sinks) of matter, $G(x)$ is a flow of matter at the boundary $S$, $\mathbf{n}$ is a unit external normal to the $S$ at the point $x$. The turbulent diffusion coefficient $k(x)$ is a small function.

In the approximation of an incompressible medium, under the certain physical state of the atmospheric surface layer, the concentration distribution in the atmospheric surface layer is described by the stationary problem:

$$\varepsilon^2 \Delta u - \varepsilon(\mathbf{A}(x), \nabla u) - B(u,x) = 0, \quad x = (x_1, x_2) \in D \subset R^2,$$
$$\left.\frac{\partial u}{\partial n}\right|_S = g(x), \quad x \in S, \tag{1}$$

where $\varepsilon > 0$ is a small parameter, the components of the vector $\mathbf{A}(x) := \{A_1(x), A_2(x)\}$, the functions $B(u,x)$, $g(x)$ and the boundary $S$ are assumed to be sufficiently smooth. We neglect the dependence of the coefficient $k(x)$ on the vertical variable.

In the present paper, the problem (1) is investigated by use of the methods of asymptotic analysis [2] and [3–6], developing the classical methods of the asymptotic construction [7] for the case of the stationary multidimensional reaction-diffusion-advection problems.

## 2    Solutions with Boundary Layers

We assume the validity of the following condition.

**Condition 1.1.** The degenerate equation $B(u,x) = 0$ has isolated solution $u = \varphi(x)$, with $B_u(\varphi(x), x) > 0$ for $x \in D \cup S$.

We define a sufficiently small neighborhood $\partial S := \{x \in D : dist(x, S) < \delta\}$, $\delta > 0$, of the curve $S$ and the local coordinates in this neighborhood: $x \in \partial S \mapsto (y,r) \in S \times [0; \delta]$, where $y = (y_1, y_2) \in S$, $r = dist(x, S) = dist(x, y)$.

Let the curve $S$ be described by the equations: $y_1 = \kappa(\eta)cos\eta$, $y_2 = \kappa(\eta)sin\eta$, $0 \le \eta < 2\pi$. Then the one-to-one correspondence between the coordinates $(x_1, x_2)$ and $(r, \eta)$ is given by the expressions:

$$x_1 = \kappa(\eta)cos\eta - r(\kappa'(\eta)sin\eta + \kappa(\eta)cos\eta)\psi(\kappa(\eta)),$$
$$x_2 = \kappa(\eta)sin\eta + r(\kappa'(\eta)cos\eta - \kappa(\eta)sin\eta)\psi(\kappa(\eta)), \tag{2}$$

where $\psi(\kappa(\eta)) := \left((\kappa'(\eta))^2 + \kappa^2(\eta)\right)^{-1/2}$.

We search for the asymptotics of a solution of boundary layer type in the form:

$$u(x,\varepsilon) = \overline{u}(x,\varepsilon) + \Pi u(\rho,\eta,\varepsilon), \tag{3}$$

where $\overline{u}(x,\varepsilon) = \varphi(x) + \varepsilon \overline{u}_1(x) + \ldots$ is a regular series, $\Pi u(\rho,\eta,\varepsilon) = \varepsilon \Pi_1 u(\rho,\eta) + \varepsilon^2 \Pi_2 u(\rho,\eta) + \ldots$ is the boundary series, describing the boundary layer in a neighborhood of the boundary $S$ and $\rho = r/\varepsilon$. Using the boundary-function method [4,7], we obtain the problems for the terms of the series (3).

For $n \geq 1$, we have the linear problems:

$$\frac{\partial^2 \Pi_n u}{\partial \rho^2} - N(\eta)\frac{\partial \Pi_n u}{\partial \rho} - B_u(\varphi(0,\eta),0,\eta)\Pi_n u = \overline{H}_n(\rho,\eta),$$

$$\frac{\partial \Pi_n u}{\partial \rho}\bigg|_{\rho=0} = q_n(\eta), \quad \Pi_n u(+\infty,\eta) = 0,$$

whose solutions can be represented in the explicit form by analogy with [8]. Here the functions $\overline{H}_n(\rho,\eta)$, $q_n(\eta)$ are known, $N(\eta) := -\psi(\kappa(\eta))\left[(\kappa' sin\eta + \kappa cos\eta)A_1(0,\eta) - (\kappa' cos\eta - \kappa sin\eta)A_2(0,\eta)\right]$. For convenience, the notation for the functions in variables $(r,\eta)$ is retained as before.

Using the explicit form of the functions $\Pi_n u(\rho,\eta)$, we obtain the following estimates by analogy with [8]: $|\Pi_n u| \leq C_n exp[-\chi_n \rho]$, $\chi_n > 0$, $C_n > 0$, $n > 0$.

## 3    Existence and Stability of Boundary Layer Solutions

We denote by $U_n(x,\varepsilon) := \varphi(x) + \varepsilon \overline{u}_1(x) + \varepsilon \Pi_1 u(\rho,\eta) + \ldots + \varepsilon^n \overline{u}_n(x) + \varepsilon^n \Pi_n u(\rho,\eta) + \varepsilon^{n+1}\Pi_{n+1} u(\rho,\eta)$ the partial sums of the $n$-th order of asymptotic series (3).

**Theorem 1.** *Suppose the assumption 1.1 to be valid. Then, for sufficiently small $\varepsilon > 0$, there exists a solution $u(x,\varepsilon)$ of problem (1) with the asymptotics (3), which satisfies the uniform estimate: $|u(x,\varepsilon) - U_n(x,\varepsilon)| \leq C\varepsilon^{n+1}$, $x \in D \cup S$ the constant $C$ is independent of $\varepsilon$.*

The proof of the existence theorem is based on the technique of lower and upper solutions [6]. By analogy with [8], it is easy to show, that the functions $\alpha_n^{(\mp)}(x,\varepsilon) = U_n(x,\varepsilon) \mp \gamma\varepsilon^n$, where $\gamma$ is a sufficiently small independent of $\varepsilon$ number, are lower and upper solutions of problem (1).

The solution $u(x,\varepsilon)$ with the asymptotics (3) is a stationary solution of the following parabolic problem:

$$L_\varepsilon v := \varepsilon^2(\Delta v - v_t) - \varepsilon(\mathbf{A}(x),\nabla v) - B(v,x) = 0, \quad x \in D, \quad t > 0,$$
$$\frac{\partial v}{\partial n}\bigg|_S = g(x), \quad x \in S, \quad t \geq 0, \quad v(x,0) = u^0(x), \quad x \in D \cup S, \tag{4}$$

for $u^0(x) = u(x,\varepsilon)$. The presence of a small parameter at the derivative $v_t$ is due to the choice of the time scale. Let the following condition be satisfied: $(\partial u^0/\partial n) = g(x)$, $x \in S$.

By analogy with the paper [8], we can define the lower and upper solutions, respectively, by the expressions: $\overline{\alpha}_0^{(\mp)}(x,t,\varepsilon) = u(x,\varepsilon) - \left(u(x,\varepsilon) - \alpha_0^{(\mp)}(x,\varepsilon)\right) e^{-\lambda(\varepsilon)t}$, where $\lambda(\varepsilon) > 0$. Taking into account, that $\left(u - \alpha_0^{(\mp)}\right) = O(\gamma)$, we have: $L_\varepsilon \overline{\alpha}_0^{(-)} > 0$, $L_\varepsilon \overline{\alpha}_0^{(+)} < 0$. Occurs

**Theorem 2.** *Under Condition 1.1 the stationary solution $u(x,\varepsilon)$ of problem (4) with the asymptotics (3) is asymptotically Lyapunov-stable with the domain of attraction at least $\left[\alpha_0^{(-)}(x,\varepsilon), \alpha_0^{(+)}(x,\varepsilon)\right]$.*

## 4  Contrast Structures

We assume the validity of the following condition.

**Condition 2.1.** The degenerate equation $B(u,x) = 0$ has isolated solutions $u = \varphi_i(x)$, $i = \overline{1,3}$ satisfying conditions $\varphi_1(x) < \varphi_2(x) < \varphi_3(x)$, $B_u(\varphi_i(x),x) > 0$, $i = 1,3$, $B_u(\varphi_2(x),x) < 0$ for $x \in D \cup S$.

In accordance with the algorithm of the papers [2] and [5], we introduce the set $\{\overline{C}\}$ of sufficiently smooth simple closed curves with the local coordinates: $x \in \delta\overline{C} \mapsto (y,r) \in \overline{C} \times [-\delta, \delta]$, $y \in \overline{C}$, $r = \pm dist(x,y)$ in the small neighborhood $\delta\overline{C}$ of each curve. Further, we select the following local coordinates: $(r, \theta) \in [-\delta, \delta] \times [0, 2\pi)$, where $\theta$ is a polar angle, corresponding to the point $y(\theta) \in \overline{C}$. For each curve from the set $\{\overline{C}\}$, we introduce the associated equation:

$$\frac{\partial^2 \tilde{u}}{\partial \xi^2} + V(\theta)\frac{\partial \tilde{u}}{\partial \xi} = B(\tilde{u},0,\theta), \quad -\infty < \xi < +\infty, \tag{5}$$

where $\theta$ is treated as parameter, $V(\theta) = -\sum\limits_{k=1}^{2} A_k(0,\theta)l^k(0,\theta,\overline{C})$, $l^k(0,\theta,\overline{C})$ are known functions, which depend on the curve $\overline{C}$; $A_k(0,\theta)$ are expressions for the functions $A_k(x)$ in the variables $(r,\theta)$ for $r = 0$.

Note, that the rest points $(\varphi_1(0,\theta),0)$ and $(\varphi_3(0,\theta),0)$ of the Eq. (5) are the rest points of the saddle type.

It is known [9], that for any value $\theta \in [0, 2\pi)$ there exists a unique value $V_0(\theta)$ such that the Eq. (5) with additional conditions

$$\tilde{u}(-\infty,\theta) = \varphi_1(0,\theta), \quad \tilde{u}(+\infty,\theta) = \varphi_3(0,\theta) \tag{6}$$

has a unique solution $\tilde{u}(\xi,\theta)$, satisfying the estimate $|\tilde{u}(\xi,\theta) - \varphi_i(0,\theta)| \leq$ $\leq Ce^{-K_0|\xi|}$, $i = 1,3$ where $C$ and $K_0$ are positive constants, and

$$V_0(\theta) = \left(\int\limits_{-\infty}^{-\infty}(u_\xi'(\xi,\theta))^2 d\xi\right)^{-1} \int\limits_{\varphi_1(0,\theta)}^{\varphi_3(0,\theta)} B(u,0,\theta)du.$$

Hence, if there exists a curve $C_0 \in \{\overline{C}\}$ for which

$$V_0(\theta) = -\sum_{k=1}^{2} A_k(0,\theta)l^k(0,\theta,C_0),$$ then on the phase plane of the Eq. (5) there

is a separatrix, connecting the saddles $(\varphi_1(0,\theta),0)$ and $(\varphi_3(0,\theta),0)$. This means the fulfilling one of the basic conditions of existence of contrast structures [5] in the problem (1).

Suppose that a curve $C_0$ exists and is described by equations: $y_1 = \sigma_0(\theta)cos\theta$, $y_2 = \sigma_0(\theta)sin\theta$, $0 \le \theta < 2\pi$, $(y_1,y_2) \in C_0$. Then the one-to-one correspondence between the coordinates $(x_1,x_2)$ and $(r,\theta)$ is given by the expressions analogous to the expressions (2) and

$$l^1(0,\theta,C_0) = \frac{\sigma_0' sin\theta + \sigma_0 cos\theta}{\sqrt{(\sigma_0')^2 + \sigma_0^2}}, \quad l^2(0,\theta,C_0) = \frac{\sigma_0 sin\theta - \sigma_0' cos\theta}{\sqrt{(\sigma_0')^2 + \sigma_0^2}}. \quad (7)$$

To ensure the existence of a curve $C_0$, we accept the following condition.

**Condition 2.2.** There exists a $2\pi$−periodic solution $\sigma = \sigma_0(\theta)$ of the equation:

$$V_0(\theta)\sqrt{(\sigma')^2 + \sigma^2} + A_1(0,\theta)(\sigma' sin\theta + \sigma cos\theta) + A_2(0,\theta)(\sigma sin\theta - \sigma' cos\theta) = 0. \quad (8)$$

We obtain the asymptotic approximation of a contrast structure by the $C^1$-matching of the boundary-layer asymptotics

$$\begin{aligned} u^-(x,\varepsilon) &= \overline{u}^-(x,\varepsilon) + Qu^-(\xi,\lambda^*,\theta,\varepsilon), \\ u^+(x,\varepsilon) &= \overline{u}^+(x,\varepsilon) + \Pi u(\rho,\eta,\varepsilon) + Qu^+(\xi,\lambda^*,\theta,\varepsilon), \end{aligned} \quad (9)$$

on the transition curve $C$. Here $\overline{u}^-(x,\varepsilon) = \varphi_1(x) + \varepsilon\overline{u}_1^-(x) + \dots$, $\overline{u}^+(x,\varepsilon) = \varphi_3(x) + \varepsilon\overline{u}_1^+(x) + \dots$ are regular series, $\Pi u(\rho,\eta,\varepsilon) = \varepsilon\Pi_1 u(\rho,\eta) + \dots$ is a boundary series, describing the boundary layer in a neighborhood of the boundary $S$, and $Qu^\pm(\xi,\lambda^*,\theta,\varepsilon) = Q_0 u^\pm(\xi,\lambda^*,\theta) + \varepsilon Q_1 u^\pm(\xi,\lambda^*,\theta) + \dots$ are series, describing the boundary layers localized in a neighborhood of the transition curve, whose position is defined by the condition

$$u(x,\varepsilon) = \varphi_2(x), \quad x \in C. \quad (10)$$

Next, in accordance with the algorithms of the method of boundary functions [7], developed in [4] and expounded in Sect. 2, we construct boundary-layer asymptotics (9).

We seek the equation of the curve $C$ in the form of a series in powers of $\varepsilon$; moreover, for the leading term we take the curve $C_0$; thus the equation of the curve $C$ in the local coordinate system, defined in the neighborhood of the curve $C_0$, has the form

$$r = \lambda^*(\theta,\varepsilon) = \varepsilon(\lambda_1(\theta) + \varepsilon\lambda_2(\theta) + \dots). \quad (11)$$

The variable $\xi$ in (9) is defined as follows: $\xi = (r - \lambda^*(\theta))\varepsilon^{-1}$. The equations for the coefficients of the expansion (11) are found from the condition of $C^1$-matching of the asymptotics (9) on the curve $C$ [5] and have the form:

$$\lambda_n(\theta)\overline{H}(\theta) - \Phi_n(\theta) = 0, \quad \theta \in [0,2\pi),$$

where $\Phi_n(\theta)$ are known functions. The function $\overline{H}(\theta)$ is defined by equality [10]:

$$
\begin{aligned}
\overline{H}(\theta) = &-\left(\tilde{u}'_\xi(0,\theta)\right)^{-1} \int_{-\infty}^{+\infty} \left(\sum_{k=1}^{2} \frac{\partial A_k}{\partial r}(0,\theta)l^k(0,\theta,C_0)\tilde{u}'_\xi(\xi,\theta)\right. \\
&\left. + \frac{\partial B}{\partial r}(\tilde{u},0,\theta)\right) \tilde{u}'_\xi(\xi,\theta)e^{-\xi\sum_{k=1}^{2} A_k(0,\theta)l^k(0,\theta,C_0)} \, d\xi := -(\tilde{u}'_\xi(0,\theta))^{-1}I(\theta),
\end{aligned}
\tag{12}
$$

where the functions $l^k(0,\theta,C_0)$ are defined by (7). Suppose that the following condition fulfilled:

**Condition 2.3.** The inequality holds: $\overline{H}(\theta) > 0$, $\theta \in [0,2\pi)$.

## 5  Existence and Stability of Contrast Structures

The proof of this assertion is carried out using the asymptotic method of differential inequalities [6]. We shall construct the upper $\beta_{n+2}(x,\varepsilon)$ and lower $\alpha_{n+2}(x,\varepsilon)$ solutions of problem (1) by modifying the terms of the asymptotic series (9), similarly to the way in which it was done in [5]. For example, for the upper solution, one can take the function, which on the different sides of the curve with the equation $r = \lambda_\beta(\theta,\varepsilon) = \varepsilon\lambda_1(\theta) + \varepsilon^2\lambda_2(\theta) + \ldots + \varepsilon^{n+2}(\lambda_{n\,|\,2} - \overline{\delta}(\theta))$ has the form:

$$
\begin{aligned}
\beta_{n+2}^{(-)} &= \varphi_1(x) + Q_0 u_\beta^-(\xi_\beta,\theta) + \ldots + \varepsilon^{n+2}\left(\overline{u}_{n+2}^-(x) + \gamma + Q_{n+2}u_\beta^-(\xi_\beta,\theta)\right) \\
&\quad + \varepsilon^{n+3}Q^* u_\beta^-(\xi_\beta,\theta), \\
\beta_{n+2}^{(+)} &= \varphi_3(x) + Q_0 u_\beta^+(\xi_\beta,\theta) + \ldots + \varepsilon^{n+2}\left(\overline{u}_{n+2}^+(x) + \gamma + Q_{n+2}u_\beta^+(\xi_\beta,\theta)\right) \\
&\quad + \varepsilon^{n+3}\left(\Pi_{n+3}u(\rho,\eta) + Q^* u_\beta^+(\xi_\beta,\theta)\right),
\end{aligned}
$$

where $\gamma > 0$ is a constant ensuring the validity of the necessary differential inequality, and $\overline{\delta}(\theta) > 0$ is a function ensuring the validity of the inequality for the jump of derivatives on the curve of transition layer. The functions $Q_n u_\beta^\pm$ are obtained by the scheme of [5] by modifying the expressions for $Q_n u^\pm$ in which $\xi$ is replaced by $\xi_\beta = (r - \lambda_\beta(\theta,\varepsilon))\varepsilon^{-1}$. A lower solution has a similar structure. All necessary inequalities are verified by the immediate calculations. The following theorem holds.

**Theorem 3.** *If Conditions 2.1–2.3 hold, then, for sufficiently small $\varepsilon > 0$, there is a solution $u(x,\varepsilon)$ of problem (1) with the asymptotics (9), and the following estimate holds: $|u(x,\varepsilon) - U_n(x,\varepsilon)| \le C\varepsilon^{n+1}$, $x \in D \cup S$, where $U_n(x,\varepsilon)$ is a partial sum of the asymptotic series (9), the constant $C$ is independent of $\varepsilon$.*

The proof of the asymptotic Lyapunov-stability of solution $u(x,\varepsilon)$ with the asymptotics (9), as a stationary solution of the problem (4), is based on the use of the following functions as the lower and upper solutions of problem (4): $\overline{\alpha}_{n+2}(x,t,\varepsilon) = u(x,\varepsilon) - (u(x,\varepsilon) - \alpha_{n+2}(x,\varepsilon))\,e^{-\lambda(\varepsilon)t}$, $\overline{\beta}_{n+2}(x,t,\varepsilon) = u(x,\varepsilon) - (u(x,\varepsilon) - \beta_{n+2}(x,\varepsilon))\,e^{-\lambda(\varepsilon)t}$, where $\lambda(\varepsilon)$ is sufficiently small. It can be shown that $L_\varepsilon\overline{\alpha}_{n+2} > 0$, $L_\varepsilon\overline{\beta}_{n+2} < 0$ for $n \ge 0$, since $(\overline{\alpha}_{n+2} - u) = O(\varepsilon^{n+1})$, $(\overline{\beta}_{n+2} - u) = O(\varepsilon^{n+1})$. Thus, the above constructed solution is stable with the influence domain being at least $[\alpha_2(x,\varepsilon), \beta_2(x,\varepsilon)]$.

**Theorem 4.** *Under Condition 2.1–2.3 the stationary solution $u(x, \varepsilon)$ of problem (4) with the asymptotics (9) is asymptotically Lyapunov-stable with the domain of attraction at least $[\alpha_2(x, \varepsilon), \beta_2(x, \varepsilon)]$.*

## 6  Example

We consider the most important particular case of the problem (1) with a cubic nonlinearity:

$$B(u, x) = (u - \varphi_1(x))(u - \varphi_2(x))(u - \varphi_3(x)). \tag{13}$$

It is known [11], that for any $\theta \in [0, 2\pi)$, two values exist:

$$V_0^{(1,2)}(\theta) = \pm 2^{-1/2}(\varphi_1(0, \theta) + \varphi_3(0, \theta) - 2\varphi_2(0, \theta)) \tag{14}$$

for which the Eq. (5) has solutions, satisfying the conditions (6). Therefore, two contrast structures can exist in the problem (1), (13): with the internal transition from the root $u = \varphi_1(x)$ to the root $u = \varphi_3(x)$ and with the reverse transition. In the each case, the position of the transition curve in the zeroth approximation determines by Eq. (8), (14).

Using the differential equations

$$v\frac{dv}{d\tilde{u}} + V_0^{(k)}(\theta)v - (\tilde{u} - \varphi_1(0, \theta))(\tilde{u} - \varphi_2(0, \theta))(\tilde{u} - \varphi_3(0, \theta)), \quad k = 1, 2 \tag{15}$$

describing the trajectories on the phase plane of the Eq. (5) for $V(\theta) = V_0^{(k)}(\theta)$, by the direct substitution we check out, that the following solutions of the Eq. (15)

$$v^{(1,2)}(\theta) = \pm 2^{-1/2}(\tilde{u} - \varphi_1(0, \theta))(\tilde{u} - \varphi_3(0, \theta)), \tag{16}$$

exist and satisfy the conditions: $v(\varphi_1(0, \theta)) = v(\varphi_3(0, \theta)) = 0$. The functions (16) describe the connecting separatrices, lying in the different half-planes of the phase plane of Eq. (5) for $V(\theta) = V_0^{(k)}(\theta)$. Since $v(\tilde{u}, \theta) = \tilde{u}'_\xi$, then using the corresponding Eq. (16), it is possible to describe the internal layer of the contrast structure in the explicit form.

For example, if the requirement 2.2, where $V(\theta) = V_0^{(2)}(\theta)$, and the condition $I(\theta) < 0$, $\theta \in [0, 2\pi)$, where $\tilde{u}'_\xi = v^{(2)}(\theta)$ (see Eq. (12)), fulfilled, then a contrast structure with the asymptotics (9) exists in the problem (1), (13). One such case was considered in [3].

In particular, if $\varphi_1(x) = a$, $\varphi_3(x) = b$ where $a$ and $b$ are constants, then the sufficient conditions of the existence of contrast structure with the asymptotics (9) can be formulated in the terms of the function

$$G(r, \theta) := \sum_{k=1}^{2} A_k(r, \theta)l^k(r, \theta, C_0) - 2^{-1/2}(a + b - 2\varphi_2(r, \theta)). \text{ Really, if we go from}$$

the integration variable $\xi$ to the integration variable $\tilde{u}$ in the integral (12) and, further, perform another substitution $\tilde{u} = u + 2^{-1}(a + b)$, then the expression for the function $I(\theta)$ is transformed to the form:

$$I(\theta) = -\left( \frac{1}{\sqrt{2}} \sum_{k=1}^{2} \frac{\partial A_k}{\partial r}(0,\theta) l^k(0,\theta,C_0) + \frac{\partial \varphi_2}{\partial r}(0,\theta) \right) \overline{I}(\theta),$$

where $\overline{I}(\theta) < 0$ for $\theta \in [0, 2\pi)$. If we take into account, that $(\partial l^k(0,\theta)/\partial r) = 0$, then we have the corresponding conditions: $G(0,\theta) = 0$, $G_r(0,\theta) < 0$ on the curve $C_0$.

The results of the study of problem (1) can be used to create a numerical algorithm that uses asymptotic analysis to construct spatially inhomogeneous mashes when describing the internal layer of contrast structure [12], and also for the purposes of constructing the test examples.

**Acknowledgements.** This work was supported by the Russian Science Foundation (project no. 18-11-00042).

# References

1. Elansky, N.F., Ponomarev, N.A., Verevkin, Y.M.: Air quality and pollutant emissions in the Moscow megacity in 2005–2014. Atmos. Environ. **175**, 54–64 (2018)
2. Nefedov, N.N., Sakamoto, K.: Multi-dimensional stationary internal layers for spatially inhomogeneous reaction-diffusion equations with balanced nonlinearity. Hiroshima Math. J. **33**(3), 391–432 (2003)
3. Davydova, M.A., Nefedov, N.N.: Existance and stability of contrast structures in multidimensional singularly perturbed reaction-diffusion-advection problems. In: Dimov, I., Faragó, I., Vulkov, L. (eds.) Numerical Analysis and Its Applications, vol. 10187, pp. 277–285. Springer, Cham (2017). https://doi.org/10.1007/978-3-319-57099-0_29
4. Davydova, M.A.: Existance and stability of solutions with boundary layers in multidimensional singularly perturbed reaction-diffusion-advection problems. Math. Notes **98**(6), 909–919 (2015)
5. Nefedov, N.N., Davydova, M.A.: Contrast structures in singularly perturbed quasilinear reaction-diffusion-advection equations. Differ. Equ. **49**(6), 688–706 (2013)
6. Nefedov, N.: Comparison principle for reaction-diffusion-advection problems with boundary and internal layers. In: Dimov, I., Faragó, I., Vulkov, L. (eds.) NAA 2012. LNCS, vol. 8236, pp. 62–72. Springer, Heidelberg (2013). https://doi.org/10.1007/978-3-642-41515-9_6
7. Vasil'yeva, A.B., Butuzov, V.F.: Asimptoticheskiye metody v teorii singulyarnykh vozmushcheniy. Vyssh. shkola, Moscow (1990)
8. Davydova, M.A., Zakharova, S.A., Levashova, N.T.: On the model problem for the reaction-diffusion-advection equation. Comput. Math. Math. Phys. **57**(9), 1528–1539 (2017)
9. Fife, P.C., Hsiao, L.: The generation and propagation of internal layers. Nonlinear Anal. Theory Methods Appl. **12**(1), 19–41 (1998)
10. Nefedov, N.N., Davydova, M.A.: Contrast structures in multidimensional singularly perturbed reaction-diffusion-advection equations. Differ. Equ. **48**(5), 745–755 (2012)

11. Romanovskiy, Y.U.M., Stepanova, N.V., Chernavskiy, D.S.: Matematicheskaya biofizika. Nauka, Moscow (1984)
12. Volkov, V., Nefedov, N.: Asymptotic-numerical investigation of generation and motion of fronts in phase transition models. In: Dimov, I., Faragó, I., Vulkov, L. (eds.) NAA 2012. LNCS, vol. 8236, pp. 524–531. Springer, Heidelberg (2013). https://doi.org/10.1007/978-3-642-41515-9_60

# Numerical Modeling of the Charge Transfer Along 1D Molecular Chain "Donor-Bridge-Acceptor" at T = 300 K

Nadezhda Fialko[(✉)] [iD] and Victor D. Lakhno[iD]

Institute of Mathematical Problems of Biology - The Branch of KIAM RAS,
142290 Pushchino, Russia
fialka@impb.ru

**Abstract.** We consider charge transfer along homogeneous chain of sites (such as DNA fragment) with the ends which imitate a donor and an acceptor. We performed direct numerical experiments based on the semi-classical Holstein model. To take into account the temperature, Langevin thermostat is used. Recently it has been shown that in homogeneous chains the charge distribution in thermodynamic equilibrium state (TDE) depends on the thermal energy of the lattice subsystem. Here, we have calculated dynamics of the system from the initial state "the charge is localized at the donor" over time intervals to the attainment of the TDE. The time intervals dependence on the length of the chain at fixed temperature is estimated. Part of parameter values are chosen as for DNA fragments of the GA...AGGG type. The results of the calculations are compared with the data of biophysical experiments on the hole transfer in DNA sequences.

**Keywords:** Holstein model · Langevin equation ·
Time to the attainment of the thermodynamic equilibrium state ·
DNA fragment

## 1 Introduction

At present, the attention of researchers is attracted to the possible mechanisms of charge transfer in quasy1D biomacromolecules, such as DNA, in connection with the potential using of this nano-objects in nanobioelectronics [1,2]. Studies of charge transfer in DNA also are of importance for biophysics; e.g., the propagation of excited charges along DNA is part of replication and reparation of DNA; movement of radicals over DNA molecules plays an important role in mutagenesis and cancerogenesis processes [3,4].

Biophysical experiments on the hole transfer from guanine G (donor) to guanine triplet GGG (acceptor), separated by adenine-thymine (A-T) bridges of various lengths [5–7], demonstrate that the rate of charge transfer between donor and acceptor decreases with increasing separation only if the guanines are separated by no more than three base pairs; if more bridging base pairs are present, the transfer rates exhibit only a weak distance dependence.

In this work we use a direct numerical experiment on the charge transfer from donor along bridge, consisting of homogeneous sites, to the acceptor. The model is based on the semi-classical Holstein Hamiltonian. The Holstein polaron model is simple and actual for explaining charge transfer in DNA [8–10]. To take into account the temperature, Langevin thermostat is used. Recently it has been shown that the charge distribution in the thermodynamic equilibrium state depends not only on the temperature, but also on the length of the chain, for homogeneous chains [11] and for the chain with a defect site in the center, which plays the role of a trap for the charge [12]. Here we have calculated dynamics of system from the initial state "the charge is localized at the donor" over time intervals to the attainment of the thermodynamic equilibrium. The time intervals dependence on the chain length $N$ is estimated at fixed temperature $T = 300\,\mathrm{K}$.

## 2    Mathematical Model

We have considered the semiclassical Holstein model: quantum particle (an electron or a hole) moves along a chain of classical sites [13]. In the case of DNA a site – oscillator – corresponds to complementary base pair, its frequency $\omega$ is related to hydrogen bonds in the base pair. Choosing the wave function $\Psi$ in the form $\Psi = \sum_{n=1}^{N} b_n |n\rangle$, where $b_n$ is the amplitude of the probability of the charge occurrence at the $n$-th site ($n = 1, \ldots, N$, $N$ is the chain length), we write the averaged Hamiltonian:

$$\langle \Psi | \hat{H} | \Psi \rangle = \sum_{m,n} \nu_{nm} b_m b_n^* + \frac{1}{2} \sum_n M \dot{\tilde{u}}_n^2 + \frac{1}{2} \sum_n K \tilde{u}_n^2 + \sum_n \alpha' \tilde{u}_n b_n b_n^*. \quad (1)$$

Here $\nu_{nn}$ is the electron energy at the $n$-th site, $\nu_{mn}$ ($m \neq n$) are matrix elements of the transition between the $m$-th and the $n$-th sites (depending on overlapping integrals). We use the nearest neighbor approximation, i.e. $\nu_{mn} = 0$ if $m \neq n \pm 1$; the intrasite fluctuations $\tilde{u}_n$ are assumed to be small and can be considered to be harmonic; the probability $|b_n|^2$ of charge occurrence at the sites depends linearly on the sites displacements $\tilde{u}_n$, $\alpha'$ is the constant of coupling between quantum and classical subsystems, $M$ is the effective mass of the site, $K$ is the elastic constant.

Equations of motion obtained from Hamiltonian (1) and written in the dimensionless form are

$$i \frac{db_n}{dt} = \eta_{n,n-1} b_{n-1} + \eta_{n,n+1} b_{n+1} + \eta_{n,n} b_n + \chi u_n b_n, \quad (2)$$

$$\frac{d^2 u_n}{dt^2} = -\omega^2 u_n - \chi |b_n|^2 - \gamma \dot{u}_n + \xi Z_n(t). \quad (3)$$

Dimensionless values are related to dimensional parameters as follows. The matrix elements $\eta_{nk} = \nu_{nk} \tau / \hbar$ ($\tau$ is characteristic time, $\tilde{t} = \tau t$). Site frequency

$\omega = \tau\sqrt{K/M}$, coupling constant $\chi = \alpha'\sqrt{\tau^3/\hbar M}$. In subsystem (3) we added the term with friction ($\gamma$ is a friction coefficient) and random force $Z_n(t)$ with the properties $\langle Z_n(t) \rangle = 0$, $\langle Z_n(t)Z_m(t+s) \rangle = \delta_{mn}\delta(s)$, $\xi^2 = (2k_B T^*\tau/\hbar)\gamma\mathrm{T}$ ($T^*$ is characteristic temperature). This way of imitating the environmental temperature with the use of Langevin equations is well known [14,15].

To model the charge transfer dynamics, we computed a set of samples. Each sample is a trajectory of system (2, 3) with its own initial data and pseudorandom time series $\{Z_n(t)\}$, i.e. ODE system with random right-hand side. It is integrated by 2o2s1g-method [16]. Having obtained a number of samples we find averaged "by ensemble" functions of time, e.g., probability of charge localization on the $n$-th site $\langle P_n \rangle = \langle |b_n|^2 \rangle$ and delocalization parameter $\langle R \rangle = \langle 1/\sum_n |b_n|^4 \rangle$. If charge is localized at one $k$-th site, i.e. charge probability $|b_k|^2 \sim 1$, then $R \sim 1$. For homogeneous $N$-site chain, if charge is in delocalized state then $R \approx N/2$ [11].

## 3  On the Model Parameter Values

The characteristic time was chosen to be $\tau = 10^{-14}$ sec, the characteristic temperature $T^* = 1\,\mathrm{K}$.

For the quantum subsystem (2) the parameter values corresponding to nucleotide pairs are chosen as follows [17–20]. The matrix elements $\eta_{n,n\pm1}$ of the transition between sites are: $\eta_{GA} = 1.352$ ($\nu_{GA} = 0.089\,\mathrm{eV}$), $\eta_{AA} = 0.456$ ($\nu_{AA} = 0.030\,\mathrm{eV}$), $\eta_{AG} = 0.744$ ($\nu_{AG} = 0.049\,\mathrm{eV}$), $\eta_{GG} = 1.276$ ($\nu_{GG} = 0.084\,\mathrm{eV}$). The electron energy on the site $\eta_{n,n}$: $\eta_G = 0$ ($\nu_G = 1.24\,\mathrm{eV}$), $\eta_A = 6.84$ ($\nu_A = 1.69\,\mathrm{eV}$). For DNA the effective mass of a site is $M = 10^{-21}$ g; the rigidity $K = 0.062\,\mathrm{eV/\mathring{A}^2}$ and coupling constant $\alpha' = 0.13\,\mathrm{eV/\mathring{A}}$, i.e. $\omega = 0.01$ and $\chi = 0.02$. When performing calculations, for the classical subsystem (3) we chose the adapted values of the parameters that speed up the system's movement to the thermodynamic equilibrium state: $\omega = 0.5$ and $\chi = 1$. We suggest that not only the average values in the TDE are the same for two systems with the same ratio of parameters $\chi/\omega$ [11], but the processes of reaching the TDE state from the same initial state (when the charge is localized at the first donor site) will be qualitatively similar.

The initial classical displacements $u_n$ and velocities $\dot{u}_n$ of sites in the samples were chosen random from the TDE distribution of 1D oscillator at a given temperature; the charge is localized at the 1st site G: $b_1 = 1$, $b_n = 0$ ($n = 2, \ldots, N$).

Thermostat temperature $T = 300\,\mathrm{K}$. The length of the chain $N$ varies from 5 (i.e. GAGGG – the bridge of one A) to 100 (i.e. GA...AGGG with 96 adenines).

## 4  On the Computational Method

For each $N$ we have calculated the average probabilities $\langle P_n \rangle$ (for 100 samples) on time interval of reaching the TDE state. The time during which the system attains the TDE is estimated by the delocalization parameter $\langle R \rangle$ and by the probability values at the G sites. Figure 1 shows the $\langle R(t) \rangle$ curve and the average

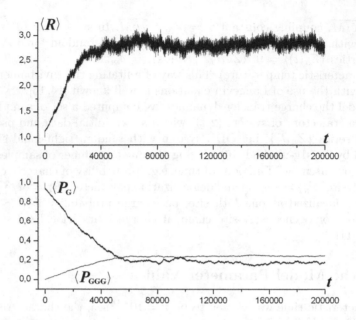

**Fig. 1.** Top: $\langle R(t) \rangle$. Bottom: average probability on donor (black curve) and on acceptor site (grey curve). The friction coefficient $\gamma = 4\omega$.

probabilities on donor $\langle P_G(t) \rangle$ and on acceptor $\langle P_{GGG}(t) \rangle = (1/3) \times$ {sum of probabilities on the sites of acceptor} for GAAGGG chain ($N = 6$).

Figure 1 demonstrates that at time $t \sim 60000$ the graphs become approximately constant, we take $t_{\text{TDE}} = 60000$. In the TDE state the probability on the donor $\langle P_G \rangle \sim 0.18$ and the sum of probabilities on the acceptor sites $\sim 0.74$. Increase in $N$ leads to an increase in the time interval $t_{\text{TDE}}$. The time during which the system attains the state of TDE can be very long which raises the question of accumulation of a computing error. To clarify this question, we made test computations.

The system (2, 3) has the first integral: the total probability $\Sigma = \sum_{n=1}^{N} |b_n|^2$ must be equal to 1. As result of numerical integration $\Sigma$ is not kept exactly. One of the ways that allows us to make computation intervals longer is forced normalization, when the variables $b_n$ will be "corrected". After the integration step we obtain values $b_n[\text{old}]$, calculate $\Sigma[\text{old}]$ and $b_n[\text{new}] = b_n[\text{old}]/\sqrt{\Sigma[\text{old}]}$, so new $\Sigma$ will again be equal to 1. In doing so we do not come back to the same trajectory, but "overjump" to a new one (corresponding to the trajectory from other initial conditions), but individual trajectories are irrelevant when we are interested in the average by a lot of samples. The forced normalization means projection of all $b_n$ onto a unit sphere in $\mathbb{C}^N$. This routine yields the same 2-nd order of accuracy as the scheme of integration.

In the programme this procedure is not performed at each step of integration. A "normalization error" $\varepsilon_{norm}$ is specified. If at any moment $t$ $|\Sigma - 1| > \varepsilon_{norm}$, then $b_n = b_n/\sqrt{\Sigma}$. Similar methods are used in molecular dynamics modeling, e.g. for forced recover to the surface with a predetermined energy in microcanonical system [21].

For $N = 6, 8, 10$ we carried out some tests by integrating the system for large time intervals with different steps $h$ and different values of $\varepsilon_{norm}$. For $h \leq 0.0005$ and $\varepsilon_{norm} \leq 0.0005$, the averaged curves $\langle R(t) \rangle$ are closed, and for $\langle P_n(t) \rangle$ we obtained the same result.

So, the problem of attaining the thermodynamic equilibrium can require huge integration time but we can expect that the averaged calculation values will not deviate greatly from the "ideal mean trajectory".

## 5  Simulation Results

Numerical experiments demonstrate that: for short chains ($N < 8$) the time interval $t_{TDE}$ increases very fast (exponentially) with the increasing $N$. For $N \geq 8$, $t_{TDE}$ is almost the same, and even decreases for $N > 20$. The calculations were performed for two friction values $\gamma = 1.8\omega = 0.9$, $\gamma = 4\omega = 2$ (periodic and aperiodic mode). In the aperiodic mode the time to reach the TDE decreases, but the quality picture is the same. The simulation results show that friction coefficient increasing accelerates the yield to the TDE of up to $\gamma \sim 4-5\omega$; with further growth of the $\gamma$ $t_{TDE}$ does not decrease.

In biophysical experiments for GA...AGGG fragments it was obtained [7], that the relative rate constants decrease exponentially for $N < 9$, and with further increase in $N$ it remains almost unchanged. If we assume that the transfer rate is a reciprocal of the time $t_{TDE}$, then a similar picture is obtained for modeling results. Figure 2 (left) shows the dependencies of $1/t_{TDE}$ on the length of the chain $N$.

For the classical subsystem we chose the specific values of the parameters $\omega = 0.5$, $\chi = 1.$, that speed up the system's output to the TDE. We assume that not only the average values in the TDE are the same, but the processes of reaching the TDE from the same initial state (the charge is localized at the first donor site) will be qualitatively similar. We performed several test calculations with "more slow" parameters $\omega = 0.1$, $\chi = 0.2$ for chains with $N = 5 - 10$ sites (see Fig. 2, right), and the results of the tests do not contradict the assumption.

Figure 2 (right) shows that the time $t_{TDE}$ of reaching the TDE increases with decrease of parameter values for classical subsystem (3); the ratio of classical frequencies is 5 and for $N \geq 8$ $t_{TDE}$ increases hundreds of times. Qualitative picture of the time dependence on the length of the bridge is the same. Also this results demonstrate that for "DNA parameter values" ($\omega = 0.01$, $\chi = 0.02$) computational experiment will require a huge operating time.

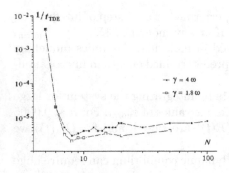 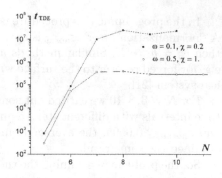

**Fig. 2.** Left: Dependencies of the transfer rate $1/t_{\text{TDE}}$ on the chain length $N$ for different friction coefficients, $\omega = 0.5$. Right: Dependence of the $t_{\text{TDE}}$ on the $N$ for two systems with the same ratio $\chi/\omega = 2$.

# 6  Discussion

In biophysical experiments [5–7], the transfer rate was not measured directly, the results are relative rate constants for different types of oligonucleotides. The authors obtained relative values of the reaction constants for hole transport from G to GGG, separated by bridges of different length. They found that for the chains with $N < 9$, lengthening the bridge on one site reduces the relative transfer rate by an order of magnitude. These values depend on oxidative damage at the G and GGG sites; the damage correlates with probability of the charge localization on the site.

We have modeled the system dynamics from the initial state "the charge is localized at the donor" over time intervals to the attainment of the thermodynamic equilibrium state and obtained dependence of the time intervals on the chain length. This dependence has the same properties as relative rate constants. Note that in the numerical experiments the values of probabilities of charge localization on the G and GGG sites in the TDE state are close. There is no direct correspondence between the data of biophysical experiments and simulation results, but we think that the results of numerical experiments are of interest.

In the thermodynamic equilibrium state, the value of $\langle R \rangle \sim N/2$ for chains $N \geq 20$, i.e. the charge is almost uniformly distributed along the chain. Based on the simulation results, it can be assumed that (similarly to the case of homogeneous chains) in short chains the charge is in the polaron state, and in long chains the charge is delocalized, and the charge transfer from the donor to the acceptor occurs by different mechanisms.

**Acknowledgements.** The work is partially supported by Russian Foundation for Basic Research, grants 16-07-00305 and 17-07-00801, and Russian Science Foundation, grant 16-11-10163.

# References

1. Lakhno, V.: DNA nanobioelectronics. Int. J. Quantum Chem. **108**(11), 1970–1981 (2008)
2. Offenhousser, A., Rinaldi, R. (eds.): Nanobioelectronics - for Electronics, Biology, and Medicine. Springer, New York (2009). https://doi.org/10.1007/978-0-387-09459-5
3. Demple, B., Harrison, L.: Repair of oxidative damage to DNA: enzymology and biology. Annu. Rev. Biochem. **63**, 915–948 (1994)
4. Loft, S., Poulsen, H.: Cancer risk and oxidative DNA damage in man. J. Mol. Med. **74**(6), 297–312 (1996)
5. Meggers, E., Michel-Beyerle, M., Giese, B.: Sequence dependent long range hole transport in DNA. J. Am. Chem. Soc. **120**(49), 12950–12955 (1998)
6. Giese, B., Wessely, S., Spormann, M., Lindemann, U., Meggers, E., Michel-Beyerle, M.: On the mechanism of long-range electron transfer through DNA. Angew. Chem. Int. Ed. Engl. **38**(7), 996–998 (1999)
7. Giese, B., Amaudrut, J., Kohler, A.-K., Spormann, M., Wessely, S.: Direct observation of hole transfer through DNA by hopping between adenine bases and by tunnelling. Nature **412**, 318–320 (2001). https://doi.org/10.1038/35085542
8. Grozema, F., Berlin, Y., Siebbeles, L.: Mechanism of charge migration through DNA: molecular wire behavior, single-step tunneling or hopping? J. Am. Chem. Soc. **122**(44), 10903–10909 (2000)
9. Henderson, P., Jones, D., Hampikian, G., Kan, Y., Schuster, G.: Long-distance charge transport in duplex DNA: the phonon-assisted polaron-like hopping mechanism. PNAS USA **96**, 8353–8358 (1999)
10. Lakhno, V., Fialko, N.: Hole mobility in a homogeneous nucleotide chain. JETP Lett. **78**(5), 336–338 (2003)
11. Fialko, N., Sobolev, E., Lakhno, V.: On the calculation of thermodynamic quantities in the Holstein model for homogeneous polynucleotides. JETP **124**(4), 635–642 (2017)
12. Fialko, N., Pyatkov, M., Lakhno, V.: On the thermodynamic equilibrium distribution of a charge in a homogeneous chain with a defect. In: Adam, G., Busa, J., Hnatic M., Podgainy D. (eds.) Mathematical Modeling and Computational Physics 2017 (MMCP 2017). EPJ Web of Conferences, vol. 173, p. 06004-4 (2018)
13. Holstein, T.: Studies of polaron motion: Part I. The molecular-crystal model. Ann. Phys. **8**(3), 325–342 (1959)
14. Lomdahl, P., Kerr, W.: Do Davydov solitons exist at 300K? Phys. Rev. Lett. **55**(11), 1235–1238 (1985)
15. Helfand, E.: Brownian dynamics study of transitions in a polymer chain of bistable oscillators. J. Chem. Phys. **69**(3), 1010–1018 (1978)
16. Greenside, H., Helfand, E.: Numerical integration of stochastic differential equations - II. Bell System Tech. J. **60**(8), 1927–1940 (1981)
17. Seidel, C., Schulz, A., Sauer, M.: Nucleobase-specific quenching of fluorescent dyes. 1. Nucleobase one-electron redox potentials and their correlation with static and dynamic quenching efficiencies. J. Phys. Chem. **100**(13), 5541–5553 (1996)
18. Lewis, F., Wu, Y.: Dynamics of superexchange photoinduced electron transfer in duplex DNA. J. Photochem. Photobiol. C **2**(1), 1–16 (2001)
19. Voityuk, A., Rosch, N., Bixon, M., Jortner, J.: Electronic coupling for charge transfer and transport in DNA. J. Phys. Chem. B **104**(41), 9740–9745 (2000)

20. Jortner, J., Bixon, M., Voityuk, A., Roesh, N.: Superexchange mediated charge hopping in DNA. J. Phys. Chem. A **106**(33), 7599–7606 (2002)
21. Frenkel, D., Smit, B.: Understanding Molecular Simulation: From Algorithms to Applications. Academic Press, London (2001)

# Simulation of Influence of Special Regimes of Horizontal Flare Systems on Permafrost

M. Yu. Filimonov[1,2]([✉]) [ID] and N. A. Vaganova[1,2] [ID]

[1] Ural Federal University, Yekaterinburg, Russia
[2] Krasovskii Institute of Mathematics and Mechanics, Yekaterinburg, Russia
{fmy,vna}@imm.uran.ru

**Abstract.** Permafrost takes place approximately 35 million $km^2$ of the globe land. In Russia, it is most widely distributed in Eastern Siberia and Baikal region with oil and gas fields. Exploitation of the fields promotes the permafrost melting because of different technical devices affect on the dynamics of thawing. The permafrost thawing due to various human-generated impacts will be accompanied by subsidence of the earth's surface around engineering facilities and development of dangerous permafrost geological processes, called thermokarst, which can lead to accidents in oil and gas fields with great damage to the environment. Therefore, an important problem for computer simulation is prediction of dynamics of the permafrost boundaries changes under long-term thermal impact of technical systems operating. In the paper a model and an algorithm for solving the problem of propagation of thermal fields in frozen ground from horizontal flare systems operated under a special regime are proposed. The maximum number of climatic and technical parameters is taken into account in the simulations. The calculations allow to choose an optimal thermal insulation of the ground surface under the flare system.

**Keywords:** Mathematical modelling · Heat transfer · Permafrost

## 1 Introduction

Permafrost occupies about 60% of the Russia territory, in which there are produced oil and gas. This production generates a number of various sources of heat, affected on degradation of permafrost [1]. The sources of heat are various technical systems, operated in oil and gas fields. Such heat sources may be production wells [2,3], since heat from hot oil moving through production wells affects on permafrost. Other engineering facilities located in the northern oil and gas fields also change the boundaries of permafrost. For example, the horizontal and vertical flare systems [4,5], on which the associated gas is burned taking into account the possible emergency discharge of gas, or various tanks filled with hot oil. In this case, the tanks can be located above the surface, and on the surface

© Springer Nature Switzerland AG 2019
I. Dimov et al. (Eds.): FDM 2018, LNCS 11386, pp. 233–240, 2019.
https://doi.org/10.1007/978-3-030-11539-5_25

of the ground, and also be buried in frozen ground. The process of filling and emptying these containers takes place according to a certain law in accordance with the technological requirements [6].

The thawing of permafrost due to various sources of heat can lead to the development of thermokarst and major accidents, which pose a particular danger to areas with the spread of permafrost. Various scenarios of climate change show that global warming will intensify in the polar regions of the Arctic and can lead to significant degradation of permafrost, which is aggravated by the technogenic impacts of [7,8]. Therefore, the actual problem is to decrease the intensity of thermal interaction in the system "technical systems (heat sources) – permafrost". To heat the soil during the operation of engineering facilities located in the permafrost spreading zone, various cooling devices and thermal insulation materials are used [9,10].

In this work, when modeling the propagation of heat in permafrost from a horizontal flare unit, technological features of the operation of these units with their periodic operation, climatic conditions and soil lithology peculiar to a specific geographical location are taken into account. The algorithms and computer programs developed on the basis of the proposed model were used to solve an applied problem related to the determination of the optimal design of a base for a horizontal flare system for a particular oil and gas field.

## 2    Statement of a Problem and Mathematical Model

To simulate non-stationary thermal fields in the permafrost zone, various climatic, physical and technological factors should be taken into account when different technical systems operate.

The first group of factors includes the consideration of solar radiation, seasonal changes in air temperature, leading to periodic thawing (freezing) of the soil, snow cover, etc.

The second group of factors includes thermophysical parameters of soils, varying from humidity, their heterogeneity and temperature. For the complete modeling of thermal fields during the construction of various engineering facilities in the permafrost zone, it is additionally necessary to take into account the possible heat sources from them and the thermophysical parameters of the applied thermal insulation.

The third group of factors includes the technological features of specific technical systems (in our case these are horizontal flare systems) and the mode of their operation for a particular field. For example, one can consider the operating mode of a horizontal flare system (HFS) when it works continuously, or when HFS runs periodically with the specified mode of switching.

Thus, it is necessary to calculate the predicted change in the temperature regime of the foundation soils under HFS, taking into account the adopted design of the heat-insulating screen, which ensures the preservation of permafrost in the base for the period of operation of the HFS.

To solve the task at computer modeling it is necessary to consider the following factors:

- solar radiation, which leads to nonlinear boundary conditions on the upper surface;
- various thermophysical parameters of soils that form permafrost rocks that vary not only in the vertical plane;
- seasonal variations in air temperatures in the area under investigation (in summer seasonal thawing of the upper layer of the soil takes place, in winter the reverse process is observed);
- possible phase transitions in the soil;
- the structure of the base thermal insulation under the flare system;
- technological features of HFS.

As a result, a boundary condition is defined dynamically over time on the surface of the soil (at the upper boundary of the calculated region) allowing to take into account all types of heat transfer inherent in the simulated conditions on this surface, taking into account the thermal resistance of the snow and vegetation cover, as well as other possible conditions.

Let us first consider heat exchange on the upper surface of the soil (without HFS), directly illuminated by the sun. Suppose at the initial time moment $t_0 = 0$ the soil occupies a given parallelepiped $\Omega$ and has a temperature $T_0(x, y, z)$. The computational domain is a three-dimensional parallelepiped, in which the axes $x$ and $y$ are parallel to the ground surface, and the $z$ axis is directed downward. We assume that the size of the region $\Omega$ is determined by the positive numbers $L_x$, $L_y$, $L_z$. Let $T = T(t, x, y, z)$ be a soil temperature at $(x, y, z)$ for the time moment $t$. Thus, the modeling of thawing in the soil is reduced to the solution in $\Omega$ of the following heat equation [11]:

$$\rho \big( c_\nu(T) + k\delta(T - T^*) \big) \frac{\partial T}{\partial t} = \nabla \left( \lambda(T) \Delta T \right), \tag{1}$$

where $\rho$ is the density $[kg/m^3]$, $T^*$ is temperature of phase transition $[K]$,

$$c_\nu(T) = \begin{cases} c_1(x, y, z), \ T < T^*, \\ c_2(x, y, z), \ T > T^*, \end{cases} \text{is specific heat [J/kg K],}$$

$$\lambda(T) = \begin{cases} \lambda_1(x, y, z), \ T < T^*, \\ \lambda_2(x, y, z), \ T > T^*, \end{cases} \text{is thermal conductivity coefficient [W/m K ],}$$

$k = k(x, y, z)$ is specific heat of phase transition, $\delta$ is Dirac delta function.

Thus, it is necessary to solve Eq. (1) in the area $\Omega$ with initial condition

$$T(0, x, y, z) = T_0(x, y, z) \tag{2}$$

and boundary conditions [2,12]

$$f(x, y, t) + \gamma q + b(T_{air} - T(x, y, 0, t)) = \varepsilon \sigma (T^4(x, y, 0, t) - T_{air}^4) + \lambda \frac{\partial T(x, y, 0, t)}{\partial z}, \tag{3}$$

$$\frac{\partial T}{\partial x} \bigg|_{x = \pm L_x} = \frac{\partial T}{\partial y} \bigg|_{y = \pm L_y} = \frac{\partial T}{\partial z} \bigg|_{z = -L_z} = 0, \tag{4}$$

where $f(x, y, t)$ is the density of heat flow from the flame, $T_{air} = T_{air}(t)$ denotes the temperature in the surface layer of air, which varies from time to time in accordance with the annual cycle of temperature, $\sigma = 5, 67 \cdot 10^{-8}$ W/(m$^2$K$^4$) is Stefan–Boltzmann constant, $b = b(t, x, y)$ is heat transfer coefficient, $\lambda = \lambda(T)$ is thermal conductivity coefficient, $\varepsilon = \varepsilon(t, x, y)$ is the coefficient of emissivity. The total solar radiation $q(t)$ contains the sum of direct solar radiation and scattered radiation.

## 3  Numerical Results

Search for exact solutions for the Eq. (1) by the method [13], or by using special series, when the exact solution [14] is taken as the zeroth term of the series, only known exact solutions are obtained, on which the developed numerical method was also tested. In this case, the nonlinear boundary condition (3) can not be satisfied even with the use of numerical-analytical methods [15–17].

A large number of works are devoted to development of numerical methods for solving boundary value problems of heat conduction. Basics of finite difference methods are detailed in the work [18].

To solve the problem (1)–(4), a finite-difference method is developed for solving problems of Stefan type [11, 19], which allows using the splitting method with respect to the space variables. At the same time, the algorithm is "adapted" to the geographic coordinates of a particular oil and gas field [2, 12].

On the surface $z = 0$ an algebraic equation of the fourth degree is generated, to be solved by Newton method. The solvability of the corresponding difference equations approximating the original problem was proved in the paper [20].

Calculations based on the proposed models and developed programs were used for 12 northern oil and gas fields. When the source of heat in the frozen ground was a well, a comparison was made between numerical data on the distribution of the boundary of the melting of frozen ground (zero isotherm) and experimental data. The comparison showed that the numerical data coincide with the experimental data with accuracy of 5%.

To simulate the thermal trace from the HFS on the surface $z = 0$, there was used the temperature or the heat flux. In accordance with the technical characteristics of the HFS, at the surface $z = 0$ the temperature was set in accordance with Fig. 1.

The right figure in Fig. 1 shows the size of the calculation area to be chosen so that the influence of the boundary conditions 4 on its side surfaces is not affected.

For the foundation under the HFS it is proposed to use heat-resistant concrete 150 mm slab, heat-insulating material (vermiculite) of 200 mm layer, geomembrane with the thick of 2 mm, which can be operated from $-60\,°$C to $+150\,°$C, and stony ground of 1300 mm thick. The soil temperature below the influence of seasonal temperature changes is $-0.3\,°$C.

Based on the numerical calculations, it is required to determine the optimum thickness of the heat-insulating material under the special operating mode of

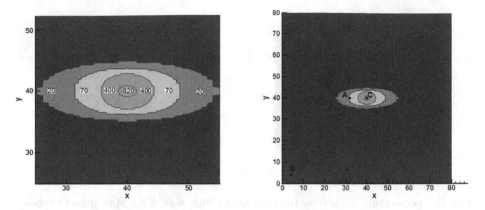

**Fig. 1.** The temperature °C from the HFS flare and the position of the heating zones from the torch on the surface $z = 0$.

the HFS. The size of the calculation area of $\Omega$ was $80\,\text{m} * 80\,\text{m} * 50\,\text{m}$, to control the accuracy of the results, calculations were made on grids $101 * 101 * 101$, $201 * 201 * 201$ and $201 * 201 * 601$ nodes. Step by time is 24 h with disabled HFS, simulation period is 30 years. When describing the mode of operation of HFS, the time step decreases. Control over the accuracy of numerical calculations was carried out by refinement of the computational grid. In this case, the size of the grid was selected, in which the results of the refined calculation are in general unchanged. An orthogonal grid was used, condensed to the boundaries of various layers of soil in accordance with the accepted vertical layout of the different layers under HFS. The article presents calculations for the thickness of vermiculite be 400 mm. Figures 2 and 3 show the temperature fields in the first and fifth years of operation of the HFS, respectively, in the $xz$ plane passing through the center of the flare trace in Fig. 1.

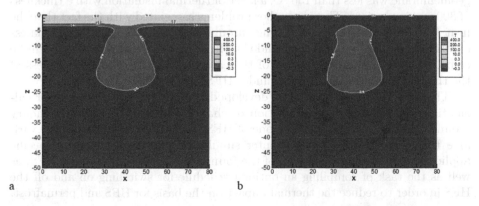

**Fig. 2.** Temperature fields from the HFS in the first year: a for May, b for October.

238     M. Yu. Filimonov and N. A. Vaganova

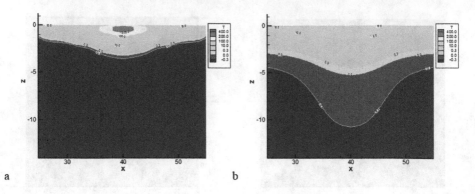

a                                          b

**Fig. 3.** Temperature fields from the HFS in the fifth year: a for May, b for October.

Numerical calculations have shown that after five years of operation of HFS, the temperature fields in the ground stabilize and, therefore, calculations can be carried out only up to five years of operation. Calculations were also carried out for the option of starting the operation of HFS in September.

## 4   Conclusion

Numerical calculations have shown that the first year of operation of HFS, when the longest thermal impact on the riprap and soil (28 calendar days) takes place, is the most significant in terms of impact, because starting from the second year, when during the year HFS works only 7 days, the impact on the riprap and the soil is not so significant and winter temperatures can reduce the effects of heat stress. Note that a thermal insulation of 200 mm thickness will not protect the geomembrane from high temperatures in any mode of operation of the HFS. Temperature on the surface of the geomembrane will be more than 150 °C. In order that in the first year of HFS operation the temperature on the surface of the geomembrane was less than 150 °C, a layer of thermal insulation with a thickness of 800 mm is required. There are other problems associated with the fact that the heat flow from the HFS, penetrating through the thermal insulation thickness of 800 mm, dissipates quite poor, and thus freezing of a ground in winter will be complicated. Based on the results of the calculations, the optimal thickness of the thermal insulation in the base under HFS is of 400 mm thickness.

Thus, the proposed model and developed computer programs allow for calculations on the long-term prediction of changes in the permafrost boundary around the HFS. The operating mode of HFS can be described by using a variable time step. Carried out computer simulation allows one to solve specific applications related to the design of the foundation construction under HFS, as well as the task of compiling an optimal schedule for switching on and off the HFS in order to reduce the thermal impact on the basis for HFS and permafrost.

**Acknowledgments.** The work is supported by Russian Foundation for Basic Research 16–01–00401.

# References

1. Vaganova, N., Filimonov, M.Yu.: Simulation of freezing and thawing of soil in Arctic regions. In: 2nd International Conference on Sustainable, ICSC 2017, IOP Conference Series: Earth and Environmental Science, vol. 72, p. 012005. IOP Publishing, UK (2017). https://doi.org/10.1088/1755-1315/72/1/012005
2. Filimonov, M.Y., Vaganova, N.A.: Simulation of technogenic and climatic influences in permafrost for northern oil fields exploitation. In: Dimov, I., Faragó, I., Vulkov, L. (eds.) FDM 2014. LNCS, vol. 9045, pp. 185–192. Springer, Cham (2015). https://doi.org/10.1007/978-3-319-20239-6_18
3. Filimonov, M.Yu., Vaganova, N.: Permafrost thawing from different technical systems in Arctic regions. In: 2nd International Conference on Sustainable, ICSC 2017, IOP Conference Series: Earth and Environmental Science, vol. 72, p. 012006. IOP Publishing, UK (2017). https://doi.org/10.1088/1755-1315/72/1/012006
4. Vaganova, N.A.: Simulation of long-term influence from technical systems on permafrost with various short-scale and hourly operation modes in Arctic region. In: Venkov, G., Pasheva, V., Popivanov, N. (eds.) 43rd International Conference Applications of Mathematics in Engineering and Economics, AMEE 2017, AIP Conference Proceedings, vol. 1970, p. 020006. American Institute of Physics, USA (2017). https://doi.org/10.1063/1.5013943
5. Filimonov, M.Yu., Vaganova, N.A.: Flare systems exploitation and impact on permafrost. In: 2nd All-Russian Scientific Conference Thermophysics and Physical Hydrodynamics with the School for Young Scientists, TPH 2017. Journal of Physics: Conference Series, vol. 899, p. 092004. IOP Publishing, UK (2017). https://doi.org/10.1088/1742-6596/899/9/092004
6. Vaganova, N., Filimonov, M.Yu.: Different shapes of constructions and their effects on permafrost. In: Pasheva, V., Venkov, G., Popivanov, N. (eds.) 42nd International Conference on Applications of Mathematics in Engineering and Economics, AMEE 2016, AIP Conference Proceedings, vol. 1789, p. 020019. American Institute of Physics, USA (2016). https://doi.org/10.1063/1.4968440
7. Muskett, R.R.: L-Band InSAR penetration depth experiment, North Slope Alaska. J. Geosci. Environ. Prot. 5(3), 14–30 (2017)
8. Nelson, F.E., Anisimov, O.A., Shiklomanov, N.I.: Subsidence risk from thawing permafrost. Nature 410, 889–890 (2001)
9. Filimonov, M.Yu., Vaganova, N.A.: Simulation of thermal stabilization of soil around various technical systems operating in permafrost. Appl. Math. Sci. 7(144), 7151–7160 (2013)
10. Vaganova, N.A., Filimonov, M.Yu.: Computer simulation of nonstationary thermal fields in design and operation of northern oil and gas fields. In: Pasheva, V., Popivanov, N., Venkov, G. (eds.) 41st International Conference Applications of Mathematics in Engineering and Economics, AMEE 2015, AIP Conference Proceedings, vol. 1690, p. 020016. American Institute of Physics, USA (2015). https://doi.org/10.1063/1.4936694
11. Samarsky, A.A., Vabishchevich, P.N.: Computational Heat Transfer, Volume 2, The Finite Difference Methodology. Wiley, Chichester (1995)
12. Filimonov, M.Yu., Vaganova N.: Numerical simulation of technogenic and climatic influence on permafrost. In: Advances in Environmental Research, vol. 54, Chap. 5. Nova Science Publishers, New York (2017)
13. Bashurov, V.V., Vaganova, N.A., Kropotov, A.I., Pchelintsev, M.V., Skorkin, N.A., Filimonov, M.Yu.: Nonlinear model of a pipeline in a gravity field with an ideal fluid moving through it. J. Appl. Mech. Tech. Phys. 53(1), 43–48 (2012)

14. Filimonov, M.Yu.: Application of special coordinated series to the solution of nonlinear partial differential equations in bounded domain. Differ. Equ. **36**(11), 1685–1691 (2000)

15. Filimonov, M.Yu.: Representation of solutions of initial-boundary value problems for nonlinear partial differential equations by the method of special series. Differ. Equ. **39**(8), 1159–1166 (2003)

16. Filimonov, M.Yu.: Application of method of special series for solution of nonlinear partial differential equations. In: Venkov, G., Pasheva, V. (eds.) 40st International Conference Applications of Mathematics in Engineering and Economics, AMEE 2014, AIP Conference Proceedings, vol. 1631, pp. 218–223. American Institute of Physics, USA (2014). https://doi.org/10.1063/1.4902479

17. Filimonov, M.Yu.: On the justification of the applicability of the Fourier method to the solution of nonlinear partial differential equations. Russ. J. Numer. Anal. Math. Model. **11**(1), 27–39 (1996)

18. Patankar, S.V.: Numerical Heat Transfer and Fluid Flow. Hemisphere, New York (1980)

19. Samarskii, A.A., Moiseyenko, B.D.: An economic continuous calculation scheme for the Stefan multidimensional problem. USSR Comput. Math. Math. Phys. **5**(5), 43–58 (1965)

20. Vaganova, N.A.: Existence of a solution of an initial-boundary value difference problem for a linear heat equation with a nonlinear boundary condition. Proc. Steklov Inst. Math. **261**(1), 260–271 (2008)

# The Method of Fundamental Solutions Combined with a Multi-level Technique

Csaba Gáspár[✉]

Széchenyi István University, Egyetem tér 1, Győr 9026, Hungary
gasparcs@sze.hu

**Abstract.** A traditional idea of the Method of Fundamental Solutions is to use some external source points where the fundamental solution should be shifted to. However, the proper definition of the locations of the sources can hardly be performed in an automated way. To circumvent this difficulty, in this paper, the source points defined along the boundary, and the collocation points are shifted to the interior of the domain together with a proper modification of the boundary conditions. Thus, the problem of singularity is avoided. The modified boundary conditions are defined on the basis of the tools of the classical finite difference methods. Several schemes are presented. The schemes can be embedded in a multi-level context in a natural way. The proposed method avoids the computational difficulties due to ill-conditioned matrices and also reduces the computational complexity of the Method of Fundamental Solutions.

**Keywords:** Method of Fundamental Solutions · Multi-level methods · Least squares · Inner collocation

## 1 Introduction

The Method of Fundamental Solutions (MFS, see e.g. [1,7]) is now a popular truly meshfree method. Moreover, it is a boundary-only technique, i.e. it requires a discretization of the boundary only (but no grid or mesh structure is needed). Recall that, for the sake of simplicity, if we restrict ourselves to a simple Laplace equation:

$$\Delta u = 0 \tag{1}$$

required in a 2D bounded domain $\Omega$ and supplied with pure Dirichlet boundary condition:

$$u|_\Gamma = u_0, \tag{2}$$

($\Gamma$ denotes the boundary of the domain $\Omega$), then the Method of Fundamental Solution provides the following approximate solution of (1)–(2):

$$u(x) := \sum_{j=1}^{N} \alpha_j \Phi(x - \tilde{x}_j) \tag{3}$$

© Springer Nature Switzerland AG 2019
I. Dimov et al. (Eds.): FDM 2018, LNCS 11386, pp. 241–249, 2019.
https://doi.org/10.1007/978-3-030-11539-5_26

Here $\tilde{x}_1$, $\tilde{x}_2$, ..., $\tilde{x}_N$ are predefined external points (source points), and the function $\Phi$ is the fundamental solution of the Laplacian, i. e., apart from a multiplicative constant:

$$\Phi(x) = \log||x||,$$

where $||.||$ denotes the Euclidean norm in $\mathbf{R}^2$. The function defined by Eq. (3) automatically satisfies the Laplace equation (1) in the domain $\Omega$. The unknown coefficients $\alpha_j$ can be determined by taking into account the boundary conditions at some boundary collocation points $x_1, x_2, ..., x_M \in \Gamma$. This leads to the following linear system of equations:

$$\sum_{j=1}^{N} \alpha_j \Phi(x_k - \tilde{x}_j) = u_k := u_0(k) \qquad (k = 1, 2, ..., M) \qquad (4)$$

The numbers of sources $(N)$ and the number of collocation points $(M)$ need not be equal. If they differ, the system (4) should be solved in a generalized sense i.e. using the singular value decomposition, or the simple technique of the least squares combined with an efficient iterative method.

For simplicity, $N$ and $M$ are often defined to be equal. In this case, the matrix of the system (4) is generally regular. However, it is often severely ill-conditioned, which may cause serious computational difficulties.

Though the MFS is simply programmable and is essentially dimension-independent and exhibits quite good accuracy, it requires defining the external source points, which is not a trivial task (see [2,3]), and the method is quite sensitive to the proper choice of the source point locations. If they are too far from the boundary, the resulting algebraic system becomes extremely ill-conditioned. If they are too close to the boundary, numerical singularities occur, and the accuracy decreases. In addition to it, the proper definition of the source locations can hardly be automated in an efficient way for sufficiently general domains.

A popular method to damp these drawbacks is to allow the source and boundary collocation points to coincide. This solves the problem of automated definition of sources, but due to the singularity of the fundamental solution at the origin, an extra problem arises, namely, the proper evaluation of the appearing singular terms in the system (4). To treat this problem, a number of special techniques have been developed (regularization and desingularization methods, see [4,5,8–10] for details).

In this paper, the sources are defined always on the boundary $\Gamma$, which can be performed easily in an automated way. To avoid the problem of singularity, the collocation points are redefined in the interior of the domain (sufficiently close to the boundary). The boundary condition should also be redefined, in which simple finite difference tools are utilized. A first-order and two second-order schemes are presented. The method can be embedded in a multi-level context in a straightforward way, which decreases further the computational cost of the technique. At the same time, the problem of severely ill-conditioned systems can be completely avoided.

## 2   Transforming the Boundary Condition to the Interior of the Domain

The main idea of the technique is similar to that of the method in [6]. For each boundary point $x \in \Gamma$, define two points $\overline{x}$ and $\overline{\overline{x}}$ by shifting $x$ in the direction $n$ (where $n$ is the outward normal unit vector at $x$): $\overline{x} := x - \delta \cdot n, \overline{\overline{x}} := x - 2\delta \cdot n$. (They are internal points if $\delta > 0$ is sufficiently small.) Performing a Taylor series expansion around $\overline{x}$, we have:

$$u(x) = u(\overline{x}) + \delta \cdot \frac{\partial u}{\partial n}(\overline{x}) + \mathcal{O}(\delta^2) \tag{5}$$

This equality provides three different schemes.

*Scheme 1:* The boundary condition $u(x) = u_0(x)$ is shifted to $\overline{x}$ without change:

$$u(\overline{x}) = u_0(x) \tag{6}$$

*Scheme 2:* The boundary condition $u(x) = u_0(x)$ is shifted to $\overline{x}$ with the first order correction:

$$u(\overline{x}) + \delta \cdot \frac{\partial u}{\partial n}(\overline{x}) = u_0(x) \tag{7}$$

*Scheme 3:* The same as Scheme 2, but the normal derivative is approximated by a central difference scheme:

$$u(\overline{x}) + \delta \cdot \frac{u_0(x) - u(\overline{\overline{x}})}{2\delta} = u_0(x), \qquad \text{i.e.} \qquad 2u(\overline{x}) - u(\overline{\overline{x}}) = u_0(x) \tag{8}$$

Let us combine these schemes with the MFS in the following way. Let $\tilde{x}_1, ..., \tilde{x}_N \in \Gamma$ be source points located on the boundary $\Gamma$, and let $x_1, ..., x_M \in \Gamma$ be additional boundary points. Denote by $n_k$ the outward normal unit vector at $x_k$. The inner points $x_k - \delta n_k$ are regarded as inner collocation points. The approximate solution is sought in the form (3) (note that the sources are located along $\Gamma$). The coefficients $\alpha_1, ... \alpha_N$ can be calculated by solving one of the following systems of equations:

$$\sum_{j=1}^{N} \alpha_j \cdot \Phi(x_k - \delta n_k - \tilde{x}_j) = u_0(x_k) \qquad \text{(Scheme 1)}$$

$$\sum_{j=1}^{N} \alpha_j \cdot [\Phi(x_k - \delta n_k - \tilde{x}_j) + \delta \cdot \frac{\partial \Phi}{\partial n_k}(x_k - \delta n_k - \tilde{x}_j)] = u_0(x_k) \qquad \text{(Scheme 2)}$$

$$\sum_{j=1}^{N} \alpha_j \cdot [2\Phi(x_k - \delta n_k - \tilde{x}_j) - \Phi(x_k - 2\delta n_k - \tilde{x}_j)] = u_0(x_k) \qquad \text{(Scheme 3)}$$

$(k = 1, 2, ..., M)$. In short:

$$A\underline{\alpha} = \underline{u}_0, \tag{9}$$

where the matrix $A$ is $M$-by-$N$, and often $M \gg N$. The solution is meant in the sense of least squares, i.e. as the solution of the Gaussian normal equations:

$$A^* A \underline{\alpha} = A^* \underline{u}_0, \tag{10}$$

The method circumvents the problem of singularity (no singular terms appear in the schemes) and also the problem of the automated definition of source points. However, the Gaussian normal equations may be ill-conditioned and the convergence may be slow, if some iterative technique is applied. This can be improved by using a multi-level technique, see later.

Note that since the sources are located close to the collocation points, the accuracy of the method is expected to be less than that of the traditional MFS based on far sources, even if the solution is an entire function.

## 3    Error Estimations

Assume, for simplicity that $\Omega$ is a circle centered at the origin, with radius $R$. Denote by $\Gamma_R$ the boundary of $\Omega$. Let the boundary condition function $u_0$ be expressed in terms of complex Fourier series:

$$u_0(t) = \sum_k \beta_k e^{ikt}$$

Then the exact solution of the Dirichlet problem (1)–(2) is as follows (written in polar coordinates):

$$u^*(r,t) = \sum_k \beta_k \left(\frac{r}{R}\right)^{|k|} e^{ikt},$$

while the approximate solution obtained by shifting the boundary to $\Gamma_{R-\delta}$ (which is located in the interior of $\Omega$) can be expressed as:

$$u(r,t) = \sum_k \gamma_k \left(\frac{r}{R}\right)^{|k|} e^{ikt}$$

The coefficients $\gamma_k$ depend on the applied scheme:

*Scheme 1:*    $u|_{\Gamma_{R-\delta}} = u_0$, which implies:

$$\gamma_k = \left(1 - \frac{\delta}{R}\right)^{-|k|} \cdot \beta_k \tag{11}$$

*Scheme 2:*    $u|_{\Gamma_{R-\delta}} + \delta \cdot \frac{\partial u}{\partial n}|_{\Gamma_{R-\delta}} = u_0$, whence:

$$\gamma_k = \frac{\left(1 - \frac{\delta}{R}\right)^{-|k|}}{1 + \frac{\delta |k|}{R-\delta}} \cdot \beta_k \tag{12}$$

*Scheme 3:*     $2 \cdot u|_{\Gamma_{R-\delta}} - u|_{\Gamma_{R-2\delta}} = u_0$, therefore:

$$\gamma_k = \frac{1}{2 \cdot \left(1 - \frac{\delta}{R}\right)^{|k|} - \left(1 - \frac{2\delta}{R}\right)^{|k|}} \cdot \beta_k \tag{13}$$

In the right-hand sides of the schemes, the boundary condition function $u_0$ is meant to be redefined on the circle $\Gamma_{R-\delta}$; the original and the redefined functions, however, have the same Fourier series expansion with respect to $t$.

Using Parseval's formula, we can express the Sobolev norm of the boundary error:

$$\|u - u^*\|^2_{H^s(\Gamma_{R-\delta})} = 2\pi \cdot \sum_k |\gamma_k - \beta_k|^2 \left(1 - \frac{\delta}{R}\right)^{2|k|} (1 + |k|)^{2s}$$

from which the following estimations are obtained:

*Scheme 1:*

$$\|u - u^*\|^2_{H^s(\Gamma_{R-\delta})} = 2\pi\delta^2 \cdot \sum_k |\beta_k|^2 (1 + |k|)^{2(s+1)} \cdot C_1(\delta, k, R) \tag{14}$$

*Scheme 2:*

$$\|u - u^*\|^2_{H^s(\Gamma_{R-\delta})} = 2\pi\delta^4 \cdot \sum_k |\beta_k|^2 (1 + |k|)^{2(s+2)} \cdot C_2(\delta, k, R) \tag{15}$$

*Scheme 3:*

$$\|u - u^*\|^2_{H^s(\Gamma_{R-\delta})} = 2\pi\delta^4 \cdot \sum_k |\beta_k|^2 (1 + |k|)^{2(s+2)} \cdot C_3(\delta, k, R) \tag{16}$$

where

$$C_1(\delta, k, R) := \frac{\left(1 - \frac{\delta}{R}\right)^{2|k|}}{(1 + |k|)^2 \delta^2} \cdot \left(\left(1 - \frac{\delta}{R}\right)^{-|k|} - 1\right)^2$$

$$C_2(\delta, k, R) := \frac{\left(1 - \frac{\delta}{R}\right)^{2|k|}}{(1 + |k|)^4 \delta^4} \cdot \left(\frac{\left(1 - \frac{\delta}{R}\right)^{-|k|}}{1 + \frac{\delta|k|}{R-\delta}} - 1\right)^2$$

$$C_3(\delta, k, R) := \frac{\left(1 - \frac{\delta}{R}\right)^{2|k|}}{(1 + |k|)^4 \delta^4} \cdot \left(\frac{1 - 2\left(1 - \frac{\delta}{R}\right)^{|k|} + \left(1 - \frac{2\delta}{R}\right)^{|k|}}{2\left(1 - \frac{\delta}{R}\right)^{|k|} - \left(1 - \frac{2\delta}{R}\right)^{|k|}}\right)^2$$

A thorough but essentially elementary analysis of the functions defined by:

$$f_1(x) := \frac{1}{R \cdot |k| \cdot x} \cdot \left(1 - (1 - x)^{|k|}\right) \qquad (0 < x \le 1)$$

$$f_2(x) := \frac{1}{R^2 k^2 x^2} \cdot \left(\frac{1}{1 + \frac{|k|x}{1-x}} - (1 - x)^{|k|}\right) \qquad (0 < x \le 1)$$

$$f_3(x) := \frac{(1-x)^{|k|}}{R^2 k^2 x^2} \cdot \frac{1 - 2 \cdot (1-x)^{|k|} + (1-2x)^{|k|}}{2 \cdot (1-x)^{|k|} - (1-2x)^{|k|}} \qquad (0 < x \le \frac{1}{2})$$

shows that $C_1(\delta, k, R)$, $C_2(\delta, k, R)$, $C_3(\delta, k, R)$ are uniformly bounded (provided that $\delta < R$ (Schemes 1, 2), resp. $\delta < \frac{R}{2}$ (Scheme 3)). The upper bound depends only on $R$ but not on $k$ and $\delta$. Thus, we have concluded that, as expected:

*Proposition 1:* Scheme 1 is of first order, Schemes 2 and 3 are of second order with respect to $\delta$ (provided that the boundary condition $u_0$ is sufficiently regular).

## 4   A Multi-level Solution Technique

First, consider two groups of sources along the boundary: $\tilde{\tilde{x}}_1, \ldots, \tilde{\tilde{x}}_{N/2}$ ('coarse level') and $\tilde{x}_1, \ldots, \tilde{x}_N$ ('fine level'); some coarse and fine source points may coincide. Define a two-level approximate solution:

$$u(x) := \sum_{j=1}^{N/2} \alpha_j^C \Phi(x - \tilde{\tilde{x}}_j) + \sum_{j=1}^{N} \alpha_j^F \Phi(x - \tilde{x}_j) \qquad (17)$$

The boundary condition can be enforced by any of the above defined schemes. In short, this leads to the equation:

$$A^C \underline{\alpha}^C + A^F \underline{\alpha}^F = \underline{u}_0 \qquad (18)$$

This equation can be solved by the following iteration

$$A^C \underline{\alpha}^C = \underline{u}_0 - A^F \underline{\alpha}^F$$

(coarse level problem: this should be solved exactly), and:

$$A^F \underline{\alpha}^F = \underline{u}_0 - A^C \underline{\alpha}^C$$

(fine level problem: this can be solved by a traditional (conjugate) gradient iteration applied to the corresponding Gaussian normal equations), and the above two steps should be repeated until convergence.

Without going into deep details, we note that the (conjugate) gradient iteration damps the high-frequency error components quite efficiently, independently of the fineness of the discretization. Consequently, the above two-level method significantly reduces the number of the necessary algebraic operations, despite the condition number of $A^F$ may be high.

This two-level method can be extended to a real multi-level method in a straightforward way by introducing even coarser levels to solve the coarse level problem.

## 4.1    A Numerical Example

The method is illustrated through the following example. Let $\Omega$ be a smooth, star-like domain, the boundary of which is parametrized by:

$$x(t) = R \cdot (1 + b \cdot \sin(ct)) \cdot \cos(t), \qquad y(t) = R \cdot (1 + b \cdot \sin(ct)) \cdot \sin(t)$$

$(0 \le t < 2\pi)$, with $b := 0.2$, $c := 5$, $R := 1.5$. Consider the harmonic test solution:

$$u^*(x, y) := \cos \frac{\pi x}{4} \sinh \frac{\pi y}{4} \tag{19}$$

and let the Dirichlet boundary condition be consistent with the test solution.

The number of the applied collocation points was always $M = 512$. The distance of the original and shifted boundary was $\delta := \frac{2R\pi}{M}$. Table 1 shows the relative $L_2$-errors along the shifted boundary when applying the two-level method with the schemes 1, 2 and 3 with different numbers of boundary source points. Here $N_{coarse}$ ($N_{fine}$, respectively) denotes the number of sources in the coarse (resp. fine) level. On both levels, a direct solution was applied.

**Table 1.** Two-level method, accuracy. Direct solution on both levels

| $N_{coarse}/N_{fine}$ | 16/32 | 32/64 | 64/128 | 128/256 | 256/512 |
|---|---|---|---|---|---|
| Relative $L_2$-errors (%) | | | | | |
| Scheme 1 | 5.72391 | 2.44171 | 1.5656 | 1.4778 | 1.474 |
| Scheme 2 | 5.56544 | 1.95703 | 0.530328 | 0.108651 | 0.021317 |
| Scheme 3 | 5.55128 | 1.95101 | 0.528485 | 0.107627 | 0.018601 |

Table 2 shows the relative $L_2$-errors along the shifted boundary, when the coarse level problem was solved directly, while at the fine level problem, 16 conjugate gradient steps were applied. It can clearly be seen that the errors are approximately the same as if a direct solver were applied on the fine level. However, the computational complexity is significantly reduced in the latter case. Note that the computational complexity can be reduced further by applying conjugate gradient steps on both levels instead of a direct solver. Experience shows that the accuracy does not decrease significantly also in this case.

**Table 2.** Two-level method, accuracy. Coarse level: direct solution, fine level: 16 conjugate gradient steps

| $N_{coarse}/N_{fine}$ | 16/32 | 32/64 | 64/128 | 128/256 | 256/512 |
|---|---|---|---|---|---|
| Relative $L_2$-errors (%) | | | | | |
| Scheme 1 | 5.72391 | 2.44171 | 1.5656 | 1.4778 | 1.47406 |
| Scheme 2 | 5.56544 | 1.95703 | 0.530328 | 0.108651 | 0.0213531 |
| Scheme 3 | 5.55128 | 1.95101 | 0.528485 | 0.107625 | 0.0185893 |

# 5    Summary and Conclusions

A version of the Method of Fundamental Solution has been presented. Unlike the traditional approach, the sources are not external points but they are located along the real boundary of the domain. The collocation points are shifted to the interior of the domain. Therefore the boundary conditions are to be adjusted in a proper way. To do this, classical finite difference tools were used. Three different schemes have been constructed: one of them is of first order, the others are of second order with respect to the distance of the shifted collocation points and the boundary. The number of the applied collocation points can (and should) be much higher than that of the source points. The discretized equations are solved in the sense of least squares, practically, by solving the corresponding Gaussian normal equations by an efficient iterative technique. To speed up the convergence, a multi-level technique has been also developed, which completely avoids the problem of severely ill-conditioned linear systems, and significantly reduces the computational complexity. Moreover, the problem of singularity is also circumvented. The accuracy of the method is less than that of the traditional Method of Fundamental Solutions based on sources located far from the boundary, but is still acceptable, and the method remains applicable also in the cases when no entire solution is available.

**Acknowledgments.** The research was partly supported by the European Union and the Hungarian Government from the project 'FIEK - Center for cooperation between higher education and the industries at the Széchenyi István University' under grant number GINOP-2.3.4-15-2016-00003.

# References

1. Alves, C.J.S., Chen, C.S., Šarler, B.: The method of fundamental solutions for solving Poisson problems. Int. Ser. Adv. Bound. Elem. **13**, 67–76 (2002)
2. Chen, C.S., Karageorghis, A., Li, Y.: On choosing the location of the sources in the MFS. Numer. Algorithms **72**, 107–130 (2016)
3. Fam, G.S.A., Rashed, Y.F.: A study on the source points locations in the method of fundamental solutions. Int. Ser. Adv. Bound. Elem. **13**, 297–312 (2002)
4. Gáspár, C.: Some variants of the method of fundamental solutions: regularization using radial and nearly radial basis functions. Cent. Eur. J. Math. **11**(8), 1429–1440 (2013)
5. Gáspár, C.: A regularized multi-level technique for solving potential problems by the method of fundamental solutions. Eng. Anal. Bound. Elem. **57**, 66–71 (2015)
6. Gáspár, C.: Fast meshless techniques based on the regularized method of fundamental solutions. In: Dimov, I., Faragó, I., Vulkov, L. (eds.) Numerical Analysis and Its Applications, NAA 2016. LNCS, vol. 10187, pp. 326–333. Springer, Cham (2017). https://doi.org/10.1007/978-3-319-57099-0_36
7. Golberg, M.A.: The method of fundamental solutions for Poisson's equation. Eng. Anal. Bound. Elem. **16**, 205–213 (1995)
8. Gu, Y., Chen, W., Zhang, J.: Investigation on near-boundary solutions by singular boundary method. Eng. Anal. Bound. Elem. **36**, 1173–1182 (2012)

9. Šarler, B.: Solution of potential flow problems by the modified method of fundamental solutions: formulations with the single layer and the double layer fundamental solutions. Eng. Anal. Bound. Elem. **33**, 1374–1382 (2009)

10. Young, D.L., Chen, K.H., Lee, C.W.: Novel meshless method for solving the potential problems with arbitrary domain. J. Comput. Phys. **209**, 290–321 (2005)

# Generalized Multiscale Discontinuous Galerkin Method for Helmholtz Problem in Fractured Media

U. Gavrileva[1]([✉])[ID], V. Alekseev[1][ID], M. Vasilyeva[1,2][ID], J. D. De Basabe[3][ID], Y. Efendiev[4][ID], and R. L. Gibson Jr.[5][ID]

[1] Multiscale Model Reduction Laboratory, North-Eastern Federal University, Yakutsk, Russia
lanasemna@mail.ru
[2] Institute for Scientific Computation, Texas A&M University, College Station, TX, USA
vasilyevadotmdotv@gmail.com
[3] Department of Seismology, Earth Sciences Division, CICESE, Baja California, Mexico
[4] Department of Mathematics and Institute for Scientific Computation (ISC), Texas A&M University, College Station, TX, USA
[5] Department of Geology and Geophysics, Texas A&M University, College Station, TX, USA

**Abstract.** In this work, we consider wave propagation in fractured media. The mathematical model is described by Helmholtz problem related to wave propagation with specific interface conditions on the fracture in the frequency domain. We use a discontinuous Galerkin method for the approximation by space that help to weakly impose interface conditions on fractures. Such approximations lead to the large system of equations and computationally expensive. In this work, we construct a coarse grid approximation for effective solution using Generalized Multiscale Discontinuous Galerkin Method (GMsDGM). In this method, we construct a multiscale space using solution of the local spectral problems in each coarse elements. The results of the numerical solution for the two-dimensional problem are presented for model problems of the wave propagation in fractured media.

**Keywords:** Fractured media · Wave propagation · Helmholtz equation · Discontinuous Galerkin method · Multiscale method · GMsFEM

## 1 Problem Formulation

We consider the Helmholtz equation for the elastic waves propagation in the computational domain $\Omega$ [1,2]

$$- \operatorname{div} \sigma - \omega^2 \rho u = f, \quad x \in \Omega \tag{1.1}$$

I. Dimov et al. (Eds.): FDM 2018, LNCS 11386, pp. 250–257, 2019.
https://doi.org/10.1007/978-3-030-11539-5_27

where $\omega$ is frequency, $\rho$ is density and $f$ is the source function.

Equation (1.1) is supplemented by the relation between the stress tensor $\sigma$ and strain tensor $\varepsilon$

$$\varepsilon(u) = \frac{1}{2}(\nabla u + (\nabla u)^T), \quad \sigma(u) = 2\mu\varepsilon(u) + \lambda \operatorname{div} u\, E \qquad (1.2)$$

where $E$ is unit tensor, $\lambda$ and $\mu$ are Lame parameters.

The problem is considered in fractured media. For numerical simulations of the elastic wave equation in the fractured media, we apply the linear-slip model (LSM) [3,4]. Specifically, we assume that the fractures have a vanishing width across which the tractions are taken to be continuous. Following the linear-slip model, we have a linear relation between traction vector and the magnitude of the discontinuity in the displacement field as follows

$$[u] = Z\sigma \cdot n, \qquad (1.3)$$

where $[u]$ is the jump of the displacement field at the fracture, $\sigma \cdot n$ is the traction vector at the surface of the fracture and $Z$ is the fracture compliance matrix.

The compliance matrix is characterized in terms of two parameters. For the fracture with up-down symmetry and rotational symmetry about the normal, the fracture compliance matrix is diagonal

$$Z = \begin{bmatrix} z_1 & 0 \\ 0 & z_2 \end{bmatrix}$$

where $z_1 = k_1^{-1}$ and $z_2 = k_2^{-1}$ are the normal and tangential compliances, respectively.

In the computations, the energy of waves needs to be absorbed in artificial boundaries in order to avoid spurious reflections caused by the finite computational domain [5]. We use a first order absorbing boundary condition

$$i\rho\omega A u = -\sigma(u)n, \quad x \in \partial\Omega, \qquad (1.4)$$

where

$$A = \begin{bmatrix} n_1 & n_2 \\ -n_2 & n_1 \end{bmatrix} \begin{bmatrix} c_p & 0 \\ 0 & c_s \end{bmatrix} \begin{bmatrix} n_1 & -n_2 \\ n_2 & n_1 \end{bmatrix}.$$

Here $n = (n_1, n_2)$ is the outward normal to the boundary and

$$c_p = \sqrt{\frac{\lambda + 2\mu}{\rho}}, \quad c_s = \sqrt{\frac{\mu}{\rho}} \qquad (1.5)$$

where $c_s$, $c_p$ are the S- and P-wave velocities.

## 2    Fine Grid Approximation

For the fine-grid approximation, we use the Interior Penalty Discontinuous Galerkin (IPDG) finite element method that allows for discontinuities in the displacement field to simulate fractures with the linear-slip model [6–8]. Let $E$ be the edge between the elements $K_1$ and $K_2$, then the average and jump of a vector function $u$ on $E$ are given by

$$\{u\} = \frac{u\mid_{K_1} + u\mid_{K_2}}{2}, \quad [u] = u\mid_{K_1} - u\mid_{K_2}.$$

Let $\mathcal{T}_h$ denote a finite element partition of the domain $\Omega$ and $\Gamma_h$ the set of all the interior faces between the elements $\mathcal{T}_h$. Let $\Gamma_c \subset \Gamma_h$ be the subset of all faces where the displacement field is continuous and $\Gamma_f \subset \Gamma_h$ be the subset of facet that represent fractures.

The weak formulation of the elastic wave equation for the interior penalty discontinuous Galerkin method in fractured media is given by

$$\sum_{K\in\mathcal{T}_h} \int_K (\sigma(u), \varepsilon(v)) dx - \sum_{K\in\mathcal{T}_h} \int_K \rho\omega^2 uv dx + \sum_{E\in\Gamma_b} \int_E i\rho\omega A uv ds$$

$$- \sum_{E\in\Gamma_c} \int_E \{\tau(u)\}\, [v]\, ds - \sum_{E\in\Gamma_c} \int_E \{\tau(v)\}\, [u]\, ds + i \sum_{E\in\Gamma_c} \frac{\overline{\omega}}{h} \int_E (\lambda + 2\mu)\, [u]\, [v]\, ds$$

$$+ \sum_{E\in\Gamma_f} \int_E Z^{-1}\, [u]\, [v]\, ds = \sum_{K\in\mathcal{T}_h} \int_K fv dx,$$

where $\overline{\omega}$ is the penalty parameter, $\tau(u) = \sigma(u)n$ is the traction vector, $\Gamma_b$ is a subset of faces on the boundary. Here

$$u = Re(u) + iIm(u),$$

and $u = \sum_j u_j\phi_j$, $\phi_j$ are linear basis functions for the fine scale approximation. We can write the complex valued problem of 2.1 in matrix form

$$(K_h + B_h - \omega^2 M_h)U = F_h, \tag{2.1}$$

where $U = Re(U) + iIm(U)$.

## 3    Multiscale Method

In this section, we describe construction of the multiscale basis functions and coarse grid approximation [9–14]. In the GMsFEM, we have following computational algorithm: (i) the construction of the multiscale basis functions by the solution of the local eigenvalue problem in local domain $K$ and (ii) the construction and solution of the coarse grid approximation on multiscale space.

We define coarse grid $\mathcal{T}_H$ in domain $\Omega$, $\mathcal{T}_H = \cup_{i=1}^{N_c} K_i$ with mesh sizes $H \gg h > 0$ where $K_i$ is coarse cell (local domain), $N_c$ is number of coarse cells. Let $V_H$ be a finite dimensional function space, which consists of functions that are smooth on each local domain related to the coarse mesh cell. We construct two local multiscale spaces $V_H^K = V_H^{K,b} + V_H^{K,i}$ (boundary and interior) by solution of the local eigenvalue problems on each coarse grid cell $K \in \mathcal{T}_H$. Furthermore, eigenfunctions corresponding to the dominant eigenvalues are used as the final multiscale basis functions. The resulting space is called the multiscale (offline) space for the coarse-grid block $K$. The global multiscale space $V_H$ is then defined as the linear span of all these $V_H^K$, $K \in \mathcal{T}_H$ and will be used as the approximation space of IPDG coupling, which can be formulated as follows: find $u_H \in V_H$:

$$a_{DG}(u_H, v) - \omega^2 m(u_H, v) + ib(u_H, v) = l(v), \quad v \in V_H \qquad (3.1)$$

where

$$a_{DG}(u, v) = \sum_{K \in \mathcal{T}_H} \int_K (\sigma(u), \varepsilon(v)) dx - \sum_{E \in \Gamma_c} \int_E \{\tau(u)\} [v] \, ds - \sum_{E \in \Gamma_c} \int_E \{\tau(v)\} [u] \, ds$$

$$+ i \sum_{E \in \Gamma_c} \frac{\overline{\omega}}{h} \int_E (\lambda + 2\mu) [u] [v] \, ds + \sum_{E \in \Gamma_f} \int_E Z^{-1} [u] [v] \, ds,$$

$$m(u, v) = \sum_{K \in \mathcal{T}_H} \int_K \rho u v dx, \quad b(u, v) = \sum_{E \in \Gamma_b} \int_E \rho \omega A u v ds$$

$$l(v) = \sum_{K \in \mathcal{T}_H} \int_K f v dx$$

and $u = Re(u) + iIm(u)$.

For the construction of the boundary space, we solve following spectral problem in $K$

$$a_{DG}(\phi^{K,b}, v) = \lambda^b s(\phi^{K,b}, v), \quad v \in V_h(K), \qquad (3.2)$$

where

$$s(\phi^{K,b}, v) = \int_{\partial K} \rho \phi^{K,b} v ds$$

To construct a multiscale space $V_H^{K,b}$, we select the first $M^b$ eigenvectors $\phi_1^{K,b}$, $\phi_2^{K,b}$, ... , $\phi_{M^b}^{K,b}$ corresponding to the first $M^b$ smallest eigenvalues $\lambda_1^b \leq \lambda_2^b \leq$ ... $\leq \lambda_{M^b}^b$, and define the space $V_b^{K,H}$ by

$$V_H^{K,b} = \text{span} \left\{ \phi_1^{K,b}, \phi_2^{K,b}, ..., \phi_{M^b}^{K,b} \right\}. \qquad (3.3)$$

Interior space $V_H^{K,i}$ is defined to capture interior eigenmodes for $K$ and use following spectral problem with homogeneous Dirichlet boundary conditions to identify the important modes

$$a_{DG}(\phi^{K,i}, v) = \lambda^i s(\phi^{K,i}, v), \quad v \in V_h(K),$$
$$\phi^i = 0, \quad x \in \partial K. \qquad (3.4)$$

We select the first $M^i$ eigenvectors $\phi_1^{K,i}, \phi_2^{K,i}, ..., \phi_{M^i}^{K,i}$ corresponding to the first $M^i$ smallest eigenvalues $\lambda_1^i \leq \lambda_2^i \leq ... \leq \lambda_{M^i}^i$. The space $V_H^{K,i}$ is spanned by the functions

$$V_H^{K,i} = \text{span} \left\{ \phi_1^{K,i}, \phi_2^{K,i}, ..., \phi_{M^i}^{K,i} \right\}. \qquad (3.5)$$

The coarse-scale system can be calculated by projecting the fine-scale matrices onto the coarse grid with the global projection matrix assembled from the calculated multiscale basis functions

$$R = (R_1, R_2, ..., R_N)^T, \quad R_j = \left[ \phi_1^{K,i}, \phi_2^{K,i}, ..., \phi_{M^i}^{K,i}, \phi_1^{K,b}, \phi_2^{K,b}, ..., \phi_{M^b}^{K,b} \right]. \quad (3.6)$$

where $R_j$ is the local projection matrix in a coarse element $K_j$ and $N$ is the number of coarse grid elements. In the numerical implementation, we first assemble the global matrices $M_h$, $K_h$, and vector $F_h$ in (2.1). Then using the global projection matrix $R$, we can define the coarse-scale system

$$(K_H + B_H - \omega^2 M_H)V_H = F_H, \quad (3.7)$$

where $V_H = Re(V_H) + iIm(V_H)$, $M_H$ and $K_H$ are the coarse-scale mass stiffness matrices and $B_H$ is the coarse-scale boundary mass matrix

$$M_H = RM_hR^T, \quad K_H = RK_hR^T, \quad B_H = RB_hR^T, \quad F_H = RF_h. \quad (3.8)$$

After calculation of the coarse-scale solution, we can recover the fine-scale solution, $V_{ms} = R^T V_H$.

## 4   Numerical Results

We present the results for the fine-scale solution and the coarse-scale solution using GMsDGM. The basis functions of the offline space are constructed by the procedure described above. For numerical simulations, we use the following parameters. Computational domains with different length of fracture are presented in Fig. 1 and have dimensions $\Omega = [0, L_x] \times [0, L_y]$ with $L_x = L_y = 500[\text{m}]$. On the left of Fig. 1, we have a fracture length of 10[m] (Case 1), and on the right of Fig. 1, we have a fracture length of 20[m] (Case 2). In all cases, we consider the fracture orientations to be the same.

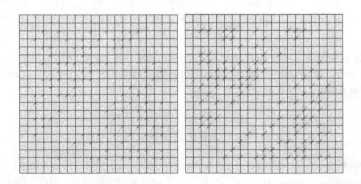

**Fig. 1.** Computational grids with different length of fractures (10[m] and 20[m]). Left: Case 1. Right: Case 2.

We set the source term $f(x) = G(x)P(\theta)$, where $P(\theta) = (cos\theta, sin\theta)$ is the polar angle of the source force vector with $\theta = 0$ and the spatial function $G(x)$ is defined as point source, $G(x) = \delta(x - x_0)$ with $x_0 = (250, 250)$ assigned as the center of the computational domain. We take penalty parameters $\overline{\omega} = 4$ and run simulations for $\omega = 2\pi f_0$ with $f_0 = 15$.

For numerical simulation, we set following parameters

$$K = \frac{E}{3(1 - 2\nu)}, \quad \mu = \frac{E}{2(1 + \nu)}, \tag{4.1}$$

with $E = 40 \cdot 10^9 [\text{Pa}]$, $\nu = 0.3$, $\rho = 2300 [kg/m^3]$. For fracture compliance matrix $Z$ in (1.3), we use $z_1 = z_2 = 10^7 [\text{m/Pa}]$.

For the numerical solution, we construct structured coarse grids with 400 cells. The fine grids are unstructured grids that resolve the fractures.

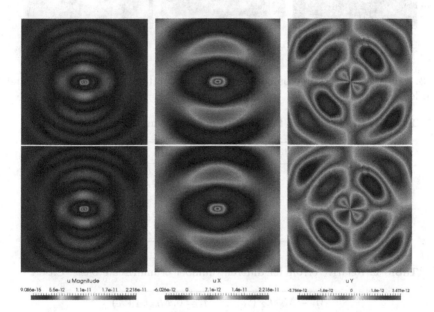

u Magnitude    u X    u Y

**Fig. 2.** Fine-scale (top) and coarse-scale (bottom) solution for magnitude, X and Y (from left to right) for Case 1.

For the Case 1, fine grid contains 16077 vertices and 31752 triangle elements. Fine grid for Case 2 contains 16509 vertices and 32616 triangle elements (see Fig. 1). Coarse grid contains 441 vertices and 400 rectangular elements are the same for Case 1 and Case 2.

In Fig. 2, we present fine-scale and multiscale solutions for Case 1. Calculations was performed using 25 boundary and 25 interior multiscale basis functions. Size of coarse grid system is $DOF = 40000$. In Fig. 3, we present fine-scale and multiscale solutions for Case 2. Calculations was performed in the same way as

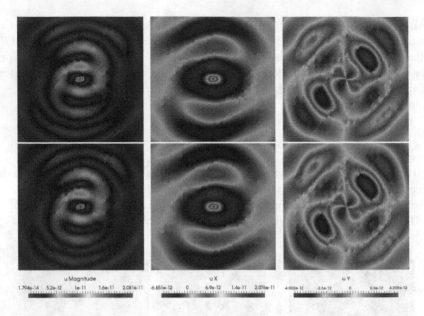

**Fig. 3.** Fine-scale (top) and coarse-scale (bottom) solution for magnitude, X and Y (from left to right) for Case 2.

**Table 1.** Relative errors for different number of multiscale basis. Case 1

| $M_i + M_b$ | $L_2$ (%) | $M_i = 0, M_b$ | $L_2$ (%) |
|---|---|---|---|
| 5+ 5 | 92.4309 | 5 | 92.0936 |
| 10+10 | 20.0548 | 10 | 22.5863 |
| 15+15 | 7.58553 | 15 | 10.1345 |
| 20+20 | 4.06103 | 20 | 5.94387 |
| 25+25 | 3.48032 | 25 | 4.15289 |

**Table 2.** Relative errors for different number of multiscale basis. Case 2

| $M_i + M_b$ | $L_2$ (%) | $M_i = 0, M_b$ | $L_2$ (%) |
|---|---|---|---|
| 5+ 5 | 98.481 | 5 | 98.4668 |
| 10+10 | 26.5191 | 10 | 26.7304 |
| 15+15 | 9.20703 | 15 | 8.55055 |
| 20+20 | 7.42222 | 20 | 8.68984 |
| 25+25 | 6.32786 | 25 | 9.39684 |

in Case 1 using 25 boundary and 25 interior multiscale basis functions. Size of coarse grid system is $DOF = 40000$. We obtain good solution for reduced order model using GMsDGM.

Relative $L_2$ errors are shown in Table 1 and 2, where $M_i$ and $M_b$ are the numbers of the multiscale interior and boundary basis functions. When we have zero interior bases $M_i = 0$ (in Table 1 and 2, right table) the errors increase. From the both tables we observe that the interior multiscale basis functions can reduce errors.

We present numerical results for two geometries with different size of fractures (10 m and 20 m). We construct reduced order model using Generalized Multiscale Discontinuous Galerkin Finite Element Method. Our results show that the presented method give good approximation of the solution and reduce size of system.

**Acknowledgments.** Work is supported by the mega-grant of the Russian Federation Government (N 14.Y26.31.0013).

# References

1. De Basabe, J.D., Sen, M.K., Wheeler, M.F.: Seismic wave propagation in fractured media: a discontinuous Galerkin approach, SEG Expanded Abstr 30 (2011)
2. De Basabe, J.D., Sen, M.K., Wheeler, M.F.: Elastic wave propagation in fractured media using the discontinuous Galerkin method. Geophysics **81**(4), T163–T174 (2016)
3. Zhang, J.: Elastic wave modeling in fractured media with an explicit approach. Geophysics **70**(5), T75–T85 (2005)
4. Schoenberg, M.: Elastic wave behavior across linear slip interfaces. J. Acoust. Soc. Am. **68**(5), 1516–1521 (1980)
5. Engquist, B., Majda, A.: Absorbing boundary conditions for numerical simulation of waves. Proc. Nat. Acad. Sci. **74**(5), 1765–1766 (1977)
6. Grote, M.J., Schneebeli, A., Schötzau, D.: Discontinuous Galerkin finite element method for the wave equation. SIAM J. Numer. Anal. **44**(6), 2408–2430 (2006)
7. Arnold, D.N.: An interior penalty finite element method with discontinuous elements. SIAM J. Numer. Anal. **19**(4), 742–760 (1982)
8. Lahivaara, T.: Discontinuous Galerkin Method for Time-Domain Wave Problems. University of Eastern Finland, Joensuu (2010)
9. Efendiev, Y., Galvis, J., Hou, T.Y.: Generalized multiscale finite element methods (GMsFEM). J. Comput. Phys. **215**, 116–135 (2013)
10. Chung, E.T., Efendiev, Y., Leung, W.T.: An online generalized multiscale discontinuous Galerkin method (GMsDGM) for flows in heterogeneous media. Commun. Comput. Phys. **21**(2), 401–422 (2017)
11. Chung, E.T., Efendiev, Y., Leung, W.T.: Generalized multiscale finite element methods for wave propagation in heterogeneous media. Multiscale Model. Simul. **12**(4), 1641–1721 (2014)
12. Gao, K., Fu, S., Gibson, R.L., Chung, E.T., Efendiev, Y.: Generalized multiscale finite-element method (GMsFEM) for elastic wave propagation in heterogeneous, anisotropic media. J. Comput. Phys. **295**, 161–188 (2015)
13. Chung, E.T., Efendiev, Y., Gibson, R.L., Vasilyeva, M.: A generalized multiscale finite element method for elastic wave propagation in fractured media. GEM-Int. J. Geomath. **7**(2), 163–182 (2016)
14. Chung, E.T., Efendiev, Y., Fu, S.: Generalized multiscale finite element method for elasticity equations. GEM-Int. J. Geomath. **5**(2), 225–251 (2014)

# Difference Schemes for the Nonlinear Equations in Partial Derivatives with Heredity

T. V. Gorbova[1], V. G. Pimenov[1,2], and S. I. Solodushkin[1,2(✉)]

[1] Ural Federal University, Ekaterinburg, Russia
solodushkin_s@mail.ru
[2] Institute of Mathematics and Mechanics, Ural Branch of the RAS,
Ekaterinburg, Russia

**Abstract.** We consider the initial-boundary value problem for nonlinear partial differential equations, the source of which are population models. Nonlinearity is contained both in the differential operator and in the inhomogeneity function. We construct a nonlinear implicit difference scheme, which requires the use of iterative solution methods on each time layer. Stability and convergence of the proposed numerical method were proved. Numerical experiments have been carried out, both on test examples and on the example of the biological model of the population.

**Keywords:** Nonlinear difference scheme ·
Convergence of the difference scheme · Partial differential equation

## 1 First Section

Partial differential equations with non-linearity in the differentiation operator are used in many mathematical models, for example, in population development models. In addition, these models could be complicated by hereditary effects, fractional derivatives, the multidimensionality of spatial variables. In this paper we consider an equation of the following form

$$\frac{\partial p(x,t)}{\partial t} = \frac{\partial^2 \phi(p(x,t))}{\partial x^2} + g(x,t,p_t(x,\cdot)), \tag{1}$$

where $t$ and $x$, $0 \leqslant t \leqslant T, 0 \leqslant x \leqslant X$, are independent variables, $p(x,t)$ is an unknown function, $p_t(x,\cdot) = \{p(x,t+s), -\tau \leqslant s \leqslant 0\}$ heredity (prehistory) of the unknown function to the moment $t$.

Initial and boundary conditions are specified $p(0,t) = p_0(t)$, $p(X,t) = p_1(t)$, $0 \leqslant t \leqslant T$, and $p(x,s) = \varphi(x,s)$, $0 \leqslant x \leqslant X$, $-\tau \leqslant s \leqslant 0$.

In [1], as in most similar works, numerical methods are not considered, but an attempts are made to find an exact solution in the form of series. However, the analytical solution in such problems can be found only in exceptional cases,

© Springer Nature Switzerland AG 2019
I. Dimov et al. (Eds.): FDM 2018, LNCS 11386, pp. 258–265, 2019.
https://doi.org/10.1007/978-3-030-11539-5_28

therefore, the development, justification of the stability and convergence, as well as the software implementation of numerical algorithms, seems to be a relevant problem. The elaboration of difference schemes is associated with a number of difficulties: a numerical experiment showed that the use of an explicit scheme leads to instability. The direct application of the implicit scheme to Eq. (1) leads to the necessity to solve nonlinear systems of large dimension.

In this paper the technique described in [2] is used: by means of the change of variables, the nonlinearity in the differentiation operator with respect to the spatial variable is transmitted to the time differentiation operator. Then an implicit difference scheme is constructed, the appeared nonlinear system is solved by the Newton method. The hereditary effect in this work is accounted by using the technique [4,5]. The main result consists in proving the stability and convergence of the constructed algorithm. To do this we elaborated a nonlinear modification of the general difference scheme for systems with heredity [3-7].

## 2  Implicit Difference Method

Assuming the single-valued invertibility of $\phi(p)$ on the domain of our interest, we make the substitution $u = \phi(p)$, $p = \omega(u)$, then (1) is transformed to the form

$$\frac{\partial \omega(u)}{\partial t} = \frac{\partial^2 u}{\partial x^2} + f(x,t,u_t(x,\cdot)),\ u_t(x,\cdot) = \{u(x,t+s), -\tau \leqslant s \leqslant 0\}. \tag{2}$$

The initial and boundary conditions are changed correspondingly:

$$u(0,t) = \phi(p_0(t)) = \mu_0(t),\ u(X,t) = \phi(p_1(t)) = \mu_1(t),\ 0 \leqslant t \leqslant T, \tag{3}$$

$$u(x,s) = \phi(\varphi(x,s)),\ 0 \leqslant x \leqslant X,\ -\tau \leqslant s \leqslant 0. \tag{4}$$

We shall assume that the problem (2)–(4) has a unique solution, understood in the classical sense, and this solution has continuous derivatives with respect to state variables $x$ up to fourth order, continuous derivatives with respect to time $t$ up to second order. Also we assume that functional $f(x,t,u_t(x,\cdot))$ is Lipschitz with respect to the last argument on the set of continuous functions, function $\omega(u)$ is twice continuously differentiable in a bounded domain containing the solution $u(x,t)$ and the following condition holds

$$\omega'(u) \geqslant \hat{\omega} > 0. \tag{5}$$

We consider an equidistant partition of $[0,X]$ into parts with step size $h = X/N$ and define the grid $x_i = ih$, $i = 0,\ldots,N$. We also split the time interval $[-\tau, T]$ into parts with step size $\Delta = T/M$ (without loss of generality $\tau/\Delta = m$ is integer) and define the grid $t_j = j\Delta$, $j = -m,\ldots,M$.

Denote by $u_j^i$ the approximation of the function value $u(x_i, t_j)$, $i = 0, 1, \ldots N$, $j = 0, \ldots M$, at the respective node.

For every fixed node $(x_i, t_j)$, $i = 0, 1, \ldots N$, $j = 0, \ldots M$, we introduce its discrete prehistory as $\{u_k^i\}_j = \{u_k^i, \max\{0, j - m\} \leqslant k \leqslant j\}$.

A mapping $I$ defined on the set of all admissible discrete prehistories and acting by the rule $\{u_k^i\}_j \to v_j^i(\cdot) = v_j^i(t_j + \xi)$, where $v_j^i(\cdot)$ is defined on $[t_j - \tau, t_j]$, is called an interpolation operator for the discrete history.

In what follows we will use piecewise linear interpolation

$$u_j^i(t_j + s) = \frac{1}{\Delta}((t_k - t_j - s)u_{k-1}^i + (t_j + s - t_{k-1})u_k^i),\ t_{k-1} \leqslant t_j + s \leqslant t_k.$$

Consider a nonlinear implicit difference scheme, $j = 0, 1, \ldots, M - 1$,

$$\frac{\omega(u_{j+1}^i) - \omega(u_j^i)}{\Delta} = \frac{u_{j+1}^{i-1} - 2u_{j+1}^i + u_{j+1}^{i+1}}{h^2} + f(x_i, t_j, u_j^i(\cdot)),\ i = 1, \ldots, N - 1,$$

$$u_{j+1}^0 = \mu_0(t_{j+1}),\ u_{j+1}^N = \mu_1(t_{j+1}),$$

$$(6)$$

with initial conditions $u_j^i = \phi(\varphi(x_i, t_j))$, $i = 0, \ldots, N$, $j = -m, \ldots, 0$.

For each fixed $j$ the (6) is a system of equations that are nonlinear with respect to $u_{j+1}^i$, $i = 1, \ldots, N - 1$. To solve the (6) at each time step $j$ we apply Newton's method [2], pp. 444–454,

$$\omega(u_{j+1}^i[k]) + \omega'(u_{j+1}^i[k])(u_{j+1}^i[k+1] - u_{j+1}^i[k])$$

$$- \Delta \frac{u_{j+1}^{i-1}[k+1] - 2u_{j+1}^i[k+1] + u_{j+1}^{i+1}[k+1]}{h^2} = \omega(u_j^i) + \Delta f(x_i, t_j, u_j^i(\cdot)),\quad (7)$$

where $k$ is an iteration number, $k = 0, 1, \ldots$, and $u_{j+1}^i[k]$ is $k$-th approximation by the Newton's method to $u_{j+1}^i$, $i = 1, \ldots, N - 1$. Note, when we search $u_{j+1}^i[k+1]$ in (7) we use $u_j^i$ which represent not the exact solution obtained at the $j$-th time layer (it is actually unknown) but its approximation in Newton's method.

The system (7) is a tridiagonal system of linear equations. Since from (5) it follows the diagonal predominance the system (7) could be effectively solved using the sweep algorithm.

Note that if condition (5) is satisfied, the method (7) can be rewritten as

$$u_{j+1}^i[k+1] - \Delta \frac{u_{j+1}^{i-1}[k+1] - 2u_{j+1}^i[k+1] + u_{j+1}^{i+1}[k+1]}{\omega'(u_{j+1}^i[k])h^2}$$

$$= u_{j+1}^i[k] + \frac{1}{\omega'(u_{j+1}^i[k])}(\omega(u_j^i) - \omega(u_{j+1}^i[k]) + \Delta f(x_i, t_j, u_j^i(\cdot))).\quad (8)$$

## 3    Elements of the General Theory of Nonlinear Difference Schemes

We will embed the proposed method into the general scheme suggested earlier for linear difference problems with heredity [3–5].

Let a segment $[-\tau, T]$ be given, divide the segment into parts by step $\Delta = T/M$ and define nodes $t_j = j\Delta$, $j = -m, \ldots, M$; without loss of generality $\tau/\Delta = m$ is integer.

A discrete model is defined as a grid function $y_j = y(t_j) \in Y$, $j = -m, \ldots, M$, where $Y$ is $q$-dimensional normed space with norm $\| \cdot \|_Y$. We will assume that the dimension $q$ of the space $Y$ depends on a number $h > 0$.

The set $\{y_i\}_j = \{y_i \in Y, \ i = j - m, \ldots, n\}$ will be called the prehistory of the discrete model by the time $t_j$, $j \geqslant 0$.

Let $V$ be a linear normed space with norm $\| \cdot \|_V$, so-called interpolation space. A mapping $I: I(\{y_i\}_j) = v \in V$ is, by definition, an operator of the interpolation of the discrete prehistory. We will assume that the interpolation operator satisfies the Lipschitz condition, i. e. there exists a constant $L_I$ such that, for all prehistories of the discrete model $\{y_i^1\}_j$ and $\{y_i^2\}_j$ the following inequality holds:

$$\| v^1 - v^2 \|_V \leqslant L_I \max_{j-m \leqslant i \leqslant j} \| y_i^1 - y_i^2 \|_Y . \tag{9}$$

Starting values of the model are defined as follow

$$y(t_i) = y_i, \ i = -m, \ldots, 0. \tag{10}$$

The formula of advance of the model by a step is, by definition, the relation

$$y_{j+1} = S(y_j) + \Delta\Phi(I(\{y_i\}_j)), \tag{11}$$

where the transition operator $S(y_j) = S(y_j, t_j, \Delta, h)$ is a nonlinear operator which is Lipschitz with a constant $L_S = L_S(\Delta, h)$; the function of advance by a step $\Phi(v) = \Phi(v, t_j, \Delta, h)$ is a nonlinear mapping into the $Y$ which is Lipschitz with respect to the first argument with a constant $L_\Phi$.

The function of exact values is, by definition, the mapping

$$Z(t_j, \Delta, h) = z_j \in Y, \ j = -m, \ldots, M. \tag{12}$$

Knowledge of the function of exact values is equivalent to knowing the exact solution of the original problem in the nodes. In what follows, for simplicity, we assume that the starting values coincide with the initial value of the function of the exact values

$$y_i = z_i, i = -m, \ldots, 0. \tag{13}$$

We will say that the method (11) converges if there exists a constant $C$ and a function $q(\Delta, h)$, $\lim\limits_{\Delta \to 0, h \to 0} q(\Delta, h) = 0$, such that the following inequality holds:

$$\| z_j - y_j \|_Y \leqslant Cq(\Delta, h) \tag{14}$$

for all $j = 0, \ldots, M$. Function $q(\Delta, h)$ defines the order of convergence.

The order of convergence depends on the approximation error, the starting error (in our case it is absent) and the stability properties of the method. An error of approximation (a residual) is, by definition, the grid function

$$d_j = (z_{j+1} - S(z_j))/\Delta - \Phi(I(\{y_j\}_j)), \ j = 0, \ldots, M - 1. \tag{15}$$

Note that for systems with heredity, there are two types of residuals for any method: a residual with interpolation and a residual without interpolation. The (15) defines a discrepancy with interpolation.

We will say that the error of approximation in method (11) has an order of $q(\Delta, h)$, if there exists a constant $C$ independent of $\Delta$ and $h$ such that for all $j = 1, \ldots, M$ the following inequality holds:

$$\| d_j \|_Y \leqslant Cq(\Delta, h).$$

The method (11) is said to be stable, if

$$L_S = L_S(\Delta, h) \leqslant 1. \tag{16}$$

**Theorem 1.** *Let method* (11) *is stable, the error of approximation has an order* $q(\Delta, h)$, $\lim\limits_{\Delta \to 0, h \to 0} q(\Delta, h) = 0$, *then method* (11) *converges with order* $q(\Delta, h)$.

*Proof.* Let us denote $\delta_j = \| z_j - y_j \|_Y$, $j = 0, \ldots, M$. Since $S$ and $\Phi$ are Lipschitz ones for all $j = 0, \ldots, M - 1$ we have

$$\delta_{j+1} = \| S(z_j) + \Delta\Phi(z_j) + \Delta d_j - S(y_j) - \Delta\Phi(y_j) \|_Y \leqslant L_S\delta_j + \Delta L_\Phi\delta_j + \Delta \| d_j \|_Y. \tag{17}$$

Denote as $\hat{\delta}_j = \max_{j-m \leqslant i \leqslant j} \delta_i$, then using the stability condition we obtain

$$\hat{\delta}_{j+1} \leqslant L_S\hat{\delta}_j + \Delta L_\Phi L_I \hat{\delta}_j + \Delta \| d_j \|_Y. \tag{18}$$

Using the stability condition, we obtain

$$\hat{\delta}_{j+1} \leqslant (1 + \Delta L_\Phi L_I)\hat{\delta}_j + \Delta \| d_j \|_Y. \tag{19}$$

From this estimate the following estimate could be deduced by standard methods

$$\hat{\delta}_j \leqslant \hat{\delta}_0 exp(TL_\Phi L_I) + \| d_j \|_Y \frac{exp(TL_\Phi L_I)}{L_\Phi L_I}. \tag{20}$$

The last estimate is right for all $j = 0, \ldots, M$. This implies the conclusion of the theorem.

Transformation of the nonlinear difference scheme (11) to the explicit form is often a difficult task. Consider the approximation of this scheme in the form of an iterative process

$$y_{j+1}[k] = S_k(y_{j+1}[k-1]) + \Delta\Phi_k(y_{j+1}[k-1], I(\{y_i[K]\}_j)), \quad k = 1, \ldots, K. \tag{21}$$

On each time layer, as the initial approximation of this iterative process, we can take $y_{j+1}[0] = y_j[K]$ (number $K$ is fixed) then the iterative process (21) is reduced to the form

$$y_{j+1}[K] = \hat{S}_K(y_j[K]) + \Delta\hat{\Phi}_K(I(\{y_i[K]\}_j)). \tag{22}$$

The definitions introduced above for the general nonlinear scheme (11) could be transformed in an obvious way for the *approximation scheme* (22).

We will say that the method (22) converges if there exists a constant $C$ and a function $q(\Delta, h, K)$,

$$\lim_{\Delta \to 0, h \to 0, K \to \infty} q(\Delta, h, K) = 0,$$

such that the following inequality holds:

$$\| z_j - y_j[K] \|_Y \leqslant Cq(\Delta, h, K)$$

for all $j = 0, \ldots, M$.

The method (22) is said to be stable, if the operator $\hat{S}_K$ is Lipschitz with the constant $L_{\hat{S}_K}$ such that

$$L_{\hat{S}_K} = L_{\hat{S}_K}(\Delta, h, K) \leqslant 1. \tag{23}$$

The error of approximation of the method (22) is defined in a similar way.

**Theorem 2.** *Let the method (22) is stable, the approximation error is of the order* $q(\Delta, h, K)$, $\lim_{\Delta \to 0, h \to 0, K \to \infty} q(\Delta, h, K) = 0$, *then the method converges with the order* $q(\Delta, h, K)$.

## 4  Embedding an Implicit Method in a General Nonlinear Scheme

Let us embed method (7) in the scheme described in the previous section. Without loss of generality, we consider homogeneous boundary conditions (3)

$$u(0, t) = 0, \ u(X, t) = 0, \quad 0 \leqslant t \leqslant T.$$

Let us denote $y_j = (u_j^1, u_j^2, \ldots, u_j^{N-1})^T \in Y$, where $Y$ is a vector space of dimension $N - 1$, $T$ is a transpose sign.

In the space $Y$ we define an operator $A$ as follow

$$Au_j^i = -a^2 \frac{u_j^{i-1} - 2u_j^i + u_j^{i+1}}{h^2};$$

we also define vector functions $\omega(y_j)$ and $f_j(I(\{y_i\}_j))$ as vectors with components $\omega(u_j^i)$ and $f(x_i, t_j, I(\{y_i\}_j))$ respectively, then system (6) could be represented as follow

$$\omega(y_{j+1}) + \Delta A y_{j+1} = \omega(y_j) + \Delta f_j(I(\{y_i\}_j)). \tag{24}$$

In a similar way we denote $y_j[k] = (u_j^1[k], u_j^2[k], \ldots, u_j^{N-1}[k])^T \in Y$, also we denote by $\omega'(y_j)$ the diagonal matrix with $\omega'(u_j^i)$ on the main diagonal in

$i$-th row. Then the iteration process (7) with the number of iterations $K$ can be written in the form

$$(\omega'(y_{j+1}[k-1]) + \Delta A)y_{j+1}[k] = \omega'(y_{j+1}[k-1])y_{j+1}[k-1]$$
$$+ \omega(y_j[K]) - \omega(y_{j+1}[k-1]) + \Delta f_j(I(\{y_i\}_j[K])), \quad k = 1, \cdots, K. \tag{25}$$

$$y_{j+1}[0] = y_j[K] \tag{26}$$

The iterative process could also be written in vector form

$$(E + \Delta \tilde{A})y_{j+1}[k] = y_{j+1}[k-1] + \Delta \tilde{F}(y_{j+1}[k-1], I(\{y_i\}_j[K])), \tag{27}$$

where

$$\tilde{A}u^i_{j+1}[k-1] = -a^2 \frac{u^{i-1}_{j+1}[k-1] - 2u^i_{j+1}[k-1] + u^{i+1}_{j+1}[k-1]}{\omega'(u^i_{j+1}[k-1])h^2},$$

and $\tilde{F}(y_{j+1}[k-1], I(\{y_i\}_j[K]))$ is a vector with components

$$\frac{1}{\omega'(u^i_{j+1}[k])} \left( \frac{\omega(u^i_j) - \omega(u^i_{j+1}[k])}{\Delta} + f(x_i, t_j, u^i_j(\cdot)) \right)$$

Since the matrix $E + \Delta \tilde{A}$ is non degenerate, the method (27) could be rewritten in explicit form (21), where $S_k(y_{j+1}[k-1]) = (E + \Delta \tilde{A})^{-1}$, $\Phi_k(y_{j+1}[k-1]) = = (E + \Delta \tilde{A})^{-1}\tilde{F}(y_{j+1}[k-1], I(\{y_i\}_j[K]))$.

We rewrite the system (6) (or Eq. (24)) in the form

$$F(y_{j+1}) = \omega(y_{j+1}) + \Delta A y_{j+1} - B = 0, \quad B = \omega(y_j) + \Delta f_j(I(\{y_i\}_j)). \tag{28}$$

Then Newton's method (7) (or (25)) could be written in the form

$$y_{j+1}[k+1] = y_{j+1}[k] - S^{-1}(y_{j+1}[k])F(y_{j+1}[k]),$$

$$S(y_{j+1}[k]) = (\omega'(y_{j+1}[k]) + \Delta A) = F'(y_{j+1}[k]), \quad k = 0, \dots, K-1. \tag{29}$$

Let us denote $\Psi(y) = y - S^{-1}(y)F(y)$; it could be shown that for small $\Delta$ operator the operator $\Psi$ is contractive.

Method (22) could be rewritten in the form

$$y_{j+1}[K] = S_K(\Psi(\Psi(\cdots\Psi(y_{j+1}[0])))) + \Delta\Phi_K(\Psi(\Psi(\cdots\Psi(y_{j+1}[0]))), I(\{y_i[K]\}_j)),$$
$$\tag{30}$$

This implies the stability of the method. Taking into account (26), method (30) could also be written in the form

$$y_{j+1}[K] = S_K(\Psi(\Psi(\cdots\Psi(y_j[K])))) + \Delta\Phi_K(\Psi(\Psi(\cdots\Psi(y_j[K]))), I(\{y_i[K]\}_j))$$
$$\tag{31}$$

Analyzing the order of the residual (including the quality of interpolation), we obtain the assertion of the theorem.

**Theorem 3.** *The implicit approximation method* (7) *written in the form* (22) *or* (31) *converges and has the order* $\Delta + h^2 + \lambda^{2^K}$, $0 < \lambda < 1$.

## 5   Numerical Examples

On the domain $x \in (0, \pi)$, $t \in (0, 4\pi)$ we consider the initial boundary value problem

$$\frac{\partial e^u}{\partial t} = \frac{\partial^2 u}{\partial x^2} - 0.5\, e^{\sin x \, \cos t} \int_{-\pi}^{0} u(x, t + s)\, ds + u, \qquad (32)$$

with initial and boundary conditions

$$u(x, s) = \sin x \, \cos s, \quad 0 \le x \le \pi, \ -\pi \le s \le 0,$$

$$u(0, t) = 0, \ u(\pi, t) = 0, \quad 0 \le t \le 4\pi.$$

Problem (32) has an exact solution $u(x, t) = \sin x \, \cos t$.

To improve the convergence rate the nonlinear analog of Crank-Nicolson method was used instead of method (7). The accuracy of the Newton method is chosen to be $\epsilon = 10^{-5}$. To calculate the integrals in the distributed delay, the trapezoid method is used. Linear extrapolation is used.

In the table below we report the deviations $\mathbf{diff} = \max_{i,j} |u_j^i - u(x_i, t_j)|$ of the approximate solution calculated by method from the exact one. Here $N$ and $M$ are the number of segments with respect to space and time.

| No. | 1 | 2 | 3 | 4 | 5 | 7 | 8 | 9 |
|-----|-----|-----|-----|-----|-----|-----|-----|-----|
| N | 32 | 32 | 32 | 32 | 8 | 8 | 8 | 8 |
| M | 32 | 64 | 128 | 256 | 16 | 32 | 64 | 128 |
| **diff** | 0.4150 | 0.1034 | 0.0274 | 0.0096 | 1.4828 | 0.4793 | 0.1613 | 0.0867 |

**Acknowledgments.** We acknowledge the support by the program 02.A03.21.0006 on 27.08.2013.

## References

1. Srivastava, V.K., Kumar, S., et al.: Two-dimensional time fractional-order biological population model and its analytical solution. Egypt J. Basic Appl. Sci. **1**, 71–76 (2014)
2. Samarskii, A.A.: Theory of Difference Schemes. Nauka, Moscow (1989). (in Russian)
3. Pimenov, V.G.: General linear methods for the numerical solution of functional-differential equations. Differ. Equ. **37**(1), 116–127 (2001)
4. Pimenov, V.G., Lozhnikov, A.B.: Difference schemes for the numerical solution of the heat conduction equation with aftereffect. Proc. Steklov Inst. Math. **275**, 137–148 (2011)
5. Pimenov, V.G.: Difference Methods for Solving Partial Differential Equations with Heredity. Publishing House of the Ural University, Ekaterinburg (2014). (in Russian)
6. Solodushkin, S.I.: A difference scheme for the numerical solution of an advection equation with aftereffect. Russ. Math. **57**, 65–70 (2013)
7. Solodushkin, S.I., Yumanova, I.F., De Staelen, R.H.: First order partial differential equations with time delay and retardation of a state variable. J. Comput. Appl. Math. **289**, 322–330 (2015)

# Conservative Finite-Difference Algorithm for the Thermodiffusion Stefan Problem

A. Gusev[1,2]([✉]), O. Mazhorova[1], and O. Shcheritsa[1,2]

[1] Keldysh Institute of Applied Mathematics,
Miusskaya sq., 4, Moscow 125047, Russia
aogus@mail.ru
[2] Bauman Moscow State Technical University,
Baumanskaya 2-ya st., 5, Moscow 105005, Russia

**Abstract.** The paper presents practically unconditionally stable procedure for numerical study of phase transition in multi-component alloys. The algorithm is based on a conservative fully implicit finite-volume scheme and coupled solution of the governing equations of heat and mass transfer with respect to the vector of all unknowns. The proposed method is implemented for computer simulation of phase changes in a ternary alloy under industrially relevant conditions.

**Keywords:** Stefan problem · Phase transition

## 1 Introduction

Phase transition driven by coupled heat and mass transport is vital to the semiconductor manufacturing. The growth of a crystal from a ternary solution is an example of such a process. For the numerical simulations of phase transition it is necessary to solve the special modification of the Stefan problem. The main difficulties are associated with the moving boundary, the strong coupling between the temperature and the composition of the liquid and solid phases, the wide range of time-scales introduced by interface motion, diffusion in solid and liquid phases. In general technological conditions, initial composition of solid and liquid phase determine a nonequilibrium situation. Such inconsistency implies infinite initial composition gradients at the surface and singular growth rate. That is why, for most algorithms [1–3] the initial guess of interface concentration is crucial for the successful simulation. A poor guess would lead to non-convergence in the interface condition iteration [4, p. 155]. Nothing of the kind is necessary for numerical procedure presented in this paper. The algorithm is fully implicit and based on simultaneous determination of concentration and temperature distribution and interface position. The conservative properties of the algorithm allow us to model regimes of growth with the preliminary dissolution of the solid phase.

© Springer Nature Switzerland AG 2019
I. Dimov et al. (Eds.): FDM 2018, LNCS 11386, pp. 266–273, 2019.
https://doi.org/10.1007/978-3-030-11539-5_29

## 2    Mathematical Model

We consider the growth of a ternary alloy $\mathcal{A}_x\mathcal{B}_{1-x}\mathcal{C}$ in cylindrical ampoule. The ampoule length $L$ is much larger than its radius $R$, $R << L$. At the solid/liquid interface the compositions of the two phases are related by the phase diagram of the system. The interfacial kinetic processes are so fast that local equilibrium may be assumed. The model accounts for diffusion heat and matter transport in the solid and liquid phases and interface motion. The interface can be assumed flat due to the size of the ampoule and the high thermal diffusivity in both phases.

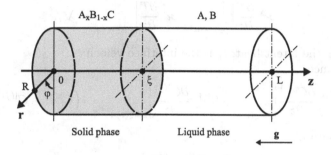

**Fig. 1.** Cylindrical ampule

There are three variables that define the composition in the liquid phase: $x^{l(\mathcal{A})}$, $x^{l(\mathcal{B})}$ and $x^{l(\mathcal{C})}$ which, however, must satisfy $x^{l(\mathcal{A})} + x^{l(\mathcal{B})} + x^{l(\mathcal{C})} = 1$, where $x^{l(\mathcal{A})}$, $x^{l(\mathcal{B})}$ and $x^{l(\mathcal{C})}$ are respectively the mole fractions of $\mathcal{A}$, $\mathcal{B}$ and $\mathcal{C}$ in the liquid. It means that only two variables, let it be $x^{l(\mathcal{A})}$, $x^{l(\mathcal{B})}$, are independent. In the solid phase $\mathcal{A}_x\mathcal{B}_{1-x}\mathcal{C}$, only one compositional variable $x$ is needed to define the composition, since $x^{s(\mathcal{A}\mathcal{C})} = x$, $x^{s(\mathcal{B}\mathcal{C})} = 1 - x$. Here $x^{s(\mathcal{A}\mathcal{C})}$ and $x^{s(\mathcal{B}\mathcal{C})}$ are the mole fractions of compounds $\mathcal{A}\mathcal{C}$ and $\mathcal{B}\mathcal{C}$ in the solid.

The domain $\Omega_s = [0, \xi) \times (0, R]$, denoted as solid phase, consists of the seed and grown crystal. The domain $\Omega_l = (\xi, L] \times (0, R]$ corresponds to the liquid phase (Fig. 1). $\Omega = \Omega_s \bigcup \Omega_l$.

The heat and matter transport equations in either bulk phases are

$$\frac{\partial C^{s(\mathcal{B})}}{\partial t} = \frac{1}{r}\frac{\partial}{\partial r}\left(rD^{s(\mathcal{B})}\frac{\partial C^{s(\mathcal{B})}}{\partial r}\right) + \frac{\partial}{\partial z}\left(D^{s(\mathcal{B})}\frac{\partial C^{s(\mathcal{B})}}{\partial z}\right), \quad (z, r) \in \Omega_s, \quad (1)$$

$$\frac{\partial C^{l(j)}}{\partial t} = \frac{1}{r}\frac{\partial}{\partial r}\left(rD^{l(j)}\frac{\partial C^{l(j)}}{\partial r}\right) + \frac{\partial}{\partial z}\left(D^{l(j)}\frac{\partial C^{l(j)}}{\partial z}\right), \quad (z, r) \in \Omega_l, \quad (2)$$

$$c_p^\gamma\frac{\partial T}{\partial t} = \frac{1}{r}\frac{\partial}{\partial r}\left(r\varkappa^\gamma\frac{\partial T}{\partial r}\right) + \frac{\partial}{\partial z}\left(\varkappa^\gamma\frac{\partial T}{\partial z}\right), \quad (z, r) \in \Omega, \quad (3)$$

where $r$, $z$ are cylindrical coordinates, t is the time, $C^{\gamma(j)}$ are the volume concentrations component $j$ in the solid and liquid phases, $D^{\gamma(j)}$ are the diffusion coefficient, $j = \mathcal{A}, \mathcal{B}$, $c_p^\gamma$ is the specific heat, $\varkappa^\gamma = k^\gamma/\rho$, $k^\gamma$ is the thermal conductivity, $\rho$ is the density. Here $\gamma = $ s if $z < \xi$ and $\gamma = $ l if $z > \xi$. The governing equation for the concentration of the component $\mathcal{B}$ in the solid phase is

$$(a_A + a_C)C^{\mathrm{s}(A)} + (a_B + a_C)C^{\mathrm{s}(B)} = \rho, \tag{4}$$

where $a_A$, $a_B$, $a_C$ are the atomic weights of corresponding components.

The conditions at the moving interface represent the internal energy balance, the mass balance and the phase diagram. Since the interface remains flat during growth this conditions can be written as

1. Stefan condition

$$\varkappa^{\mathrm{s}} \left.\frac{\partial T}{\partial z}\right|_{z=\xi(t)-} - \varkappa^{\mathrm{l}} \left.\frac{\partial T}{\partial z}\right|_{z=\xi(t)+} = \lambda \xi_t, \tag{5}$$

where $\lambda$ is the latent heat, $\xi_t$ is the interface velocity.

2. Mass balance

$$D^{\mathrm{s}(j)} \left.\frac{\partial C^{\mathrm{s}(j)}}{\partial z}\right|_{z=\xi(t)-} - D^{\mathrm{l}(j)} \left.\frac{\partial C^{\mathrm{l}(j)}}{\partial z}\right|_{z=\xi(t)+} = -\left(C^{\mathrm{s}(j)} - C^{\mathrm{l}(j)}\right) \xi_t. \tag{6}$$

3. Phase diagram

$$F_n(C^{\mathrm{s}(A)}, C^{\mathrm{s}(B)}, C^{\mathrm{l}(A)}, C^{\mathrm{l}(B)}, T) = 0, \quad n = 1, 2, 3. \tag{7}$$

The specific form of (7) depends on the ternary alloy properties.

The temperature of the ampule faces changes according to the law $T|_{z=0} = T_b(t)$, $T|_{z=L} = T_t(t)$. Boundary conditions at the cylinder lateral surface $r = R$ are

$$\varkappa^\gamma \left.\frac{\partial T}{\partial r}\right|_{r=R} = \frac{\tilde{\alpha}}{\rho} \left(T^* - T|_{r=R}\right), \tag{8}$$

where $\tilde{\alpha}$ is the heat transfer coefficient, $T^*$ is the temperature of the surrounding.

The ampoule walls are impermeable

$$D^{\gamma(j)} \left.\frac{\partial C^{\gamma(j)}}{\partial n}\right|_{\partial\Omega} = 0, \quad j = \mathcal{A}, \mathcal{B}, \quad \gamma = \mathrm{s}, \mathrm{l}. \tag{9}$$

As it has been mentioned above, the interface can be assumed flat due to the size of the ampoule and the high thermal diffusivity in both phases. So that system can be described by the average composition $\tilde{C}(z,t)$ and the temperature $\tilde{T}(z,t)$ in the cross-section [8]

$$\tilde{f}(z,t) = \frac{2}{R^2} \int_0^R f(r,z,t)r\,dr, \quad f = T, C^{\mathrm{s}(j)}, C^{\mathrm{l}(j)}, \quad j = \mathcal{A}, \mathcal{B}. \tag{10}$$

Using (10) we can rewrite (1)–(3) as

$$c_p^\gamma \frac{\partial \tilde{T}}{\partial t} = \frac{\partial}{\partial z}\left(\varkappa^\gamma \frac{\partial \tilde{T}}{\partial z}\right) + \alpha\tilde{T} + \beta, \quad \gamma = \mathrm{s}, \mathrm{l}, \tag{11}$$

$$\frac{\partial \widetilde{C}^{\gamma(j)}}{\partial t} = \frac{\partial}{\partial z} \left( D^{\gamma} \frac{\partial \widetilde{C}^{\gamma(j)}}{\partial z} \right), \quad \gamma = \mathrm{s,l}, \quad j = \mathcal{A}, \mathcal{B}. \tag{12}$$

Here $\alpha = -2\widetilde{\alpha}/(R\rho)$, $\beta = 2\widetilde{\alpha}T^*/(R\rho)$. Thus, axially symmetric problem (1)–(3) is reduced to the one-dimensional problem (11) and (12). The heat source term in the Eq. (11) represents the heat exchange with the surrounding in the original model.

We use the Landau transformation $(t, z) \to (t, y)$ in order to track the interface position [5,9,10]. The solid phase is mapped to the interval $[0, 1)$ and liquid phase is mapped to the interval $(1, 2]$. For any time the interface position is fixed to the point $y = 1$ in the new coordinate system $(t, y)$. The change of variables is given by $t = t$, $z = \varphi(y, t)$. We denote the length of the corresponding zone in the original coordinate system by $l = l(t)$; $l = l^{\mathrm{s}} = \xi(t)$ in the solid phase and $l = l^{\mathrm{l}} = L - \xi(t)$ in the liquid phase. The governing Eqs. (11) and (12) in variables $(t, y)$ can be written as

$$c_{\mathrm{p}} \left[ \frac{\partial}{\partial t} \left( l\widetilde{T} \right) - \frac{\partial}{\partial y} \left( \varphi_t \widetilde{T} \right) \right] = \frac{\partial}{\partial y} \left( \frac{\varkappa}{l} \frac{\partial \widetilde{T}}{\partial y} \right) + l \left( \alpha \widetilde{T} + \beta \right), \tag{13}$$

$$\frac{\partial}{\partial t} \left( l\widetilde{C}^{\gamma(j)} \right) - \frac{\partial}{\partial y} \left( \varphi_t \widetilde{C}^{\gamma(j)} \right) = \frac{\partial}{\partial y} \left( \frac{D^{\gamma(j)}}{l} \frac{\partial \widetilde{C}^{\gamma(j)}}{\partial y} \right), \quad y \in (0,1) \cup (1,2). \tag{14}$$

Here $c_{\mathrm{p}}$, $\varkappa$, $D$, and $l$ are piecewise constant functions such that $c_{\mathrm{p}} = c_{\mathrm{p}}^{\mathrm{s}}$, $\varkappa = \varkappa^{\mathrm{s}}$, $D = D^{\mathrm{s}}$, $l = l^{\mathrm{s}}$ if $y \in [0, 1)$ and $c_{\mathrm{p}} = c_{\mathrm{p}}^{\mathrm{l}}$, $\varkappa = \varkappa^{\mathrm{l}}$, $D = D^{\mathrm{l}}$, $l = l^{\mathrm{l}}$ if $y \in (1, 2]$.

At the point $y = 1$ we obtain

$$\left( \frac{\varkappa^{\mathrm{s}}}{l^{\mathrm{s}}} \frac{\partial \widetilde{T}}{\partial y} \right) \bigg|_{y=1-} - \left( \frac{\varkappa^{\mathrm{l}}}{l^{\mathrm{l}}} \frac{\partial \widetilde{T}}{\partial y} \right) \bigg|_{y=1+} = \lambda \xi_t, \tag{15}$$

$$\left( \frac{D^{(j)}}{l} \frac{\partial \widetilde{C}^{(j)}}{\partial y} \right) \bigg|_{y=1-}^{y=1+} = - \left( \widetilde{C}^{\mathrm{s}(j)} - \widetilde{C}^{\mathrm{l}(j)} \right) \xi_t, \quad j = \mathcal{A}, \mathcal{B}. \tag{16}$$

The set of Eqs. (13)–(16) is nonlinear, since the functions $\varphi_t$, $l$ depend on the unknown moving interface position $\xi$. In the later discussion the tilde character over $T$ and $C$ is omitted for the sake of simplicity.

## 3    Numerical Solution Technique

We introduce the spatial grid $\omega^y = \{y_i, 0 \le i \le N, y_0 = 0, y_{i^*} = 1, y_N = 2\}$ in the domain $\Omega_y = [0, 2]$, where $h_{i+1/2} = y_{i+1} - y_i$. The interface point $y = 1$ coincides with the grid point denoted by $y_{i^*}$. Also we introduce the semi–integer nodes $y_{i+1/2} = (y_i + y_{i+1})/2$, $i = 0, 1, \ldots, N - 1$. The finite-difference grid in time is denoted by $\omega^t = \{t_0 = 0, t_{k+1} = t_k + \tau, k = 0, 1, \ldots\}$. Here $\tau$ is the time step. We denote the finite-difference time derivative by $f_{t,i} = f_t = (\widehat{f}_i - f_i)/\tau$, where $\widehat{f}_i = f(y_i, t_j + \tau)$. The finite-difference approximation of spatial derivative is denoted by $f_y(i) = (f_{i+1} - f_i)/h_{i+1/2}$, $f_{\bar{y}}(i) = f_y(i-1)$.

The fully implicit conservative finite-difference scheme is obtained with finite-volume method [7,11]. The functions $T$, $C^{\gamma(j)}$ are associated with the grid nodes. We assume that the solution in each control volume is constant i.e. $f(y_i, t_k) = f_i$ for $y \in (y_{i-1/2}, y_{i+1/2})$, where $f = T, C^{\gamma(j)}$. At the interface $(i = i^*)$ concentration takes two values $C^{(j)}_{i^*-0} = C^{s(j)}$ and $C^{(j)}_{i^*+0} = C^{l(j)}$. Here $C^{s(j)}$, $C^{l(j)}$ are equilibrium concentrations satisfying the phase diagram of the system. Thus, $C^{(j)}(y, t_k) = \widetilde{C}^{(j)}_{i^*-0}$, if $y \in (y_{i^*-1/2}, y_{i^*})$ and $C^{(j)}(y, t_k) = C^{(j)}_{i^*+0}$, if $y \in (y_{i^*}, y_{i^*+1/2})$. The specific heat, the diffusivity, the thermal diffusivity and the zone length are referred to the semi–integer nodes. These functions are constant at the intervals between the grid nodes. The equation coefficients corresponding to coordinate system transformation $\varphi_t$ and $l$ are approximated in implicit way.

The approximation of heat transport Eq. (13) is

$$
\frac{h^y_{i-1/2}}{2} c_{p,i-1/2}\left(l_{i-1/2}T_i\right)_t + \frac{h^y_{i+1/2}}{2} c_{p,i+1/2}\left(l_{i+1/2}T_i\right)_t
$$
$$
- \left[c_{p,i+1/2}\left(\varphi_t\widehat{T}\right)(y_{i+1/2}) - c_{p,i-1/2}\left(\varphi_t\widehat{T}\right)(y_{i-1/2})\right] =
$$
$$
\left[\varkappa_{i+1/2}\left(\frac{\widehat{T}_y}{\widehat{l}_{i+1/2}}\right) - \varkappa_{i-1/2}\left(\frac{\widehat{T}_{\bar{y}}}{\widehat{l}_{i-1/2}}\right)\right] + \xi_t\left[\lambda\delta_{ii^*} + \left(c_{p,i-1/2} - c_{p,i+1/2}\right)\widehat{T}_{i^*}\right]
$$
$$
+ \left[\frac{h_{i-1/2}}{2}\widehat{l}_{i-1/2}(\alpha_{i-1/2}\widehat{T}_i + \beta_{i-1/2}) + \frac{h_{i+1/2}}{2}\widehat{l}_{i+1/2}(\alpha_{i+1/2}\widehat{T}_i + \beta_{i+1/2})\right]. \quad (17)
$$

Here $\delta_{ii^*}$ is the Kronecker delta. At a regular points relation (17) represents the scheme with central differences for one-dimensional convection-diffusion equation with a heat source. On the solid/liquid interface $(i = i^*)$ it is supplemented by terms representing a jump of thermodynamical parameters and heat generation or absorbtion due to the phase transition. The approximation of mass transport Eq. (14) is given by

$$
\frac{h^y_{i-1/2}}{2}\left(l_{i-1/2}C^{\gamma(j)}_{i-0}\right)_t + \frac{h^y_{i+1/2}}{2}\left(l_{i+1/2}C^{\gamma(j)}_{i+0}\right)_t
$$
$$
- \left[\left(\varphi_t\widehat{C}^{\gamma(j)}\right)(y_{i+1/2}) - \left(\varphi_t\widehat{C}^{\gamma(j)}\right)(y_{i-1/2})\right]
$$
$$
= D_{i+1/2}\left(\frac{\widehat{C}^{\gamma(j)}_y}{\widehat{l}_{i+1/2}}\right) - D_{i-1/2}\left(\frac{\widehat{C}^{\gamma(j)}_{\bar{y}}}{\widehat{l}_{i-1/2}}\right), \quad \gamma = s,l, \quad j = A, B. \quad (18)
$$

The set of nonlinear algebraic Eqs. (17) and (18) is solved by the Newton's method for unknown vector, which components are concentrations of all species, temperature and interface rate. Thus, proposed numerical method is based on simultaneous determination of concentration and temperature distribution and interface position. Fully implicit coupled approach allows to overcome

computational difficulties caused by the combination of super- or undersatura-
tion together with local surface equilibrium. Such inconsistency implies infinite
initial composition gradients at the surface and singular growth rate. Without
any preliminary adjustment of interface concentration, a convergent solution is
obtained after 6 Newton iterations at the beginning of the process, 2–3 iterations
are required at the developed stage. At the first time step initial data is used as
starting estimate for iteration, and the solution from the previous time level is
used at the next time levels.

## 4   Results and Discussion

Let us suppose that initial data determine a nonequilibrium situation: $\mathcal{AC}$ crystal
introduced to an undersaturated ternary $\mathcal{A}$–$\mathcal{B}$–$\mathcal{C}$ solution. The computer simulations are done for realistic phase diagram and experimentally used growth condi-
tions. The calculations were carried out with the following parameters: $T_t = 793\text{K}$,
$T_b = 783\text{K}$, $T_{sat} = 773\text{K}$, $\delta T = 20\text{K}$, $R = 0.5\,\text{cm}$, $L = 15\,\text{cm}$, $\xi(t = 0) = 30$
$\mu\text{m}$, $D^{s(\mathcal{AC})} = D^{s(\mathcal{BC})} = 5 \cdot 10^{-12}\,\text{sm}^2/s$, $D^{l(\mathcal{A})} = D^{l(\mathcal{B})} = 5 \cdot 10^{-5}\,\text{sm}^2/s$,
$\tilde{\alpha} = 10^{-5}\,\text{W}/(\text{K} \cdot \text{sm}^2)$, $\lambda = 463\,\text{J/g}$, $c_p^s = 0.441\,\text{J}/(\text{g} \cdot \text{K})$, $c_p^s = 0.393\,\text{J}/(\text{g} \cdot \text{K})$,
$\varkappa^s = 0.02\,\text{sm}^2 \cdot \text{W}/(\text{g} \cdot \text{K})$, $\varkappa^l = 0.08\,\text{sm}^2 \cdot \text{W}/(\text{g} \cdot \text{K})$. The time step is $\tau = 0.1\,t_D$,
$t_D = l_0^2 c_p^l / \varkappa^l \approx 5\text{s}$, $l_0 = 1\,\text{sm}$. The space grids are chosen sufficiently fine to
resolve sharp compositional profiles ($i^* = 5000$, $N - i^* = 1000$). The detailed
description of phase diagram and physical parameters can be found in [6].

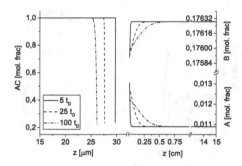

**Fig. 2.** Composition distribution in the solid and liquid phases for different processing
times (dissolution)

The initial distribution of the temperature in the ampule is $T(0, z) = T_b + (T_t - T_b)\, z/L$. The composition of the liquid phase is in equilibrium with the
solid phase $\mathcal{A}_{0.207}\mathcal{B}_{0.783}\mathcal{C}$ at temperature $T_{sat} < T(0, \xi)$. Thus, $\mathcal{A}$–$\mathcal{B}$–$\mathcal{C}$ solution
is undersaturated, the dissolution of initial crystal starts. To initiate the crys-
tal growth the system is cooled. The temperature of the ampule faces changes
according to the law $T_{b/t}(t) = T_{b/t} - \mu t$, where $\mu = 0.5\,\text{K/min}$ while $t < t^* = 20$
min, $\mu = 0$ while $t > t^*$. The temperature of the environment is given by

$T^*(z) = T(0, z) - \delta T$. The interaction of the liquid and solid phases at the beginning of the process is given in Fig. 2. The dissolution of crystal at the initial stage feeds the solution with the component $\mathcal{A}$. We observe the formation of $\mathcal{A}_x \mathcal{B}_{1-x}\mathcal{C}$ solid skin with simultaneous dissolution of the solid. Therefore, the component $\mathcal{B}$ is supplied from the solution to the solid phase and the concentration of $\mathcal{B}$ in the liquid phase decreases. The diffusion layer in the solid is very thin and dissolves before diffusion increases the penetration distance noticeably. Then in the vicinity of the solid phase solution becomes saturated due to programmed cooling and ongoing feeding with the dissolved crystal. Further cooling causes transition from dissolution to growth. It is important to note that transition occurs automatically since the proposed numerical algorithm is self–consistent.

Figure 3 displays the crystal growth process. Large amount of component $\mathcal{A}$ supplies the liquid phase during the dissolution of the $\mathcal{A}\mathcal{C}$ crystal. The nonmonotonic concentration distribution of the component $\mathcal{A}$ occurs in the liquid phase since the diffusion in the solution is not sufficiently strong and ampoule is rather long. The components $\mathcal{A}$ and $\mathcal{B}$ are supplied from the solution into the solid phase. When the programmed cooling is over growth takes place under isothermal conditions. The concentration distributions in the solution are homogenized by diffusion transport. The concentration of the component $\mathcal{B}$ in the vicinity of the solid/liquid interface is increasing, while the concentration of the component $\mathcal{A}$ is decreasing. The graphs of the temperature are not presented, since the thermal diffusivity of the proposed ternary compound is rather high and distributions of temperature in the solid and solution are close to the linear.

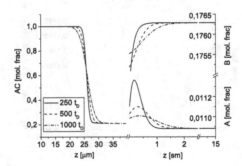

**Fig. 3.** Composition distribution in the solid and liquid phases for different processing times (growth)

## 5    Conclusions

New numerical procedure for computer simulations of multicomponent crystallization process is presented. Mathematical model accounts for the heat and matter transport in both solid and liquid phases and the interface motion. Proposed algorithm is based on coupled solution of governing equations. The corresponding system of nonlinear algebraic equations is solved by Newton's method. The conservative properties of algorithm allow us to model regimes of growth with preliminary dissolution of the solid. Computer simulations show that relatively small value of time step $\tau = 0.1\ t_D$ should be used only at the beginning of the process. After 3–10 time levels, the time step can be increased up to $t_D$ and more. Usually, we use $\tau = 0.5\ t_D$. This value is chosen for the accuracy and is not dictated by stability consideration. That's why we characterize the proposed algorithm as practically unconditionally stable. It allows to perform full–scale numerical study of phase transition in multicomponent solutions for a wide range of phase diagrams.

**Acknowledgments.** The study was supported by the Russian Foundation for Basic Research under grant 18-01-00436.

# References

1. Small, M.B., Ghez, R.: Growth and dissolution kinetics of III-V heterostructures formed by LPE. J. Appl. Phys. **50**, 5322 (1979)
2. Small, M.B., Ghez, R.: Growth and dissolution kinetics of III-V heterostructures formed by LPE. II. Comparison between thermodynamic and kinetic models. J. Appl. Phys. **51**, 1589 (1980)
3. Kimura, M., Qin, Z., Dost, S.: A solid-liquid diffusion model for growth and dissolution of ternary alloys by liquid phase epitaxy. J. Crystal Growth **158**, 231–240 (1996)
4. Dost, S., Lent, B.: Single Crystal Growth of Semiconductors from Metallic Solutions. Elsevier, Amsterdam (2007)
5. Shcheritsa, O.V., Mazhorova, O.S., Denisov, I.A., Popov, Y.P., Elyutin, A.V.: Numerical study for diffusion processes in dissolution and growth of $Cd_xHg_{1-x}Te/CdTe$ heterostructures formed by LPE. Part I. Isothermal conditions. J. Crystal Growth **290**, 350–356 (2006)
6. Gusev, A.O., Shcheritsa, O.V., Mazhorova, O.S.: Numerical study of ternary alloy crystallization in the cylindrical ampule. Keldysh Institute of Applied Mathematics, Preprint 215 (2016)
7. Shcheritsa, O.V., Gusev, A.O., Mazhorova, O.S.: On solution to phase transition problem in multicomponent alloy in the cylindrical ampule. Herald Bauman Moscow State Tech. Univ. Nat. Sci. **5**, 118–138 (2017)
8. Bakirova, O.I.: Numerical simulation of zone melting processes on the basis of the solution of the problem of phase transition to binary systems. Mathematical modeling. Producing of crystals and semiconductor structures, pp. 142–158. Moscow, Nauka Publisher (1986). [publication in Russian]
9. Landau, H.G.: Heat conduction in a melting solid. J. Appl. Math. **8**, 81–94 (1950)
10. Illingworth, T.C., Golosnoy, I.O.: Numerical solutions of diffusion controlled moving boundary problems which conserve solute. J. Comput. Phys. **209**, 207–225 (2005)
11. Mazhorova, O.S., Popov, Y.P., Shcheritsa, O.V.: Conservative scheme for the thermodiffusion Stefan problem. Differ. Equ. **49**(7), 869–882 (2013)

# On One Method for Solving
# of a Non-stationary Fluid Flows
# with Free Surface

Valentin A. Gushchin[1($\boxtimes$)] and Vasilii G. Kondakov[2]

[1] Institute for Computer Aided Design Russian Academy of Sciences, Moscow, Russia
gushshin@icad.org.ru
[2] Nuclear Safety Institute of Russian Academy of Sciences, Moscow, Russia
kondakov@ibrae.ac.ru

**Abstract.** The flow of an incompressible Newtonian fluid with a free boundary in contact with air is considered. In this paper, there will be problems with a flat bottom for different initial and boundary conditions. The choice of these tasks is based on the principle of "from simple to complex". This stage is considered as the initial and "debugging" in the development of the CABARET technique in application to the problems of incompressible fluid flows with a free surface. The system of equations describing such a model of the medium is a transformed Navier-Stokes system in a curvilinear coordinate system, such that at any instant the curvilinear transformation conforms the computational domain into a rectangle of unit height. The free surface is described by the kinematic boundary condition, which is obtained from the assumption that the liquid particles located on the interface between the two media remain on this boundary all the time. The comparison of numerical results with some theoretical data is discussed.

**Keywords:** CABARET method · Free surface · Numerical simulation

## 1 Introduction

Let's consider two-dimensional flow of incompressible medium with free surface. Earlier [1] the flow of a liquid past an obstacle with breaking of the wave front was discussed. A whole series of methods has been developed which enable one to model numerically the non-stationary flows of a viscous liquid [2]. Two approaches are generally used in these methods: the Eulerian and Lagrangian approaches, each of which has its own advantages and disadvantages. The following cases were calculated and compared with theoretical, experimental and numerical data: wave motion in a container [2], undulate (wave) bora [1], simulation of hydraulic jump [2].

A new method for solving of similar problems was proposed in [3], where the dynamics of vortex pairs and its interaction with a free surface was considered. To simplify the task, we assumed that fluid is non-viscous and homogeneous.

© Springer Nature Switzerland AG 2019
I. Dimov et al. (Eds.): FDM 2018, LNCS 11386, pp. 274–280, 2019.
https://doi.org/10.1007/978-3-030-11539-5_30

Note that if a viscosity is not zero the vortexes can jump off from a free surface and move in to the fluid back [4,5]. The obtained results for test problems were in a good agreement with the theoretical, experimental and numerical data. The results can be used for the developing of the expert systems for the study of spectral characteristics of the surface waves and their anomalies.

This paper proposes a new approach for the solution of problems of interaction of vortex structures with a free surface. The second-order accuracy finite-difference scheme based on the famous CABARET scheme [6–8] is suggested for incompressible viscous fluid with a free surface. In the case of incompressible fluid the CABARET technique should solve the elliptical task for pressure using incompressibility condition. After that the calculated pressure gradient is used in the equations of motion.

The aim of this work is extension of proposed method [3] on multiple problems, for example hydraulic jump and undulate (wave) bora. The authors are planning to extend this new methodology to three-dimensional model with solving advection-diffusion equations for modeling sea water (thermal and salinity equations) in future.

## 2    Equations and Boundary Conditions

The equations describing the motion of a viscous incompressible fluid with a free surface can be written as following:

$$
\begin{cases}
\dfrac{\partial \mathbf{v}}{\partial t} + (\mathbf{v} \cdot \nabla)\,\mathbf{v} + \dfrac{1}{\rho_0}\,\nabla p = \nu\,\triangle\,\mathbf{v} + \mathbf{g}, \\
\qquad\qquad\qquad\qquad \nabla \cdot \mathbf{v} = 0.
\end{cases}
\tag{1}
$$

here $z = \eta(x,t)$ - free surface level, $\mathbf{v} = (u,w)$ - velocity vector, p - pressure, $\nu$ - is the coefficient of kinematic viscosity of the fluid, $\mathbf{g}$ - gravitation acceleration, $\rho_0$ - density of fluid.

The boundary conditions: on the bottom boundary (solid impermeable wall) $w = 0$.

On the free surface: (a) the dynamic boundary condition $p = P_a$, where $P_a$ - is a atmosphere pressure; (b) the kinematic boundary condition:

$$
\frac{\partial \eta}{\partial t} + u \frac{\partial \eta}{\partial x} - w = 0.
\tag{2}
$$

Let $H_0$ and $U_0$ are length and velocity scales, then non-dimensional variables will be:

$$
\tilde{x} = \frac{x}{H_0}, \tilde{z} = \frac{z}{H_0}, \tilde{\eta} = \frac{\eta}{H_0}, \tilde{t} = \frac{tU_0}{H_0},
$$

$$
\tilde{u} = \frac{u}{U_0}, \tilde{w} = \frac{w}{U_0}, \tilde{p} = \frac{p}{\rho_0 U_0^2}, Re = \frac{U_0 H_0}{\nu}, Fr = \frac{U_0^2}{gH_0},
\tag{3}
$$

here $Re$ - Reynolds number, $Fr$ - number of Froude. Let omit tilde sign and rewrite the Eq. (1):

$$
\begin{cases}
\dfrac{\partial u}{\partial t} + u\dfrac{\partial u}{\partial x} + w\dfrac{\partial u}{\partial z} + \dfrac{\partial p}{\partial x} = \dfrac{1}{Re}\Delta u \\[2mm]
\dfrac{\partial w}{\partial t} + u\dfrac{\partial w}{\partial x} + w\dfrac{\partial w}{\partial z} + \dfrac{\partial p}{\partial z} = \dfrac{1}{Re}\Delta w - \dfrac{1}{Fr}, \\[2mm]
\dfrac{\partial u}{\partial x} + \dfrac{\partial w}{\partial z} = 0.
\end{cases}
\tag{4}
$$

The main difficulty arising in numerical problem solving with a free surface is a necessity to carry out the calculation in the areas of complex shapes with changing in time boundaries. To overcome this difficulty, it is proposed to use the moving coordinate system:

$$
\begin{cases}
x_1 = x, \\[2mm]
x_2 = \dfrac{z + H(x)}{\eta(x,t) + H(x)},
\end{cases}
\tag{5}
$$

which allows to transform a domain to a rectangle in the plane $(x_1, x_2)$ per each time moment. The equation of continuity takes the form:

$$
\frac{1}{J}\left\{ \frac{\partial}{\partial x_1}\left( u\frac{\partial z}{\partial x_2} - w\frac{\partial x}{\partial x_2}\right) + \frac{\partial}{\partial x_2}\left( w\frac{\partial x}{\partial x_1} - u\frac{z}{x_1}\right)\right\} = 0,
$$

$$
\frac{1}{\eta + H}\left\{ \frac{\partial}{\partial x_1}\left( u(\eta + H)\right) + \frac{\partial}{\partial x_2}\left( w - u\left[ x_2(\eta_x + H_x) - H_x\right]\right)\right\} = 0,
$$

$$
\frac{\partial}{\partial x_1}\left( u(\eta + H)\right) + \frac{\partial}{\partial x_2}\left( w - u\left[ x_2(\eta_x + H_x) - H_x\right]\right) = 0.
$$

Here J - Jacobian of the transformation (5). With next replacements $x_1 \to x, x_2 \to \xi, R = \eta + H, A = 1/R, B = -AR_x, C = AH_x$ the previous equation is converted into:

$$
\frac{\partial}{\partial x}(uR) + \frac{\partial}{\partial \xi}\left( w + uR\left[\xi B + C\right]\right) = 0.
\tag{6}
$$

Let's convert the equations of motion:

$$
\begin{cases}
\dfrac{\partial u}{\partial t} + \dfrac{\partial u^2 + p}{\partial x} + \dfrac{\partial uw}{\partial z} = \dfrac{1}{Re}\Delta u \\[2mm]
\dfrac{\partial w}{\partial t} + \dfrac{\partial uw}{\partial x} + \dfrac{\partial w^2 + p}{\partial z} = \dfrac{1}{Re}\Delta w - \dfrac{1}{Fr},
\end{cases}
$$

$$
\begin{cases}
R\dfrac{\partial u}{\partial t} + \dfrac{\partial}{\partial x}\left( (u^2 + p)R\right) + \dfrac{\partial}{\partial \xi}\left( uw - (u^2 + p)R[\xi B + C]\right) = \dfrac{1}{Re}\Delta u \\[2mm]
R\dfrac{\partial w}{\partial t} + \dfrac{\partial}{\partial x}\left( uwR\right) + \dfrac{\partial}{\partial \xi}\left( w^2 + p - uwR[\xi B + C]\right) = \dfrac{1}{Re}\Delta w - \dfrac{1}{Fr},
\end{cases}
\tag{7}
$$

## 3   Method CABARET

Let's consider the curvilinear transformation:

$$x = x, \xi = \frac{z + H(x)}{\eta(x,t) + H(x)}.$$

New system of Navier-Stokes equations from (6), (7)is:

$$\frac{\partial \mathbf{U}}{\partial t} + \frac{\partial \mathbf{F}}{\partial x} + \frac{\partial \mathbf{G}}{\partial \xi} = \mathbf{Q},$$

$$\mathbf{U} = \begin{pmatrix} Ru \\ Rw \end{pmatrix}, \mathbf{F} = \begin{pmatrix} R(u^2 + p) \\ Ruw \end{pmatrix}, \quad (8)$$

$$\mathbf{G} = \begin{pmatrix} uw + (u^2 + p)R(\xi B + C) + \xi DRu \\ w^2 + p + uwR(\xi B + C) + \xi DRw \end{pmatrix}, \mathbf{Q} = \begin{pmatrix} \frac{R}{Re}\triangle u \\ \frac{R}{Re}\triangle w - \frac{R}{Fr} \end{pmatrix}.$$

Here $R = \eta + H, A = 1/R, B = -AR_x, C = AH_x, D = -AR_t$. CABARET scheme for system (8) in implicit form will be:

$$\frac{\mathbf{U}^{n+1} - \mathbf{U}^n}{\tau} + \frac{\frac{\mathbf{F}_{i+1}^n + \mathbf{F}_{i+1}^{n+1}}{2} - \frac{\mathbf{F}_i^n + \mathbf{F}_i^{n+1}}{2}}{x_{i+1} - x_i} + \frac{\frac{\mathbf{G}_{j+1}^n + \mathbf{G}_{j+1}^{n+1}}{2} - \frac{\mathbf{G}_j^n + \mathbf{G}_j^{n+1}}{2}}{\xi_{j+1} - \xi_j} = \mathbf{Q}^{n+1/2}. \quad (9)$$

To simulate an incompressible medium, the CABARET technique uses the classical approach of the procedure that makes the velocity vector field solenoidal, starting from the incompressibility condition and setting the boundary conditions for the pressure. That is, at the first step of the algorithm, we calculate the velocity field don't taking into account the work of the pressure forces, then the pressure equation is calculated taking into account the distribution of the velocity divergence in the entire design area. Since we are dealing with two types of variables, the same procedure also occurs separately for "conservative" and "flux" variables.

In an arbitrary curvilinear coordinate system in one space-time cell, in the case of an incompressible medium, two velocity components are the invariants: the normal component of the velocity of the face and the tangential component of the velocity tangential to it. A rigorous proof of this circumstance is given in the monograph [9]. The characteristic system is written as:

$$\begin{cases} \dfrac{\partial u_n}{\partial t} + u_n \dfrac{\partial u_n}{\partial \mathbf{n}} = 0, \\[2mm] \dfrac{\partial u_\tau}{\partial t} + u_n \dfrac{\partial u_\tau}{\partial \mathbf{n}} = 0, \\[2mm] u_n = (\mathbf{u}, \mathbf{n}), u_\tau = (\mathbf{u}, \tau), \\[2mm] (\mathbf{n}, \tau) = 0. \end{cases} \quad (10)$$

where $\mathbf{n}$ - external normal to side, $\tau$ - tangential to $\mathbf{n}$ unit vector.

# 4   The Test Problems

## 4.1   Fluctuations of Liquid in a Container

Let the rectangular container be partially filled with a liquid at rest. At the initial time t = 0, an instantaneous pressure pulse $\Phi = A\sigma(t)cos(kx)$ is applied to the free surface, after that the free surface starts to vibrate. From a linear analysis the shape of a free surface will be:

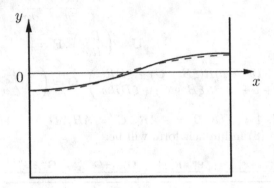

**Fig. 1.** The fluctuation of liquid in container.

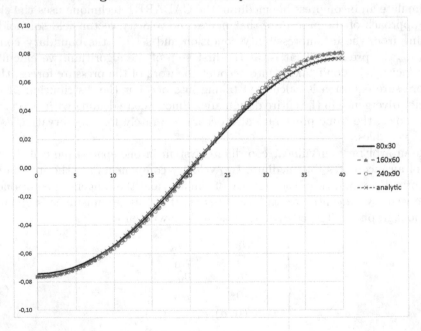

**Fig. 2.** Levels of fluctuations in container. The comparison with analytic solution.

$$\eta(x,t) = -A\sqrt{\frac{\xi}{gH}} sin\left(\sqrt{\frac{\xi}{gH}}t\right) cos\left(kx\right),$$

(11)

$$\xi = kHth\left(kH\right),$$

here $A = 11.78$, $k = 2\pi/\lambda$, the perturbation wavelength $\lambda = 80$, the number $Re = 10$, and the initial depth of the basin $H = 15$ in the considered variant. The results shown in Fig. 1 correspond to grid $40 \times 15$. Here solid line - analytical solution, dashed line - our numerical results. The comparison is good enough. The calculations were carried out for different numbers of grid points (Fig. 2).

## 5   Conclusion

The new approach for numerical modeling of incompressible fluid flows with free surface was suggested. This approach is based on CABARET method possessing by second order accuracy in space and time variables. Special transformation of coordinate system is used, which allows to transform calculated area to rectangle per each time moment. Such approach may be considered as Lagranjian-Eulerian. To solve the kinematic boundary equation, two approaches were considered: one-dimensional CABARET [9–11] scheme with internal interpolation of variable deviations into nodes and one-dimensional SMIF [1] scheme (hybrid scheme with switching by criterion). Both approaches give the same results and monotonic solution over the entire time interval. The basic scheme remains universal - the CABARET method for an incompressible medium, with boundary input/output conditions, non-flow with no slip, and a boundary condition for a free surface.

The methodical calculations for different number of grid points demonstrate the internal convergence. The test calculations of oscillating free surface in open container and undulate (wave) bora were performed. The results are in a good agreement with theoretical data.

## References

1. Gushchin, V.A., Konshin, V.N.: Computational aspects of the splitting method for incompressible flow with free surface. J. Comput. Fluids **21**(3), 345–353 (1992). https://doi.org/10.1016/0045-7930(92)90043-U
2. Belotserkovskii, O.M., Gushchin, V.A., Kon'shin, V.N.: The splitting method for investigating flows of a stratified liquid with a free surface. USSR Comput. Math. Math. Phys. **27**(2), 181–191 (1987). https://doi.org/10.1016/0041-5553(87)90175-3
3. Gushchin, V.A., Kondakov, V.G.: On cabaret scheme for incompressible fluid flow problems with free surface. In: International Multidisciplinary Scientific GeoConference Surveying Geology and Mining Ecology Management, SGEM 2017, no. 21, pp. 485–492 (2017)
4. Barker, S.J., Crow, S.C.: The motion of two-dimensional vortex pairs in a ground effect. J. Fluid Mech. **82**(4), 659–671 (1977). https://doi.org/10.1017/S0022112077000913

5. Peace, A.J., Riley, N.: A viscous vortex pair in ground effect. J. Fluid Mech. **129**, 409–426 (1983). https://doi.org/10.1017/S002211208300083X
6. Karabasov, S.A., Goloviznin, V.M.: Compact accurately boundary-adjusting high-resolution technique for fluid dynamics. J. Comput. Phys. **228**(19), 7426–7451 (2009). https://doi.org/10.1016/j.jcp.2009.06.037
7. Faranosov, G.A., Goloviznin, V.M., Karabasov, S.A., Kondakov, V.G., Kopiev, V.F., Zaitsev, M.A.: CABARET method on unstructured hexahedral grids for jet noise computation. Comput. Fluids **88**, 165–179 (2013). https://doi.org/10.1016/j.compfluid.2013.08.011
8. Goloviznin, V.M., Karabasov, S.A., Kondakov, V.G.: Generalization of the CABARET scheme to two-dimensional orthogonal computational grids. Math. Models Comput. Simul. **6**(1), 56–79 (2014). https://doi.org/10.1134/S2070048214010050
9. Goloviznin, V.M., Zaitsev, M.A., Karabasov, S.A., Korotkin, I.A.: New CFD Algorithms for Multiprocessor Computer Systems. Mosk. Gos. Univ, Moscow (2013). [in Russian]
10. Goloviznin, V.M., Samarskii, A.A.: Finite difference approximation of convective transport equation with space splitting time derivative. Matematicheskoe Modelirovanie **10**(1), 86–100 (1998). [in Russian]
11. Goloviznin, V.M., Karabasov, S.A.: Nonlinear correction of Cabaret scheme. Matematicheskoe Modelirovanie **10**(12), 107–123 (1998). [in Russian]

# The Third Order Iterative Method for Solving Nonlinear Parabolic Equations and Its Application to the Biological Tissues Models

I. F. Iumanova[1] and S. I. Solodushkin[1,2(✉)]

[1] Ural Federal University, Yekaterinburg, Russia
solodushkin_s@mail.ru
[2] Institute of Mathematics and Mechanics, Ural Branch of the RAS,
Yekaterinburg, Russia

**Abstract.** We supply an iterative method for solving nonlinear difference schemes appearing after discretization of evolution partial differential equations. Proposed method is a kind of two stage iterative process, which does not use derivatives. The theorem of third order convergence is proven. Results of numerical experiments with test equations, which sources are equation from echocardiography, are presented.

**Keywords:** Nonlinear difference scheme ·
Acceleration of convergence · Iterative method · Echocardiography

## 1  Introduction

In [1] it was presented a mathematical model for processing the three-dimensional (3D) + time echocardiography. The model, which could be formalized as second-order (degenerate) parabolic partial differential equation (PDE)

$$\frac{\partial u}{\partial t} = \mathcal{F}\left(t, u, \frac{\partial u}{\partial x}, \frac{\partial^2 u}{\partial x^2}\right),\tag{1}$$

is applied to initially given image sequence. Non-linear partial differential equations for the processing of this kind of data have been used also in [2].

Since the analytical solution in such nonlinear problems could be found in some exceptional cases only, the elaboration of numerical algorithms and acceleration of their convergence as well as their software implementation are relevant problems. Using of explicit difference schemes for (1) suffer from instability and lead to unbounded growth of error. On the other hand using of implicit schemes for (1) leads to the necessity of solving nonlinear systems of high dimension, and iterative methods play a major role here.

In [1] nonlinear PDEs are discretized by semi-implicit finite volume method. In [4] implicit difference scheme are considered and the corresponding nonlinear systems are solved by Newton's method.

© Springer Nature Switzerland AG 2019
I. Dimov et al. (Eds.): FDM 2018, LNCS 11386, pp. 281–288, 2019.
https://doi.org/10.1007/978-3-030-11539-5_31

## 2   Two-Step Iterative Method

In this section we consider a one-dimensional equation and focus our efforts on two main difficulties: the nonlinearity and the convergence acceleration. Exactly, we consider a nonlinear initial boundary value problem with nonlinear differential operator

$$\frac{\partial u}{\partial t} = \mathcal{F}\left(t, u, \frac{\partial u}{\partial x}, \frac{\partial^2 u}{\partial x^2}\right), \tag{2}$$

here $\mathcal{F}$ is a nonlinear function. Initial and boundary conditions are specified as follow

$$u(x, 0) = u_0(x), \ x \in [0, X], \tag{3}$$
$$u(0, t) = 0, \ u(X, t) = 0, \ t \in [0, T]. \tag{4}$$

Question of existence and uniqueness are discussed of (2)–(4) in [3].

To solve (1)–(3) a uniform grid on the domain $[0, X] \times [0, T]$ is constructed. We consider an equidistant partition of $[0, X]$ into parts with step size $h = X/N$ and define the grid $x_i = ih$, $i = 0, \ldots, N$. We also split the time interval $[0, T]$ into parts with step size $\Delta = T/M$ and define the grid $t_j = j\Delta$, $j = 0, \ldots, M$. The derivatives are approximated as follow

$$\frac{\partial u}{\partial t} = \frac{u_{j+1}^i - u_j^i}{\Delta} + O(\Delta), \qquad \frac{\partial u}{\partial x} = \frac{u_{j+1}^{i+1} - u_{j+1}^{i-1}}{2h} + O(h^2),$$
$$\frac{\partial^2 u}{\partial x^2} = \frac{u_{j+1}^{i+1} - 2u_{j+1}^i + u_{j+1}^{i-1}}{h^2} + O(h^2), \tag{5}$$

where $j = 0, \ldots, M - 1$, and $i = 1, \ldots, N - 1$. To find a solution $u_{j+1} = (u_{j+1}^1, u_{j+1}^2, \ldots, u_{j+1}^{N-1})$ on each next time layer it is necessary to solve a corresponding nonlinear three-diagonal system, which can be written in the form $F(u_{j+1}) = 0$, $F : \mathbb{R}^{N-1} \to \mathbb{R}^{N-1}$. To deal with it we consider a two-step iterative method

$$u_{j+1}^{(k+1)} = \widetilde{u}_{j+1}^{(k)} - \mu \left[ F\left(u_{j+1}^{(k)}, \Phi\left(u_{j+1}^{(k)}\right)\right) \right]^{-1} F\left(\widetilde{u}_{j+1}^{(k)}\right), \tag{6}$$

$$\widetilde{u}_{j+1}^{(k)} = u_{j+1}^{(k)} - \left[ F\left(u_{j+1}^{(k)}, \Phi\left(u_{j+1}^{(k)}\right)\right) \right]^{-1} F\left(u_{j+1}^{(k)}\right), \tag{7}$$

where $u_{j+1}^{(k)}$ is the $k$-th iteration of a solution $u_{j+1}$, and $\widetilde{u}_{j+1}^{(k)}$ is an auxiliary sequence, $F(u', u'') = \frac{1}{\lambda}(E - \Phi(u', u''))$ is a first divided difference (see Definition 1), $\lambda \in (0, 1]$, $\mu \in (0, 1]$, $k = 0, 1, 2, \ldots$.

**Definition 1** (see [5]). *Let $F$ be a continuous nonlinear mapping from a Banach space $X$ to $Y$. Linear operator $F(u', u'')$ is called the first divided difference of operator $F$ if the following requirements hold:*

*1. for each fixed $u', u'' \in X$ operator $F(u', u'')$ is such that $F(u', u'')(u' - u'') = F(u') - F(u'')$;*

2. *if the Frechet derivative $F'(u)$ exists then $F(u, u) = F'(u)$.*

Note, that functional $F(u', u'')$ is obviously nonlinear with respect to $u', u''$. Definition 1 does not defines a unique operator $F(u', u'')$, indeed the first divided difference could be specified in a numerous of ways. One of such specification is given in

**Statement 1** *(see [6]). If there exists a continuous derivative $\Phi'(u)$ in the Gateau sense on the interval $[u', u'']$, then we can define the first divided difference $\Phi(u', u'')$ as a Riemann's abstract integral[1] [7]:*

$$\Phi(u', u'') = \int_0^1 \Phi'(u'' + t(u' - u'')) \, dt.$$

*Using this specification of the definition, we can obtain the estimate*

$$\|\Phi(u', u'')\| = \left\| \int_0^1 \Phi'(u'' + t(u' - u'')) \, dt \right\|$$

$$\leq \int_0^1 \|\Phi'(u'' + t(u' - u''))\| \, dt \leq \int_0^1 M_1 \, dt = M_1, \tag{8}$$

*where $M_1$ is the maximum of the norm of the Gateau derivative $\Phi'$ on the segment $[u', u'']$.*

Let us consider how to apply method (6)–(7) to nonlinear systems of two equations

$$f_1(u_1, u_2) = 0, \quad f_2(u_1, u_2) = 0.$$

For convenience, the subscript $j$ is eliminated, $\mu = 1$.
We use the following notation:

$$u^{(k)} = \left( u_1^{(k)}, u_2^{(k)} \right), \quad \tilde{u}^{(k)} = \left( \tilde{u}_1^{(k)}, \tilde{u}_2^{(k)} \right),$$

$$\Phi(u^{(k)}) = \left( \varphi_1(u_1^{(k)}, u_2^{(k)}), \varphi_2(u_1^{(k)}, u_2^{(k)}) \right),$$

$$\Phi(\tilde{u}^{(k)}) = \left( \varphi_1(\tilde{u}_1^{(k)}, \tilde{u}_2^{(k)}), \varphi_2(\tilde{u}_1^{(k)}, \tilde{u}_2^{(k)}) \right),$$

$$\Delta u^{(k)} = \left( u_1^{(k)} - \tilde{u}_1^{(k)}, u_2^{(k)} - \tilde{u}_2^{(k)} \right) = \left( \Delta u_1^{(k)}, \Delta u_2^{(k)} \right),$$

$$\Delta \tilde{u}^{(k)} = \left( \tilde{u}_1^{(k+1)} - \tilde{u}_1^{(k)}, \tilde{u}_2^{(k+1)} - \tilde{u}_2^{(k)} \right) = \left( \Delta \tilde{u}_1^{(k)}, \Delta \tilde{u}_2^{(k)} \right),$$

$$u' = (u_1', u_2'), \quad u'' = (u_1'', u_2'').$$

---

[1] Here we use well-established terminology, see [8,9]. The Riemann integral of a function of a real argument with values in an abstract Banach space.

Let us define

$$\Phi\left(u', u''\right) = \begin{pmatrix} \dfrac{\varphi_1(u_1', u_2') - \varphi_1(u_1'', u_2')}{u_1' - u_1''} & \dfrac{\varphi_1(u_1'', u_2') - \varphi_1(u_1'', u_2'')}{u_2' - u_2''} \\[2mm] \dfrac{\varphi_2(u_1', u_2') - \varphi_2(u_1'', u_2')}{u_1' - u_1''} & \dfrac{\varphi_2(u_1'', u_2') - \varphi_2(u_1'', u_2'')}{u_2' - u_2''} \end{pmatrix}. \tag{9}$$

From (6)–(7) we obtain the following system of equations:

$$\frac{f_1\left(u_1^{(k)}, u_2^{(k)}\right) - f_1\left(\varphi_1\left(u_1^{(k)}, u_2^{(k)}\right), u_2^{(k)}\right)}{u_1^{(k)} - \varphi_1\left(u_1^{(k)}, u_2^{(k)}\right)}\Delta u_1^{(k)}$$

$$+\frac{f_1\left(\varphi_1\left(u_1^{(k)}, u_2^{(k)}\right), \tilde{u}_2^{(k)}\right) - f_1\left(\varphi_1\left(u_1^{(k)}, u_2^{(k)}\right), \varphi_2\left(u_1^{(k)}, x_2^{(k)}\right)\right)}{u_2^{(k)} - \varphi_2\left(u_1^{(k)}, u_2^{(k)}\right)}\Delta u_2^{(k)}$$

$$= -f_1\left(\tilde{u}_1^{(k)}, \tilde{u}_2^{(k)}\right) - f_1\left(u_1^{(k)}, u_2^{(k)}\right),$$

$$\frac{f_2\left(u_1^{(k)}, u_2^{(k)}\right) - f_2\left(\varphi_1\left(u_1^{(k)}, u_2^{(k)}\right), u_2^{(k)}\right)}{u_1^{(k)} - \varphi_1\left(u_1^{(k)}, u_2^{(k)}\right)}\Delta u_1^{(k)}$$

$$+\frac{f_2\left(\varphi_1\left(u_1^{(k)}, u_2^{(k)}\right), u_2^{(k)}\right) - f_2\left(\varphi_1\left(u_1^{(k)}, u_2^{(k)}\right), \varphi_2\left(u_1^{(k)}, u_2^{(k)}\right)\right)}{u_2^{(k)} - \varphi_2\left(u_1^{(k)}, u_2^{(k)}\right)}\Delta u_2^{(k)}$$

$$= -f_2\left(\tilde{u}_1^{(k)}, \tilde{u}_2^{(k)}\right) - f_2\left(u_1^{(k)}, u_2^{(k)}\right),$$

$$\frac{f_1\left(u_1^{(k)}, u_2^{(k)}\right) - f_1\left(\varphi_1\left(u_1^{(k)}, u_2^{(k)}\right), u_2^{(k)}\right)}{u_1^{(k)} - \varphi_1\left(u_1^{(k)}, u_2^{(k)}\right)}\tilde{\Delta}u_1^{(k)}$$

$$+\frac{f_1\left(\varphi_1\left(u_1^{(k)}, u_2^{(k)}\right), \tilde{u}_2^{(k)}\right) - f_1\left(\varphi_1\left(u_1^{(k)}, u_2^{(k)}\right), \varphi_2\left(u_1^{(k)}, u_2^{(k)}\right)\right)}{u_2^{(k)} - f_2\left(u_1^{(k)}, u_2^{(k)}\right)}\Delta\tilde{u}_2^{(k)}$$

$$= -f_1\left(u_1^{(k)}, u_2^{(k)}\right),$$

$$\frac{f_2\left(u_1^{(k)}, u_2^{(k)}\right) - f_2\left(\varphi_1\left(u_1^{(k)}, u_2^{(k)}\right), u_2^{(k)}\right)}{u_1^{(k)} - \varphi_1\left(u_1^{(k)}, u_2^{(k)}\right)}\tilde{\Delta}u_1^{(k)}$$

$$+\frac{f_2\left(\varphi_1\left(u_1^{(k)}, u_2^{(k)}\right), u_2^{(k)}\right) - f_2\left(\varphi_1\left(u_1^{(k)}, u_2^{(k)}\right), \varphi_2\left(u_1^{(k)}, u_2^{(k)}\right)\right)}{u_2^{(k)} - \varphi_2\left(u_1^{(k)}, u_2^{(k)}\right)}\tilde{\Delta}u_2^{(k)} = -f_2\left(u_1^{(k)}, u_2^{(k)}\right).$$

Similarly, it is easy to generalize to nonlinear systems from any number of equations.

The proposed method (6)–(7) is a development of Aitken–Steffensen method (also known as Steffensen's method) which is one of the effective methods for approximate solving of nonlinear equations. Steffensen's method is a kind of chord method, but in comparison with classical chord method it has some advantages, e.g. it does not require two initial approximations, it has quadratic convergence like Newton's method, moreover in particular cases it converges faster than Newton method.

# 3     Study of Convergence of a Two-Step Iterative Method

Let $F(u)$ be a continues, $\mu = \lambda = 1$, we also assume that the operator of the first divided difference is invertible, i.e. in the domain of our interest there exist $[F(u', u'')]^{-1} = [E - \Phi(u', u'')]^{-1}$.

**Theorem 1** *(proof could be found in [10]). Let the following conditions hold:*

1. $\left\| F(u^{(0)}) \right\| = \left\| u^{(0)} - \Phi\left(u^{(0)}\right) \right\| \leq \eta$;
2. *there exists an open domain* $\Omega \subseteq X$ *such that for each* $u', u'', u''' \in \Omega$ *the following three estimations hold*
   (a) $\left\| [F(u', u'')]^{-1} \right\| = \left\| [E - \Phi(u', u'')]^{-1} \right\| \leq B$;
   (b) $\left\| \Phi(u', u'') \right\| \leq M$;
   (c) $\left\| \Phi(u', u'') - \Phi(u'', u''') \right\| \leq K \left\| u' - u''' \right\|$,

   *where* $B, M, K$ *are constants;*
3. $h = C_2 B^2 K M \eta < 1$, *where* $C_2$ *is a constant which is not small;*
4. *the set* $\Omega$ *fully contains a closed ball*

$$\left\| u - u^{(0)} \right\| \leq R, \tag{10}$$

*where* $R = \dfrac{C_1 S_0}{C_2 B K M}$, $C_1$ *is a constant, and* $S_k = \displaystyle\sum_{n=k}^{\infty} h^{2^n}$.

*Then (a) all elements of the sequence* $\left( u^{(k)} \right)$, *defined by the method (6)–(7) which starts from the certain* $u^{(0)}$, *lie in the ball (10), (b) the sequence* $\left( u^{(k)} \right)$ *has a limit* $u^*$ *in the ball (10), and (c) the estimation takes place*

$$\left\| u^* - u^{(k)} \right\| \leq \frac{C_1}{C_2 B K M} S_k \quad (k = 0, 1, 2, \dots).$$

*Remark 1.* If parameters $\mu \in (0, 1), \lambda \in (0, 1)$, in method (6)–(7) then condition (3) of Theorem 1 is rewritten in the following form: (3) $h = C_2 B^2 K M \dfrac{\mu}{\lambda} \eta < 1$, where $C_2$ is positive constant.
The technique of proof is preserved.

**Theorem 2.** *Assume that*

1. *equation* $F(u) = 0$ *has a solution in the ball*

$$\left\| u - u^{(0)} \right\| \leq \rho; \tag{11}$$

2. *for each* $u', u'', u'''$ *from the ball* $\left\| u - u^{(0)} \right\| \leq (1 + \alpha) \rho$ *the following three estimations hold*
   (a) $\left\| [F(u', u'')]^{-1} \right\| = \left\| [E - \Phi(u', u'')]^{-1} \right\| \leq B$;

*(b)* $\|\Phi(u', u'')\| \le M$;

*(c)* $\|\Phi(u', u'') - \Phi(u'', u''')\| \le K \|u' - u'''\|$,

*where* $B, M, K$ *are constants, at that* $\alpha = \max\{l^2\rho^2, M\}$, *where* $l = \sqrt{2C}BKM$, $C$ *is a constant;*

*3.* $l\rho < 1$.

*Then* *(a)* *the solution* $u^*$ *of equation* $F(u) = 0$ *is unique in the ball (11),* *(b)* *sequence* $(u^{(k)})$, *defined by the method (6)–(7), converges to* $u^*$ *with the third order, i.e. the following estimation of the convergence rate holds*

$$\left\| u^* - u^{(k)} \right\| \le \frac{1}{l} \left( l\rho \right)^{3^k} \quad (k = 0, 1, 2, \dots).$$

## 4    Numerical Examples

### 4.1    Example 1

In order to confirm the theoretical results we find the computational order of convergence (see Definition 2) of the method (6)–(7).

**Definition 2** *(see [12]). The computational order of convergence (COC) of iteration method is*

$$COC \approx \frac{\ln \left( \|u^{(k+1)} - u^*\| / \|u^{(k)} - u^*\| \right)}{\ln \left( \|u^{(k)} - u^*\| / \|u^{(k-1)} - u^*\| \right)},$$

*where* $u^{(k-1)}, u^{(k)}, u^{(k+1)}$ *is members of the iteration sequences near the root* $u^*$ *of the equation* $F(u) = 0$, $k \in \mathbb{N}$, $F : \mathbb{R}^{N-1} \to \mathbb{R}^{N-1}$, $u^{(k)} = \left( u_1^{(k)}, \dots, u_{N-1}^{(k)} \right)$.

We consider system of $N - 1$ nonlinear equations which appear after discretization of boundary value problem [11]

$$3u_i \left( u_{i+1} - 2u_i + u_{i-1} \right) + \left( u_{i+1} - u_{i-1} \right)^2 / 4 = 0, \tag{12}$$
$$u_0 = 0, \quad u_N = 20,$$

where $i = 1, \dots, N - 1$. The parameter $N$ can be changed. According to [11] the initial approximation was chosen as follows $u^{(0)} = (10, \dots, 10)$.

Calculations for system (12) were carried out for $N = 10$, $N = 30$ and $N = 50$. In all cases, the computational order of convergence for the method (6)–(7) is 2.9999. The results of numerical experiments confirmed the theoretical results obtained for method (6)–(7). In order to carry out numerical experiments long arithmetic was used. Calculations were made with 300 decimal places with accuracy $\varepsilon = 10^{-50}$.

## 4.2    Example 2

On the domain $x \in (0,1)$, $t \in (0, 4\pi)$ we consider the initial boundary value problem

$$\frac{\partial u}{\partial t} = \sqrt{1 + (u_x)^2} \, \frac{\partial}{\partial x} \left( \frac{u_x}{\sqrt{1 + (u_x)^2}} \right) + \frac{2}{1 + 4x^2} - \sin t, \qquad (13)$$

where $u_x = \frac{\partial u}{\partial x}$, with initial and boundary conditions

$$u(x, 0) = x(1 - x) + 1, \quad 0 \le x \le 4,$$

$$u(0, t) = \cos t, \ u(1, t) = \cos t, \quad 0 \le t \le 4\pi.$$

Problem (13) has an exact solution $u(x, t) = x(1 - x) + \cos t$.

The source of this initial boundary value problem (13) is [1], but the heterogeneous function is added in order to the problem has an exact solution.

An implicit difference scheme is used to reduce (13) to a nonlinear system. Method (6)–(7) as well as Newton's method were used for comparison. The accuracy of the Newton's method is chosen to be $\epsilon = 10^{-5}$, i.e. stop criterion is $\|u^{k+1} - u^k\| < \epsilon$.

In the table below we report the deviations $\mathbf{diff} = \max_{i,j} |u_j^i - u(x_i, t_j)|$ of the approximate solution calculated by method from the exact one and the average (over the time layers) number of iterations. Here $N$ and $M$ are the number of segments with respect to space and time (Table 1).

Table 1. Average number of iterations and errors norms.

| | Method (6)–(7) | | | | Newton's method | | | |
|---|---|---|---|---|---|---|---|---|
| N | 8 | 8 | 8 | 8 | 8 | 8 | 8 | 8 |
| M | 16 | 32 | 64 | 128 | 16 | 32 | 64 | 128 |
| diff | 0.3945 | 0.1908 | 0.0961 | 0.0490 | 0.3864 | 0.1934 | 0.0984 | 0.0498 |
| Iter. number | 1.63 | 1.50 | 1.06 | 1.03 | 5.125 | 5.125 | 5.03 | 4.53 |

**Acknowledgments.** We acknowledge the support by the program 02.A03.21.0006 on 27.08.2013 and the project Development of a personalized computer model of electrotherapy in heart failure patients with risk of sudden death MK-6328.2018.7.

# References

1. Sarti, A., Mikula, K., Sgallari, F., et al.: Evolutionary partial differential equations for biomedical image processing. J. Biomed. Inf. **5**, 77–91 (2002)
2. Bagheri, B., Ezzati, R.: Partial differential equations applied to medical image segmentation. Int. J. Ind. Math. **6**(4), 345–350 (2014). Article ID IJIM-00569
3. Alvarez, L., Guichard, F., Lions, P.L., et al.: Axioms and fundamental equations of image processing. Arch. Rat. Mech. Anal. **123**, 200–57 (1993)
4. Samarskii, A.A.: The Theory of Difference Schemes. CRC Press, Boca Raton (2001)
5. Ulm, S.Y.: On generalized divided differences. I. Izv. AN EstSSR **16**, 13–26 (1967)
6. Sergeev, A.S.: On the method of chords. Sibirskij matematiceskij zurnal **2**(2), 282–289 (1961). (in Russian)
7. Ulm, S.Y.: On generalized divided differences. II, Izv. AN EstSSR **16**(2), 146–156 (1967). (in Russian)
8. Sergeev, A.S.: On the convergence of certain variants of the method of chords in normed spaces. Sbornik nauchnykh trudov Permskogo politekhnicheskogo instituta **13**, 43–54 (1963). (in Russian)
9. Bel'tyukov, B.A.: On the perturbed analog of the Aitken-Steffensen method for solving nonlinear operator equations. Sibirskij matematiceskij zurnal **12**(5), 983–1000 (1974). (in Russian)
10. Yumanova, I.F.: One specification of Steffensen's method for solving nonlinear operator equations. Vestn. Udmurtsk. Univ. Mat. Mekh. Komp. Nauki **26**(4), 579–590 (2016)
11. Rump, S.M.: Solution of linear and nonlinear algebraic problems with sharp, guaranteed bounds. In: Bhmer, K., Stetter, H.J. (eds.) Defect Correction Methods. Computing Supplementum, vol. 5, pp. 147–168. Springer, Vienna (1984). https://doi.org/10.1007/978-3-7091-7023-6_9
12. Cordero, A.: Variants of Newton's method using fifth-order quadrature formulas. Appl. Math. Comput. **190**, 686–698 (2007)

# Iterative Process for Numerical Recovering the Lowest Order Space-Wise Coefficient in Parabolic Equations

D. Kh. Ivanov[1(✉)] and P. N. Vabishchevich[1,2]

[1] Institute of Mathematics and Information Science,
North-Eastern Federal University, 58, Belinskogo, Yakutsk, Russia
djulus.ivanov@yandex.ru
[2] Nuclear Safety Institute, Russian Academy of Sciences,
52, B. Tulskaya, Moscow, Russia

**Abstract.** In this work we suggest an iterative process for coefficient inverse problem. A parabolic equation in a bounded area supplied with initial condition and monotonic nondecreasing on time Dirichlet condition on a boundary is considered. We state a problem to recover the lowest order coefficient that depends only on spatial variables under an additional information as the observation of a solution taken at the final point of time. For numerical recovering of the coefficient we build the iterative process, at each iteration we perform finite-element approximation in space and fully implicit two-level discretization in time. For capabilities of given iterative process we present computational test for a model problem.

**Keywords:** Inverse problem · Parabolic equation ·
Finite element method · Implicit scheme

## 1 Introduction

Inverse problems for partial differential equations of the second order were studied in a number of papers [1,2,14], and the interest in them remains strong. Theoretical studies of such problems generally treat with the questions of the well-posedness of the problem, the uniqueness of the solution and its stability.

Among inverse problems for parabolic equations one can give an attention to the problems with unknown coefficients and/or the right-hand side of an equation, in this case we have coefficient inverse problem [7]. For identification the unknowns some additional information is needed, for instance, one can take the information about a state of the solution at the final point of observation time, known as final overdetermination, or in more general case integral overdetermination [6,8–11,13].

In this work we present an iterative method to recover the lowest order coefficient in a non-stationary parabolic equation supplemented by final overdetermination, in our case the lowest order coefficient depends only on spatial

I. Dimov et al. (Eds.): FDM 2018, LNCS 11386, pp. 289–296, 2019.
https://doi.org/10.1007/978-3-030-11539-5_32

variables. The method monotonically approximates the exact solution of the inverse problem, such approach was considered in works [7,8].

## 2  Problem Statement

Primarily, we consider a 2-D problem, the generalization to the 3-D case can be made analogically. Let $\Omega$ be a bounded polygon with a boundary $\partial\Omega$, $\boldsymbol{x} = (x_1, x_2) \in \Omega$. We search a function $u(\boldsymbol{x}, t)$ in $\Omega \times (0, T]$, $T > 0$ such that it is the solution of the initial-boundary value problem for the parabolic equation of second order

$$\frac{\partial u}{\partial t} - \operatorname{div}(k(\boldsymbol{x}) \operatorname{grad} u) + c(\boldsymbol{x})u = f(\boldsymbol{x}, t), \quad \boldsymbol{x} \in \Omega, \quad 0 < t \le T \tag{1}$$

with coefficients $0 < k_1 < k(\boldsymbol{x}) < k_2$, $c(\boldsymbol{x}) \ge 0$. The Dirichlet boundary condition is specified as

$$u(\boldsymbol{x}, t) = g(\boldsymbol{x}, t), \quad \boldsymbol{x} \in \partial\Omega, \quad 0 < t \le T, \tag{2}$$

and the initial condition is taken as

$$u(\boldsymbol{x}, 0) = u_0(\boldsymbol{x}), \quad \boldsymbol{x} \in \Omega, \tag{3}$$

where $u_0(\boldsymbol{x}) \ge 0$. The system (1)–(3) presents a direct problem for unknown function $u(\boldsymbol{x}, t)$ under the consideration that input data as the right-hand side $(f(\boldsymbol{x}, t))$, the coefficients $(k(\boldsymbol{x}), c(\boldsymbol{x}))$ as well as the boundary $(g(\boldsymbol{x}, t))$ and initial $(u_0(\boldsymbol{x}))$ conditions are specified.

We consider the inverse problem for finding a pair of functions $\{u(\boldsymbol{x}, t), c(\boldsymbol{x})\}$ satisfying (1)–(3) and additional condition taken as an observation at the final point of time

$$u(\boldsymbol{x}, T) = u_T(\boldsymbol{x}), \quad \boldsymbol{x} \in \Omega, \tag{4}$$

known as the final overdetermination.

Let us introduce an elliptic operator as follows

$$\mathcal{A}u = -\operatorname{div}(k(\boldsymbol{x}) \operatorname{grad} u), \tag{5}$$

and the Eq. (1) has a simpler form

$$\frac{\partial u}{\partial t} + \mathcal{A}u + c(\boldsymbol{x})u = f(\boldsymbol{x}, t), \quad \boldsymbol{x} \in \Omega, \quad 0 < t \le T. \tag{6}$$

*Remark 1.* The input data should satisfies matching conditions first type

$$g(\boldsymbol{x}, 0) = u_0(\boldsymbol{x}), \quad g(\boldsymbol{x}, T) = u_T(\boldsymbol{x}), \quad \boldsymbol{x} \in \partial\Omega,$$

and second type

$$\frac{\partial g}{\partial t}(\boldsymbol{x}, 0) + \mathcal{A}u_0 + c(\boldsymbol{x})u_0 = f(\boldsymbol{x}, 0),$$

$$\frac{\partial g}{\partial t}(\boldsymbol{x}, T) + \mathcal{A}u_T + c(\boldsymbol{x})u_T = f(\boldsymbol{x}, T), \quad \boldsymbol{x} \in \partial\Omega.$$

*Remark 2.* The unknown coefficient $c(\boldsymbol{x})$ is constrained such that, see [9],

$$c(\boldsymbol{x}) \equiv 0, \quad \boldsymbol{x} \in \partial\Omega, \tag{7}$$

if this condition is violated then we introduce a new function $\eta(\boldsymbol{x})$ satisfying

$$\frac{\partial g}{\partial t}(\boldsymbol{x}, T) + \mathcal{A}u_T + \eta(\boldsymbol{x})u_T = f(\boldsymbol{x}, T), \quad \boldsymbol{x} \in \partial\Omega.$$

Appropriately extending it into the $\Omega$ we define new variables for unknown coefficient and for elliptic operator (5) as follows

$$\tilde{c}(\boldsymbol{x}) = c(\boldsymbol{x}) - \eta(\boldsymbol{x}), \quad \boldsymbol{x} \in \Omega,$$

$$\tilde{\mathcal{A}}u = -\mathrm{div}(k(\boldsymbol{x})\,\mathrm{grad}\,u) + \eta(\boldsymbol{x})u.$$

and the Eq. (6) becomes as follows

$$\frac{\partial u}{\partial t} + \tilde{\mathcal{A}}u + \tilde{c}(\boldsymbol{x})u = f(\boldsymbol{x}, t), \quad \boldsymbol{x} \in \Omega, \quad 0 < t \le T.$$

Hence, new variable satisfies the relation (7), $\tilde{c}(\boldsymbol{x}) \equiv 0$, $\boldsymbol{x} \in \partial\Omega$.

Let us supplement the right-hand side function and the function in boundary condition with next relations

$$
\begin{aligned}
f(\boldsymbol{x}, t) \ge 0, \quad f_t(\boldsymbol{x}, t) \ge 0, \quad \boldsymbol{x} \in \Omega, \quad 0 < t \le T, \\
g(\boldsymbol{x}, t) \ge 0, \quad g_t(\boldsymbol{x}, t) \ge 0, \quad \boldsymbol{x} \in \partial\Omega, \quad 0 < t \le T.
\end{aligned}
\tag{8}
$$

Then taking into account all above given remarks and relations (8) the coefficient inverse problem is to find a pair $\{u(\boldsymbol{x}, t), c(\boldsymbol{x})\}$ satisfying (2)–(4) and (6). The existence and uniqueness of solution such problem can be found in works [7,10, 11]. In addition, for forward problem (2), (3) and (6) we have

$$u_T(\boldsymbol{x}) = u(\boldsymbol{x}, T) > 0, \quad \boldsymbol{x} \in \Omega.$$

# 3   Iterative Process for Identification the Lowest Order Coefficient

Let us consider an interative process as follows: we set an approximation of the lowest order coefficient with initial value; at each iteration we solve (2), (3) and (6); at next iteration coefficient approximation is found from the equation (6) at the final point of time with the relations (4) and (7).

In contrast to the works [7,8], we set initial value for the coefficient as a solution of next equation with (7)

$$u_T(\boldsymbol{x})\,c^0 = -\mathcal{A}u_T(\boldsymbol{x}) + f(\boldsymbol{x}, T), \quad \boldsymbol{x} \in \Omega. \tag{9}$$

Next, for given coefficient $c^k(\boldsymbol{x})$, $k = 0, 1, \ldots$, where $k$ is the iteration number, we solve the following forward problem:

$$\frac{\partial u^k}{\partial t} + \mathcal{A}u^k + c^k u^k = f \quad \text{on } \Omega \times (0, T], \tag{10}$$

$$u^k = g \quad \text{on } \partial\Omega \times (0, T], \tag{11}$$

$$u^k = u_0 \quad \text{on } \overline{\Omega} \times \{0\}. \tag{12}$$

A new approximation of coefficient is found at the final time with the final overdetermination and the relation (7)

$$u_T\, c^{k+1} = -\frac{\partial u^k}{\partial t} - \mathcal{A}u_T + f \quad \text{on } \Omega \times \{T\}. \tag{13}$$

**Theorem 1.** *Iterative process (9)–(13) has monotonicity such that*

$$u(\boldsymbol{x}, t) \geq u^{k+1}(\boldsymbol{x}, t) \geq u^k(\boldsymbol{x}, t),$$
$$c(\boldsymbol{x}) \leq c^{k+1}(\boldsymbol{x}) \leq c^k(\boldsymbol{x}), \quad \boldsymbol{x} \in \Omega, \quad 0 < t \leq T$$

*for every $k = 0, 1, \ldots$.*

A proof comes from the fact that the nonnegativity of the solution (2)–(4) and (6), therefore at each iteration of the process (9)–(13), in turn comes from the use of the maximum principle and considering the relations (8), specifically nonnegativity of right-hand function and nonnegativity of the initial state. For a solution's derivative we have analogical case on the base of nonnegativity of the derivative of the right-hand side function.

## 4 Computational Implementation

As generally we make a common preparations for the use the finite element method to discretization in space. We introduce Hilbert space $H = L_2(\Omega)$ with traditional scalar product and norm

$$(u, v) = \int_\Omega uv\, dx, \qquad \|u\| = \sqrt{(u, u)}.$$

We define bilinear form as follows

$$a(u, v) = (\mathcal{A}u, v) = \int_\Omega (\operatorname{grad} u, \operatorname{grad} v)\, dx, \quad a(u, u) \geq \hat{\delta}\|u\|^2, \quad \hat{\delta} > 0.$$

Also we introduce a Sobolev subspace $V^h \subset H^1(\Omega)$, for example, it is Lagrange finite elements of the first degree. For this case a function in $V^h$ has the next form

$$u(\boldsymbol{x}) \approx \sum_{i=1}^{M} a_i \chi_i(\boldsymbol{x}),$$

where $a_i = u(\boldsymbol{x}_i)$ and $\chi_i(\boldsymbol{x})$ is piece-wise pyramid basis-function

$$\chi_i(\boldsymbol{x}_j) = \begin{cases} 1, & \text{if } i = j, \\ 0, & \text{if } i \neq j. \end{cases}$$

Here, $\boldsymbol{x}_i$, $i = 1, 2, \ldots, M$, are nodes of Delaunay-type triangulation $\omega$ constructed over $\Omega$.

For time discretization we divide the time interval $[0, T]$ uniformly into $N$ subintervals, $t_n = n\tau$, $\tau = T/N$, $n = 0, 1, \ldots, N$. Numerically solving the inverse problem (2)–(4), (6) by using fully implicit two-level scheme in time, [12], we deal with functions $w_n(\boldsymbol{x}) \in V^h$, $n = 0, 1, \ldots, N$ and $s(\boldsymbol{x}) \in V^h$ such that they satisfy next variational problem formulated as follows

$$((w_{n+1} - w_n)/\tau, v) + a(w_{n+1}, v) + (sw_{n+1}, v) = (\phi_{n+1}, v), \tag{14}$$

$$w_{n+1} = \theta \quad on \ \partial\omega, \quad n = 0, 1, \ldots, N - 1, \tag{15}$$

$$w_0 = \varphi \quad on \ \bar{\omega}, \tag{16}$$

and final overdetermination

$$(w_N, v) = (\psi, v), \tag{17}$$

where $\phi_n = Pf(., t_n)$, $n = 0, 1, \ldots, N$, $\theta = Pg$, $\varphi = Pu_0$, $\psi = Pu_T$, $P$ is $L_2$-projection onto $V^h$.

Iterative process (9)–(13) has discrete analogue as follows, initial value for coefficient

$$(s^0\psi, v) = -a(\psi, v) + (\phi_N, v), \tag{18}$$

At $k$-th iteration we find a function $w^k \in V^h$ such that

$$((w_{n+1}^k - w_n^k)/\tau, v) + a(w_{n+1}^k, v) + (s^k w_{n+1}^k, v) = (\phi_{n+1}, v), \tag{19}$$

$$w_{n+1}^k = \theta \quad on \ \partial\omega, \quad n = 0, 1, \ldots, N - 1, \tag{20}$$

$$w_0^k = \varphi \quad on \ \bar{\omega}. \tag{21}$$

New value for coefficient at next iteration satisfies

$$(s^{k+1}\psi, v) = - ((w_N^k - w_{N-1}^k)/\tau, v) - a(\psi, v) + (\phi_N, v), \tag{22}$$

and

$$s^{k+1} = 0 \quad on \ \partial\omega. \tag{23}$$

The discrete analogue for Theorem 1 takes place due to the discrete maximum principle for elliptic operators in parabolic/hyperbolic PDEs, which puts a necessary condition for the grid geometry as Delaunay triangulation [3–5,12].

## 5  Model Problem

Let us test the iterative process (18)–(23). We consider next model problem in a domain $\Omega = (0, 1) \times (0, 1)$, $\boldsymbol{x} = (x_1, x_2)$

$$\frac{\partial u}{\partial t} - \Delta u + c(\boldsymbol{x})u = 0, \quad \boldsymbol{x} \in \Omega, \quad 0 < t \leq T,$$

$$u(\boldsymbol{x}, t) = 100t^2 e^{-x_1}, \quad \boldsymbol{x} \in \partial\Omega, \quad 0 < t \leq T,$$

$$u(\boldsymbol{x}, 0) = 0, \quad \boldsymbol{x} \in \overline{\Omega}.$$

We take the exact coefficient of the lowest order as follows

$$c(\boldsymbol{x}) = \begin{cases} 5, & \text{if } \sqrt{(x_1 - 0.6)^2 + (x_2 - 0.4)^2} \leq 0.3, \\ 1, & \text{if } |x_1 - 0.3| \leq 0.1 \text{ and } |x_2 - 0.8| \leq 0.1, \\ 0, & \text{else.} \end{cases} \quad (24)$$

Solution at the final point of time $t_N = T$, which is set to 0.25 for the base case, is taken by solving the forward problem (14)–(16) with a coefficient (24) and a time step $\tau = 1.0 \cdot 10^{-5}$, and it is used for final overdetermination function $\psi$ (17). The mesh $\omega$ is constructed like uniform right-triangled grid that in both direction the division number is 50.

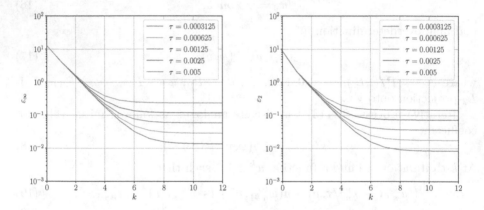

**Fig. 1.** The dependence of the iterative process (18)–(23) on different time step $\tau$.

**Fig. 2.** The solution $c^k(\boldsymbol{x})$ of the iterative process (18)–(23) at $k = 1$ (left), $k = 2$ (middle), $k = 3$ (right).

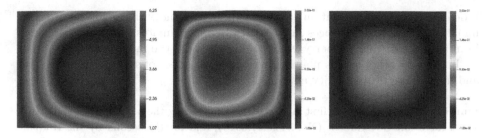

**Fig. 3.** The solution $u(x, T)$ of the iterative process (18)–(23) at $k = 1$ (left), the difference between solutions at iteration $k = 2$ and $k = 1$ (middle), at iteration $k = 3$ and $k = 2$ (right).

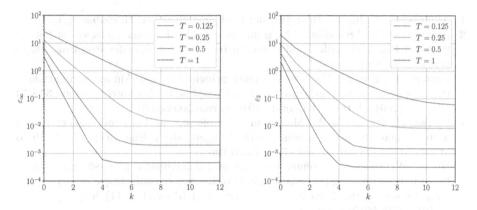

**Fig. 4.** The dependence of the iterative process (18)–(23) on the final point of time $T$ with fixed time step $\tau = 3.125 \cdot 10^{-4}$.

The inverse problem for unknown lowest order coefficient is solved by (18)–(23). We perform the evaluation of the effect of computational errors on the basis of calculations on different time grids, when using the input data derived from the solution of the forward problem on more detailed time grid and with more accurate approximations in time. Figure 1 shows a dependence on the time step $\tau$ of approximation accuracy at each iteration calculated in standard $L_\infty$- and $L_2$-norm. The monotonicity of the approximation, specifically convergence from above for coefficient solution of the iterative process (18)–(23), is presented in Figs. 2 and 3. In Fig. 4 there is a dependence of the approximation on the final point of time $T$.

## 6   Conclusions

In this article we suggest the iterative process for the identification of space-wise dependent coefficient in lowest term for parabolic equation, which serves the initial-boundary value problem. At each iteration the coefficient approximation is defined by the state equation at the final point of time with the final overdetermination. The iterative process monotonically approximates the exact coefficient from above, and that property is very well tested for the model problem.

In future work, authors plan to investigate an effect of space grid properties and make numerical stability analysis.

**Funding.** The work for the first author was supported by the mega-grant of the Russian Federation Government (N 14.Y26.31.0013), and for the second by the Russian Foundation for Basic Research (project 17-01-00689).

# References

1. Alifanov, O.M.: Inverse Heat Transfer Problems. Springer, Heidelberg (2012)
2. Alifanov, O.M., Artiukhin, E.A., Rumiantsev, S.V.: Extreme Methods for Solving Ill-Posed Problems with Applications to Inverse Heat Transfer Problems. Begell House, New York (1995)
3. Burman, E., Ern, A.: Discrete maximum principle for galerkin approximations of the laplace operator on arbitrary meshes. Comptes Rendus Mathematique **338**(8), 641–646 (2004). https://doi.org/10.1016/j.crma.2004.02.010
4. Ciarlet, P., Raviart, P.A.: Maximum principle and uniform convergence for the finite element method. Comput. Methods Appl. Mech. Eng. **2**(1), 17–31 (1973). https://doi.org/10.1016/0045-7825(73)90019-4
5. Huang, W.: Discrete maximum principle and a delaunay-type mesh condition for linear finite element approximations of two-dimensional anisotropic diffusion problems. Numer. Math.: Theory Methods Appl. **4**(3), 319–334 (2011). https://doi.org/10.1017/S1004897900000374
6. Isakov, V.: Inverse parabolic problems with the final overdetermination. Commun. Pure Appl. Math. **44**(2), 185–209 (1991)
7. Isakov, V.: Inverse Problems for Partial Differential Equations, vol. 127, 3rd edn. Springer, Cham (2017). https://doi.org/10.1007/978-3-319-51658-5
8. Kostin, A.B.: Inverse problem of finding the coefficient of u in a parabolic equation on the basis of a nonlocal observation condition. Differ. Equ. **51**(5), 605–619 (2015)
9. Prilepko, A.I., Solov'ev, V.V.: On the solvability of inverse boundary value problems for the determination of the coefficient preceding the lower derivative in a parabolic equation. Differentsial'nye Uravneniya **23**, 136–143 (1987)
10. Prilepko, A.I., Kostin, A.B.: Inverse problems of the determination of the coefficient in parabolic equations. I. Siberian Math. J. **33**(3), 489–496 (1992)
11. Prilepko, A.I., Kostin, A.B.: On inverse problems of determining a coefficient in a parabolic equation. II. Siberian Math. J. **34**(5), 923–937 (1993)
12. Samarskii, A.A.: The Theory of Difference Schemes. Marcel Dekker Inc., New York (2001)
13. Vabishchevich, P.N.: Iterative computational identification of a space-wise dependent source in parabolic equation. Inverse Probl. Sci. Eng. **25**(8), 1168–1190 (2017). https://doi.org/10.1080/17415977.2016.1230611
14. Vasil'ev, F.P.: Solution Methods for Incorrect Extremal Problems. Nauka, Moscow (1981). In Russian

# P-Thinned Gamma Process and Corresponding Random Walk

Pavlina Jordanova[1]([✉]) and Milan Stehlík[2,3]

[1] Faculty of Mathematics and Informatics, Shumen University,
115 Universitetska str., 9712 Shumen, Bulgaria
pavlina_kj@abv.bg
[2] Institute of Statistics, Universidad de Valparaíso, Valparaíso, Chile
[3] Linz Institute of Technology and Department of Applied Statistics,
Johannes Kepler University, Altenbergerstrasse 69, 4040 Linz, Austria
mlnstehlik@gmail.com

**Abstract.** P-thinned Gamma processes could be considered as a particular case of renewal processes which inter-renewal times are zero-inflated Gamma distributed. This paper considers also the difference between two, not obligatory identically distributed, processes which time intersections coincide in distribution with convolutions of zero-inflated Gamma distributed random variables. The idea comes from the Variance-Gamma model which is defined as Gamma time changed Wiener processes and is stochastically equivalent to a difference between two independent Gamma processes. The main properties and numerical characteristics of the resulting process are obtained. Simulation illustrates the theoretical results.

**Keywords:** Random walk · Gamma process · Mixed distribution ·
Convolutions · Zero-inflation

## 1 Introduction

In 1980 Cinlar [1] pointed out the importance of continuous time Gamma processes for analysing different types of subordinators (increasing processes with independent additive increments). In 1982 Lawrance [8] introduced first order auto-regressive process with Gamma innovations. Lewis, McKenzie and Hugus [9] used the multiplicative relation between Gamma and Beta distributions and investigated the properties of Beta-Gamma auto-regressive model with random coefficients. A very good presentation of Gamma processes, as a particular case of a stationary process with almost sure infinite number of jumps in any time interval is done in Dufresne, Gerber and Shiu [3]. The authors explain that "the majority of jumps are very small in some sense" and therefore their sum in a fixed time interval is finite and reveal the meaning of the corresponding Lévy measure in the terms of insurance. In 1993 Dickson and Waters [2] use Gamma processes in survival analysis. Seneta [13] proposes a procedure for fitting these

© Springer Nature Switzerland AG 2019
I. Dimov et al. (Eds.): FDM 2018, LNCS 11386, pp. 297–304, 2019.
https://doi.org/10.1007/978-3-030-11539-5_33

processes. Madan, Carr, and Chang [11] and Kaishev and Dimitrova [6] use the variance gamma model for option pricing. Kumar, Wylomanska, Poloczanski and Sundar [7] obtain the probability density function of the first hitting time of the Gamma process.

The idea about the results in this paper comes from the Variance-Gamma models, introduced by Madan and Seneta [10] in 1990. They define it as Gamma time changed Wiener process and show that it is stochastically equivalent to a difference between two independent Gamma processes. These kinds of processes seems to be very useful in practice because the Gamma distribution is one of the most easy to handle and in the same time relatively general for modeling of unbounded, absolutely continuous and positive random variables (r.vs.). Their distributional tails, in the sense of p-mild and extreme measures of heavy-tailedness, are heavier than the normal one, and lighter than the Pareto one.[1] In Sect. 2 the general properties and numerical characteristics of p-thinned gamma sequences are derived. Their time intersections coincide in distribution with convolutions of zero-inflated Gamma r.vs. They are stochastically equivalent to a corresponding compound binomial random variable with Gamma summands. We should note that our concept about tinning does not coincide with those considered in Wolpert [15]. In Sect. 3 a very specific particular case of a six parametric random walk is considered. The summands are difference between two zero-inflated and independent mixed gamma r.vs., not obligatory identically distributed. The time intersections of these processes coincide in distribution with the difference between two compound binomial random variables with Gamma summands. More general results about random walk can be seen in Spitzer [14].

The paper finishes with a brief simulation study. We decompose the time series into two new time series. One with the positive and another with the negative jumps. The presence of the Multinomial r.vs. allows us to consider also consecutive constant values.

Denote by $X \sim Gamma(\alpha, \beta)$ the fact that the random variable (r.v.) $X$ belongs to Gamma distribution with parameters $\alpha > 0$ and $\beta > 0$.

Along the paper $C_i$ means "condition $i$" and $\Gamma(\alpha, t) = \int_t^\infty x^{\alpha-1} e^{-x} dx, t \geq 0$, is the upper incomplete Gamma function. $\Gamma(1, t) = e^{-t}$. $\overset{d}{=}$ means coincidence in distribution. $I\{A\}$ is for the indicator of the event $A$ or it is a Bernoulli r.v. with $p = P(I\{A\} = 1) = P(A)$. $Bi(n, p)$ is the Binomial distribution with parameters $n \in N$ and $p \in (0, 1)$. By assumption $Bi(0, p) = 0$. The geometric distribution concentrated on $0, 1, \ldots$ will be denoted by $G(p)$, and $NBi(n; p)$ is the Negative binomial distribution on $0, 1, \ldots$, with parameters $n \in N$ and $p \in (0, 1)$. $Mn$ means Multinomial distribution. The cumulative distribution function (c.d.f.) of a random variable (r.v.) $\xi$ will be denoted by $F_\xi(x) = P(\xi \leq x)$.

---

[1] These measures of heavy-tailedness were introduced and partially investigated in Jordanova and Petkova [4].

## 2   P-Thinned Gamma Process

Here we consider the following discrete time p-thinned Gamma process

$$\tilde{S}(n) = \tilde{S}(0) + \sum_{i=1}^{n} I_i X_i, \ n = 1, 2, \ldots \tag{1}$$

where:

$C_1 : \tilde{S}(0) = \tilde{S}(0) \geq 0$ is a constant.

$C_2 : I_1, I_2, \ldots, I_n$ are i.i.d. Bernoulli r.vs. with $P(I_1 = 1) = p \in (0, 1)$, and

$C_3 : X_1, X_2, \ldots, X_n$ are i.i.d. $Gamma(\alpha, \beta)$ distributed r.vs.

$C_4 :$ By assumption $I_1, I_2, \ldots, I_n$ and $X_1, X_2, \ldots, X_n$ are independent.

Briefly we will denote the time intersections of these processes by $\tilde{S}(n) \sim CZIG(S(0); n; p; \alpha, \beta)$.

For $\alpha = 1$ one obtains the innovations in the Lawrance [8] model.

The family of these distributions is closed with respect to convolution. More precisely if $\tilde{S}(n) \sim CZIG(\tilde{S}_1(0); n; p; \alpha, \beta)$ and $\tilde{S}(m) \sim CZIG(\tilde{S}_2(0); m; p; \alpha, \beta)$ are independent, then $\tilde{S}(n) + \tilde{S}(m) \sim CZIG(\tilde{S}_1(0) + \tilde{S}_2(0); n + m; p; \alpha, \beta)$.

For fixed $n = 1, 2, \ldots$ denote the number of summands, which are different from zero by $N_n = I_1 + I_2 + \ldots + I_n$. It is clear that $N_n \sim Bi(n, p)$. If $S(n) = \sum_{k=1}^{n} X_k$ and $S(0) = \tilde{S}(0)$, then conditions $C_1$ and $C_4$ imply:

$$\tilde{S}(n) = \tilde{S}(0) + I\{N_n > 0\} \sum_{k=1}^{N_n} X_k = S(0) + I\{N_n > 0\}S(N_n), \ n = 1, 2, \ldots \tag{2}$$

is (possibly shifted) compound Binomially distributed with Gamma compounding distribution. The additive increments of this sequence are homogeneous in time. Due to the fact that these sequences have independent additive increments, all their finite dimensional distributions(f.d.ds.) are determined via the univariate one.

Denote by $T_1 = min\{n \in N : I_n = 1\}$, $T_i = min\{n > T_{i-1} : I_n = 1\}$, $i = 2, 3, \ldots$. It is clear that $T_1 - 1 \sim G(p)$, for $i = 2, 3, \ldots, T_i - T_{i-1} - 1 \sim G(p)$ are independent and $T_n - n \sim NBi(n, p)$, $n = 1, 2, \ldots$.

Assume $T(z) := inf\{n \geq 0 : \tilde{S}(n) \geq z\}$ is the first hitting time of level $z > 0$.

In the next theorem we summarise the properties of these sequences. The results (1–4) follow immediately by the total probability formula and the formula for double expectations and therefore we will skip their proves.

**Theorem 1.** The sequence $\tilde{S}(n) \sim CZIG(S(0); n; p; \alpha, \beta)$, $n = 1, 2, \ldots$ defined in (1) possesses the following properties.

1. For $x \geq \tilde{S}(0)$, $k = 0, 1, 2, \ldots$, and $n = k + 1, k + 2, \ldots$,
$F_{\tilde{S}(n)}(y|\tilde{S}(k) = x) = 0$ if $y < x$ and

$$F_{\tilde{S}(n)}(y|\tilde{S}(k) = x) = 1 - (1 - p)^{n-k} \sum_{s=1}^{n-k} \frac{\Gamma(s\alpha, \beta(y - x))}{\Gamma(s\alpha)} \binom{n - k}{s} \left(\frac{p}{1 - p}\right)^s$$

if $y \geq x$.

$$P_{\tilde{S}(n)}(y|\tilde{S}(k) = x) = \begin{cases} 0 & , y < x \\ \frac{(n-k)!(1-p)^{n-k}e^{-\beta(y-x)}}{y-x}\sum_{s=1}^{n-k}\frac{\left(\frac{p\beta^{\alpha}(y-x)^{\alpha}}{1-p}\right)^{s}}{\Gamma(s\alpha)s!(n-k-s)!} & , y > x. \end{cases}$$

and $P(\tilde{S}(n) = x|\tilde{S}(k) = x) = (1-p)^{n-k}$.

The mean square regression of $\tilde{S}(n)$ with respect to some previous value is linear.

$$E[\tilde{S}(n)|\tilde{S}(k) = x] = x + (n-k)\frac{p\alpha}{\beta},$$

$$Var[\tilde{S}(n)|\tilde{S}(k) = x] = (n-k)\frac{p\alpha}{\beta^2}[1 + \alpha(1-p)].$$

$$E[e^{-z\tilde{S}(n)}|\tilde{S}(k) = x] = e^{-zx}(1 - p + \frac{p\beta^{\alpha}}{(\beta+z)^{\alpha}})^{n-k}, \quad s \geq 0.$$

2. For any $k = 0, 1, 2, ...$, $\tilde{S}(n) \overset{d}{=} [\tilde{S}(n+k)|\tilde{S}(k) = \tilde{S}(0)]$ is a Mixed Gamma distributed and

$$\tilde{S}(n) \overset{d}{=} \tilde{S}(0) + I\{N_n > 0\}Gamma(\alpha N_n, \beta), \quad N_n \sim Bi(n, p).$$

3. The auto-correlation function does not depend on $p$, $\alpha$ and $\beta$.

$$cov(\tilde{S}(n), \tilde{S}(k)) = min(k, n)\frac{p\alpha}{\beta^2}[1 + \alpha(1-p)], \quad k, n = 1, 2, ...$$

$$cor(\tilde{S}(n), \tilde{S}(k)) = \sqrt{\frac{min(k, n)}{max(k, n)}}, \quad k, n = 1, 2, ...$$

4. $P(\tilde{S}(n) = \tilde{S}(0)|N_n = 0) = 1$ and for $k = 1, 2, ...$,

$$[\tilde{S}(n)|N_n = k] \overset{d}{=} \tilde{S}(0) + \eta, \quad where \quad \eta \sim Gamma(k\alpha, \beta).$$

More precisely for $x \leq \tilde{S}(0)$ the p.d.f. $P_{\tilde{S}(n)}(x|N_n > 0) = 0$, and for $x \geq \tilde{S}(0)$,

$$P_{\tilde{S}(n)}(x|N_n > 0) = \frac{e^{-\beta(x-\tilde{S}(0))}}{1 - (1-p)^n}\sum_{k=1}^{n}\frac{n!}{k!(n-k)!}\frac{(p\beta^{\alpha})^k}{\Gamma(\alpha k)}(x-S(0))^{\alpha k-1}(1-p)^{n-k}.$$

5. For $n = 1, 2, ...$ and $m = 1, 2, ..., n$

$$P(T_{N_n} = j|N_n = m) = \frac{C_{j-1}^{m-1}}{C_n^m}, \quad j = m, m+1, ..., n.$$

$$E(T_{N_n}|N_n = m) = \frac{m(n+1)}{m+1}, \quad Var(T_{N_n}|N_n = m) = \frac{m(n+1)(n-m)}{(m+1)^2(m+2)}.$$

$$P(T_{N_n} = s|N_n > 0) = \frac{p(1-p)^{n-s}}{1 - (1-p)^n}, \quad s = 1, 2, ..., n$$

$$E(z^{T_{N_n}}|N_n > 0) = \frac{zp}{[1-(1-p)^n]} \frac{(1-p)^n - z^n}{1-p-z},$$

$$E(T_{N_n}|N_n > 0) = \frac{np - (1-p)[1-(1-p)^n]}{p[1-(1-p)^n]} \tag{3}$$

6. The first passage time. For $t > S(0)$, and $k = 1, 2, \ldots$

$$P(T(t) = k) = \sum_{m=1}^{k} \frac{\Gamma(\alpha m, \beta(t - S(0)))}{\Gamma(\alpha m)} \frac{k!}{m!(k-m)!} p^m (1-p)^{k-m-1} \left(\frac{m}{k} - p\right).$$

**Note:** 1. Property (5) shows that $n + 1 - T_{N_n}|N_n > 0$ is truncated geometrically distributed on $1, 2, \ldots, n$ with probability for success $1 - p$. The definition for the truncated geometrically distribution can be seen e.g. in Kapadia and Thomasson [5].

2. According to the Central Limit Theorem the time intersections of these processes are asymptotically normal.

## 3 Zero-Inflated Gamma Random Walk

In this section we investigate the substraction between two independent convolutions of zero-inflated gamma distributions. We call it Zero-inflated Gamma random walk. It is defined by the relation

$$R(n) = R(0) + \sum_{i=1}^{n} I_i^+ X_i^+ - \sum_{i=1}^{n} I_i^- X_i^- = R(n-1) + I_n^+ X_n^+ - I_n^- X_n^-, \ n = 1, 2, \ldots \tag{4}$$

where

$C_1$ : $(I_i^+, I_i^-, 1 - I_i^+ - I_i^-) \sim Mn(1; p^+, p^-, p_0)$, $i = 1, 2, \ldots$ are i.i.d. with $p^+, p^- \in (0, 1)$ and $p_0 = 1 - p^+ - p^- \in (0, 1]$.

$C_2$ : $X_1^+, X_2^+, \ldots, X_n^+$ are i.i.d. $Gamma(\alpha^+, \beta^+)$ distributed r.vs. $X_1^-, X_2^-, \ldots, X_n^-$ are i.i.d. $Gamma(\alpha^-, \beta^-)$ distributed r.vs.

$C_3$ : For fixed $n = 1, 2, \ldots$ denote by $N_n^+ = I_1^+ + I_2^+ + \ldots + I_n^+$, $N_n^- = I_1^- + I_2^- + \ldots + I_n^-$, and by $N_n^{(0)} = n - N_n^+ - N_n^-$.

$C_4$ : By assumption $(N_1^+, N_1^-, N_1^{(0)})$, $(N_2^+, N_2^-, N_2^{(0)})$, $\ldots$, $(N_n^+, N_n^-, N_n^{(0)})$, $X_1^+, X_2^+, \ldots, X_n^+$ and $X_1^-, X_2^-, \ldots, X_n^-$ are independent.

$C_5$ : $R(0) \geq 0$ is a constant.

As a consequence of the conditions $C_1 - C_5$:

$$R(n) = R(0) + I\{N_n^+ > 0\} \sum_{k=1}^{N_n^+} X_k^+ - I\{N_n^- > 0\} \sum_{k=1}^{N_n^-} X_k^-, \ n = 0, 1, 2, \ldots \tag{5}$$

Denote by $T_1^+ = min\{n \in N : I_n^+ = 1\}$, $T_i^+ = min\{n > T_{i-1}^+ : I_n^+ = 1\}$, $T_0^+ = 0$, $T_1^- = min\{n \in N : I_n^- = 1\}$, $T_i^- = min\{n > T_{i-1}^- : I_n^- = 1\}$, $T_0^- = 0$, $T_1^{(0)} = min\{n \in N : I_n^+ = 0 \ and \ I_n^- = 0\}$, $T_i^{(0)} = min\{n > T_{i-1}^{(0)} : I_n^+ = 0 \ and \ I_n^- = 0\}$, $i = 2, 3, \ldots, T_0^{(0)} = 0$.

It is clear that $(N_n^+, N_n^-, N_n^{(0)}) \sim Mn(n; p^+, p_-, p_0)$, $T_{i+1}^{(0)} - T_i^{(0)} - 1 \sim G(p_0)$, $T_{i+1}^- - T_i^- - 1 \sim G(p^-)$ and $T_{i+1}^+ - T_i^+ - 1 \sim G(p^+)$ $i = 1, 2, \ldots$.

The sequence, defined by (4) possesses the following properties.

**Theorem 2.** Assume the sequence $R$ is defined by (4). Then

1. Univariate marginals. Their Laplace transform is:

$$Ee^{-sR(n)} = e^{-sR(0)} \left[ p_0 + p^+ \left( \frac{\beta^+}{\beta^+ + s} \right)^{\alpha^+} + p^- \left( \frac{\beta^-}{\beta^- - s} \right)^{\alpha^-} \right]^n, \quad s \geq 0.$$

The corresponding mean and variances are:

$$E[R(n)] = R(0) + n \left( \frac{p^+\alpha^+}{\beta^+} - \frac{p^-\alpha^-}{\beta^-} \right),$$

$$Var[R(1)] = \frac{p^+\alpha^+}{(\beta^+)^2} [1 + \alpha^+(1 - p^+)] + \frac{p^-\alpha^-}{(\beta^-)^2} [1 + \alpha^-(1 - p^-)] + p^-p^+ \frac{\alpha^-\alpha^+}{\beta^-\beta^+}$$

$$Var[R(n)] = nVar[R(1)]$$

2. Conditional distributions. For $x \geq R(0)$,

$$F_{R(n)}(y|R(n-1) = x) = p^- \frac{\Gamma(\alpha^-, x - y)}{\Gamma(\alpha^-)}, \quad y < x$$

$$P(R(n) = x|R(n-1) = x) = p_0$$

$$F_{R(n)}(y|R(n-1) = x) = 1 - p^+ \frac{\Gamma(\alpha^+, y - x)}{\Gamma(\alpha^+)}, \quad y > x.$$

The mean square regression of $R(n)$ with respect to $R(n-1)$ is linear

$$E[R(n)|R(n-1) = x] = x + \frac{p^+\alpha^+}{\beta^+} - \frac{p^-\alpha^-}{\beta^-},$$

$$Var[R(n)|R(n-1) = x] = Var[I_1^- X_1^- - I_1^+ X_1^+] = Var[R(1)].$$

3. The sequence $\{R(n) - R(0) - n \left( \frac{p^+\alpha^+}{\beta^+} - \frac{p^-\alpha^-}{\beta^-} \right), n = 1, 2, ...\}$ is a martingale.
   - ✓ If $\frac{p^+\alpha^+}{\beta^+} = \frac{p^-\alpha^-}{\beta^-}$, then $\{R(n), n = 1, 2, ...\}$ is a martingale.
   - ✓ If $\frac{p^+\alpha^+}{\beta^+} > \frac{p^-\alpha^-}{\beta^-}$, then $\{R(n), n = 1, 2, ...\}$ is a sub-martingale.
   - ✓ If $\frac{p^+\alpha^+}{\beta^+} < \frac{p^-\alpha^-}{\beta^-}$, then $\{R(n), n = 1, 2, ...\}$ is a super-martingale.
4. Bivariate distributions. For $0 \leq n_1 \leq n_2$, $s_1 \geq 0$, $s_2 \geq 0$

$$Ee^{-s_2 R(n_2) - s_1 R(n_1)}$$

$$= e^{-(s_1+s_2)R(0)} \left\{ p_0 + p^+ \left( \frac{\beta^+}{\beta^+ + s_1 + s_2} \right)^{\alpha^+} + p^- \left( \frac{\beta^-}{\beta^- - s_1 - s_2} \right)^{\alpha^-} \right\}^{n_1}$$

$$\times e^{-s_2} \left\{ p_0 + p^+ \left( \frac{\beta^+}{\beta^+ + s_2} \right)^{\alpha^+} + p^- \left( \frac{\beta^-}{\beta^- - s_2} \right)^{\alpha^-} \right\}^{n_2-n_1}$$

5. $cov(R(n), R(k)) = kVar[R(1)]$, $cor(R(n), R(k)) = \sqrt{\frac{k}{n}}$, $1 \leq k \leq n$.

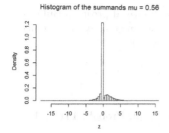

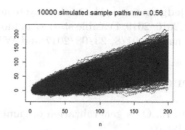

**Fig. 1.** The histogram of the summands of $R(n)$ for $p^+ = 0.5$, $p^- = 0.44$, $\alpha^+ = 2$, $\alpha^- = 1$, $\beta^+ = 1$, $\beta^- = 1$.

**Fig. 2.** 10000 sample paths of $R(n)$ for parameters $p^+ = 0.5$, $p^- = 0.44$, $\alpha^+ = 2$, $\alpha^- = 1$, $\beta^+ = 1$, $\beta^- = 1$.

## 4  Simulation Study

In this section we have denoted by $\mu = \frac{p^+ \alpha^+}{\beta^+} - \frac{p^- \alpha^-}{\beta^-}$ and consider separately 3 particular cases $\mu < 0$, $\mu = 0$, and $\mu > 0$. In any of these cases the corresponding $\{R(n), n = 0, 1, 2, ...\}$ is super-martingale, martingale or sub-martingale. All plots in this paper are plotted using standard functions in R software [12]. To obtain Figs. 1–4 we have simulated $m = 10000$ sample paths of the process $\{R(n), n = 0, 1, 2, ...\}$ for time horizon $1, 2, ..., 200$ and $S(0) = 0$.

Case 1. $\mu > 0$. Figure 1 visualises the histogram of the simulated values of $I_i^+ X_i^+ - I_i^- X_i^-$ for $p^+ = 0.5$, $p^- = 0.44$, $\alpha^+ = 2$, $\alpha^- = 1$, $\beta^+ = 1$, $\beta^- = 1$. The image on Fig. 2 confirms the well known results from the theory of random walks (see [14]) that in this case, and in infinite time horizon the process goes to infinity and it has almost sure finite global infimum.

Case 2. $\mu = 0$. The histogram on Fig. 3 summarizes the simulated values of $I_i^+ X_i^+ - I_i^- X_i^-$ for $p^+ = 0.45$, $p^- = 0.5$, $\alpha^+ = 0.1$, $\alpha^- = 0.9$, $\beta^+ = 0.2$, $\beta^- = 2$. The plot on Fig. 4 depicts the fact that in this case in infinite time horizon the process fluctuates unboundedly. See [14].

Case 3. For $\mu < 0$ analogous simulation study shows that in infinite horizon the process goes a.s. to minus infinity and has a.s. finite supremum.

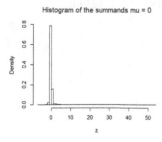

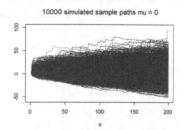

**Fig. 3.** The histogram of the summands of $R(n)$ for $p^+ = 0.45$, $p^- = 0.5$, $\alpha^+ = 0.1$, $\alpha^- = 0.9$, $\beta^+ = 0.2$, $\beta^- = 2$.

**Fig. 4.** 10000 sample paths of $R(n)$ for parameters $p^+ = 0.45$, $p^- = 0.5$, $\alpha^+ = 0.1$, $\alpha^- = 0.9$, $\beta^+ = 0.2$, $\beta^- = 2$.

**Acknowledgements.** The authors were supported by the bilateral projects Bulgaria - Austria, 2016–2019, Feasible statistical modelling for extremes in ecology and finance, Contract number 01/8, 23/08/2017 and WTZ Project No. BG 09/2017.

# References

1. Çinlar, E.: On a generalization of gamma processes. J. Appl. Probab. **17**(2), 467–480 (1980)
2. Dickson, D.C.M., Waters, H.R.: Gamma processes and finite time survival probabilities. ASTIN Bull.: J. IAA **23**(2), 259–272 (1993)
3. Dufresne, F., Gerber, H.U., Shiu, E.S.W.: Risk theory with the gamma process. ASTIN Bull.: J. IAA **21**(2), 177–192 (1991)
4. Jordanova, P.K., Petkova, M.P.: Measuring heavy-tailedness of distributions. In: AIP Conference Proceedings, vol. 1910, no. 1, pp. 0600021–0600028 (2017)
5. Kapadia, C.H., Thomasson, R.L.: On estimating the parameter of a truncated geometric distribution by the method of moments. Ann. Inst. Stat. Math. **27**(1), 269–272 (1975)
6. Kaishev, V.K., Dimitrova, D.S.: Dirichlet bridge sampling for the variance gamma process: pricing path-dependent options. Manag. Sci. **55**(3), 483–496 (2009)
7. Kumar, A., Wylomanska, A., Poloczanski, R., Sundar, S.: Fractional Brownian motion time-changed by gamma and inverse gamma process. Phys. A: Stat. Mech. Appl. **468**, 648–667 (2017)
8. Lawrance, A.J.: The innovation distribution of a gamma distributed autoregressive process. Scand. J. Stat. JSTOR **9**, 234–236 (1982)
9. Lewis, P.A.W., McKenzie, E., Hugus, D.K.: Gamma processes. Stoch. Models **5**(1), 1–30 (1989)
10. Madan, D.B., Seneta, E.: The variance gamma model for share market returns. J. Bus. **63**, 511–524 (1990)
11. Madan, D.B., Carr, P.P., Chang, E.C.: The variance gamma process and option pricing. Rev. Financ. **2**(1), 79–105 (1998)
12. R Development Core Team. R: a language and environment for statistical computing. R Foundation for Statistical Computing, Vienna, Austria
13. Seneta, E.: Fitting the variance-gamma model to financial data. J. Appl. Probab. **41**(A), 177–187 (2004)
14. Spitzer, F.: Principles of Random Walk. Graduate Texts in Mathematics, vol. 34. Springer, New York (1976). https://doi.org/10.1007/978-1-4757-4229-9
15. Wolpert, R.L.: Stationary gamma processes (2016)

# Analytic Theory of Singular Perturbations and Lomov's Regularization Method

Vasiliy I. Kachalov[✉]

National Research University "MPEI", Moscow, Russia
vikachalov@rambler.ru

**Abstract.** The paper contains results related to the so-called analytic theory of singular perturbations. The main of them are sufficient conditions for ordinary convergence of series in powers of a small parameter representing solutions of singularly perturbed problems.

**Keywords:** Method of regularization · Pseudoholomorphic solutions · Analytic family of homomorphisms

## 1 Introduction

At present, the theory of singular perturbations is represented by a large number of different asymptotic methods. The foundations of the theory of asymptotic integration are laid by the works of Prandtl, Birkhoff, and Schlesinger. Fundamental importance in the development of the theory of singular perturbations is the work of A.N. Tikhonov, where classical theorems on the passage to the limit in nonlinear singularly perturbed problems are proved. Deep development of this theory was obtained in the works of V. Vazov, M.I. Vishik, L.A. Lyusternik, A.B. Vasil'eva, V.F. Butuzov, N.N. Nefedov, Yu.A. Mitropol'skii, S.A. Lomov, A.N. Filatov, N.Kh. Rozov, N.I. Shkil, M.V. Fedoryuk, V.P. Maslov, V.A. Trenogin, N.N. Moiseev, M.M. Khapaev, V.F. Safonov, A.A. Bobodzhanov, A.V. Nesterov, V.G. Zadorozhniy, M.G. Dmitriev, M.I. Imanaliev, K.A. Kasymov and other researchers.

In contrast to regular perturbation theory, the theory of singular perturbations is an asymptotic theory (therefore, it is called the theory of asymptotic integration) [1–3].

The regularization method of Lomov [4] makes it possible to construct solutions of singularly perturbed problems in the form of series in powers of a small parameter that converge in the usual sense (not asymptotically), under certain restrictions on the data of the problem [5,7]. Such solutions are called pseudo-analytic (pseudoholomorphic) and were obtained, in the main, for linear problems using the theory of spaces of vectors of exponential type [5]. The aim of our research is to construct, based on the regularization method of S.A. Lomov and his generalizations, the analytic theory of singular perturbations, a section of mathematics designed to equalize the rights of regular and singular theories.

© Springer Nature Switzerland AG 2019
I. Dimov et al. (Eds.): FDM 2018, LNCS 11386, pp. 305–312, 2019.
https://doi.org/10.1007/978-3-030-11539-5_34

## 2    Functional Analysis in the Theory of Singular Perturbations. Linear Problems

The first step in constructing the theory of ordinary convergence of series in powers of a small parameter was the study of the conditions for the existence of solutions of the equation analytic in the small parameter

$$\varepsilon \frac{dy}{dt} - A(t)y = h(t), t \in [0, T], \tag{1}$$

in which $A(t) \in C([0, t]; \mathcal{L}(E))$—algebra of continuous functions on the interval $[0, T]$ operator-valued functions with values in the space of bounded operators acting in a Banach space $E$.

Denote by $C_E = C([0, T]; E)$ the Banach space of functions that are continuous on the interval $[0, T]$ and with values $E$, and let $A(t)$ be continuously invertible for each $t$ from the specified interval. Denote by $F = A^{-1}(t)\frac{d}{dt}$ a closed unbounded operator acting in $C_E$. We take an arbitrary $c > 0$ and denote by the set of elements $u(t) \in C_E$, such that

$$\sup_k \frac{\|F^k u\|_{C_E}}{c^k} < +\infty, \tag{2}$$

with the norm defined by the formula (2), and call it the space of vectors of exponential type $\leq c$. One can prove that $Y^C$ is a Banach space [4,5]. Further, we unite such spaces over all positive $C$, we define the topology of the inductive limit there and call the resulting space the space of vectors of exponential type: $\exp_F E = \lim_{c \to +\infty} \operatorname{ind} Y^c$. In [5] it is proved that any vector $u(t) \in \exp_F E$ has the representation

$$
\begin{aligned}
u(t) = u(0) &+ \int_0^t A(t_1)dt_1 \cdot Fu(0) + \int_0^t A(t_1)dt_1 \int_0^{t_1} A(t_2)dt_2 \cdot F^2 u(0) \\
&+ \ldots + \int_0^t A(t_1)dt_1 \int_0^{t_1} A(t_2)dt_2 \ldots \int_0^{t_{k-1}} A(t_k)dt_k \cdot F^k u(0) + \ldots,
\end{aligned} \tag{3}
$$

and here the power series $u(0) + \varepsilon Fu(0) + \ldots + \varepsilon^k F^k u(0) + \ldots$ converges in a certain neighborhood of the value $\varepsilon = 0$.

**Theorem 1** [5]. *In order that there exist a solution of the Eq. (1), analytic at the point $\varepsilon = 0$, it is necessary and sufficient that*

$$\int_0^t h(\tau)d\tau \in \exp_F E, \tag{4}$$

*and such a solution is unique.*

If we give the initial condition for the Eq. (1), then the resulting problem will be singularly perturbed and the corresponding space of vectors of exponential type is constructed for the so-called basic operator of the regularization method [4] (in any case, (4) is necessary).

We consider the question of the existence of an analytic solution at the point $\varepsilon = 0$ of the equation

$$\varepsilon \frac{dy}{dt} - A(t)y = h(t), \quad t \in [0, T], \tag{5}$$

when the operator $A(t)$ for each $t \in [0, T]$ is closed unbounded.

S.A. Lomov formulated the problem of describing such classes of unbounded operators $A(t)$ so that the space $\exp_F E$ is maximally simple, for example, coinciding with a functional space well studied in the analysis.

**Theorem 2.** *Let $A(t)$ be a closed unbounded operator acting for each $t \in [0, T]$ in a Banach space $E$ having a domain of definition independent of $t$, and the inverse operator $A^{-1}(t) \in \mathcal{L}(E) \ \forall t \in [0, T]$ admits an analytic continuation to the disk $|t| \leq R$, where $R > T$. Then, if there exists a constant closed unbounded operator $B$ such that for all natural $n$*

$$\max_{\substack{|\eta_i| = R \\ i = \overline{1,n}}} \| A^{-1}(\eta_1) \dots A^{-1}(\eta_n) \| \leq \| B^{-n} \|,$$

*and the resolvent of this operator $R(\lambda; B)$ is an entire function of exponential type, then any vector-valued function $u(t)$ with values in $E$ analytic in the indicated disk belongs to $\exp_F E$.*

In the proof of this theorem, the integral Cauchy formula for functions of several complex variables was used [6].

We note that for $A(t)$ we can take an ordinary linear differential operator with zero initial conditions, that is, consider equations of the form

$$\varepsilon u_t - \sum_{m=1}^{n} a_m(t, x) \partial_x^m u = h(t, x),$$

in which the functions $h(t, x)$ and $a_m(t, x)$ $(m = \overline{1, n})$ are analytic on the rectangle $[0, T] \times [0, X]$. For example, the equation $\varepsilon u_t = (t + 1)u_x + tx^3$, $t \geq 0$, where $D((t + 1)\partial_x) = \{y(x) : \ y(x) \in C^1[0, X], \ y(0) = 0\}$, has a unique analytic at the point $\varepsilon = 0$ solution

$$u(t, x, \varepsilon) = -\frac{tx^3}{3(t + 1)} + 2 \sum_{k=1}^{\infty} \varepsilon^k \frac{(-1)^k (2k - 1)!! x^{k+3}}{(k + 3)!(t + 1)^{2k+1}}.$$

This series converges as $|\varepsilon| < 1/2X$ uniformly in the domain $\{(t, x) : \ t \geq 0, \ 0 \leq x \leq X\}$.

## 3 Algebraic Foundations of the Theory of Singular Perturbations and Holomorphic Regularization

The well-known Poincare decomposition theorem guarantees the existence and uniqueness of the solution $y(t, \varepsilon)$ of the Cauchy problem

$$\frac{dy}{dt} = f(t, y, \varepsilon), \quad y(t_0, \varepsilon) = y_0,$$

analytic at the point $\varepsilon = 0$ if $f(t, y, \varepsilon)$ is analytic at the point $(t_0, y_0, 0)$ as a function of three variables. It is clear that for a singularly perturbed problem

$$\varepsilon \frac{dy}{dt} = f(t, y), y(t_0, \varepsilon) = y_0 \tag{6}$$

it is not so (in the general case). It is proved (and this is the main statement of the method of holomorphic regularization) that instead of solving the analytic dependence on $\varepsilon$ of the left side of the Eq. (6), inherit the integrals of this equation [8]. The algebraic basis of the method is the commutation relations and equivalent homomorphisms of the algebras of analytic functions of a different number of variables [9]. We denote by $\mathcal{A}_{t_0}$ the algebra of functions of $t$ analytic on the interval $[t_0, T]$, and through $\mathcal{A}_{t_0 y_0}$—algebra of functions of variables $t$ and $y$ analytic on the rectangle $\Pi = [t_0, T] \times [y_0 - \gamma, y_0 + \gamma]$, where $\gamma$ is a sufficiently large positive number.

To describe the algorithm of the method of holomorphic regularization, we shall consider the right-hand side of Eq. (6) is scalar function for which condition ($\alpha$): $f(t, y) \in \mathcal{A}_{t_0 y_0}$ and $f(t, y) \neq 0 \; \forall (t, y) \in \Pi$ is satisfied. From the nonlinear Eq. (6) we pass to the linear equation of its integrals

$$\varepsilon U_t + f(t, y)U_y = 0, \tag{7}$$

whose solution, assuming the operator $\partial_t$ to be subordinate to the operator $f\partial_y$, we look for a series in powers of $\varepsilon$:

$$U(t, y, \varepsilon) = U_0(t, y) + \varepsilon U_1(t, y) + \ldots + \varepsilon^n U_n(t, y) + \ldots. \tag{8}$$

In accordance with the method of undetermined coefficients, we obtain a series of problems for determining the coefficients of this series:

$$
\begin{aligned}
fU_{0,y} &= 0, \\
fU_{1,y} &= -U_{0,t}, \\
fU_{2,y} &= -U_{1,t}, \\
&\ldots\ldots\ldots\ldots\ldots \\
fU_{n,y} &= -U_{n-1,t}, \\
&\ldots\ldots\ldots\ldots\ldots
\end{aligned}
\tag{9}
$$

As a solution of the first equation of the series, we can take an arbitrary function $\varphi(t) \in \mathcal{A}_{t_0}$. We now introduce the following notation: $g_k \equiv \frac{1}{f(t, y_k)}$, $k = 1, 2, \ldots$; if $h(t, y)$ is some function, then

$$J_k h \equiv \int_{y_0}^{y_k} h(t, y_{k+1}) dy_{k+1}, \quad k = 1, 2, \ldots; \quad J_0 h \equiv \int_{y_0}^{y} h(t, y_1) dy_1.$$

Then we have:

$$
\begin{aligned}
U(t, y, \varepsilon) = \varphi &- \varepsilon J_0(g_1 \partial_t \varphi) + \varepsilon^2 J_0(g_1 \partial_t J_1(g_2 \partial_t \varphi)) \\
&- \varepsilon^3 J_0(g_1 \partial_t J_1(g_2 \partial_t J_2(g_3 \partial_t \varphi))) + \ldots.
\end{aligned}
\tag{10}
$$

To prove the convergence of the power series (10), we use the lemma, which is proved by the method of mathematical induction.

**Lemma 1.** *If in expression $\partial_t(b_1(\partial_t(b_2(\partial_t \ldots (\partial_t b_n)) \ldots)$, in which $b_1, \ldots, b_n$ — functions of the variable $t$, expand the brackets by the formula of the derived product, and replace $\partial_t^s b_r$, $1 \le r \le n$, $0 \le s \le n$ $s!$, then the obtained sum will be $(2n-1)!!$.*

We represent the coefficient $U_n$ of the series (10) as follows:

$$U_n = (-1)^n (J_0 J_1 \ldots J_{n-1})[g_1(\partial_t(g_2(\partial_t \ldots \partial_t(g_n \partial_t \varphi)) \ldots)], \qquad (11)$$

where

$$(J_0 J_1 \ldots J_{n-1})[H(t, y_1, \ldots, y_n)] \equiv \int_{y_0}^{y} dy_1 \int_{y_0}^{y_1} dy_2 \ldots \int_{y_0}^{y_{n-1}} H(t, y_1, \ldots, y_n) dy_n.$$

The functions $\varphi, g_1, \ldots, g_n$ are analytic in $t$ on the segment $[0, T]$ uniformly in $y \in [y_0 - \gamma, y_0 + \gamma]$, so $\exists C > 0$: $|\partial_t^s \varphi| \le C^s s!$, $|\partial_t^s g_r| \le C^s s!$ on a rectangle $\Pi$. Thus,

$$|U_n(t, y)| \le \left| \int_{y_0}^{y} dy_1 \int_{y_0}^{y_1} dy_2 \ldots \int_{y_0}^{y_{n-1}} dy_n \right| C^n (2n-1)!!,$$

from which it follows that $|U_n(t, y)| \le \frac{\gamma^n C^n (2n-1)!!}{n!}$ $\forall (t, y) \in \Pi$.

The convergence of the series (10) in a neighborhood of the point $\varepsilon = 0$ uniformly on the rectangle $\Pi$ follows from the d'Alembert test.

Since $\varphi(t)$ enters the expression (10) linearly, we can look at $U(t, y, \varepsilon)$ for each fixed sufficiently small $\varepsilon$ as an image of some linear operator: $U(t, y, \varepsilon) = A_f^\varepsilon[\varphi]$.

**Theorem 3.** *The mappings $\{A_f^\varepsilon\}$ form an analytic at the point $\varepsilon = 0$ family of homomorphisms from algebra $\mathcal{A}_{t_0}$ to algebra $\mathcal{A}_{t_0 y_0}$.*

**Proof.** First we establish the commutation relation for the operator $A_f^\varepsilon$. Since for any $\varphi(t)$ the function $U(t, y, \varepsilon)$ is an integral of the Eq. (6), then there exists a function $\Phi$ of one variable such that $A_f^\varepsilon[\varphi(t)] = \Phi(A_f^\varepsilon[t])$. If we put $y = y_0$ in the left and right sides of this equality, then $\varphi(t) \equiv \Phi(t)$, i.e. it can be represented as a commutation relation

$$A_f^\varepsilon[\varphi(t)] = \varphi(A_f^\varepsilon[t]). \qquad (12)$$

Further, $A_f^\varepsilon[\varphi_1 \varphi_2] = (\varphi_1 \varphi_2)(A_f^\varepsilon[t]) = \varphi_1(A_f^\varepsilon[t])\varphi_2(A_f^\varepsilon[t]) = A_f^\varepsilon[\varphi_1]A_f^\varepsilon[\varphi_2]$, i.e. $A_f^\varepsilon : \mathcal{A}_{t_0} \to \mathcal{A}_{t_0 y_0}$—homomorphism. $\square$

Thus, it is proved that the images of the homomorphisms $\{A_f^\varepsilon\}$ are analytic in the parameter integrals of Eq. (6).

## 4   Pseudoholomorphic Solutions of Singularly Perturbed Problems

**Definition 1.** *The solution $y(t, \varepsilon)$ of the initial problem (6) is called pseudo-holomorphic at $\varepsilon = 0$ if there exists a function $Y(t, \eta, \varepsilon)$, analytic in the third variable in the neighborhood of the value $\varepsilon = 0$ for each $t \in [t_0, T]$ and every $\eta$ from some unbounded set $G$ such that for some analytic function $\varphi(t)$ the equality $y(t, \varepsilon) = Y(t, \varphi(t)/\varepsilon, \varepsilon)$ $\forall t \in [t_0, T]$.*

The following theorem gives sufficient conditions for the existence of a pseu-doholomorphic solution $y$ of the Cauchy problem (6).

**Theorem 4.** *Suppose that the function $\varphi(t)$, analytic on the interval $[t_0, T]$, is such that $\varphi(t_0) = 0$, and equation*

$$\varphi'(t) \int_{y_0}^{y} \frac{dy_1}{f(t, y_1)} = \frac{\varphi(t)}{\varepsilon} \tag{13}$$

*has a solution of the form*

$$y = Y_0 \left( t, \Psi \left( \frac{\varphi(t)}{\varepsilon} \right) \right), \tag{14}$$

*in which $q = \Psi(\eta)$ is an entire function with an asymptotic value equal to $p_0$, and the function $Y_0(t, q)$ is analytic on the rectangle $\Pi_{tq}^0 = [t_0, T] \times Q$, where $Q$ is a segment containing the points $p_0$ and $\Psi(0)$. Then the solution $y(t, \varepsilon)$ of the Cauchy problem (6) is pseudoholomorphic at the point $\varepsilon = 0$.*

**Proof.** We write the general integral of the Eq. (6) in the form

$$\widetilde{U}(t, y, \varepsilon) = \frac{\varphi(t)}{\varepsilon}, \tag{15}$$

where

$$\begin{aligned}
\widetilde{U}(t, y, \varepsilon) &= J_0(g_1 \partial_t \varphi) - \varepsilon J_0(g_1 \partial_t J_1(g_2 \partial_t \varphi)) \\
&\quad + \varepsilon^2 J_0(g_1 \partial_t J_1(g_2 \partial_t J_2(g_3 \partial_t \varphi))) + \dots.
\end{aligned} \tag{16}$$

We calculate the values of the function $\Psi$ from the left and right sides of (15)

$$\Psi(\widetilde{U}(t, y, \varepsilon)) = \Psi \left( \frac{\varphi(t)}{\varepsilon} \right).$$

We denote the right-hand side of the resulting equality by $q$ and select the principal term on the left-hand side:

$$\Psi(J_0(g_1 \partial_t \varphi)) + \varepsilon F(t, y, \varepsilon) = q. \tag{17}$$

Next, we take $p > p_0$ very close to $p_0$ and, assuming that $p_0 < \Psi(0)$, we construct a rectangle $\Pi_{tq} = [t_0, T] \times [p, \Psi(0)]$. It is obvious that for the Eq. (17) all the conditions of the implicit function theorem on the rectangle $\Pi_{tq}$, $\varepsilon = 0$   $y = Y_0(t, q)$, and the estimate of the modulus of this function does not depend on $p$ because of the analyticity on the rectangle $Pi_{tq}^0$. Consequently, in a neighborhood $\sigma_{tq}$ of each point $(t, q) \in \Pi_{tq}$ there exists a solution $y = Y(t, q, \varepsilon)$ of the Eq. (17), analytic in some neighborhood of the value $\varepsilon = 0$. We choose a finite subcovering from the covering $\{\sigma_{ta}\}$ of the rectangle $\Pi_{tq}$. Then the function $Y(t, q, \varepsilon)$ is analytic, uniformly on this rectangle, in the smallest neighborhood $0 < \varepsilon < \varepsilon_0$ corresponding to the finite subcovering.

Let the parameter $\varepsilon$ in the Eq. (6) satisfy the inequality $0 < \varepsilon < \varepsilon_0$ and a curve $\Gamma : q = \Psi(\varphi(t)/\varepsilon)$ entirely belongs to a rectangle $\Pi_{tq}$. Then representation

$$y(t,\varepsilon) = \sum_{n=0}^{\infty} \varepsilon^n Y_n(t, \varphi(t)/\varepsilon), \tag{18}$$

takes place and this series converges uniformly on the segment $[t_0, T]$. If the rectangle $\Pi_{tq}$ contains only part of the curve $\Gamma$, then the series (18) converges uniformly on some interval $[t_0, t_1] \subset [t_0, T]$ and a pseudoholomorphic extension $y(t, \varepsilon)$ to the right is required.     □

## 5   Generalizations and Examples

The method of holomorphic regularization admits a generalization to the case of equations of higher orders and systems [10, 12, 13]. In this section of the paper we give examples covering a sufficiently wide range of singularly perturbed initial problems.

$1°$. $\varepsilon y' = y^2 - e^{2t}$,   $y(0, \varepsilon) = 0$,

$$y(t, \varepsilon) = e^t \text{th} \frac{1 - e^t}{\varepsilon} + \frac{\varepsilon}{2} \text{th}^2 \frac{1 - e^t}{\varepsilon} + \ldots.$$

$2°$. $\varepsilon y' = e^{-ye^t} - 10$,   $y(0, \varepsilon) = 0$,

$$y(t, \varepsilon) = e^{-t}\left(1 + \frac{\varepsilon e^{-t}}{10}\right) \ln \frac{1 + 9e^{10(1-e^t)}}{10} + \ldots.$$

$3°$. $\varepsilon y'' + yy' - y = 0$,   $y(t_0, \varepsilon) = y_0 > 0$,   $y'(t_0, \varepsilon) = v_0 \neq 1$,

$$y(t, \varepsilon) = t - t_0 + y_0 + \frac{\varepsilon}{y_0}(v_0 - 1)\left(e^{-\frac{y_0(t-t_0)}{\varepsilon}} - 1\right) + \ldots.$$

$4°$. $\varepsilon y'' = e^{2t} - e^{2y}(y')^2$,   $y(0, \varepsilon) = y'(0, \varepsilon) = 0$,

$$y(t, \varepsilon) = t - \varepsilon e^t \ln\left(1 - \text{th}\frac{1 - e^t}{\varepsilon}\right) + \ldots.$$

$5°$. A mixed problem for a nonlinear parabolic equation:

$$\varepsilon u_t = u_{xx} - u^3,$$
$$u(0, \varepsilon) = \sin x,$$
$$u(t, 0) = u(t, \pi) = 0.$$

Second-order Galerkin approximation:

$$u_2(t, x, \varepsilon) = \frac{2e^{-t/\varepsilon}\sin x}{\sqrt{7 - 3e^{-t/\varepsilon}}}.$$

$6°$. The Tikhonov type system

$$\begin{cases} y' = v^2, \\ \varepsilon v' = v^2 - y^2 e^{2t}, \end{cases}$$
$$y(0, \varepsilon) = -2, \quad v(0, \varepsilon) = 0$$

has the following pseudoholomorphic solution:

$$\begin{cases} y(t,\varepsilon) = -2e^{-2t} - 2\varepsilon e^{-t}\text{th}\frac{2(e^{-t}-1)}{\varepsilon} + \dots, \\ v(t,\varepsilon) = 2e^{-t}\text{th}\frac{2(e^{-t}-1)}{\varepsilon} + \dots. \end{cases}$$

In the paper [14] an example of a weakly nonlinear singularly perturbed system is given:

$$\begin{cases} \varepsilon y_1' = -(e^t + 1)y_1 + 2e^{-t}y_2 + \varepsilon e^{ty_2}, \\ \varepsilon y_2' = e^{2t}y_1 - 3y_2, \\ y_1(0,\varepsilon) = 1, \quad y_2(0,\varepsilon) = 0. \end{cases}$$

In these examples it is easy to see the existence of a limit transition characteristic of singularly perturbed problems satisfying the conditions of the theorem of A.N. Tikhonov [11].

# References

1. Vasilyeva, A.B., Butuzov, V.F.: Asymptotic expansion of solutions of singularly perturbed problems. Nauka, Moscow (1973)
2. Butuzov, V.F., Vasilyeva, A.B., Nefedov, N.N.: Asymptotic theory of contrast structures. Autom. Telemech. **7**, 4–42 (1997)
3. Vasilyeva, A.B., Butuzov, V.F., Nefedov, N.N.: Singularly perturbed problems with boundary and inner layers. Proc. Steklov Math. Inst. **268**, 268–283 (2010)
4. Lomov, S.A., Lomov, I.S.: Fundamentals of the mathematical theory of the boundary layer. Michigan State University, Michigan (2011)
5. Kachalov, V.I., Lomov, S.A.: Smoothness of solutions of differential equations with respect to a singularly incoming parameter. DAN SSSR **299**(4), 805–808 (1988)
6. Kachalov, V.I.: On the smoothness of solutions of differential equations containing a parameter. Differ. Equ. **26**(10), 1711–1716 (1990)
7. Kachalov, V.I., Lomov, S.A.: Pseudoanalytic solutions of singularly perturbed problems. Rep. Russ. Acad. Sci. **334**(6), 694–695 (1994)
8. Kachalov, V.I.: Holomorphic regularization of singularly perturbed problems. Bull. MPEI **6**, 54–62 (2010)
9. Kachalov, V.I.: Commutation relations, homomorphisms, and differential equations. Diff. Equ. **50**(1), 10–16 (2014)
10. Kachalov, V.I.: Holomorphic in the parameter of the integrals of singularly perturbed second-order equations and limit theorems. Sci. Tech. Bull. St. Petersburg GPU. Phys. Math. **194**(2), 103–109 (2014)
11. Kachalov, V.I.: Tikhonov's theorem on the passage to the limit and pseudoholomorphic solutions of singularly perturbed problems. Rep. Russ. Acad. Sci. **458**(6), 630–632 (2014)
12. Kachalov, V.I.: Holomorphic regularization of singularly perturbed systems of differential equations. J. Comput. Math. Math. Phys. **57**(4), 64–71 (2017)
13. Kachalov, V.I.: On the method of holomorphic regularization of singularly perturbed problems. Proc. High Sch. Math. **6**, 52–59 (2017)
14. Kachalov, V.I., Fedorov, Yu.S.: Holomorphic regularization of weakly nonlinear singularly perturbed problems. Differ. Equ. Control Process. **3**, 17–30 (2016)

# On Inverse Problem of Determination of the Coefficient in the Black-Scholes Type Equation

Vitaly L. Kamynin$^{(\boxtimes)}$ and Tatiana I. Bukharova

National Research Nuclear University MEPhI
(Moscow Engineering Physics Institute),
31, Kashirskoe shosse, 115409 Moscow, Russia
vlkamynin2008@yandex.ru, bukharova_t@mail.ru

**Abstract.** We prove the existence and uniqueness theorems for inverse problem of determination of the lower coefficient in the Black-Scholes type equation with additional condition of integral observation. These results are based on the investigation of unique solvability of corresponding direct problem which is of independent interest. We give the example of the inverse problem for which the conditions of the theorems proved are fulfilled.

## 1 Introduction

In present work we study the questions of unique solvability of the inverse problem of determination of a pair of functions $\{u(t,x), \gamma(t)\}$ satisfying in the rectangle $Q \equiv [0,T] \times [0,l]$ the parabolic equation

$$u_t - a(t,x)u_{xx} + b(t,x)u_x + d(t,x)u + \gamma(t)u = f(t,x), \tag{1}$$

the initial and boundary conditions

$$u(0,x) = u_0(x), \ x \in [0,l]; \ u(t,l) = 0, \ t \in [0,T]; \tag{2}$$

and the additional condition of integral observation

$$\int_0^l u(t,x)\omega(x)dx = \varphi(t), \quad t \in [0,T]. \tag{3}$$

In our investigation we suppose that the leading coefficient $a(t,x)$ in the Eq. (1) is strongly degenerate for $x = 0$, namely, we assume that

$$a(t,x) \sim a_0 x^2, \ x \to 0, \quad a_0 = \text{const} > 0.$$

In what follows we use Lebesgue, Sobolev and Hölder spaces with corresponding norms in usual sense (see, for example, [1]). The derivatives are understood in generalized sense, the equalities and inequalities are satisfied almost everywhere.

© Springer Nature Switzerland AG 2019
I. Dimov et al. (Eds.): FDM 2018, LNCS 11386, pp. 313–320, 2019.
https://doi.org/10.1007/978-3-030-11539-5_35

We introduce notation

$$Q_\tau^\delta = [0,\tau] \times [\delta, l],\ 0 \le \delta < l,\ 0 \le \tau \le T;\ Q_\tau^0 \equiv Q_\tau,\ Q_T^\delta \equiv Q^\delta,\ Q_T \equiv Q;$$

$$L_\infty^+(0,T) = \{\gamma(t) \in L_\infty(0,T) : \gamma(t) \ge 0\}.$$

We assume that the functions occurring in the input data of the problem (1)–(3) are measurable and satisfy the following conditions:

(A) $x^2 a_1 \le a(t,x) \le a_2$, $a_1, a_2 = \text{const} > 0$, $a(t,0) = 0$,
 $a_x, a_{xx}, a_x^2/a \in L_\infty(Q)$, $|a_x^2/a| \le K_a, |a_{xx}| \le K_a^{**}$;
(B) $b^2/a, d^2/a, f^2/a, b_x \in L_\infty(Q)$,
 $|b^2/a| \le K_{b,a}, |d^2/a| \le K_{d,a}, |f^2/a| \le K_{f,a}, |b_x| \le K_b^*$;
(C) $\omega(x), \omega'(x), \omega''(x) \in L_\infty(0,l),\ \omega(l) = 0,\ |\omega| \le K_\omega,\ |\omega'| \le K_\omega^*,\ |\omega''| \le K_\omega^{**}$;
(D) $u_0(x) \in \overset{0}{W}_2^1(0,l),\ |u_0(x)| \le M_0,\ x \in [0,l]$;
(E) $\varphi(x) \in W_\infty^1(0,T),\ \varphi(t) \ge \varphi_0 > 0$;
(F) $-\varphi_1^* \le \varphi'(t) \le -\varphi_0^* < 0,\quad F_1 \ge \int_0^l f(t,x)\omega(x)dx \ge F_0 \ge 0$;

(G) $\int_0^l u_0(x)\omega(x)dx = \varphi(0)$.

Here $a_1, a_2, K_a, K_a^{**}, K_\omega, K_\omega^*, \varphi_0, \varphi_0^*, \varphi_1^*, M_0 = \text{const} > 0$, $K_{b,a}, K_{d,a}, K_{f,a}$, $K_\omega^{**} = \text{const} \ge 0$.

**Remark 1.** From conditions $(A)$–$(C)$ we obtain the estimates

$$|b| \le \sqrt{K_{b,a}\, a_2} \equiv K_b,\quad |d| \le \sqrt{K_{d,a}\, a_2} \equiv K_d,\quad |f| \le \sqrt{K_{f,a}\, a_2} \equiv K_f,$$

$$|a_x| \le \sqrt{K_a a_2} \equiv K_a^*,\quad |(a\omega)_{xx}| \le K_a^{**} K_\omega + 2K_a^* K_\omega^* + a_2 K_\omega^{**} \equiv K_{a,\omega},$$

$$|(b\omega)_x| \le K_b^* K_\omega + K_b K_\omega \equiv K_{b,\omega}.$$

**Definition 1.** Generalized solution of the problem (1)–(3) is a pair of functions $\{u(t,x), \gamma(t)\}$,

$$u(t,x) \in L_\infty(0,T; W_2^1(0,l)) \bigcap C^{0,\beta}(Q), \beta \in (0,1),\ u_t(t,x) \in L_2(Q),$$

$$a u_{xx}^2 \in L_1(Q),\ u_{xx}(t,x) \in L_2(Q^\delta)\ \forall \delta > 0,\ \gamma(t) \in L_\infty^+(0,T),$$

satisfying Eq. (1) almost everywhere in $Q$ and such that the function $u(t,x)$ satisfies the conditions (2), (3) in classical sense.

**Remark 2.** Conditions $(A)$ and $(B)$ are satisfied for the well-known Black-Scholes equation (see [2,3]). Hence, this equation can be considered in our investigation.

**Remark 3.** Equation (1) degenerates on the boundary of $Q$ for $x = 0$. From the well-known Fichera's theory it follows that whether or not boundary conditions should be given at the particular part of the boundary where the equation degenerates, depends on the sign of so-called Fichera function on that part of the boundary (see [4], [5, p. 20]). We impose such conditions on the input data that it is not necessary to specify the value of $u(t, x)$ for $x = 0$ (this value cannot be arbitrary). But in our existence theorem we prove that the function $u(t, x)$ obtained in this theorem automatically satisfies the condition $u(t, 0) = 0$.

Inverse problems for degenerate parabolic equations of the type (1) are of great interest, for example, in financial mathematics (see Remark 2). In various settings but different from those considered in present paper they were investigated (including numerical calculation) in [6–11], etc.

Method of proving the existence and uniqueness theorems for inverse problem (1)–(3) in our paper is based on the study of the unique solvability of the direct problem (1)–(2) (when the function $\gamma(t)$ is assumed to be known) and on the estimates of its solution. We note that the investigation of the direct problem (1)–(2) is of independent interest and the results obtained here for it are also new.

## 2   Unique Solvability and Estimates for the Direct Problem

Suppose that $\gamma(t) \in L_\infty^+(0, T)$ is a known function and consider the direct problem (1)–(2) with this function $\gamma(t)$ in the Eq. (1). A generalized solution $u(t, x)$ of this problem will be understood in a sense of Definition 1.

Using the technique similar to [11] and the existence theorem for uniformly parabolic equation with two independent variables from [1] we prove the following existence and uniqueness theorems.

**Theorem 1.** Let the conditions $(A), (B), (D)$ hold, $\gamma(t) \in L_\infty^+(0, T)$. Then the generalized solution of the direct problem (1)–(2) is unique.

**Theorem 2.** Let the conditions $(A), (B), (D)$ hold, $\gamma(t) \in L_\infty^+(0, T)$. Then there exists a generalized solution $u(t, x)$ of the direct problem (1)–(2). For this solution we have $u(t, 0) = 0$, $t \in [0, T]$, and it satisfies the estimates

$$\sup_{0 \le t \le T} \|u_x(t, \cdot)\|_{L_2(0, l)}^2 + \|au_{xx}^2\|_{L_1(Q)}$$

$$\le e^{\lambda^* T} \|u_0'\|_{L_2(0, l)}^2 + \frac{3l}{\lambda^*} \left( e^{\lambda^* T} - 1 \right) K_{f, a}, \quad (4)$$

$$\|u_{xx}\|_{L_2(Q^\delta)}^2 \le \frac{1}{a_1 \delta^2} \left[ e^{\lambda^* T} \|u_0'\|_{L_2(0, l)}^2 + \frac{3l}{\lambda^*} \left( e^{\lambda^* T} - 1 \right) K_{f, a} \right] \forall \delta \in (0, l), \quad (5)$$

$$|u(t, x)| \le [M_0 + T\sqrt{K_{f, a}(a_2 + 1)}]e^{K_d T}, \quad (t, x) \in Q, \quad (6)$$

$$\|u_t\|_{L_2(Q)}^2 \le c_1(\|u_0'\|_{L_2(0,l)}^2 + K_{f,a}), \tag{7}$$

$$|u(t_1,x_1) - u(t_2,x_2)| \le c_2(|x_1 - x_2|^{1/2} + |t_1 - t_2|^{1/6}), \ (t_1,x_1),(t_2,x_2) \in Q, \tag{8}$$

where

$$\lambda^* = 3\left(K_{b,a} + \frac{l^2}{\pi^2}K_{d,a}\right) \tag{9}$$

and $c_1, c_2 = \mathrm{const} > 0$ depend only on $l, T, a_1, a_2, K_{b,a}, K_{d,a}, K_{f,a}, \|u_0'\|_{L_2(0,l)}$ and $\|\gamma\|_{L_\infty(0,T)}$.

## 3  Investigation of the Inverse Problem

Now we consider the inverse problem (1)–(3). Denote

$$F(t) = \int_0^l u(t,x)\omega(x)\,dx, \quad B_R^+ = \{\gamma(t) \in L_\infty(0,T)\colon 0 \le \gamma(t) \le R\}.$$

**Theorem 3.** Let the conditions $(A)$–$(E), (G)$ hold. Then there are no two different generalized solutions of the problem (1)–(3).

**Sketch of the Proof.** Assume that there are two different solutions $\{u^{(1)}, \gamma^{(1)}\}$ and $\{u^{(2)}, \gamma^{(2)}\}$ of inverse problem (1)–(3). We put $v(t,x) = u^{(1)}(t,x) - u^{(2)}(t,x)$, $\nu(t) = \gamma^{(1)}(t) - \gamma^{(2)}(t)$.

Taking into account condition (3) and assumptions $(A)$–$(C), (E)$, after a series of calculations analogous to those carried out in the proof of the corresponding uniqueness theorem in [12] we obtain that the function $v(t,x)$ satisfies in $Q$ the integro-differential equation

$$v_t - a(t,x)v_{xx} + b(t,x)v_x + d(t,x)v + \gamma^{(1)}(t)v$$

$$= -\frac{1}{\varphi(t)}\int_0^l [(a\omega)_{yy} - (b\omega)_y - d\omega]v\,dy \cdot u^{(2)}(t,x) \tag{10}$$

and the function $\nu(t)$ satisfies the relation

$$\nu(t) = \frac{1}{\varphi(t)}\int_0^l [(a\omega)_{xx} - (b\omega)_x - d\omega]v(t,x)\,dx. \tag{11}$$

Using the well-known technique of energy estimates applied to the relation (10) we proof that $v(t,x) \equiv 0$ in $Q$. Then from (11) we obtain that $\nu(t) \equiv 0$ in $[0,T]$. Theorem 3 is proved.

In order to prove the existence of generalized solution of the inverse problem (1)–(3) we in addition to conditions $(A)$–$(G)$ assume that

$$F_0 + \varphi_0^* \geq l[K_{a,w} + K_{b,w} + K_d K_w][M_0 + T\sqrt{K_{f,a}(a_2 + 1)}]e^{K_d T}. \quad (12)$$

We derive the operator equation for the unknown function $\gamma(t) \in L_\infty^+(0, T)$.

Let pair of functions $\{u(t, x), \gamma(t)\}$ be a generalized solution of the inverse problem (1)–(3). We multiply Eq. (1) by $w(x)$ and integrate the result over the segment $[0, l]$. Taking into account conditions (2), (3) and also $(A)$–$(E)$ we after integration by parts obtain the relation

$$\gamma(t) = \frac{1}{\varphi(t)} \left\{ F(t) - \varphi'(t) + \int_0^l [(aw)_{xx} + (bw)_x - dw]u\, dx \right\}. \quad (13)$$

In view of this relation let us introduce the operator $\mathcal{A} : L_\infty^+(0, T) \to L_\infty(0, T)$ by the formula

$$\mathcal{A}(\gamma) = \frac{1}{\varphi(t)} \left\{ F(t) - \varphi'(t) + \int_0^l [(aw)_{xx} + (bw)_x - dw]u\, dx \right\}, \quad (14)$$

where $\gamma(t)$ is an arbitrary function in $L_\infty^+(0, T)$ and $u \equiv u(t, x; \gamma)$ is a solution of direct problem (1)–(2) with given coefficient $\gamma(t)$ in the Eq. (1). Then the relation (13) can be written as

$$\gamma = \mathcal{A}(\gamma). \quad (15)$$

**Remark 4.** By virtue of Theorems 1 and 2 and assumptions $(A)$–$(C), (E)$ the operator $\mathcal{A}$ is well defined.

**Lemma 1.** Let conditions $(A)$–$(G)$ hold. Then the operator Eq. (15) is equivalent to the inverse problem (1)–(3) in the following sense. If pair $\{u(t, x), \gamma(t)\}$ is a generalized solution of the inverse problem, then $\gamma(t)$ satisfies (15). Conversely, if $\gamma(t) \in L_\infty^+(0, T))$ is a solution of operator Eq. (15), and $u = u(t, x; \gamma)$ is a solution of direct problem (1)–(2) with this $\gamma$ in the Eq. (1), then the pair $\{u(t, x; \gamma); \gamma(t)\}$ is a generalized solution of inverse problem (1)–(3).

The proof of this Lemma is standard (see, for example [11]).

Further we establish a number of properties of the operator $\mathcal{A}$ given by (14). We set

$$R_0 = F_1 + \varphi_1^* + l[K_{a,w} + K_{b,w} + K_d K_w][M_0 + T\sqrt{K_{f,a}(a_2 + 1)}]e^{K_d T}. \quad (16)$$

**Lemma 2.** Let conditions $(A)$–$(G)$, (12) hold. Then the operator $\mathcal{A}$ maps the set $B_{R_0}^+$ into itself.

**Proof.** By virtue of estimate (6), assumption $(B)$ and Remark 1 we obtain

$$\left| \int_0^l [(a\omega)_{xx} + (b\omega)_x - d\omega] u\, dx \right|$$

$$\leq l[K_{a,\omega} + K_{b,\omega} + K_d K_\omega][M_0 + T\sqrt{K_{f,a}(a_2 + 1)}]e^{K_d T}.$$

Then due to the assumption (12), the definition of $R_0$ in (16) and the definition of $\mathcal{A}$ in (14) we have for any $\gamma(t) \in B_{R_0}^+$ that $0 \leq \mathcal{A}(\gamma) \leq R_0$ and so $\mathcal{A}(\gamma) \in B_{R_0}^+$. Lemma is proved.

**Lemma 3.** Let conditions $(A)$–$(G)$ hold. Then the operator $\mathcal{A}$ is continuous operator on $B_{R_0}^+$.

**Proof.** Let $\gamma^{(i)}(t) \in B_{R_0}^+$ and $u^{(i)}(t,x)$ are the corresponding solutions of the direct problem (1)–(2) with $\gamma^{(i)}(t)$ in the Eq. (1), $i = 1, 2$. Define $v(t,x) = u^{(1)}(t,x) - u^{(2)}(t,x)$. It is easy to see that $v(t,x)$ satisfies in $Q$ the equation

$$v_t - a(t,x)v_{xx} + b(t,x)v_x + d(t,x)v + \gamma^{(1)}(t)v = -[\gamma^{(1)}(t) - \gamma^{(2)}(t)]u^{(2)}(t,x)$$

and homogeneous initial and boundary conditions. Then applying the estimate (4) for the function $v(t,x)$ we have

$$\sup_{0 \leq t \leq T} \|v(t,\cdot)\|_{L_2(0,l)} \leq c_1 \|\gamma^{(1)} - \gamma^{(2)}\|_{L_\infty(0,T)}, \tag{17}$$

where $c_1 = \text{const} > 0$.

On the other hand by virtue of definition of $\mathcal{A}$ we obtain the estimate

$$\|\mathcal{A}\gamma^{(1)} - \mathcal{A}\gamma^{(2)}\|_{L_\infty(0,T)}$$

$$\leq \frac{1}{\varphi_0} l^{1/2}(K_{a,\omega} + K_{b,\omega} + K_d K_\omega) \sup_{0 \leq t \leq T} \|v(t,\cdot)\|_{L_2(0,l)}. \tag{18}$$

From estimates (17) and (18) and the assumptions $(A)$–$(C)$ we obviously obtain that the operator $\mathcal{A}$ is continuous operator on $B_{R_0}^+$.

**Lemma 4.** Let conditions $(A)$–$(G)$ hold. Then the operator $\mathcal{A}$ is compact operator on $B_{R_0}^+$.

**Proof.** The assertion of this lemma is a consequence of the estimates (6) and (8), the compactness of the imbedding of the space $C^{0,1/3}(Q)$ into the space $C(Q)$ and the definition of the operator $\mathcal{A}$.

Using Lemmas 1–4 we obtain the following existence theorem.

**Theorem 4.** Let conditions $(A)$–$(G)$, (12) hold. Then there exists a generalized solution $\{u(t,x), \gamma(t)\}$ of the inverse problem (1)–(3) wherein $u(t,0) = 0$, $t \in [0,T]$. Moreover we have the estimate

$$0 \leq \gamma(t) \leq R_0, \tag{19}$$

where $R_0$ is from (16), and the estimates (4)–(8) where the constants $c_1, c_2$ depend only on $l, T, a_1, a_2, K_{b,a}, K_{d,a}, K_{f,a}, \|u_0'\|_{L_2(0,l)}$ and $R_0$.

**Proof.** By Lemmas 2–4 operator $\mathcal{A}$ is a completely continuous operator mapping the bounded convex closed set $B^+_{R_0}$ into itself. By the Schauder fixed point theorem (e.g. see [13, p. 193]) in this case Eq. (15) has a solution $\gamma(t)$ from $B^+_{R_0}$, so the estimate (19) hold. Now the assertion of the theorem follows from Lemma 1.

Let us show that the conditions of our uniqueness and existence Theorems 3 and 4 are valid for the inverse problem for Black-Scholes equation.

**Example 1.** In the rectangle $Q$ we consider the following inverse problem:

$$u_t - \frac{1}{2}\sigma x^2 u_{xx} + \mu x u_x + \gamma(t)u = 0, \tag{20}$$

$$u(0,x) = u_0(x) \equiv x(l-x), \quad u(t,l) = 0, \tag{21}$$

$$\int_0^l u(t,x)(l-x)dx = \varphi(t) \equiv \frac{l^4}{12} - \kappa t l^m. \tag{22}$$

Here $\sigma, m, \kappa > 0$, $\mu \geq 0$ – are some given constants.

It is easy to check that the conditions $(A)$–$(D), (F), (G)$ are fulfilled for any $l, T, \sigma, m, \kappa, \mu$. The constant $\varphi_0$ from the condition $(E)$ is equal to $l^4/24$. Therefore the inequality $\varphi(t) \geq \varphi_0$ is true if

$$\kappa T l^m < \frac{l^4}{24}. \tag{23}$$

In this case the uniqueness Theorem 3 is valid for the problem (20)–(22).

Inequality (16) is valid if

$$\kappa l^{m-4} \geq \frac{2\sigma + \mu}{2}. \tag{24}$$

Thus if the conditions (23) and (24) are fulfilled simultaneously then the solution of the inverse problem (19)–(21) exists and is unique.

Let us consider some special cases.

Case 1: $m = 4, \kappa \geq \frac{2\sigma+\mu}{2}$. In this case condition (24) is valid for arbitrary $T$ and $l$ and condition (23) is fulfilled if $T < \frac{1}{24\kappa}$. Thus in this case the solution of the problem (20)–(22) exists and is unique for sufficiently small $T$ and arbitrary $l$.

Case 2: $m > 4$. In this case conditions (23) and (24) are valid if $l$ is sufficiently large and $T$ is small (and depends on the choice of $l$).

Case 3: $m > 3, \kappa = l$. In this case conditions (23) and (24) are also valid if $l$ is sufficiently large and $T$ is small (and depends on the choice of $l$).

In both these cases the solution of the problem (20)–(22) also exists and is unique.

**Acknowledgements.** This work was partially supported by the Program of competitiveness increase of the National Research Nuclear University MEPhI (Moscow Engineering Physics Institute); contract No. 02.a03.21.0005, 27.08.2013.

# References

1. Kruzhkov, S.N.: Quasilinear parabolic equations and systems with two independent variables. Trudy Sem. im. I.G. Petrovskogo **5**, 217–272 (1979)
2. Black, F., Scholes, M.: The pricing of options and corporate liabilities. J. Polit. Econ. **81**, 637–659 (1973)
3. Hull, J.: Options, Futures and Other Derivatives. Prentice Hall, Upper Saddle River (2005)
4. Fichera, G.: Sulle equazioni differenziali lineari ellitico-paraboliche del secondo ordine. Atti Accad. Nazionale dei Lincei. Mem. Cl. Sci. Fis. Mat. Natur. Ser. I(8) **5**, 1–30 (1956)
5. Oleĭnik, O.A., Radkevič, E.A.: Second Order Differential Equations with Nonnegative Characteristic Form. AMS, Rhode Island and Plenum Press, New York (1973)
6. Deng, Z.C., Yang, L.: An inverse problem of identifying the coefficient of first-order in a degenerate parabolic equation. J. Comput. Appl. Math. **235**, 4404–4417 (2011)
7. Deng, Z.C., Yang, L.: An inverse problem of identifying the radiative coefficient in a degenerate parabolic equation. Chin. Ann. Math. Ser. B. **35B**(3), 355–382 (2014)
8. Bouchouev, I., Isakov, V.: Uniqueness, stability and numerical methods for the inverse problem that arises in financial markets. Inverse Prob. **15**(3), 95–116 (1999)
9. Lishang, J., Yourshan, T.: Identifying the volatility of underlying assets from option prices. Inverse Prob. **17**(1), 137–155 (2001)
10. Lishang, J., Qihong, C., Lijun, W., Zhang, J.E.: A new well-posed algorithm to recover implied local volatility. Quant. Financ. **3**(6), 451–457 (2003)
11. Prilepko, A.I., Kamynin, V.L., Kostin, A.B.: Inverse source problem for parabolic equation with the condition of integral observation in time. J. Inverse III-posed Prob. **26**(4), 523–539 (2018)
12. Bukharova, T.I., Kamynin, V.L.: Inverse problem of determining the absorption coefficient in the multidimensional heat equation with unlimited minor coefficients. Comput. Math. Math. Phys. **55**(7), 1183–1195 (2015)
13. Lyusternik, L.A., Sobolev, V.I.: Kratkii Kurs Functcional'nogo Analiza (Brief Course of Functional Analysis). Vysshaya Shkola, Moscow (1982)

# Finite Difference Schemes on Locally Refined Cartesian Grids for the Solution of Gas Dynamic Problems on the Basis of Quasigasdynamics Equations

Yury N. Karamzin[1], Tatiana A. Kudryashova[1], Sergey V. Polyakov[1,2], and Viktoriia O. Podryga[1,3(✉)]

[1] Keldysh Institute of Applied Mathematics, 4 Miusskaya sq., 125047 Moscow, Russia
polyakov@imamod.ru, pvictoria@list.ru
[2] National Research Nuclear University MEPhI,
(Moscow Engineering Physics Institute), 31 Kashirskoe sh., 115409 Moscow, Russia
[3] Moscow Automobile and Road Construction State Technical University,
64 Leningradsky prospect, 125319 Moscow, Russia

**Abstract.** The paper is devoted to the numerical solution of gas dynamic problems on the basis of a system of quasigasdynamic equations in domains of complex shape. One possible grid approach to solving this class of problems is used. An approach is applying to the locally refined Cartesian (LRC) grids, consisting of rectangles (parallelepipeds) of various sizes. In this paper some variants of the construction of finite difference schemes in the two-dimensional case are considered. Their order of approximation is investigated. The analysis of the schemes is carried out numerically on the example of two-dimensional problem of gas flow under conditions of the real equation of state.

**Keywords:** Initial boundary value problems for quasigasdynamic equation system · Finite difference schemes ·
Locally refined Cartesian grids

## 1 Introduction

In the modeling of technological processes, the need for numerical calculation of liquid and gas flows in regions of complex shape often occurs. One of the problems of such calculations is the strong anisotropy of the elements of the computational domain boundary, which causes various physical and computational effects, for example, the exponential growth of the solution near the boundary, the gradient catastrophe of the solution, numerical instability, etc. (see also Fig. 1). In this connection, it becomes necessary to resolve boundary layers with a predetermined accuracy near the external or internal boundaries of the domain. The solution of this problem is done by developing special algorithms for refining the calculated grids and methods for calculating the fluxes on such grids. As

© Springer Nature Switzerland AG 2019
I. Dimov et al. (Eds.): FDM 2018, LNCS 11386, pp. 321–328, 2019.
https://doi.org/10.1007/978-3-030-11539-5_36

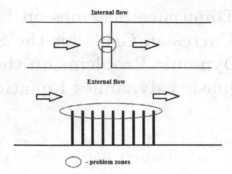

**Fig. 1.** Illustration to the problem of boundary anisotropy.

an alternative to grid methods, various boundary layer differential models also appear (for example, models based on the apparatus of wall functions).

This work is aimed at the analysis of the grid approach, oriented to the use of locally refined Cartesian (LRC) grids. The motivation for this choice is the development of a multilevel multiscale approach [1–3], that combines models of continuous medium and molecular dynamics (MD). The use of LRC grids has long been known in the literature (see, for example, [4,5]). However, the questions of their use in complex heterogeneous models are not sufficiently studied. Therefore, within the framework of this paper, some variants of the construction of LRC grids and numerical schemes based on them were considered once again.

As an example, a system of quasigasdynamic (QGD) equations is chosen [6–8], which is used in the framework of the aforementioned multiscale approach for modeling flows in microchannels of technical systems. A particular feature of this system is the parabolic type of its equations. The latter allows considering numerical schemes on LRC grids on the example of a single convection-diffusion equation. With the help of this approach, in the spatially two-dimensional case, different variants of the approximation of this equation on LRC grids of various configurations were investigated. In particular, both cases are considered when the grid has interface layers between cells of different sizes, and when they are absent. Grids of nodal and cellular type were also considered. For all four variants, the order of approximation of numerical schemes was investigated. Example of calculation of gas flow using the developed technique is also given.

## 2  Problem Formulation

The QGD equations [6–8] are used to describe the model problem. These equations are written for the case of a real viscous heat-conducting gas in the absence of volume external forces and sources of mass and heat. In the form invariant relative to the coordinate system in dimensional variables (the SI unit system is used below), in the three-dimensional formulation, together with the equations of constraints and state, we have the following system:

$$\begin{cases} \dfrac{\partial \rho}{\partial t} + \operatorname{div} \mathbf{j} = 0, \quad \mathbf{j} = \rho \left( \mathbf{u} - \mathbf{w} \right), \\[2mm] \mathbf{w} = \tau \left[ \dfrac{1}{\rho} \operatorname{div} \left( \rho \mathbf{u} \right) \right] \mathbf{u} + \mathbf{w}^*, \quad \mathbf{w}^* = \tau \left[ (\mathbf{u} \operatorname{grad}) \mathbf{u} + \dfrac{1}{\rho} \operatorname{grad} p \right], \end{cases} \quad (1)$$

$$\begin{cases} \dfrac{\partial}{\partial t} \left( \rho u_k \right) + \operatorname{div} \left( \mathbf{j} u_k + \mathbf{e}_k p - \mathbf{\Pi}_k \right) = 0, \quad \mathbf{\Pi}_k = \left( \Pi_{j,k} \right), \\[2mm] \Pi_{k,j} = \Pi_{k,j}^{NS} + \Pi_{k,j}^{AD}, \quad \Pi_{k,j}^{NS} = \mu \left( \dfrac{\partial u_k}{\partial x_j} + \dfrac{\partial u_j}{\partial x_k} + \delta_{k,j} \left( \zeta - \dfrac{2}{3} \right) \operatorname{div} \mathbf{u} \right), \\[2mm] \Pi_{k,j}^{AD} = \rho u_k w_j^* + \delta_{k,j} \tau \left[ (\mathbf{u} \operatorname{grad}) p \right], \end{cases} \quad (2)$$

$$\begin{cases} \dfrac{\partial E}{\partial t} + \operatorname{div} \left( \mathbf{j} H + \mathbf{q} - \mathbf{\Pi} \mathbf{u} \right) = 0, \\[2mm] \mathbf{q} = -\kappa \operatorname{grad} T - \tau \left[ (\mathbf{u} \operatorname{grad}) \varepsilon + p \left( \mathbf{u} \operatorname{grad} \right) \left( \dfrac{1}{\rho} \right) \right] \rho \mathbf{u}, \end{cases} \quad (3)$$

$$\begin{cases} E = \rho \left( \dfrac{1}{2} |u_l|^2 + \varepsilon \right), \quad H = \dfrac{E + p}{\rho}, \quad p = Z \rho \mathfrak{R} T, \quad \varepsilon = c_V T, \\[2mm] \gamma = \dfrac{c_p}{c_V}, \quad \tau = \dfrac{\mu}{p \operatorname{Sc}}, \quad \operatorname{Pr} = \dfrac{\mu c_p}{\kappa}, \quad \operatorname{Sc} = \dfrac{\mu}{\rho D}, \quad \operatorname{Ma} = \dfrac{|\mathbf{u}|}{a}, \\[2mm] \operatorname{Re} = \dfrac{\rho |\mathbf{u}| \lambda}{\mu}, \quad \operatorname{Kn} = \dfrac{\lambda}{L}, \quad a^2 = \left( \dfrac{\partial p}{\partial \rho} \right) \approx \gamma \dfrac{p}{\rho}, \quad \lambda = \dfrac{A \mu}{\rho \sqrt{Z \mathfrak{R} T}}. \end{cases} \quad (4)$$

Here it is assumed that the gas is characterized by four main parameters: mass density $\rho = m \cdot n$ ($m$ is the mass of gas molecules, $n$ is the numerical density or concentration), macroscopic velocity $\mathbf{u}$, pressure $p$ and temperature $T$. Other gas parameters: $E$, $H$ and $\varepsilon$ are the total energy density, the enthalpy and the internal energy; $\mu = \mu(T)$, $\zeta = \zeta(T)$, $\kappa = \kappa(T)$ and $\tau = \tau(T, \rho)$ are the kinetic coefficients, namely, the coefficients of dynamic viscosity, relative bulk viscosity, thermal conductivity, and the gas relaxation time to a quasiequilibrium state; $Z = Z(T, \rho)$, $\gamma = \gamma(T, \rho)$, $c_v = c_V(T)$, $c_p = c_p(T)$ are the compressibility coefficient, the adiabatic exponent, the specific heat at constant volume and the pressure; $\mathfrak{R} = k_B/m$ is the individual gas constant ($k_B$ is the Boltzmann constant); Sc and Pr are the Schmidt and Prandtl numbers. In Eqs. (1)–(3) $\mathbf{j}$, $\mathbf{q}$ and $\mathbf{\Pi}$ are the modified fluxes of density, heat, and the viscous stress tensor, $\rho \mathbf{w}$ and $\rho \mathbf{w}^*$ are the vectors of the QGD corrections to the density flux $\rho \mathbf{u}$; $\mathbf{e}_k$ ($k = 1, 2, 3$) are the unit vectors, div and grad are the divergence and gradient operators. Also here additional parameters are introduced, where $a$ is the local velocity of sound in gas, $\lambda$ is the average mean free path of gas molecules; Ma, Re and Kn are the Mach, Reynolds and Knudsen numbers, $A$ is the constant characterizing the collision model of molecules, $L$ is the characteristic size of the problem.

The system of Eqs. (1)–(3) is closed by the initial and boundary conditions. The initial conditions correspond to the equilibrium state of the gaseous medium in the absence of interaction with external factors. In the calculations discussed below, the case of a stationary gas medium was considered:

$$\rho = \rho_0, \quad \mathbf{u} = 0, \quad p = p_0, \quad T = T_0. \quad (5)$$

Here $\rho_0$ and $p_0$ are the initial density and pressure of gas medium, $T_0$ is the initial temperature.

At the entrance to the medium, the gas parameters are set, which depend on the specific formulation of the problem and are generally defined as follows:

$$(\mathbf{u}, \mathbf{n}) = u_{in}, \quad \rho = \rho_{in}, \quad T = T_{in}. \tag{6}$$

where the parameters with the index $in$ correspond to the specified values of the parameters at the entrance to the medium.

On the free surfaces of the computational domain, the so-called "soft" boundary conditions are given (see, for example, [7]):

$$\frac{\partial \rho}{\partial n} = 0, \quad \frac{\partial (\rho \mathbf{u})}{\partial n} = 0, \quad \frac{\partial p}{\partial n} = 0. \tag{7}$$

## 3  Numerical Methods and Analysis

Analysis of Eqs. (1)–(3) shows that they all have a parabolic type. Therefore, in order to consider the space-time approximation on LRC grids, each of these equations can be represented in the following general dimensionless form:

$$\frac{\partial u}{\partial t} = Lu + f, \quad t > 0, \quad \mathbf{x} = (x_1, \ldots, x_m) \in \Omega \subset R^m; \tag{8}$$

$$\begin{cases} Lu = \sum_{\alpha=1}^{m} L_\alpha u - qu, \quad L_\alpha u = \dfrac{\partial w_\alpha}{\partial x_\alpha} + r_\alpha \dfrac{\partial u}{\partial x_\alpha}, \\[4mm] w_\alpha = \sum_{\beta=1}^{m} k_{\alpha\beta} \dfrac{\partial u}{\partial x_\beta} + p_\alpha u, \quad \alpha = 1, \ldots, m. \end{cases} \tag{9}$$

Here, by function $u$ one of the functions $\rho, u_1, \ldots, u_m, E$ is meant, $m = 1, 2, 3$ is the dimensionality of Euclidean space $R^m$, $\Omega$ is the computational domain, $k_{\alpha\beta}$, $p_\alpha$, $r_\alpha$, and $q$ are the coefficients of the corresponding equations.

For Eq. (8) with operator (9), the following initial-boundary value problem is posed:

$$u = u_0(\mathbf{x}), \quad t = 0, \quad \mathbf{x} \in \Omega; \tag{10}$$

$$u = u_G(\mathbf{x}, t), \quad t > 0, \quad \mathbf{x} \in \partial \Omega_1; \tag{11}$$

$$(\mathbf{w}, \mathbf{n}) = w_G(\mathbf{x}, t), \quad t > 0, \quad \mathbf{x} \in \partial \Omega_2. \tag{12}$$

Here $\partial \Omega_1$ and $\partial \Omega_2$ are the parts of the boundary $\partial \Omega$ of the region where different boundary conditions are given; $\mathbf{w} = \{w_\alpha\}$ is the already used flux vector; $\mathbf{n}$ is the vector of the outward normal to the boundary of the domain, $u_G$ and $w_G$ are given boundary functions.

Problem (8)–(12) is solved by the finite volume method [9]. As a spatial grid, the LRC grid $\bar{\Omega}_h$ is used, in the general case an approximating region $\Omega$ together with its boundary $\partial \Omega$. By time, an ordinary uniform grid $\bar{\omega}_t$ with a step $\Delta t$ is taken. Let us first consider the possible arrangement of a spatial grid. First, we will use only such LRC grids, in which the refinement of neighboring cells is made exactly 2 times. Secondly, let's single out two classes of LRC grids, which are differed in the way of transition from large cells to smaller ones (that is, differing in the type of interface). The first class is characterized by the absence of any interface between cells and allows for the presence of so-called hanging nodes (denoted as a classical interface). The second class on the contrary prohibits the presence of hanging nodes. For its realization in the two-dimensional (three-dimensional) case, partitioning of rectangular (parallelepiped) cells into triangles (tetrahedrons) is used. An illustration of the classes of LRC grids is shown in Fig. 2 for the two-dimensional case.

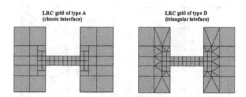

**Fig. 2.** Classes of LRC grids.

In addition to the use of different types of grids in calculating a single function, two classes of numerical schemes are usually used in the control volume method; these classes are nodal and cellular schemes. They differ in that the required function is determined respectively at the grid nodes, or at the centers of its cells (if there are several unknown functions, then numerical schemes on spaced grids can also be used). This division will also be taken into account below.

For convenience of the following exposition, the two-dimensional version of problem (8)–(12) will be considered. Its approximation on the space-time grid $\bar{\Omega}_h \times \bar{\omega}_t$ can be written in the form of the following well-known scheme with weights [10]:

$$\frac{\hat{u}_h - u_h}{\Delta t} = \sigma \left( \hat{\Lambda}_h \hat{u}_h + \hat{f}_h \right) + (1 - \sigma) \left( \Lambda_h u_h + f_h \right), \quad t > 0, \quad \mathbf{x} \in \bar{\Omega}_h \backslash \partial \Omega_1, \tag{13}$$

$$u = u_0 (\mathbf{x}), \quad t = 0, \quad \mathbf{x} \in \bar{\Omega}_h, \tag{14}$$

$$u = u_G (\mathbf{x}, t), \quad t \in \bar{\omega}_t, \quad \mathbf{x} \in \bar{\Omega}_h \cap \partial \Omega_1. \tag{15}$$

Here $\sigma \in [0, 1]$ is the weight of the scheme, the values $\Lambda_h$, $u_h$, $f_h$ and $\hat{\Lambda}_h$, $\hat{u}_h$, $\hat{f}_h$ refer to layers on $t$ and $t + \Delta t$. Obviously, the operator view of the scheme does not depend on the type of grid and the method of determining the function (at nodes or in cell centers). Both of these questions are addressed to ways of approximating the operator $\Lambda_h$. The definition of $\Lambda_h$ also includes the question of approximating the boundary conditions on a part of the boundary $\partial \Omega_2$.

To clarify the approximation of the operator $\Lambda_h$ on the grid $\bar{\Omega}_h$ let us recall the algorithm of the control volume method. In the general case, it consists in integrating the Eq. (8) over the control volume $V_i$, where the index $i$ denotes the number of the next unknown value of the function $u_i \equiv u(\mathbf{x}_i, t)$. In the case of nodal schemes, a grid node with an index $i$ is used as the center of the control volume; in the case of cellular schemes, the center of the cell with the index $i$ is used. In both cases, the baricentric control volumes used. Then, in the construction of nodal schemes, the volumes $V_i$ consist of rectangles and triangles, and when constructing cellular schemes, the volumes $V_i$ coincide with the grid cells $C_i$ (see Fig. 3).

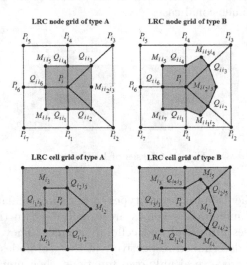

**Fig. 3.** Control volumes $V_i$ (shaded) in case of nodal (top) and cellular (bottom) schemes for two types of LRC grids (left and right). The points $M$ coincide with the centers of the cells, the points $Q$ are on the intersections of the lines of the gradient and the boundaries of the control volume.

The construction of the approximations themselves is carried out analogously to the techniques detailed in [11] for the case of exponential schemes with a double integral transformation. As a result, we can write the following general form of the operator $\Lambda_h$ in the internal nodes of the grid:

$$\Lambda_h u_h = \sum_{k=1}^{N_i} \left( \sum_{\alpha=1}^{m} \frac{1}{g_{\alpha,i}^k} \frac{\left(g_\alpha^k w_\alpha^k\right)_{i_\alpha+1/2} - \left(g_\alpha^k w_\alpha^k\right)_{i_\alpha-1/2}}{l_{\alpha,i_\alpha}^k} \right) - \tilde{q}_i u_i. \qquad (16)$$

Here it is meant that $|V_i|$ is the size of the control volume; $N_i$ is the number of its constituent rectangular or triangular parts; $l_{\alpha,i_\alpha}^k$ are the distances along the coordinates $x_\alpha$ between the point $P_i$ and its neighbors $P_{i_k}$ in nodal schemes and between the centers of neighboring cells $M_{i_k}$ in cellular schemes.

An important correction of (16) in interface zones is the implementation of balance expressions at the boundaries of the control volumes. The general formula for such a correction is quite difficult to write out. However, in the form of an algorithm, it reduces to the fact that the final flux of desired value ($g$w), coming to a larger cell, consisted of the sum of the fluxes leaving the cells of a smaller size.

On the boundaries of the computational domain, the approximation of the operator $\Lambda_h$ does not differ from the case of the ordinary Cartesian grid (see [11]). Therefore here we omit it.

Concerning the accuracy of proposed numerical scheme, we note that by analogy with [11] the second order of local approximation in space and the first or second order of approximation in time (depending on the weight of the scheme) can be achieved. A detailed analysis of this issue requires additional research.

## 4    Computer Experiment and Conclusions

An example of the proposed method's application is the calculation of the flow near a blunt body, carried out by the authors earlier [12] and shown in Fig. 4. The calculation is carried out within the framework of a nodal scheme with a triangular interface. In the framework of this paper, the calculations were repeated using a cellular scheme with a triangular interface. In order to achieve the same accuracy, the grid had to be refined 2 times more.

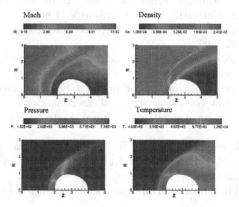

**Fig. 4.** Stationary distributions of the flow parameters of a viscous heat-conducting gas near a blunt body.

In order to solve problems on the basis of a multiscale approach, combining QGD and MD models, the most effective from the point of view of the homogeneity of the algorithm is the use of cellular schemes without an interface. However, a stronger refinement of the grid will have to be used. Also a similar problem can be solved in a combined way. In its framework, to calculate the QGD equations, it is necessary to use a node scheme on an LRC grid with a triangular interface (this will help improve the accuracy of calculations). To calculate the motion of particles based on the MD model, it is necessary to use the LRC grid cells without an interface. In this case, it is necessary to use logarithmic interpolation to interpolate gas-dynamic quantities into cell centers.

**Acknowledgment.** The work was supported by the Russian Foundation for Basic Research (projects No. 18-07-01292-a, 18-51-18004-bolg-a, 16-29-15095-ofi_m).

# References

1. Podryga, V.O., Karamzin, Y.N., Kudryashova, T.A., Polyakov, S.V.: Multiscale simulation of three-dimensional unsteady gas flows in microchannels of technical systems. In: Proceedings of the VII European Congress on Computational Methods in Applied Sciences and Engineering (ECCOMAS Congress 2016), Crete Island, Greece, 5–10 June 2016, vol. 2, pp. 2331–2345 (2016)
2. Kudryashova, T., Karamzin, Y., Podryga, V., Polyakov, S.: Two-scale computation of N2–H2 jet flow based on QGD and MMD on heterogeneous multi-core hardware. Adv. Eng. Softw. **120**, 79–87 (2018)
3. Kudryashova, T., Podryga, V., Polyakov, S.: HPC-simulation of gasdynamic flows on macroscopic and molecular levels. In: Uvarova, L.A., Nadykto, A.B., Latyshev, A.V. (eds.) Nonlinearity. Problems, Solutions and Applications, vol. I, chap. 26, pp. 543–556. Nova Science Publishers, New York (2017)
4. Sethian, J.A.: Level Set Methods and Fast Marching Methods: Evolving Interfaces in Computational Geometry, Fluid Mechanics, Computer Vision, and Materials Science. Cambridge University Press, Cambridge (1999)
5. Osher, S.J., Fedkiw, R.P.: Level Set Methods and Dynamic Implicit Surfaces. Springer, New York (2002)
6. Chetverushkin, B.N.: Kinetic Schemes and Quasi-Gasdynamic System of Equations. CIMNE, Barcelona (2008)
7. Elizarova, T.G.: Quasi-Gas Dynamic Equations. Springer, Berlin (2009)
8. Zlotnik, A.A.: Entropy-conservative spatial discretization of the multidimensional quasi-gasdynamic system of equations. Comput. Math. Math. Phys. **57**(4), 706–725 (2017)
9. Eymard, R., Gallouet, T.R., Herbin, R.: The finite volume method. In: Ciarlet, P.G., Lions, J.L. (eds.) Handbook of Numerical Analysis, vol. 7, pp. 713–1020. North Holland, Amsterdam (2000)
10. Samarskii, A.A.: The Theory of Difference Schemes. Marcel Dekker Inc., New York (2001)
11. Polyakov, S.V., Karamzin, Y.N., Kudryashova, T.A., Tsybulin, I.V.: Exponential difference schemes for solving boundary-value problems for Diffusion-Convection-type equations. Math. Models Comput. Simul. **9**(1), 71–82 (2017)
12. Kudryashova, T.A., Polyakov, S.V., Sverdlin, A.A.: Calculation of gas flow parameters around a reentry vehicle. Math. Models Comput. Simul. **1**(4), 445–452 (2009)

# Time to Start a Crowded Period in a Finite-Buffer Queue with Poisson Input Flow and General Processing Times

Wojciech M. Kempa[(✉)]

Institute of Mathematics, Silesian University of Technology, 23 Kaszubska Street,
44-100 Gliwice, Poland
wojciech.kempa@polsl.pl

**Abstract.** A finite-capacity queueing model with Poisson input flow and generally distributed processing times of jobs is considered. An idea of a crowded period is introduced, namely the time period in the system operation during which the number of jobs present in the system is continually greater than or equal the fixed level $M > 0$. A system of integral equations for the tail cumulative distribution function of the time to start a crowded period is derived, conditioned by the number of jobs present in the accumulating buffer before the start moment. A solution of the equivalent system written for Laplace transforms is found using the linear algebraic approach.

**Keywords:** Crowded period · Finite buffer · Poisson process · Single-server queue · Transient state

## 1 Introduction

Queueing models with limited capacities of buffers accumulating jobs waiting for processing have a lot of applications in solving different-type problems of technical (packet switches in telecommunications networks, issues of production engineering, road traffic) and economic nature (optimal number of cash-desks in hypermarkets, organization of freight transport). One of the common problems that appear in these types of systems is the phenomenon of a buffer overflow, which results either in the loss of some of the incoming jobs and the need of resending them for processing (as, for example, in packet networks), or the need of redirecting the excess traffic to another service channel/channels (as in the case of, for example, customer service in a hypermarket, in which, at high load, an additional cash machine is activated, but also in the case of an inefficient production line). A "warning signal" announcing the risk of a possible buffer overflow is the growing number of jobs accumulated in the buffer. The period in which the number is permanently equal to or greater than a certain fixed level $M > 0$ will be called the crowded period in the system evolution (the period with at most $M - 1$ jobs is called a quiet period). Knowledge of the probability

© Springer Nature Switzerland AG 2019
I. Dimov et al. (Eds.): FDM 2018, LNCS 11386, pp. 329–336, 2019.
https://doi.org/10.1007/978-3-030-11539-5_37

distribution of time to the occurrence of a crowded period is of key importance from the point of view of optimizing the system operation: it allows to prepare an additional service channel (or channels) which can be used during the crowded period.

In the paper we deal with the $M/G/1/N$-type queueing model in which the incoming jobs arrive according to a Poisson process with rate $\lambda > 0$ and are processed in generally distributed service times with a cumulative distribution function $F(\cdot)$. The maximal number of jobs simultaneously present in the system equals $N$, i.e. we have a buffer with $N - 1$ places and one place in service facility. Identifying Markov moments in the evolution of the considered system and applying the continuous version of the total probability law, we construct a system of equations for conditional tail distributions of the time to start a crowded period. The solution of the corresponding system written for Laplace transforms is found using the algebraic approach.

Steady-state analysis of finite-capacity systems can be found e.g. in monograph of Takagi [13]. One can find analytical results on buffer overflow period and the loss process in [5] and [12]. In particular, in [5] the joint transform of the busy period and numbers of jobs being processes and lost during the busy period is obtained for the system with phase-type distributions of interarrival and service times. In [12] the representation for the distribution of the number of buffer overflows during a single busy period is found in the case of exponential interarrival times. Cumulative distribution functions of buffer overflow durations in the finite- and infinite-buffer $M/G/1$-type models are investigated e.g. in [1,2] and [4]. In [10] the closed-form representation for the distribution of the time to buffer overflow in the $M/G/1/N$-type system with warm up and closedown times is obtained. In [9] the distribution of the time to buffer overflow in the $M/G/1/N$ queue with multiple vacation policy is considered. In [7] and [8] analytical results for the probability distribution of the buffer overflow period duration are obtained for models with vacation-type disciplines.

Results for the $GI/M/1/N$ queueing system can be found in [6], where the explicit formula for the Laplace transform of the distribution of the time to the first buffer overflow, conditioned by the initial buffer state, is derived and next used to get the corresponding formula for the time to reach the $k$th buffer overflow, where $k \geq 2$.

## 2    Transient Equations for Time to Start a Crowded Period

In this section we derive time-dependent equations for conditional tail distributions of the time to start a crowded period in the considered queueing model, identifying Markov moments in the system operation and applying the law of total probability. Next, we transform the obtained system of equations to the equivalent one written for Laplace transforms.

Let us denote by $C_M$ the time to start the crowded period in the system operation, determined by the queue level $0 < M \leq N$, i.e. the crowded period begins if the number of jobs present in the system reaches $M$. Let also $\xi_0$ be the number of jobs accumulating in the buffer before the opening time of the system.

Introduce the following notation:

$$C_n(t) = C_{n,M}(t) \overset{def}{=} \mathbf{P}\{C_M > t \,|\, \xi_0 = n\}, \quad 0 \le n \le N, t > 0, \tag{1}$$

so $C_n(t)$ stands for the conditional tail cumulative distribution function of the time to start a crowded period, where the condition is the initial buffer state.

Assume, firstly, that the accumulating buffer is empty before the starting moment $(n = 0)$. Let us note that then the following equation is true:

$$C_0(t) = \lambda \int_0^t e^{-\lambda x} C_1(t - x)dx + e^{-\lambda t}. \tag{2}$$

Indeed, if the first arrival coincides with time $0 < x < t$, then at this time the state of the system changes into 1, the service process begins immediately and the time to start the crowded period will shortened to $t - x$. If the first job enters the system after time $t$ (with probability $e^{-\lambda t}$), then $C > t$ with probability 1.

Consider now the case in that $1 \le n \le M - 1$. Due to exponential intrearrival times, successive departure epochs are Markov moments in the evolution of the system (see e.g. [3]). Hence, applying the law of total probability with respect to the first departure moment after the starting time, we get

$$C_n(t) = \sum_{k=0}^{M-n-1} \int_0^t \frac{(\lambda y)^k}{k!} e^{-\lambda y} C_{n+k-1}(t - y)dF(y)$$

$$+ [1 - F(t)] \sum_{k=0}^{M-n-1} \frac{(\lambda t)^k}{k!} e^{-\lambda t}, \quad 1 \le n \le M - 1. \tag{3}$$

Let us comment (3) briefly. Indeed, if the number of jobs arriving before the first departure moment $0 < y < t$ does not exceed $M - n - 1$, then the system renews its operation with $n + k - 1$ jobs present after a departure at time $y$. In the case the number of jobs arriving during $(0, y)$ is greater than $M - n - 1$, a crowded period begins before time $t$. If the first departure occurs after time $t$ (with probability $1 - F(t)$), then the time to start a crowded period exceeds $t$ if and only if the number of arrivals up to $t$ equals at most $M - n - 1$ (the second summand on the right side of (3)).

Evidently, we have

$$C_n(t) = 0, \quad M \le n \le N. \tag{4}$$

Let us transform the equations of the system (2)–(4) to the equivalent forms written for Laplace transforms.

Define for $\mathrm{Re}(s) > 0$

$$\widehat{c}_n(s) \overset{def}{=} \int_0^\infty e^{-st} C_n(t)\, dt, \tag{5}$$

$$a_k(s) \overset{def}{=} \int_0^\infty e^{-(\lambda+s)t} \frac{(\lambda t)^k}{k!}\, dF(t) \tag{6}$$

and

$$b_k(s) \overset{def}{=} [1 - F(t)] \sum_{i=0}^{M-k-1} \int_0^\infty e^{-(\lambda+s)t} \frac{(\lambda t)^k}{k!}\, dt. \tag{7}$$

Now, the system (2)–(4) can be rewritten as follows:

$$\widehat{c}_0(s) = \frac{1}{\lambda+s}[\lambda\widehat{c}_1(s) + 1], \tag{8}$$

$$\widehat{c}_n(s) = \sum_{k=0}^{M-n-1} a_k(s)\widehat{c}_{n+k-1}(s) + b_n(s), \quad 1 \le n \le M - 1, \tag{9}$$

$$\widehat{c}_n(s) = 0, \quad M \le n \le N. \tag{10}$$

Let us apply to the system (8)–(10) the following substitution:

$$\widehat{d}_n(s) \overset{def}{=} \widehat{c}_{N-n}(s), \quad 0 \le n \le N. \tag{11}$$

As a consequence we obtain

$$\widehat{d}_n(s) = 0, \quad 0 \le n \le N - M, \tag{12}$$

$$\widehat{d}_n(s) = \sum_{k=0}^{M-N+n-1} a_k(s)\widehat{d}_{n-k+1}(s) + b_{N-n}(s), \quad N - M + 1 \le n \le N - 1, \tag{13}$$

$$\widehat{d}_N(s) = \frac{1}{\lambda+s}[\lambda\widehat{d}_{N-1}(s) + 1]. \tag{14}$$

## 3   Exact Solution for Transforms

In the present section, applying the linear algebraic approach, we obtain the solution of the system (12)–(14) in the closed form.

Let us firstly note that, defining

$$\phi_n(s) \overset{def}{=} -b_{N-n}(s), \tag{15}$$

we can write (13) in the following equivalent form:

$$\sum_{k=-1}^{n-(N-M+2)} a_{k+1}(s)\widehat{d}_{n-k}(s) - \widehat{d}_n(s) = \phi_n(s), \quad N - M + 1 \le n \le N - 1. \tag{16}$$

In [11] it is proved that each solution of the system of the form

$$\sum_{k=-1}^{n} \alpha_{k+1}x_{n-k} - x_n = \beta_n, \quad n \geq 0, \tag{17}$$

with unknowns $x'_k s$ and with $\alpha'_k s$ and $\beta'_k s$ being known sequences (complex-valued, in general), can be written as follows:

$$x_n = AR_{n+1} + \sum_{k=1}^{n} R_{n-k}\beta_k, \quad n \geq 0, \tag{18}$$

where the sequence $(R_k)$ is defined via the generating function of the sequence $(\alpha_k)$, namely

$$\sum_{k=0}^{\infty} R_k z^k = \frac{1}{P_\alpha(z) - 1}, \quad |z| < 1, \tag{19}$$

where $P_\alpha(z) \stackrel{def}{=} \sum_{k=-1}^{\infty} z^k \alpha_{k+1}$ and $A$ is a constant.

Successive terms of the sequence $(R_k)$ can also be obtained using the following recursion:

$$R_0 = 0, \quad R_1 = \frac{1}{\alpha_0}, \quad R_{k+1} = R_1\Big(R_k - \sum_{i=0}^{k} \alpha_{i+1}R_{k-i}\Big), \tag{20}$$

where $k \geq 1$.

Basing on this result, let us prove the following lemma:

**Lemma 1.** *Each solution of the linear system of the form*

$$\sum_{k=-1}^{n-r} \alpha_{k+1}x_{n-k} - x_n = \beta_n, \quad n \geq r, \tag{21}$$

*with unknowns $x'_k s$ and with $\alpha'_k s$ and $\beta'_k s$ being known sequences (in general complex-valued), and where $r \geq 0$, can be written as follows:*

$$x_n = AR_{n+1-r} + \sum_{k=r}^{n} R_{n-k}\beta_k, \quad n \geq r, \tag{22}$$

*where $A$ is a constant.*

*Proof.* Substituting $m = n - r$ in (21), we obtain

$$\sum_{k=-1}^{m} \alpha_{k+1}x_{m+r-k} - x_{m+r} = \beta_{m+r}, \quad m \geq 0. \tag{23}$$

Now, taking $y_m = x_{m+r}$ and $\gamma_m = \beta_{m+r}$, we get

$$\sum_{k=-1}^{m} \alpha_{k+1} y_{m-k} - y_m = \gamma_m, \quad m \geq 0. \tag{24}$$

Since (24) has the same form as (17), then it follows from (18) that the solution of (24) can be written as follows:

$$y_m = A R_{m+1} + \sum_{k=0}^{m} R_{m-k} \gamma_k, \quad m \geq 0, \tag{25}$$

where the sequence $(R_k)$ is given by the equality (19).

Returning in (25) to sequences $x'_k s$ and $\beta'_k s$, we have

$$x_{m+r} = A R_{m+1} + \sum_{k=0}^{m} R_{m-k} \beta_{k+r}, \quad m \geq 0, \tag{26}$$

and utilizing the fact that $n = m + r$, we obtain, finally,

$$x_n = A R_{n+1-r} + \sum_{k=0}^{n-r} R_{n-r-k} \beta_{k+r} = A R_{n+1-r} + \sum_{k=r}^{n} R_{n-k} \beta_k, \quad n \geq r, \tag{27}$$

that ends the proof.

Let us note that (16) has the same form as (21) with $r = N - M + 2$. Evidently, (16) is valid only for $N - M + 1 \leq n \leq N - 1$. From Lemma 1 (compare (22)) it follows that

$$d_n(s) = A(s) R_{n+1-(N-M+2)}(s) + \sum_{k=N-M+2}^{n} R_{n-k}(s) \phi_k(s), \quad n \geq N - M + 2, \tag{28}$$

where now $A(s)$ and the sequences $(R_k(s))$ and $(\phi_k(s))$ depend on the argument $s$.

The generating function of the $(R_k(s))$ is defined as follows (see (19))

$$\sum_{k=0}^{\infty} R_k(s) z^k = \frac{1}{P_a(s,z) - 1}, \quad |z| < 1, \tag{29}$$

where $P_a(s,z) \stackrel{def}{=} \sum_{k=-1}^{\infty} z^k a_{k+1}(s)$.

Due to the fact that the number of equations in (16) is finite, we can utilize the Eq. (14) as a boundary condition and find the explicit representation for $A(s)$.

Introducing (28) into the right side of (14), we get

$$\hat{d}_N(s) = \frac{1}{\lambda + s} \left\{ \lambda \left[ A(s) R_{M-2}(s) + \sum_{k=N-M+2}^{N-1} R_{N-1-k}(s) \phi_k(s) \right] + 1 \right\}. \tag{30}$$

Similarly, writing (28) for $n = N$, we obtain

$$\widehat{d}_N(s) = A(s)R_{M-1}(s) + \sum_{k=N-M+2}^{N} R_{N-k}(s)\phi_k(s). \tag{31}$$

Comparing the right sides of (30) and (31), we eliminate $A(s)$ as follows:

$$A(s) = \frac{\sum_{k=N-M+2}^{N-1}\left[R_{N-k}(s) - \lambda(\lambda+s)^{-1}R_{N-1-k}(s)\right]\phi_k(s) - (\lambda+s)^{-1}}{\lambda(\lambda+s)^{-1}R_{M-2}(s) - R_{M-1}(s)}. \tag{32}$$

Referring to (11), (28) and (32), we have

$$\widehat{c}_n(s) = \frac{\sum_{k=N-M+2}^{N-1}\left[R_{N-k}(s) - \lambda(\lambda+s)^{-1}R_{N-1-k}(s)\right]\phi_k(s) - (\lambda+s)^{-1}}{\lambda(\lambda+s)^{-1}R_{M-2}(s) - R_{M-1}(s)}$$
$$\cdot R_{M-n-1}(s) + \sum_{k=N-M+2}^{N-n} R_{N-n-k}(s)\phi_k(s), \quad 0 \le n \le M-2. \tag{33}$$

The representation for $\widehat{c}_{M-1}(s)$ we obtain directly from (13). Indeed, substituting $n = N - M + 1$, we have

$$\widehat{d}_{N-M+1}(s) = a_0(s)\widehat{d}_{N-M+2}(s) + b_{M-1}(s) \tag{34}$$

and hence

$$\widehat{c}_{M-1}(s) = a_0(s)\widehat{c}_{M-2}(s) + b_{M-1}(s). \tag{35}$$

Collecting now the formulae (33) and (35) and having in mind (4) and (15), we can formulate the following main result:

**Theorem 1.** *The Laplace transform of the time to start the crowded period in the M/G/1/N-type finite-capacity queue with n jobs present in the accumulating buffer at the starting epoch of the system, where $0 \le n \le N$, can be represented as follows:*

$$\widehat{c}_n(s) = \frac{\sum_{k=N-M+2}^{N-1}\left[\lambda(\lambda+s)^{-1}R_{N-1-k}(s) - R_{N-k}(s)\right]b_{N-k}(s) - (\lambda+s)^{-1}}{\lambda(\lambda+s)^{-1}R_{M-2}(s) - R_{M-1}(s)}$$
$$\cdot R_{M-1-n}(s) - \sum_{k=N-M+2}^{N-n} R_{N-n-k}(s)b_{N-k}(s), \quad 0 \le n \le M-2, \tag{36}$$

$$\widehat{c}_{M-1}(s) = a_0(s)\widehat{c}_{M-2}(s) + b_{M-1}(s) \tag{37}$$

*and*

$$\widehat{c}_n(s) = 0, \quad M \le n \le N, \tag{38}$$

*where $\mathrm{Re}(s) > 0$ and $0 < M \le N$ is the queue level (a threshold) at which the crowded period begins. Moreover, the functional sequences $\big(a_k(s)\big)$, $\big(b_k(s)\big)$ and $\big(R_k(s)\big)$ are defined in (6), (7) and (29), respectively.*

# References

1. de Boer, P.T., Nicola, V.F., van Ommeren, J.V.C.: The remaining service time upon reaching a high level in M/G/1 queues. Queueing Syst. **39**, 55–78 (2001)
2. Chae, K.C., Kim, K., Kim, N.K.: Remarks on the remaining service time upon reaching a target level in the M/G/1 queue. Oper. Res. Lett. **35**, 308–310 (2007)
3. Cohen, J.W.: The Single Server Queue. North-Holand Publishing Company, Amsterdam, New York, Oxford (1982)
4. Fakinos, D.: The expected remaining service time in a single server queue. Oper. Res. **30**, 1014–1018 (1982)
5. Al Hanbali, A.: Busy period analysis of the level dependent PH/PH/1/K queue. Queueing Syst. **67**, 221–249 (2011)
6. Kempa, W.M.: On the distribution of the time to buffer overflow in a queueing system with a general-type input stream. In: Proceedings of the 35th International Conference on Telecommunication and Signal Processing (TSP 2012), Prague, 3–4 July 2012, pp. 207–211 (2012)
7. Kempa, W.M.: On buffer overflow duration in WSN with a vacation-type power saving mechanism. In: Proceedings of 2017 International Conference on Systems, Signals and Image Processing (IWSSIP), Poznań, Poland, 22–24 May 2017, pp. 87–90. Institute of Electrical and Electronics Engineers, Piscataway (2017)
8. Kempa, W.M.: On buffer overflow duration in a finite-capacity queueing system with multiple vacation policy. In: Pasheva, V., Popivanov, N., Venkov, G. (eds.) Proceedings of the 43th International Conference Applications of Mathematics in Engineering and Economics, AMEE 2017, Sozopol, Bulgaria, 8–13 June 2017, pp. 1–6. American Institute of Physics, Melville (2017) (AIP Conference Proceedings, vol. 1910, iss. 1)
9. Kempa, W.M., Marjasz, R.: Distribution of the time to buffer overflow in the single-server queueing model with multiple vacation policy. In: Wituła, R., et al. (eds.) Selected Problems on Experimental Mathematics, pp. 113–127. Silesian University of Technology Press, Gliwice (2017)
10. Kempa, W.M., Paprocka, I.: Time to buffer overflow in a finite-capacity queueing model with setup and closedown times. In: Świątek, J., Wilimowska, Z., Borzemski, L., Grzech, A. (eds.) Information Systems Architecture and Technology: Proceedings of 37th International Conference on Information Systems Architecture and Technology—ISAT 2016—Part III. AISC, vol. 523, pp. 215–224. Springer, Cham (2017). https://doi.org/10.1007/978-3-319-46589-0_17
11. Korolyuk, V.S.: Boundary-Value Problems for Compound Poisson Processes. Naukova Dumka, Kiev (1975)
12. Lee, E.Y., Kinateder, K.K.J.: The expected wet period of finite dam with exponential inputs. Stoch. Anal. Appl. **90**, 175–180 (2000)
13. Takagi: Queueing Analysis, vol. 2: Finite Systems. North-Holland (1993)

# Two-Grid Newton Algorithms
# for a System of Heat Conducting
# Gas Equations

Miglena N. Koleva$^{(\boxtimes)}$ and Lubin G. Vulkov

University of Ruse, 8 Studentska str., 7017 Ruse, Bulgaria
{mkoleva,lvalkov}@uni-ruse.bg

**Abstract.** We present two-grid Newton's algorithms for solving gas dynamics coupled with non-linear heat conduction problems. The algorithms involve solving one small, non-linear coarse mesh system and one linear problem on a fine mesh. We also examine a two-grid method combined with Richardson extrapolation. Test examples compare the efficiency of the proposed algorithms.

**Keywords:** Gas dynamics · Conservative difference schemes · Newton linearization · Two-grid method · Richardson extrapolation

## 1 Introduction

The theory of difference schemes has two basic aspects: (i) methods for construction of the difference schemes; (ii) validation of the chosen difference scheme, i.e. accuracy and convergence investigation. With respect to the applications, the computational cost of the algorithms for achieving the necessary accuracy is very important.

Applying Newton's method (NM), numerical methods for 1D system of gas dynamics are developed in [10,11] and for non-linear viscoelasticity in [13]. Computational results show the advantage of the NM in comparison with other iterative methods [10], since NM works well on coarse space meshes and large time step.

The goal of the present paper is to construct and study numerically two-grid interpolation algorithms for the implementation of conservative difference schemes for gas dynamics equations with heat conduction [11]. To improve the computational efficiency we implement the two-grid idea, combined with Richardson extrapolation.

The two-grid FEM was originally proposed by Axelsson and Xu [1,15] to semi-linear elliptic equations and later to deal with the asymmetry and indefiniteness, localization and parallelization for PDEs. It was shown it's efficiency for some coupled systems [3,4,14] as well as for problems of physics and mechanics, see e.g. [5–8,16].

© Springer Nature Switzerland AG 2019
I. Dimov et al. (Eds.): FDM 2018, LNCS 11386, pp. 337–345, 2019.
https://doi.org/10.1007/978-3-030-11539-5_38

We consider the following model problem (in Lagrange variables) [12]:

$$v_t = -p_s, \quad \eta_t = v_s, \quad \varepsilon_t + pv_s = [k(T, \eta)T_s]_s, \\ p = \mathcal{P}(T, \eta), \quad \varepsilon = E(T, \eta), \quad \eta = 1/\rho, \tag{1}$$

where $(t, s) \in \Omega = (0, t^*) \times (0, S)$, $ds = \rho dx$ is the Lagrange variable, $x$ is the the the Euler variable, $\rho = \rho(s, t)$ is the density, $v = v(s, t)$ is the velocity, $p = p(s, t)$ is the pressure, $T = T(s, t)$ is the temperature, $\varepsilon = \varepsilon(s, t)$ is the gas energy, $E(T, \eta)$ is the internal energy and $k(T, \eta)$ is the heat coefficient. We solve the problem (1) with boundary conditions:

$$v(0, t) = \varphi(t), \quad \eta(S, t) = \psi(t), \quad T(0, t) = \theta_1(t), \quad T(S, t) = \theta_2(t), \quad t > 0, \tag{2}$$

initial medium status and and state inequalities:

$$v(s, 0) = v^0(s), \; p(s, 0) = p^0(s), \; \eta(s, 0) = \eta^0(s), \; T(s, 0) = T^0(s), \; \varepsilon(s, 0) = \varepsilon^0(s), \tag{3}$$
$$\mathcal{P}_\eta > 0, \quad E_t > 0, \quad k(T, \eta) > 0. \tag{4}$$

If in (1), we eliminate $p$ and $\varepsilon$, we obtain a system of three differential equations for unknowns $v$, $\eta$ and $T$

$$v_t + \mathcal{P}_t = 0, \quad \eta_t - v_s = 0, \quad E_t + \mathcal{P}v_s = [k(T, \eta)T_s]_s. \tag{5}$$

In this work we consider the system (5), associated with the boundary and initial conditions (2), (3).

The remaining part of the paper is organized as follows. In Sect. 2, we introduce a basic for the two-grid algorithms, finite difference scheme. In Sect. 3 we construct different two-grid algorithms with and without Richardson extrapolation. The computational results are presented and discussed in Sect. 4. Finally we give some concluding remarks.

## 2    Finite Difference Scheme and Newton's Linearization

In this section we present the underlying difference scheme, which is a base of the two-grid approach.

In the computational domain $(0, S) \times (0, t^*)$ we consider uniform mesh $\overline{\omega}_{h\tau} = \overline{\omega}_h \times \overline{\omega}_\tau$, defined by

$$\overline{\omega}_h = \{s_i = ih, \; i = 0, 1, \dots, n, \; nh = S\}, \\ \overline{\omega}_\tau = \{t_j = j\tau, \; j = 0, 1, \dots, J, \; J\tau = t^*\}.$$

Let $y$ be a mesh function, defined in $\overline{\omega}_{h\tau}$. Further, we use the notations

$$\begin{aligned} &y(s_i, t_j) = y_i^j = y, \quad y(s_i \pm h) = y_{i\pm 1}^j = y(\pm 1), \quad \widehat{y} = y_i^{j+1} = y(s_i, t_{j+1}), \\ &y_t = (\widehat{y} - y)/\tau, \quad y_s = (y(+1) - y)/h, \quad y_{\overline{s}} = y_s(-1), \\ &y^{(\sigma)} = \sigma\widehat{y} + (1 - \sigma)y, \; 0 \le \sigma \le 1, \quad \overline{y} = 0.5[y + y(-1)]. \end{aligned}$$

We study the following conservative finite difference scheme, written only on the integer grid nodes of $\overline{\omega}_{h\tau}$ [9,12]

$$\overline{v}_t + \mathcal{P}_{\overline{s}}^{(\sigma_1)} = 0, \quad \overline{\eta}_t(+1) - v_s^{(0.5)} = 0, \tag{6}$$

$$E_t + 0.5[\overline{\mathcal{P}}^{(\sigma_1)}(+1)v_s^{(0.5)} + \overline{\mathcal{P}}^{(\sigma_1)}v_{\overline{s}}^{(0.5)}] = (\overline{k}T_{\overline{s}})_s^{(\sigma_2)}, \tag{7}$$

$$v_i^0 = v^0(s_i), \quad \eta_i^0 = \eta^0(s_i), \quad T_i^0 = T^0(s_i), \quad i = 0, 1, \ldots, n,$$
$$v_0^j = \varphi(t_j), \quad \eta_n^j = \psi(t_j), \quad T_0^j = \theta_1(t_j), \quad T_n^j = \theta_2(t_j), \quad j = 0, 1, \ldots, J, \tag{8}$$

where $0 \leq \sigma_1, \sigma_2 \leq 1$ are weights.

The non-linear system of algebraic Eqs. (6)–(8) can be solved by Newton's method. For $m = 0, 1, \ldots$ starting with $\overset{0}{v} = v^0, \overset{0}{\eta} = \eta^0, \overset{0}{T} = T^0$, at each time level, we iterate

$$\overset{m+1}{v_t} + \left(\overset{m+1}{\overline{\mathcal{P}}^{(\sigma_1)}}\right)_{\overline{s}} = 0, \quad \overset{m+1}{v_0} = \varphi(t_j), \quad \overset{m+1}{\overline{\eta}_t}(+1) - \overset{m+1}{v_s^{(0.5)}} = 0, \quad \overset{m+1}{\eta_n} = \psi(t_j), \tag{9}$$

$$\left(\frac{\partial \overset{m}{E}}{\partial T} \overset{m+1}{T} + \frac{\partial \overset{m}{E}}{\partial \eta} \overset{m+1}{\eta}\right)_t + \left(\overset{m}{E} - \frac{\partial \overset{m}{E}}{\partial T} \overset{m}{\eta} - \frac{\partial \overset{m}{E}}{\partial \eta} \overset{m}{\eta}\right)_t$$

$$+ 0.5\left(\overset{m+1}{\overline{\mathcal{P}}^{(\sigma_1)}}(+1) \overset{m}{v_s^{(0.5)}} + \overset{m+1}{\overline{\mathcal{P}}^{(\sigma_1)}} \overset{m}{v_{\overline{s}}^{(0.5)}}\right) +$$

$$0.5\left(\overset{m}{\overline{\mathcal{P}}^{(\sigma_1)}}(+1)\left(\overset{m+1}{v_s} - \overset{m}{v_s}\right) + \overset{m}{\overline{\mathcal{P}}^{(\sigma_1)}}\left(\overset{m+1}{v_{\overline{s}}} - \overset{m}{v_{\overline{s}}}\right)\right) \tag{10}$$

$$= \left(\overset{m}{\overline{k}} \overset{m+1}{T_{\overline{s}}}\right)_s + 0.5\,\overset{m}{T_s}\left(\frac{\partial \overset{m}{k}(+1)}{\partial T(+1)}\left(\overset{m+1}{T}(+1) - \overset{m}{T}(+1)\right)\frac{\partial \overset{m}{k}}{\partial T}(\overset{m+1}{T} - \overset{m}{T})\right)$$

$$- 0.5\,\overset{m}{T_{\overline{s}}}\left(\frac{\partial \overset{m}{k}}{\partial T}(\overset{m+1}{T} - \overset{m}{T}) + \frac{\partial \overset{m}{k}(-1)}{\partial T(-1)}\left(\overset{m+1}{T}(-1) - \overset{m}{T}(-1)\right)\right),$$

$$\overset{m+1}{T_0} = \theta_1(t_j), \quad \overset{m+1}{T_n} = \theta_2(t_j), \quad j = 0, 1, \ldots, J,$$

where

$$\overset{m+1}{\mathcal{P}^{(\sigma_1)}} := \frac{\partial \overset{m}{\mathcal{P}^{(\sigma_1)}}}{\partial \eta} \overset{m+1}{\eta} + \frac{\partial \overset{m}{\mathcal{P}^{(\sigma_1)}}}{\partial T} \overset{m+1}{T} + \overset{m}{\overline{\mathcal{P}}^{(\sigma_1)}} - \frac{\partial \overset{m}{\mathcal{P}^{(\sigma_1)}}}{\partial \eta} \overset{m}{\eta} - \frac{\partial \overset{m}{\mathcal{P}}}{\partial T} \overset{m}{T}.$$

The iteration process terminates, when the maximal discrete norm of the difference between the solutions, obtained on two subsequent iterations becomes less than a given tolerance (*tol*).

Following the methodology in [12] one can prove the following assertion.

**Theorem 1 (Convergence).** *Let the conditions* (4) *hold and* $\eta$, $V$, $p$, $T \in C_l^r$, *where* $l$ *and* $r$ *are the differentiation with respect to* $t$ *and* $s$, *respectively. Then, for sufficiently small* $\tau$ *and* $h$, *the solution obtained by Newton's linearization* (9)–(10) *is convergent to the solution of problem* (1) *with rate of convergence* $\mathcal{O}(\tau + h^2)$, *if* $\sigma_1 = \sigma_2 = 1$ *and* $\mathcal{O}(\tau^2 + h^2)$, *if* $\sigma_1 = \sigma_2 = 0.5$.

## 3    Two-Grid Algorithms

In order to accelerate the computational efficiency, we develop two-grid algorithms. The basic idea of the two-grid method [1,15] mesh $(\overline{\omega}_H)$ and the linearized (around the interpolant of the coarse-grid solution) problem on the fine mesh $(\overline{\omega}_h)$.

Let us define a primal coarse mesh $\overline{\omega}_H$

$$\overline{\omega}_H = \{s_i = iH, \ i = 0, 1, \ldots, N, \ NH = S\}, \quad H >> h.$$

Let $y = (v, \eta, T)$ and denote by $y^H$ the solution, obtained on the coarse mesh $\overline{\omega}_H$. For the interpolated solution of $y^H$ on the fine mesh $\overline{\omega}_h$ we use the notation $\mathcal{I}(y^H)$.

To solve the problem (6)–(8), at each time level we apply one of the following algorithms.

**Algorithm 1** (*Two-grid approach*)

**step 1.** Solve the non-linear system (6)–(8) on the coarse mesh $\overline{\omega}_H$, using Newton iteration process (9)–(10) to find $v^H$, $\eta^H$, $T^H$.

**step 2.** Perform only one Newton iteration (9)–(10) on the fine mesh $\overline{\omega}_h$, using $\mathcal{I}(v^H)$, $\mathcal{I}(\eta^H)$ and $\mathcal{I}(T^H)$ as initial guesses.

Another way to improve the accuracy of the approximated solution is the Richardson extrapolation [2]. We implement this idea in the two-grid procedure. Richardson extrapolation is executed on the fine mesh at final time, i.e. we perform passive Richardson extrapolation, which posses better stability properties the the active approach.

Let $\sigma_1 = \sigma_2 = 0.5$.

**Algorithm 2** (*Two-grid approach with Richardson extrapolation*)

**step 1.** Execute Algorithm 1 on the coarse mesh $\overline{\omega}_H$ and fine mesh $\overline{\omega}_h$, with time step $\tau$. Denote the solution by $y_1$;

**step 2.** Execute Algorithm 1 on the coarse mesh $\overline{\omega}_{H_1}$ and fine mesh $\overline{\omega}_{h_1}$, $h_1 = h/2$ with time step $\tau/2$. Denote the solution by $y_2$;

**step 3.** Find the solution $y = \dfrac{1}{3}(4y_2 - y_1)$.

In view of Theorem 1, the expected order of convergence in maximal discrete norm ($\| \cdot \|$) for the two-grid Algorithm 1 is

$$\mathcal{O}(\tau^2 + H^4 + h^2) \text{ for } \sigma_1 = \sigma_2 = 0.5 \text{ and } \mathcal{O}(\tau + H^4 + h^2) \text{ otherwise.}$$

Regarding to Algorithm 2, as the Richardson extrapolation is applied in time and fine spatial space, in view of [2], for $\sigma_1 = \sigma_2 = 0.5$ we expect the third order of convergence both in time and fine spatial space.

## 4  Computational Results

In this section we will verify the theoretical suggestions and efficiency of Algorithms 1–2 for solving (2)–(5). The test example is [11]

$$\mathcal{P} = \rho RT, \quad E = \frac{RT}{\gamma - 1}, \quad k(T, \eta) = k_0 T^\alpha \text{ for } R = 1, \; \gamma = 5/3, \; k_0 = 0.5, \; \alpha = 2.$$

We chose appropriate initial and boundary conditions (3), (2), and add residual function in the right-hand side of the third equation in (5), such that the exact solution of the system (2)–(5) is

$$v = e^{-t} \cos(s), \quad \eta = e^{-t} \sin(s) + 1, \quad T = (e^{-t} \sin(s) + 1)^2.$$

The computations are performed for $S = 1$ and $t^* = 0.5$. We will give CPU time (in seconds), the errors $(\mathcal{E}_v, \mathcal{E}_\eta, \mathcal{E}_T)$ in maximal discrete norm and the order of convergence $(CR_v, CR_\eta, CR_T)$ of the numerical solution, computed by Algorithms 1–2 and standard one grid approach, as a benchmark test. The convergence rate is computed by a double mesh principle, i.e. as a $\log 2$ of the ratios of the errors (in maximal discrete norm), obtained on two consequently refined space grids. For the transition of the solution between the two meshes-coarse and fine, we use a shape-preserving piecewise cubic Hermite interpolation and for termination of the Newton's iteration process we take $tol = 1.0e{-}6$. The results are given at final time $t^*$, for $\sigma_1 = \sigma_2 = 0.5$ and mesh parameters:

$$h \lesssim H^2 \quad \text{and} \quad \tau \lesssim h \text{ for Algorithm 1,}$$
$$h \lesssim H^q \quad \text{and} \quad \tau \lesssim h \text{ for Algorithm 2, step 1,}$$
$$H_1 \lesssim 2^{-1/q} H, \quad h_1 \lesssim H_1^q \quad \text{and} \quad \tau \lesssim h_1 = h/2 \text{ for Algorithm 2, step 2.}$$

In Table 1 we illustrate the performance of the one-grid computations. We observe second order convergence rate in space and also in time (as the ratio $\tau/h$ is fixed).

In Table 2 we give the results, obtained by Algorithm 1. It is obvious (see Table 2) that the order of convergence (in maximal discrete norm) on the coarse mesh of the numerical solution, computed by Algorithm 1 is 4. Because of the choice of the mesh step sizes ($h \lesssim H^2$ and $\tau \lesssim h$), we may conclude that the accuracy of Algorithm 1 for $\sigma_1 = \sigma_2 = 0.5$ is $\mathcal{O}(\tau^2 + H^4 + h^2)$. We observe significant improvement of the computational efficiency - one and the same precision can be obtained by one and two-grid approach, but the CPU time of the two-grid Algorithm 1 is much smaller than the standard one-grid method.

If we use a linear interpolation for the transition of the solution from the coarse to the fine mesh and in the opposite direction - from the fine to the coarse

**Table 1.** Standard one-grid computations

| $n$ | $\mathcal{E}_v$ | $CR_v$ | $\mathcal{E}_\eta$ | $CR_\eta$ | $\mathcal{E}_T$ | $CR_T$ | CPU |
|---|---|---|---|---|---|---|---|
| 100 | 1.913988e−5 | | 2.233231e−5 | | 2.216652e−5 | | 0.198 |
| 200 | 4.784809e−6 | 2.0000 | 5.576770e−6 | 2.0016 | 5.541508e−6 | 2.0000 | 0.440 |
| 400 | 1.196205e−6 | 2.0000 | 1.394039e−6 | 2.0002 | 1.385369e−6 | 2.0000 | 1.201 |
| 800 | 2.990506e−7 | 2.0000 | 3.484875e−7 | 2.0001 | 3.463442e−7 | 2.0000 | 5.118 |
| 1600 | 7.476259e−8 | 2.0000 | 8.712124e−8 | 2.0000 | 8.658599e−8 | 2.0000 | 23.252 |
| 3200 | 1.869086e−8 | 2.0000 | 2.178033e−8 | 2.0000 | 2.164699e−8 | 2.0000 | 114.984 |
| 6400 | 4.672549e−9 | 2.0001 | 5.445147e−9 | 2.0000 | 5.411516e−9 | 2.0001 | 625.895 |
| 12800 | 1.168309e−9 | 1.9999 | 1.361154e−9 | 2.0001 | 1.353928e−9 | 1.9989 | 3884.682 |
| 25600 | 2.914108e−10 | 2.0033 | 3.403875e−10 | 1.9996 | 3.351179e−10 | 2.0144 | 26380.783 |

**Table 2.** Two-grid computations with Algorithm 1

| $N$ | $n$ | $\mathcal{E}_v$ | $CR_v$ | $\mathcal{E}_\eta$ | $CR_\eta$ | $\mathcal{E}_T$ | $CR_T$ | CPU |
|---|---|---|---|---|---|---|---|---|
| 10 | 100 | 1.914314e−5 | | 2.232033e−5 | | 2.216511e−5 | | 0.276 |
| 20 | 400 | 1.196281e−6 | 4.0002 | 1.393908e−6 | 4.0011 | 1.385376e−6 | 3.9999 | 1.222 |
| 40 | 1600 | 7.476247e−8 | 4.0002 | 8.711991e−8 | 4.0000 | 8.658604e−8 | 4.0000 | 14.635 |
| 80 | 6400 | 4.672825e−9 | 3.9999 | 5.445022e−9 | 4.0000 | 5.411919e−9 | 3.9999 | 335.446 |
| 160 | 25600 | 2.930938e−10 | 3.9949 | 3.399767e−10 | 4.0014 | 3.391980e−10 | 3.9959 | 13307.035 |

mesh, the error is a little bit bigger, but the computational time is slightly less. For example, if $N = 80$, $n = 6400$, the computational results with Algorithm 1 are $\mathcal{E}_v = 4.695109e - 9$, $\mathcal{E}_\eta = 5.624833e - 9$, $\mathcal{E}_T = 5.415666e - 9$, CPU $= 325.844$. The order of convergence in the coarse space is again 4.

The computations with Algorithm 1 for $\sigma_1 = \sigma_2 = 1$ with $\tau \lesssim h^2$, $h \lesssim H^2$ confirm $O(\tau + H^4 + h^2)$ order of convergence in maximal discrete norm.

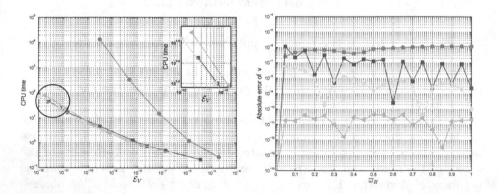

**Fig. 1.** Error in maximal discrete norm vs CPU time (*left*) and absolute error on the coarse mesh, $N = 20$ (*right*) of the numerical solution $v$, computed by Algorithm 1 (line with circles); Algorithm 2, $q = 2$ (line with triangles); Algorithm 2, $q = 5/3$ (line with diamonds); Algorithm 2, $q = 4/3$ (line with squares)

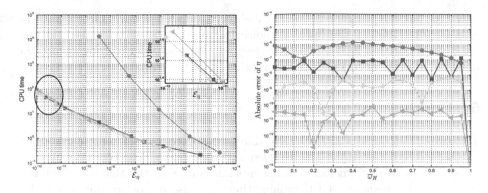

**Fig. 2.** Error in maximal discrete norm vs CPU time (*left*) and absolute error on the coarse mesh, $N = 20$ (*right*) of the numerical solution $\eta$, computed by Algorithm 1 (line with circles); Algorithm 2, $q = 2$ (line with triangles); Algorithm 2, $q = 5/3$ (line with diamonds); Algorithm 2, $q = 4/3$ (line with squares)

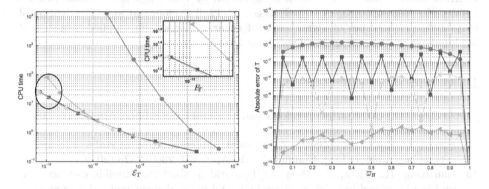

**Fig. 3.** Error in maximal discrete norm vs CPU time (*left*) and absolute error on the coarse mesh, $N = 20$ (*right*) of the numerical solution $T$, computed by Algorithm 1 (line with circles); Algorithm 2, $q = 2$ (line with triangles); Algorithm 2, $q = 5/3$ (line with diamonds); Algorithm 2, $q = 4/3$ (line with squares)

The order of convergence in $L_2$ norm is also investigated numerically for the same test example. The computations showed bigger convergence order in space, namely $O(\tau + H^5 + h^{2.5})$ for $\sigma_1 = \sigma_2 = 1$ and $O(\tau^2 + H^5 + h^{2.5})$ for $\sigma_1 = \sigma_2 = 0.5$.

On Figs. 1, 2 and 3 we depict the error in maximal discrete norm vs CPU time and absolute error on the coarse mesh $\overline{\omega}_H$, $N = 20$ of the solutions $v$, $\eta$ and $T$, computed with Algorithm 1 and Algorithm 2 for different values of $q$. It is obvious that the Algorithm 2 is more efficient than Algorithm 1. Despite the fact that for one and the same number of space grid nodes, Algorithm 2, $q = 2$ reaches high precision of the solution in comparison with the other computational methods (see Figs. 1, 2 and 3 (*right*)), the CPU time is bigger than the cases $q = 4/3$ and $q = 5/3$ (see Figs. 1, 2 and 3 (*left*)). On the other side, Algorithm 2, $q = 4/3$ is very fast, but the precision is lower. We observe à better performance Algorithm 2, $q = 5/3$. Thus, as the optimal choice for relation between coarse and fine mesh (in Algorithm 2) is $q = 5/3$, we suggest that the order of convergence (in maximal discrete norm) of Algorithm 2 is $\mathcal{O}(\tau + H^5 + h^3)$.

# 5   Conclusions

The two-grid (coarse and fine) algorithms, based on the Newton's method for solving gas dynamic equations, coupled with non-linear heat conduction problems are presented. Numerical experiments illustrate that the algorithms can not only save a large amount CPU running time, but also reproduce the same order of approximation accuracy that provides the fine mesh.

Combination of the Richardson extrapolation with the two-grid technique significantly increase the computational efficiency.

**Acknowledgments.** This research is supported by the Bulgarian National Science Fund under Bilateral Project DNTS/Russia 02/12 "Development and investigation of finite-difference schemes of higher order of accuracy for solving applied problems of fluid and gas mechanics, and ecology" from 2018.

# References

1. Axelsson, O.: On mesh independence and newton-type methods. Appl. Math. **38**(4–5), 249–265 (1983)
2. Faragó, I., Havasi, A., Zlatev, Z.: Efficient implementation of stable Richardson extrapolation algorithms. Comput. Math. Appl. **60**, 2309–2325 (2010)
3. Hanzhang, H., Chen, Y., Zhou, J.: Two-grid method for miscible displacement problem by mixed finite element methods and finite element method of characteristics. Comput. Math. Appl. **72**(11), 2694–2715 (2016)
4. Jovanovic, B., Koleva, M.N., Vulkov, L.G.: Convergence of a FEM and two-grid algorithms for elliptic problems on disjoint domains. J. Comput. Appl. Math. **236**(3), 364–374 (2011)
5. Jovanovic, B., Koleva, M.N., Vulkov, L.G.: Application of the two-grid method to a heat radiation problem. In: AIP CP, vol. 1186, pp. 352–360 (2009)
6. Koleva, M.N.: A two-grid algorithm for implementation of fully conservative difference schemes of the gas dynamics equations. In: AIP CP, vol. 1067, pp. 262–270 (2008)
7. Koleva, M.N., Vulkov, L.G.: A two-grid approximation of an interface problem for the nonlinear Poisson-Boltzmann equation. In: Margenov, S., Vulkov, L.G., Waśniewski, J. (eds.) NAA 2008. LNCS, vol. 5434, pp. 369–376. Springer, Heidelberg (2009). https://doi.org/10.1007/978-3-642-00464-3_41
8. Koleva, M.N., Vulkov, L.G.: Two-grid quasilinearization approach to ODEs with applications to model problems in physics and mechanics. Comput. Phys. Commun. **181**(3), 663–670 (2010)
9. Moskalkov, M.N.: A fully conservative difference scheme in gas dynamics. Zh. Vychisl. Mat. Mat. Fiz. **20**(1), 162–170 (1980). (in Russian)

10. Povechenko, Y.A., Popov, Y.P., Samarskaya, E.A.: Iterative methods for solving difference schmes for equations of gas dynamics with heat conductivity. Zh. Vychisl. Mat. Mat. Fiz. **22**(4), 903–912 (1992)
11. Samarskii, A., Popov, Y.: Difference Methods for Solution of Problems of Gas Dynamics. Nauka, Moscow (1st edn. 1980, 2nd edn. 1992) (in Russian)
12. Vakulchik, P., Schorec, A.: Convergence of the Newton method for finite difference schemes for gas dynamics equations with heat transfer. Diff. Eqns. **11**, 1149–1155 (1986). (in Russian)
13. Vulkov, L.: On the realization and convergence of difference schemes for the nonlinear viscoelasticity equations. Diff. Uravnenia **27**(1), 1123–1132 (1991). (in Russian)
14. Wu, L., Allen, M.: Two-grid methods for mixed finite element solution of coupled reaction-diffusion systems. Numer. Meth. PDE **15**(5), 589–604 (1999)
15. Xu, J.: A novel two-grid method for semilinear elliptic equations. SIAM J. Sci. Comput. **15**, 231–237 (1994)
16. Zadorin, A.I., Tikhovskaya, S.V., Zadorin, N.A.: A two-grid method for elliptic problem with boundary layers. Appl. Numer. Math. **93**, 270–278 (2015)

# On Convergence of Difference Approximations of Extremum Problems Described by Elliptic Equations with Unbounded Nonlinearity

Aigul Manapova[✉]

Bashkir State University,
Zaki Validi Street, 32, Ufa, Republic of Bashkortostan, Russia
aygulrm@yahoo.com

**Abstract.** The present work is devoted to investigation of an optimization problem for elliptic equations with mixed derivatives and unbounded nonlinearity. The coefficients multiplying partial derivatives of the second order in the state equation are used as a control function. We develop finite difference approximations of extremum problems, study their well-posedness and estimate the approximation accuracy with respect to the state.

**Keywords:** Optimal control problem · Nonlinear elliptic equations ·
Unbounded nonlinearity · Mixed derivatives

## 1 Introduction and Setting of the Problem

During the last forty years, such a trend as the construction of difference approximations for problems of optimal control of mathematical physics equations has emerged and is intensively developing in the theory of finite differences. Note that the first results regarding discrete methods for optimization problems with solutions of the state equations from $W_2^k$ were obtained in the theory of finite element methods. However, techniques for constructing difference schemes and deriving estimates in grid methods differ from those used in the finite element method.

In the paper we study an optimal control problem (OCP) of processes described by nonlinear elliptic equations with mixed derivatives and unbounded nonlinearity. The controls are contained in the coefficients multiplying the second order partial derivatives. In the case of unbounded nonlinearity the coefficients of the state equation in the optimization problem, depending on the exact solution, are assumed to satisfy the required properties (positive definiteness, boundedness of $u$-derivatives) only in the range of the exact solution $u$ or in its small neighborhood (see, for example, [1–7] and references therein). Nonlinearity of such type allows to consider a serious of interesting applied problems (problems of heat conductivity, gas dynamics, etc.)

© Springer Nature Switzerland AG 2019
I. Dimov et al. (Eds.): FDM 2018, LNCS 11386, pp. 346–353, 2019.
https://doi.org/10.1007/978-3-030-11539-5_39

The numerical solving of optimal control problems (NSOCP) is in a wide sense related with studying the following issues:

1. The formulation of optimization problem ensuring the existence of solution on the set of admissible controls being a set of some infinite-dimensional vector space;
2. The reduction of optimal control problem to a sequence of finite-dimensional problem guaranteeing the convergence in some sense of solutions to the finite-dimensional problems to the solutions of original optimal control problems;
3. Numerical solving of the finite-dimensional problems.

The present paper is aimed on solving the first two steps of NSOCP, namely, constructing of difference approximations for extremum problems, studying well-posedness of the initial problems and their approximations, and convergence analysis of the approximations with respect to the state. Studies concerning difference schemes for this class of problems have shown that the convergence analysis in this case is a rather complicated technical problem even for smooth solutions, since the problem for the error of the grid method with respect to the state is now nonlinear. Moreover, in this case, we need to prove that the solution of the discrete scheme for the boundary value problem belongs to the domain (or its small neighborhood) of values of the corresponding exact solution, which, in turn, requires that the convergence rate of the grid method with respect to the state for the optimization problem be analyzed in the norm $C(\overline{\omega})$.

It should be noted that such result for the optimal control problems is obtained for the first time.

Now consider the following problem. Let $\Omega = \left\{ x = (x_1, x_2) \in \mathbf{R}^2 : 0 < x_\alpha < l_\alpha, \alpha = 1, 2 \right\}$ be a rectangle in $\mathbf{R}^2$ with boundary $\Gamma = \partial \Omega$. Consider the following optimal control problem: to minimize an objective functional $J : U \to \mathbf{R}^1$ of the form

$$g \to J(g) = \int_\Omega |u(r; g) - u_0^{(1)}(r)|^2 d\Omega, \tag{1}$$

on solutions $u(g)$ to the first boundary value problem for a second-order nonlinear differential equation:

$$-\sum_{\alpha,\beta=1}^2 k_{\alpha\beta}(x) \frac{\partial^2 u(x)}{\partial x_\alpha \partial x_\beta} + q(u)u = f(u), \quad x \in \Omega, \tag{2}$$

$$u(x) = 0, \qquad x \in \partial \Omega = \Gamma; \quad k_{12}(x) = k_{21}(x), \tag{3}$$

obeying all admissible controls

$$g = (k_{11}, k_{22}) \in U = \left\{ k_{\alpha\alpha} \equiv g_{\alpha\alpha} \in W_\infty^1(\Omega), \alpha = 1, 2 : \right.$$

$$\nu \sum_{\alpha=1}^2 \xi_\alpha^2 \le g_{11}(x) \xi_1^2 + g_{22}(x)\xi_2^2 + 2k_{12}(x) \xi_1\xi_2 \le \mu \sum_{\alpha=1}^2 \xi_\alpha^2, \tag{4}$$

$$\left. \forall x \in \Omega, \xi \ne 0, \xi \in \mathbf{R}^2, \left| \frac{\partial g_{\alpha\alpha}}{\partial x_1} \right| \le R_1, \left| \frac{\partial g_{\alpha\alpha}}{\partial x_2} \right| \le R_2, \alpha = 1, 2 \right\},$$

where $u_0 \in W_2^1(\Omega)$, $k_{12} \in W_2^1(\Omega)$ are given functions, $q(\eta)$, $f(\eta)$ are given functions of $\eta$.

It is a priori assumed that the problem (2)–(3) is uniquely solvable in the class $W_{2,0}^m(\Omega) = W_2^m(\Omega) \cap \overset{\circ}{W}_2^1(\Omega)$, $3 < m \le 4$ (see [7]) and the following estimate

$$\sup_{g \in U} \|u(g)\|_{W_2^2(\Omega)} \le M, \quad M = Const > 0$$

holds. We denote by $M_u = \{u : M_1 \le u(x) \le M_2, x \in \Omega\}$ a range of values of the exact solution to the problem (2)–(3) (which is a bounded set by the assumption that the solution of the original problem is smooth). We define a neighborhood $D_u$ ($\delta$ – neighborhood) of the exact solution values domain $M_u$ as

$$D_u = \left\{ \overline{u} : \overline{M}_1 = M_1 - \delta \le \overline{u}(x) \le M_2 + \delta = \overline{M}_2, \quad x \in K \subseteq \overline{\Omega}, \delta > 0 \right\},$$

where $\delta > 0$ is an arbitrary constant and can be quite small.

Additionally, let the coefficients of Eq. (2) satisfy the following smoothness conditions on the solution of problem (2)–(3):

$$0 \le q_0 \le q(\eta) \le \overline{q}_0, \qquad |f(\eta)| \le f_0, \quad \forall \eta \in D_u;$$

$$|q(\eta_1) - q(\eta_2)| \le L_q |\eta_1 - \eta_2|, \quad |f(\eta_1) - f(\eta_2)| \le L_f |\eta_1 - \eta_2|, \forall \eta_1, \eta_2 \in D_u;$$

$$\frac{2\mu(\max l_\alpha)^2}{\nu^2 - \mu \overline{q}_0 (\max l_\alpha)^2} \left\{ L_f + \frac{\mu f_0 L_q (\max l_\alpha)^2}{\nu^2 - \mu \overline{q}_0 (\max l_\alpha)^2} \right\} = q_0^*,$$

$$\nu^2 - \mu \overline{q}_0 (\max l_\alpha)^2 > 0, \qquad q_1^* = \frac{q_0^*}{2} < 1.$$

Here, $\nu, \mu, R_\alpha, \alpha = 1, 2, q_0, \overline{q}_0, L_q, L_f, f_0, 0 < \rho < 1$ are given positive constants. Note that the conditions imposed on the coefficients of the state equation (2) are satisfied only in the vicinity of the exact solution values that indicates the presence of nonlinearities of the unbounded growth.

Note that the condition of uniform ellipticity

$$\nu \sum_{\alpha=1}^{2} \xi_\alpha^2 \le \sum_{\alpha,\beta=1}^{2} k_{\alpha\beta}(x)\xi_\alpha\xi_\beta \le \mu \sum_{\alpha=1}^{2} \xi_\alpha^2, \quad \alpha, \beta = 1, 2, \forall x \in \Omega, \xi \ne 0, \xi \in \mathbf{R}^2,$$

which characterize the ellipticity of (2) and the boundedness of the coefficients $k_{\alpha\beta}(x)$, $x \in \Omega$ on the solution $u = u(x)$ to problem (2), (3) implies the estimates:

$$0 < \nu \le k_{\alpha\alpha}(x) \le \mu, \ \alpha = 1, 2, \qquad k_{12}^2(x) = k_{21}^2(x) < k_{11}(x)k_{22}(x), \quad \forall x \in \Omega.$$

In a more detailed form, Eq. (2) can be written as

$$-\sum_{\alpha=1}^{2} k_{\alpha\alpha}(x)\frac{\partial^2 u}{\partial x_\alpha^2} - 2k_{12}(x)\frac{\partial^2 u}{\partial x_1 \partial x_2} + q(u)u = f(u), \quad x \in \Omega,$$

$$k_{12}(x) = k_{21}(x).$$

*Remark 1.* While setting a nonlinear boundary value problem for the function $u = u(x)$, describing a particular physical process, we a priori assumed that the problem is uniquely solvable in the class $W_{2,0}^m(\Omega)$, $3 < m \le 4$. In connection with this and taking into account the embedding $W_2^m(\Omega) \to C^2(\overline{\Omega})$, under the solution of the boundary value problem (2)–(3) we mean a function $u = u(x)$, $x = (x_1, x_2) \in \Omega \subset \mathbf{R}^2$, $u \in W_{2,0}^m(\Omega)$, $m > 3$, satisfying the Eq. (2) in the classical sense.

The solvability of optimal control problem (1)–(4) is guaranteed by the following result.

**Theorem 1.** *There exists at least one optimal control $g_* \in U$ in problem (1)–(4); i.e., $J_* = \inf\{J(g) : g \in U\} > -\infty$ and $U_* = \{g_* \in U : J(g_*) = J_*\} \ne \emptyset$. The set $U_*$ of minimizers of the cost functional $J(g)$ in optimization problem (1)–(4) is weakly compact in $H = W_2^1(\Omega) \times W_2^1(\Omega)$. Any minimizing sequence $\{g^{(n)}\}_{n=1}^{\infty} \subset U$ of $J(g)$ converges weakly in $H$ to the set $U_*$.*

Proof. One can show that $U$ is a convex, closed, and bounded subset of $H$. Consequently, $U$ is weakly compact in $H$. Moreover, the functional $J(g)$ is weakly semicontinuous on U. Combining these two facts with the results of [8], p. 505, Theorem 4, we obtain all the assertions of the Theorem 1.

## 2   Difference Approximation of the OCPs

To approximate problem (1)–(4) and analyze the convergence of difference approximations, we introduce the following grids on $[0, l_\alpha]$, $\alpha = 1, 2$ and in $\overline{\Omega}$: $\overline{\omega}_\alpha = \{x_\alpha = x_\alpha^{(i_\alpha)} = i_\alpha h_\alpha \in [0, l_\alpha] : i_\alpha = \overline{0, N_\alpha}, N_\alpha h_\alpha = l_\alpha\}$, $\alpha = 1, 2$; $\omega_\alpha = \overline{\omega}_\alpha \cap (0, l_\alpha)$; $\omega_\alpha^+ = \overline{\omega}_\alpha \cap (0, l_\alpha]$, $\omega_\alpha^- = \overline{\omega}_\alpha \cap [0, l_\alpha)$, $\alpha = 1, 2$; $\overline{\omega} = \overline{\omega}_1 \times \overline{\omega}_2$; $\omega^{(\pm 1)} = \omega_1^{\pm} \times \omega_2$, $\omega^{(\pm 2)} = \omega_1 \times \omega_2^{\pm}$; $\gamma = \overline{\omega} \setminus \omega$, $\omega^{-1} = [\omega_1 \setminus l_1 - h_1] \times \omega_2$, $\omega^{-2} = \omega_1 \times [\omega_2 \setminus l_2 - h_2]$; $|h|^2 = h_1^2 + h_2^2$.

Let $V$ be the set of grid functions defined on the grid $\overline{\omega} = \overline{\omega}_1 \times \overline{\omega}_2$, and $\overset{\circ}{V}$ be its subset consisting of all grid functions vanishing on $\gamma = \overline{\omega} \setminus \omega$. For grid functions from $\overset{\circ}{V}$, the inner products, norms, and seminorms are defined as [9–12]:

$$(y, v)_{L_2(\omega)} = \sum_{\omega} h_1 h_2 y(x) v(x), \qquad \|y\|_{L_2(\omega)}^2 = (y, y)_{L_2(\omega)} = \sum_{\omega} h_1 h_2 y^2(x),$$

$$\|y\|_{\overset{\circ}{W}_2^1(\omega)}^2 = \|y_{\overline{x}_1}\|_{L_2(\omega_1^+ \times \omega_2)}^2 + \|y_{\overline{x}_2}\|_{L_2(\omega_1 \times \omega_2^+)}^2$$

$$= |y|_{W_2^1(\omega)}^2 = \sum_{\alpha=1}^{2} \sum_{\omega(+\alpha)} h_1 h_2 y_{\overline{x}_\alpha}^2 = \|\nabla y\|_{L_2(\omega)}^2,$$

$$\|y\|_{W_{2,0}^2(\omega)}^2 = \|y_{\overline{x}_1 x_1}\|_{L_2(\omega)}^2 + \|y_{\overline{x}_2 x_2}\|_{L_2(\omega)}^2 + 2\|y_{\overline{x}_1 \overline{x}_2}\|_{L_2(\omega_1^+ \times \omega_2^+)}^2$$

$$= \sum_{\alpha=1}^{2} \sum_{\omega} h_1 h_2 y_{\overline{x}_\alpha x_\alpha}^2 + 2 \sum_{\omega_1^+ \times \omega_2^+} h_1 h_2 y_{\overline{x}_1 \overline{x}_2}^2 = |y|_{W_2^2(\omega)}^2,$$

$$\|y\|_{C(\overline{\omega})} = \|y\|_{L_\infty(\overline{\omega})} = \max_{\overline{\omega}} |y(x)|.$$

Here, $|\cdot|_{W_2^1(\omega)}$ and $|\cdot|_{W_2^2(\omega)}$ denote the seminorms in $W_2^1(\omega)$ and $W_2^2(\omega)$, respectively.

To optimal control problem (1)–(4) we associate the following difference approximations: to minimize the grid functional

$$J_h(\Phi_h) = \sum_{x \in \overline{\omega}} |y(x; \Phi_h) - u_{0h}(x)|^2 \hbar_1 \hbar_2 = \|y(\Phi_h) - u_{0h}\|_{L_2(\overline{\omega})}^2, \qquad (5)$$

under the conditions that grid function $y \equiv y(\Phi_h) = (y_1(\Phi_h), y_2(\Phi_h)) \in \overset{\circ}{V}$ is the solution of the difference boundary value problem for problem (2)–(3):

$$A(u)\,u \equiv -\sum_{\alpha=1}^{2} \Phi_{\alpha\alpha}(x) y_{\overline{x}_\alpha x_\alpha} - 2k_{12}^h(x) Q(y) + q(y)y = f(y), \quad x \in \omega, \qquad (6)$$

$$y(x) = 0, \qquad x \in \gamma,$$

where

$$Q(y)(x) = \frac{y_{x_1 x_2}(x) + y_{\overline{x}_1 x_2}(x) + y_{x_1 \overline{x}_2}(x) + y_{\overline{x}_1 \overline{x}_2}(x)}{4}$$

$$= \frac{y(x_1 + h_1, x_2 + h_2) - y(x_1 - h_1, x_2 + h_2) - y(x_1 + h_1, x_2 - h_2)}{4h_1 h_2}$$

$$+ \frac{y(x_1 - h_1, x_2 - h_2)}{4h_1 h_2} = y_{\overset{\circ}{x}_1 \overset{\circ}{x}_2}(x_1, x_2), \qquad (7)$$

and the grid controls $\Phi_h$ belong to the set of admissible grid controls

$$\Phi_h = (\Phi_{11}, \Phi_{22}) \in U_h = \{\Phi_{\alpha\alpha} \in W_\infty^1(\omega), \alpha = 1, 2 :$$

$$\nu \sum_{\alpha=1}^{2} \xi_\alpha^2 \le \sum_{\alpha=1}^{2} \Phi_{\alpha\alpha}(x)\xi_\alpha^2 + 2k_{12}^h(x)\xi_\alpha\xi_\beta \le \mu \sum_{\alpha=1}^{2} \xi_\alpha^2, \ \forall x \in \omega; \ \xi \in \mathbf{R}^2, \xi \ne 0,$$

$$|\Phi_{\alpha\alpha h x_1}(x)| \le R_1, \ x \in \omega^{-1}, \quad |\Phi_{\alpha\alpha h x_2}(x)| \le R_2, \ x \in \omega^{-2}\}. \qquad (8)$$

Here, $k_{12}^h$ is a grid approximation of the function $k_{12}$ defined by Steklov average. The problem (6)–(7) is a system of nonlinear equations for the grid function $y = y(x)$, $x \in \overline{\omega} = \overline{\omega}_1 \times \overline{\omega}_2$. In this context, a natural question to ask is concerned with the existence and uniqueness of the solution to the nonlinear problem. It is well known that a major approach to finding solutions of nonlinear equations is based on iterative methods. In some cases, they make it possible to analyze the existence and uniqueness of solutions to nonlinear equations.

To find an approximate solution to the nonlinear difference equations (6)–(7), we construct an iterative process associated with successive approximations with respect to the nonlinearities, see [7]:

$$A(y^{(s)})y^{(s+1)} \equiv -\sum_{\alpha=1}^{2} \Phi_{\alpha\alpha}(x) y_{\overline{x}_\alpha x_\alpha}^{(s+1)} - 2k_{12}^h(x)\,Q(y^{(s+1)})$$

$$+ q(y^{(s)})y^{(s+1)} = f(y^{(s)}), \quad x \in \omega, \qquad (9)$$

$$y^{(s+1)}(x) = 0, \qquad x \in \gamma_h. \tag{10}$$

Specifically, the coefficients of the grid operator $A(y)$ are taken from the preceding iteration, so that the new approximation $y^{(s+1)}(x)$, $x \in \overline{\omega}_h$, is found by solving the linear problem.

In the following section sufficient conditions of the iterative method convergence will be specified, and theorems on difference boundary value and optimization problems resolvability and the accuracy estimate of the approximations with respect to the state are provided.

## 3   Well-Posedness of the Approximations. A Priory Error Estimate for the Method with Respect to the State

By the $\delta$-neighborhood of the exact solution to differential problem (2)–(3), we mean the set $S_u^\delta = \{v : \|v - u\|_C \le \delta\}$. Obviously, if $v \in S_u^\delta$, then $v \in D_u$.

The following theorems on the solution existence to problems (6)–(7) and (5)–(8), correspondingly, hold.

**Theorem 2.** *Suppose that problem (2)–(3) satisfies the conditions from Sect. 1 and the initial approximation in the iterative process (9)–(10) for the nonlinear grid boundary value problem (6)–(7) obeys the conditions:*

$$y^{(0)} \in S_u^*, \qquad \frac{\beta}{1 - q_1^*} < \delta, \qquad q_1^* = \frac{q_0^*}{2} < 1,$$

*where*

$$S_u^* = \left\{ v : \|v - u\|_C \le \frac{1 - q_1^*}{1 + q_1^*} \cdot \frac{\delta}{2} \right\},$$

*and $\beta$ is defined by the relation*

$$\beta = (1 + q_1^*) \|z^{(0)}\|_{C(\overline{\omega})} + C_0 C_2 \|\psi\|_{L_2(\omega)}, \qquad z^{(0)} = y^{(0)} - u,$$

*and $\psi(x)$, $x \in \omega$, is the residual of difference equation (6):*

$$\psi(x) = f(u) - \left[ -\sum_{\alpha=1}^{2} \Phi_{\alpha\alpha}(x) u_{\overline{x}_\alpha x_\alpha} - 2k_{12}^h(x) Q(u) + q(u)u \right] = f(u) - A(u)u.$$

*Then, in the neighborhood $S_u^\delta$ of the exact solution $u$ to differential problem (2)–(3), the nonlinear difference problem (6)–(7) has a unique solution $y$, which can be obtained as the limit of the sequence $\{y^{(s)}\}_{s=1}^{\infty}$ generated by the iterative process (9)–(10). Moreover, the convergence rate of the $\{y^{(s)}\}_{s=1}^{\infty}$ to the solution of problem (6)–(7) is estimated by inequality*

$$\|y^{(s)} - y\|_{C(\overline{\omega})} \le \frac{\beta}{1 - q_1^*} (q_1^*)^s.$$

*Additionally, a priori estimates for the grid solution to scheme (6)–(7) are established in various grid norms:*

$$\|y\|_{W_{2,0}^2(\omega)} \le C_5, \qquad \|y\|_{C(\overline{\omega})} \le C_2 C_5,$$
$$\|y\|_{L_2(\omega)} \le C_3 C_5, \qquad \|y\|_{\overset{\circ}{W_2^1}(\omega)} \le C_4 C_5,$$

*where $C_k$, $k = 2, 3, 4, 5$ are positive constants:*

$$C_2 = \frac{(\max l_\alpha)^2}{2(l_1 l_2)^{1/2}}, \quad C_3 = \frac{(\max l_\alpha)^2}{2}, \quad C_4 = \left(\frac{l_1^2 + l_2^2}{32}\right)^{1/2},$$
$$C_5 = C_0 f_0 (l_1 l_2)^{1/2}, \quad C_0 = \frac{2\mu}{\nu^2 - \mu \overline{q}_0 (\max l_\alpha)^2}.$$

**Theorem 3.** *For every $h > 0$, there exists at least one optimal control $\Phi_{h*} \in U_h$ in the sequence of grid (difference) optimization problems (5)–(8), i.e.,*

$$J_{h*} = \inf\{J_h(\Phi_h) : \Phi_h \in U_h\} > -\infty, \ U_{h*} = \{\Phi_{h*} \in U_h : J_h(\Phi_{h*}) = J_{h*}\} \ne \emptyset.$$

This theorem is proved using the same technique as in the proof of Theorem 1 and taking into account that (5)–(8) is a finite-dimensional problem of minimizing the functional (see also [8], p. 502, Theorem 1 and the comments on pp. 565–566]).

The following theorem states the result concerning the accuracy of the approximations with respect to the state.

**Theorem 4.** *Suppose that the solution of differential problem (2)–(3) belongs to the class $W_2^m(\Omega)$, where $m$ is any number from the interval $3 < m \le 4$, and let the initial approximation $y^{(0)}$ of the iterative process (9)–(10) be in $S_u^*$. Additionally, assume that problem (1)–(4) satisfies the conditions from Sect. 1. Then, for sufficiently small $h < h_0$, the convergence rate of the grid method with respect to the state for optimization problem (1)–(4) satisfies the estimate:*

$$\|z\|_{W_{2,0}^2(\omega)} = \|y - u\|_{W_{2,0}^2(\omega)} \le C_6 |h|^{m-2} \|u\|_{W_2^m(\Omega)}$$
$$+ C_7 \sum_{\alpha=1}^{2} \|k_{\alpha\alpha} - \Phi_{\alpha\alpha}\|_{L_\infty(\omega)} \|u\|_{W_{2,0}^2(\Omega)}, \quad 3 < m \le 4,$$

*where $C_6$ and $C_7$ do not depend on $h$ and $u(x)$.*

The proofs of Theorems 2 and 4 are based on applying the iterative process (9)–(10), which can be regarded as an efficient technique for implementing nonlinear difference scheme (6)–(7). Note that the rigorous substantiation of the convergence of iterative methods in the case of nonlinearities of unlimited growth is a nontrivial problem of interest in itself.

The results obtained will be heavily used later in studying issues, such as the convergence of the approximations with respect to the functional and control, and a regularization of the approximations.

**Acknowledgments.** The author thanks Prof. F. V. Lubyshev for his important remarks and useful discussions.

# References

1. Abrashin, V.N., Asmolik, V.A.: Locally one-dimensional difference schemes for multidimensional quasilinear hyperbolic equations. Differ. Uravn. **18**(7), 294–308 (1982)
2. Matus, P.P., Moskal'kov, M.N., Shcheglik, V.S.: Consistent estimates for rate of convergence of the grid method for a second-order nonlinear equation with generalized solutions. Differ. Uravn. **31**(7), 1198–1206 (1995)
3. Jovanovic, B.S.: Finite difference scheme for partial differential equations with weak solutions and irregular coefficients. Comput. Methods Appl. Math. **4**(1), 48–65 (2004)
4. Jovanovic, B.S., Suli, E.: Analysis of Finite Difference Schemes. Springer Series in Computational Mathematics, vol. 46. Springer, London (2014). https://doi.org/10.1007/978-1-4471-5460-0
5. Matus, P.: On convergence of difference schemes for IBVP for quasilinear parabolic equations with generalized solutions. Comp. Meth. Appl. Math. **14**(3), 361–371 (2014)
6. Lubyshev, F.V., Fairuzov, M.E.: Consistent convergence rate estimates in the grid $W_{2,0}^2(\omega)$ norm for difference schemes approximating nonlinear elliptic equations with mixed derivatives and solutions from $W_{2,0}^m(\Omega)$, $3 < m \leq 4$. Comput. Math. Math. Phys. **57**(9), 1427–1452 (2017)
7. Lubyshev, F.V., Fairuzov, M.E., Manapova, A.R.: Accuracy of difference schemes for nonlinear elliptic equations with non-restricted nonlinearity. Zhurnal SVMO **19**(3), 41–52 (2017). (in Russian)
8. Vasil'ev, F.P.: Optimization Methods. Faktorial, Moscow (2002). (in Russian)
9. Samarskii, A.A.: The Theory of Difference Schemes. Nauka/Marcel Dekker, Moscow/New York (1989, 2001)
10. Samarskii, A.A., Andreev, V.B.: Difference Methods for Elliptic Equations. Nauka, Moscow (1976). (in Russian)
11. Samarskii, A.A., Vabishchevich, P.N.: Computational Heat Transfer. Wiley/Librokom, Moscow/New York (1996, 2009)
12. Samarskii, A.A., Lazarov, R.D., Makarov, V.L.: Difference Schemes for Differential Equations with Weak Solutions. Vysshaya Shkola, Moscow (1987). (in Russian)

# Spijker's Example and Its Extension

Miklós E. Mincsovics[1,2(✉)]

[1] MTA-ELTE Numerical Analysis and Large Networks Research Group,
Pázmány Péter sétány 1/C, Budapest 1117, Hungary
[2] Department of Differential Equations, Budapest University of Technology
and Economics, Egry József utca 1, Budapest 1111, Hungary
mincso@math.bme.hu

**Abstract.** Strongly and weakly stable linear multistep methods can behave very differently. The latter class can produce spurious oscillations in some of the cases for which the former class works flawlessly. The main question is if we can find a well defined property which clearly tells the difference between them. There are many explanations from different viewpoints. We cite Spijker's example which shows that the explicit two step midpoint method is unstable with respect to the Spijker norm. We show that this result can be extended for the general weakly stable case.

**Keywords:** Linear multistep methods · Stability · Spijker-norm

## 1  Introduction

This paper focuses on the stability and instability of linear multistep methods. When we introduce linear multistep methods it is unavoidable to talk about the root-condition and usually about the two types of it, which divide these methods into two classes: the weakly and strongly stable linear multistep methods. This can be found in almost every textbook about the numerical solution of ordinary differential equations, see e.g. [1, Sect. 5.2.3]. The root-condition is closely related to the stability of linear multistep methods. As it is well-known stability together with consistency implies the convergence of the method. This result can be obtained in different setups, we follow the book [6] where stability, consistency and convergence are defined in a general sense forming the base of a beautiful theoretical framework. This book also gives a detailed description how to use this framework for nonlinear ODEs. Our intention is to clarify the relation of the strongly/weakly stable linear multistep methods and stability of linear multistep methods in the above mentioned setting. Spijker's example [5] gave the first (negative) result about this relation. The example shows that the explicit two step midpoint method is unstable with respect to the general notion of stability if we use an unusual norm. We extend this example to the whole weakly stable class.

We organized the paper as follows. We introduce linear multistep methods and their basic notions which are important for us, including the definition of

I. Dimov et al. (Eds.): FDM 2018, LNCS 11386, pp. 354–361, 2019.
https://doi.org/10.1007/978-3-030-11539-5_40

weakly/strongly stable linear multistep methods. Then we reformulate linear multistep methods and define stability in the general sense. After this preparation we recall Spijker's example and finally we present the new result which generalizes Spijker's example. We conclude the paper with a critical remark.

## 2    Stability of Linear Multistep Methods

Without loss of generality we consider the scalar autonomous *initial value problem* (IVP)

$$\begin{cases} u(0) = u^0 \,, \\ \dot{u}(t) = f(u(t)) \,, \end{cases} \tag{1}$$

where $t \in [0, T]$, $u^0 \in \mathbb{R}$ is the initial value, $u : [0, T] \to \mathbb{R}$ is the unknown function and we assume that $f$ is Lipschitz continuous.

In practice we have to use a numerical method to approximate the solution of (1) since finding the solution analytically is impossible in most of the cases. There are many possible choices, one is the application of a linear multistep method (LMM).

*Linear multistep method*s can be given in the following way:

$$\begin{cases} u_i = c^i \,, & i = 0, \dots, k-1 \\ \dfrac{1}{h} \sum_{j=0}^{k} \alpha_j u_{i-j} = \sum_{j=0}^{k} \beta_j f(u_{i-j}) \,, & i = k, \dots, n+k-1 = N \,, \end{cases} \tag{2}$$

where $h = T/N$ is the step size, $\alpha_j$, $\beta_j \in \mathbb{R}$, $\alpha_0 \neq 0$ are the coefficients of the method and the constants $c^i$ are some approximation of the solution on the first $k$ time levels. When these latter are known (here we do not go into the details how to determine these values since this is irrelevant to the results of the paper) the method can "run", we can calculate the next approximation and so on. To get $u_i$ which approximates the solution at the $i$-th time level $u(i \cdot h)$ we only need to know the previous $k$ approximations. Thus the formula represents a $k$-step method. Note that while $k$ is fixed for the method $n$, $N = k + n - 1$ and $h$ can vary as the grid gets finer. For shorthand notation later we will use $f_{i-j}$ for $f(u_{i-j})$. As an example consider the explicit two step midpoint method (sometimes called leapfrog scheme in the context of parabolic PDEs)

$$\begin{cases} u_i = c^i \,, & i = 0, 1 \\ \dfrac{1}{h} \left( \tfrac{1}{2} u_i - \tfrac{1}{2} u_{i-2} \right) = f_{i-1} \,, & i = 2, \dots, n+k-1 = N \end{cases} \tag{3}$$

which plays the main role in Spijker's example.

The *first characteristic polynomial* associated to (2) is defined as

$$\varrho(z) = \sum_{j=0}^{k} \alpha_j z^{k-j} \,.$$

Usually, two types of root-conditions are defined. These are presented below.

**Definition 1.** *The method is said to be* strongly stable *if for every root* $\xi_i \in \mathbb{C}$ *of the first characteristic polynomial* $|\xi_i| < 1$ *holds except* $\xi_1 = 1$, *which is a simple root.*

*A not strongly stable method is said to be* weakly stable *if for every root* $\xi_i \in \mathbb{C}$ *of the first characteristic polynomial* $|\xi_i| \leq 1$ *holds and if* $|\xi_i| = 1$ *then it is a simple root, moreover* $\xi_1 = 1$.

We note that sometimes these are defined slightly differently. The two main possible differences are the following. First, not requiring that $\xi_1 = 1$ holds. Second, the weakly stable class containing the strongly stable class. Our reason not to vote for this option is that we want to distinguish clearly between the two.

Roughly speaking being weakly (or strongly) stable means that applying a method for $\dot{u}(t) = 0$ the approximation remains bounded which is an understandable requirement.

The explicit two step midpoint method is weakly stable since its first characteristic polynomial is $\frac{1}{2}\left(z^2 - 1\right)$ with roots $z = \pm 1$.

In the weakly stable case we have another root at the boundary of the unit circle which could cause problems in some of the cases. One type of explanation about the difference between weakly and strongly stable LMMs tries to exploit this fact directly, see e.g. [1, Example 5.7]. Our approach is different.

In the following we rewrite LMMs (2) into the form for which we can define stability in the general sense. A method can be represented with a sequence of operators $F_N : \mathcal{X}_N \to \mathcal{Y}_N$, where $\mathcal{X}_N, \mathcal{Y}_N$ are $k + n$ dimensional normed spaces with norms $\|\cdot\|_{\mathcal{X}_N}, \|\cdot\|_{\mathcal{Y}_N}$ respectively and

$$(F_N(\mathbf{u}_N))_i = \begin{cases} u_i - c^i, & i = 0, \ldots, k-1 \\ \dfrac{1}{h}\sum_{j=0}^{k} \alpha_j u_{i-j} - \sum_{j=0}^{k} \beta_j f(u_{i-j}), & i = k, \ldots, n+k-1 = N. \end{cases}$$

Finding the approximating solution means that we have to solve the non-linear system of equations $F_N(\mathbf{u}_N) = \mathbf{0}$. $F_N$ can be represented in the following way:

$$F_N(\mathbf{u}_N) = \mathbf{A}_N \mathbf{u}_N - \mathbf{B}_N f(\mathbf{u}_N) - \mathbf{c}_N,$$

where $\mathbf{u}_N = (\mathbf{u}_k, \mathbf{u}_n)^T = (u_0, \ldots, u_{k-1}, u_k, \ldots, u_{n+k-1})^T \in \mathbb{R}^{k+n}$, $\mathbf{u}_k \in \mathbb{R}^k$, $\mathbf{u}_n \in \mathbb{R}^n$, $f(\mathbf{u}_N) = (f(u_0), f(u_1), \ldots, f(u_{n+k-1}))^T \in \mathbb{R}^{k+n}$, $\mathbf{c}_n = (c^0, c^1, \ldots, c^{k-1}, 0, \ldots, 0)^T \in \mathbb{R}^{k+n}$,

$$\mathbf{A}_N = \begin{pmatrix} \mathbf{I} & \mathbf{0} \\ \mathbf{A}_k & \mathbf{A}_n \end{pmatrix}, \qquad \mathbf{B}_N = \begin{pmatrix} \mathbf{0} & \mathbf{0} \\ \mathbf{B}_k & \mathbf{B}_n \end{pmatrix},$$

where $\mathbf{I} \in \mathbb{R}^{k \times k}$ is the identity matrix, $\mathbf{A}_k, \mathbf{B}_k \in \mathbb{R}^{n \times k}$, $\mathbf{A}_n, \mathbf{B}_n \in \mathbb{R}^{n \times n}$,

$$
\mathbf{A}_k = \frac{1}{h}
\begin{pmatrix}
\alpha_k & \cdots & \alpha_2 & \alpha_1 \\
0 & \alpha_k & \cdots & \alpha_2 \\
\vdots & \ddots & \ddots & \vdots \\
0 & \cdots\cdots & & \alpha_k \\
0 & \cdots\cdots & & 0 \\
\vdots & \ddots & \ddots & \vdots \\
0 & \cdots\cdots & & 0
\end{pmatrix}
\qquad
\mathbf{A}_n = \frac{1}{h}
\begin{pmatrix}
\alpha_0 & 0 & \cdots\cdots\cdots\cdots & 0 \\
\alpha_1 & \alpha_0 & 0 & \cdots\cdots & 0 \\
\alpha_2 & \alpha_1 & \alpha_0 & 0 & \cdots & 0 \\
\vdots & \ddots & \ddots & \ddots & \ddots & \vdots \\
\vdots & \ddots & \ddots & \ddots & \ddots & \vdots \\
0 & \cdots & 0 & \alpha_k & \cdots & \alpha_0
\end{pmatrix}
$$

and $\mathbf{B}_k$, $\mathbf{B}_n$ are the same as $\mathbf{A}_k$, $\mathbf{A}_n$, except that we have to omit the $\frac{1}{h}$ factor and the $\alpha$-s have to be changed to $\beta$-s.

**Definition 2.** *We call a method* stable *in the norm pair* $\left( \|\cdot\|_{\mathcal{X}_n}, \|\cdot\|_{\mathcal{Y}_n} \right)$ *if for all IVP* (1) $\exists S \in \mathbb{R}$ *and* $\exists N_0 \in \mathbb{N}$ *such that* $\forall N \geq N_0$, $\forall \mathbf{u}_N, \mathbf{v}_N \in \mathbb{R}^{k+n}$ *the estimate*

$$
\|\mathbf{u}_N - \mathbf{v}_N\|_{\mathcal{X}_N} \leq S \, \|F_N(\mathbf{u}_N) - F_N(\mathbf{v}_N)\|_{\mathcal{Y}_N} \tag{4}
$$

*holds.*

To define stability in this way has a definite profit. It is general in the sense that it works for almost every type of numerical method approximating the solution of ODEs and PDEs as well. Convergence can be proved by the popular recipe "consistency + stability = convergence"

$$
\|\varphi_N(\bar{u}) - \bar{\mathbf{u}}_N\|_{\mathcal{X}_N} \leq S \, \|F_N(\varphi_N(\bar{u})) - F_N(\bar{\mathbf{u}}_N)\|_{\mathcal{Y}_N} = S \, \|F_N(\varphi_N(\bar{u}))\|_{\mathcal{Y}_N} \to 0,
$$

where $\bar{u}$, $\bar{\mathbf{u}}_N$ denote the solution of the original problem (1) and the approximating problem $F_N(\mathbf{u}_N) = \mathbf{0}$ respectively, $\varphi_N : \mathcal{X} \to \mathcal{X}_N$ are projections from the normed space where the original problem is set, thus $\varphi_N(\bar{u}) - \bar{\mathbf{u}}_N$ represents the error (measured in $\mathcal{X}_N$). Finally, $\|F_N(\varphi_N(\bar{u}))\|_{\mathcal{Y}_N} \to 0$ is exactly the definition of consistency in this framework. We note that the existence of $\bar{\mathbf{u}}_N$ (from some index) is also the consequence of stability, see [2, Lemma 24. and 25.], cf. [6, Lemma 1.2.1]. There are many versions of Definition 2 which are requiring the stability estimate only in some neighbourhood, see [2], but as we defined it is satisfactory for the IVP (1).

In the following we introduce norm pairs which are interesting for us. We start with some norm notations: for $k \in \mathbb{N}$ fixed, $\mathbf{u}_N \in \mathbb{R}^{k+n}$ the $k\infty$ norm is defined as

$$
\|\mathbf{u}_N\|_{k\infty} = \max_{0 \leq i \leq k-1} |u_i| + \max_{k \leq i \leq N} |u_i|,
$$

thus $\|\mathbf{u}_N\|_{k\infty} = \|\mathbf{u}_k\|_{\infty} + \|\mathbf{u}_n\|_{\infty}$. While the $k$–Spijker-norm is defined as

$$
\|\mathbf{u}_N\|_{k\$} = \max_{0 \leq i \leq k-1} |u_i| + h \max_{k \leq l \leq N} \left| \sum_{i=k}^{l} u_i \right|.
$$

Using the notation $\|\mathbf{u}_n\|_{\$} = h \max_{k \le l \le N} \left| \sum_{i=k}^{l} u_i \right|$ the $k$–Spijker-norm can be expressed as $\|\mathbf{u}_N\|_{k\$} = \|\mathbf{u}_k\|_{\infty} + \|\mathbf{u}_n\|_{\$}$. Introducing another notation, the Spijker-norm can be given in a useful way which will be presented in the following.

First, we introduce $\mathbf{E}_n \in \mathbb{R}^{n \times n}$

$$
\mathbf{E}_n = \frac{1}{h}
\begin{pmatrix}
1 & 0 & \dots & \dots & 0 \\
-1 & 1 & 0 & \dots & 0 \\
0 & -1 & 1 & 0 & \vdots \\
\vdots & \ddots & \ddots & \ddots & \vdots \\
0 & \dots & 0 & -1 & 1
\end{pmatrix}
\quad \text{for which} \quad
\mathbf{E}_n^{-1} =
\begin{pmatrix}
h & 0 & \dots & \dots & 0 \\
h & h & 0 & \dots & 0 \\
h & h & h & 0 & \vdots \\
\vdots & \ddots & \ddots & \ddots & \vdots \\
h & h & \dots & h & h
\end{pmatrix}.
$$

Note that $\mathbf{E}_n$ represents the linear part of the explicit Euler method (without the initial step) and its inverse can be interpreted as the simplest numerical integration. Second, if $\mathbf{A}$ is a regular matrix and $\|\cdot\|_{\star}$ is a norm then $\|\mathbf{u}\|_{\mathbf{A},\star} = \|\mathbf{A}\mathbf{u}\|_{\star}$ defines a norm. Then clearly

$$
\|\mathbf{u}_n\|_{\$} = \left\| \mathbf{E}_n^{-1} \mathbf{u}_n \right\|_{\infty} = \|\mathbf{u}_n\|_{\mathbf{E}_n^{-1},\infty}.
$$

It is known that weakly and strongly stable linear multistep methods are stable in the norm pair $(\|\cdot\|_{k\infty}, \|\cdot\|_{k\infty})$, cf. [3]. Moreover, strongly stable methods are stable in the $(\|\cdot\|_{k\infty}, \|\cdot\|_{k\$})$ norm pair, see [4]. These are positive results and there is a natural question: are weakly stable methods stable in the $(\|\cdot\|_{k\infty}, \|\cdot\|_{k\$})$ norm pair or not? The following section is devoted to answer this question.

## 3   Spijker's Example and Its Extension

First we recall Spijker's example, see [6, Example 2 in Sect. 2.2.4] for the proof.

**Theorem 1.** *The explicit two-step midpoint method* (3) *is not stable in the* $(\|\cdot\|_{2\infty}, \|\cdot\|_{2\$})$ *norm pair.*

In the following we present the extension of this result.

**Theorem 2.** *Weakly stable methods are not stable in the* $(\|\cdot\|_{k\infty}, \|\cdot\|_{k\$})$ *norm pair.*

*Proof.* We assume that the method is weakly stable, thus we assume that $|\xi_2| = 1$, $\xi_2 \ne 1$. We set $f \equiv 0$, $\mathbf{v}_N = \mathbf{0}$ and $\mathbf{u}_N = (0, \dots, 0, u_k, \dots, u_{n+k-1})^T$. For this setting stability (4) is simplified to

$$
\|\mathbf{u}_n\|_{\infty} \le S \|\mathbf{A}_n \mathbf{u}_n\|_{\$}.
$$

For all $S$ and for all $n_0$ we will present a vector $\mathbf{u}_n$, $n > n_0$ for which

$$
\|\mathbf{u}_n\|_{\infty} > S \|\mathbf{A}_n \mathbf{u}_n\|_{\$}. \tag{5}
$$

Note that

$$h\mathbf{A}_n = \alpha_0 \prod_{i=1}^{k} (\mathbf{I} - \xi_i \mathbf{H}_n) \,,$$

where $\mathbf{I} \in \mathbb{R}^{n \times n}$ stands for the identity matrix, $\mathbf{H}_n \in \mathbb{R}^{n \times n}$ is defined as

$$\mathbf{H}_n = \begin{pmatrix} 0 & 0 & \dots\dots & 0 \\ 1 & 0 & 0 & \dots & 0 \\ 0 & 1 & 0 & 0 & \dots \\ \vdots & \ddots & \ddots & \ddots & \vdots \\ 0 & \dots & 0 & 1 & 0 \end{pmatrix} ,$$

and $\xi_i$, $i = 1, \dots, k$ are the roots of the first characteristic polynomial. This comes from the following calculation.

$$h\mathbf{A}_n = \alpha_0 \mathbf{I} + \alpha_1 \mathbf{H}_n + \alpha_2 \mathbf{H}_n^2 + \dots + \alpha_k \mathbf{H}_n^k = \alpha_k \prod_{i=1}^{k} (\mathbf{H}_n - \nu_i \mathbf{I}) =$$

$$\alpha_k (-1)^k \left( \prod_{i=1}^{k} \nu_i \right) \prod_{i=1}^{k} \left( \mathbf{I} - \frac{1}{\nu_i} \mathbf{H}_n \right) = \alpha_0 \prod_{i=1}^{k} \left( \mathbf{I} - \frac{1}{\nu_i} \mathbf{H}_n \right) = \alpha_0 \prod_{i=1}^{k} (\mathbf{I} - \xi_i \mathbf{H}_n) \,,$$

where we exploited the commutativity of the terms $(\mathbf{I} - \xi_i \mathbf{H}_n)$ and that $\xi_i = \frac{1}{\nu_i}$ since $\alpha_0 + \alpha_1 z + \alpha_2 z^2 + \dots + \alpha_k z^k$ is the reciprocal polynomial of $\varrho$. This covers the case when $\forall \xi_i \neq 0$. If $\exists \xi_i = 0$ the modification of the calculation is straightforward.

Let us introduce $\mathbf{w}_n = (w_1, \dots, w_n)^T \in \mathbb{R}^n$, $\mathbf{w}_n = \mathbf{E}_n \mathbf{u}_n$. With this (5) is equivalent to

$$\|\mathbf{w}_n\|_\$ = \left\| \mathbf{E}_n^{-1} \mathbf{w}_n \right\|_\infty = \|\mathbf{u}_n\|_\infty > S \left\| \alpha_0 \prod_{i=2}^{k} (\mathbf{I} - \xi_i \mathbf{H}_n) \mathbf{w}_n \right\|_\$ .$$

If $\xi_2 = -1$ then

$$\left\| \prod_{i=2}^{k} (\mathbf{I} - \xi_i \mathbf{H}_n) \mathbf{w}_n \right\|_\$ \leq \left( \prod_{i=3}^{k} \| (\mathbf{I} - \xi_i \mathbf{H}_n) \|_\$ \right) \| (\mathbf{I} - \xi_2 \mathbf{H}_n) \mathbf{w}_n \|_\$ \leq$$

$$2^{k-1} \| (\mathbf{I} - \xi_2 \mathbf{H}_n) \mathbf{w}_n \|_\$ \,,$$

since

$$\| (\mathbf{I} - \xi_i \mathbf{H}_n) \|_\$ = \max_{\|\mathbf{u}\|_\$ = 1} \| (\mathbf{I} - \xi_i \mathbf{H}_n) \mathbf{u} \|_\$ \leq 1 + \max_{\|\mathbf{u}\|_\$ = 1} \| \mathbf{H}_n \mathbf{u} \|_\$ \leq 2 \,.$$

Now, let us choose $w_m = m \, \xi_2^m = m(-1)^m$.

$$((\mathbf{I} - \xi_2 \mathbf{H}_n) \mathbf{w}_n)_m = \xi_2^m = (-1)^m \,,$$

thus its norm

$$h \max_{1 \le l \le n} \left| \sum_{i=1}^{l} (-1)^m \right| \to 0,$$

as $h \to 0$, while

$$\|\mathbf{w}_n\|_\$ = h \max_{1 \le l \le n} \left| \frac{l\xi_2^{l+1}}{\xi_2 - 1} - \frac{\xi_2^{l+1} - \xi_2}{(\xi_2 - 1)^2} \right| = h \max_{1 \le l \le n} \left| \frac{l(-1)^l}{2} - \frac{(-1)^{l+1} + 1}{4} \right| \ge$$

$$\frac{h(n-1)}{2} \to \frac{1}{2}.$$

Else $\xi_2 = e^{i\varphi}$ with $0 < \varphi < \pi$ and then $\xi_3 = e^{-i\varphi}$. The right side can be estimated similarly as before:

$$\left\| \prod_{i=2}^{k} (\mathbf{I} - \xi_i \mathbf{H}_n) \, \mathbf{w}_n \right\|_\$ \le \left( \prod_{i=4}^{k} \|(\mathbf{I} - \xi_i \mathbf{H}_n)\|_\$ \right) \|(\mathbf{I} - \xi_2 \mathbf{H}_n)(\mathbf{I} - \xi_3 \mathbf{H}_n) \, \mathbf{w}_n\|_\$ \le$$

$$2^{k-2} \left\| \left( \mathbf{I} - 2\cos\varphi \mathbf{H}_n + \mathbf{H}_n^2 \right) \mathbf{w}_n \right\|_\$.$$

Now, let us choose $w_m = m \, \Re \xi_2^m$, where $\Re$ is the notation for the real part.

$$\left( \left( \mathbf{I} - 2\cos\varphi \mathbf{H}_n + \mathbf{H}_n^2 \right) \mathbf{w}_n \right)_m = \begin{cases} \cos\varphi, & \text{if } m = 1 \\ \cos m\varphi - \cos(m-2)\varphi, & \text{if } m \ge 2 \end{cases}$$

thus its norm

$$h \max_{1 \le l \le n} \left\{ |\cos\varphi|, |\cos l\varphi + \cos(l-1)\varphi - 1| \right\} \to 0,$$

as $h \to 0$, while $\xi_2^l$ is either periodic with period $\ge 3$ or dense on the unit circle which means that $\exists c > 0$ such that

$$\|\mathbf{w}_n\|_\$ = h \max_{1 \le l \le n} \left| \Re \left( \frac{l\xi_2^{l+1}}{\xi_2 - 1} - \frac{\xi_2^{l+1} - \xi_2}{(\xi_2 - 1)^2} \right) \right| \ge$$

$$h \max_{1 \le l \le n} l \left| \Re \left( \frac{\xi_2^{l+1}}{\xi_2 - 1} \right) \right| - \frac{2h}{|\xi_2 - 1|^2} > c,$$

if $n$ is large enough. This proves the statement.

## 4     Concluding Discussion

We conclude the paper adding a critical remark. Although Theorem 2 clearly showed the difference between weakly and strongly stable LMMs the practical side of this result is not clear at all. Stability is only a partial achievement, no doubt an important one, however, we are mostly interested in the convergence of methods.

Simply speaking the problem is the following. A weakly stable method is stable in the norm pair $(\|\cdot\|_{k\infty}, \|\cdot\|_{k\infty})$ resulting convergence in the norm $\|\cdot\|_{k\infty}$. For a strongly stable method we can obtain convergence in the same norm not depending on which type of stability $((\|\cdot\|_{k\infty}, \|\cdot\|_{k\infty})$ or $(\|\cdot\|_{k\infty}, \|\cdot\|_{k\$}))$ we use.

The profit is shifted to the consistency check. Note that consistency in the norm $\|\cdot\|_{k\infty}$ with order $m$ implies consistency in the norm $\|\cdot\|_{k\$}$ with the same order $m$ or higher. This means that for strongly stable methods we have the freedom to check consistency in the $\|\cdot\|_{k\$}$ norm. It is a technical gain, see the tricky example [6, Example 1 in Sect. 2.2.4]:

$$
(F_n(\mathbf{u}_n))_i = \left\{
\begin{array}{ll}
u_0 - c_0 & , \text{if } i = 0, \\
\dfrac{u_i - u_{i-1}}{h} - f_{i-1} & , \text{if } 1 \leq i \leq n \quad \text{odd}, \\
\dfrac{u_i - u_{i-1}}{h} - f_i & , \text{if } 2 \leq i \leq n \quad \text{even}.
\end{array}
\right.
$$

This one-step method is consistent of order 2 with respect to the $\|\cdot\|_{1\$}$ norm. To get consistency of order 2 with respect to the $\|\cdot\|_{1\infty}$ norm is less straightforward (however, it is possible).

Consequently, this freedom could be a technical gain. Unfortunately, not more, we can not win an order this way.

# References

1. Ascher, U.M., Petzold, L.R.: Computer Methods for Ordinary Differential Equations and Differential-Algebraic Equations. SIAM, Philadelphia (1998)
2. Faragó, I., Mincsovics, M.E., Fekete, I.: Notes on the basic notions in nonlinear numerical analysis. Electron. J. Qual. Theory Diff. Equat. **6**, 1–22 (2012). Proc. 9th Coll. Qualitative Theory of Diff. Equ. 2011
3. Mincsovics, M.E.: Stability of one-step and linear multistep methods - a matrix technique approach. Electron. J. Qual. Theory Diff. Equat. **15**, 1–10 (2016). https://doi.org/10.14232/ejqtde.2016.8.15. Proc. 10th Coll. Qualitative Theory of Diff. Equ. 2015
4. Mincsovics, M.E.: Note on the stability of strongly stable linear multistep methods. In: AIP Conference Proceedings, vol. 1895, no. 1, p. 110006 (2017). https://doi.org/10.1063/1.5007412
5. Spijker, M.N.: Stability and convergence of finite-difference methods. Thesis, University of Leiden (1968)
6. Stetter, H.J.: Analysis of Discretization Methods for Ordinary Differential Equations. Springer, Heidelberg (1973). https://doi.org/10.1007/978-3-642-65471-8

# Numerical Estimation of Seismic Wave Attenuation in Fractured Porous Fluid-Saturated Media

Mikhail Novikov$^{(\boxtimes)}$, Vadim Lisitsa⬨, and Tatiana Khachkova

IPGG SB RAS, Koptug ave. 3, Novosibirsk 630090, Russia
NovikovMA@ipgg.sbras.ru
http://www.ipgg.sbras.ru/ru

**Abstract.** Study of wave-induced fluid flow (WIFF) became actual in geophysics last years, because attenuation caused by this effect can serve as indicator of fractured highly-permeable reservoirs. In our work we model two-scale fractured domains with small scale fractures forming percolating clusters. Statistical geometry analysis and numerical wave propagation simulations using finite-difference approximation of Biot's dynamic equations were done to estimate seismic attenuation and investigate the dependence of attenuation due to WIFF on percolation length. Theoretical predictions of at tenuation due to scattering are also provided. Obtained estimations demonstrate sufficient correlation between fracture connectivity and attenuation of waves propagating in considered fractured media.

**Keywords:** Wave-induced fluid flow · Seismic attenuation ·
Finite difference method · Biot model

## 1 Introduction

Seismic attenuation in porous fractured fluid-saturated media is under active study in modern geophysics. One of the attenuation mechanisms is the so-called wave-induced fluid flow (WIFF), caused by pressure gradients at the interface between highly permeable fracture-filling material and background rock when seismic waves propagate in the media. There are two types of such phenomena considered. The first type is fluid flow between background rock and fracture-filling material (fracture-to-background WIFF, FB-WIFF), which is mainly characterized by properties of fluid and background as well as fracture geometry and orientation. The second type is flow within connected fractures (fracture-to-fracture WIFF, FF-WIFF). FF-WIFF mostly depends on fracture connectivity and fracture-filling material properties. Estimations of frequency-dependent attenuation can provide information about transport properties of fractured media and fluid mobility and thus serve for detection of fluid-filled highly permeable reservoirs.

I. Dimov et al. (Eds.): FDM 2018, LNCS 11386, pp. 362–369, 2019.
https://doi.org/10.1007/978-3-030-11539-5_41

Recent WIFF studies involves includes different approaches such as theoretical investigations [4], numerical upscaling with quasi-static oscillatory tests [10] and full waveform numerical simulation [9]. One principal interest in WIFF study is link between fracture connectivity and frequency-dependent anisotropy at seismic and acoustic wavefields scale. Unfortunately, most studies involves quite simple fractured media models. For example, connectivity of fractures can be provided by pairwise intersection of fractures from two differently aligned sets. This structure, obviously, provides only local interconnection of fractures without forming secondary scale intersected fracture clusters, strongly limiting their ability to study FF-WIFF influence. We are only aware of one attempt reported in [5] to extend FF-WIFF investigations by using a homogeneous distribution of the fracture orientation and study of attenuation dependence of both fractures length and number of intersections.

In our study we generate models of fractured media using a simulated annealing approach, including percolation in the functional under maximization. Starting from two perpendicularly oriented sets of aligned small-scale fractures distributed uniformly in the domain, we apply simulated annealing to form fracture clusters with prescribed percolation length. We do statistical geometrical and topological analysis of several obtained percolation stages. Also we use constructed models for numerical wave propagation modelling. Wave propagation simulations are performed using finite-difference approximation of dynamic Biot's poroelasticity system. The resulting wavefields are used to estimate seismic attenuation due to wave-induced fluid flow. Estimations show that the mesoscale structure of percolating clusters has the most impact on the attenuation, and small-scale fractures mostly affect scattering.

## 2    Generation of Discrete Fracture Network

For statistical model generation a simulated annealing technique has been applied. We considered two sets of fractures, parallel and perpendicular to wave propagation direction. The fractures' length is 30 mm, and width is 4 mm. In our initial model fractures have uniform distribution. As objective function we consider the connectivity index [12] - i.e., probability of percolation within a $250 \times 250$ mm window in the whole fractured domain. Several constructed models with different percolation probabilities ("stages") are presented in Fig. 1. One can observe the formation of mesoscale percolating structures while percolation increases. We generated 10 realisations in total. Each realisation consist of 6 different percolation stages. Note that percolation probability varies with equidistant step between stages.

We then applied a technique similar to the one used for handwriting recognition to recover the medial axis of fracture clusters. Medial axis was divided in a set of simple polygonal chains ("branches"). Geological and topological analysis was performed for model structure characterization. We obtained average number of branches, length of branches, distance between its endpoints, integral curvature and tortuosity of branches in dependence of branch orientation. Results are showed in Fig. 2.

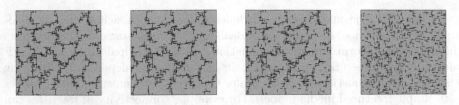

**Fig. 1.** Fragments of fractured domain models. Percolation decreases from left to right.

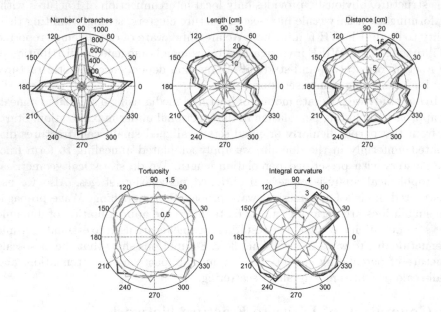

**Fig. 2.** The statistical parameters of the mesoscale structures in the following order: number of branches oriented at particular direction, mean length of the branches, mean distance between the branch endpoints, mean tortuosity and mean integral curvature with respect to branch orientation. Colors correspond to different percolation stages in the descending order: black, blue, green, red, yellow, pink. (Color figure online)

Most branches are oriented along coordinate directions, which is provided by orientation of small-scale fractures. Nonetheless, length distribution makes it clear that the longest branches are oriented at ±45°. These branches define main directions in which percolation is delivered. Note, that the increase of the considered geometrical parameters, except tortuosity, is more intense at low percolation stages, while stages with relatively high percolation result in weak variation of these parameters. Finally, tortuosity remains almost the same for all considered stages.

# 3    Numerical Modeling of Wave Propagation

## 3.1    Numerical Setup

To investigate the fracture connectivity effect on seismic attenuation we perform plane fast P-wave propagation simulations in our fractured media. We consider both background and fractured-filling materials to be porous and fluid-saturated. The dynamic Biot poroelastic equations [1,2] were solved numerically using an explicit second-order finite-difference scheme in a staggered grid [3,6–8]. Rock within fractures is much more permeable than background rock and also boasts a higher porosity. The solid materials' physical properties as well as fluid properties are shown at Table 1.

**Table 1.** Material properties.

| Parameter | Background | Fractures |
|---|---|---|
| Fluid properties | | |
| Bulk modulus $[GPa]$ | 2.25 | |
| Density $[kg/m^3]$ | 1090 | |
| Viscosity $[Pa \cdot s]$ | 0.001 | |
| Mineral phase | | |
| Bulk modulus $[GPa]$ | 37 | |
| Shear modulus $[GPa]$ | 44 | |
| Density $[kg/m^3]$ | 2650 | |
| Dry rock | | |
| Bulk modulus $[GPa]$ | 26 | 0.02 |
| Shear modulus $[GPa]$ | 31 | 0.01 |
| Density $[kg/m^3]$ | 2385 | 1325 |
| Porosity [%] | 10 | 50 |
| Fluid-saturated rock | | |
| Bulk modulus $[GPa]$ | 27.8 | 4.3 |
| Shear modulus $[GPa]$ | 31 | 0.01 |
| Density $[kg/m^3]$ | 2494 | 1870 |

We consider a rectangular computational domain containing fractured layers as schematically showed in Fig. 3. Top and bottom boundary conditions are periodic. At the sides we construct perfectly matched layers to eliminate reflections. Input signal is a Ricker wavelet. We simulate propagation of waves with dominant frequencies varying from 1 to 10 kHz with 1 kHz step. Waves start propagating from a source line, and traces are recorded at two receiver lines before and after the fractured layer. Note, that simulations were done using all 10 realisations of fracture models including 6 percolation stages each.

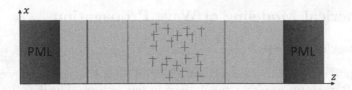

**Fig. 3.** Schematic illustration of the computational domain. The red line corresponds to the source positions, the green lines correspond to the receiver positions. (Color figure online)

## 3.2 Results

Figure 4 shows resulting fluid pressure snapshots for different percolation stages and wave peak frequency of 3 kHz. At all stages fluid flow between background rock and fracture-filling material occurs, caused mostly by high permeability contrast. However, fluid flow intensity clearly depends on the fracture connectivity. An increase in connectivity causes pressure gradients to grow between fractures and background, which results in a stronger fracture-to-background fluid flow. Thus, more intense attenuation of propagating seismic waves is observed for models with higher connectivity. Besides, one can see high fluid pressure at the fractures for highly connected ones. Comparing to background rock, fluid mobility at the fractures due to material permeability is enough to distribute freely, so fracture-to-fracture flows within mesoscale fracture clusters occur.

Also signals were recorded at two receiver lines for each of 6 considered connectivity stages. Traces for each realisation and also averaged ones are shown in Fig. 5. Here we also can clearly see a significant attenuation increase with connectivity for highly connected fractures. Higher structure connectivity causes much stronger attenuation, especially in comparison with almost nonintersecting fractures at the first stage and the highest connectivity case. Moreover, the difference between realisations is almost negligible, which testifies that the considered size of domain is representative.

Next, the spectral ratio technique was applied to recorded signals to estimate frequency-dependent wave attenuation. By this approach one can obtain estimates for both dispersion and frequency-dependent attenuation within the frequency range of approximately $[0.5\nu_0, 2\nu_0]$, where $\nu_0$ is the peak frequency of initial signal. As we performed wave propagation simulation considering central frequencies from 1 to 10 kHz with 1 kHz step, our considered range of frequencies is fully covered. Moreover, averaged scattering attenuation estimates for all considered stages were obtained using effective cross-section evaluation [11].

Figure 6 illustrates both attenuation estimations obtained for numerical experiments and theoretical estimates of attenuation due to scattering in our considered frequency range. Results show, that a percolation length increase results in an attenuation increase. Again, one can readily see that this attenuation increase between different connectivity stages is more intense when the percolation length in our domain is relatively small. For highly connected

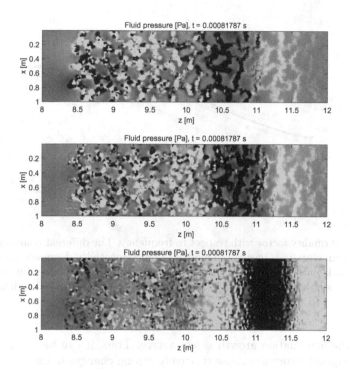

**Fig. 4.** Snapshots illustrating fluid pressure for different stages of fracture connectivity. Fracture connectivity decreases from top to bottom.

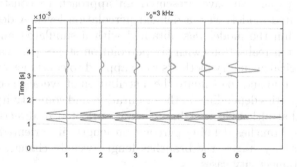

**Fig. 5.** Signals recorded at two receiver arrays. The Blue line represents the signal before interaction with fractures, the red lines correspond to the signal after interaction with the fractures. The connectivity of the structures decreases from the left to the right. (Color figure online)

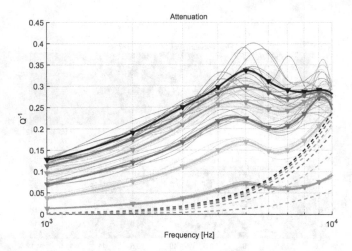

**Fig. 6.** Inverse quality factor with respect to frequency. The different colors correspond to different connectivity stages of the models. The connectivity decreases from black to pink. The thick lines correspond to average over realisations, the thin lines represent estimation for single realisation. The dashed lines represent theoretical estimates of the attenuation due to the scattering. (Color figure online)

fractures, the attenuation growth is less active. Thus, it can be concluded, that the small original structures cause relatively strong changes in the average structure. For already highly percolated models the relative changes in fractured structure geometry are smaller.

## 4    Conclusion

In the present paper, we have presented an approach to construct complex mesoscale fractured models with a defined percolation level. A desired connectivity index within the model was obtained with a simulated annealing technique. In total, ten realisations with six percolation stages per realisation were modeled. Numerical geometry methods were applied to construct medial axis for each fracture system and to estimate the distribution of average geometrical and topological properties depending on the orientation and connectivity. The results demonstrated that although most branches are oriented in directions of small-scale structures, branches defining percolation length are oriented along ±45°. At high connectivity stages, the branches length weakly changes compared to relatively low connectivity cases.

Wave propagation numerical simulations were performed to investigate the fracture connectivity influence on seismic attenuation for fractured porous fluid-saturated media. Results of numerical modeling and theoretical predictions show that mesoscale structures size is the main parameter affecting the attenuation due to WIFF and scattering. An additional statistical analysis of fracture systems compared to numerical results demonstrates strong correlation between percolation length and seismic wave attenuation. Wavefield snapshots and registered traces show, that the increase of percolation length causes pressure gradient growth between fractures and background rock, resulting in an increase of attenuation due to FB-WIFF and FF-WIFF. However, further study of relations between percolation and geometrical characteristics of the fracture network is needed in order to be performed to make seismic attenuation approaches applicable to the estimation of fractured reservoir transport properties.

**Acknowledgments.** This research was supported by Russian Foundation for Basic Research grants no. 18-05-00031, 18-01-00579, 16-05-00800. The computations were performed using supercomputer "Lomonosov" of Moscow State University and cluster NKS-30T+GPU of the Siberian supercomputer center.

# References

1. Biot, M.A.: Theory of propagation of elastic waves in fluid-saturated porous solid. I. Low-frequency range. J. Acoust. Soc. Am. **28**, 168–178 (1956)
2. Biot, M.A.: Theory of propagation of elastic waves in fluid-saturated porous solid. II. Higher frequency range. J. Acoust. Soc. Am. **28**, 179–191 (1956)
3. Carcione, J.M.: Computational poroelasticity - a review. Geophysics **75**(5), 1264–1276 (2010)
4. Guo, J., et al.: Effects of fracture intersections on seismic dispersion: theoretical predictions versus numerical simulations. Geophys. Prospect. **65**(5), 1264–1276 (2017)
5. Hunziker, J., et al.: Seismic attenuation and stiffness modulus dispersion in porous rocks containing stochastic fracture networks. J. Geophys. Res.: Solid Earth **123**(1), 125–143 (2018)
6. Masson, Y.J., Pride, S.R., Nihei, K.T.: Finite difference modeling of Biot's poroelastic equations at seismic frequencies. J. Geophys. Res.: Solid Earth **111**(B10), 305 (2006)
7. Masson, Y.J., Pride, S.R.: Poroelastic finite difference modeling of seismic attenuation and dispersion due to mesoscopic-scale heterogeneity. J. Geophys. Res.: Solid Earth **112**(B03), 204 (2007)
8. Masson, Y.J., Pride, S.R.: Finite-difference modeling of Biot's poroelastic equations across all frequencies. Geophysics **75**(2), N33–N41 (2010)
9. Novikov, M., et al.: Numerical study of fracture connectivity response in seismic wavefields. In: SEG Technical Program Expanded Abstracts 2017, pp. 3786–3790. Society of Exploration Geophysicists, Tulsa (2017)
10. Rubino, J.G., et al.: Seismoacoustic signatures of fracture connectivity. J. Geophys. Res.: Solid Earth **119**(3), 2252–2271 (2014)
11. Rytov, S.M., Kravtsov, Y.A., Tatarskii, V.I.: Principles of Statistical Radiophysics 2. Correlation Theory of Random Processes. Springer, Heidelberg (1988)
12. Xu, C., et al.: A connectivity index for discrete fracture networks. Math. Geol. **38**(5), 611–634 (2006)

# Relativistic Burgers Models on Curved Background Geometries

Baver Okutmustur[✉]

Department of Mathematics, Middle East Technical University (METU),
06800 Ankara, Turkey
baver@metu.edu.tr

**Abstract.** Relativistic Burgers model and its generalization to various spacetime geometries are recently studied both theoretically and numerically. The numeric implementation is based on finite volume and finite difference approximation techniques designed for the corresponding model on the related geometry. In this work, we provide a summary of several versions of these models on the Schwarzschild, de Sitter, Schwarzschild-de Sitter, FLRW and Reissner-Nordström spacetime geometries with their particular properties.

## 1 Introduction

The theory of derivation of relativistic type Burgers models on curved spacetimes was started and improved by LeFloch and collaborators [2–5]. The model was firstly derived on a flat spacetime both from the Euler system and from a hyperbolic balance law satisfying the Lorentz invariance property [5]. This analysis has recently been extended to the Schwarzschild, de Sitter (dS), Schwarzschild-de Sitter (SdS), FLRW and Reissner-Nordström (RN) spacetimes and examined numerically by means of finite volume and finite difference approximations. The current work provides a summary of [2–5,10] on different spacetime geometries. We are interested in compressible fluids developing on a curved background. The fluid under consideration may include shock/rarefaction waves and we study a class of weak solutions into the Euler system on the given geometry. The Levi-Civita connection is denoted by covariant derivative $\nabla$. It follows that, the Euler equations for a compressible fluid on a curved spacetime is

$$\nabla_\alpha\big(T^{\alpha\beta}(\rho,u)\big) = \nabla_\alpha\big(\rho c^2 u^\alpha u^\beta + p(\rho)\big(u^\alpha u^\beta + g^{\alpha\beta}\big)\big) = 0, \qquad (1)$$

where $T^{\alpha\beta}(\rho,u)$ is the energy-momentum, $\rho$ is the mass-energy density, $u = (u^\alpha)$ is its unit velocity field, $c > 0$ is the light speed, $p$ is the pressure. Further details on energy-momentum tensor and perfect fluids can be found in the articles [1,5,11]. The simplest form of these equations so called the Euler system of compressible fluids on the flat background reads

$$\partial_t \rho + \partial_x(\rho v) = 0, \qquad (2)$$
$$\partial_t(\rho v) + \partial_x(\rho v^2 + p(\rho)) = 0. \qquad (3)$$

© Springer Nature Switzerland AG 2019
I. Dimov et al. (Eds.): FDM 2018, LNCS 11386, pp. 370–377, 2019.
https://doi.org/10.1007/978-3-030-11539-5_42

It can easily be checked that, imposing a vanishing pressure to this system and taking a combination of these equations, one can derive the inviscid Burgers equation We refer [2–5] for this derivation and for further details.

A relativistic generalization of the Burgers model on flat and Schwarzschild spacetimes are proposed in [5] and a numerical scheme was designed by finite volume approximation resulting weak solutions with shock waves. This analysis was extended to SdS spacetime in [2], dS spacetime in [3], FLRW spacetime in [4] and RN spacetime in [10]. The metric elements of each of these geometries describe a solution to the Einstein's field equations. We cite [1,6,8] for convergence and geometric formulation of finite volume methods on Lorentzian spacetimes. More details on the RN metric and its properties can be found in [7,9]. The reader can find general instruction for the general relativity theory and related topics in [11].

The outlook of this work is as follow. Firstly, we give some basic information about spacetime geometry. Then Schwarzschild, dS, SdS, FLRW and RN spacetime metrics and their particular properties are presented, respectively. Relativistic Burgers models for each background are also given in this part. We next describe a general approach for derivation of the model on any background. The final part is dedicated to description of a general geometric finite volume scheme on a curved background and it ends up with some concluding remarks.

## 2   Spacetime and Metric

In general relativity, different than Newton's theory, space and time are a single continuum as spacetime. By the relativistic point of view, there is no well-defined construct of two distinct events happening at the same time. For this reason, there is a light cone defined at any event that is the location of paths through spacetime. A spacetime is illustrated by an $(n+1)$ dimensional Lorentzian geometry, where $n$ describes dimensions of space and 1 refers dimension of time. The sign of its spherically symmetric metric is denoted by $(-, +, \cdots, +)$. If we restrict the dimension $(n+1)$ to a particular dimension $(3+1)$, then a general spherically symmetric metric dimension will be of the form

$$g = g_{00} \ dt^2 + g_{11} \ dr^2 + g_{22} \ d\theta^2 + g_{33} \ d\phi^2.$$

It follows that the line element for a $(3+1)$ dimensional form in terms of time $t$, the radial $r$ and angular coordinates $\theta$ and $\varphi$ can be written by the formula

$$g = -A(t,r) \ dt^2 + B(t,r) \ dr^2 + r^2(d\theta^2 + \sin^2\theta \ d\phi^2) \qquad (4)$$

with nonzero covariant elements $g_{00} = -A(t,r)$, $g_{11} = B(t,r)$, $g_{22} = r^2$, $g_{33} = r^2\sin^2\theta$; and the corresponding contravariant elements $g^{00} = -\frac{1}{A(t,r)}$, $g^{11} = -\frac{1}{B(t,r)}$, $g^{22} = \frac{1}{r^2}$, $g^{33} = \frac{1}{r^2(\sin^2\theta)}$. Here $A(t,r)$ and $B(t,r)$ are functions depending on $t$ and $r$ variables. In the following, we introduce some of the well-known spacetime geometries having spherically symmetric metric elements and corresponding relativistic Burgers models on these geometries.

## 2.1  Minkowski Metric

This spacetime is also called as flat spacetime. We consider a $(3+1)$ dimensional coordinate system given by $(x^0, x^1, x^2, x^3) = (t, x, y, z)$ where $(x^0, x^1, x^2) = (x, y, z)$ and $(x^0) = (t)$ are spatial and time components, respectively. It follows that, the metric of a $(3+1)$ dimensional Minkowski spacetime is

$$g = -c^2 dt^2 + dx^2 + dy^2 + dz^2$$

where $c$ is the light speed. In usual spherical coordinates $r, \theta, \varphi$, it becomes

$$g = -c^2 dt^2 + dr^2 + r^2 d\theta^2 + r^2 \sin^2 \theta d\varphi^2. \tag{5}$$

**Burgers Model on Flat Geometry.** The relativistic Burgers equation on flat spacetime can be derived either by Lorentz invariance property or by the Euler system on a curved spacetime. The proofs of both derivation methods can be found in [5]. The relativistic Burgers equation on flat background is

$$\partial_t v + \partial_r \big(1/\epsilon^2 \big(-1 + \sqrt{1 + \epsilon^2 v^2}\big)\big) = 0, \quad \epsilon = 1/c. \tag{6}$$

## 2.2  Schwarzschild Metric

The Schwarzschild spacetime describes the gravitational field of the universe and defines a spherically symmetric black hole solution to the Einstein's field equations. Its metric is represented by

$$g = -\Big(1 - \frac{2M}{r}\Big)c^2 dt^2 + \Big(1 - \frac{2M}{r}\Big)^{-1} dr^2 + r^2(d\theta^2 + \sin^2 d\varphi^2) \tag{7}$$

where $M > 0$ is the mass parameter, $r$ is the radial spatial coordinate. The sign of the quantity $M$ and $r$ have significant effects on the geometry.

- If $M = 0$, the Schwarzschild metric reduces to the Minkowski metric.
- If $M \neq 0$, it is singular for $r = 0$ and has a coordinate singularity for $r = 2M$. That is, $g_{00}$ vanishes and $g_{11}$ becomes infinite, so the equation is not valid at that point. For $r < 2M$, $r \neq 0$, the Schwarzschild metric is a regular Lorentzian metric, but the timelike and spacelike behaviors of the coordinates $t$ and $r$ are interchanged, i.e., the Schwarzschild metric in standard coordinates is again a smooth Lorentzian metric, but $t$ is a space coordinate while $r$ is a time coordinate. If $r > 2M > 0$, the metric is a regular Lorentzian metric with $t$ timelike and $r$ spacelike. If $r = 2M$, the Schwarzschild metric with $M > 0$ is no more a smooth Lorentzian metric.

**Burgers Model on Schwarzschild Geometry.** According to the paper [5], relativistic Burgers model on Schwarzschild background is

$$\partial_t(v) + \Big(1 - \frac{2M}{r}\Big)\partial_r\big(\frac{v^2}{2}\big) - \frac{M}{r^2}(v^2 - c^2) = 0. \tag{8}$$

It can easily be observed that, if $M = 0$, we get the inviscid Burgers equation.

## 2.3  De Sitter Metric

The de Sitter (dS) spacetime is a particular background of the Lorentzian spacetime and its metric describes a cosmological solution to the Einstein's field equations. The metric element contains a cosmological constant $\Lambda$. If $\Lambda > 0$, the
background geometry is called the de Sitter spacetime; if $\Lambda < 0$, the background
geometry is called the Anti-de Sitter spacetime. Particularly, if $\Lambda = 0$, the metric turns to be a Minkowski metric and hence we get a flat geometry. In $(3 + 1)$
dimension, this metric is

$$g = -(1 - \Lambda r^2)dt^2 + \frac{1}{1 - \Lambda r^2}dr^2 + r^2(d\theta^2 + \sin^2\theta d\varphi^2). \qquad (9)$$

It is easy to verify that for $\Lambda = 0$, it becomes the Minkowski metric.

**Burgers Model on the dS Geometry.** Following the paper [3], the relativistic
Burgers equation on a dS background is

$$\partial_t v + (1 - \Lambda r^2)\,\partial_r\left(\frac{v^2}{2}\right) + \Lambda r(v - c^2 - 2v^2) = 0. \qquad (10)$$

Substituting $\Lambda = 0$ in this equation, we recover the inviscid Burgers equation.

## 2.4  Schwarzschild-de Sitter (SdS) Metric

The SdS geometry is a spherically symmetric solution to the Einstein's field
equations. Its metric is a composition of Schwarzschild and dS metrics. In a
$(3 + 1)$ dimensional spherical coordinates, this metric is given by

$$g = -\left(1 - \frac{2M}{r} - \frac{\Lambda r^2}{3}\right)dt^2 + \left(1 - \frac{2M}{r} - \frac{\Lambda r^2}{3}\right)^{-1} dr^2 + r^2(d\theta^2 + \sin^2\theta d\varphi^2), \quad (11)$$

where $M > 0$ is the mass parameter, $\Lambda$ is the cosmological constant, and $c$ is the
light speed. It can be observed that whenever $\Lambda = 0$, the SdS metric reduces to
the Schwarzschild metric. If the mass parameter $M = 0$, the metric reduces to
the dS metric. If both $\Lambda = M = 0$ then it turns to be the Minkowski metric.

**Burgers Model on SdS Geometry.** According to the article [2], the relativistic Burgers equation on SdS geometry is

$$\partial_t v + (1 - \frac{2M}{r} - \frac{\Lambda r^2}{3})\partial_r\left(\frac{v^2}{2}\right) = \frac{Mv^2}{r^2} - \frac{\Lambda rv^2}{3} - \frac{mc^2}{r^2} + \frac{\Lambda rc^2}{3}. \qquad (12)$$

Substituting $\Lambda = 0$ in this model yields the Burgers model on the Schwarzschild
background. Moreover, if $M = 0$ the model reduces to the model on the dS
spacetime. Finally, if $\Lambda = M = 0$, then it gives the inviscid Burgers equation.

## 2.5   Friedmann–Lemaitre–Robertson–Walker (FLRW) Metric

The FLRW metric is a solution to the Einstein's field equations and is given by

$$g = -c^2 dt^2 + a(t)^2 \left( \frac{dr^2}{1 - kr^2} + r^2 d\theta^2 + r^2 \sin^2 \theta d\varphi^2 \right) \tag{13}$$

where $t$ is the time, $c$ is the light speed, $k = \{-1, 0, 1\}$ is the curvature, $a(t)$ is the cosmic expansion factor and $r$, $\theta$, $\varphi$ are the spherical coordinates.

**Burgers Model on a FLRW Geometry.** Following the paper [4], the relativistic Burgers equation on a FLRW background is

$$a\, v_t + \left(1 - kr^2\right)^{1/2} \partial_r \left( \frac{v^2}{2} \right) + v \left( 1 - \frac{v^2}{c^2} \right) a_t = 0. \tag{14}$$

For $k = -1, 0, 1$, we get three different models of interest on FLRW background.

## 2.6   Reissner-Nordström (RN) Metric

The RN spacetime is a spherically symmetric solution to the Einstein's field equations. The main difference between RN and Schwarzschild blackholes is that RN spacetime is electrically charged with an electrically charge term $Q$. The corresponding metric is described by

$$g = -\left(1 - \frac{2M}{r} + \frac{Q^2}{r^2}\right) dt^2 + \left(1 - \frac{2M}{r} + \frac{Q^2}{r^2}\right)^{-1} dr^2 + r^2 (d\theta^2 + \sin^2 \theta d\varphi^2). \tag{15}$$

RN metric becomes Minkowski metric in polar coordinates at very large radius $r$. On the other hand, if $Q = 0$, it turns to be a Schwarzschild metric. RN metric is smooth and Lorentzian under the condition that

$$\frac{2M}{r} - \frac{Q^2}{r^2} < 1.$$

Event horizon is located at where $g^{11} = 0$, that is, $1 - \frac{2M}{r} + \frac{Q^2}{r^2} = 0 \Rightarrow r^2 - 2Mr + Q^2 = 0$ with roots $r_\pm = M \pm \sqrt{M^2 - Q^2}$. Depending on the relation between $M$ and $Q$, we get more information about the event horizon and the geometry.

- If $M^2 < Q^2$, $r_\pm$ are not real and $g^{11}$ is positive except at $r = 0$ where there is a singularity. As $g^{11} > 0$, $r$ is spacelike coordinate. At $r = 0$ we have timelike line. There is no event horizon for this case and this solution is non-physical.
- If $M^2 > Q^2$, there are three regions
  **1st region** ($r_+ < r < \infty$): In this region $g^{11} > 0$. Event horizon is the surface defined by $r = r_+$. The singularity at $r = 0$ is timelike line.
  **2nd region** ($r_- < r < r_+$): If we set $r = r_+ - \delta$ then with the condition $r_+ > M$, we get $g^{11} = 1 - \frac{2M}{r_+ - \delta} + \frac{Q^2}{(r_+ - \delta)^2} \Rightarrow 2\delta(-Mr_+ + Q^2) < 0$.
  **3rd region** ($0 < r < r_-$): If we set $r = r_- - \delta$ then with the condition $r_- < M$, we get $g^{11} = 1 - \frac{2M}{r_- - \delta} + \frac{Q^2}{(r_- - \delta)^2} \Rightarrow 2\delta(-Mr_- + Q^2) > 0$.

– If $M^2 = Q^2$: This case is known as the extreme Reissner-Nordström solution. Event horizon is $r_+ = r_- = M$ and $g^{11} = 0$ at $r = r_\pm$.

We address the reader to the articles [7,9,10] for further detail.

**Burgers Model on RN Geometry.** According to the paper [10], the derived model is

$$\partial_t(v) + \partial_r\left(\frac{v^2}{2}(1 - \frac{2M}{r} + \frac{Q^2}{r^2})\right) = \left(2v^2 - 1\right)\left(\frac{M}{r^2} - \frac{Q^2}{r^3}\right). \qquad (16)$$

Note that if $Q = 0$, it yields the Burgers model on the Schwarzschild geometry. Moreover, if both $M = Q = 0$, then the classical Burgers equation is recovered.

### 2.7   Derivation of the Burgers Models

A general approach to derive the Burgers model for a given metric element is briefly introduced in this part. We take a metric of the general form (4) and consider the Christoffel symbols given by

$$\Gamma^\mu_{\alpha\beta} = \frac{1}{2}g^{\mu\nu}(-\partial_\nu g_{\alpha\beta} + \partial_\beta g_{\alpha\nu} + \partial_\alpha g_{\beta\nu}). \qquad (17)$$

We substitute $\alpha, \beta, \mu, \nu \in \{0, 1, 2, 3\}$ in (17) to obtain all the terms of $\Gamma^\mu_{\alpha\beta}$. Then by using the unit vector property of $u^\alpha$, we find a relation between $u^0, u^1$ and a velocity component $v$ depending on $u^0$ and $u^1$. It follows to substitute all these values into the energy momentum tensor for perfect fluids relation given by

$$T^{\alpha\beta} = (\rho c^2 + p)\, u^\alpha\, u^\beta + p\, g^{\alpha\beta}. \qquad (18)$$

We then obtain all the terms $T^{\alpha\beta}$ for $\alpha, \beta = 0, 1, 2, 3$. These values are plugged into the Euler system given by

$$\nabla_\alpha T^{\alpha\beta} = \partial_\alpha T^{\alpha\beta} + \Gamma^\alpha_{\alpha\gamma}T^{\gamma\beta} + \Gamma^\beta_{\alpha\gamma}T^{\alpha\gamma} = 0. \qquad (19)$$

As a result of this calculation, we obtain a system of two equations. Next we impose vanishing pressure to this system and take a suitable combination of both equations in order to write it in one equation form. This final equation is the desired Burgers model depending on the given geometry [2–5,10].

## 3   Finite Volume Method (FVM) Formulation

This part is based on the papers [1–5]. We consider an $(n + 1)$–dimensional spacetime $M$ and a hyperbolic balance law given by

$$div\big(T(v)\big) = S(v), \quad v : M \to \mathbb{R}, \qquad (20)$$

where $v$ is a scalar field, $div(\cdot)$ is the divergence operator, $T(v)$ is the flux vector field and $S(v)$ is the scalar field. We establish the FVM by averaging (20) over each element of the constructed triangulation. For convergence of the scheme on curved manifolds and assumptions on triangulation we refer the articles [1,5].

In local coordinates, we suppose that the spacetime is described in coordinates $(t, r)$ and consider equally spaced cells $I_j = [r_{j-1/2}, r_{j+1/2}]$ of size $\Delta r$, centred at $r_j$, with $r_{j+1/2} = r_{j-1/2} + \Delta r$, and $r_{j-1/2} = j\Delta r$, $r_j = (j+1/2)\Delta r$. Next we rewrite (20) in $(1+1)$ dimension

$$\partial_t T^0(t, r) + \partial_r T^1(t, r) = S(t, r), \tag{21}$$

with $T^0, T^1$ are flux fields, and $S$ is the source term. Integrating (21) over each grid cell $[t_n, t_{n+1}] \times [r_{j-1/2}, r_{j+1/2}]$ yields

$$\int_{r_{j-1/2}}^{r_{j+1/2}} T^0(t_{n+1}, r)\, dr = \int_{r_{j-1/2}}^{r_{j+1/2}} T^0(t_n, r)\, dr$$

$$- \int_{t_n}^{t_{n+1}} (T^1(t, r_{j+1/2}) - T^1(t, r_{j-1/2}))\, dt$$

$$+ \int_{[t_n, t_{n+1}] \times [r_{j-1/2}, r_{j+1/2}]} S(t, r)\, dt\, dr.$$

Approximate these terms by numerical fluxes

$$\widetilde{T}_j^n \approx \frac{1}{\Delta r} \int_{r_{j-1/2}}^{r_{j+1/2}} T^0(t_n, r)\, dr, \quad \widetilde{Q}_{j\pm 1/2}^n \approx \frac{1}{\Delta t} \int_{t_n}^{t_{n+1}} T^1(t, r_{j\pm 1/2})\, dt,$$

$$\widetilde{S}_j^n \approx \frac{1}{\Delta r\, \Delta t} \int_{[t_n, t_{n+1}] \times [r_{j-1/2}, r_{j+1/2}]} S(t, r)\, dt\, dr,$$

the scheme takes the form

$$\widetilde{T}_j^{n+1} = \widetilde{T}_j^n - \frac{\Delta t}{\Delta r} (\widetilde{Q}_{j+1/2}^n - \widetilde{Q}_{j-1/2}^n) + \Delta t \widetilde{S}_j^n. \tag{22}$$

The numerical implementation of the Burgers models via FVM are analyzed in the papers [2–5,10]. We address the reader to these works for further detail.

## 3.1   Concluding Remarks

- In [2–4,10], nonlinear Burgers models describing the propagation and interaction of shock waves on flat, Schwarzschild, dS, SdS, FLRW and RN spacetimes are studied and examined. Here we provide a review summary of these works.
- Depending on the geometry and the derived relativistic Burgers model equation, the finite volume/difference schemes are redesigned.
- The schemes are consistent with the conservative form of the Burgers models which results correct computations of weak solutions with shock/rarefaction waves.
- The convergence, efficiency and robustness of these schemes are numerically analyzed for each spacetime geometry of interest.

– One of the most obvious findings emerging from this study is that it allows us to make a comparison of the relativistic Burger models and numerical results on different spacetimes.

**Acknowledgments.** Supported by METU-GAP Project, Project no: GAP-101-2018-2767.

# References

1. Amorim, P., LeFloch, P.G., Okutmustur, B.: Finite volume schemes on Lorentzian manifolds. Commun. Math. Sci. **6**(4), 1059–1086 (2008)
2. Ceylan, T., Okutmustur, B.: Finite volume approximation of the relativistic Burgers equation on a Schwarzschild-(anti-)de Sitter spacetime. Turk. J. Math. **41**, 1027–1041 (2017)
3. Ceylan, T., Okutmustur, B.: Finite volume method for the relativistic Burgers model on a (1+1)-dimensional de Sitter spacetime. Math. Comput. Appl. **21**(2), 16 (2016)
4. Ceylan, T., LeFloch, P.G., Okutmustur, B.: A finite volume method for the relativistic Burgers equation on a FLRW background spacetime. Commun. Comput. Phys. **23**, 500–519 (2018)
5. LeFloch, P.G., Makhlof, H., Okutmustur, B.: Relativistic Burgers equations on a curved spacetime. Derivation and finite volume approximation. SIAM J. Numer. Anal. **50**(4), 2136–2158 (2012)
6. LeFloch, P.G., Okutmustur, B.: Hyperbolic conservation laws on spacetimes. A finite volume scheme based on differential forms. Far East J. Math. Sci. **31**, 49–83 (2008)
7. Nordebo, J.: The Reissner-Nordström metric. M.S. Dissertion, Department of Physics, Umea University, Switzerland (2016)
8. LeVeque, R.J.: Finite Volume Methods for Hyperbolic Problems, 1st edn. Cambridge University Press, Cambridge (2002)
9. Nashed, G.G.L.: Stability of Reissner-Nordström black hole. Acta Phys. Pol. **112**, 13–19 (2007)
10. Okutmustur, B.: Propagations of shock and rarefaction waves on the Reissner-Nordström spacetimes for Burgers models. SDU J. Nat. Appl. Sci. **22**(Spec. Issue), 448–459 (2018)
11. Wald, R.M.: General Relativity, 1st edn. The University of Chicago Press, Chicago (1984)

# Consistent Discrete-Analytical Schemes for the Solution of the Inverse Source Problems for Atmospheric Chemistry Models with Image-Type Measurement Data

Alexey Penenko[1,2]($\boxtimes$) , Vladimir Penenko[1,2] , Elena Tsvetova[1] , and Zhadyra Mukatova[1,2]

[1] Institute of Computational Mathematics and Mathematical Geophysics SB RAS, pr. Akademika Lavrentjeva 6, 630090 Novosibirsk, Russia
a.penenko@yandex.ru
[2] Novosibirsk State University, Pirogova str. 1, 630090 Novosibirsk, Russia

**Abstract.** The inverse source problems for nonlinear chemical transport models with image-type measurement data are considered. The use of the sensitivity operators, constructed of the ensemble of adjoint problem solutions, allows transforming the inverse problems stated as the systems of nonlinear ODE or PDE to a family of operator equations depending on the given set of functions in the space of measurement results. In the paper, the set of consistent discrete analytical schemes for 1D diffusion-reaction model is presented. The operator equations are solved with the relevant methods for nonlinear operator equations.

## 1 Introduction

The inverse source problems for nonlinear chemical transformation models with image-type measurement data are considered. These inverse problems arise when atmospheric composition monitoring data obtained as vertical profiles are assimilated in the forecast models.

The use of the sensitivity operators, constructed of the ensembles of adjoint problem solutions, allows transforming the inverse problems initially stated as the systems of nonlinear ODE or PDE to a family of operator equations depending on the given set of functions in the space of measurement results. The operator equation can be solved with the relevant methods for nonlinear ill-posed operator equations (e.g. the methods based on the truncated SVD or Newton-type methods).

For numerical solution, the discrete-analytical schemes for transport and transformation processes are applied. The schemes are constructed with the use of the locally-adjoint problems. Multidimensional problems are treated according to the splitting technique with respect to spatial dimensions and physical processes. The accuracy of the sensitivity operator calculation is assured by the

I. Dimov et al. (Eds.): FDM 2018, LNCS 11386, pp. 378–386, 2019.
https://doi.org/10.1007/978-3-030-11539-5_43

consistency of the numerical schemes for the direct and adjoint problems in the sense of Lagrange-type identities. The results of the numerical experiments are presented. The objective of the paper is to present the developed consistent original numerical schemes for the solution of the inverse problems as well as the results of their application.

## 2    The Inverse Problem Statement

A diffusion-reaction model for $l = 1, \ldots, N_c$ is considered in a domain $\Omega_T = \Omega \times (0, T)$, $\Omega = [0, X]$, $T > 0$, bounded by $\partial \Omega_T = \partial \Omega \times [0, T]$:

$$\frac{\partial \varphi_l}{\partial t} - \frac{\partial}{\partial x} \left( \mu_l \frac{\partial \varphi_l}{\partial x} \right) + P_l(t, \varphi) \varphi_l = \Pi_l(t, \varphi) + r_l, \quad (x, t) \in \Omega_T, \tag{1}$$

$$\varphi_l(0, t) = \alpha_l^L(t), \quad \varphi_l(X, t) = \alpha_l^R(t), \quad t \in (0, T), \tag{2}$$

$$\varphi_l = \varphi_l^0, \quad x \in \Omega, \ t = 0, \tag{3}$$

where $N_c$ is the number of considered substances, $\varphi_l(x, t)$ denotes the concentration of the $l^{th}$ substance at a point $(x, t) \in \Omega_T$, $\varphi$ is the vector of $\varphi_l$ for $l = 1, \ldots, N_c$. The functions $\mu_l$ correspond to the diffusion coefficients, $\alpha_l^L$, $\alpha_l^R$, $\varphi_l^0$ are boundary conditions and initial conditions, correspondingly, $r_l$ is a source function. Let $r \in F$, where $F$ is the set of admissible sources. Destruction and production operator elements $P_l, \Pi_l : [0, T] \times \mathbb{R}_+^{N_c} \to \mathbb{R}_+$ are defined by the chemical kinetics system. We suppose all the functions and model parameters to be smooth enough for the solutions to exist and the further transformations to make sense.

We introduce the discrete grid in time $\omega_t = \{t^j\}_{j=0}^{N_t-1} \subset [0, T]$ with intervals of length $\tau^{j+0.5}$, where $t^0 = 0$ and $t^{N_t-1} = T$. The spatial grid is $\omega_x = \{x_i\}_{i=0}^{N_x-1} \subset [0, X]$, $x_0 = 0$, $x_{N_x-1} = X$. To solve the problem (1)–(3) we will use the additive-averaged splitting scheme with respect to processes. Let $\sum_{\xi \in \{x, c\}} \gamma_\xi = 1$, $\gamma_\xi \geq 0$ and $\phi_l^0(x_i) = \varphi_l^0(x_i)$, $i = 0, \ldots, N_x - 1$. By recursion, let $\phi_l^{j-1}$ be the solution of the direct problem at the time-step $j - 1$.

At the transport stage of the splitting scheme ($\xi = x$), the differential equation is approximated with:

$$\gamma_x \frac{\left( \phi_{[x]} \right)_l^j - \phi_l^{j-1}}{\tau^{j-0.5}} - \frac{\partial}{\partial x} \left( \mu_l \frac{\partial \left( \phi_{[x]} \right)_l^j}{\partial x} \right) = \gamma_x r_l^{j-0.5}, \quad x \in \Omega, \tag{4}$$

$$\left( \phi_{[x]} \right)_l^j (0) = \alpha_l^L(t^j), \quad \left( \phi_{[x]} \right)_l^j (X) = \alpha_l^R(t^j). \tag{5}$$

Let $[a(x)]_{x=x_i} = a(x_i+) - a(x_i-)$, $a(x_i\pm)$ are right and left limits.

**Lemma 1 (Analogous to [6]).** *Let*

$$\left[\mu(x, t^j)\frac{\partial \phi^j(x)}{\partial x}\right]_{x=x_i} = \left[\phi^j(x)\right]_{x=x_i} = 0, \ i = 1, ..., N_x - 2, \quad (6)$$

*then the solution $(\phi_{[x]})_l^j$ of the direct problem (4)–(5) in the gridpoints $\omega_x$ is given by the solution of the system*

$$-\hat{L}_i\phi^j(x_{i-1}) + \hat{C}_i\phi^j(x_i) - \hat{R}_i\phi^j(x_{i+1}) = F_i, \ i = 0, ..., N_x - 1. \quad (7)$$

$$\hat{L}_i = \begin{cases} 0 & i = 0 \vee N_x - 1, \\ \tau \ \mu(x_{i-1}+, t^j)\frac{\partial \psi_L(x_{i-1})}{\partial x} & i = 1, ..., N_x - 2, \end{cases} \quad (8)$$

$$\hat{C}_i = \begin{cases} 1 & i = 0 \vee N_x - 1, \\ \tau \left(\left(\mu(x_i-, t^j)\frac{\partial \psi_L(x_i)}{\partial x}\right) - \left(\mu(x_i+, t^j)\frac{\partial \psi_R(x_i)}{\partial x}\right)\right) & i = 1, ..., N_x - 2, \end{cases} \quad (9)$$

$$\hat{R}_i = \begin{cases} -\tau \ \mu(x_{i+1}-, t^j)\frac{\partial \psi_R(x_{i+1})}{\partial x} & i = 1, ..., N_x - 2, \\ 0 & i = 0 \vee N_x - 1, \end{cases} \quad (10)$$

$$F_i = \begin{cases} (\alpha_l^L)^j & i = 0, \\ \int_{x_{i-1}}^{x_i} \Upsilon^j(x)\psi_L(x)dx + \int_{x_i}^{x_{i+1}} \Upsilon^j(x)\psi_R(x)dx & i = 1, ..., N_x - 2, \\ (\alpha_l^R)^j & i = N_x - 1 \end{cases} \quad (11)$$

$$\Upsilon^j(x) = \phi_l^{j-1}(x) + \tau\gamma_x r_l^{j-0.5}(x), \quad (12)$$

*where $\tau = \tau^{j-0.5}/\gamma_x$, $\mu = \mu_l$, $\psi_L$ and $\psi_R$ are the solutions of*

$$-\frac{\partial}{\partial x}\left(\tau \ \mu(x, t^j)\frac{\partial \psi(x)}{\partial x}\right) + \psi(x) = 0 \quad (13)$$

*in the intervals $[x_{i-1}, x_i]$ and $[x_i, x_{i+1}]$ with boundary conditions*

$$\psi(x_{i-1}) = 0, \ \psi(x_i) = 1, \quad (14)$$
$$\psi(x_i) = 1, \ \psi(x_{i+1}) = 0, \quad (15)$$

*correspondingly.*

The solution of (13) is then approximated with the solution for piece-wise-constant coefficients to obtain the numerical scheme.

At the transformation stage of the splitting scheme ($\xi = c$), the differential equation is approximated with

$$(\phi_{[c]})_l^j = \phi_l^{j-1}L^j(P_l(t^{j-1}, \phi^{j-1}))$$
$$+ G^j(P_l(t^{j-1}, \phi^{j-1}))(\Pi_l(t^{j-1}, \phi^{j-1}) + \gamma_c r_l^{j-0.5}), \quad (16)$$

where $l = 1, ..., N_c$, $\tau = \tau^{j-0.5}/\gamma_c$,

$$L^j(P) = \begin{cases} 0 & P\tau > \bar{\varepsilon} \\ e^{-P\tau} & P\tau \in [\underline{\varepsilon}, \bar{\varepsilon}] \\ 1 - P\tau & 0 < P\tau < \underline{\varepsilon} < 1 \end{cases} , \quad G^j(P) = \begin{cases} 1/P & P\tau > \varepsilon \\ \frac{1-e^{-P\tau}}{P\tau}\tau & P\tau \in [\underline{\varepsilon}, \varepsilon] \\ \tau & 0 < P\tau < \underline{\varepsilon} < 1 \end{cases} ,$$

then the results of calculations at the both stages are averaged

$$\phi_l^j = \sum_{\xi \in \{x,c\}} \gamma_\xi \left(\phi_{[\xi]}\right)_l^j . \tag{17}$$

**Direct Problem:** Given $r$, $\mu_l$, $\alpha_l^L$, $\alpha_l^R$, $\varphi_l^0$, determine $\phi$ from (7)–(12), (16), (17). Let $\phi[r]$ denote the solution of the direct problem.

Suppose that the source $r$ is unknown but there are measurements of the state function elements with $l \in L_{mes}$. Let us define the space $U_{mes} \subset \mathbb{R}^{N_t \times N_c \times N_x}$, a set of $\Xi$ functions $U = \{u_\xi\}_{\xi=1}^{\Xi} \subset U_{mes}$ and the Fourier-transform based operator:

$$U_{mes} = \left\{ \left\{ \begin{cases} h_l, & l \in L_{mes} \\ 0, & l \notin L_{mes} \end{cases} \right\}_{l=1}^{N_c} \middle| h_l \in \mathbb{R}^{N_t \times N_x} \right\}, \quad \bar{A}_U : \begin{cases} \mathbb{R}^{N_t-1 \times N_c \times N_x} \to \mathbb{R}^{\Xi} \\ r \mapsto \sum_{\xi=1}^{\Xi} \langle u_\xi, \phi[r] \rangle_3 \, e_\xi \end{cases},$$

where for some weights $\rho_l, \delta t^j, \delta x_i > 0$:

$$\langle \cdot, \cdot \rangle_3 : \begin{cases} \mathbb{R}^{N_t-1 \times N_c \times N_x} \times \mathbb{R}^{N_t-1 \times N_c \times N_x} \to \mathbb{R} \\ \{h, \phi\} \mapsto \sum_{j=1}^{N_t-1} \langle \phi^j, h^j \rangle \, \delta t^j \end{cases}, \tag{18}$$

$$\langle \cdot, \cdot \rangle : \begin{cases} \mathbb{R}^{N_c \times N_x} \times \mathbb{R}^{N_c \times N_x} \to \mathbb{R} \\ \{a, b\} \mapsto \sum_{i=0}^{N_x-1} \sum_{l=1}^{N_c} (a_l)_i (b_l)_i \, \rho_l \delta x_i \end{cases}, \tag{19}$$

Let us choose an $r^{(*)} \in F$, which will be called the "exact solution", and define a measurement data vector:

$$I_U = \bar{A}_U \left( r^{(*)} \right). \tag{20}$$

**Inverse Problem:** Given $\mu_l$, $\alpha_l^L$, $\alpha_l^R$, $\varphi_l^0$, $I_U$ determine $r^{(*)}$ from (20).

## 3  Solution with the Sensitivity Operator

**Lemma 2.** Let $h^j \in \mathbb{R}^{N_c \times N_x}$, $j = 1, ..., N_t - 1$,

$$\delta\phi^j = 0, \ j = 0, \tag{21}$$

$$A_\xi^{j-0.5} \delta\phi_{[\xi]}^j = B_\xi^{j-0.5} \delta\phi^{j-1} + C_\xi^{j-0.5} \gamma_\xi \delta r^{j-0.5}, \ j = 1, ..., N_t - 1, \ \xi \in \Xi, \tag{22}$$

$$\delta\phi^j = \sum_{\xi \in \{x,c\}} \gamma_\xi \delta\phi_{[\xi]}^j, \ j = 1, ..., N_t - 1, \ \sum_{\xi \in \{x,c\}} \gamma_\xi = 1, \ \gamma_\xi \geq 0. \tag{23}$$

*If*

$$\left(A_\xi^{j-0.5}\right)^* \psi_\xi^j = \frac{1}{\delta t^j}\Psi^{j+1} + h^j, \ j = 1, \ldots N_t - 1, \tag{24}$$

$$\Psi^j = \begin{cases} \sum_{\xi \in \{x,c\}} \gamma_\xi \left(B_\xi^{j-0.5}\right)^* \psi_\xi^j \delta t^j, \ 1 \le j \le N_t - 2 \\ 0, \ j = N_t - 1 \end{cases}, \ j = 1, \ldots N_t - 1, \tag{25}$$

*then*

$$\sum_{j=1}^{N_t-1} \left\langle \delta\phi^j, h^j \right\rangle \delta t^j = \sum_{j=1}^{N_t-1} \left\langle \delta r^j, \sum_{\xi \in \{x,c\}} \gamma_\xi^2 \left(C_\xi^j\right)^* \psi_\xi^j \right\rangle \delta t^j. \tag{26}$$

*Proof. Let's multiply (23) by $\gamma_\xi \psi_\xi^j$ and (22) by $\psi^{j+1}$ and sum the results:*

$$0 = \left\langle \delta\phi^j - \sum_\xi \gamma_\xi \delta\phi_{[\xi]}^j, \psi^{j+1} \right\rangle + \sum_\xi \left\langle A_\xi^j \delta\phi_{[\xi]}^j - B_\xi^j \delta\phi^{j-1} - C_\xi^j \gamma_\xi \delta r^j, \gamma_\xi \psi_\xi^j \right\rangle =$$

$$\left\langle \delta\phi^j, \psi^{j+1} \right\rangle + \sum_\xi \gamma_\xi \left\langle \delta\phi_{[\xi]}^j, \left(A_\xi^j\right)^* \psi_\xi^j - \psi^{j+1} \right\rangle - \left\langle \delta\phi^{j-1}, \sum_\xi \gamma_\xi \left(B_\xi^j\right)^* \psi_\xi^j \right\rangle$$

$$- \left\langle \delta r^j, \sum_\xi \gamma_\xi \left(\gamma_\xi \left(C_\xi^j\right)^* \psi_\xi^j\right) \right\rangle.$$

*If $\left(A_\xi^j\right)^* \psi_\xi^j - \psi^{j+1} = 0$, then*

$$\left\langle \delta\phi^j, \psi^{j+1} \right\rangle = \left\langle \delta\phi^{j-1}, \sum_\xi \gamma_\xi \left(B_\xi^j\right)^* \psi_\xi^j \right\rangle + \left\langle \delta r^j, \sum_\xi \gamma_\xi \left(\gamma_\xi \left(C_\xi^j\right)^* \psi_\xi^j\right) \right\rangle.$$

*Summing the previous relation for $j = 1, \ldots, N_t - 1$ and using the discrete analog of the integration by parts, we have*

$$\sum_{j=1}^{N_t-1} \left\langle \delta\phi^j, \psi^{j+1} - \sum_\xi \gamma_\xi \left(B_\xi^{j+1}\right)_\xi^* \psi_\xi^{j+1} \frac{\delta t^{j+1}}{\delta t^j} \right\rangle \delta t^j$$

$$+ \left\langle \delta\phi^{N_t-1}, \sum_\xi \gamma_\xi \left(B_\xi^{N_t}\right)^* \Psi_\xi^{N_t} \frac{\delta t^{N_t}}{\delta t^{N_t-1}} \right\rangle \delta t^{N_t-1}$$

$$= \sum_{j=1}^{N_t-1} \left\langle \delta r^j, \sum_\xi \gamma_\xi^2 \left(C_\xi^j\right)^* \psi_\xi^j \right\rangle \delta t^j.$$

*Let*

$$\psi^{j+1} = h^j + \frac{1}{\delta t^j} \begin{cases} \sum_\xi \gamma_\xi \left(B_\xi^{j+1}\right)_\xi^* \psi_\xi^{j+1} \delta t^{j+1}, \ j = 1, \ldots N_t - 2 \\ 0, \ j = N_t - 1 \end{cases},$$

*then using notations (25) we obtain the needed result.*

Due to the linearity of the transport process model, the equation in variations has the similar coefficients. The variational equation for the discrete-analytical scheme for the transformation step (16) can be found in [4]. In the terms of Lemma 2

$$A_c^{j-0.5} = Id, \quad B_c^{j-0.5} = w^j(t^j, \phi^j, r^j, \delta\phi^j, \delta r^j), \quad C_c^{j-0.5} = R^j(t^j, \phi^j),$$

$$w^j(t, \phi, r, \delta\phi, \delta r) = \underset{l=1,\ldots,N_c}{diag} \; L^j\left(p_l^2\right) + \left\{ S_l^j(t, \phi, r, \delta\phi, \delta r) \right\}_{l=1}^{N_c},$$

$$R^j(t, \phi) = \underset{l=1,\ldots,N_c}{diag} \; G^j\left(p_l^1\right), \; p_l^2 = P_l(t^j, \phi^j + \delta\phi^j), \; p_l^1 = P_l(t^j, \phi^j),$$

where $\{S_l\}_{l=1}^{N_c}$ denotes a matrix with rows $S_l$. If $p_l^2 = p_l^1$, then

$$S_l^j(t, \phi, r, \delta\phi, \delta r) = G^j\left(p_l^1\right) \bar{\nabla}\Pi_l(t, \phi + \delta\phi, \phi).$$

Otherwise

$$S_l^j(t, \phi, r, \delta\phi, \delta r) = G^j\left(p_l^1\right) \bar{\nabla}\Pi_l(t, \phi + \delta\phi, \phi) +$$
$$\left(\phi_l \bar{\nabla}L^j(p_l^2, p_l^1) + \bar{\nabla}G^j(p_l^2, p_l^1)(\Pi_l(t, \phi + \delta\phi) + r_l + \delta r_l)\right) \bar{\nabla}P_l(t, \phi + \delta\phi, \phi).$$

Let us denote the sensitivity function for the source from (26) as

$$m[r, \delta r; h]^j = \sum_{\xi \in \{x,c\}} \gamma_\xi^2 \left(C_\xi^j\right)^* \psi_\xi^j, \quad j = 1, \ldots, N_t - 1.$$

**Theorem 1 (Analogous to [4,5]).** *For any* $r + \delta r, r \in \mathbb{R}^{N_t-1 \times N_c \times N_x}$ *and* $U \subset U_{mes}$, *the identities hold*

$$\bar{A}_U(r + \delta r) - \bar{A}_U(r) = m_U[r, \delta r]\delta r, \quad m_U[r, \delta r] : \left\{ \begin{array}{l} \mathbb{R}^{N_t-1 \times N_c \times N_x} \to \mathbb{R}^\Xi \\ z \mapsto \sum\limits_{\xi=1}^\Xi \langle m[r, \delta r; u_\xi], z \rangle_3 \, e_\xi \end{array} \right.,$$

$$I_U - \bar{A}_U(r) = m_U[r, 0]\left(r^{(*)} - r\right) + \left(m_U[r, r^{(*)} - r] - m_U[r, 0]\right)\left(r^{(*)} - r\right). \tag{27}$$

As the set $U$, we use the trigonometric basis:

$$U_\Theta = \{e_{\eta\theta_x\theta_t} | \; 0 \le \theta_x \le \Theta_x, 0 \le \theta_t \le \Theta_t, \; \eta \in L_{mes}\},$$

$$(e_{\eta\theta_x\theta_t})_i^j = \left\{ \left\{ \begin{array}{l} \frac{1}{\sqrt{\rho_\eta}}C(T, \theta_t, t^j)C(X, \theta_x, x_i), \; l = \eta \\ 0, \; l \ne \eta \end{array} \right. \right\}_{l=1}^{N_c},$$

$$C(T, \theta, t) = \frac{1}{\sqrt{T}} \left\{ \begin{array}{l} \sqrt{2}\cos\left(\frac{\pi\theta t}{T}\right), \; \theta > 0 \\ 1, \; \theta = 0 \end{array} \right..$$

In this case $\Xi = \Theta_x \Theta_t |L_{mes}|$, where $|L_{mes}|$ is the number of elements $L_{mes}$. Suppose that it is known that only a given set of substances are emitted. We denote their indices by $L_{src}$. For the solution of operator equations (27), we consider Newton-Kantorovich type method based on the truncated SVD with the use of right inverse matrices [1–3,5,8] :

1. Let $r^{(0)}$ be an initial approximation, $\Delta p = \Xi/N_p$, $cond_{max}$ is the maximal considered conditional number.
2. Iterate by $p$, starting with $p = \Delta p$:
   (a) We compute $m^{(k)}$, the matrix of the sensitivity operator $m_U[r^{(k)}, 0]$ and the singular value decomposition of $m^{(k)} (m^{(k)})^T$. Let $\sigma_l$ be the singular values of $m^{(k)}$. If $\sigma_1/\sigma_p > cond_{max}$, then the algorithm is stopped.
   (b) If not, we define the Newton-type increment

$$\delta r^{(k)} = \Pr_{src} \left(m^{(k)}\right)^T \left[m^{(k)} \left(m^{(k)}\right)^T\right]_p^+ \left(I_U - \bar{A}_U(r^{(k)})\right),$$

$$\Pr_{src} z = \left\{ \begin{cases} z_l, \ l \in L_{src} \\ 0, \ l \notin L_{src} \end{cases} \right\}_{l=1}^{N_c}, \quad \left[m^{(k)} \left(m^{(k)}\right)^T\right]_p^+ = \sum_{l=1}^{p} \frac{U_l}{\sigma_l^2} \langle ., U_l \rangle_{\mathbb{R}^\Xi},$$

where $\langle ., . \rangle_{\mathbb{R}^\Xi}$ is the Euclidean scalar product in $\mathbb{R}^\Xi$, $\{U_l\}_{l=1}^{rank(m^{(k)})}$ is the orthonormal system of the left singular vectors of $m^{(k)}$.
   (c) We choose $r^{(k+1)} = r^{(k)} + \gamma^{(k)} \delta r^{(k)}$ to satisfy the inequality (dividing $\gamma^{(k)}$ several times by 2, if necessary)

$$\left\| I_U - \bar{A}_U(r^{(k)} + \gamma^{(k)} \delta r^{(k)}) \right\|_{\mathbb{R}^\Xi} < \left\| I_U - \bar{A}_U(r^{(k)}) \right\|_{\mathbb{R}^\Xi}.$$

   (d) The algorithm continues to work until the iterations are stabilized. Then, the number $p$ increases, i.e. $p := p + \Delta p$.

## 4    Numerical Experiments

As an example, consider the modified atmospheric chemistry mechanism from [7]. The reaction rates depend on time of day. The system can be rewritten in the form of a system of ordinary differential equations of the production-destruction type. Suppose that the sources emit $NO$ only (i.e. $L_{src} = \{NO\}$). $O_3$ are available for measurement (i.e. $L_{mes} = \{O_3\}$). The grid parameters are $T = 24 \times 3600$, $N_t = 3000$, $X = 6000$, $N_x = 100$, $N_c = 22$, $\Theta_x = \Theta_t = 15$. Initial approximation to the sources $r^{(0)}$ is zero. The exact source and the results of the source reconstruction are presented in Fig. 1(a) and (b). The corresponding convergence parameter dynamics is shown in Fig. 1(c) and (d).

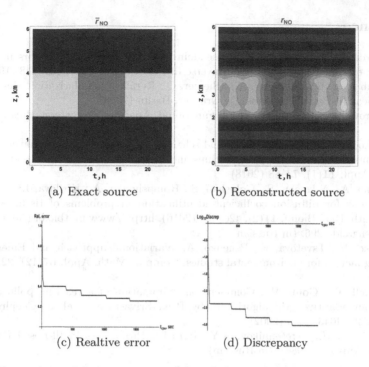

(a) Exact source    (b) Reconstructed source

(c) Realtive error    (d) Discrepancy

**Fig. 1.** Comparison of the exact source (a) and the result of the reconstruction (b). Error (c) and discrepancy (d) dynamics with respect to computation time.

# 5    Conclusions

The inverse source problem for 1D diffusion-reaction model is considered. The measurement data is given as the images of the concentration fields dynamics for the subset of the interacting species. The sensitivity operator, composed of the ensemble of the independent adjoint problem solutions, allow to transform the inverse problem to the family of nonlinear integral equations. Consistent discrete-analytical approximations for the solution of the direct and adjoint problems are presented. An equation from the family is solved with the Newton-Kantorovich-type algorithm based on the truncated SVD.

**Acknowledgments.** The algorithm development was supported by the Russian Science Foundation project 17-71-10184. Vectorization and optimization of the codes were implemented with the support of the Ministry of education and science of the Russian Federation (4.1.3 Joint laboratories of NSU-NSC SB RAS). Siberian Supercomputer Center is gratefully acknowledged for providing the supercomputer facilities.

# References

1. Cheverda, V.A., Kostin, V.I.: R-pseudoinverses for compact operators in Hilbert spaces: existence and stability. J. Inverse Ill-Posed Probl. **3**(2), 131–148 (1995)
2. Kaltenbacher, B., Neubauer, A., Scherzer, O.: Iterative Regularization Methods for Nonlinear Ill-Posed Problems. De Gruyter, Berlin (2008)
3. Kantorovich, L.V., Akilov, G.P.: Functional Analysis. Pergamon Press, Oxford (1982)
4. Penenko, A.V.: Consistent numerical schemes for solving nonlinear inverse source problems with gradient-type algorithms and Newton-Kantorovich methods. Numer. Anal. Appl. **11**(1), 73–88 (2018)
5. Penenko, A., Nikolaev, S., Golushko, S., Romashenko, A., Kirilova, I.: Numerical algorithms for diffusion coefficient identification in problems of tissue engineering. Math. Biol. Bioinf. **11**(2), 426–444 (2016). http://www.matbio.org/article.php?lang=eng&id=296. (in Russian)
6. Penenko, V., Tsvetova, E., Penenko, A.: Variational approach and Euler's integrating factors for environmental studies. Comput. Math. Appl. **67**(12), 2240–2256 (2014)
7. Stockwell, W., Goliff, W.: Comment on "Simulation of a reacting pollutant puff using an adaptive grid algorithm" by R.K. Srivastava et al. J. Geophys. Res. **107**(D22), 4643–4650 (2002)
8. Vainikko, G.M., Veretennikov, A.Y.: Iterative Procedures in Ill-Posed Problems. Nauka, Moscow (1986). (in Russian)

# Multiscale Mathematical Modeling of the Metal Nanoparticles Motion in a Gas Flow

Viktoriia Podryga[1,2](✉) and Sergey Polyakov[1,3]

[1] Keldysh Institute of Applied Mathematics of RAS,
4 Miusskya sq., 125047 Moscow, Russia
pvictoria@list.ru, polyakov@imamod.ru
[2] Moscow Automobile and Road Construction State Technical University,
64, Leningradsky prospect, 125319 Moscow, Russia
[3] National Research Nuclear University MEPhI,
31 Kashirskoe sh., 115409 Moscow, Russia

**Abstract.** The paper is devoted to modeling the motion of metal nanoparticles in a gas flow. Two different ways are proposed to solve the problem. The first way is based on the use of the molecular dynamics method, the second method is based on the use of the multiscale approach combining the quasigasdynamic equations system and the molecular dynamics method. A cluster of nickel atoms is considered as a metallic nanoparticle, a flow of nitrogen molecules is considered as a gas flow. In numerical experiments, the conditions for matching the applied mathematical models and the behavior of the nanoparticle are determined.

**Keywords:** Multiscale modeling · Metallic nanoparticles ·
Fluid dynamics with nanoinclusions · Molecular dynamics

## 1 Introduction

Nanotechnology is an important factor in modern development. One of the intensively developing scientific directions in nanotechnology is studying the properties of various materials surfaces on micro- and nanoscales. The creation of micro- and nanostructures and systems with specified parameters and properties gives the widest possibilities. Within the framework of this direction, one of the important tasks is study and development of various methods for creating surfaces with desired properties using gas-dynamic spraying technique [1,2].

One of the reasons for attention to such physical processes is their technological simplicity and wide field of specific industrial applications. At the present moment, the stage of creating continuous micro- and nanocoatings for the purpose of hardening the structural materials has been practically completed.

In contrast to methods for creating a continuous nanocoating (where only a uniform flux of nanoparticles is required to be created), the technology of forming a certain pattern or mask consisting of several dozen nanoparticles on the

© Springer Nature Switzerland AG 2019
I. Dimov et al. (Eds.): FDM 2018, LNCS 11386, pp. 387–394, 2019.
https://doi.org/10.1007/978-3-030-11539-5_44

substrate is considered here. Such a point-like sputtering technology requires from the appropriate nanoimprinting installations a high process controllability and high accuracy of mask formation. This technological task in its entirety has not been investigated. One of the reasons for this situation is the high complexity and cost of field studies. Theoretical analysis of this problem also experiences difficulties both at the level of the formation of adequate mathematical models and at the level of their computer implementations. However, recently the computer complexity of program realizations began to decline due to the use of modern supercomputers. As a result, it became possible to approbate complex mathematical models and to test certain mathematical approaches.

Various aspects of computer of jet flows of gases, flows of gas mixtures in microchannels of technical systems, etc. were analyzed in [3,4] on the basis of quasigasdynamic models of continuum mechanics and [5–7] on the basis of a new multiscale approach combining solution of the macroscopic equations of gas dynamics with Newton's equations of dynamics for individual particles. In this paper, the processes of motion of metallic nanoparticles in a gas flow are studied. A rarefied gas accelerated to velocities belonging to the transonic and supersonic ranges is considered at low and room temperatures (100–350 K). To solve the problem, two ways are proposed, the application of which depends on the size of the system under study and the required level of the processes detail.

The first way is based on the method of molecular dynamics (MD) [8]. This method is applicable for nano- and microsize systems, when the total number of particles of the system under investigation does not exceed 1 billion, for a larger number of particles the method is possible, but it becomes unprofitable in view of the high cost of computational resources and the large computing time. For rarefied gas systems the MD method is usually used with the systems size of the order of 10 mean free paths and for metal systems it is used with a size not exceeding 100–1000 unit cell lengths. The optimal system sizes for calculations are much smaller discussed.

The second way is based on the multiscale approach combining a quasigasdynamic (QGD) [4] equations system and a MD method. At the molecular level, it allows to determine the parameters of the equation of state of a real gas [9], calculate the kinetic properties of the gas [10], and determine the boundary conditions on the surfaces of solids [5,7]. This approach includes two levels of modeling, which are macroscopic and microscopic levels. It uses the system of QGD equations as a model at the macrolevel, and the MD method is used as a microlevel model. The implementation of the approach is based on splitting by physical processes. Method is applicable for systems of various sizes; in the case of small system sizes, it turns into the first way of solving. In this paper, in view of the specific nature of the calculation of gas media with solid inclusions, this approach will be considered only as a combination without going over to pure MD solutions. The approach is not limited from above either by the size of the systems or by the number of particles. For the lower limit of the application of the approach, the limit of MD application, described above, is adopted.

As a model problem, the motion of a nickel nanocluster in a nitrogen flow was considered. The first way was used to solve the problem. In this case, it was interesting to obtain information about the modeled process on the basis of the first principles. The purpose of the modeling was to reproduce the actual acceleration process taking into account the internal state of the nanocluster. A specific algorithm for applying the second way was also determined on the basis of calculations within the first way. In particular, the best algorithm for conjugation of the QGD and MD models was of interest. As a result, it was suggested to select one or several grid cells inside the QGD model, within which a nanocluster is currently located, and set the dynamic boundary conditions on the boundary of this volume, and move the volume as the nanoparticle moves in the same direction. In this case, it is necessary to monitor the implementation of grid analogs of conservation laws (mass, momentum, and energy). A similar solution is used in macroscopic models for the motion of large bodies in fluid.

## 2   Statement of the Problem

The first solution way, based on MD calculations, is a special case of the second way, representing a multiscale two-level approach. For this reason, the mathematical formulation for the second way will be described.

The mathematical model includes two components corresponding to the scale levels. At the macroscopic level, the QGD equations are used. In the case of a pure gas of the one type, the system of QGD equations in the three-dimensional case in the form invariant with respect to the coordinate system in dimensional variables, together with the equations of state, is written as:

$$\frac{\partial \rho}{\partial t} + div\mathbf{W}^{(\rho)} = 0, \quad \frac{\partial E}{\partial t} + div\mathbf{W}^{(E)} = 0, \tag{1}$$

$$\frac{\partial (\rho u_k)}{\partial t} + div\mathbf{W}^{(\rho u_k)} = 0, \quad k = x, y, z,$$

$$E = \frac{1}{2}\rho|\mathbf{u}|^2 + \rho\varepsilon, \quad \varepsilon = c_V T, \quad p = Z\rho RT. \tag{2}$$

Here $\rho = mn$ is the mass density ($m$ is the mass of molecule of gas, $n$ is the concentration), $T$ is the temperature and $\mathbf{u}$ is the macroscopic velocity. Other parameters: $p$ is the partial pressure of gas; $E$ and $\varepsilon$ are the total energy density and internal energy. Variables $Z = Z(T, \rho)$, $c_V = c_V(T)$ and $R = k_B/m$ are the compressibility coefficient, specific heat capacity and individual gas constant ($k_B$ is the Boltzmann constant); vectors $\mathbf{W}^{(\rho)}$, $\mathbf{W}^{(\rho u_k)}$, $\mathbf{W}^{(E)}$ coincide, up to sign, with the fluxes of mass density, momentum density of the corresponding components, and energy density.

The system of Eqs. (1)–(2) is closed by the corresponding initial and boundary conditions. The initial conditions are taken in accordance with the equilibrium state of the gas medium in the absence of interaction with external factors. The boundary conditions for the QGD equations near the solid surfaces can be defined by determining the fluxes of mass density, momentum density, and energy density at the boundary according to Newton's conditions or their normal component near the surfaces by the MD method. On the free surfaces of the computational domain, the so-called "soft" boundary conditions are set [4].

To describe the motions of gas molecules and nanocluster atoms, the equations of two-component molecular dynamics are used:

$$m_{l,i}\frac{d\mathbf{v}_{l,i}}{dt} = \mathbf{F}_{l,i}, \quad \mathbf{v}_{l,i} = \frac{d\mathbf{r}_{l,i}}{dt}, \quad i = 1, ..., N_l, \tag{3}$$

where $i$ is the particle number, $l$ is the particle type ($a$ are the atoms of nanocluster, $b$ are the molecules of gas), $N_l$ is the total number of particles of type $l$. Particle of type $l$ with number $i$ has own mass $m_{l,i}$, the position vector $\mathbf{r}_{l,i}$, the velocity vector $\mathbf{v}_{l,i}$ and the resultant force $\mathbf{F}_{l,i}$, acting on this particle.

The forces depend on choosing the potential of particle interaction. The choice of a certain type of potential is based on comparing the mechanical properties of the potential computer model and the real material. In this work, nickel was used as the nanocluster material, and nitrogen was used as the gas. For these specific substances, the corresponding potentials were chosen [5,6].

The initial conditions at the microlevel are determined by the thermodynamic equilibrium state of the system at given temperature, pressure, and average momentum. The boundary conditions at the molecular level depend on the simulated situation.

To implement the second way of solving the problem, it is necessary to coordinate the above-mentioned models in the closeness of the nanocluster. An algorithm for such an agreement is proposed in Sect. 3.

## 3   Numerical Methods and Program Implementation

The QGD equations are solved by means of an explicit on-time grid numerical algorithm, which is based on the finite volume method on grids of arbitrary type. For the convenience of solving problems in areas of complex geometry, hybrid block meshes consisting of cells of various shapes and sizes can be used. In the two-dimensional case, it is proposed to use quadrangles and triangles, in the three-dimensional case, polyhedrons with a number of vertices from four to eight. All the parameters of gas components (density, pressure, temperature, velocity vector components, etc.) refer to the centers of mass of the cells. Flux variables refer to the centers of the faces of cells. Spatial approximations of the basic terms of the QGD equations are performed by standard methods. Computational two-stage (predictor–corrector) scheme is chosen.

The MD equations are solved according to the Verlet scheme either in each cell of grid independently or in groups of coupled cells. The system of MD equations is used in the case of the first calculation way alone, or in the case of the second way as a subgrid algorithm applied inside each control volume.

The algorithm for matching the QGD and MD calculations for the case of Cartesian grids consists in the following. The dimensions of the grid cells in solving the QGD equations are determined by the Knudsen number ($Kn = \lambda/L$,

$\lambda$ is the mean free path of gas molecules, $L$ is the characteristic size of the problem). In solving this problem, in addition to $\lambda$ and $L$, there is also the size of the nanocluster $L_c$ and the characteristic length of the motion of the gaseous medium $L_u = |\mathbf{u}|\Delta t$ over time $\Delta t$. Therefore, the QGD equations are solved provided that all the above dimensions are consistent with the parameters of the computational grid (corresponding to the space step $h_{QGD}$ and the time step $\Delta t_{QGD}$), namely: $\lambda \leq h_{QGD} \ll L$, $L_c \ll L$, $L_u = |\mathbf{u}|\Delta t_{QGD} \leq h_{QGD}$.

The linear size of the selected area in which the nanocluster is located may be smaller, or may exceed the size of a single cell. Therefore, the volume $\Omega_c$ surrounding the cluster is introduced with a linear dimension $h_c = k \cdot h_{QGD}$ and the center located near the center of the cluster; here $k = 2 + [L_c/h_{QGD}]$ is the whole parameter. The digit ?two? in the expression for the parameter is selected in order to track the movement of the cluster in time across the boundaries of cells and shift the center of the volume to the desired direction.

At the boundaries of the volume $\Omega_c$ conditions are set that correspond to the flow parameters at the macroscopic level. In this case, the total mass, momentum and energy enclosed in the volume $\Omega_c$ are frozen, taking into account the gas and the nanocluster content. As a result, the combined calculation algorithm looks like this. At a particular time step, the total gas flow moves the extracted gas volume as a single unit, which is described by the QGD equations. Then the configuration $\Omega_c$ (that is, the state of the gas and nanocluster) is recalculated taking into account external boundary conditions. Then, if it is necessary, the volume $\Omega_c$ changes its shape and position.

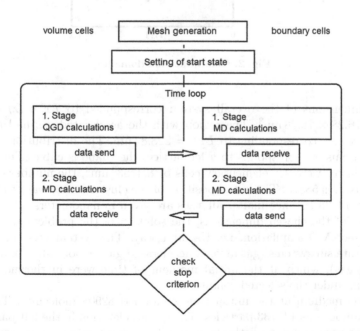

**Fig. 1.** QGD + MD algorithm block scheme.

To carry out a correct calculation the QGD model is supplemented by real equations of gas state, transport coefficients and other accompanying parameters (enthalpies, average mean free paths, etc.), as well as real boundary conditions that are obtained with the help of MD calculations. A block scheme of the algorithm for the second calculation way is shown in Fig. 1.

The parallelization method is partition into domains of equal power. Each domain is divided into "boxes of interaction". Topology of distribution on domains and boxes is a three-dimensional lattice. Topology exchanges are a three-dimensional torus. Realization was carried out by MPI + OpenMP technologies.

## 4   Results of Modeling

In this section, the motion of a metallic nanocluster in a gas flow is investigated. Nanocluster consists of nickel atoms, nitrogen is considered as a gas. Initially, the system is under normal conditions ($T = 273.15\,\mathrm{K}$, $p = 101325\,\mathrm{Pa}$), the nanocluster and the gas around it are at rest. After, the left along the x-axis (Fig. 2), a gas flow with a supersonic velocity is incident on the nanoparticle.

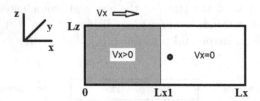

**Fig. 2.** Computational domain.

The dimensions of the overall system corresponded to $Lx \times Ly \times Lz = 533.96 \times 76.28 \times 76.28\,\mathrm{nm}^3$. The region with the accelerated gas had a length along the x axis corresponding to $Lx1 = 228.84\,\mathrm{nm}$. The distribution of nickel atoms in a cluster corresponds to a fcc lattice, the shape is cubic. The sizes of the nanocluster were $24^3$ elementary cells or $8,4753^3\,\mathrm{nm}^3$, the number of atoms in nanocluster is 58825. The initial position of the cluster center was at the point $(266.98, 38.14, 38.14)$, hereinafter all sizes are given in nanometers.

In view of the chosen dimensions, the solution of the problem was carried out by direct MD simulation, i.e. the first way. The system was constructed from separate subsystems (gas at rest, accelerated gas, nanocluster, nanocluster in resting gas), which at the initial moment of time were in thermodynamic equilibrium under the selected conditions.

The gas medium at the initial time contained 82908 molecules. The total particle number was 141733 particles. The gas acceleration in the left part of the system was carried out with the help of a Langevin thermostat by sending to the gas a nonzero longitudinal velocity $V_x$. In calculations, approximately half of the nitrogen molecules in the left-hand part of the calculated region had at the start an average velocity $V_x \approx (2.8; 8.4; 14.0)$ Ma. The sound velocity in nitrogen under normal conditions is $357$ nm/ns, so the value $V_x$ was about $1; 3; 5$ nm/ps.

Calculations were carried out on supercomputer K-60 (KIAM RAS, Moscow, Russia), which has 78 nodes, each includes $2 * \mathrm{CPU}$ Intel Xeon E5-2690 v4; 2,6 GHz microprocessor with 28 cores. Total performance is 74 TFlops, transfer is FDR InfiniBand 56 Gbit/s.

The result is the accelerating the entire gas (and the nanocluster together with it). Most of the gas at rest begins to move in the same direction as the supersonic flow that has struck it. For the same reason, the gas surrounding the cluster quickly "flies" forward. Nanocluster moves much more slowly. It eventually accelerates evenly to a certain limited velocity (see Fig. 3). The magnitude of this velocity is related to the amount of gas colliding with the cluster and its total momentum. To accelerate the cluster to a certain average velocity, it will take either a very long time (if the input flow is continuous but has a low speed), or a high speed and/or density (if the flow acts on the cluster in the pulsed mode). Figure 4 illustrates the change in the position of the nanocluster at different instants of time. These displacements will be used later to calibrate the second multiscale calculation way.

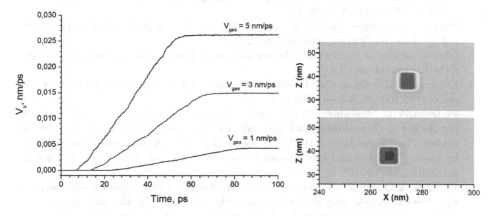

**Fig. 3.** The resulting longitudinal velocities of the nanocluster, obtained by the acceleration for three values of the gas velocity.

**Fig. 4.** Position of nanocluster at initial time (below) and at time point 350 ps (above)

## 5   Conclusion

The paper deals with the problem of modeling the motion of individual metallic nanoparticles in a gas flow. For its numerical solution, two ways are proposed. The first way is based on the application of the molecular dynamics method in the whole computational domain. The second way consists in applying a multiscale approach combining a quasigasdynamical equations system and a method of molecular dynamics. The advantages of the first way are in the uniformity of the general algorithm and the possibility of analyzing the problem at the micro- and macrolevels without introducing additional errors. However, this technique is strongly limited by the size of the computational domain. The second way allows to overcome dimensional constraints, but requires additional research in terms of correctness and accuracy. In this paper, a specific implementation of the second way is proposed and the results of calculations of computing the motion of a nickel nanoparticle in a nitrogen flow are presented. In numerical experiments based on the first way, the flow parameters and the dynamics of the nanoparticle motion were obtained, as well as the conditions for matching the applied mathematical models in the case of the second solution way.

**Acknowledgment.** The work was funded by Russian Foundation for Basic Research, projects no. 18-07-01292-a, 18-51-18004-bolg_a, 16-07-00206-a.

# References

1. Irissou, E., Legoux, J.-G., Ryabinin, A.N., Jodoin, B., Moreau, C.: Review on cold spray process and technology: part I - intellectual property. J. Therm. Spray Technol. **17**(4), 495–516 (2008)
2. Papyrin, A., Kosarev, V., Klinkov, S., Alkhimov, A., Fomin, V.: Cold Spray Technology. Elsevier Science, Amsterdam (2007)
3. Chetverushkin, B.N.: Kinetic schemes and quasi-gasdynamic system of equations. CIMNE, Barcelona (2008)
4. Elizarova, T.G.: Quasi-Gas Dynamic Equations. Springer, Berlin (2009). https://doi.org/10.1007/978-3-642-00292-2
5. Podryga, V.O., Karamzin, Y.N., Kudryashova, T.A., Polyakov, S.V.: Multiscale simulation of three-dimensional unsteady gas flows in microchannels of technical systems. In: Proceedings of the VII European Congress on Computational Methods in Applied Sciences and Engineering, Crete Island, Greece, 5–10 June 2016, vol. 2, pp. 2331–2345 (2016)
6. Podryga, V.O.: Multiscale approach to computation of three-dimensional gas mixture flows in engineering microchannels. Dokl. Math. **94**(1), 458–460 (2016)
7. Podryga, V.O.: Computational technology of multiscale modeling the gas flows in microchannels. In: IOP Conference Series: Materials Science and Engineering, vol. 158, p. 6 (2016). Paper 012078
8. Rapaport, D.C.: The Art of Molecular Dynamics Simulation. Cambridge University Press, Cambridge (2004)
9. Podryga, V., Polyakov, S.: Correction of the gas flow parameters by molecular dynamics. In: Proceedings of the Fourth International Conference on Particle-Based Methods-Fundamentals and Applications, pp. 779–788. CIMNE, Barcelona (2015)
10. Podryga, V.O.: Calculation of kinetic coefficients for real gases on example of nitrogen. In: Dimov, I., Farago, I., Vulkov, L. (eds.) Numerical Analysis and Its Applications. NAA 2016, LNCS, vol. 10187, pp. 542–549. Springer, Cham (2017). https://doi.org/10.1007/978-3-319-57099-0_61

# Construction of Higher-Order Approximation Difference Scheme for Nonlinear Convection-Diffusion Equation Using Adaptive Artificial Viscosity in Case of Two-Phase Filtering Problems

I. V. Popov[1,2]([⊠]), Yu. A. Poveshchenko[1,2], and S. V. Polyakov[1,2]

[1] Keldysh Institute of Applied Mathematics of RAS, 4 Miusskaya square, 125047 Moscow, Russia
piv2964@mail.ru
[2] National Research Nuclear University MEPhI (Moscow Engineering Physics Institute), 31, Kashirskoe shosse, 115409 Moscow, Russia

**Abstract.** The method of adaptive artificial viscosity is used to model the process of one-dimensional nonlinear convection-diffusion equation. For this purpose, a finite difference scheme (FDS) of the second order of time and space approximation has been developed. The scheme was tested using a numerical solution of the problem on formation of a gradient catastrophe. The process of two-phase filtration was analyzed with the help of constructed FDS. Numerical calculations showed that the proposed method, and in this case reliably tracks the discontinuities of the solution.

**Keywords:** Two-phase filtering problem · Finite difference schemes · Computer simulations

## 1 Formulation of the Problem

In article we will consider the problem of two-phase filtration. As the first phase, incompressible water is taken $S_W$, and as the second phase the ideal gas to be compressed $S_G$. The process is considered in the isothermal approximation [1]. For both phases, the normalization condition is satisfied $S_W + S_G = 1$. The equation of continuity for phases has the form

$$\frac{\partial}{\partial t}(m\rho_i S_i) + \frac{\partial}{\partial x}(\rho_i u_i) + q_i = 0, \tag{1}$$

where $m$ is the volumetric porosity of the layer (which we will consider as a constant, that means, the skeleton is not deformable), $\rho_i$ is density of $i$-th fluid, $i = W$ (or $G$) is the water or gas component, $u_i$ is filtration rate of $i$-th fluid, $q_i$

© Springer Nature Switzerland AG 2019
I. Dimov et al. (Eds.): FDM 2018, LNCS 11386, pp. 395–402, 2019.
https://doi.org/10.1007/978-3-030-11539-5_45

is the volumetric flow rate of $i$-th fluid. For this model only gas is taken away $q_G \neq 0$, and water remains in the reservoir $q_W = 0$.

The filtration rate is determined from the Darcy's law of motion for each of the phases, which have the form

$$u_i = -k \frac{k_i}{\eta_i} \nabla p_i \qquad (2)$$

where $k$ is coefficient of permeability, $k_i = k(S_i)$ is the relative phase permeability of $i$-th fluid, $\eta_i$ is the dynamic viscosity of i-th fluid, $p_i$ is the pressure of i-th fluid. In future we can notice, that is $p_W = p_G = p$, there is no capillary pressure jump, as in the Backley-Leverett model, but unlike this model $\rho_W = $ const and $\rho_G \neq$ const. Also in the model there is no gravity. Finally, we obtained the following system of equations:

$$\begin{cases} m \dfrac{\partial S_W}{\partial t} + \dfrac{\partial u_W}{\partial x} = 0, \\[2mm] m \dfrac{\partial}{\partial t} (\rho_G S_G) + \dfrac{\partial}{\partial x} (\rho_G u_G) + q_G = 0, \\[2mm] u_i = -k \dfrac{k_i}{\eta_i} \nabla p_i, i = W, G. \end{cases} \qquad (3)$$

The system of equations closing by the equation of state of the gas, which has the form

$$\rho_G = \frac{pM}{z(p, T) RT}, \qquad (4)$$

where $M$ is the molar gas mass, $z(p, T)$ is the coefficient of supercompressibility (for an ideal gas $z = 1$), $R$ is the gas constant, $T$ is the absolute temperature for a given model ($T = $ const). Further, we denote by $C_G = \dfrac{M}{z(p, T) RT}$, then the equation of state for the gas (in the isothermic approximation) has the form $\rho_G = C_G p$.

The volume source $q_G$ will be modeled by the difference of pressure in the reservoir $p$ and the atmospheric one $p_{\text{atm}}$, then the model source of the gas phase selection getting form $q_G = \alpha (p - p_{\text{atm}})$, where $\alpha$ is the parameter.

In models construction based that at the limiting saturation $S_W^* = S^*$ and $S_G^* = 1 - S^*$, the first or second phases do not form a connected system. It follows that phases are immobile:

$$k_W = 0, \quad \text{for} \ S_W \leq S_W^* = S^*,$$

$$k_G = 0, \quad \text{for} \ S_G \geq S_G^* = 1 - S^*.$$

When the saturation becoming reached, the gas withdrawal stops and $p_{\text{atm}} = 0$. The proposed model will consider using initial and boundary conditions, which will be defined below. The model is solved in the field $\Omega = \left\{ 0 \leq t_0 \leq t \leq T; 0 \leq \right.$

$x \leq L\}$. In the task set, it is required to determine the pressure $p$ and moisture saturation $S_i$, where $i = W, G$. A similar building solution can be founded in articles [2,3].

## 2    Steps to Solve the Filtration Problem

At the first stage, the equation of piezoconductivity is solved,

$$mS_G C_G \frac{\partial p}{\partial t} = -\frac{\partial}{\partial x}\left(-\rho_G k \frac{k_G}{\eta_G}\frac{\partial p}{\partial x}\right) - \rho_G \frac{\partial}{\partial x}\left(-k \frac{k_W}{\eta_W}\frac{\partial p}{\partial x}\right) - q_G. \qquad (5)$$

This equation is derived from system of Eq. (3), an implicit difference scheme is used for the numerical solution, which will be given below. The scheme is implicit, in order to avoid a strict limitation on the Courant number, since the piezoconductivity equation is an equation of parabolic type. The source on the right side of the equation is also taken in implicit form $q_G = \alpha\left(p^{n+1} - p_{atm}\right)$. The implicit assignment of the source makes it possible to increase the stability of the circuit because the diagonal predominance of the coefficients in the matrix increases.

At the second stage, the equation of continuity is solved for the volume fraction of the pore space of water. For this phase there is no source on the right-hand side and it is therefore simpler to solve, although one can take the equation of continuity for the gas. The equation of continuity is an equation of hyperbolic type. It will be solved by an explicit difference scheme with a second order of approximation in time and space. It should also be noted that the continuity equation for filtration problems is a nonlinear transport equation. The method for constructing of the difference scheme for the nonlinear transport equation is described in detail in [4], so in this paper only the difference scheme and the test calculation according to the proposed scheme will be presented.

The second stage is completed, when the pressure is found and the fraction of the volume of the pore space of the water is determined by all the remaining unknowns, after which we proceed to calculate the new time layer.

The time step for calculating the layers is determined by the continuity equation, since for the piezoconductivity equation there is no restriction on the time step.

## 3    Construction of Difference Schemes

With respect to the spatial variable $x$, for simplicity, we introduce a uniform grid with step $h = \frac{L}{N}$, where $N$ is the number of nodes: $x_0 = 0 < x_1 < ... < x_N = L$. We introduce the time grid $t_n = n\tau$, where $n = 0, 1, ... , \tau$ is the grid step, which is determined by the formula $\tau = \frac{\gamma h}{M_u}$ $(M_u = \max\left(||u_G||_C, ||u_W||_C\right), || \cdot ||_C$ is grid $C$-norm), $\gamma$ is the Courant number.

We write out the difference scheme for the piezoconductivity equation

$$(mS_G C_G)_k^n \frac{p_k^{n+1} - p_k^n}{\tau} =$$

$$= \frac{1}{h} \left[ \left( \rho_G k \frac{k_G}{\eta_G} \right)_{k+1/2}^n \nabla_h^+ p_k^{n+1} - \left( \rho_G k \frac{k_G}{\eta_G} \right)_{k-1/2}^n \nabla_h^- p_k^{n+1} \right]$$

$$+ \frac{(\rho_G)_k^n}{h} \left[ \left( k \frac{k_W}{\eta_W} \right)_{k+1/2}^n \nabla_h^+ p_k^{n+1} - \left( k \frac{k_W}{\eta_W} \right)_{k-1/2}^n \nabla_h^- p_k^{n+1} \right] \qquad (6)$$

$$- \alpha \left( p_k^{n+1} - p_{\mathrm{atm}} \right),$$

where

$$\nabla_h^+ p_k^{n+1} = \frac{p_{k+1}^{n+1} - p_k^{n+1}}{h}, \quad \nabla_h^- p_k^{n+1} = \frac{p_k^{n+1} - p_{k-1}^{n+1}}{h}.$$

We rewrite the resulting finite difference Eq. (6) in the three-point form

$$A_k p_{k-1}^{n+1} - C_k p_k^{n+1} + B_k p_{k+1}^{n+1} = -F_k^n, \qquad (7)$$

where

$$A_k = \frac{\tau}{h^2} \frac{1}{(mS_G C_G)_k^n} \left[ \left( \rho_G k \frac{k_G}{\eta_G} \right)_{k-1/2}^n + (\rho_G)_k^n \left( k \frac{k_W}{\eta_W} \right)_{k-1/2}^n \right],$$

$$B_k = \frac{\tau}{h^2} \frac{1}{(mS_G C_G)_k^n} \left[ \left( \rho_G k \frac{k_G}{\eta_G} \right)_{k+1/2}^n + (\rho_G)_k^n \left( k \frac{k_W}{\eta_W} \right)_{k+1/2}^n \right],$$

$$C_k = -(1 + A_k + B_k), \quad F_k = -p_k^n.$$

At the points where the gas is extracted the coefficients have the form:

$$C_k = -\left( 1 + A_k + B_k + \frac{\tau \alpha}{(mS_G C_G)_k^n} \right), \quad F_k = -\left( p_k^n + \frac{\tau \alpha}{(mS_G C_G)_k^n} p_{\mathrm{atm}} \right).$$

The system of linear equations is solved by the tridiagonal matrix algorithm [5,6], after which the pressure on the new time layer becomes known.

The proposed difference scheme for the piezoconductivity equation is well known (see [5]). Note that the difference scheme for this equation is stable for any Courant number. We write out the difference scheme for the continuity equation for the first phase (water), in stream form:

$$\frac{(S_W)_k^{n+1} - (S_W)_k^n}{\tau} + \frac{k}{m\eta_W} \frac{1}{h} \left[ W_{k+1/2}^n - W_{k-1/2}^n \right] = 0, \qquad (8)$$

where

$$W_{k+1/2}^n = -k_W (S_W)_{k+1/2}^n \nabla_h^+ p_k^{n+1}$$

$$-\frac{\tau}{2}\left[\left(\frac{\partial k_W}{\partial S_W}\left[(S_W)^n_{k+1/2}\right]\right)^2\left(\nabla^+_h p^{n+1}_k\right)^2\nabla^+_h(S_W)^n_k\right]-\mu^n_{k+1/2}\nabla^+_h(S_W)^n_k,$$

$\mu^n_{k+1/2}$ is the adaptive artificial viscosity, which is used to ensure monotonicity and increase the stability of the difference scheme.

The proposed difference scheme for the continuity equation was studied using the classical equation of nonlinear transport. In flow form, the equation of nonlinear transport has the form:

$$u^{n+1}_k = u^n_k - \frac{\tau}{h}\left[W^n_{k+1/2} - W^n_{k-1/2}\right], \quad k = 1, \ldots, N_x - 1, \quad n = 0, 1, \ldots; \quad (9)$$

where

$$W^n_{k\pm1/2} = f\left(u^n_{k\pm1/2}\right) - \frac{\tau}{2}g\left(u^n_{k\pm1/2}\right)\nabla^\pm_h u^n_k,$$

$$u^n_{k\pm1/2} = \frac{1}{2}\left(u^n_{k\pm1} + u^n_k\right), \quad \nabla^+_h u^n_k = \frac{u^n_{k+1} - u^n_k}{h}, \quad \nabla^-_h u^n_k = \frac{u^n_k - u^n_{k-1}}{h}.$$

The following initial and boundary conditions are added to the scheme (9):

$$u^0_k = \Psi(x_k), k = 0, \ldots, N_x; \quad (10)$$

$$u^n_0 = \varphi(t_n), \quad u^n_{N_x} = u^n_{N_x-1}, \quad n = 0, \ldots, N_t. \quad (11)$$

We note that in the linear case, when $f(u) = au$, the scheme (9)–(11) becomes the Lax-Wendroff difference scheme:

$$u^{n+1}_k = u^n_k - \frac{\tau a}{2h}\left(u^n_{k+1} - u^n_{k-1}\right) + \frac{\tau^2 a^2}{2h^2}\left(u^n_{k+1} - 2u^n_k - u^n_{k-1}\right).$$

The constructed finite difference scheme was tested on various problems. In particular, we give an example of a numerical calculation of the problem of gradient catastrophe. For this problem, the initial pulse had the form given $u(0, x) = \sin(\pi x)$ on the segment $[0, 1]$. Outside this area, the value of the function was zero. Calculations were carried out on a segment $[0, 2]$ with Courant number equal to 0.5. The numerical solution is shown on Fig. 1. On this figure the solution curves for 50, 100, 150, ... time steps are proposed. From these figures it can be seen the following. In the case of formation of gradient catastrophe the finite difference scheme without the use of artificial viscosity gives the oscillations of solution (Fig. 1, left). The finite difference scheme with the using of artificial viscosity gives stable monotonic numerical solution (Fig. 1, right).

A detailed description of the construction and properties of the finite difference scheme for the continuity equation is given in [4]. Here we list only the basic properties for the proposed difference scheme:

1. This scheme has a second order in time, since the second time derivatives are taken into account in the scheme, and also has a second order in the spatial variable, since the derivative with respect to the spatial variable is approximated by the central-difference.

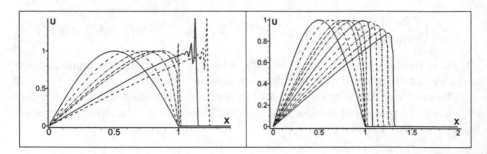

**Fig. 1.** Numerical solution at different time moments before and after of gradient catastrophe. The left figure was obtained without artificial viscousity, the right figure was obtained with using artificial viscousity.

2. The scheme is monotone and stable under the following conditions:

$$0 < \gamma < \frac{\sqrt{2}}{2}, \quad \max\left(0, \frac{1}{2} - \gamma\right) \le \frac{\mu}{h\|u\|_C} \le \frac{1}{2} - \gamma^2.$$

It should be noted that the piezoconductivity equation can also be solved by an explicit scheme. In this case, the time steps for calculating the layers in time will be determined from the piezoconductivity equation.

## 4   The Numerical Experiment

The following data were taken for the proposed filtration model: the reservoir volume porosity is equal $m = 0.1$, the formation temperature $T = 320\,K$, the initial pressure in the formation $p = 7 * 10^6\,Pa$, the pressure at the point of collection $p_{atm} = 1 * 10^5\,Pa$, the gas density $\rho_G = 42.118\,\frac{kg}{m^3}$, the permeability coefficient $k = 0.2\,darcy = 2 * 10^{-13}\,m^2$, the relative phase permeability of water and gas, respectively:

$$k_W = k_W\,(S_W) = \begin{cases} 0, & S_W \le 0.2, \\ 2.78 * (S_W - 0.2)^2, & 0.2 < S_W < 0.8, \\ 1.0008, & S_W \ge 0.8. \end{cases}$$

$$k_G = k_G\,(S_W) = \begin{cases} 0.798, & S_W \le 0.2 \\ 1.33 * (0.8 - S_W)^2, & 0.2 < S_W < 0.8 \\ 0, & S_W \ge 0.8. \end{cases}$$

where $\eta_W = 0.4 * 10^2\,\text{Pois} \equiv 0.4 * 10^{-3}\dfrac{N \cdot s}{m^2}$ is the dynamic viscosity of the 1st

fluid (water), $\eta_G = 0.0155 * 10^2\,\text{Pois} \equiv 1.55 * 10^{-4}\dfrac{N \cdot s}{m^2}$ is the dynamic viscosity

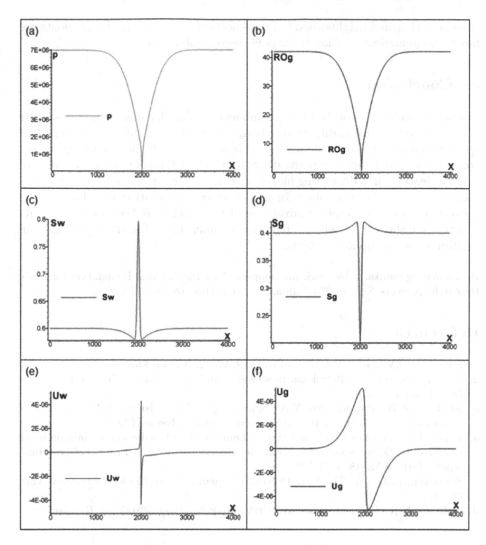

**Fig. 2.** Steady state distributions: pressure (a), gas density (b), water (c) and gas (d) saturations, water (e) and gas (f) velocities.

of the 2nd fluid (gas), the initial fractions of the volume of the pore space for water $S_W = 0.6$, respectively for the gas $S_G = 0.4$. The size of the calculation area $L = 4\,\mathrm{km}$, the sampling point is located in the middle of the calculation area. Numerical results ara proposed in Fig. 2.

The results shown in Fig. 2 correspond to the state of gas well environment after 296 h from the beginning of gas extraction $l = 2\,\mathrm{km}$. These figure show the distributions of pressure (a), gas density (b), moisture saturation (c) and gas saturation (d), as well as water (e) and gas (f) velocities. It can be seen from the figure that in a neighborhood of the selection point, where the gradients are large, the numerical solution behaves without oscillations.

## 5  Conclusion

Within the framework of the filtration model considered, a new difference scheme is constructed for the continuity equation, which has a second order of time and space approximation. It is shown that this difference scheme has monotonicity, conservatism, and well traces the discontinuities of the solution in the form of singular features in a depressing funnel near the gas selection zone. The proposed numerical method can be used for modeling and forecasting the state of gas-saturated reservoirs. Another advantage of the method is the possibility of its effective parallelization when solving multidimensional filtration problems on multiprocessor computer systems.

**Acknowledgments.** This work was supported by the Russian Foundation for Basic Research (projects Nos. 16-07-00519-a, 18-07-00841-a, 16-29-15095-ofi-m).

## References

1. Sivukhin, D.V.: General Course of Physics, vol. 2. Nauka, Moscow (1979)
2. Aziz, K., Settari, A.: Petroleum Reservoir Simulation. Applied Science Publishers Ltd., London (1979)
3. Koldoba, A.B., Poveschenko, Y.A., Samarskaya, E.A., Tishkin, V.F.: Methods of mathematical modeling of the environment. Nauka, Moscow (2000)
4. Popov, I.V.: Construction of difference scheme with high order approximation using adaptive artificial viscosity for nonlinear advection equation. Prepr. Keldysh Inst. Appl. Math. RAS **68**, 1–21 (2017)
5. Samarskii, A.A.: The Theory of Difference Schemes. Marcel Dekker Inc., New York (2001)
6. Kalitkin, N.N.: Chislennye metody. BHV, St.-Petersburg (2011). (in Russian)

# About Free-Volumetric Approximation of a Piezoconductive Medium with Gas Hydrate Inclusions

Yury Poveshchenko[1,2], Viktoriia Podryga[1,3(✉)], and Parvin Rahimly[4]

[1] Keldysh Institute of Applied Mathematics of RAS,
4 Miusskya sq., 125047 Moscow, Russia
`hecon@mail.ru, pvictoria@list.ru`
[2] National Research Nuclear University MEPhI,
31 Kashirskoe sh., 115409 Moscow, Russia
[3] Moscow Automobile and Road Construction State Technical University,
64 Leningradsky Prospect, 125319 Moscow, Russia
[4] Moscow Institute of Physics and Technology (State University),
Dolgoprudny 141701, Moscow Region, Russia
`pervin@rehimli.info`

**Abstract.** The paper deals with the thermodynamically equilibrium model of the splitting by physical processes of a two-component three-phase filtration fluid dynamics with gas hydrate inclusions, for which a family of two-layer completely conservative difference schemes based on support operators method with space-time temporal scales is constructed.

**Keywords:** Gas hydrates · Filtration · Mathematical modeling · Support operators

## 1  Introduction

Many porous materials, saturated with one or several fluids, undergo physico-chemical transformations of the solid phase. For example, gas hydrates, which are compounds of gas and water molecules, can be mentioned [1–3]. In the initial state, the pore space is completely or partially filled with hydrates in the solid state at low temperature and high pressure. When the temperature rises and the compression level decreases, the hydrate decomposes into gas and water. A decrease in temperature and an increase in pressure lead to the formation of hydrates. Huge hydrocarbon reserves in gas hydrate deposits can be extracted quite fully only with the use of mining technologies based on physicochemical

Supported by the Academic Excellence Project of the NRNU MEPhI under contract with the Ministry of Education and Science of the Russian Federation No. 02.A03.21.0005.

transformations of the porous medium, which indicates the relevance of studying such transformations.

Filtration fluid dynamics of free water and gas in the presence of solid hydrate inclusions in a porous medium has a number of specific features.

First of all, according to the Gibbs phase rule, this thermodynamically equilibrium two-component ($H_2O$, $CH_4$) three-phase (hydrate, free water and gas) system has only one thermodynamic degree of freedom. Those there is a thermobaric relationship $T_{dis} = f(P)$ between pressure and temperature of gas hydrate dissociation. It is clear that in this situation, for any basic thermodynamic variable, one can choose any (for example, the internal energies of water or gas), if through them the temperature and pressure are expressed. In what follows we will need this to represent the fluid dynamics of the hydrated medium by Darcy's laws in a discrete energy formulation (see Sect. 3). However, it must be borne in mind that in calculating in a thawed free-hydrate zone in a medium, there are two independent thermodynamic parameters $(P, T)$. In practice, the dynamics of the joint behavior of spatially separated thawed and hydrate-containing zones is of interest.

## 2   Formulation of the Problem

In a spatial domain $O$ with a boundary $\partial O$, we consider thermodynamically equilibrium two-component (water, methane) three-phase equations of filtration fluid dynamics with gas hydrate inclusions [4]:

$$\frac{\partial}{\partial t}\{m\left[S_\nu S_w \rho_w + (1 - S_\nu)\rho_\nu \beta_w\right]\} + div[\rho_w \mathbf{V}_w] + q_w = 0, \tag{1}$$

$$\frac{\partial}{\partial t}\{m\left[S_\nu (1 - S_w)\rho_g + (1 - S_\nu)\rho_\nu (1 - \beta_w))\right]\} + div[\rho_g \mathbf{V}_g] + q_g = 0, \tag{2}$$

$$\mathbf{V}_w = -\frac{k \cdot k_{rw}}{\mu_w}(\nabla P - g\rho_w \mathbf{k}), \tag{3}$$

$$\mathbf{V}_g = -\frac{k \cdot k_{rg}}{\mu_g}(\nabla P - g\rho_g \mathbf{k}), \tag{4}$$

$$\frac{\partial}{\partial t}\{m\left[S_\nu(S_w \rho_w \varepsilon_w + (1 - S_w)\rho_g \varepsilon_g) + (1 - S_\nu)\rho_\nu \varepsilon_\nu\right] + (1 - m)\rho_s \varepsilon_s\}$$
$$+ div\{\rho_w \varepsilon_w \mathbf{V}_w + \rho_g \varepsilon_g \mathbf{V}_g + [P(\mathbf{V}_w + \mathbf{V}_g)]\} + div\mathbf{W} + q_\varepsilon = 0, \tag{5}$$

$$\mathbf{W} = -\{m\left[S_\nu(S_w \lambda_w + (1 - S_w)\lambda_g) + (1 - S_\nu)\lambda_\nu\right] + (1 - m)\lambda_s\}\nabla T, \tag{6}$$

$$T_{dis} = f(P). \tag{7}$$

The indices $g$, $w$, $\nu$, $s$ refer to gas, water, hydrate, skeleton of porous medium; $l$ is the index indicating the phase; $P$ is the pressure, $T$ is the temperature, $S_w$ is the water saturation, $\nu$ is the hydrate saturation, $S_\nu = 1 - \nu$ is the thawing, $\rho_l(P, T)$ and $\varepsilon_l(P, T)$ are the densities and internal energies of the unit mass of the phases, $\mathbf{V}_l$ is the filtration rate of the corresponding phase; $\beta_w$ is the mass fraction of water in the hydrate, $t$ is the time, $q_w$, $q_g$ and $q_\varepsilon$ are the respective sources densities depending on the parameters $(t, \mathbf{r}, S_w, S_\nu, P)$, $\mathbf{r}$ is the radius vector. $\mathbf{W}$ is the heat flux, $\lambda_l(P, T)$ are the coefficients of thermal conductivity. $g\mathbf{k}$ is the vector of acceleration of gravity, directed vertically downwards, $k(r, S_\nu, P)$ is the absolute permeability, $k_{rl}(S_w)$ are the related phase permeabilities, $\mu_l$ are the viscosities. According to the Gibbs phase rule, the three-phase two-component hydrate system is multivariant, i.e. has one degree of freedom (temperature or pressure) [5]. Consequently, for a gas hydrate that is in equilibrium with liquid water or ice, the dependence $T_{dis} = f(P)$ is unambiguous.

The enthalpies $i_l = \varepsilon_l + P/\rho_l$ of hydrate, free water and gas are thermodynamically consistent in the sense of the following relation:

$$\beta_w i_w + (1 - \beta_w) i_g = i_\nu + h, \tag{8}$$

where $h$ is the latent heat of the phase transition of a unit of hydrate mass. For a gas phase the equation of state is

$$\rho_g = \frac{PM}{z_g RT} \tag{9}$$

with a coefficient of supercompressibility $z_g$, $M$ is the molar mass of gas.

From Eqs. (1), (2), (5), an equation for the three-phase piezoconductive of fluid dynamics with hydrate inclusions can be obtained in the following form [4]:

$$m\delta_\varepsilon \left\{ S_\nu \left[ S_w \frac{(\rho_w)_t}{\rho_w} + (1 - S_w) \frac{(\rho_g)_t}{\rho_g} \right] + (1 - S_\nu) \frac{(\rho_\nu)_t}{\rho_\nu} + \frac{(m)_t}{m} \right\}$$
$$+ \frac{\psi}{m\rho_\nu} \left\{ m \left\{ S_\nu \left[ S_w \rho_w (\varepsilon_w)_t + (1 - S_w) \rho_g (\varepsilon_g)_t \right] + (1 - S_\nu) \rho_\nu (\varepsilon_\nu)_t \right\} \right. \tag{10}$$
$$+ \left[ (1 - m) \rho_s \varepsilon_s \right]_t \right\} + \delta_\varepsilon DIG + \frac{\psi}{m\rho_\nu} DIG_\varepsilon = 0,$$

$$DIG = \frac{1}{\rho_w} div[\rho_w \mathbf{V}_w] + \frac{1}{\rho_g} div[\rho_g \mathbf{V}_g] + \frac{q_w}{\rho_w} + \frac{q_g}{\rho_g}, \tag{11}$$

$$DIG_\varepsilon = [div(\rho_w \varepsilon_w \mathbf{V}_w) - \varepsilon_w div(\rho_w \mathbf{V}_w)] + [div(\rho_g \varepsilon_g \mathbf{V}_g) - \varepsilon_g div(\rho_g \mathbf{V}_g)]$$
$$+ div[P(\mathbf{V}_w + \mathbf{V}_g)] + div\mathbf{W} + (q_\varepsilon - \varepsilon_w q_w - \varepsilon_g q_g) = \rho_w \mathbf{V}_w \nabla \varepsilon_w + \rho_g \mathbf{V}_g \nabla \varepsilon_g$$
$$+ div[P(\mathbf{V}_w + \mathbf{V}_g)] + div\mathbf{W} + (q_\varepsilon - \varepsilon_w q_w - \varepsilon_g q_g). \tag{12}$$

Here

$$\frac{\psi}{m\rho_\nu} = \left( \phi - \frac{1}{\rho_\nu} \right) \geq 0, \qquad \phi = \frac{\beta_w}{\rho_w} + \frac{(1 - \beta_w)}{\rho_g}, \tag{13}$$

$$\delta_\varepsilon = \beta_w \varepsilon_w + (1 - \beta_w) \varepsilon_g - \varepsilon_\nu \geq 0 \tag{14}$$

are respectively specific jumps (per unit mass) for the phase transition of the volume and internal energy. The notation is also used hereinafter

$$( )_t = \frac{\partial}{\partial t}, \qquad ( )_p = \frac{\partial}{\partial P}.$$

This is the main piezoconductive-dissipative thermodynamically equilibrium equation of three-phase two-component fluid dynamics with hydrate inclusions physically split with the saturation block (1)–(4) possessing mainly hyperbolic properties on the background of the fixed thermodynamic parameters of the medium [4]. Let's introduce a new value - the intensity of the hydrate system $D_p$ by the relation:

$$D_p = m\delta_\varepsilon \{ S_\nu [S_w \frac{(\rho_w)_p}{\rho_w} + (1 - S_w)\frac{(\rho_g)_p}{\rho_g}] + (1 - S_\nu)\frac{(\rho_\nu)_p}{\rho_\nu} + \frac{(m)_p}{m} \}$$

$$+ \frac{\psi}{m\rho_\nu} \{ m[S_\nu S_w \rho_w (\varepsilon_w)_p + S_\nu (1 - S_w)\rho_g (\varepsilon_g)_p] + (1 - S_\nu)\rho_\nu (\varepsilon_\nu)_p \} \qquad (15)$$

$$+ [(1 - m)\rho_s \varepsilon_s]_p$$

and rewrite Eq. (14) in a more compact form:

$$D_p \frac{\partial P}{\partial t} + \delta_\varepsilon DIG + \frac{\psi}{m\rho_\nu} DIG_\varepsilon = 0. \qquad (16)$$

In (15) we take the total derivative with respect to pressure, taking into account the dependence (7). Using this dependence (7) and choosing the corresponding internal energies $\varepsilon_w$ and $\varepsilon_g$ as the unique thermodynamic degree of freedom in Eqs. (3), (4), (10)–(12), we obtain the Eq. (10) in the energy representation.

## 3    Free-Volume Approximation of Divergent-Piezoconductive Difference Schemes for Filtration Fluid Dynamics Problems with Gas Hydrate Inclusions

Let's introduce some notations for the grid functions of the support operators method [6] (see Fig. 1).

We will refer to its nodes $\omega$ the quantities:

$$\overline{m}, S_\nu, S_w, \rho_\nu, \rho_w, \rho_g, \rho_s, P, T, \varepsilon_\nu, \varepsilon_w, \varepsilon_g, \varepsilon_s, \mu_w, \mu_g, k_{rw}, k_{rg}, q_w, q_g, q_\varepsilon.$$

To the grid bases $\phi$ [6], we classify the vector functions:

$$\mathbf{V}_w, \mathbf{V}_g, \nabla P, \nabla T, \mathbf{W}.$$

We assign grid functions that represent the discontinuous material properties of substances to cells $\Omega$

$$m, k, \lambda_\nu, \lambda_w, \lambda_g, \lambda_s.$$

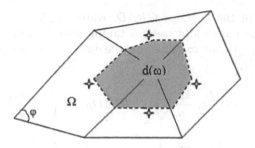

**Fig. 1.** Support operators grid: domains – $d(\omega)$, nodes – $\omega$, cells – $(\Omega)$, faces – $\sigma$ and edges – $\lambda$, bases – $(\phi)$.

The relations are obvious

$$\overline{m_\omega} = \sum_{\phi(\omega)} V_\phi m_{\Omega(\phi)}, \quad \overline{(1-m)_\omega} = \sum_{\phi(\omega)} V_\phi(1 - m_{\Omega(\phi)}) = V_\omega - \overline{m_\omega}, \quad V_\omega = \sum_{\phi(\omega)} V_\phi,$$
(17)

i.e. $\overline{m_\omega}$ and $\overline{(1-m)_\omega}$ represent the volume of the pore domain $d(\omega)$ (see Fig. 1) and its frame part, respectively.

Then on the temporary layers $t$ and $\widehat{t} = t + \tau$ ($\tau > 0$ is the time step), we introduce the difference derivatives with respect to time and to interpolations in mesh nodes $\omega$: $a_t = (\widehat{a} - a)/\tau$, $a^\delta = \delta \widehat{a} + (1 - \delta)a$. Here the interpolation weight $\delta$ may depend on the spatial grid node $\omega$.

Under the value

$$\delta_\nu = \sqrt{(\overline{m}S_\nu)\widehat{}} \Big/ \left( \sqrt{(\overline{m}S_\nu)\widehat{}} + \sqrt{(\overline{m}S_\nu)} \right), \qquad 0 < S_\nu < 1 \tag{18}$$

we mean the free-volume time approximation of the grid functions given at the nodes $\omega$, i.e. interpolation weight $\delta_\nu$ is determined by the proportion of the pore volume, intended for free movement of the liquid and gas. The choice of such an approximation will allow us in the future to produce discrete transformations of equations related to their splitting by physical processes, close to continual ones. Other arbitrary interpolations with respect to time will be denoted by $[\,]\widetilde{}$. They can relate not only to grid nodes $\omega$, but also to its other elements (bases $\phi$ and etc.).

Under the difference analysis operations of vector analysis $DIN$ and $GRAD$ we mean the approximation of their continual analogues $div$ and $grad$. Using the methods of support operators [6], we write the discrete analogue of Eqs. (1), (2) and (5) in the following form.

Equations of continuity, representing the balance of the mass of the water and gas components are

$$\{\overline{m}[S_\nu S_w \rho_w + (1 - S_\nu)\rho_\nu \beta_w]\}_t + DIN(\rho_w \mathbf{V}_w)\widetilde{} + \widetilde{q_w} = 0, \tag{19}$$

$$\{\overline{m}[S_\nu(1 - S_w)\rho_g + (1 - S_\nu)\rho_\nu(1 - \beta_w)]\}_t + DIN(\rho_g \mathbf{V}_g)\widetilde{} + \widetilde{q_g} = 0. \tag{20}$$

With the help of the operator $GRAD$, water $(\rho_w \mathbf{V}_w)\tilde{\ }$ and gas $(\rho_g \mathbf{V}_g)\tilde{\ }$ fluxes are approximated in grid bases $\phi$ taking into account the discretization of Darcy's law (3), (4), for example, on the implicit time layer by any of the standard methods [7,8].

$$(\rho_w \mathbf{V}_w)_\phi^{p\tilde{\ }} = -\left(\rho_w \frac{kk_{rw}}{\mu_w}\right)_{\underline{\Delta\phi}}^{\tilde{\ }} GRADp\tilde{\ } + \left(\rho_w^2 \frac{kk_{rw}}{\mu_w}\right)_{\underline{\Delta\phi}}^{\tilde{\ }} g\mathbf{k}, \qquad (21)$$

$$(\rho_g \mathbf{V}_g)_\phi^{p\tilde{\ }} = -\left(\rho_g \frac{kk_{rg}}{\mu_g}\right)_{\underline{\Delta\phi}}^{\tilde{\ }} GRADp\tilde{\ } + \left(\rho_g^2 \frac{kk_{rg}}{\mu_g}\right)_{\underline{\Delta\phi}}^{\tilde{\ }} g\mathbf{k}. \qquad (22)$$

Here under $(\ )_{\underline{\Delta\phi}}^{\tilde{\ }}$ we mean approximations of the corresponding expressions in grid bases $\phi$ with some interpolation on time.

However, in the presence of a thermobaric dependence of the form (7) for preserving the continual properties of the sign-definiteness of the quadratic forms of the gradients of thermodynamic quantities of the form $\int \varepsilon div(\rho \mathbf{V}) dV$ (see also (33) below), the form of the Darcy's law in the energy formulation is more preferable. We obtain it from the following considerations.

Taking into account the thermobaric dependence (7) in the three-phase equilibrium zone, hydrate-water-gas we can write:

$$d\varepsilon_w = \varepsilon_{wp}' dP, \qquad d\varepsilon_g = \varepsilon_{gp}' dP,$$

where $\varepsilon_{wp}'$ and $\varepsilon_{gp}'$ are the total derivatives of the internal energy with respect to pressure, taking into account (7). Then the Darcy's law (3), (4) in the grid bases (formed by the nodes in which the thermobaric relation (7) is satisfied) can be represented in the energy form:

$$(\rho_w \mathbf{V}_w)_\phi^{\varepsilon\tilde{\ }} = -\left(\rho_w \frac{kk_{rw}}{\mu_w \varepsilon_{wp}'}\right)_{\underline{\Delta\phi}}^{\tilde{\ }} GRAD\varepsilon_w^{(\delta_\nu)} + \left(\rho_w^2 \frac{kk_{rw}}{\mu_w}\right)_{\underline{\Delta\phi}}^{\tilde{\ }} g\mathbf{k}, \qquad (23)$$

$$(\rho_g \mathbf{V}_g)_\phi^{\varepsilon\tilde{\ }} = -\left(\rho_g \frac{kk_{rg}}{\mu_g \varepsilon_{gp}'}\right)_{\underline{\Delta\phi}}^{\tilde{\ }} GRAD\varepsilon_g^{(\delta_\nu)} + \left(\rho_g^2 \frac{kk_{rg}}{\mu_g}\right)_{\underline{\Delta\phi}}^{\tilde{\ }} g\mathbf{k}. \qquad (24)$$

In this way

$$(\rho_w \mathbf{V}_w)_\phi^{\tilde{\ }} = -\left\{(\rho_w \mathbf{V}_w)_\phi^{p\tilde{\ }} \,|\,(\rho_w \mathbf{V}_w)_\phi^{\varepsilon\tilde{\ }}\right\}, \qquad (25)$$

$$(\rho_g \mathbf{V}_g)_\phi^{\tilde{\ }} = -\left\{(\rho_g \mathbf{V}_g)_\phi^{p\tilde{\ }} \,|\,(\rho_g \mathbf{V}_g)_\phi^{\varepsilon\tilde{\ }}\right\}. \qquad (26)$$

The equation for the balance of internal energy approximating (5) has the form:

$$\left\{\overline{m}[S_\nu(S_w \rho_w \varepsilon_w + (1 - S_w)\rho_g \varepsilon_g) + (1 - S_\nu)\rho_\nu \varepsilon_\nu] + \overline{(1 - m)}\rho_s \varepsilon_s\right\}_t$$

$$+DIN\left[\left(\varepsilon_w^{(\delta_\nu)}\right)_{up}(\rho_w \mathbf{V}_w)\tilde{\ }\right] + DIN\left[\left(\varepsilon_g^{(\delta_\nu)}\right)_{up}(\rho_g \mathbf{V}_g)\tilde{\ }\right] \qquad (27)$$

$$+ DIN\left\{[P(\mathbf{V}_w + \mathbf{V}_g)]\tilde{\ }\right\} + DIN\,\mathbf{W}\tilde{\ } + q_\varepsilon\tilde{\ } = 0.$$

Index $up$ in the expression for the energy of water $(\varepsilon_w^{(\delta_\nu)})_{up}$ indicates that the respective values are taken up (upwind) by the water flow $(\rho_w \mathbf{V}_w)\tilde{\ }$ in a previously defined divergence $DIN(\rho_w \mathbf{V}_w)\tilde{\ }$. Similarly, the index $up$ is understood in the expression for the energy of the gas $(\varepsilon_g^{(\delta_\nu)})_{up}$.

Work of pressure forces $[P(\mathbf{V}_w + \mathbf{V}_g)]\tilde{\ }$ and the total heat flux $\mathbf{W}\tilde{\ }$ in the medium are approximated in grid bases $\phi$, for example, on the implicit time layer standard manner [7,8]:

$$[P(\mathbf{V}_w + \mathbf{V}_g)]_\phi^{\tilde{\ }} = \left(\frac{P}{\rho_w}\right)_\phi^{\tilde{\ }} (\rho_w \mathbf{V}_w)_\phi^{p\,\tilde{\ }} + \left(\frac{P}{\rho_g}\right)_\phi^{\tilde{\ }} (\rho_g \mathbf{V}_g)_\phi^{p\,\tilde{\ }} . \tag{28}$$

Further, the discrete analogue of the piezoconductive Eqs. (10)–(14) physically split with the saturation block (19), (20), but difference-equivalent to the system of first principles of the model (19), (20), (21) has the form:

$$\delta_\varepsilon^{(\delta_\nu)} \{ [(\overline{m}S_\nu)S_w]^{(1-\delta_\nu)} \frac{(\rho_w)_t}{(\rho_w)^{(\delta_\nu)}}$$

$$+ [(\overline{m}S_\nu)(1 - S_w)]^{(1-\delta_\nu)} \frac{(\rho_g)_t}{(\rho_g)^{(\delta_\nu)}} + [\overline{m}(1 - S_\nu)]^{(1-\delta_\nu)} \frac{(\rho_\nu)_t}{(\rho_\nu)^{(\delta_\nu)}} + (\overline{m})_t \}$$

$$+ [\psi/(m\rho_\nu)]\tilde{\ } \{ [(\overline{m}S_\nu)S_w\rho_w]^{(1-\delta_\nu)}(\varepsilon_w)_t + [(\overline{m}S_\nu)(1 - S_w)\rho_g]^{(1-\delta_\nu)}(\varepsilon_g)_t \tag{29}$$

$$+ [\overline{m}(1 - S_\nu)\rho_\nu]^{(1-\delta_\nu)}(\varepsilon_\nu)_t + \left[\overline{(1 - m)}\rho_s\varepsilon_s\right]_t \}$$

$$+ \delta_\varepsilon^{(\delta_\nu)} DIG\tilde{\ } + [\psi/(m\rho_\nu)]\tilde{\ } DIG_\varepsilon\tilde{\ } = 0,$$

$$\delta_\varepsilon = [\beta_w\varepsilon_w + (1 - \beta_w)\varepsilon_g] - \varepsilon_\nu, \tag{30}$$

$$[\psi/(m\rho_\nu)]\tilde{\ } = \left[\beta_w/(\rho_w)^{(\delta_\nu)} + (1 - \beta_w)/(\rho_g)^{(\delta_\nu)}\right] - 1/(\rho_\nu)^{(\delta_\nu)}, \tag{31}$$

$$DIG\tilde{\ } = DIN(\rho_w \mathbf{V}_w)\tilde{\ }/(\rho_w)^{(\delta_\nu)} + DIN(\rho_g \mathbf{V}_g)\tilde{\ }/(\rho_g)^{(\delta_\nu)}$$

$$+ q_w\tilde{\ }/(\rho_w)^{(\delta_\nu)} + q_w\tilde{\ }/(\rho_g)^{(\delta_\nu)}, \tag{32}$$

$$DIG_\varepsilon\tilde{\ } = \left[DIN\left\{(\varepsilon_w^{(\delta_\nu)})_{up}(\rho_w \mathbf{V}_w)\tilde{\ }\right\} - (\varepsilon_w)^{(\delta_\nu)}DIN(\rho_w \mathbf{V}_w)\tilde{\ }\right]$$

$$+ \left[DIN\left\{(\varepsilon_g^{(\delta_\nu)})_{up}(\rho_g \mathbf{V}_g)\tilde{\ }\right\} - (\varepsilon_g)^{(\delta_\nu)}DIN(\rho_g \mathbf{V}_g)\tilde{\ }\right] \tag{33}$$

$$+ DIN\left\{[P(\mathbf{V}_w + \mathbf{V}_g)]\tilde{\ }\right\} + DIN\mathbf{W}\tilde{\ } + \left(q_\varepsilon\tilde{\ } - \varepsilon_w^{(\delta_\nu)}q_w\tilde{\ } - \varepsilon_g^{(\delta_\nu)}q_g\tilde{\ }\right).$$

In the expression $DIG_\varepsilon\tilde{\ }$, which appears in (29), there is a monotonic (upwind) energy approximation for $\varepsilon_w^{(\delta_\nu)}$ and $\varepsilon_g^{(\delta_\nu)}$ in the corresponding combinations of divergent expressions containing these quantities.

# 4   Conclusion

In the work on metric grids of the theory of the method of support operators applied to the processes of a two-component three-phase filtration fluid dynamics with gas hydrate inclusions, a family of discrete algorithms split by physical processes with free-volumetric approximation of the piezoconductivity equation in a hydrated medium was constructed. These algorithms allow to make applied calculations with a large time step and are difference-equivalent to the system of the first balanced principles of the model. Also, they allow qualitatively approximate the gradient changes in thermodynamic quantities in the depression wells, the Joule-Thomson effect, and other physical processes inherent in the theory of nonisothermal filtration.

**Acknowledgments.** The work was funded by Russian Foundation for Basic Research, projects no. 18-07-00841-a, 16-29-15081-ofi_m, 16-29-15095-ofi_m.

# References

1. Bondarev, E.A., Babe, G.D., Groisman, A.G., Kanibolotskii, M.A.: Mechanics of hydrate formation in gas flows. Nauka, Novosibirsk (1976). (in Russian)
2. Bogatyrenko, R.S.: Features of the development and operation of gas hydrate deposits (on the example of the Messoyakhsky deposit). Dissertatsiia, Moscow (1979). (in Russian)
3. Dubrovskii, D.A.: Creation of methods for analysis and forecasting of indices of development of gas hydrate deposits. Dissertatsiia, Moscow (1989). (in Russian)
4. Poveshchenko, Y.A., Kazakevich, G.I.: Mathematical modeling of gas hydrate processes. Math. Mach. Syst. **3**, 105–110 (2011). (in Russian)
5. Byk, S.S., Makogon, Y.F., Fomina, V.I.: Gas hydrates. Khimiia, Moscow (1980). (in Russian)
6. Poveshchenko, Y.A., Podryga, V.O., Rahimly, P.I.: About one approach to free volumetric approximation of a piezoconductive medium with gas hydrate inclusions. Math. Montisnigri **40**, 68–89 (2017). (in Russian)
7. Koldoba, A.V., Poveshchenko, Y.A., Gasilova, I.V., Dorofeeva, E.Y.: Numerical schemes of the support operators method for elasticity theory equations. Matem. Modelirovanie **24**(12), 86–96 (2012). (in Russian)
8. Koldoba, A.V., Poveshchenko, Y.A., Samarskaya, E.A., Tishkin, V.F.: Methods of mathematical modeling of the environment. Nauka, Moscow (2000). (in Russian)

# Modeling of Fluidodynamic Processes in a Porous Medium with Gashydrate Deposits

Yu. A. Poveshchenko[1,2], P. I. Rahimly[3], I. V. Gasilova[1], G. I. Kazakevich[4], Yu. S. Sharova[1], V. O. Podryga[1,5($\boxtimes$)], and S. B. Popov[1]

[1] Keldysh Institute of Applied Mathematics of RAS, Moscow, Russia
pvictoria@list.ru
[2] National research nuclear university MEPhI, Moscow, Russia
[3] Moscow Institute of Physics and Technology (State University), Dolgoprudny, Moscow Region, Russia
[4] P.P. Shirshov Institute of Oceanology of RAS, Moscow, Russia
[5] Moscow Automobile and Road Construction State Technical University, Moscow, Russia

**Abstract.** In this paper, the calculations of water saturation, thermal expansion hydrate thaw and thermodynamic parameters (pressure and temperature) were made using the proposed two-block mathematical model of the dissociation of gas hydrates in a porous medium. The numerical model allows discretizing the task in a one-dimensional case and implementing unconditionally stable difference scheme. The obtained results demonstrate the applicability of the proposed model for the solution of typical problems of gas hydrate fluid dynamics, including the studies of the complex dynamics of water and hydrate saturations of the formation in respect to adiabatic expansion of the gas in the collector space. This model will help in the research and modeling of the problems of the phase transformations of gas hydrate inclusions in the porous media.

**Keywords:** Gas hydrates · Filtration · Mathematical modeling

## 1 Introduction

The presence of water vapor in hydrocarbon gases under higher pressure and low temperature leads to the condensation of water vapor and to the formation of gas-containing ice plugs (gas hydrates), which complicates both the processes of transportation and refinement of hydrocarbons. The studies showed that the formation of gas hydrates is also possible in natural conditions, moreover, the volumes of natural deposits of gas hydrates are extremely large [1]. Possibility of extraction of hydrocarbons from gas hydrates makes these compounds an extremely promising target of research in the capacity of the alternative source of hydrocarbons. This requires detailed studies focused on the development of

© Springer Nature Switzerland AG 2019
I. Dimov et al. (Eds.): FDM 2018, LNCS 11386, pp. 411–418, 2019.
https://doi.org/10.1007/978-3-030-11539-5_47

production techniques and solution of possible ecological problems such as the release of gas into the atmosphere.

For the studies of filtration processes with respect to gas hydrates dissociation, the equations of continuum mechanics expressing the laws of conservation of mass, momentum, and energy [2] are used. Mathematical modeling of subterranean gas hydrates has a long history and is carried out in different directions [3–5].

## 2    Mathematical Model

The mathematical model of filtration consists of two blocks: the saturation transfer and the equation of the piezoconductivity for the pressure.

The saturation transfer is written in the following form:

$$\frac{\partial}{\partial t}\left\{m(S_\nu S_w \rho_w + (1 - S_\nu)\rho_\nu \beta_w)\right\} + div[\rho_w \boldsymbol{V}_w] + q_w = 0, \qquad (1)$$

$$\frac{\partial}{\partial t}\left\{m(S_\nu(1 - S_w)\rho_g + (1 - S_\nu)\rho_\nu(1 - \beta_w))\right\} + div[\rho_g \boldsymbol{V}_g] + q_g = 0. \qquad (2)$$

The equation of the piezoconductivity is given by:

$$D_p\frac{\partial P}{\partial t} + \delta_\varepsilon DIG + \frac{\psi}{m\rho_\nu}DIG_\varepsilon = 0. \qquad (3)$$

Material coefficients of the Eq. (3) are given by the following expressions:

$$D_p = m\delta_\varepsilon\left\{S_\nu\left[S_w\frac{(\rho_w)_p}{\rho_w} + (1 - S_w)\frac{(\rho_g)_p}{\rho_g}\right] + (1 - S_\nu)\frac{(\rho_\nu)_p}{\rho_\nu} + \frac{(m)_p}{m}\right\}$$

$$+ \frac{\psi}{m\rho_\nu}\left\{m[S_\nu S_w \rho_w(\varepsilon_w)_p + S_\nu(1 - S_w)\rho_g(\varepsilon_g)_p\right.$$

$$\left. + (1 - S_\nu)\rho_\nu(\varepsilon_\nu)_p] + [(1 - m)\rho_s\varepsilon_s]_p\right\}, \qquad (4)$$

$$\delta_\varepsilon = \beta_w\varepsilon_w + (1 - \beta_w)\varepsilon_g - \varepsilon_\nu,$$

$$\frac{\psi}{m\rho_\nu} = \frac{\beta_w}{\rho_w} + \frac{(1 - \beta_w)}{\rho_g} - \frac{1}{\rho_\nu},$$

$$DIG = \frac{1}{\rho_w}div[\rho_w \boldsymbol{V}_w] + \frac{1}{\rho_g}div[\rho_g \boldsymbol{V}_g] + \frac{q_w}{\rho_w} + \frac{q_g}{\rho_g},$$

$$DIG_\varepsilon = \rho_w \boldsymbol{V}_w \nabla \varepsilon_w + \rho_g \boldsymbol{V}_g \nabla \varepsilon_g + div[P(\boldsymbol{V}_w + \boldsymbol{V}_g)]$$

$$+ divW + (q_\varepsilon - \varepsilon_w q_w - \varepsilon_g q_g) \qquad (5)$$

$$W = -\left\{m[S_\nu(S_w\lambda_w + (1 - S_w)\lambda_g) + (1 - S_\nu)\lambda_\nu] + (1 - m)\lambda_s\right\} \bigtriangledown T.$$

Here $D_p$ is the pressure coefficient, $\delta_\varepsilon$ is the internal energy jump per mass unit during the phase transition, $\frac{\psi}{m\rho_\nu}$ is the specific volume jump. Indices $g$, $w$, $\nu$, $s$ refer to gas, water, hydrate and the skeleton of a porous medium; $l$ is an index referred to a phase; $P$ is the pressure, $T$ is the temperature, $S_w$ is the water saturation, $S_g = 1 - S_w$ is the gas saturation, $\nu$ is the hydration saturation, $S_\nu = 1 - \nu$ is the hydrate thaw, $\rho_l = (P,T)$ refers to a phase density, $V_l$ is the filtration velocity of the corresponding phase; $\beta_w$ is the mass fraction of water in hydrate, $m(r, P)$ is the porosity, $r$ is the radius vector, $t$ is the time, $q_l(t, r, S_w, S_\nu, P)$ are the phase source densities, $\varepsilon_l(P,T)$ are the phase internal energies, $\lambda_l(P,T)$ are the thermal conductivity coefficients included in the thermal flux $W$.

The filtration velocities are given by the following expressions (Darcy's law respect of gravity in a medium with total pressure):

$$V_w = -\frac{k \cdot k_{rw}}{\mu_w}(\bigtriangledown P - g\rho_w k), \qquad (6)$$

$$V_g = -\frac{k \cdot k_{rg}}{\mu_g}(\bigtriangledown P - g\rho_g k), \qquad (7)$$

where $k$ is vertical coordinate axis, $g$ is gravity acceleration, $k(r, S_\nu, P)$ is absolute permeability, $k_{rw}(S_w)$, $k_{rg}(S_w)$ are the phase permeabilities, $\mu_w$, $\mu_g$ are the viscosities of water and gas.

The internal energy of the hydrate is expressed through the energies of its compounds (gas and water):

$$\beta_w i_w + (1 - \beta_w)i_g = i_\nu + h, \qquad (8)$$

where $h$ is the latent heat of a phase transition of a hydrate mass unit, $i_l = \varepsilon_l + \frac{P}{\rho_l}$ are the enthalpies.

Hydrate state is described by the phase equilibrium formula:

$$T = AlnP + B, \qquad (9)$$

where $A$, $B$ are the constants.

Due to this formula, the temperature dependence in all the expressions can be converted to the pressure dependence.

Gas state equation is given in the following form:

$$\rho_g = \frac{PM}{z(P,T)RT}, \qquad (10)$$

where $M$ is the molar mass of the gas, $R$ is the universal gas constant, $z(P,T)$ is the gas supercompressibility coefficient.

Thus, four given above Eqs. (1)–(3) with (9) contain three unknown independent variables: water saturation $S_w$, thaw $S_\nu$ and pressure $P$ are the unknowns for the three Eqs. (1)–(3) of model. All the Eqs. (1)–(3) are nonlinear.

## 3 Formulation of the Problem

We consider the one-dimensional formulation of the problem. It is assumed that the thermal conductivity deal in the total heat transfer balance is negligible compared to convection, i.e. the conductive component in the energy equation is taken to be equal to zero ($div\mathbf{W} = 0$). The gravity acceleration is also not counted ($g = 0$). As a result, the system (1)–(3) results to the following form:

$$\frac{\partial}{\partial t}\left\{ m(S_\nu S_w \rho_w + (1 - S_\nu)\rho_\nu \beta_w) \right\} + \frac{\partial}{\partial x}\left[ \rho_w V_w \right] + q_w = 0, \tag{11}$$

$$\frac{\partial}{\partial t}\left\{ m(S_\nu(1 - S_w)\rho_g + (1 - S_\nu)\rho_\nu(1 - \beta_w)) \right\} + \frac{\partial}{\partial x}\left[ \rho_g V_g \right] + q_g = 0. \tag{12}$$

$$D_p \frac{\partial P}{\partial t} + \delta_\varepsilon \left( \frac{1}{\rho_w}\frac{\partial}{\partial x}\left[ \rho_w V_w \right] + \frac{1}{\rho_g}\frac{\partial}{\partial x}\left[ \rho_g V_g \right] + \frac{q_w}{\rho_w} + \frac{q_g}{\rho_g} \right)$$
$$+ \frac{\psi}{m\rho_\nu}\left( \rho_w V_w \frac{\partial \varepsilon_w}{\partial x} + \rho_g V_g \frac{\partial \varepsilon_g}{\partial x} + \frac{\partial}{\partial x}\left[ P(V_w + V_g) \right] \right.$$
$$\left. + (q_\varepsilon - \varepsilon_w q_w - \varepsilon_g q_g) \right) = 0. \tag{13}$$

$$V_w = -\frac{k \cdot k_{rw}}{\mu_w}\left( \frac{\partial P}{\partial x} \right), \; V_g = -\frac{k \cdot k_{rg}}{\mu_g}\left( \frac{\partial P}{\partial x} \right). \tag{14}$$

The boundaries of the computational domain are assumed impermeable solid "walls", i.e. they have zero flux through them:

$$V_w\big|_{x=0} = 0, \; V_g\big|_{x=0} = 0, \; V_w\big|_{x=l} = 0, \; V_g\big|_{x=l} = 0, \; t > 0. \tag{15}$$

The fact of utmost importance is that the initial problem, stated in the form of conservation laws (masses of $H_2O$, $CH_4$ and total energy of the medium) with a total matrix of the system written for the functions $S_\nu$, $S_w$, pressure $P$ and temperature $T$, possesses both hyperbolic and parabolic properties. The direct using of this system for determining of the dynamics of variables $S_\nu$, $S_w$, $P$, $T$ and for building implicit difference scheme required for the calculation of parabolic equations with a big time step is rather difficult.

In the developed numerical method the splitting of the system into the saturation transport block of the fluids alongside with given velocities (possessing mainly hyperbolic properties) and the piezoconductivity block of the system with hydrate inclusions (determining the dissipative evolution of thermodynamic parameters of the steady-state fluid-hydrate model) is performed. Such splitting of the initial task by physical processes allows creating applied algorithms for the system matrix of difference equations only for pressure $P$, which gives the possibility of performing calculations with the least amount of unknown parameters.

For solution of the given initial-boundary value problem finite difference method is used. For this, a grid with a uniform spatial step $h$ and time step $\tau_n$, where $n$ is the number of the time step, therefore their grid counterparts replace the original equations, boundary and initial conditions. When constructing schemes the UPWIND approximation is used for saturation and the DOWNWIND approximation is used for thaw. This procedure follows from the analysis of the hyperbolic properties of the system of equations with respect to $S_\nu$, $S_w$ with fixed velocities, defined by the Darcy's law.

## 4   Initial Data Definition

Let us consider a process, in which at the start time the pressure is distributed by the following law:

$$P(x,0) = 2 \cdot 10^7 \, \text{Pa}, \tag{16}$$

and water saturation and thaw are also homogeneously distributed in space:

$$S_w(x,0) = S_w^*, \ S_\nu(x,0) = S_\nu^*, \tag{17}$$

where $0 < S_w^* < 1$ and $0 < S_\nu^* < 1$ are constants, $x \in [0, l]$, $l$ is the length of the computational domain.

Two cases of source operation in a reservoir environment are considered:

$$q_w = \alpha(P - P^*), \ q_g = \beta(P - P^*), \ q_\varepsilon = \varepsilon_w q_w + \varepsilon_g q_g. \tag{18}$$

As well as energy sources taking into account adiabatic expansion of the gas

$$q_\varepsilon = \varepsilon_w q_w + \varepsilon_g q_g + \frac{q_g}{\rho_g} P, \tag{19}$$

where $P^*$ is a constant, $\alpha$ and $\beta$ are the constant parameters describing certain characteristics of the source (discharge) in the borehole zone due to the pressure drop inside the bore hole and in the shelf. The following values of the parameters, peculiar for the Messoyakhsky gas hydrate deposit, were selected:

$$\rho_w = 10^3 \, \frac{kg}{m^3}, \ \rho_\nu = 910 \, \frac{kg}{m^3}, \rho_s = 2800 \, \frac{kg}{m^3}, \ \beta_w = 0.9,$$

$$A = 7.28 \, K, \ B = 169.7 \, K,$$

$$\mu_w = 10^{-3} \, Pa \cdot s, \ \mu_g = 0.014 \cdot 10^{-3} \, Pa \cdot s,$$

$$c_w = 4165 \, \frac{J}{kg \cdot K}, \ c_g = 2500 \, \frac{J}{kg \cdot K}, \ c_s = 873 \, \frac{J}{kg \cdot K},$$

$$M = 0.016 \, \frac{kg}{mol}, \ h = 514810 \, \frac{J}{kg}, \ R = 8.31 \, \frac{J}{mol \cdot K}, \ m = 0.35,$$

$$S_w^* = 0.6, \ S_\nu^* = 0.5, \ k(S_\nu) = k_0(S_\nu)^3, \ k_0 = 10^{-14} \, m^2, \ P^* = 2 \, MPa,$$

$$k_{rw}(S_w) = 1.477 S_w^5 - 1.587 S_w^6 + 1.11 S_w^7 - 0.0473,$$

$$k_{rg}(S_w) = 1.044 - 1.7 S_w + 0.6 S_w^2.$$

The minimum value of water saturation $S_{w\ min} = 0.55$.
$k_{rw}(S_w) = 0$, $k_{rg}(S_w) = k_{rg}(S_{w\ min})$ at $S_w \leq S_{w\ min}$.
The maximum value of water saturation $S_{w\ max} = 0.9$.
$k_{rw}(S_w) = k_{rw}(S_{w\ max})$, $k_{rg}(S_w) = 0$ at $S_w \geq S_{w\ max}$.
The length of the model tube is assumed to be equal to $l = 1$ m, the step along the spatial coordinate $h = 0.01$ m. The parameters of source are $\alpha = 0$, $\beta = 10^{-5}\ kg/(Pa \cdot s \cdot m^3)$ The calculations are performed for the time moments $t = 1, 10, 50$ s. Area of action of the sources has the boundaries $x \in [0.4,\ 0.6]$. The computational domain is symmetric with respect to $x = 0.5$ and is not halved for the purpose of checking the symmetry properties of the computational algorithm (the sweep coefficients are calculated from left to right, the grid functions – from right to left).

## 5   Results of Calculations

Figure 1 shows the spatial distribution of thaw taking into account adiabatic expansion of gas (19) for time points 1, 10, 50 s, respectively. Figure 2 also shows the distribution for the case (18) when there is no adiabatic expansion of gas. In the first case, when adiabatic expansion of gas is taken into account, there is no additional thawing of the hydrated medium on the boundaries of the are of action of sources ($x \in [0.4,\ 0.6]$). Other parameters ($S_w$, $P$, $T$) show the similar behavior in both cases, their spatial distributions for the same time points are presented in Figs. 3, 4, and 5, respectively.

Additional analysis and testing of the mechanisms of appearance of thaw peaks on the boundary of the area of action of sources ($x \in [0.4,\ 0.6]$) were carried out in the case of absence of adiabatic expansion of gas (i.e., when selecting sources in the collector medium in the form (18)). Appearance of peaks can be explained in the following way: gas drawing is set up in the form (18) with $\alpha = 0$, the water generated during hydrate decomposition is incompressible, is not drained and cumulates on the boundaries of the drain zone. This situation leads to the corresponding thaw peaks when continuing gas drawing on the boundaries of the drain zone area of action. The results of calculations with different phase permeabilities showed that the nonlinearity of the relative permeabilities in water and gas, as well as absolute permeability, does not affect the appearance of peaks. If to switch off gas drawing after the first second, the peaks do not appear.

Calculations were performed on a personal computer with clock speed and processor: 2.7 GHz Intel Core i5. A typical calculation on a grid of 100 cells took several minutes.

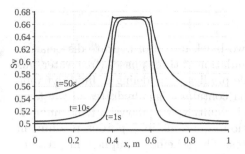

**Fig. 1.** Distribution of thaw for time steps 1, 10, 50 s taking into account adiabatic expansion of gas

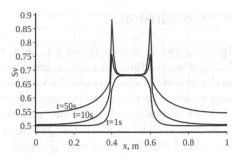

**Fig. 2.** Distribution of thaw for time steps 1, 10, 50 s without adiabatic expansion of gas

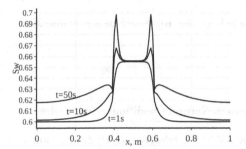

**Fig. 3.** Distribution of water saturation for time steps 1, 10, 50 s taking into account adiabatic expansion of gas

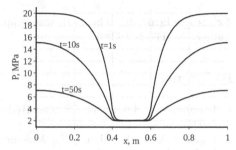

**Fig. 4.** Distribution of pressure for time steps 1, 10, 50 s taking into account adiabatic expansion of gas

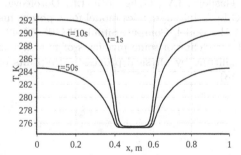

**Fig. 5.** Distribution of temperature for time steps 1, 10, 50 s taking into account adiabatic expansion of gas

# 6    Conclusion

In the paper it is formulated a general two-block model of hydrate dissociation in a porous medium for heat exchange simulation of three phases (free water, gas and gas hydrates) with regard to full-scale predictive modeling of fluid dynamic and dissipative processes in porous media containing gas hydrates in the form of solid inclusions. A difference scheme is built, which allows solving numerically the system of fluid and gas filtration equations in a porous medium in respect of the dissociation of gas hydrates. The scheme is implicit in pressure and explicit in water saturation and thaw. The results of performed numerical experiments, including the ones with adiabatic expansion of gas in the collector space, demonstrate the applicability of developed methods for the calculations of real problems, related to gas hydrates deposits, as well as for the study of complex processes of water and hydrate saturation dynamics in the reservoir.

**Acknowledgments.** The work was supported by the Russian Science Foundation (project No 17-71-20118).

# References

1. Istomin, V.A., Iakushev, V.S.: Gas Hydrates in Natural Conditions. Nedra, Moscow (1992)
2. Basniev, K.S., Kochina, I.N., Maksimov, V.M.: Underground Hydromechanics. Nedra, Moscow (1993)
3. Poveshchenko, Yu.A., Podryga, V.O., Rahimly, P.I., Sharova, Yu.S.: About one discrete model of splitting by the physical processes of a piezoconductive medium with gas hydrate inclusions. J. Phys.: Conf. Ser **946**, paper 012077 (16 p.) (2018). https://doi.org/10.1088/1742-6596/946/1/012077
4. Poveschenko, O.J., Gasilova, I.V., Galiguzova, I.I., Dorofeeva, E.J., Olkhovskaya, O.G., Kazakevich, G.I.: A fluid dynamics model for a porous media, containing gas hydrate deposits. Math. Model. Comput. Simul. **25**(10), 32–42 (2013)
5. Rahimly, P.I., et al.: Modeling of some problems of fluid dynamics containing gas hydrates based on splitting by physical processes. Preprints of the Keldysh Inst. Appl. Math. **39** (2018)

# Nonlinear Differential Equation of the Surface Section of Gas-Liquid

L. A. Prokudina$^{(\boxtimes)}$

South Ural State University, pr. Lenina 76, Chelyabinsk, Russia
prokudina-la@mail.ru

**Abstract.** Study of flow of thin layers of viscous liquids (liquid films) is of great theoretical and practical importance. For liquid films is the basis of technological processes in many industries: petrochemical, thermal power, food and other. The development of nonlinear mathematical models that adequately reflect the real flow of a liquid film, the contact and interaction between fluids (gas-liquid), the calculation characteristics of the film flow is an important task.

Presents the nonlinear mathematical model of the state of free surface liquid film, accounting for the interaction with the gas flow is a nonlinear differential equation of fourth order for deflection of the free liquid surface from the unperturbed state $\psi(x,t)$, where $x$ – the spatial variable, $t$ – time.

The transition to the finite-difference equation is completed. Computational algorithms for the calculation of wave characteristics and deviation of free surface liquid film in contact with the gas stream are developed. The results of computational experiments for the vertical film of water at moderate Reynolds numbers are presented.

**Keywords:** Liquid film · Instability · Finite-difference equation · Reynolds number · Shear stress

## 1 Introduction

The first studies of liquid films were carried out by Kapitsa [1–3]. The study of thin liquid films is of considerable interest to many industries such as chemical, petrochemical, thermal power, etc. [4–8]. For liquid film is realized in numerous systems (columns with parallel-sided nozzle, the tubular film column, absorbers etc.). A feature of liquid films is the presence of a large free surface with a small thickness of the film, which allows to accelerated chemical, diffusion, chemical processes on the surface of the gas-liquid section. The flows of the thin liquid layer in contact with the gas flow of the opposite direction are of interest. Such trends occur in the extraction of oil from reservoirs saturated with gas. The liquid is pushed aside to the walls of the pipe and forms a thin film.

For the theoretical study of film is regarded as a model system. To study the phenomena in the liquid film, it is necessary to have an equation that establishes a relationship between the main characteristics of its wave flow. This paper

I. Dimov et al. (Eds.): FDM 2018, LNCS 11386, pp. 419–426, 2019.
https://doi.org/10.1007/978-3-030-11539-5_48

presents a model equation, a nonlinear differential equation of the liquid-gas interface. Difference methods are widely used to solve partial differential equations describing unsteady physical processes [9–11]. In this paper, the nonlinear partial differential equation was solved by finite difference method on a uniform rectangular grid.

The aim of this work is to study the wave parameters of liquid films and the state of the gas-liquid interface with the combined action of gravity and shear stress.

## 2    Problem Statement

Consider the flow of a thin layer of liquid (liquid film) by gravity ($g$) on a solid impermeable surface (Fig. 1) for a range of Reynolds numbers $Re < 15$. A gas flow moves over the liquid film in counter flow mode, creating a constant tangential stress on its free surface ($\tau$).

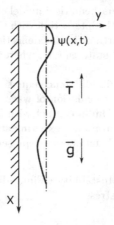

**Fig. 1.** Flow pattern of a liquid film.

The system of Navier-Stokes equations and continuity equation is written in projections on the coordinate axis:

$$\begin{cases} \frac{\partial u}{\partial t} + u\frac{\partial u}{\partial x} + v\frac{\partial u}{\partial y} = -\frac{\partial P}{\partial x} + F_x + \frac{1}{Re}\left(\frac{\partial^2 u}{\partial y^2} + \frac{\partial^2 u}{\partial x^2}\right), \\ \frac{\partial v}{\partial t} + u\frac{\partial v}{\partial x} + v\frac{\partial v}{\partial y} = -\frac{\partial P}{\partial y} + \frac{1}{Re}\left(\frac{\partial^2 v}{\partial y^2} + \frac{\partial^2 v}{\partial x^2}\right), \\ \frac{\partial u}{\partial x} + u\frac{\partial v}{\partial y} = 0; \end{cases} \tag{1}$$

The boundary conditions on the liquid film surface express the balance of normal and tangential forces acting on the gas-liquid interface. Nonlinear boundary conditions are include:

(1) Surface tension;
(2) The gas flow, creating on the free surface of the film constant shear stress;
(3) Presence of free surface films of insoluble surfactants

$$y = 0: \quad u = 0, \ v = 0 \tag{2}$$

$$y = \delta: \ \frac{1}{Re}\left[2\frac{\partial u}{\partial x}\frac{\partial \delta}{\partial x} - 2\frac{\partial v}{\partial y}\frac{\partial \delta}{\partial x} - \left(\frac{\partial v}{\partial x} + \frac{\partial u}{\partial y}\right)\right] + N\frac{\partial^2 u}{\partial x^2} + \tau_x = 0 \tag{3}$$

$$P = \frac{2}{Re}\left[\frac{\partial v}{\partial y} - \left(\frac{\partial u}{\partial y} + \frac{\partial v}{\partial x}\right)\frac{\partial \delta}{\partial x}\right] - \sigma\frac{\partial^2 \delta}{\partial x^2} + P_0^+ \tag{4}$$

The kinematic boundary condition

$$\frac{\partial \delta}{\partial t} = v - u\frac{\partial \delta}{\partial x} \tag{5}$$

Dimensionless quantities in formulas (1–5): $u$, $v$ are projections of velocity; $x$, $y$ are variables; $t$ is time; $\delta$ is the thickness of the liquid film; $Re$ is the Reynolds number; $F_x$ is the projection of the Froude number; $P$ is pressure; $\tau_x$ is shear stress; $\sigma$ is the surface tension parameter; $N$ is the surface viscosity parameter.

## 3  Nonlinear Model Equation of Surface of Section

To solve the flow problem of thin liquid films (1–4), the method of narrow strips (small parameter) [8–13] was used. Given that the thickness of the liquid film is much less than the wavelength, we introduce a small parameter

$$\varepsilon = \frac{\delta}{\lambda}$$

$$\begin{cases} \varepsilon\frac{\partial u}{\partial t} + \varepsilon u\frac{\partial u}{\partial x} + \varepsilon v\frac{\partial u}{\partial y} = -\varepsilon\frac{\partial P}{\partial x} + F_x + \frac{1}{Re}\left(\frac{\partial^2 u}{\partial y^2} + \varepsilon^2\frac{\partial^2 u}{\partial x^2}\right), \\ \varepsilon^2\frac{\partial v}{\partial t} + \varepsilon^2 u\frac{\partial v}{\partial x} + \varepsilon^2 v\frac{\partial v}{\partial y} = -\frac{\partial P}{\partial y} + \frac{1}{Re}\left(\varepsilon^3\frac{\partial^2 v}{\partial y^2} + \varepsilon\frac{\partial^2 v}{\partial x^2}\right), \\ \frac{\partial u}{\partial x} + u\frac{\partial v}{\partial y} = 0; \end{cases}$$

$$y = 0: \quad u = 0, \ v = 0$$

$$y = \delta: \ \frac{1}{Re}\left[2\varepsilon^2\frac{\partial u}{\partial x}\frac{\partial \delta}{\partial x} - 2\varepsilon^2\frac{\partial v}{\partial y}\frac{\partial \delta}{\partial x} - \left(\varepsilon^2\frac{\partial v}{\partial x} + \frac{\partial u}{\partial y}\right)\right] + \varepsilon\overline{N}\frac{\partial^2 u}{\partial x^2} + \tau_x = 0$$

$$P = \frac{2}{Re}\left[\varepsilon\frac{\partial v}{\partial y} - \left(\frac{\partial u}{\partial y} + \varepsilon^2\frac{\partial v}{\partial x}\right)\varepsilon\frac{\partial \delta}{\partial x}\right] - \overline{\sigma}\frac{\partial^2 \delta}{\partial x^2} + P_0^+$$

where $\overline{N} = \varepsilon N$; $\overline{\sigma} = \varepsilon^2\sigma$

Unknown functions $u$, $v$, $P$ we decompose into a series by a small parameter $\varepsilon$[14]:

$$\begin{cases} u = u_0 + \varepsilon u_1 + \varepsilon^2 u_2 + \dots, \\ v = v_0 + \varepsilon v_1 + \varepsilon^2 v_2 + \dots, \\ P = P_0 + \varepsilon P_1 + \varepsilon^2 P_2 + \dots \end{cases}$$

The wide application of asymptotic perturbation methods in the study of physic-chemical hydrodynamic systems is justified by a good coincidence of the obtained data with the experiment or with the results of numerical methods.

Believing that the thickness of the liquid film

$$\delta(x,\, t) = 1 + \psi(x,\, t),$$

and substituting by the obtained solution $u_0$, $v_0$, $u_1$, $v_1$ in Eq. (5) we get a nonlinear differential equation for the deviation $\psi(x,\, t)$. We write this equation in the form:

$$\frac{\partial \psi}{\partial t} = b_1 \frac{\partial \psi}{\partial x} + b_2 \frac{\partial^2 \psi}{\partial x^2} + b_3 \frac{\partial^3 \psi}{\partial x^3} + b_4 \frac{\partial^4 \psi}{\partial x^4} + b_5 \psi \frac{\partial \psi}{\partial x} + b_6 \psi \frac{\partial^2 \psi}{\partial x^2} +$$
$$+ b_7 \left( \frac{\partial \psi}{\partial x} \right)^2 + b_8 \frac{\partial \psi}{\partial x} \frac{\partial^2 \psi}{\partial x^2} + b_9 \psi \left( \frac{\partial \psi}{\partial x} \right)^2, \tag{6}$$

where the coefficients of Eq. (6) include physical-chemical factors in the original mathematical model of a liquid film

$$b_1 = -ReF_x - Re\tau_x, \quad b_2 = \frac{3}{40} Re^3 F_x(\tau_x + F_x), \quad b_3 = -\frac{Re^2 F_x N}{2}, \quad b_4 = -\frac{Re\sigma}{3},$$

$$b_5 = -2ReF_x - Re\tau_x, \quad b_6 = b_7 = \frac{3}{8} Re^3 F_x \tau_x + \frac{9}{20} Re^3 F_x^2, \quad b_8 = -\frac{1}{2} Re^2 F_x N,$$

$$b_9 = 2\left( \frac{3}{4} Re^3 F_x \tau_x + \frac{9}{8} Re^3 F_x^2 \right).$$

## 4    Finite Difference Method

We write the Eq. (6) in finite differences. Replace derivatives (the subscript $i$ indicates the step in spatial coordinate, and the index $j$ - at the time), as follows

$$\frac{\partial \psi}{\partial x} \rightarrow \frac{\psi_{i+1}^j - \psi_{i-1}^j}{2\Delta x}, \quad \frac{\partial^2 \psi}{\partial x^2} \rightarrow \frac{\psi_{i+1}^j - 2\psi_i^j + \psi_{i-1}^j)}{(\Delta x)^2}, \quad \frac{\partial \psi}{\partial t} \rightarrow \frac{\psi_i^{j+1} - \psi_i^j}{\Delta t}$$

$$\frac{\partial^3 \psi}{\partial x^3} \rightarrow \frac{\psi_{i+2}^j - 2\psi_{i+1}^j + 2\psi_{i-1}^j - \psi_{i-2}^j}{4(\Delta x)^3},$$

$$\frac{\partial^4 \psi}{\partial x^4} \rightarrow \frac{\psi_{i+2}^j - 4\psi_{i+1}^j + 6\psi_i^j - 4\psi_{i-1}^j + \psi_{i-2}^j}{(\Delta x)^4}.$$

We introduce the parameter $\lambda = \frac{\Delta t}{(\Delta x)^4}$.

Then the equation for the deviation of the free surface of the film in finite differences can be written as

$$\psi_i^{j+1} = \psi_i^j + \lambda \left( b_1 (\Delta x)^3 \frac{\psi_{i+1}^j - \psi_{i-1}^j}{2} + b_2 (\Delta x)^2 (\psi_{i+1}^j - 2\psi_i^j + \psi_{i-1}^j) + \right.$$

$$+ b_3 \frac{\Delta x}{4} (\psi_{i+2}^j - 2\psi_{i+1}^j + 2\psi_{i-1}^j - \psi_{i-2}^j) +$$

$$+ b_4 (\psi_{i+2}^j - 4\psi_{i+1}^j + 6\psi_i^j + \psi_{i-2}^j) +$$

$$+ b_5 \frac{(\Delta x)^3}{2} \psi_i^j (\psi_{i+1}^j - \psi_{i-1}^j) + b_6 (\Delta x)^2 \psi_i^j (\psi_{i+1}^j - 2\psi_i^j + \psi_{i-1}^j) + \qquad (7)$$

$$+ b_7 \frac{(\Delta x)^2}{4} \psi_{i+1}^j (\psi_{i+1}^j - \psi_{i-1}^j) - b_7 \frac{(\Delta x)^2}{4} \psi_{i-1}^j (\psi_{i+1}^j - \psi_{i-1}^j)$$

$$+ b_8 \frac{\Delta x}{2} \psi_{i+1}^j (\psi_{i+1}^j - 2\psi_i^j + \psi_{i-1}^j) - b_8 \frac{\Delta x}{2} \psi_{i-1}^j (\psi_{i+1}^j - 2\psi_i^j + \psi_{i-1}^j) +$$

$$\left. + b_9 \frac{(\Delta x)^2}{4} \psi_i^j (\psi_{i+1}^j - \psi_{i-1}^j)^2 \right).$$

## 5   Computational Experiments

### 5.1   Instability of Liquid Films

We consider the linear part of the free surface equation of a liquid film (7)

$$\frac{\partial \psi}{\partial t} = b_1 \frac{\partial \psi}{\partial x} + b_2 \frac{\partial^2 \psi}{\partial x^2} + b_3 \frac{\partial^3 \psi}{\partial x^3} + b_4 \frac{\partial^4 \psi}{\partial x^4}$$

Believing in the formula $\psi = A e^{i(kx - \omega t)}$, where $\omega = \omega_r + i\omega_i$, k – wave number, we get dependencies for wave parameters: $\omega_r$ – frequency, $\omega_i$ – increment.

The criterion of instability of liquid films is $\omega_i > 0$. Computational experiments were carried out for the vertical water film. Physical factors are related by relationships

$$\sigma = 4887 Re^{-\frac{5}{3}}, \quad F_x = \frac{3}{Re} - \frac{3}{2} \tau_x.$$

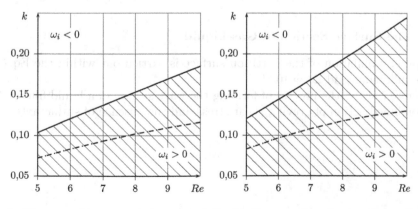

**Fig. 2.** The unstable area of the vertical liquid film: a – free flowing ($\tau = 0$); b – countercurrent mode ($\tau = -0.1$); intermittent curve is the maximum values of the increment

Figure 2 shows the instability regions of the vertical liquid film. Gas flow in counter-flow mode has a destabilizing effect on the film.

Figures 3 and 4 shows the increment for free flowing and countercurrent mode.

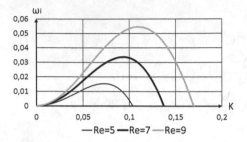

**Fig. 3.** Increment $\tau = 0$                         **Fig. 4.** Increment $\tau = -0.1$

Figures 5 and 6 shows the phase velocity for free flowing and countercurrent mode.

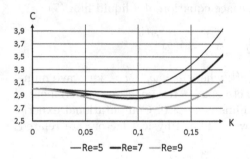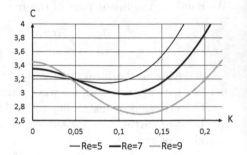

**Fig. 5.** Phase velocity $\tau = 0$                 **Fig. 6.** Phase velocity $\tau = -0.1$

## 5.2   The Surface Section of Gas-Liquid

Computer simulation of the partition surface is carried out within the Eq. (7). Finite-difference grid $\Delta t = 10^{-9}$

Figure 7 shows the process of forming the wave surface of a liquid film under the combined action of gravity, shear stress and the absence of surfactants.

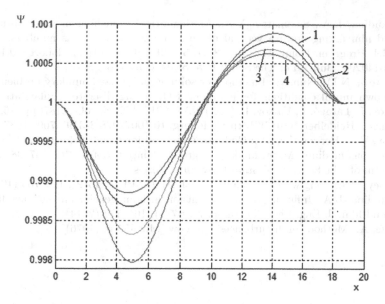

**Fig. 7.** The state of a free surface $Re = 9$, $1 - t = 0.001$, $2 - t = 0.003$, $3 - t = 0.005$, $4 - t = 0.008$

**Acknowledgments.** Supported by Ministry of Education and Science of the Russian Federation within the framework of the basic part of the State task "Development, research and implementation of data processing algorithms for dynamic measurements of spatially distributed objects", Terms of Reference 8.9692.2017/8.9 from 17.02.2017.

# References

1. Kapitsa, P.L.: Wave flow of thin layers of viscous fluid. Free Flow ZhETF **18**(1), 3–18 (1948)
2. Kapitsa, P.L.: Wave flow of thin layers of viscous fluid. Curr. Contact Gas Flow Heat Transf. ZhETF **18**(1), 19–28 (1948)
3. Kapitza, P.L., Kapitza, S.P.: Wave flow of thin layers of viscous fluid. Experimental study of wave flow regime. ZhETF **19**(2), 105–120 (1949)
4. Ganchev, B. G.: Cooling elements of nuclear reactors with flowing films, Moscow. Energoatomizdat, 192 p. (1987, in Russia)
5. Olevsky, V.M., Sweet, V.R., Kashnikov, A.M., Chernyshev, V.I.: Film heat and mass transfer equipment, Moscow. Chemistry, 240 p. (1988, in Russia)
6. Kholpanov, L.P., Shkadov, V.Y.: Fluid flow and heat and mass transfer with the surface of section, Moscow. Science, 271 p. (1990, in Russia)
7. Alekseenko, S.V., Nakoryakov, V.E., Pokusaev, B.G.: Wave flow of liquid films, Moscow. Science, 256 p. (1992, in Russia)
8. Ivanilov, Y.P.: Roll waves in an inclined channel. J. Calculate. Math. Math. Phys. **1**(6), 1061–1076 (1961)

9. Prokudina, L.A., Salamatov, Y.A.: Mathematical modelling of wavy surface of liquid film falling down a vertical plane at moderate Reynolds numbers. Math. Model. Program. Comput. Softw. **8**(4), 30–39 (2015). https://doi.org/10.14529/mmp140208. Bulletin of The South Ural State University. Series
10. Yaparova, N.M.: Numerical method for solving an inverse boundary problem with unknown initial conditions for parabolic PDE using discrete regularization. In: Dimov, I., Faragó, I., Vulkov, L. (eds.) NAA 2016. LNCS, vol. 10187, pp. 752–759. Springer, Heidelberg (2017). https://doi.org/10.1007/978-3-319-57099-0_87
11. Surov, V.S., Berezansky, I.V.: Godunov's method for a multivelocity model of heterogeneous medium. Math. Model. Program. Comput. Softw. **7**(2), 87–98 (2014). Bulletin of The South Ural State University, Series
12. Benney, D.J.: Long waves on liquid film. J. Math. Phys. **45**(2), 150–155 (1966)
13. Prokudina, L.A.: Influence of surface tension in homogeneity on the wave flow of a liquid film. J. Eng. Phys. Thermophys. **87**(1), 165–173 (2014)
14. Nayfe, A.: Methods of disturbances, Moscow. Mir 455 p. (1976)

# Completely Conservative Difference Schemes for Simultaneous Calculations of Thawed Hydrated Zone and Piezoconductive Medium with Gas Hydrate Inclusions

Parvin Rahimly[1], Yury Poveshchenko[2,3], Viktoriia Podryga[2,4(✉)],
and Orkhan Rahimly[1]

[1] Moscow Institute of Physics and Technology (State University),
Dolgoprudny 141701, Moscow Region, Russia
{pervin,orxan}@rehimli.info
[2] Keldysh Institute of Applied Mathematics of RAS,
4 Miusskya sq., 125047 Moscow, Russia
hecon@mail.ru,pvictoria@list.ru
[3] National Research Nuclear University MEPhI,
31 Kashirskoe sh., 115409 Moscow, Russia
[4] Moscow Automobile and Road Construction State Technical University,
64 Leningradsky Prospect, 125319 Moscow, Russia

**Abstract.** In the paper the thermodynamically equilibrium joint discrete model of a two-component three-phase (water, methane, hydrate) filtration fluid dynamics and two-phase processes in a melted zone with absence of gas hydrates is considered, for which the splitting by physical processes is performed. In the numerical calculations the direct unsplit using of the system being studied is difficult. Thus the splitting by physical processes is important for the purposes of determining the dynamics of variables and constructing the implicit difference scheme required for calculations of filtering processes with large steps in time is difficult.

**Keywords:** Gas hydrates · Filtration · Melted zone ·
Support operators · Completely conservative difference schemes

## 1 Introduction

The initial problem, formulated in the form of conservation laws (mass $H_2O$, $CH_4$ and total energy of the medium), with the general matrix of the system with respect to a function $S_\nu$ characterizing the volume fraction in the pores that is attributable to free water and gas in the hydrate-saturated part of the porous medium, and functions of saturation $S_w$, pressure $P$ and temperature $T$ has mixed hyperbolic and parabolic properties. Unsplit utilization of the studying system for the purposes of determining the dynamics of variables and

© Springer Nature Switzerland AG 2019
I. Dimov et al. (Eds.): FDM 2018, LNCS 11386, pp. 427–434, 2019.
https://doi.org/10.1007/978-3-030-11539-5_49

constructing the implicit difference scheme required for calculations of filtering processes with large steps in time is difficult. Filtration fluid dynamics of free water and gas in the presence of solid hydrate inclusions in a porous medium has a number of specific features. First of all, according to the Gibbs phase rule, this thermodynamically equilibrium two-component ($H_2O$, $CH_4$) three-phase (hydrate, free water and gas) system has only one thermodynamic degree of freedom. Those there is a thermobaric relationship $T_{dis} = f(P)$ between the temperature of gas hydrate dissociation and pressure. However, it must be borne in mind that in calculating in a thawed free-hydrate zone in a medium, there are two independent thermodynamic parameters $(P, T)$. In practice, the dynamics of the joint behavior of spatially separated thawed and hydrate-containing zones is of interest.

## 2    Formulation of the Problem

### 2.1    Three-Phase Medium with Hydrate Inclusions

In a spatial domain $O$ with a boundary $\partial O$, we consider thermodynamically equilibrium two-component (water, methane) three-phase equations of filtration fluid dynamics with gas hydrate inclusions [1]:

$$\frac{\partial}{\partial t}\left\{m\left[S_\nu S_w \rho_w + (1 - S_\nu)\rho_\nu \beta_w\right]\right\} + div[\rho_w \mathbf{V}_w] + q_w = 0, \tag{1}$$

$$\frac{\partial}{\partial t}\left\{m\left[S_\nu(1 - S_w)\rho_g + (1 - S_\nu)\rho_\nu(1 - \beta_w)\right]\right\} + div[\rho_g \mathbf{V}_g] + q_g = 0, \tag{2}$$

$$\mathbf{V}_w = -\frac{k \cdot k_{rw}}{\mu_w}(\nabla P - g\rho_w \mathbf{k}), \tag{3}$$

$$\mathbf{V}_g = -\frac{k \cdot k_{rg}}{\mu_g}(\nabla P - g\rho_g \mathbf{k}), \tag{4}$$

$$\frac{\partial}{\partial t}\left\{m\left[S_\nu(S_w \rho_w \varepsilon_w + (1 - S_w)\rho_g \varepsilon_g) + (1 - S_\nu)\rho_\nu \varepsilon_\nu\right] + (1 - m)\rho_s \varepsilon_s\right\}$$
$$+ div\left\{\rho_w \varepsilon_w \mathbf{V}_w + \rho_g \varepsilon_g \mathbf{V}_g + [P(\mathbf{V}_w + \mathbf{V}_g)]\right\} + div\mathbf{W} + q_\varepsilon = 0, \tag{5}$$

$$\mathbf{W} = -\left\{m\left[S_\nu(S_w \lambda_w + (1 - S_w)\lambda_g) + (1 - S_\nu)\lambda_\nu\right] + (1 - m)\lambda_s\right\}\nabla T, \tag{6}$$

$$T_{dis} = f(P) = AlnP + B. \tag{7}$$

The indices $g$, $w$, $\nu$, $s$ refer to gas, water, hydrate, skeleton of porous medium; $l$ is the index indicating the phase; $P$ is the pressure, $T$ is the temperature, $S_w$ is the water saturation, $\nu$ is the hydrate saturation, $S_\nu = 1 - \nu$ is the thaw, $\rho_l(P, T)$ and $\varepsilon_l(P, T)$ are the densities and internal energies of the unit mass of the phases, $\mathbf{V}_l$ is the filtration rate of the corresponding phase; $\beta_w$ is the mass fraction of water in the hydrate, $t$ is the time, $q_w$, $q_g$ and $q_\varepsilon$ are the respective sources densities depending on the parameters $(t, \mathbf{r}, S_w, S_\nu, P)$, $\mathbf{r}$ is the radius vector.

$\mathbf{W}$ is the heat flux, $\lambda_l(P,\,T)$ are the coefficients of thermal conductivity. $g\mathbf{k}$ is the vector of acceleration of gravity, directed vertically downwards, $k(r, S_\nu, P)$ is the absolute permeability, $k_{rl}(S_w)$ are the related phase permeabilities, $\mu_l$ are the viscosities. According to the Gibbs phase rule, the three-phase two-component hydrate system is multivariant, i.e. has one degree of freedom (temperature or pressure) [2]. Consequently, for a gas hydrate that is in equilibrium with liquid water or ice, the dependence $T_{dis} = f(P)$ is unambiguous.

The enthalpies $i_l = \varepsilon_l + P/\rho_l$ of hydrate, free water and gas are thermodynamically consistent in the sense of the following relation

$$\beta_w i_w + (1 - \beta_w) i_g = i_\nu + h, \tag{8}$$

where $h$ is the latent heat of the phase transition of a unit of hydrate mass. For a gas phase the equation of state is

$$\rho_g = \frac{PM}{z_g RT}. \tag{9}$$

with a coefficient of supercompressibility $z_g$, $M$ is the molar mass of gas.

From Eqs. (1), (2), (5), an equation for the three-phase piezoconductive of fluid dynamics with hydrate inclusions can be obtained in the following form [1]:

$$m\delta_\varepsilon \left\{ S_\nu \left[ S_w \frac{(\rho_w)_t}{\rho_w} + (1 - S_w) \frac{(\rho_g)_t}{\rho_g} \right] + (1 - S_\nu) \frac{(\rho_\nu)_t}{\rho_\nu} + \frac{(m)_t}{m} \right\}$$
$$+ \frac{\psi}{m\rho_\nu} \left\{ m \left\{ S_\nu \left[ S_w \rho_w (\varepsilon_w)_t + (1 - S_w) \rho_g (\varepsilon_g)_t \right] + (1 - S_\nu) \rho_\nu (\varepsilon_\nu)_t \right\} \right. \tag{10}$$
$$+ [(1 - m)\rho_s \varepsilon_s]_t \} + \delta_\varepsilon DIG + \frac{\psi}{m\rho_\nu} DIG_\varepsilon = 0,$$

$$DIG = \frac{1}{\rho_w} div[\rho_w \mathbf{V}_w] + \frac{1}{\rho_g} div[\rho_g \mathbf{V}_g] + \frac{q_w}{\rho_w} + \frac{q_g}{\rho_g}, \tag{11}$$

$$DIG_\varepsilon = [div(\rho_w \varepsilon_w \mathbf{V}_w) - \varepsilon_w div(\rho_w \mathbf{V}_w)] + [div(\rho_g \varepsilon_g \mathbf{V}_g) - \varepsilon_g div(\rho_g \mathbf{V}_g)]$$
$$+ div[P(\mathbf{V}_w + \mathbf{V}_g)] + div\mathbf{W} + (q_\varepsilon - \varepsilon_w q_w - \varepsilon_g q_g) = \rho_w \mathbf{V}_w \nabla \varepsilon_w + \rho_g \mathbf{V}_g \nabla \varepsilon_g$$
$$+ div[P(\mathbf{V}_w + \mathbf{V}_g)] + div\mathbf{W} + (q_\varepsilon - \varepsilon_w q_w - \varepsilon_g q_g). \tag{12}$$

Here

$$\frac{\psi}{m\rho_\nu} = \left( \phi - \frac{1}{\rho_\nu} \right) \geq 0, \qquad \phi = \frac{\beta_w}{\rho_w} + \frac{(1 - \beta_w)}{\rho_g}, \tag{13}$$

$$\delta_\varepsilon = \beta_w \varepsilon_w + (1 - \beta_w)\varepsilon_g - \varepsilon_\nu \geq 0 \tag{14}$$

are respectively specific jumps (per unit mass) for the phase transition of the volume and internal energy. The notation is also used hereinafter

$$( )_t = \frac{\partial}{\partial t}, \qquad ( )_p = \frac{\partial}{\partial P}.$$

Equation (10) is the main piezoconductive-dissipative thermodynamically equilibrium equation of three-phase two-component fluid dynamics with hydrate inclusions physically split with the saturation block (1)–(4) possessing mainly hyperbolic properties on the background of the fixed thermodynamic parameters of the medium [1]. Let's introduce a new value - the intensity of the hydrate system $D_p$ by the relation

$$D_p = m\delta_\varepsilon \{ S_\nu [S_w \frac{(\rho_w)_p}{\rho_w} + (1 - S_w) \frac{(\rho_g)_p}{\rho_g}] + (1 - S_\nu) \frac{(\rho_\nu)_p}{\rho_\nu} + \frac{m_p}{m} \}$$
$$+ \frac{\psi}{m\rho_\nu} \{ m[S_\nu S_w \rho_w (\varepsilon_w)_p + S_\nu (1 - S_w) \rho_g (\varepsilon_g)_p] + (1 - S_\nu) \rho_\nu (\varepsilon_\nu)_p \} \qquad (15)$$
$$+ [(1 - m)\rho_s \varepsilon_s]_p$$

and rewrite Eq. (14) in a more compact form:

$$D_p \frac{\partial P}{\partial t} + \delta_\varepsilon DIG + \frac{\psi}{m\rho_\nu} DIG_\varepsilon = 0. \qquad (16)$$

In (15) we take the total derivative with respect to pressure, taking into account the dependence (7). Using this dependence (7) and choosing the corresponding internal energies $\varepsilon_w$ and $\varepsilon_g$ as the unique thermodynamic degree of freedom in equations (3), (4), (10)–(12), we obtain the equation (10) in the energy representation.

## 2.2   Two-Phase Melt Zone

Analogously to Sect. 2.1, assuming thaw $S_\nu = 1$, we obtain a two-phase equation in the thawed zone:

$$\frac{\partial}{\partial t} \{ mS_w \rho_w \} + div[\rho_w \mathbf{V}_w] + q_w = 0, \qquad (17)$$

$$\frac{\partial}{\partial t} \{ m(1 - S_w)\rho_g \} + div[\rho_g \mathbf{V}_g] + q_g = 0, \qquad (18)$$

$$\frac{\partial}{\partial t} \{ m[(S_w \rho_w \varepsilon_w + (1 - S_w)\rho_g \varepsilon_g)] + (1 - m)\rho_s \varepsilon_s \}$$
$$+ div \{ \rho_w \varepsilon_w \mathbf{V}_w + \rho_g \varepsilon_g \mathbf{V}_g + [P(\mathbf{V}_w + \mathbf{V}_g)] \} + div \mathbf{W} + q_\varepsilon = 0, \qquad (19)$$

$$\mathbf{W} = - \{ m[(S_w \lambda_w + (1 - S_w)\lambda_g)] + (1 - m)\lambda_s \} \nabla T. \qquad (20)$$

Further, from Eqs. (17)–(19), excluding the function from under the sign of the time derivative, we obtain the equations that determine the nonisothermal process of piezoconductivity in the thawed zone:

$$\frac{S_w}{\rho_w} \frac{\partial(m\rho_w)}{\partial t} + \frac{1 - S_w}{\rho_g} \frac{\partial(m\rho_g)}{\partial t} + DIG = 0, \qquad (21)$$

$$m \{ S_w \rho_w \frac{\partial \varepsilon_w}{\partial t} + (1 - S_w)\rho_g \frac{\partial \varepsilon_g}{\partial t} \} + \frac{\partial[(1 - m)\rho_s \varepsilon_s]}{\partial t} + DIG_\varepsilon = 0. \qquad (22)$$

Here the combination of mass $(DIG)$ and energy $(DIG_\varepsilon)$ divergences together with the action of the corresponding sources $(q_w, q_g, q_\varepsilon)$ are defined similarly to 10.

For numerical realization of the above differential models (1), (2), (10) and (17), (21), (22), discrete algorithms and corresponding difference schemes were developed in the works [3–6].

# 3   Results of Calculations

Consider the following spatially one-dimensional process on an interval $x \in [0, l]$, $l$ is the length of the computational domain. In it, at the initial instant of time, the pressure $P(x, 0) = 2 \cdot 10^7$ $Pa$, water saturation $S_w(x, 0) = S_w^*$ and thaw $S_\nu(x, 0) = S_\nu^*$ are uniform in space, $0 < S_w^* < 1$ and $0 < S_\nu^* < 1$ are constants quantities. The acceleration of free fall is not taken into account $(g = 0)$. The boundaries of the calculation area are assumed to be impermeable solid "walls", i.e. they have zero flux through them: $V_w\big|_{x=0} = 0$, $V_g\big|_{x=0} = 0$, $V_w\big|_{x=l} = 0$, $V_g\big|_{x=l} = 0$, $t > 0$. Suppose that in a certain narrow spatial region $x \in [0.4, \ 0.6]$ $m$ the source works:

$$q_w = \alpha(P - P^*), \ q_g = \beta(P - P^*), \ q_\varepsilon = \varepsilon_w q_w + \varepsilon_g q_g. \tag{23}$$

where $P^*$ is the constant, $\alpha$ and $\beta$ are the characteristics of the source (discharge) in the borehole zone due to the pressure drop inside the bore hole and in the shelf. The parameters of source are $\alpha = 0$, $\beta = 10^{-5}\,s/m^2$. In this case, the dimension of the pressure $[P] = Pa$. For the calculation, the following values of the parameters, characteristic for the Messoyakhsky gas hydrate deposit [7], were chosen:

$$\rho_w = 10^3\,\frac{kg}{m^3}, \ \rho_\nu = 910\,\frac{kg}{m^3}, \rho_s = 2800\,\frac{kg}{m^3}, \ \beta_w = 0.9,$$
$$A = 7.28\ K, \ B = 169.7\ K, \ z_g = 1$$
$$\mu_w = 10^{-3}\ Pa \cdot s, \ \mu_g = 0.014 \cdot 10^{-3}\ Pa \cdot s,$$
$$c_w = 4165\,\frac{J}{kg \cdot K}, \ c_g = 2500\,\frac{J}{kg \cdot K}, \ c_s = 873\,\frac{J}{kg \cdot K},$$
$$M = 0.016\,\frac{kg}{mol}, \ h = 514810\,\frac{J}{kg}, \ R = 8.31\,\frac{J}{mol \cdot K}, \ m = 0.35,$$
$$S_w^* = 0.6, \ S_\nu^* = 0.8, \ k(S_\nu) = k_0(S_\nu)^3, \ k_0 = 10^{-14}\ m^2, \ P^* = 2\ MPa,$$
$$k_{rw}(S_w) = 1.477S_w^5 - 1.587S_w^6 + 1.11S_w^7 - 0.0473,$$
$$k_{rg}(S_w) = 1.044 - 1.7S_w + 0.6S_w^2.$$

The minimum value of water saturation $S_{w\ min} = 0.55$.
$k_{rw}(S_w) = 0$, $k_{rg}(S_w) = k_{rg}(S_{w\ min})$ at $S_w \leq S_{w\ min}$.
The maximum value of water saturation $S_{w\ max} = 0.9$.
$k_{rw}(S_w) = k_{rw}(S_{w\ max})$, $k_{rg}(S_w) = 0$ at $S_w \geq S_{w\ max}$.

The model tube length is assumed to be equal to $l = 1$ m, the step along the spatial coordinate $h = 0.01$ m. The calculations are performed for the time moments $t = 1, 10, 100$ s. In Fig. 1, 2, 3 and 4, it can be seen that in the calculations a joint transphase process of pressure expansion in the emerging thawed zone and in the remaining piezoconductive medium of the reservoir with gas-hydrate inclusions is spatially expanding within the gas extraction range of $[0.4, 0.6]$. Such a transphase calculation is carried out for the first time. The temperature peaks in Fig. 4 at the boundaries of the gas extraction region ($x = 0.4$ m and $x = 0.6$ m) are related to the absence of thermobaric dependence $T_{dis} = f(P)$ in the thawed zone that has arisen here and continues to expand spatially. The hotter gas comes here from areas where there is no gas extraction, the energy of which is not wasted on the hydrates that are already absent in the thawed zone. Figures 5 and 6 show the temperature-pressure dependences obtained in the calculations for the time points 10, 100 s. The calculation data are marked in contrast to the thermobaric equilibrium curve $T_{dis} = f(P)$ depicted by the solid line. It is also seen that in the process of calculation, the region of the three-phase flow moves to the zone of reduced pressure along the thermobaric equilibrium

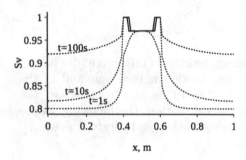

**Fig. 1.** Distribution of the hydrate thaw. The three phase zone - dash line, thawed zone- solid line

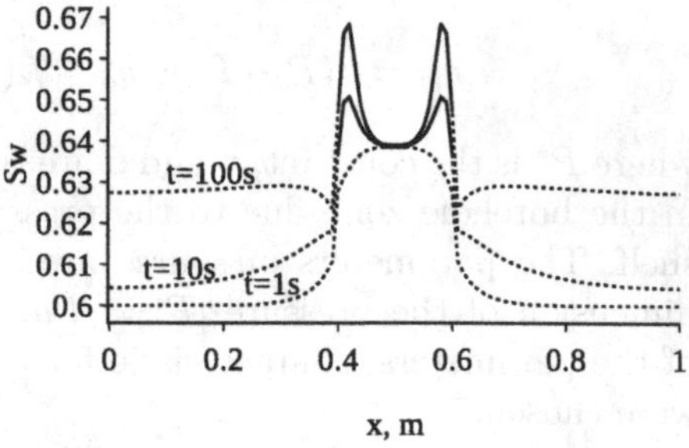

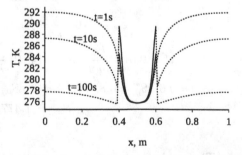

**Fig. 2.** Distribution of water saturation. The three phase zone - dash line, thawed zone- solid line

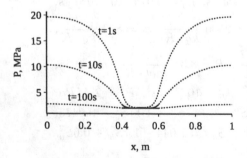

**Fig. 3.** Distribution of pressure. The three phase zone - dash line, thawed zone- solid line

**Fig. 4.** Distribution of temperature. The three phase zone - dash line, thawed zone- solid line

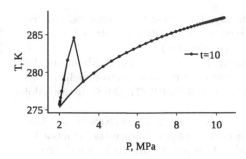

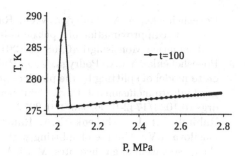

**Fig. 5.** Temperature versus pressure for time 10 s. The thermobaric equilibrium $T = f(P)$ - solid line.

**Fig. 6.** Temperature versus pressure for time 100 s. The thermobaric equilibrium $T = f(P)$ - solid line

curve. In this case, part of the region of the fluid dynamic flow near the boundaries of the gas selection region ($x = 0.4$ m and $x = 0.6$ m), corresponding to the emerging two-phase thawed zone, "slips" from the equilibrium curve into the region of higher temperatures. At the time $t = 1$ s the temperature dependence of the pressure corresponds to the thermobaric equilibrium curve $T_{dis} = f(P)$.

## 4    Conclusion

In the work with reference to both thawed zone and medium with solid-phase hydrate inclusions, a method for solving these problems is developed. The method is based on the splitting by physical processes of the transfer of water and hydrate saturations on the one hand and the extraction of piezoconductive processes taking into account the presence of a solid hydrate phase in the pores on the other hand. Algorithms are created and software implementation is developed. A joint transphase calculation with dynamically varying spatial localizations of a different number of phases is carried out for the first time. The results of calculations show the effectiveness of the developed methods for calculating real problems associated with the deposits of gas hydrates, studying the complex transphase processes, as well as the dynamics of water and hydrate saturations in the reservoir.

**Acknowledgments.** The work was supported by the Russian Science Foundation (project No 17-71-20118).

## References

1. Poveshchenko, Y.A., Kazakevich, G.I.: Mathematical modeling of gas hydrate processes. Math. Mach. Syst. **3**, 105–110 (2011). (in Russian)
2. Landau, L.D., Lifshitz, E.M.: Course of Theoretical Physics: Theory of Elasticity, vol. 7. Pergamon Press Ltd., Oxford (1970)

3. Poveshchenko, Y.A., Podryga, V.O., Rahimly, P.I.: About one approach to free volumetric approximation of a piezoconductive medium with gas hydrate inclusions. Mathematica Montisnigri **40**, 68–89 (2017). (in Russian)
4. Poveshchenko, Yu.A., Podryga, V.O., Rahimly, P.I., Sharova, Yu.S.: About one discrete model of splitting by the physical processes of a piezoconductive medium with gas hydrate inclusions. J. Phys. Conf. Ser. (JPCS) **946**, 012077 (2018). https://doi.org/10.1088/1742-6596/946/1/012077
5. Rahimly, P.I., Poveshchenko, Y.A., Rahimly, O.R., Podryga, V.O., Kazakevich, G.I., Gasilova, I.V.: The use of splitting with respect to physical processes for modeling the dissociation of gas hydrates. Math. Models Comput. Simul. **10**(1), 69–78 (2018). https://doi.org/10.1134/S2070048218010118
6. Rahimly, P.I., Poveshchenko, Yu.A., Podryga, V.O., Rahimly, O.R., Popov, S.B.: Modeling the processes of joint filtration in melted zone and piezoconductive medium with gas hydrate inclusions. Preprints IPM im. M.V. Keldysha [KIAM Preprints], Moscow, vol. 40 (2018). https://doi.org/10.20948/prepr-2018-40
7. Bogatyrenko, R.S.: Features of the development and operation of gas hydrate deposits (on the example of the Messoyakhsky deposit). Dissertatsiia, Moscow (1979). (in Russian)

# Converging Shock Wave for Ignition of a Pre-compressed Target of Laser Thermonuclear Fusion

Orkhan Rahimly[1]($\boxtimes$), Nikolai Zmitrenko[2], and Sergey Guskov[3]

[1] Moscow Institute of Physics and Technology (State University),
Dolgoprudny, 141701 Moscow Region, Russia
orxan@rehimli.info
[2] Keldysh Institute of Applied Mathematics of RAS,
4 Miusskya sq., 125047 Moscow, Russia
zmitrenko@imamod.ru
[3] Lebedev Physical Institute of RAS,
53 Leninskiy Prospekt, 119333 Moscow, Russia
guskov@sci.lebedev.ru

**Abstract.** The formation of a convergent shock wave in a homogeneous spherical target is studied, the outer layer of which is heated by a flow of monoenergetic fast electrons with a given particle energy. The ablation pressure generating the wave is formed during the spherical expansion of the layer of the heated substance, the surface density of which remains constant throughout the heating process and is equal to the product of the initial warming depth and the initial density of the target. The investigations were performed on the basis of numerical calculations on a one-dimensional hydrodynamic program with reference to one of the most promising methods of ignition of a laser thermonuclear fusion target - the ignition of a pre-compressed target by a shock wave (shock ignition).

**Keywords:** Gas dynamics · Ablation pressure ·
Convergent shock wave · Laser thermonuclear fusion

## 1 Introduction

One of the promising methods of ignition of the target of laser thermonuclear fusion (LTS) is the action on the previously compressed spherical target by an additional powerful laser pulse that initiates a convergent shock wave (SW) [1,2], which should ensure thermonuclear ignition. Such a scheme in the literature is called "ignition by a focused shock wave" or "shock ignition". This approach, with the necessary synchronization of the dynamics of the preliminary compression of the target and the propagation of the igniting shock wave, turns out to be energetically more favorable than the traditional spark ignition.

© Springer Nature Switzerland AG 2019
I. Dimov et al. (Eds.): FDM 2018, LNCS 11386, pp. 435–442, 2019.
https://doi.org/10.1007/978-3-030-11539-5_50

In paper [3] we considered a similar problem of the convergence of a spherical SW caused by the release of energy in a medium due to a volumetric heat source. This paper is devoted to the investigation of generation and dynamics of a convergent SW as applied to the conditions of "shock ignition", when the pressure of an external evaporating layer, which initiates the generation of a SW target converging toward the center, is formed as a result of heating the layer by a stream of fast electrons.

In work [4], based on the results of many numerical calculations, the requirements for the realization of ignition by a shock wave of a target in the form of a shell of DT ice calculated for the energy of a laser pulse of 130 kJ are summarized. According to the data of this work, during a time of about 10 ns the target should be compressed to a density of about its radius of about 250 μm. Generation of an igniting shock wave must be carried out under the action of radiation with a power of the order of 110 TW, which corresponds to the radiation intensity alt the surface of the pre-compressed target about $14 \times 10^{15}$ V/cm$^2$. With this power of the absorbed radiation, the velocity of the igniting shock wave was about 290 km/s. It's necessary that, to maintain the pressure of the igniting shock wave, at least until its collision with the first shock wave reflected from the center, which is observed at a radius of about 70 μm, the duration of the second powerful part of the pulse was about 600 ps. According to numerical calculations [5], the energy of fast electrons under the above-mentioned conditions for the interaction of the powerful part of the pulse with the plasma can be from 30 to 100 keV [5]. The run of fast electron, which is determined by Coulomb collisions with thermal plasma electrons, does not depend on the plasma temperature. Therefore, the surface density of the layer in which the nonrelativistic fast electron is decelerated is determined only by the initial electron energy and increases as its square. For definiteness, we shall assume that fast electrons are monoenergetic with an energy of 50 keV. The run of such electrons in a fully ionized DT plasma with a density is 3 μm [6], and the corresponding surface density of the heated layer is 0.003 g/cm$^2$.

The main goal of this paper is to elucidate the details of formation of an igniting shock wave in relation to the "shock ignition" conditions presented in [4] for the "HiPER" target: the fast electron source power is 100–200 TW, the energy is 30–50 keV, which corresponds to their mass range 0.001–0.003 cm$^2$. For this purpose, the paper investigates the temporal evolution of ablation pressure when the outer layer of a sphere with a given surface density is heated, comparison with analytical solutions of similar problems is carried out.

Numerical calculations were performed using a one-dimensional hydrodynamic FLORA code [7]. Since only the hydrodynamic aspects of the problem were investigated, the calculations were performed without taking into account the electronic thermal conductivity and in the one-temperature approximation.

## 2    Statement of the Computational Problem

The mathematical model for this problem (the system of equations) in Lagrangian mass coordinates in the spherically symmetric case is

$$\frac{\partial}{\partial t}\left(\frac{1}{\rho}\right) = \frac{\partial}{\partial x}(r^2 v), \quad \frac{\partial v}{\partial t} = -r^2 \frac{\partial p}{\partial x}, \quad \frac{\partial r}{\partial t} = v, \quad \frac{\partial r}{\partial x} = \frac{1}{\rho r^2}, \quad (1)$$

$$\frac{\partial \epsilon}{\partial t} + p\frac{\partial}{\partial x}(r^2 v) = \frac{Q}{\rho}, \quad \epsilon = \frac{R}{\gamma - 1}T, \quad p = \rho RT, \quad \frac{Q}{\rho} = \begin{cases} \frac{W}{m_Q}, & t < t_m \\ 0, & t \ge t_m \end{cases}$$

where $\gamma$ is the ratio of specific heats, $R$ is the gas constant, $r_v(t)$ is the coordinate of the boundary with the vacuum, $r_k$ is the internal coordinate of the boundary of the heated layer, which we obtain from the condition:

$$\mu = \int_{r_k}^{r_v} \rho r dr = \rho_0 \Delta_0 \qquad (2)$$

$\Delta_0 = r_v(0) - r_k(0)$ is initial depth of heating, $Q \left[\frac{V}{cm^3}\right]$ is energy release of fast electrons, $m_Q = \int_{r_k}^{r_v} \rho r^2 dr$ is mass (per 1 sr) of the energy release area. It is believed that the energy release occurs uniformly in the mass of the layer with a fixed surface density $\mu$.

The conditions for carrying out the calculations were chosen as follows: the initial radius and density of the target in the form of a homogeneous $DT$-ball are, respectively, $R_0 = 250\,\mu m$ and $\rho_0 = 10\,g/cm^3$; initial depth of heating of the ball $\Delta_0 = 3\,\mu m$; constant, not changing with time, the value of the optical thickness of the region of energy absorption is $\mu = 0.003\,g/cm^2$. The equation of state of matter is an ideal gas $p = \rho RT$. The gas constant $R = 766\,[10^5\,J/(g \cdot keV)]$, which corresponds to the substance DT, the ratio of specific heat capacities corresponds to a monatomic gas ($\gamma = 5/3$). The target mass is

$$M_0 = \frac{4\pi}{3}\rho_0 R_0^3 \cong 654.5 \times 10^{-6}(mkg)$$

Let's discuss now the results of a calculation in which the absorbed energy was $W = 110\,TW$, the action time was $t_m = 600\,ps$ and the total absorbed energy was $E = Wt_m = 66\,kJ$.

## 3    Generation of a Spherical SW During the Heating of a Sphere by Fast Electrons

Let's analyse the area of energy release. In Fig. 1 shows the density profile in the heated layer at different times. At short times, not exceeding 20 ps, an unloading plane wave propagates in the heated layer. On the right and left, unloading waves come towards each other. The pressure reaches a maximum of 10 ps and equals 2.6 Gbar, then begins to decrease, and the temperature increases monotonically throughout the region [8,9].

Moments of 20–30 ps correspond to the transition from a planar expansion of matter to a quasiplanar one. In Fig. 2 it is clearly seen how the evaporated mass during this period of time changes its dispersion law to a faster one. By that time the entire heated layer has already been covered by the motion. The density decreases markedly, with the exception of the inner boundary of the heated layer, this is due to the emerging density "peak" (Fig. 3) on the inner boundary of the energy release region that plays the role of the piston at the beginning of the process. During the process, the dimensions of the energy release region increase in proportion to the square of the outer radius (the thickness of the region $0.003\,\mathrm{g/cm^2}$ remains constant), it is observed the expansion accompanied by a decrease in pressure and an increase in temperature in this region. Let us now turn to the density and pressure profiles in the vicinity of the front of the shock wave. The value of the density on the front of SW increases by a factor of 4 (Fig. 3), as it should be for the adiabatic index 5/3. Pressure during the process falls, as in a plane wave with conservation of energy. From Fig. 4 it can be seen that the temperature in the shock wave rises only in the region of energy release, but inside the target, due to the absence of thermal conductivity, increases insignificantly only due to heating on the front of the SW.

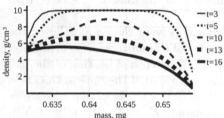

**Fig. 1.** The density profile in the heated layer at times corresponding to planar expansion (the boundaries are determined by a thickness of $0.003\,\mathrm{g/cm^2}$)

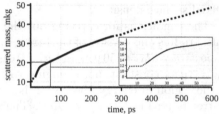

**Fig. 2.** The dot line in the enlarged tab corresponds to the planar expansion, the line in the same place is quasi-plane, the dot line on the top is spherical

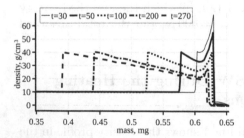

**Fig. 3.** The density profile at time instants corresponding to a quasiplane expansion

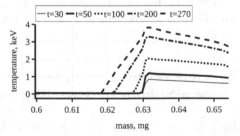

**Fig. 4.** The temperature profile at time instants corresponding to the quasi-plane expansion

After 270 ps the law of expansion is replaced by a "spherical" (see Fig. 2). By 600 ps, the pressure at the inner edge of the energy release has decreased to 600 Mbar, and the connection between the heating region and the SW is lost, and the latter propagates according to its internal laws ("breaking away" from the "piston", whose role is played by the internal boundary of the energy release region). As shown below, in order to create a convergent and combustion-initiating shock wave, there is an optimal value for the pulse duration, which no longer makes sense to heat the layer. The corresponding parameters of the pulse are determined by analyzing the dynamics of the convergent SW, performed below for a number of power values and the pulse duration.

We note that the temporal evolution of the thermodynamic parameters of the heated layer in the performed numerical calculations is well described by the analytical solution of isothermal expansion of the outer layer of a homogeneous sphere with constant surface density obtained in [8]. For the numerical calculation parameters ($W = 110$ TW, $R_0 = 250, \mu = 0.003 \, \text{g/cm}^2, \rho_0 = 10 \, \text{g/cm}^3$) solution in papers [3–5] becomes:

$$\rho = 10 \, \text{g/cm}^3 \times \begin{cases} 1, & 0 \le t \le 15.7 \\ (t/15.7)^{-3/2}, & 15.7 \le t \le 268 , \\ (17t/15.7)^{-3/4}, & 268 \le t \le 600 \end{cases} \tag{3}$$

$$T = 0.318 \, \text{keV} \times \begin{cases} t/15.7, & 0 \le t \le 268 \\ 70.5(t/15.7)^{-1/2}, & 268 \le t \le 600 \end{cases} \tag{4}$$

$$P = 2.44 \, \text{Gbar} \times \begin{cases} t/15.7, & 0 \le t \le 15.7 \\ (t/15.7)^{-1/2}, & 15.7 \le t \le 268 . \\ 8.15(t/15.7)^{-5/4}, & 268 \le t \le 600 \end{cases} \tag{5}$$

The solution describes: (1) during a period of time $0 \le t \le 15.4$ ps, while the unloading wave goes from the outer surface of the ball to the inner boundary of the heated layer, the static (without expansion of matter) heating of the layer; (2) for a period of time $15.7 \le t \le 268$ ps the plane expansion of the layer with a given non-changing mass; (3) in a time period of $268 \le t \le 600$ ps the spherical expansion of the heated layer with a given surface density and increasing mass.

In Fig. 5 it is seen a comparison of the numerical calculation results and the analytical solution describing the density evolution in the heated layer during the first 100 ps.

It is useful to discuss the obtained results by involving the solution of the problem of disintegration of a discontinuity in a configuration typical of numerical problems about the heating of a plane layer by fast electrons. One of the solutions to this problem, in particular, is shown in Fig. 3. The density profile reflects the zone behind the front SW, and also, the energy release zone, where the density tends to zero due to the expansion. The density profile in the SW region demonstrates the appearance of a density "peak" as it approaches the internal boundary of the heating region, which plays the role of a piston.

The energy release region that acts on the inner parts of the target is a hot region with an approximately initial density (to the left of the intensive expansion region) and high pressure. Such a configuration can be analyzed within the framework of a typical formulation of the problem of disintegration of a discontinuity: at the initial instant of time the parameters of the medium on the left and right of the discontinuity are constant and equal, respectively, $\rho_0, p_0, v_0 = 0$ and, $\rho_1, p_1, v_1 = 0$ where $p_1 \gg p_0$. Such a configuration leads to the appearance of a shock wave propagating towards a lower pressure and a rarefaction wave propagating in the opposite direction. Between these structures there is a region of constant flow with some intermediate pressure $p$. The construction of the solution of this idealized model leads to the consideration of the nonlinear (6):

$$A\sqrt{y} + y^B = 1 \tag{6}$$

where $y = \frac{p}{p_1}, A = \frac{\gamma-1}{2}\sqrt{\frac{2\rho_1}{\gamma(\gamma+1)\rho_0}}, B = \frac{\gamma-1}{2\gamma}$, $p$ is the required pressure. Having determined $p$, we find the velocity on the shock wave and the density $\rho_1'$ behind the rarefaction wave according to formulas.

$$v = \sqrt{\frac{2}{\gamma+1}\frac{p}{\rho_0}}, \quad \rho_1' = \rho_1\left(\frac{p}{p_1}\right)^{\frac{1}{\gamma}}$$

For our problem, the densities on the right and on the left are equal $\rho_0 = 10\,\mathrm{g/cm}^3$ and $\gamma = 5/3$.

The nonlinear Eq. (6) can be easily solved by the iteration method. For $\rho_0 = 10\,\mathrm{g/cm}^3$, $\rho_1 = 10\,\mathrm{g/cm}^3$ and $\gamma = 5/3$ we find $p \cong 0.4456\,p_1$, $\rho_1' \cong 10(0.4456)^{0.6} \approx 6.16\,\mathrm{g/cm}^3$. During the time of the 600 ps heating pulse, the pressure at the internal boundary of the energy input region reaches 220 Gbar. Considering the beginning of the process (planar expansion), 6 ps and pressure 2 Gbar, we find the intermediate pressure is 891 Mbar, as it is observed in our calculations.

The effect of the appearance of a density peak near the inner boundary of the heated layer is of interest. Calculations in the geometry of planar expansion show that the "peak" of the density is formed independently of the geometry of the problem. "Peak" is formed immediately before the inner boundary of the heated layer, its magnitude reaches a maximum of about 100 ps, and then begins to decrease. Let us follow the pressure at the inner boundary of the heated layer. In Fig. 6 it is seen that a graph of the dependence of the growth rate or pressure decrease on time. It can be seen that approaching 40–50 ps the rate of pressure growth stabilizes. Note that this moment corresponds to the time when all the substance is already involved in the movement, i.e. corresponds to the time of the meeting of the unloading wave going beyond the SW and the unloading wave from the vacuum boundary. Starting from 70 ps, the pressure growth slows down, and after 100 ps the growth rate changes its sign. It is shown in [8] that the pressure at the inner boundary of the heated layer increases linearly until the unloading wave has passed a layer with a thickness equal to the initial depth of heating, and then the pressure decreases. This explains why the "peak" begins

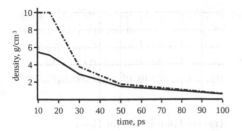

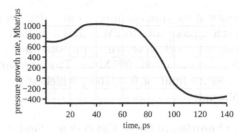

**Fig. 5.** Time evolution of density in a heated layer. The dashcurve corresponds to the analytical curve, and the continuous curve corresponds to the numerical curve.

**Fig. 6.** Pressure changes at the inner boundary of the heated layer (its thickness is $3\,\mathrm{mg/cm^2}$ from the inner to the outer boundary).

to decay after 100 ps. The unloading wave "pulls" the mass out of the "peak" area, and the pressure decreases.

The origin of the density "peak" can also be explained by the classical problem of a piston inserted into a gas. In actual calculations, the pressure at the boundary increases almost linearly up to 100 ps. Consequently, the condition on the piston can be set by the law, where is the dimensional constant, which is approximately equal to the ratio of pressure at 100 ps, divided by 100 ps.

$$P_0 \approx \frac{P(100\,\mathrm{ps})}{100\,\mathrm{ps}} = 0.09\frac{\mathrm{Gbar}}{100\,\mathrm{ps}} = 9 \times 10^5\,\mathrm{Mbar}/\mu s$$

In this case, the dependence of the piston velocity on time is $\sim t^{1/2}$. As it is well known, this problem has a self-similar solution [10]. In this solution, in the case of an accelerating piston (which is in this case under consideration), the density tends to infinity as it approaches the piston. The ideal solution, containing a feature, is realized in our calculations with the detail that the grid provides.

## 4    Conclusion

On the basis of one-dimensional hydrodynamic simulation, computational data were obtained on the generation and beginning of propagation in a homogeneous sphere of a spherical shock wave converging to the center under the action of a pulsed constant-power energy source in the outer layer. Such a heating regime corresponds to the heating of matter by a flow of fast electrons, the run of which does not depend on the temperature of the heated substance. The calculations confirmed the conclusions obtained on the basis of an analytical solution of the problem of the possibility of achieving the shock wave power necessary for ignition of a preliminarily compressed LTS target. It is shown that heating of the outer layer with a surface density of $0.003\,\mathrm{g/cm^2}$ (corresponds to the absorption of fast electrons with an energy of 50 keV) of a deuterium-tritium plasma with a mass of $0.65\,\mathrm{mg}$ compressed to a density of $10\,\mathrm{g/cm^2}$, of a pulsed energy source with a power of $110\,\mathrm{TW}$ and a duration of $600\,\mathrm{ps}$, provides generation of a convergent shock wave, whose pressure during the entire time of wave convergence to the center exceeds $900\,\mathrm{Mbar}$. This pressure exceeds the known value of the lower pressure limit of $600\,\mathrm{Mbar}$, which is necessary for ignition of a precompressed plasma.

**Acknowledgments.** The work was funded by Russian Foundation for Basic Research, project no. 16-11-10174.

# References

1. Shcherbakov, V.A.: Calculation of thermonuclear laser target ignition by focusing shock wave. Fiz. Plazmy **9**(2), 409–411 (1983). (in Russian)
2. Betti, R., Zhou, C.D., Anderson, K.S., Perkins, L.J., Theobald, W., Solodov, A.A.: Shock ignition of thermonuclear fuel with high areal density. Phys. Rev. Lett. **98**(15), paper 155001 (2007)
3. Zmitrenko, N.V., Rahimly, O.R.: Convergence of a shock wave in the presence of energy release in the environment. Preprints of the Keldysh Institute of Applied Mathematics, vol. 55 (2016). (in Russian). https://doi.org/10.20948/prepr-2016-55
4. Lafon, M., Ribeyre, H., Schurtz, G.: Gain curves and hydrodynamic modeling for shock ignition. Phys. Plasmas **17**(5), paper 052704 (2010)
5. Klimo, O., et al.: Laser plasma interaction studies in the context of shock ignition-transition from collisional to collisionless absorption. Phys. Plasmas **18**(8), paper 082709 (2011)
6. Ribeyre, X., Gus'kov, S.Yu., Feugeas, J.L., Nicolai, Ph., Tikhonchuk, V.T.: Dense plasma heating and Gbar shock formation by a high intensity flux of energetic electrons. Phys. Plasmas **20**(6), paper 062705 (2013)
7. Gayfullin, S.A., et al.: A software package for calculating the equations of one-dimensional gas dynamics with thermal conductivity. In the book packages of application programs. Computational experiment, pp. 50–70. Nauka, Moscow (1983). (in Russian)
8. Gus'kov, S.Yu.: Isothermal expansion of a spherical layer with a given areal density into vacuum. JETP Lett. **103**(7), 494–497 (2016)
9. Gus'kov, S.Yu., Zmitrenko, N.V., Rahimly, O.R.: A converging shock wave for ignition of a preliminarily compressed target of laser thermonuclear fusion. Preprints of the Keldysh Institute of Applied Mathematics, vol. 94 (2017). (in Russian). https://doi.org/10.20948/prepr-2017-94
10. Samarskiy, A.A., Popov, Yu.P.: Difference Methods for Solving Problems of Gas Dynamics. Nauka, Moscow (1992). (in Russian)

# Monte Carlo Solution of Dirichlet Problem for Semi-linear Equation

Abdujabbor Rasulov[1]([✉]), Gulnora Raimova[2], and Matyokub Bakoev[1]

[1] University of World Economy and Diplomacy, Tashkent, Uzbekistan
asrasulov@gmail.com, mbakoev@gmail.com
[2] Institute of Mathematics, Academy of Sciences, Tashkent, Uzbekistan
raimova27@gmail.com

**Abstract.** In the present work Dirichlet boundary value problem (BVR) for semi-linear elliptic equation is considered. Assuming the existence of solution BVR obtained a probabilistic representation of the solution as a mathematical expectation of some random variable. In accordance with a probabilistic representation on the trajectories of the branching random process were constructed unbiased estimator of the solution. An unbiased estimator of the solution has finite variance, based on the trajectories of a branching process with a finite average number of branching and easily simulated. Some numerical experiments are performed.

**Keywords:** Monte Carlo method · An unbiased estimator ·
Branching random process · Markov chain · Dirichlet problem

## 1 Introduction

The method of statistical modeling for the solution BVR for elliptic equations were considered in [6, 7, 11]. Recently for semi-linear parabolic equations Monte Carlo method proposed in the work [16]. Constructing unbiased and biased estimators for the solutions of linear Helmholtz equations $\Delta u - c(x)u = -g(x)$ were considered in case $c(x) = const$ by Mikhailov and Elepov [4,5], in the case $c(x)$ is variable for Dirichlet problem with mixed boundary condition by Simonov [19] and Sipin [20] investigate for Dirichlet problem in case $\Delta u + \sum_{i=1}^{n} a_i \frac{\partial u}{\partial x_i} + au = -g$. In the works [12,13] of Mikhailov and Makarov estimates of the solution of boundary value problems for the linear Helmholtz equation are constructed on the basis of a special integral-difference equation, using the process of "walks on the spheres" with reflection from the boundary.

These methods were extended for the equations $-\Delta v + 2cv = c^2 + v^2$ in the work [17], for the equation $-\Delta u + u^n = 0$ in [10] and for the equations $\Delta u(x) = \sum_{i=1}^{n} a_i(x)u^{2i}(x) + a_0(x)$ in the work [15].

© Springer Nature Switzerland AG 2019
I. Dimov et al. (Eds.): FDM 2018, LNCS 11386, pp. 443–451, 2019.
https://doi.org/10.1007/978-3-030-11539-5_51

## 2    Description of the Problem

Let $D$ be bounded domain in $R^3$ with regular boundary $\Gamma$,

$$f(x, u) = \sum_{n=0}^{\infty} a_n(x) u^n(x),$$

here the coefficients satisfy the condition $\lim_{n \to \infty} \sqrt[n]{\bar{a}_n} = 0$, where $\sup_{x \in D} |a_n(x)| \leq \bar{a}_n$. We will consider the following Dirichlet problem:

$$- \Delta u(x) + cu(x) = f(x, u), \quad x \in D, \quad u|_\Gamma = \varphi(x). \qquad (1)$$

Suppose, that functions $\varphi(x) \in C(\bar{\Gamma})$, $a_n(x) \in C(\bar{D})$ $(n = 0, 1, 2, \ldots)$ and the coefficient $c > 0$ that,there exists unique continuous solution of semi-linear problem [2]. Let

$$\sum_{n=0}^{\infty} \bar{a}_n n < \infty. \qquad (2)$$

Under the above assumptions, we construct an unbiased estimate of the solution of the finite variance problem at some arbitrary point $x \in D$.

## 3    Probabilistic Representation of the Solution of the Problem

Using Green function, we write a special integral equation, which connected the value of function $u(x)$ with its integral on the sphere and ball with the center in the point $x$ and has a maximal radius in the considered domain.

$$u(x) = \frac{R\sqrt{c}}{4\pi R^2 \mathrm{sh}(R\sqrt{c})} \int_{S_R(x)} u(y) dy + \int_{K_R(x)} \frac{\mathrm{sh}[(R - |x - y|)\sqrt{c}]}{4\pi |x - y| \mathrm{sh}(R\sqrt{c})} f(y, u(y)) dy.$$

Here $R = R(x) = \min_{y \in \Gamma} |x - y|$ is the distance from the point $x$ to the boundary $\Gamma$, $K_R(x)$ is the ball with radius $R$ and center of ball at the point $x$, $S_R(x)$ - corresponding sphere with radius $R$. Write the last expression as follows:

$$u(x) = q(x) \int_{S_R(x)} u(x + R\omega) d\omega + (1 - q(x)) \int_{K_R(x)} p(x, y) \sum_{n=0}^{\infty} \frac{a_n(y)}{c} u^n(y) dy, \quad (3)$$

where $\omega$ - uniformly distributed points on the $S_1$,

$$q(x) = \frac{R\sqrt{c}}{\mathrm{sh}(R\sqrt{c})}, \quad p(x, y) = \frac{\mathrm{sh}[(R - |x - y|)\sqrt{c}]}{4\pi |x - y|(\mathrm{sh}(R\sqrt{c}) - R\sqrt{c})},$$

$p(x, y)$ is transition density function from the point $x$ to $y$ $(x, y \in K_R(x))$.

# 4    Constructing Branching Random Processes

According to representation (3) we will construct a branching random process in $D$. Let us $M = \sum_{n=1}^{\infty} n \bar{a}_n$ and $\alpha$ some constance, $0 < \alpha < 1$. In initial moment we have one particle at the point $x_0 = x$. Let $n > 0$ and $x_n$ known.

For the unit time a particle moved to the uniformly distributed on sphere $S_R(x_n)$ to the random point $x_{n+1}$ with the probability $q(x_n)$, or with probability $1 - q(x_n)$ moved to the point $x_{n+1} \in K_R(x_n)$, which is distributed with density function $p(x_n, y)$. In the second case with probability $\pi_n = \frac{\alpha}{M} \bar{a}_n$, $(n = 1, 2, \cdots)$a particle is divided into $n$ new particles or with the probability $\pi_0 = 1 - \sum_{n=1}^{\infty} \pi_n$ will absorbed. New particles behave independently from the others (similar to the original). A random process will be terminated if all the particles absorbed in $\bar{D}$ or particles hit the boundary $\Gamma$. A parameter $\alpha$ allowed to control the quantity of new born particles. Now we will describe modelling algorithm for transition from $x_n \rightarrow x_{n+1}$. If a random point uniformly distributed on sphere $S_R(x_n)$, then $x_{n+1} = x_n + R\omega_n$, where $\omega_n$ is an isotropic vector. If a random point $x_{n+1}$ distributed by density $p(x_n, y)$ for the modeling next random point, we can use acceptance rejection method (about the acceptance rejection method see [7]).

The details of the acceptance rejection algorithm go as follows:

(A) Simulate $\alpha_n^0$, $\alpha_n^1$, $\alpha_n^2$ - uniformly distributed on the interval $(0; 1)$ random variables and calculate $\xi = -\ln\left(\alpha_n^1 \alpha_n^2\right)/\sqrt{c}$; if $\xi > R$ the process starts from the beginning, otherwise calculate $\zeta = \alpha_n^0 \, \xi \exp(-\sqrt{c}\xi)$;

(B) If $\zeta \geq \xi \frac{\mathrm{sh}\left[(R-\xi)\sqrt{c}\right]}{\mathrm{sh}\left(R\sqrt{c}\right)}$ than go to (A) otherwise $x_{n+1} = \xi$.

Let $Z_0 = (x, 1)$, $Z_1$, $Z_2$, ..., $Z_n$, ... be the trajectory of random processes, where $Z_i = (x_i^1, n_i^1; x_i^2, n_i^2; \ldots; x_i^{l_i}, n_i^{l_i})$ is a point distribution at the moment $i$. We can prove the following lemma.

**Lemma 1.** *The branching random processes $\{Z_n\}$ with probability one terminated in the domain $D$ or converges to the finite point distribution $Z = (x^1, n^1; x^2, n^2; \ldots; x^k, n^k)$, where $x^i \in \Gamma$, $i = 1, 2, \ldots k$.*

*Proof.* The process can be considered as a branching diffusion process for the particles that are diffusing in the bounded domain $\bar{D}$ with absorbing boundaries. If the average number of particles $K$, which generates new particles for the one step is $K = K(x, 1) < 1$, then the constructed process is absorbed with probability 1, and the condition $K < 1$ is a necessary and sufficient condition for the average number of the branches of the process to be finite [8, p. 104]. Let us show for the process, which was described above, $K < 1$. In accordance with our suggestions, the series $M = \sum_{k=0}^{\infty} \bar{a}_n n$ converges. We will fix $0 < \alpha < 1$. Taking into consideration that $\pi_n = \frac{\alpha}{M} \bar{a}_n$, we find: $K = q(x) + (1 - q(x))\frac{\alpha}{M} \sum_{n=1}^{\infty} n \bar{a}_n < 1$.

Thus, the current process terminates, then the general number of particles, which took part in the process, is finite [3]. If the process does not terminate inside of $D$, then starting at some time all the particles hit to the boundary $\Gamma$. There exists $n = n_0$, so that starting at $Z_n$ that has the following form $(x_n^1, 1; x_n^2, 1; \ldots; x_n^k, 1)$, where $k$ does not depend on $n$. Further, all of the $k$ particles independently are doing random walk on the spheres. For random walks on sphere from the results [8], when $n \to \infty$, $i = 1, 2, \ldots, k$, $x_n^i \to x^i \in \Gamma$ almost surely. Lemma is proved.

Furthermore on the trajectory random walk on sphere with branching we will construct unbiased and $\varepsilon$-biased estimators of the solution.

# 5   Construction of Unbiased and $\varepsilon$-biased Estimators of the Solution

In accordance with the probabilistic representation 1, a random process in whose trajectories we define an unbiased estimator for the solution of the problem. The random process of walk on sphere with branching starts at the point $x_0 = x$, an initial point we have one particle at the point $x$. For the unit time a particle moved to the uniformly distributed on sphere $S_R(x)$ random point $x_1 = y_1$ with the probability $q(x) = \frac{R\sqrt{c}}{\mathrm{sh}(R\sqrt{c})}$, or with probability $1 - q(x)$ moved to the point $x_1 = y_2 \in K_R(x)$, which is distributed with density function $p(x, y)$. In the second case with probability $\pi_n$ $(n \neq 0)$ a particle divided into $n$ new particles or with the probability $\pi_0$ absorbed. From point $x_1$, $n$ independent trajectories of the walk on sphere are constructed. New particles behave independently from each other (similar to the original). A random process terminates, if all of particles absorbed in $\bar{D}$ or particles hit $\varepsilon$-neighbourhood of boundary $\Gamma_\varepsilon = \{y \in \bar{D} : \min_{z \in \Gamma} |y - z| < \varepsilon\}$.

On the trajectory of process were constructed the estimator which defined recurrently. Let

$$\zeta_0(x) = u(x), \quad \zeta_k(x) = \Psi(\zeta_{k-1}(x)) \tag{4}$$

$$\Psi(\zeta(x)) = \begin{cases} \zeta(y_1), & \text{with probability } q(x); \\ W_n(y_2) \prod_{i=1}^{n} \zeta^{(i)}(y_2), & \text{with probability } (1 - q(x))\pi_n, n \neq 0; \\ W_0(y_2), & \text{with probability } (1 - q(x))\pi_0. \end{cases} \tag{5}$$

Here $\zeta^{(i)}(y)$ - an independent realization of random variable $\zeta(y)$, $W_n(y) = \frac{Ma_n(y)}{c\bar{a}_n\alpha}$, $W_0(y) = \frac{a_0(y)}{c\pi_0}$. Let $\Re_k$, be $\sigma$-algebra, generated by sequences $\{\omega_i\}_{i=0}^{k-1}$, $\{\alpha_i^0\}_{i=0}^{k-1}$, $\{\alpha_i^1\}_{i=0}^{k-1}$, $\{\alpha_i^2\}_{i=0}^{k-1}$. The following statement is valid.

**Theorem 1.** *The sequence $\{\zeta_k(x)\}_{k=0}^{\infty}$ forms a martingale on $\{\Re_k\}_{k=0}^{\infty}$. If $M < c$, then $\zeta_k(x)$ is a uniformly integrable martingale.*

*Proof.* It follows from definition $\Re_k$ that $\zeta_k(x)$ is $\Re_k$ measurable. From the properties of conditional expectation, it follows that:

$$\mathbf{E}(\zeta_{n+1}(x)/\Re_n) = \mathbf{E}\left(\Psi(\zeta_n(x))/\Re_n\right) = q(x)\mathbf{E}\left(\zeta_n(y_1)/\Re_n\right)$$

$$+(1-q(x))\sum_{i=1}^{\infty}\pi_n\,\mathbf{E}\left(W_i(y_2)\prod_{j=1}^{i}\zeta_n^{(j)}(y_2)/\Re_n\right) + (1-q(x))\pi_0\,\mathbf{E}\left(W_0(y_2)/\Re_n\right).$$

From the validity of the probabilistic representation (3) we get:

$$\mathbf{E}(\zeta_{n+1}(x)/\Re_n) = q(x)\int_{S_R(x)}\zeta_n(y)d\omega$$

$$+(1-q(x))\int_{K_R(x)}p(x,y)\left[\sum_{i=1}^{\infty}\pi_n\frac{Ma_n(y)}{c\bar{a}_n\alpha}\zeta_n^i(y) + \pi_0\frac{a_0(y)}{c\pi_0}\right] = \zeta_n(x).$$

Thus, the sequence $\{\zeta_k(x)\}_{k=0}^{\infty}$ forms a martingale on $\{\Re_k\}_{k=0}^{\infty}$. For proving that $\zeta_k(x)$ is uniformly integrable, so, it is enough to show that $|\zeta_k(x)| \leq C_0$, $(C_0 - const)$. Let the parameter $\alpha$ be chosen from the condition $\frac{M}{c} \leq \alpha < 1$.

Since $u(x) \in C(\bar{D}) \cap C^2(\bar{D})$ and $\bar{D}$ bounded domain, then $|u(x)| \leq const$ for $(x) \in \bar{D}$. Furthermore, under the condition of theorem $|W_n(y)| = \left|\frac{Ma_n(y)}{c\bar{a}_n\alpha}\right| \leq 1$ and therefore we have $|\zeta_n(x)| \leq C_0 < \infty$, $(C_0 = const)$. It follows that the sequence $\{\zeta_n(x)\}$ is uniformly integrable, and theorem has been proven.

Moreover, for using the estimator $\zeta_n(x)$, we will build $\varepsilon$-biased estimator which is easy implementable for coding. We will take some sufficiently small $\varepsilon$, and will consider the inner $\varepsilon$-neighborhood of the boundary $\Gamma_\varepsilon$. Let $N_1$ be the time index of the absorption of the process inside the domain, and $N_\varepsilon$ the time index of first passage of all particulars in $\Gamma_\varepsilon$. Then $N = \min\{N_1, N_\varepsilon\}$ is the stopping time of the process. Then the probability of absorption at a point $x_n$ the trajectory will be equal to:

$$g(x_n) = \begin{cases} 1, & \text{if } x_n \in \Gamma_\varepsilon, \\ (1-q(x_{n-1}))\pi_0, & \text{if } x_n \in \overline{D}\backslash\Gamma_\varepsilon. \end{cases}$$

From lemma it follows that $N < \infty$. The next statement is valid.

**Theorem 2.** *Let the conditions $M < c$ be satisfied. Then $\zeta_N(x)$ is an unbiased estimator for $u(x)$ with finite variance.*

*Proof.* According to Theorem 1, $\zeta_n(x)$ is a uniformly integrable martingale, and $N$ is a Markov time, then by Doobs optional sampling theorem (see [9,18]) for the martingale $\{\zeta_k(x)\}_{k=0}^{\infty}$ we get $E\zeta_N(x) = E\zeta_1(x)$. By definition $\zeta_1(x)$ using formulae (4)–(5) and probabilistic representation (3) we have the following representation

$$\mathbf{E}(\zeta_1(x)) = q(x)\mathbf{E}u(y_1) + (1-q(x))\mathbf{E}f(y_2, u(y_2))/c = u(x).$$

In accordance with the conditions of Theorem 1 $E(\zeta_N(x))^2 < \infty$ and thus its variance is finite. The theorem has been proven.

Further, for using $\zeta_N(x)$, we will build $\varepsilon$-biased, but practically implementable estimator $\zeta_N^*(x)$ in the following way:

Let $x^*$ be closest point of the bound $\Gamma$ to the point $x$. $\zeta_N^*(x)$ is obtained by substituting $u(x_N)$ into $\zeta_N(x)$ with $\varphi(x_N^*)$. Let us estimate the bias in $\zeta_N^*(x)$. Using the result of Theorem 1 and according to the property of the mathematical expectation we can record $|\mathbf{E}\zeta_N^*(x) - u(x)| \le \mathbf{E}|\zeta_N^*(x) - \zeta_N(x)|$. If $N = N_1$, then the process is terminated without hitting $\Gamma_\varepsilon$. In this case $\zeta_N^*(x) = \zeta_N(x)$ and the bias is equal to zero. If $N = N_\varepsilon$, then $Z_N = (x_N^1, n_N^1; x_N^2, n_N^2; \ldots; x_N^k, n_N^k)$, where $x_N^i \in \Gamma$, $i = 1, 2, \ldots, k$ and the number $k$ does not depend on $N$. Taking into account that $|W_n(y)| \le 1$ for arbitrary $n$, and $y \in \bar{D}$, we have

$$\zeta_N^*(x) - \zeta_N(x) \le \prod_{i=1}^k \left[\varphi\left(x_N^{i\,*}\right)\right]^{n_N^i} - \prod_{i=1}^k \left[u\left(x_N^{i\,*}\right)\right]^{n_N^i}.$$

Let $L(\varepsilon)$ be the absolute value of continuity of the function $u(x)$, then the following is valid: $|\mathbf{E}\zeta_N^*(x) - u(x)| \le L(\varepsilon)\mathbf{E}\left(n_N^1 + n_N^2 + \cdots + n_N^k\right)$.

Since the average number of particles in the $N$-th generation is $\mathbf{E}\left(n_N^1 + n_N^2 + \cdots + n_N^k\right) \le K^N < 1$, we will find (see [3, p. 101]), that, the bias doesn't exceed $L(\varepsilon)$. Finiteness of the variance follows from $|W_n(y)| \le 1$.

# 6    Special Cases

We can consider the cases when $f(x, u) = g\exp(u)$, $f(x, u) = g\sin(u)$, $f(x, u) = g\cos(u)$, $f(x, u) = g\sinh(u)$, $f(x, u) = g\cosh(u)$, $g$ is constant, and obtain the corresponding branching probabilities. For example, consider the following cases.

A. Let us consider Dirichlet problem for $f(x, u) = g\exp(u)$:

$$-\Delta u(x) + cu(x) = g\exp(u), \ g > 0, \ x \in D; \quad u(x) = \varphi(x), \ x \in \Gamma.$$

It is known for any $u$ one can write an expansion $\exp(u) = \sum_{n=0}^{\infty} \frac{u^n}{n!}$. In this case $|a_n| = a_n = \frac{g}{n!}$, and $M = \sum_{n=1}^{\infty} n|a_n| = \sum_{n=1}^{\infty} \frac{g}{(n-1)!} = g\exp(1)$. The condition (2) is satisfies. The parameter $\alpha$ will be chosen from the condition $g\exp(1)/c \le \alpha < 1$. The probabilities of branching defined by following recurrent formulas: $\pi_n = \frac{\alpha}{M}|a_n| = \frac{\alpha}{\exp(1)}\frac{1}{n!} = \frac{\pi_{n-1}}{n}$, $(n = 2, 3, \ldots)$, $\pi_1 = \frac{\alpha}{\exp(1)}$ and the probability of absorption: $\pi_0 = 1 - \sum_{n=1}^{\infty} \pi_n = 1 - \alpha(1 - \exp(1))$.

B. Dirichlet problem for $f(x, u) = g\sin(u)$. In this case $\sin(u) = \sum_{n=0}^{\infty} (-1)^n \frac{u^{2n+1}}{(2n+1)!}$, $a_{2n+1} = \frac{(-1)^n g}{(2n+1)!}$, $a_{2n} = 0$, $|a_{2n+1}| = \frac{g}{(2n+1)!}$, $M = g\sum_{n=0}^{\infty} \frac{1}{(2n)!} = g \, \mathrm{ch}(1)$. The condition (2) is satisfies. The parameter $\alpha$

will be chosen from the condition $g\,\mathrm{ch}(1)/c \le \alpha < 1$. The probabilities of branching defined by following recurrent formulas $(n = 1, 2, \ldots)$:
$$\pi_{2n+1} = \frac{\alpha}{M}|a_{2n+1}| = \frac{\alpha}{\mathrm{ch}(1)}\frac{1}{(2n+1)!} = \frac{\pi_{2n-1}}{2n(2n+1)}, \quad pi_1 = \frac{\alpha}{\mathrm{ch}(1)}, \quad pi_{2n} = 0. \text{ The}$$
probability of absorption: $\pi_0 = 1 - \sum_{n=1}^{\infty}\pi_n = 1 - \frac{\alpha}{\mathrm{ch}(1)}\sum_{n=0}^{\infty}\frac{1}{(2n+1)!} = 1 - \alpha\,\mathrm{sh}(1)/\mathrm{ch}(1)$.

# 7  Numerical Experiments

In the numerical experiment, 5 model problems for cases (a) $f(x, u) = g\exp(u) + f_0(x)$, (b) $f(x, u) = g\sin(u) + f_0(x)$, (c) $f(x, u) = g\cos(u) + f_0(x)$, (d) $f(x, u) = g\sinh(u) + f_0(x)$, (e) $f(x, u) = g\cosh(u) + f_0(x)$, (f) $f(x, u) = \sum_{n=0}^{\infty}a_n(x)u^n(x)$ in the cube and in the sphere were considered (Table 1).

**Table 1.** The results of the computational experiment.

| MP | $f(x,u)$ | $D$ | $R$ | $x_0$ | $U_{ex}$ | $MC$ | $err$ | $3\sigma$ | $anb$ | $anp$ |
|----|------|------|-----|---------|-----------|-----------|---------|---------|------|-------|
| 2 | (d) | sph | 0.9 | $(0.4;0.2;-0.6)$ | 0.00230 | 0.00282 | 0.00052 | 0.00080 | 1.04 | 12.30 |
| 1 | (d) | sph | 0.9 | $(-0.4;0.2;-0.6)$ | 0.89022 | 0.89553 | 0.00531 | 0.01781 | 1.04 | 12.29 |
| 1 | (a) | sph | 0.9 | $(-0.4;0.2;-0.6)$ | 0.89022 | 0.89204 | 0.00182 | 0.01908 | 1.05 | 12.32 |
| 3 | (a) | sph | 0.9 | $(-0.7;0.2;-0.1)$ | $-0.56464$ | $-0.56740$ | 0.00276 | 0.02328 | 1.06 | 12.43 |
| 3 | (c) | cube | 1.0 | $(0.7;0.2;0.1)$ | 0.84147 | 0.83611 | 0.00536 | 0.01916 | 1.01 | 11.68 |
| 5 | (e) | cube | 1.0 | $(0.9;0.2;0.1)$ | 0.65244 | 0.64132 | 0.01112 | 0.01901 | 0.99 | 10.37 |
| 2 | (b) | sph | 1.2 | $(0.9;0.2;0.1)$ | 0.00032 | $-0.00049$ | 0.00082 | 0.00259 | 1.08 | 13.18 |
| 2 | (b) | cube | 1.2 | $(0.9;0.2;0.1)$ | 0.00032 | 0.00198 | 0.00166 | 0.00225 | 1.07 | 13.22 |
| 5 | (e) | sph | 2.0 | $(0.9;0.2;0.1)$ | 0.65244 | 0.67290 | 0.02046 | 0.11216 | 0.37 | 10.56 |
| 1 | (f) | cube | 1.0 | $(0.5;0.5;0.5)$ | 0.08925 | 0.06402 | 0.02523 | 0.01320 | 1.06 | 12.36 |
| 2 | (f) | sph | 1.0 | $(0.5;-0.5;0.5)$ | 0.01563 | 0.01139 | 0.00424 | 0.00519 | 1.05 | 11.77 |
| 4 | (f) | sph | 1.0 | $(-0.1;-0.5;0.5)$ | 0.97531 | 1.00188 | 0.02657 | 0.06357 | 1.10 | 13.19 |
| 4 | (f) | sph | 1.0 | $(-0.1;-0.5;0.5)$ | 0.97531 | 0.95390 | 0.02141 | 0.02919 | 1.11 | 13.03 |

**Explanation of Table:** MP - number of the model problem; $D$ - the domain in which the problem is considered (sph - sphere); $R$ - the radius of the ball, if the ball area, the length of the face, if the cube; $x_0$ - the point at which the problem is solved; $U_{ex}$ - the exact solution; $MC$ - the estimation $S = \bar{\xi} = \frac{1}{N_t}(\xi_1 + \xi_2 + \cdots + \xi_N)$, where $N$ - the number of realisation and $\xi_i$ is an independent implementation of estimation $\zeta_N^*(x)$; $err$ - the difference between the exact solution and the estimation $err = |u(x_0) - S|$; $\sigma$ - a statistical estimation of $\sqrt{D\xi/N}$, where $D\xi$ is the sample variance of the estimation; $anp$ - the average number of branching; $anp$ - the average number of particles of the tree.

**Conclusion.** The results of computational experiments show, the constructed statistic algorithms could be used in practice. The parameter $\alpha$ which controlled the probability of branching can control the quantity of average number of particles of the tree. We remind that choosing this parameter was from the condition $M/c \leq \alpha < 1$. For the computational experiments we chose $\alpha = 0.9$, the quantity of trails was equal to $N = 5000$. From the table one can see the average number of particles of the tree $anb \approx 12$. The level of confidence equals 99.7% and the exact solutions will be in interval $(\bar{\xi} - 3\sigma, \bar{\xi} + 3\sigma)$. The future we want to apply these results to the physics (see: [14]) and to the differential geometric problems (see: [1]).

# References

1. Babich, M.V.: Real finite-gap solutions of the equation $\Delta u = \cosh(u)$. J. Math. Notes **50**(1–2), 663–667 (1991)
2. Courant, R., Hilbert, D.: Methods of Mathematical Physics: Partial Differential Equations, vol. 2. Interscience, New York (1962)
3. Harris, T.E.: The Theory of Branching Processes. Courier Corporation, Chelmsford (2002)
4. Elepov, B.S., Mikhailov, G.A.: Solution of the Dirichlet problem for the equation $\Delta u - cu = -q$ by a model of "walks on spheres". USSR Comput. Math. Math. Phys. **9**(3), 194–204 (1969)
5. Elepov, B.S., Mikhailov, G.A.: Use of the fundamental solutions of elliptic equations for constructing Monte Carlo algorithms. USSR Comput. Math. Math. Phys. **14**(3), 187–194 (1974)
6. Ermakov, S.M.: The Monte Carlo method and related problems. Nauka, Moscow (1975)
7. Ermakov, S.M., Mikhailov, G.M.: Statistical Modelling. Nauka, Moscow (1982). (in Russian)
8. Ermakov, S.M., Nekrutkin, V.V., Sipin, A.S.: Random Processes for Classical Equations of Mathematical Physics. Kluwer Academic Publishers, Dordrecht (1989)
9. Meyer, P.A.: Probability and potentials. Blaisdell Publishing Company, New York (1966)
10. Mikhailov, G.A.: Solving the Dirichlet problem for nonlinear elliptic equations by the Monte Carlo method. Sib. Math. J. **35**(5), 967–975 (1994)
11. Mikhailov, G.A., Voitishek, A.V.: Numerical statistical modelling. In: Monte Carlo Methods. Akademiya, Moscow (2006)
12. Mikhailov, G.A., Makarov, R.N.: Solution of boundary value problems of the second and third kind by Monte Carlo methods. Sib. Math. J. **38**(3), 518–527 (1997)
13. Mikhailov, G.A., Makarov, R.N.: Solution of boundary value problems by the method of "walks on spheres" with reflection from the boundary. J. Dokl. RAN **353**(6), 720–722 (1997). (in Russian)
14. Neklyudov, A.V.: Behavior of the solution to Gauss-Bieberbach-Rademacher equation on plane. Ufa Math. J. **6**(3), 85–94 (2014)
15. Rasulov, A.S.: Monte Carlo method for solving nonlinear problems. Fan, Tashkent (1992). (in Russian)

16. Rasulov, A.S., Raimova, G.M.: Monte Carlo method for solution of initial-boundary value problems for nonlinear parabolic equations. J. Math. Comput. Simul. **146**, 240–250 (2018)
17. Rasulov, A.S., Sipin, A.S.: Solution of one nonlinear equation by Monte Carlo method. In: Monte Carlo Methods in Computational Mathematics and Mathematical Physics, Novosibirsk, pp. 149–155 (2001)
18. Shiryaev, A.N.: Probability. Graduate Texts in Mathematics, vol. 95. Springer, New York (1996). https://doi.org/10.1007/978-1-4757-2539-1. Print ISBN: 978-1-4757-2541-4. Online ISBN: 978-1-4757-2539-1. Series Print ISSN: 0072-5285
19. Simonov, N.A.: Random walks on spheres algorithms for solving mixed boundary value problem and Neumann problem. Sib. J. Comput. Math. **10**(2), 209–220 (2007)
20. Sipin, A.S.: Solution of two Dirichlet boundary value problems by the Monte Carlo method. USSR Comput. Math. Math. Phys. **19**(2), 119–134 (1979)

# Parallel Numerical Method to Estimate the Effective Elastic Moduli of Rock Core Samples from 3D Tomographic Images

Galina Reshetova[1]([✉])[iD] and Tatiana Khachkova[2]

[1] Institute of Computational Mathematics and Mathematical Geophysics SB RAS, 630090 Novosibirsk, Russia
kgv@nmsf.sscc.ru
[2] Trofimuk Institute of Petroleum Geology and Geophysics SB RAS, 630090 Novosibirsk, Russia

**Abstract.** We present a new parallel numerical technique to estimate the effective elastic parameters of a rock core sample from three-dimensional Computed Tomography (CT) images. Our method is based on the energy equivalence principle and a new approach to solve 3D static elasticity problem by iterative relaxation technique. We determine the elastic moduli by the parallel computation of potential energy of the elastic deformations arising in the sample under some homogeneous strains applied to the boundary thus simulating effects occurring in laboratory measurements. The obtained numerical results are discussed. The proposed method is verified using homogeneous samples with specified properties as well as for layered materials with effective parameters obtained according to the Schoenberg method. The effective parameters of a real carbonate core sample obtained from 3D CT-image are presented.

**Keywords:** Effective parameters · Elastic moduli ·
The principle of energy equivalence · The relaxation method ·
Homogeneous boundary conditions · Representative volume ·
3D tomographic images

## 1 Introduction

In recent years, the oil and gas industry has been actively developing a new technology for the core research, based on computer simulation. The use of tomographic core images allows obtaining a digital three-dimensional reconstruction of the core sample material under study and performing a numerical experiment using modern computer technologies. The Digital rock physics includes the whole complex of digital tomographic core images research and their processing (primarily filtration and segmentation). Within the framework of these studies, there are a number of approaches for determining the effective inhomogeneous elastic moduli, including the methods based on the analysis of inclusions [1–3], the use

© Springer Nature Switzerland AG 2019
I. Dimov et al. (Eds.): FDM 2018, LNCS 11386, pp. 452–460, 2019.
https://doi.org/10.1007/978-3-030-11539-5_52

of a wide range of homogenization methods [4] and some other approaches. In this paper, we present a parallel numerical algorithm for estimating the elastic properties of a rock sample from their three-dimensional tomographic images. For solving the problem, a method based on the energy equivalence principle proposed in the paper [5] to study the properties of composite materials and a new approach to solve 3D static elasticity problem by iterative relaxation technique were chosen. This choice is due to the possibility of numerical parallelization to subtasks with linear acceleration and, as subsequently it was shown by the numerical experiments, an high accuracy. In the method proposed, the boundary conditions for solving the static elasticity problem are homogeneous kinematic or static boundary conditions. Although there is a theoretical possibility of using the static boundary conditions, but in practice, only kinematic conditions are used. This is due to both the complexity of solving the static problem of the elasticity theory in the formulation of stresses and specific feature of the problem of designing a composite material with a periodic structure for which the energy equivalence principle method was developed. However, when studying rock samples, it is more natural to apply static boundary conditions in the form of applying external strains that simulate a real physical experiment.

## 2 Statement of the Problem and Description of the Method

The effective elastic properties of a sample are determined based on the generalized Hooke's law, which expresses the relationship between the deformations and stresses averaged over a representative volume:

$$\bar{\sigma}_{ij} = c^*_{ijkl}\bar{\varepsilon}_{kl} \quad or \quad \bar{\varepsilon}_{ij} = s^*_{ijkl}\bar{\sigma}_{kl}. \tag{1}$$

The components of the stiffness tensor $c^*_{ijkl}$ and the compliance tensor $s^*_{ijkl}$ form the tensors of the fourth rank and by definition are effective stiffness $C^*$ and the compliance $S^*$ tensors. The average stresses and strains are determined by the formulas:

$$\bar{\sigma}_{ij} = \frac{1}{V}\int_V \sigma_{ij}dV, \quad \bar{\varepsilon}_{ij} = \frac{1}{V}\int_V \varepsilon_{ij}dV, \tag{2}$$

where $\sigma_{ij}$ and $\varepsilon_{ij}$ are components of the stress and strain tensors describing the stress-strain state of a sample in the representative volume and satisfying the equilibrium equations and the Saint-Venant compatibility equations.

To find the effective stiffness and compliance tensors $C^*$ and $S^*$ we use the energy equivalence principle method [5]. To do this we introduce the notion of homogeneous boundary conditions [4]. Such conditions can be either kinematic or static and are defined in such a way that when applied to the boundary S of a homogeneous elastic body with a volume V, they cause in it the uniform (constant) stresses and deformations. In particular, the homogeneous static (3a) and kinematic (3b) boundary conditions are boundary conditions with stresses

(3a) and displacements (3b) specified on the boundary in the form of the linear functions

$$(a)\ t_i(S) = \sigma_{ij}^0 n_j, \quad (b)\ u_i(S) = \varepsilon_{ij}^0 x_j, \tag{3}$$

where $\sigma_{ij}^0, \varepsilon_{ij}^0$ are some constant symmetric stress and strain tensors, and $\boldsymbol{n}$ is the vector of the outer normal to the boundary S.

The energy equivalence principle method is based on a theorem asserting that the homogeneous static (kinematic) boundary conditions applied to the boundary S of a non-homogeneous representative volume V generate such a stress field $\sigma_{ij}$ (strain $\varepsilon_{ij}$) that its averaging over the volume (2) is equal to the value of the constant stress $\sigma_{ij}^0$ (strain $\varepsilon_{ij}^0$) applied to the boundary (3):

$$\bar{\sigma}_{ij} = \sigma_{ij}^0, \quad \bar{\varepsilon}_{ij} = \varepsilon_{ij}^0. \tag{4}$$

The potential energy of deformations arising in an inhomogeneous elastic deformable volume V is expressed by the formula:

$$U = \frac{1}{2} \int_V \sigma_{ij} \varepsilon_{ij} dV. \tag{5}$$

We calculate the energy of deformations when homogeneous static boundary conditions are applied to an inhomogeneous elastic body:

$$U = \tfrac{1}{2} \int_V \sigma_{ij}\varepsilon_{ij} dV = \tfrac{1}{2} \int_S \sigma_{ij} u_i n_j dS = \tfrac{1}{2}\sigma_{ij}^0 \int_S u_i n_j dS = \tfrac{1}{2}\sigma_{ij}^0 \int_V u_{i,j} dV = \tfrac{1}{2}\sigma_{ij}^0 \int_V \varepsilon_{ij} dV$$

$$= \tfrac{1}{2}\bar{\sigma}_{ij}\bar{\varepsilon}_{ij} V = \tfrac{1}{2} s_{ijkl}^* \sigma_{kl}^0 \sigma_{ij}^0 V. \tag{6}$$

It follows that the potential energy of an inhomogeneous elastic body in a stress-strain state is represented in the following form

$$U = \frac{1}{2} s_{ijkl}^* \sigma_{kl}^0 \sigma_{ij}^0 V. \tag{7}$$

Thus, if the value of the potential energy $U$ of the stress-strain state of the elastic body in which it has been transferred under the action of homogeneous boundary conditions (static stresses) $\sigma_{ij}^0$ is known, then Eq. (7) can be used to find the components of the effective compliance tensor $S_{ijkl}^*$. If we calculate the potential energy $U_0$ for a homogeneous sample with properties determined by the effective stiffness tensor $C^*$, then we obtain the expression:

$$U_0 = \frac{1}{2} s_{ijkl}^* \sigma_{kl}^0 \sigma_{ij}^0 V. \tag{8}$$

It follows from formulas (7) and (8) that the energy method can be regarded as a method based on the equivalence principle of the potential energies for inhomogeneous and homogeneous samples, which are "equivalent":

$$U_0 = U. \tag{9}$$

It should be noted that one can find both $C^*$ and $S^*$ because of their reversibility. In this paper, $S^*$ is determined first, and then $C^*$ is calculated by inversion.

## 3    The Algorithm for Determining the Components of the Compliance Tensor $S^*$

We assume that the volume $V$ is fixed in space by a rectangular (parallelepiped in 3D) region with the sides (edges) parallel to the coordinate axes. To find the components $s^*_{ijkl}$, we seek the solution of the boundary-value problem of the static theory of elasticity

$$\sigma_{ij,j} = 0, \tag{10}$$

$$\sigma_{ij} = c_{ijkl}\varepsilon_{kl} = c_{ijkl}u_{k,l}, \quad i,j = 1,2 \tag{11}$$

with the corresponding homogeneous static boundary conditions applied to the faces of the sample.

In case of 2D sample the tensor $S^*$ is written in the form:

$$\begin{bmatrix} \varepsilon_{11} \\ \varepsilon_{22} \\ 2\varepsilon_{12} \end{bmatrix} = S^* \begin{bmatrix} \sigma_{11} \\ \sigma_{22} \\ \sigma_{12} \end{bmatrix}, \quad S^* = \begin{pmatrix} s^*_{1111} & s^*_{1122} & s^*_{1112} \\ & s^*_{2222} & s^*_{2212} \\ sym & & s^*_{1212} \end{pmatrix} \tag{12}$$

When calculating $s^*_{1111}$, $s^*_{2222}$ and $s^*_{1212}$ with the static boundary conditions (Table 1) according to (7), we obtain:

$$U^{(1)} = \frac{1}{2}s^*_{1111}V, \quad s^*_{1111} = 2U^{(1)}/V. \tag{13}$$

Here and below, the superscript in the notation indicates to number of the case under consideration. Cases 1–3 are presented in Table 1.

To determine the remaining components $s^*_{1122}, s^*_{2212}$ and $s^*_{1112}$, we use the linearity property of the elasticity problem. Since the boundary conditions for cases 4–6 are a linear combination of boundary conditions in cases 1–3 (Table 1), then

$$U^{(4)} = \frac{1}{2}\int_V \sigma^{(4)}_{ij}\varepsilon^{(4)}_{ij}dV = \frac{1}{2}\int_V (\sigma^{(1)}_{ij} + \sigma^{(2)}_{ij})(\varepsilon^{(1)}_{ij} + \varepsilon^{(2)}_{ij})dV = U^{(1)} + U^{(2)} + U^{(1,2)},$$

$$\tag{14}$$

$$U^{(k,l)} = \frac{1}{2}\int_V (\sigma^{(k)}_{ij}\varepsilon^{(l)}_{ij} + \sigma^{(l)}_{ij}\varepsilon^{(k)}_{ij})dV$$

$$= \int_V \left[ \frac{1}{E}(\sigma^{(k)}_{11}\sigma^{(l)}_{11} + \sigma^{(k)}_{22}\sigma^{(l)}_{22}) - \frac{\nu}{E}(\sigma^{(k)}_{11}\sigma^{(l)}_{22} + \sigma^{(l)}_{11}\sigma^{(k)}_{22}) + \frac{2(\nu+1)}{E}\sigma^{(k)}_{12}\sigma^{(l)}_{12} \right] dV,$$

$$\tag{15}$$

**Table 1.** The boundary conditions and formula for finding the components of $s^*$ (2D).

| Case | U | Faces a | Faces b | Faces a, b | Value $s_{ijkl}^*$ |
|------|---|---------|---------|------------|--------------------|
| 1 | $U^{(1)}$ | $\sigma_{11} = 1$ | $\sigma_{22} = 0$ | $\sigma_{12} = 0$ | $s_{1111}^* = 2U^{(1)}/V$ |
| 2 | $U^{(2)}$ | $\sigma_{11} = 0$ | $\sigma_{22} = 1$ | $\sigma_{12} = 0$ | $s_{2222}^* = 2U^{(2)}/V$ |
| 3 | $U^{(3)}$ | $\sigma_{11} = 0$ | $\sigma_{22} = 0$ | $\sigma_{12} = 1$ | $s_{1212}^* = 2U^{(3)}/V$ |
| 4 | $U^{(4)} = U^{(1)} + U^{(2)} + U^{(1,2)}$ | $\sigma_{11} = 1$ | $\sigma_{22} = 1$ | $\sigma_{12} = 0$ | $s_{1122}^* = U^{(1,2)}/V$ |
| 5 | $U^{(5)} = U^{(2)} + U^{(3)} + U^{(2,3)}$ | $\sigma_{11} = 0$ | $\sigma_{22} = 1$ | $\sigma_{12} = 1$ | $s_{2212}^* = U^{(2,3)}/V$ |
| 6 | $U^{(6)} = U^{(1)} + U^{(3)} + U^{(1,3)}$ | $\sigma_{11} = 1$ | $\sigma_{22} = 0$ | $\sigma_{12} = 1$ | $s_{1112}^* = U^{(1,3)}/V$ |

where $U^{(k)}, \sigma_{ij}^{(k)}, \varepsilon_{ij}^{(k)}, U^{(l)}, \sigma_{ij}^{(l)}, \varepsilon_{ij}^{(l)}$ are calculated energies, stresses and deformations for cases 1–3. So all components are defined by formula (Table 1).

In the three-dimensional case, the algorithm for finding the components of the compliance tensor is analogous to the two-dimensional one. The tensor $S^*$ is written in the form:

$$
\begin{bmatrix} \varepsilon_{11} \\ \varepsilon_{22} \\ \varepsilon_{33} \\ 2\varepsilon_{23} \\ 2\varepsilon_{13} \\ 2\varepsilon_{12} \end{bmatrix} = S^* \begin{bmatrix} \sigma_{11} \\ \sigma_{22} \\ \sigma_{33} \\ \sigma_{23} \\ \sigma_{13} \\ \sigma_{12} \end{bmatrix}, \quad S^* = \begin{pmatrix} s_{1111}^* & s_{1122}^* & s_{1133}^* & s_{1123}^* & s_{1113}^* & s_{1112}^* \\ & s_{2222}^* & s_{2233}^* & s_{2223}^* & s_{2213}^* & s_{2212}^* \\ & & s_{3333}^* & s_{3323}^* & s_{3313}^* & s_{3312}^* \\ & & & s_{2323}^* & s_{2313}^* & s_{2312}^* \\ & sym & & & s_{1313}^* & s_{1312}^* \\ & & & & & s_{1212}^* \end{pmatrix} \quad (16)
$$

**Table 2.** The boundary conditions for finding the diagonal components of $s_{ijkl}^*$ (3D).

| U | Faces a | Faces b | Faces c | Value $s_{ijkl}^*$ |
|---|---------|---------|---------|--------------------|
| 1 | $\sigma_{33} = \sigma_{13} = \sigma_{23} = 0$ | $\sigma_{22} = \sigma_{12} = \sigma_{23} = 0$ | $\sigma_{11} = 1, \sigma_{12} = \sigma_{13} = 0$ | $s_{1111}^* = 2U^{(1)}/V$ |
| 2 | $\sigma_{33} = \sigma_{13} = \sigma_{23} = 0$ | $\sigma_{22} = 1, \sigma_{12} = \sigma_{23} = 0$ | $\sigma_{11} = \sigma_{12} = \sigma_{13} = 0$ | $s_{2222}^* = 2U^{(2)}/V$ |
| 3 | $\sigma_{33} = 1, \sigma_{13} = \sigma_{23} = 0$ | $\sigma_{22} = \sigma_{12} = \sigma_{23} = 0$ | $\sigma_{11} = \sigma_{12} = \sigma_{13} = 0$ | $s_{3333}^* = 2U^{(3)}/V$ |
| 4 | $\sigma_{23} = 1, \sigma_{33} = \sigma_{13} = 0$ | $\sigma_{23} = 1, \sigma_{22} = \sigma_{12} = 0$ | $\sigma_{11} = \sigma_{12} = \sigma_{13} = 0$ | $s_{2323}^* = 2U^{(4)}/V$ |
| 5 | $\sigma_{13} = 1, \sigma_{33} = \sigma_{23} = 0$ | $\sigma_{22} = \sigma_{12} = \sigma_{23} = 0$ | $\sigma_{13} = 1, \sigma_{11} = \sigma_{12} = 0$ | $s_{1313}^* = 2U^{(5)}/V$ |
| 6 | $\sigma_{33} = \sigma_{13} = \sigma_{23} = 0$ | $\sigma_{12} = 1, \sigma_{22} = \sigma_{23} = 0$ | $\sigma_{12} = 1, \sigma_{11} = \sigma_{13} = 0$ | $s_{1212}^* = 2U^{(6)}/V$ |

Similar to the two-dimensional case, we calculate $U^{(1)}$ - $U^{(6)}$ (Table 2) and then use the linearity property and the results of previous considerations to calculate the remaining components of $s_{ijkl}^*$ (Table 3), where the values of $U^{(k,l)}$

are calculated by the formula:

$$U^{(k,l)} = \frac{1}{2}\int\limits_V (\sigma_{ij}^{(k)}\varepsilon_{ij}^{(l)} + \sigma_{ij}^{(k)}\varepsilon_{ij}^{(l)})dV = \int\limits_V \frac{1}{E}(\sigma_{11}^{(k)}\sigma_{11}^{(l)} + \sigma_{22}^{(k)}\sigma_{22}^{(l)} + \sigma_{33}^{(k)}\sigma_{33}^{(l)})dV$$

$$-\int\limits_V \frac{\nu}{E}(\sigma_{11}^{(k)}\sigma_{22}^{(l)} + \sigma_{11}^{(l)}\sigma_{22}^{(k)} + \sigma_{22}^{(k)}\sigma_{33}^{(l)} + \sigma_{22}^{(l)}\sigma_{33}^{(k)} + \sigma_{11}^{(k)}\sigma_{33}^{(l)} + \sigma_{11}^{(l)}\sigma_{33}^{(k)})dV$$

$$+\int\limits_V \frac{2(\nu+1)}{E}(\sigma_{12}^{(k)}\sigma_{12}^{(l)} + \sigma_{23}^{(k)}\sigma_{23}^{(l)} + \sigma_{13}^{(k)}\sigma_{13}^{(l)})dV. \tag{17}$$

**Table 3.** The formulas for computing the remaining components of $s_{ijkl}^*$ (3D case).

| | | | | |
|---|---|---|---|---|
| $s_{1122}^* = U^{(1,2)}/V$ | $s_{1123}^* = U^{(1,4)}/V$ | $s_{1113}^* = U^{(1,5)}/V$ | $s_{2313}^* = U^{(4,5)}/V$ | $s_{3312}^* = U^{(3,6)}/V$ |
| $s_{1133}^* = U^{(1,3)}/V$ | $s_{2223}^* = U^{(2,4)}/V$ | $s_{2213}^* = U^{(2,5)}/V$ | $s_{1112}^* = U^{(1,6)}/V$ | $s_{2312}^* = U^{(4,6)}/V$ |
| $s_{2233}^* = U^{(2,3)}/V$ | $s_{3323}^* = U^{(3,4)}/V$ | $s_{3313}^* = U^{(3,5)}/V$ | $s_{2212}^* = U^{(2,6)}/V$ | $s_{1312}^* = U^{(5,6)}/V$ |

## 4  Numerical Solution to the Static Problem of the Elasticity Theory

The most time-consuming component of the approach from the standpoint of computation is the solution of a series of static problems in the elasticity theory with external stresses given at the boundaries. In the final analysis, these problems are reduced to systems of linear algebraic equations, for whose it is possible to apply both direct and iterative methods. The fact is, direct methods, having certain advantages, in this case are not suitable for solving three-dimensional problems due to excessive demands for computer resources. Therefore, we have chosen iterative methods for determining effective parameters.

We propose to find solution of static problem (10), (11) with the static boundary conditions (3a) by finding the steady-state solution of the dynamic problem of the elasticity theory in the formulation of the stress/displacement velocity with additional dissipative terms to equations of motion (18):

$$\rho\dot{v}_i + \alpha v_i = \sigma_{ij,j} \tag{18}$$

$$\dot{\sigma}_{ij} = C_{ijkl}\dot{\varepsilon}_{kl} = C_{ijkl}v_{k,l} \tag{19}$$

with zero initial conditions for $t = 0$:

$$v_i = 0, \quad \sigma_{ij} = 0 \tag{20}$$

and time-invariant boundary conditions on the boundary $S$ (3a). Here $v_i = \dot{u}_i$ is the displacement velocity of the $i$-th component of the displacement vector.

In order to show the convergence of problem (18)–(20) to static problem, we use the virial theorem ([8], Sect. 10) asserting that the kinetic energy of the mechanical system $T$ averaged over an infinite time interval is equal to the virial averaged over the same time interval. If the potential energy $U$ is a homogeneous function of the first degree of inverse values of the radius vectors, then the relation

$$2T = -U \tag{21}$$

is satisfied.

It follows that if the kinetic energy of the system is reduced through an artificially introduced damping mechanism in the equation, then the rigid connection between the kinetic and potential energies provided by this theorem leads to a decrease in the potential energy up to its minimum. Then, based on the Lagrange-Dirichlet principle, for a statically stressed body (*Of all possible stress-strain states of a deformable solid, the actual stress state corresponds to a minimum of the total deformation energy*), we can conclude that the solution of the dynamic problem (18)–(20) converges to the solution of the stationary problem. For a numerical solution of the initial boundary value problem (18)–(20), we apply a finite-difference scheme on staggered grids [9], whose coefficients are modified to provide approximation in inhomogeneous media [10].

## 5    Parallel Implementation

The most time-consuming part of the algorithm, requiring in the 3-dimensional case the obligatory parallel implementation, consists of solving six problems for calculating the stress-strain state of an elastic sample. As these tasks can be solved independently, the most natural way in this case is the MPI parallelization by tasks, which allows each task to be solved independently. In addition, the solution of an individual task is parallelized using OpenMP technology.

## 6    Numerical Experiments

The representative series of numerical experiments was carried out. Firstly, the homogeneous isotropic materials samples (plexiglas, copper and steel) were considered and the calculated effective parameters were compared with the elastic moduli of the material itself (the difference is $10^{-6}$). Secondly, the elasic moduli for the samples of layered materials were calculated (Fig. 1). The size of the models was varied along the interlayers, across, the number of layers and their incline was changed and so on. The results of the proposed method were compared with the Schoenberg averaging method ones [7]. The difference decreased with increasing size along the interlayer and at the size of 500 * 500 * 30 was 4%. Finally, calculations were made for a three-dimensional segmented digital model of a carbonate core (Fig. 1c), then the seismic velocities were estimated and compared with the results of a laboratory measurement. The difference make up less than 3%.

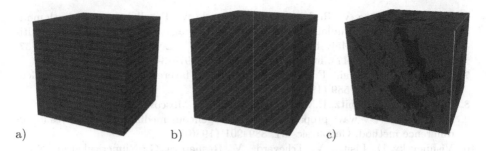

a)                          b)                          c)

**Fig. 1.** The different models of layered media (a–b) and the segmented digital model of core sample (c).

# 7   Conclusions

In our work we proposed a new method of calculating the effective elastic moduli of rock that use tomographic core images and perform a numerical experiment using modern computer technologies. Our method is based on the energy equivalence principle and the new approach to solve 3D static elasticity problem by iterative relaxation technique. We determine the elastic moduli by the parallel computation of potential energy of the elastic deformations arising in the sample under some homogeneous strains applied to the boundary thus simulating effects occurring in laboratory measurements. The performed numerical experiments showed a high accuracy of the method and an acceptable calculating time.

**Acknowledgements.** This work was supported by the Russian Foundation for Basic Research, Grant 16-05-0800, 18-05-00031, 18-01-00579, 18-41-540016. The research is carried out using the equipment of the shared research facilities of HPC computing resources at Lomonosov Moscow State University, Joint Supercomputer Center of RAS and the Siberian Supercomputer Center.

# References

1. Christensen, R.: Introduction to Mechanics of Composite Materials, 1st edn. Wiley, New York (1979)
2. Sendetski, J. (ed.): Composition Materials, vol. 2. Mechanics of Composition Meterials [Russian translation]. Mir, Moscow (1978)
3. Shermergor, T.: The Theory of Elasticity of Microinhomogeneous Media [Russian translation]. Nauka, Moscow (1977)
4. Aboudi, J.: Mechanics of Composite Materials: A Unified Micromechanical Approach. Elsevier Science, Amsterdam (1991)
5. Zhang, W., Dai, G., Wang, F., Sun, S., Bassir, H.: Using strain energy-based prediction of effective elastic properties in topology optimization of material microstructures. Acta Mechanica Sinica **23**(1), 77–89 (2007)

6. Kostin, V., Lisitsa, V., Reshetova, G., Tcheverda, V.: Parallel algorithm with modulus structure for simulation of seismic wave propagation in 3D multiscale multiphysics media. In: Malyshkin, V. (ed.) PaCT 2017. LNCS, vol. 10421, pp. 42–57. Springer, Cham (2017). https://doi.org/10.1007/978-3-319-62932-2_4
7. Schoenberg, M., Muir, F.: A calculus for finely layered anisotropic media. Geophysics **54**(5), 581–589 (1989)
8. Landau, L.D., Lifshitz, E.M.: Mechanics. Nauka, Moscow (1988)
9. Virieux, J.: P-SV wave propagation in heterogeneous media: Velocity-stress finite-difference method. Geophysics **51**, 889–901 (1986)
10. Vishnevsky, D., Lisitsa, V., Tcheverda, V., Reshetova, G.: Numerical study of the interface errors of finite-difference simulations of seismic waves. Geophysics **79**, T219–T232 (2014)

# Integral-Consistent Numerical Technique for Self-gravitating Medium Model

Yu. S. Sharova[1], Yu. A. Poveshchenko[1,2], V. A. Gasilov[1,2], N. S. Smirnova[3], and V. O. Podryga[1,4(✉)]

[1] Keldysh Institute of Applied Mathematics of RAS, Moscow, Russia
pvictoria@list.ru
[2] National Research Nuclear University MEPhI, Moscow, Russia
[3] Laboratoire d'Annecy-le-Vieux de Physique Theorique, Annecy, France
[4] Moscow Automobile and Road Construction State Technical University,
Moscow, Russia

**Abstract.** The supercompression of matter caused by gravitational coupling or self-gravitational forces leads to density growth by several orders in magnitude. Keeping in mind the importance of self-gravitation in astrophysical processes like supernovae star evolution we consider it reasonable to develop a numerical technique based on the consistent approximation to the terms describing gravitational energy transfer into the kinetic energy of a matter in the star along its life cycle. The so-called completely conservative gas-dynamics difference schemes including the gravitation effects are the proper numerical technique able to simulate correctly the problems concerning gravitational coupling effects. The accounting for gravitational forces in the construction of completely conservative difference schemes is a significant complication. In the paper, we propose an integrally-consistent difference scheme that utilizes the method of support difference operators thus providing a possibility to conform the balance between kinetic and gravitational energy increments or losses. According to this method, we use the result of the total gravitational energy varying and construct the symmetrized strain rate tensor as the base operator. The result of varying the gravitational energy of the system is a discrete convolution of the Newton gravitational tensor in the difference media under study, which exhaustively answers all the gravitational processes unfolding against the background of the hydrodynamic motion of matter. The symmetrized strain tensor governs the kinematic motion in a considered system. The conjugate operator related to the convolution of these tensors automatically gives the approximation to the gravitational forces acting in the interior of the balance volume of the difference model built via the support operator approach.

**Keywords:** Self-gravitation · Mathematical modeling

## 1 Introduction

The significant area in computational astrophysics is the development of numerical methods for modeling gravitationally coupled (self-gravitating) objects.

© Springer Nature Switzerland AG 2019
I. Dimov et al. (Eds.): FDM 2018, LNCS 11386, pp. 461–467, 2019.
https://doi.org/10.1007/978-3-030-11539-5_53

Gravitational forces should be taken into account in studies of matter evolution related to such objects like protoplanetary clouds, supernovas and other clusters of large masses.

Numerical simulation of huge gravitating masses dynamics by means of Lagrangian (moving) adaptive grids, can lead to noticeable violations of the gravitational energy balance due to significant redistribution of matter and grid distortions. Here we consider the problem of building a robust technique mitigating such effects which may lead to the loss of computational accuracy and meaningless results. We propose the construction of completely conservative difference schemes accounting for the effect of gravitational forces on the matter dynamics and energy balance in a consistent fashion [1,2].

## 2   Formulation of the Problem

We consider an ensemble of Lagrangian (moving) particles having fixed masses and occupying some volume with moving boundary surface. The volume comprises a constant total mass, however individual particles can redistribute over the volume interior. We pursue the goal to derive the equation of the gravitational energy balance of the system in the form:

$$\frac{1}{2} \int_M \hat{F} dM = \frac{1}{2} \int_M F dM + \int_t^{\hat{t}} d\tau \left\{ - \int_{O(\tau)} D_g dV + \frac{1}{8\pi\gamma} \int_{\Sigma(\tau)} \left| \begin{matrix} F & \boldsymbol{g} ds \\ \frac{dF}{dt} & \frac{d}{dt}(\boldsymbol{g} ds) \end{matrix} \right| \right\} \quad (1)$$

Here we consider a flow region $O$ with mass $M$ witch is bounded by a surface $\Sigma$ (with elements of an externally oriented surface $d\boldsymbol{s}$). $F$ is the gravitational potential, $\boldsymbol{g} = -grad\ F$ is the intensity of the gravitational field, so that the gravitational analog of the Gauss's flux theorem is satisfied

$$\int_\Sigma \boldsymbol{g} d\boldsymbol{s} = -4\pi\gamma \int_M dM, \quad (2)$$

where $\gamma$ is the gravitational constant, $d/dt$ is the material derivative. Values referred to the final point in time $\hat{t}$ are marked with a "lid". The gravitational "dissipative" function $D_g = tr(t_g t_v)$ is defined as the trace of the product of the gravitational

$$t_g = \frac{1}{4\pi\gamma} \left( \boldsymbol{g} \cdot \boldsymbol{g} - \frac{g^2}{2} \delta \right) \quad (3)$$

and kinematic

$$t_v = \frac{1}{2} \left( \frac{d\boldsymbol{v}}{dt} + grad\ \boldsymbol{v} \right) \quad (4)$$

tensors. It determines the transformation of gravitational energy into kinetic energy, accelerating the flow elements "as a whole" without the thermodynamic compression of matter. This is achieved by the work of gravitational forces

$$- \, div \, t_g = \rho g \tag{5}$$

per unit volume of the medium. Hence, the balance of the pulse in such an integral system in the presence of gravitational forces can be written in the form:

$$\int_M \hat{v} dM = \int_M v dM - \int_t^{\hat{t}} d\tau \left( \int_{\Sigma(\tau)} P ds + \int_{O(\tau)} div \, t_g dV \right) \tag{6}$$

Here $v$ is velocity, $P$ is pressure, $\rho$ is density of matter. It is also assumed that the integral identity of vector analysis is done:

$$\int_O D_g dV + \int_O v \, div \, t_g dV = \int_\Sigma t_g v ds \tag{7}$$

## 3  Metric Grids of the Support Operator Method for Describing Gravitational Processes

Grids of this type [3] consist of cells ($\Omega$) formed by nodes ($\omega$) and faces ($\sigma$) (see Fig. 1). The nodes $\omega(\Omega)$ forming the cell $\Omega$ correspond to the bases $\varphi(\Omega)$ in this cell. The bases consist of unit normals $e(\sigma)$ to the basis faces $\sigma$. Metric calibration of the difference grid consists in choosing the volumes of bases $V_\varphi > 0$ with the natural normalization condition $\sum_{\varphi(\Omega)} V_\varphi = V_\Omega$. That is the sum of the base volumes inside the cell $\Omega$ is equal to the volume of this cell $V_\Omega$. For example, for a 2d-quadrangular grid $V_\varphi = 1/2 S_\varphi$, where $S_\varphi$ is the area of the triangle corresponding to the basis $\varphi$. We will refer the components of the gravitational

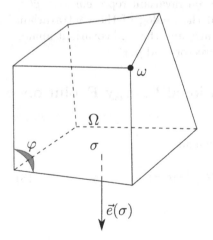

**Fig. 1.** The cell ($\Omega$)

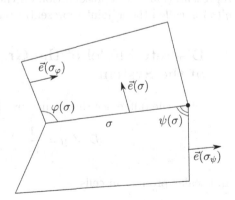

**Fig. 2.** The quasiregular grid

field intensity to a grid $(\sigma).(\varphi)$ consisting of faces $\sigma$ with their normals $e(\sigma)$ ordered into the system of local bases $(\varphi)$. We consider a quasiregular grid (see Fig. 2) with the order of regularity $h^m$ if for any index $\sigma$ and any bases adjacent to it $\varphi(\sigma)$, $\psi(\sigma)$ there exists a one-to-one correspondence $\sigma_\varphi \leftrightarrow \sigma_\psi$ of the indices entering into these bases, so that in some norm the following equality holds:

$$\|e(\sigma_\varphi) - e(\sigma_\psi)\| = O(h^m) \tag{8}$$

here $h$ is the small parameter characterizing the accuracy of differentiation at the considered computational grid.

We introduce the grid scalar product:

$$(\hbar_1, \hbar_2)_\sigma = \sum_\sigma V_\sigma \hbar_1(\sigma) \hbar_2(\sigma) \tag{9}$$

with weight $V_\sigma = \sum_{\varphi(\sigma)} V_\varphi > 0$ considered as a bordering volume.

The scalar product $\int_O (\hbar, g) dV$ in the domain $O$ is approximated by the grid analog

$$(\bar{\hbar}, g')_\sigma = \sum_\varphi V_\varphi \sum_{\sigma(\varphi), \tilde{\sigma}(\varphi)} Gr_\varphi(\sigma, \tilde{\sigma}) \bar{\hbar}(\sigma) \bar{g}(\tilde{\sigma}) \tag{10}$$

Here

$$g'(\sigma) = \frac{1}{V_\sigma} \sum_{\varphi(\sigma)} V_\varphi \sum_{\tilde{\sigma}(\varphi)} Gr_\varphi(\sigma, \tilde{\sigma}) \bar{g}(\tilde{\sigma}) \tag{11}$$

or in operator form

$$g' = H\bar{g} \tag{12}$$

Thus we introduce a self-adjoint positive-definite metric grid operator $H$ : $(\sigma) \to (\sigma)$, $H = H^* > 0$ given by the Gram matrices in local bases $Gr_\varphi(\sigma, \tilde{\sigma}) = (e(\sigma), e(\tilde{\sigma}))$ and taking the grid analog of the contravariant representation $\bar{g}(\sigma)$ to a grid covariant analogue. We understand the analog of the contravariant representation as a representation on the average, and the grid covariant analog $g'(\sigma)$ is called the adjoint representation of a vector field $g$.

## 4    Discrete Model of the Gravitational Energy Evolution of the System

We define the difference divergence by the formula:

$$DIV\, g = \frac{1}{V_\Omega} \sum_{\sigma(\Omega)} s_\sigma(\Omega) g'(\sigma) s(\sigma) \tag{13}$$

and scalar product in cells

$$(F_1, F_2)_\Omega = \sum_\Omega V_\Omega F_{1\Omega} F_{2\Omega} \tag{14}$$

Here $s(\sigma)$ is the area of the face $\sigma$, $s_\sigma(\Omega)$ is a sign function equal to 1 if the normal $e(\sigma)$ for the cell $\Omega$ is external and $-1$ to the contrary.

Further, for the domain $O$ bounded by the surface $\sigma$, modeling the integral relation

$$\int_O grad\ F \hbar dV + \int_O F\ div\ \hbar dV = \int_\Sigma F \hbar ds \tag{15}$$

we define a difference operator $GRAD : (\Omega) \to (\sigma)$ from the identity

$$(\overline{GRAD\ F},\ \hbar')_\sigma + (F,\ DIV\ \hbar)_\Omega = \sum_{\partial\sigma} s_{\partial\sigma} F_{\partial\sigma}\hbar'(\partial\sigma)s(\partial\sigma) \tag{16}$$

Here, for a boundary face $\partial\sigma$, $s(\partial\sigma)$ is its area, $s_{\partial\sigma}$ equals 1 if the normal $e(\partial\sigma)$ is external to the approximation of the region $O$. Finally, $F_{\partial\sigma}$ is the gravitational potential on this boundary face. We emphasize that $F$ and $\hbar' = H\hbar$, in general, are any grid functions. Therefore, on the face $\sigma$

$$\overline{GRAD\ F} = \frac{\Delta F}{h'} \tag{17}$$

where

$$\Delta F = -\sum_{\Omega(\sigma)} s_{\sigma(\Omega)} F_\Omega + s_{\partial\sigma} F_{\partial\sigma} \tag{18}$$

The second term in the increment of the potential across the face exists here if this boundary is a borderline

$$h' = \frac{V_\sigma}{s(\sigma)} \tag{19}$$

Modeling the gravitational analog of the Gauss's theorem (2) on the grid $(\sigma).(\varphi)$ as

$$DIV\ \boldsymbol{g} = \frac{-4\pi\gamma M}{V}, \quad \bar{g} = -\overline{GRAD\ F} \tag{20}$$

(here $M$ and $V$ are the mass and volume of the cell $\Omega$), we write the variation of the gravitational energy (1) on the grid in the form

$$\frac{1}{2}\frac{d}{dt}\left(F, \frac{M}{V}\right)_\Omega = -\sum_\varphi D_{g\varphi} V\varphi + \frac{1}{8\pi\gamma}\sum_{\partial\sigma} s_{\partial\sigma} \left| \begin{matrix} F_{\partial\sigma} & g'(\partial\sigma)s(\partial\sigma) \\ \frac{dF_{\partial\sigma}}{dt} & \frac{d}{dt}(g'(\partial\sigma)s(\partial\sigma)) \end{matrix} \right| \tag{21}$$

$$D_{g\varphi} = \frac{1}{4\pi\gamma} \sum_{\sigma(\varphi),\tilde{\sigma}(\varphi)} \bar{g}(\sigma)\bar{g}(\tilde{\sigma})(t^g_{v(-1/2)\varphi})_{\sigma\tilde{\sigma}} \tag{22}$$

$$\left(t^g_{v(-1/2)\varphi}\right)_{\sigma\tilde{\sigma}} = -\frac{h'(\sigma)h'(\tilde{\sigma})}{2V_\varphi}\frac{d}{dt}\left(\frac{Gr_\varphi(\sigma,\tilde{\sigma})}{h'(\sigma)h'(\tilde{\sigma})}V_\varphi\right) \tag{23}$$

Since

$$D_g = \frac{1}{4\pi\gamma}tr\left(\boldsymbol{g}\cdot\boldsymbol{g}\left(t_v - \frac{1}{2}tr(t_v)\delta\right)\right) \tag{24}$$

we conclude that the value $(t^g_{v(-1/2)\varphi})_{\sigma\tilde{\sigma}}$ approximates the covariant tensor $(t_v - 1/2 tr(t_v)\sigma)$ on the grid $(\sigma).(\varphi)$. Hence the gravitationally consistent approximation of the covariant strain rate tensor $t_v$ in the basis $\varphi$ has the form

$$\left(t^g_{v\varphi}\right)_{\sigma\tilde{\sigma}} = -\frac{h'(\sigma)h'(\tilde{\sigma})}{2}\frac{d}{dt}\left(\frac{Gr_\varphi(\sigma,\tilde{\sigma})}{h'(\sigma)h'(\tilde{\sigma})}\right) \qquad (25)$$

on the grid $(\sigma).(\varphi)$.

## 4.1   Approximation of the Tensor $\left(t^g_{v\varphi}\right)_{\sigma\tilde{\sigma}}$ on Parallel Shear Displacements of the Continuum Medium

For example, consider the parallel shear displacement of the continuum (see Fig. 3) along the $Oxy$ axis. The continuum is consist of a Lagrangian parallelogram grid with faces $\sigma$, $\tilde{\sigma}$ and corresponding normals $e(\sigma), e(\tilde{\sigma})$. Displacements of environmental points $u = u_x i + u_y j$ are defined as $u_x = y\ ctg\ \varphi$, $u_y = 0$. Velocities are defined as $v_y = \frac{du_y}{dt} = 0$, $v_x = \frac{du_x}{dt} = -\frac{y}{sin^2\varphi}\dot{\varphi}$. For grid parameters it is also obvious $h'(\sigma) = const$, $h'(\tilde{\sigma}) = const \cdot sin\ \varphi$. Performing non-complex continual transformations [4] for the symmetrized strain rate tensor in the covariant representation we have:

$$t_v = \frac{1}{2}\left(\frac{dv}{dr} + \nabla v\right) = -\frac{\dot{\varphi}}{2\ sin\ \varphi}\begin{pmatrix} 0 & 1 \\ 1 & -2\ cos\ \varphi \end{pmatrix} \qquad (26)$$

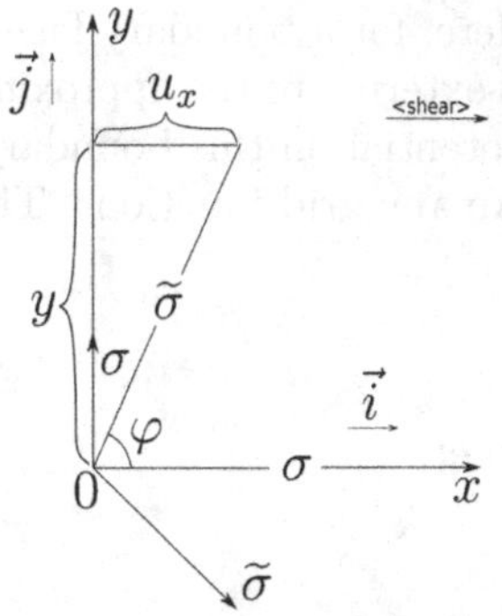

**Fig. 3.** The parallel shear displacement of the continuum along the $Oxy$ axis

The difference formula (25) for $\left(t^g_{v\varphi}\right)_{\sigma\tilde{\sigma}}$ on the grid $(\sigma).(\varphi)$ also gives the same result (26).

## 4.2   Approximation of the Gravitational Forces $-div\ t_g dV$ Consistent with the Change in the Gravitational Energy of the System

Simulating the integral relation (7) and collecting the factors of velocities $v_\omega$ in linear on these node velocities and quadratic in the gravitational field, we define the operator $-(div\ t_g dV)_\triangle : (\sigma) \to (\omega)$ which approximates the gravitational force acting on the adjacent volume, from the following identity:

$$\sum_\varphi D_{g\varphi}V_\varphi + \sum_\omega (v_\omega, (div\ t_g dV)_{\triangle\omega}) = \left(\int_\Omega t_g v ds\right)_\triangle$$

Approximating the surface $ds$ of the area $\Omega$ as the sum of surfaces $s_{\partial\varphi(\partial\sigma)}$ in boundary bases with a boundary face in accordance with

$$s_{\partial\varphi(\partial\sigma)} = \sum_{\partial\varphi(\partial\sigma)} s_{\partial\varphi(\partial\sigma)}$$

we write the approximation of the surface integral in the form

$$\left(\int_{\Sigma} t_g v ds\right)_{\triangle} = \sum_{\partial\sigma} \sum_{\partial\varphi(\partial\sigma)} (t_g)_{\partial\varphi} v_{\partial\omega(\partial\varphi)} s_{\partial\varphi(\partial\sigma)}$$

here, according to (3)

$$(t_g)_{\partial\varphi} = \frac{1}{4\pi\gamma}\left( g_{\partial\varphi} \cdot g_{\partial\varphi} - \frac{(g_{\partial\varphi})^2}{2}\delta \right)$$

$$g_{\partial\varphi} = \sum_{\tilde\sigma(\partial\varphi)} \bar{g}(\tilde\sigma) e(\tilde\sigma)$$

$\partial\omega(\partial\varphi)$ is the central node $\partial\omega$ of the basis $\partial\varphi$.

## 5    Conclusion

We have developed the finite difference approximation to the gravitational forces on Lagrangian grids with quadrilateral cells which is consistent with the difference equations describing the gravitational and kinetic energy balances evolution for the self-gravitating object. The developed approximations does not lead to the energy disbalanses which otherwise appear during the simulation of the matter supercompression performed via calculations at moving grids with distorted quadrilateral cells.

**Acknowledgments.** The work was funded by Russian Foundation for Basic Research, projects no. 18-07-00841-a, 16-29-15081-ofi_m, 16-29-15095-ofi_m.

Supported by the Academic Excellence Project of the NRNU MEPhI under contract with the Ministry of Education and Science of the Russian Federation No. 02.A03.21.0005.

## References

1. Samarskii, A.A., Popov, Yu.P.: Difference methods for solving problems of gas dynamics. Nauka, Moscow (1992)
2. Denisov, A.A., Koldoba, A.V., Poveshchenko, Yu.A., Popov, Yu.P., Chechetkin, V.M.: The role of rotation in the thermonuclear model of supernova explosion. Preprints of the Keldysh Institute of Applied Mathematics, no. 99 (1986)
3. Koldoba, A.V., Kuznetsov, O.A., Poveshchenko, Yu.A., Popov, Yu.P., Samarskii, A.A.: Completely conservative difference schemes for the equations of continuum mechanics in quasi-Lagrangian variables in the presence of gravitational and magnetohydrodynamic processes. Preprints of the Keldysh Institute of Applied Mathematics, no. 55 (1985)
4. Kochin, N.E.: Vector calculus and initiate of tensor calculus. Lenand, Moscow (2017)

# Computation of Mean-Field Equilibria with Correlated Stochastic Processes

V. Shaydurov[1,2(✉)], S. Zhang[1], and V. Kornienko[1,3]

[1] Tianjin University of Finance and Economics, 25# Zhujiang Road,
Hexi District, Tianjin 300222, China
[2] Institute of Computational Modeling of SB RAS, 50/44# Akademgorodok,
Krasnoyarsk 660036, Russia
`shaidurov04@mail.ru`
[3] Siberian Federal University, 79# Svobodnyi Prospect, Krasnoyarsk 660041, Russia

**Abstract.** The numerical algorithm is presented for solving differential problem formulated as the Mean-Field Game (MFG) with the coupled system of two parabolic partial differential equations: the Fokker-Plank-Kolmogorov equation and the Hamilton-Jacobi-Bellman one. The case is considered with correlation of the considered stochastic processes. The description focuses on the discrete semi-Lagrangian approximation of these equations and on the application of the MFG theory directly at discrete level. The constructed algorithm is implemented to the problem of carbon dioxide pollution as an illustration.

**Keywords:** Optimal control · Mean-Field Game ·
Numerical approximation · Finite differences · Carbon dioxide pollution

## 1 Introduction

The Mean-Field Game (MFG) approach is theoretically developed in papers by Lasry and Lions [1,2]. A "historical" information on the rapid development of this area of investigations is presented in the brief monograph [3] providing mathematical analysis. This approach has been adapted to many problems in physics, biology, engineering, and economics [1–6]. In the problem formulated here, the mean-field equilibrium is described by the coupled system of two parabolic partial differential equations: the Fokker-Plank-Kolmogorov (FPK) equation and the Hamilton-Jacobi-Bellman (HJB) one.

In this paper, we focus on the discrete approximation of these equations and on an application of the MFG theory directly at discrete level. Contrary to difference schemes applied by other authors, we propose the semi-Lagrangian approximation which improves some properties of a discrete problem of this type. Earlier this approximation was used for solving the same type of one-dimensional problem [7].

Partially supported by Project 17-01-00270 of Russian Foundation for Basic Research and the Major Research Plan of the National Natural Science Foundation of China (91430108), the National Natural Science Foundation of China (11771322).

© Springer Nature Switzerland AG 2019
I. Dimov et al. (Eds.): FDM 2018, LNCS 11386, pp. 468–475, 2019.
https://doi.org/10.1007/978-3-030-11539-5_54

## 2    The Mathematical Model

We shall not derive the differential statement of the MFG problem and refer the reader to comprehensive book [3] for the general description. We begin with the Kolmogorov equation (which is called Fokker-Plank one in other content) for the density $m(t, x, y)$ of "atomized" agents on the rectangle $(0, H_1) \times (0, H_2)$ at time segment $t \in [0, T]$.

Put $\Omega = (0, H_1) \times (0, H_2)$ with the boundary $\Gamma$ and the closure $\bar{\Omega} = \Omega \bigcup \Gamma$. First, we introduce the forward FPK problem [3]:

$$\frac{\partial m}{t} - \frac{\sigma_1^2}{2} \frac{\partial^2 m}{\partial x^2} - \frac{\sigma_2^2}{2} \frac{\partial^2 m}{\partial y^2} - \gamma \sigma_1 \sigma_2 \frac{\partial^2 m}{\partial x \partial y} + \frac{\partial(\alpha m)}{\partial x} + \frac{\partial(\beta m)}{\partial y} = 0 \quad \text{in } (0, T) \times \Omega, \quad (1)$$

$$m(0, x, y) = m_0(x, y) \quad \text{on} \quad \bar{\Omega}, \quad \partial m / \partial n = 0 \quad \text{on} \quad [0, T] \times \Gamma \qquad (2)$$

where $\partial m / \partial n$ means the normal derivative at boundary points of $\Gamma$. Here $\sigma_1 > 0$, $\sigma_2 > 0$ are fixed constants characterizing the probable noises produced by the Brownian motion [5] and $\gamma \in (-1, 1)$ is a coefficient of correlation for these stochastic processes. The control functions $\alpha(t, x, y)$, $\beta(t, x, y)$ reflect the efforts directed towards the decreasing of $m$. Besides, (2) defines the initial density of agents on $\bar{\Omega}$ and provides the stay of agents on a "feasible" closed domain $\bar{\Omega}$ at each moment of time and interferes their exit out of limits.

In the model used, we want to minimize the cost functional

$$J(m, \alpha, \beta) = \int_0^T \exp(-rt) \int_\Omega (d_1 \alpha^2 m / 2 + d_2 \beta^2 m / 2 + g(t, x, y, m)) \, d\Omega \, dt. \qquad (3)$$

Here $d_1, d_2$ are positive constants; nonnegative $r$ is the risk-free discount rate. For function $g(t, x, y, m)$ we demand its concavity in argument $m$: $g(t, x, y, \tilde{m}) - g(t, x, y, m) \leq (\tilde{m} - m) f(t, x, y, m)$ with $f(t, x, y, m) = \partial g / \partial m(t, x, y, m)$ for all admissible values of other arguments.

So, we get the optimization problem

$$\begin{cases} \inf_{\alpha, \beta} J(m, \alpha, \beta) = \int_0^T \exp(-rt) \int_\Omega (d_1 \alpha^2 m / 2 + d_2 \beta^2 m / 2 + g(t, x, y, m)) \, d\Omega dt \\ \textit{with conditions } (1) - (2). \end{cases}$$

$$(4)$$

Here we briefly describe a formal way to get the optimality conditions for this differential problem. The rigorous derivation can be found in [3].

Take an arbitrary function $v \in C^\infty([0, T] \times \bar{\Omega})$, multiply (1) by it, and integrate by parts with respect to $t$ and $x$:

$$-\int_0^T \int_\Omega \left( \frac{\partial v}{\partial t} + \frac{\sigma_1^2}{2} \frac{\partial^2 v}{\partial x^2} + \frac{\sigma_2^2}{2} \frac{\partial^2 v}{\partial y^2} + \gamma \sigma_1 \sigma_2 \frac{\partial^2 m}{\partial x \partial y} + \alpha \frac{\partial v}{\partial x} + \beta \frac{\partial v}{\partial y} \right) m \, d\Omega dt$$

$$+ \int_\Omega (v(T, x, y) m(T, x, y) - v(0, x, y) m_0(x, y)) \, d\Omega = 0$$

$$(5)$$

taking into account the boundary condition similar to (2) $\partial v/\partial n = 0$ on $[0, T] \times \Gamma$.

In addition to the cost functional, we also formulate the Lagrangian

$$\Im(m, \alpha, \beta, v) := J(m, \alpha, \beta) - \int_{\Omega} \left( v(T, x, y) m(T, x, y) - v(0, x, y) m_0(x, y) \right) \, d\Omega$$

$$+ \int_0^T \int_{\Omega} \left( \frac{\partial v}{\partial t} + \frac{\sigma_1^2}{2} \frac{\partial^2 v}{\partial x^2} + \frac{\sigma_2^2}{2} \frac{\partial^2 v}{\partial y^2} + \gamma \sigma_1 \sigma_2 \frac{\partial^2 v}{\partial x \partial y} + \alpha \frac{\partial v}{\partial x} + \beta \frac{\partial v}{\partial y} \right) m \, d\Omega \, dt. \quad (6)$$

Thus, the minimization problem (4) converts to the saddle point problem [3, 5]

$$\inf_{(m, \alpha, \beta)} \sup_{v} \Im(m, \alpha, \beta, v). \quad (7)$$

After "differentiation" with respect to some functions, we get the backward HJB equation with the initial and boundary conditions:

$$\frac{\partial v}{\partial t} + \frac{\sigma_1^2}{2} \frac{\partial^2 v}{\partial x^2} + \frac{\sigma_2^2}{2} \frac{\partial^2 v}{\partial y^2} + \gamma \sigma_1 \sigma_2 \frac{\partial^2 v}{\partial x \partial y} + \alpha \frac{\partial v}{\partial x} + \beta \frac{\partial v}{\partial y} \quad (8)$$
$$= - \exp(-rt) \left( f + d_1 \alpha^2/2 + d_2 \beta^2/2 \right) \quad \text{on } [0, T] \times \Omega,$$

$$v(T, x, y) = 0 \quad \text{on} \quad \bar{\Omega}, \quad \partial v/\partial n = 0 \quad \text{on } [0, T] \times \Gamma, \quad (9)$$

$$\alpha = -\frac{\exp(rt)}{d_1} \frac{\partial v}{\partial x}, \quad \beta = -\frac{\exp(rt)}{d_2} \frac{\partial v}{\partial y} \text{ on } [0, T] \times \bar{\Omega}. \quad (10)$$

## 3     The Numerical Solution of the FPK Equation

So, we have to solve problem (1)–(2) with the functions $\alpha$ and $\beta$ satisfying the property $\alpha = \beta = 0$ on $[0, T] \times \Gamma$ due to (9) and (10). Introduce discrete uniform grids in time and in space: $t_k = k\tau, k = 0, \ldots, M, \tau = T/M; x_{i+1/2} = (i + 1/2)h_1, i = -1, \ldots, N_1, h_1 = H_1/N_1; y_{j+1/2} = (j + 1/2)h_2, j = -1, \ldots, N_2, h_2 = H_2/N_2;$ for integers $M, N_1, N_2 \geq 2$. Denote $z_{i+1/2, j+1/2} = (x_{i+1/2}, y_{j+1/2})$ and put $\bar{\Omega}_h = \left\{ z_{i+1/2, j+1/2}; i = -1, \ldots, N_1, j = -1, \ldots, N_2 \right\}$ and $\Omega_h = \left\{ z_{i+1/2, j+1/2}; i = 0, \ldots, N_1 - 1, j = 0, \ldots, N_2 - 1 \right\}$. Introduce also the points $x_i = ih_1, i = 0, \ldots, N_1; y_j = jh_2, j = 0, \ldots, N_2$.

We shall find a solution to this problem as a grid function $m^h(t, x, y)$ at each time level $t_k$ on $\bar{\Omega}_h$. Split the approximation of Eq. (1) into two parts. First consider the approximation of diffusion part of (1). The most discrete approximations are not appropriate for our demands since the approximation must produce self-adjoint operator with M-property. Therefore, we take the stencil of different scheme dependent on sign of $\gamma$. Introduce the expressions

$$A_{i+1/2, j+1/2} = \begin{bmatrix} a_{i-1/2, j+3/2} & a_{i+1/2, j+3/2} & a_{i+3/2, j+3/2} \\ a_{i-1/2, j+1/2} & a_{i+1/2, j+1/2} & a_{i+3/2, j+1/2} \\ a_{i-1/2, j-1/2} & a_{i+1/2, j-1/2} & a_{i+3/2, j-1/2} \end{bmatrix},$$

$$M^h_{k,i+1/2,j+1/2} = \begin{bmatrix} m^h_{k,i-1/2,j+3/2} & m^h_{k,i+1/2,j+3/2} & m^h_{k,i+3/2,j+3/2} \\ m^h_{k,i-1/2,j+1/2} & m^h_{k,i+1/2,j+1/2} & m^h_{k,i+3/2,j+1/2} \\ m^h_{k,i-1/2,j-1/2} & m^h_{k,i+1/2,j-1/2} & m^h_{k,i+3/2,j-1/2} \end{bmatrix}$$

with the 9-point scalar product. First, treat the case $\gamma \geq 0$. Put

$$
\begin{aligned}
a_{i-1/2,j+3/2} &= a_{i+3/2,j-1/2} = 0; \\
a_{i+1/2,j+3/2} &= a_{i+1/2,j-1/2} = -\sigma_1^2/2h_1^2 + |\gamma|\sigma_1\sigma_2/2h_1h_2; \\
a_{i-1/2,j+1/2} &= a_{i+3/2,j+1/2} = -\sigma_2^2/2h_2^2 + |\gamma|\sigma_1\sigma_2/2h_1h_2; \\
a_{i+3/2,j+3/2} &= a_{i-1/2,j-1/2} = -|\gamma|\sigma_1\sigma_2/2h_1h_2; \\
a_{i+1/2,j+1/2} &= 1/\tau + \sigma_1^2/h_1^2 + \sigma_2^2/h_2^2 - |\gamma|\sigma_1\sigma_2/h_1h_2.
\end{aligned}
\tag{11}
$$

Off-diagonal coefficients will be non-positive if

$$|\gamma|\sigma_2/\sigma_1 \leq h_2/h_1 \leq \sigma_2/|\gamma|\sigma_1. \tag{12}$$

In principle, these inequalities are satisfied when $h_2 \simeq h_1\sigma_2/\sigma_1$. And the less $\gamma \in [0,1)$ the wider boundary of the ratio $h_2/h_1$. In the case $\gamma < 0$, we change only 4 coefficients with the same 5 other ones:

$$
\begin{aligned}
a_{i+3/2,j+3/2} &= a_{i-1/2,j-1/2} = 0; \\
a_{i-1/2,j+3/2} &= a_{i+3/2,j-1/2} = \gamma\sigma_1\sigma_2/2h_1h_2.
\end{aligned}
\tag{13}
$$

Combine this approximation with the semi-Lagrangian approach [7] to get

$$
\begin{aligned}
A_{i+1/2,j+1/2}M^h_{k,i+1/2,j+1/2} &= \beta^{i+1/2,j+1/2}_{k-1,1}m^h_{k-1,i-1/2,j+1/2} \\
+\beta^{i+1/2,j+1/2}_{k-1,2}m^h_{k-1,i+1/2,j+1/2} &+ \beta^{i+1/2,j+1/2}_{k-1,3}m^h_{k-1,i+3/2,j+1/2} \\
+\beta^{i+1/2,j+1/2}_{k-1,4}m^h_{k-1,i+1/2,j-1/2} &+ \beta^{i+1/2,j+1/2}_{k-1,5}m^h_{k-1,i+1/2,j+3/2}
\end{aligned}
\tag{14}
$$

$$\beta^{i+1/2,j+1/2}_{k-1,1} = \max\{0, a_{k-1,i,j+1/2}/h_1\},$$

$$\beta^{i+1/2,j+1/2}_{k-1,3} = \max\{0, -\alpha_{k-1,i+1,j+1/2}/h_1\},$$

$$\beta^{i+1/2,j+1/2}_{k-1,4} = \max\{0, \beta_{k-1,i+1/2,j}/h_2\},$$

$$\beta^{i+1/2,j+1/2}_{k-1,5} = \max\{0, -\beta_{k-1,i+1/2,j+1}/h_2\},$$

$$\beta^{i+1/2,j+1/2}_{k-1,2} = 1/\tau - \beta^{i+3/2,j+1/2}_{k-1,1} - \beta^{i-1/2,j+1/2}_{k-1,3} - \beta^{i+1/2,j+3/2}_{k-1,4} - \beta^{i+1/2,j-1/2}_{k-1,5},$$

$$\forall i = 0,\ldots,N_1-1,\ j = 0,\ldots,N_2-1,$$

with initial condition $m^h_{0,i+1/2,j+1/2} = m_0(x_{i+1/2}, y_{j+1/2}) \forall i = 0,\ldots,N_1-1, j = 0,\ldots,N_2-1$. For any node $z_{i+1/2,j+1/2} \in \bar{\Omega}^h \backslash \Omega^h$ outside the domain $\Omega$, we put $m^h_{k,i+1/2,j+1/2} = m^h_{k,i'+1/2,j'+1/2}$ with the nearest node $z_{i'+1/2,j'+1/2} \in \Omega^h$ to ensure Neumann discrete boundary condition. We will exclude values with arguments out of a rectangle and designate the remained system of the linear algebraic equations as

$$\aleph m^h_{\cdot,\cdot,\cdot} = \Im m^h_{0,\cdot,\cdot}. \tag{15}$$

**Remark.** Let inequalities (12) hold and $\tau|\alpha^h_{k,i,j+1/2}| \le h_1/8$, $\tau|\beta^h_{k,i+1/2,j}| \le h_2/8 \,\forall\, k = 0,\ldots,M-1, i = 1,\ldots,N_1-1,\ j = 1,\ldots,N_2-1$. Then, all coefficients in the right-hand side of (14) are nonnegative. Therefore, the matrix of system (15) has the property of M-matrix.

## 4   The Discrete Optimal Conditions

Instead of cost functional (3), introduce the discrete one

$$J^h(m^h,\alpha^h,\beta^h) = \sum_{k=0}^{M-1}\sum_{i=0}^{N_1-1}\sum_{j=0}^{N_2-1} \exp(-rt_k)\Big(d_1 r^h_{k,i+1/2,j+1/2} m^h_{k,i+1/2,j+1/2}/2$$

$$+d_2 s^h_{k,i+1/2,j+1/2} m^h_{k,i+1/2,j+1/2}/2 + g^h_{k,i+1/2,j+1/2}\Big) h_1 h_2 \tau \qquad (16)$$

for

$$r^h_{k,i+1/2,\cdot} = (\alpha^h_{k,i,\cdot})^2/2 + (\alpha^h_{k,i+1,\cdot})^2/2,$$
$$s^h_{k,\cdot,j+1/2} = (\beta^h_{k,\cdot,j})^2/2 + (\beta^h_{k,\cdot,j+1})^2/2, \qquad (17)$$
$$g^h_{k,i+1/2,j+1/2} = g(t_k, x_{i+1/2}, y_{j+1/2}, m^h_{k,i+1/2,j+1/2}).$$

Thus, we have the discrete problem for the minimization

$$\begin{cases} \inf_{\alpha^h,\beta^h} J^h(m^h,\alpha^h,\beta^h), \\ \aleph m^h_{\cdot,\cdot,\cdot} = \Im m^h_{0,\cdot,\cdot}. \end{cases} \qquad (18)$$

To formulate the discrete optimal control problem, introduce a discrete function $v^h_{\cdot,\cdot,\cdot} = \Big\{v^h_{k,i+1/2,j+1/2};\ k = 0,\ldots,M,\ i = -1,\ldots,N_1,\ j = -1,\ldots,N_2\Big\}$.

We omit intermediate considerations at discrete level that are analogous to the differential ones (5)–(10) and derive the system of discrete equations characterizing optimal condition with the stencil $V^h$ analogous to $M^h$:

$$A_{i+1/2,j+1/2}\cdot V^h_{k-1,i+1/2,j+1/2} = \beta^{i+3/2,j+1/2}_{k-1,1} v^h_{k,i+3/2,j+1/2}$$
$$+\beta^{i+1/2,j+1/2}_{k-1,2} v^h_{k,i+1/2,j+1/2} + \beta^{i-1/2,j+1/2}_{k-1,3} v^h_{k,i-1/2,j+1/2}$$
$$+\beta^{i+1/2,j+3/2}_{k-1,4} v^h_{k,i+1/2,j+3/2} + \beta^{i+1/2,j-1/2}_{k-1,5} v^h_{k,i+1/2,j-1/2} + b^h_{k,i+1/2,j+1/2} \qquad (19)$$
$$\forall\, k = M-1, M-2,\ldots,0,\quad i = 0,\ldots,N_1-1,\ j = 0,\ldots,N_2-1,$$

for the initial data $v^h_{M,\cdot,\cdot} = \mathbf{0}$ where $b^h_{k,i+1/2,j+1/2} = -\exp(-rt_k)\Big(f(t_k, x_{i+1/2},$ $y_{i+1/2}, m^h_{k,i+1/2,j+1/2}) + d_1 r^h_{k,i+1/2,j+1/2}/2 + d_2 s^h_{k,i+1/2,j+1/2}/2\Big).$

Again, for any node $z_{i+1/2,j+1/2} \in \bar{\Omega}^h\backslash\Omega^h$ (outside of the domain $\Omega$), we put $v^h_{k,i+1/2,j+1/2} = v^h_{k,i'+1/2,j'+1/2}$ with the nearest node $z_{i'+1/2,j'+1/2} \in \Omega^h$ in accordance with zeroth Neumann boundary condition.

Differentiating with respect to $\alpha^h_{\cdot,\cdot,\cdot}$ and $\beta^h_{\cdot,\cdot,\cdot}$ gives equalities

$$\alpha^h_{k,i,\cdot} = -\exp(rt_k)\Big(v^h_{k,i+1/2,\cdot} - v^h_{k,i-1/2,\cdot}\Big)/d_1 h_1 \quad \forall\, i = 1,\ldots,N_1-1,$$
$$\beta^h_{k,\cdot,j} = -\exp(rt_k)\Big(v^h_{k,\cdot,j+1/2} - v^h_{k,\cdot,j-1/2}\Big)/d_2 h_2 \quad \forall\, j = 1,\ldots,N_2-1. \qquad (20)$$

## 5   The Numerical Solution of the Complete Model

Suppose that some initial approximation $\alpha_{\cdot,\cdot,\cdot}^h, \beta_{\cdot,\cdot,\cdot}^h$, and $m_{\cdot,\cdot,\cdot}^h$ are given. For example, we can firstly take $\alpha_{\cdot,\cdot,\cdot}^h = 0$, $\beta_{\cdot,\cdot,\cdot}^h = 0$ and compute $m_{k,\cdot,\cdot}^h$. The better approximations of the control grid functions $\alpha_{\cdot,\cdot,\cdot}^h, \beta_{\cdot,\cdot,\cdot}^h$ may be computed by the following iterative way.

**Iterative Algorithm**

1. Solve (19) $\forall k = M - 1, M - 2, \ldots, 0$ to get $v_{\cdot,\cdot}^h$.
2. Compute $\tilde{\alpha}_{k,\cdot,\cdot}^h$ and $\tilde{\beta}_{k,\cdot,\cdot}^h$ by (20) for $k = 0, \ldots, M - 1$.
3. Compute $\tilde{m}_{\cdot,\cdot,\cdot}^h$ by (15).
4. Compute $J^h(\tilde{m}^h, \tilde{\alpha}^h, \tilde{\beta}^h)$ by (16).
5. If $\left| J^h(m^h, \alpha^h, \beta^h) - J^h(\tilde{m}^h, \tilde{\alpha}^h, \tilde{\beta}^h) \right| > Tol$ then $\{ \alpha_{\cdot,\cdot,\cdot}^h := \tilde{\alpha}_{\cdot,\cdot,\cdot}^h ; m_{\cdot,\cdot,\cdot}^h := \tilde{m}_{\cdot,\cdot,\cdot}^h$; go to 1 $\}$.
6. Take $\tilde{\alpha}_{\cdot,\cdot,\cdot}^h, \tilde{\beta}_{\cdot,\cdot,\cdot}^h$, and $\tilde{m}_{\cdot,\cdot,\cdot}^h$ as an approximate solution of (18).

Now we give the hint to demonstrate that (20) ensures the steepest descent of the above estimate for the difference between values of discrete cost functional. Consider two controls: fix $\alpha^h, \beta^h$ and vary $\tilde{\alpha}^h, \tilde{\beta}^h$. Take $r^h, \tilde{r}^h, s^h, \tilde{s}^h$ from (17) and the solutions $m^h, \tilde{m}^h$ of discrete problem (15). Find the difference between two values of the discrete functional $J^h(\tilde{m}^h, \tilde{\alpha}^h, \tilde{\beta}^h) - J^h(m^h, \alpha^h, \beta^h)$. Imply the concavity of the function $g$ with respect to $m$ and disintegrate each addend in the right-hand side into independent parts with respect to every $\tilde{\alpha}_{k,i,j+1/2}^h$ and $\tilde{\beta}_{k,i+1/2,j}^h$ like in [4]. First, the minimization of this difference independently decays into the minimization of the quadratic polynomials. Second, the coefficients of the principal term in $\tilde{\alpha}_{k,i,j+1/2}^h$, $\tilde{\beta}_{k,i+1/2,j}^h$ are strictly positive. Thus, the minimum for each of them is reached at points

$$\bar{\alpha}_{k,i,\cdot}^h = -\exp(rt_k)\left( v_{k,i+1/2,\cdot}^h - v_{k,i-1/2,\cdot}^h \right)/d_1 h_1 \, \forall i = 1, \ldots, N_1 - 1, \bar{\beta}_{k,\cdot,j}^h =$$
$$-\exp(rt_k)\left( v_{k,\cdot,j+1/2}^h - v_{k,\cdot,j-1/2}^h \right)/d_2 h_2 \quad \forall j = 1, \ldots, N_2 - 1.$$ And these minimization points indeed coincide with the computed in the iterative algorithm.

At first glance, it seems that the coincidence of equalities (20) and the condition of minimization is a natural consequence of approximation of the differential problem. However, this coincidence is not automatic for many other difference schemes (see, for example, [5,6]) for which additional inner iterations or corrections are necessary.

Taylor expansion at the point $z_{k,i+1/2,j+1/2}$ demonstrates the approximation order of $O(\tau + h)$ for schemes (14) and (19). Due to the M-matrix property, these schemes are stable and provide the accuracy of $O(\tau + h)$.

## 6   The Model

The agents of our model are the producers. Each of them is associated to a production at time $t$, denoted by $q(t)$ which results in amount of emission $e(t)$.

Generally speaking, an increase in production can result in more emissions and vice versa. So, it is reasonable to use the emission as a state variable instead of production (see, e.g., [6]). The dynamics of agent's emission $e(t)$ describes by the following controlled process:

$$de(t) = -l(t)\,dt + \sigma_1 dW_1(t) + dN_t\,(e(t)), \quad e(0) = e_0, \tag{21}$$

where $l(t)$ is the level of emission reduction and $\sigma_1 dW_1(t)$ is its stochastic disturbance. It results from technological innovations, market fluctuation, and some other uncertain factors. Here $W_1(t)$ is a standard Brownian motion and $\sigma_1$ is a noise parameter. $N_t\,(e(t))$ is the reflection part which guarantees for the process to stay in $[e_{min},\,e_{max}]$ determined by production capacity of an agent [4-6]. Another dynamic state of the system is the amount of the permitted emission $x(t)$. Let the agents try their best to negotiate between them to pursue the highest level of the permitted emission. The effort level is represented by $\mu(t)$. The permitted emission $x(t)$ follows the stochastic process

$$dx(t) = \mu(t)\,dt + \sigma_2 dW_2(t) + dN_t\,(x(t)), \quad x(0) = x_0. \tag{22}$$

Here $\sigma_2 dW_2(t)$ is the stochastic disturbance in which $W_2(t)$ is dependent on $W_1(t)$ with correlation $\gamma \in [0,1)$. $N_t\,(x(t))$ is the reflection part which guarantees for the process to stay in $[x_{min},\,x_{max}]$.

Assume that all agents have the same capacity in production and negotiation but make different decisions. The states $e$ and $x$ of the producers continuum at time $t$ are distributed in $\bar{\Omega} = [e_{min},\,e_{max}] \times [x_{min},\,x_{max}]$ with the probability density function $m(t,e,x)$. The initial density $m_0(e,x)$ is given.

The agents' net revenues include five parts: the production revenue, the cost of emission reduction, the cost of pursuing permitted emission, the cost of the carbon tax, and the carbon trading. The production revenue of agent is $Y(t,e,m) = \left(e_{max}e - e^2/2\right)/2\,(c_1 + c_2 m)$, $e \in [e_{min},\,e_{max}]$, where $c_1$ and $c_2$ are positive constants. The cost of emission reduction and negotiation for the permitted emission are $C_e(l) = d_1 l^2(t)$ and $C_n(l) = d_2 \mu^2(t)$, respectively.

The cost of carbon tax is $C_a(t,e,x) = p_a(t)\min\{e(t),x(t)\}$ where $p_a(t)$ denotes the tax rate. If the emission amount $e(t)$ is less than permission level $x(t)$, the agent should pay the carbon tax of his emission. If the emission amount $e(t)$ is more than $x(t)$, the agent should pay the carbon tax of the basic part. However, the permission of the exceed part should be brought by the carbon trading mechanism expressed by the fourth part of revenue $C_b(t,e,x) = p_b(t)\,(e(t) - x(t))^+$ where the emission permits price $p_b(t) > p_a(t)$. After summation, the net revenue of agent at time $t$ is $f(t,e,x,l,\mu,m) = Y(t,e,m(t,e,x)) - d_1 l^2 - d_2 \mu^2 - p_a(t)\min\{e,x\} - p_b(t)\,(e-x)^+$.

The agents can adjust their strategies $l(t)$ and $\mu(t)$ to maximize their expected discounted revenues over the time segment $[0,T]$. When the probability density function is not related with the control variables, we have [6]

$$\sup_{l(t),\mu(t)} \left\{ \int_0^T \exp(-rt) \int_\Omega f\left(t, e(t), x(t), l(t), \mu(t), m\left(t, e(t), x(t)\right)\right) m \, d\Omega dt \right\}.$$
(23)

In order to rewrite this problem in the form of (4), we produce linear substitution $(e - e_{\min}, x - x_{\min})$ by variables $(x, y)$ and take the minus before $f$ to change supremum in (23) for infimum in (4). Then put $l = -\alpha$ and $\mu = -\beta$.

# 7  Conclusion

Thus, we suggest the discrete approximation of the MFG problem which completely inherited the basic properties of the differential problem simultaneously with the approximation of each differential equation. The approximations of direct FPK and inverse HJB problems have adjoint operators which are monotone in corresponding adjoint vector spaces. And the discrete approximation (20) of differential connection (10) for the minimization of the cost functional gives the condition for steepest descent of the discrete cost functional.

# References

1. Lasry, J.M., Lions, P.L.: Jeux champ moyen. I. Le cas stationnaire. C. R. Math. Acad. Sci. Paris **343**(9), 619–625 (2006)
2. Lasry, J.M., Lions, P.L.: Jeux champ moyen. II. Horizon fini et contrle optimal. C. R. Math. Acad. Sci. Paris **343**(10), 679–684 (2006)
3. Bensoussan, A., Frehse, J., Yam, P.: Mean Field Games and Mean Field Type Control Theory. Springer, Berlin (2013). https://doi.org/10.1007/978-1-4614-8508-7
4. Shaydurov, V., Zhang, S., Karepova, E.: Conservative difference schemes for the computation of mean-field equilibria. In: AIP Conference Proceedings, vol. 1892, pp. 20–35 (2017)
5. Röhl, T., Röhl, C., Schuster, H.C., Traulsen, A.: Impact of fraud on the mean-field dynamics of cooperative social systems. Phys. Rev. E **76**, 026114 (2007)
6. Chang, S., Wang, X.: Modeling and computation of mean field equilibria in producers game with emission permits trading. Commun. Nonlinear Sci. Numer. Simulat. **37**, 238–248 (2016)
7. Shaydurov, V., Vyatkin, A., Kuchunova, E.: Semi-Lagrangian difference approximations with different stability requirements. Russ. J. Numer. Anal. Math. Model. **33**(2), 123–135 (2018)

# Development and Numerical Study of Robust Difference Schemes for a Singularly Perturbed Transport Equation

Lidia Shishkina$^{(\boxtimes)}$ [ID] and Grigorii Shishkin [ID]

Krasovskii Institute of Mathematics and Mechanics, UB RAS,
S. Kovalevskaya Street, 16, 620990 Yekaterinburg, Russia
Lida@convex.ru, shishkin@imm.uran.ru

**Abstract.** On the set $\overline{G} = G \cup S$, $G = (0, d] \times (0, T]$ with the boundary $S = S_0 \cup S^{\ell}$, we consider an initial-boundary value problem for the singularly perturbed transport equation with a perturbation parameter $\varepsilon$ multiplying the spatial derivative, $\varepsilon \in (0, 1]$. For small values of the perturbation parameter $\varepsilon$, the solution of such a problem has a singularity of the boundary layer type, which makes standard difference schemes unsuitable for practical computations. To solve this problem numerically, an approach to the development of a robust difference scheme is proposed, similar to that used for constructing special $\varepsilon$-uniformly convergent difference schemes for singularly perturbed elliptic and parabolic equations. In this paper, we give a technique for constructing a robust difference scheme and justifying its $\varepsilon$-uniform convergence, and we study numerically solutions of standard and special robust difference schemes for a model initial-boundary value problem for a singularly perturbed transport equation. The results of numerical experiments confirm theoretical results.

**Keywords:** Singularly perturbed transport equation ·
Boundary layer · Robust difference scheme · $\mathcal{E}$-uniform convergence ·
Maximum norm · Standard difference scheme · Uniform mesh ·
Piecewise-uniform mesh · Solution decomposition · *A priori* estimates

## 1 Introduction

Construction of highly efficient numerical methods for solving problems for regular transport equations was carried out in schools of Marchuk, Samarskii, Kalitkin (see, for example, [1–3] and the bibliography therein) for a sufficiently long time. But singularly perturbed transport equations remained for a long time outside the interests of researchers, although problems with such equations are sufficient often encountered in practice. In the present paper, on the set

This research was partially supported by the Russian Foundation for Basic Research under grant No. 16-01-00727.

I. Dimov et al. (Eds.): FDM 2018, LNCS 11386, pp. 476–483, 2019.
https://doi.org/10.1007/978-3-030-11539-5_55

$\overline{G} = G \cup S$, $G = (0, d] \times (0, T]$ with the boundary $S = S_0 \cup S^\ell$ an initial-boundary value problem is considered for a new class, namely, for a singularly perturbed transport equation with a perturbation parameter $\varepsilon$ multiplying the spatial derivative, $\varepsilon \in (0, 1]$.

For small values of the parameter $\varepsilon$, the problem solution involves a boundary layer of width $\mathcal{O}(\varepsilon)$, on which the problem solution changes to a finite value, and the derivatives of the solution in a neighborhood of the boundary layer grow unboundedly as $\varepsilon \to 0$. When using standard grid approximations of the problem, the error in the grid solution depends on the parameter $\varepsilon$ and for small values $\varepsilon$ becomes large, commensurate with the solution of the differential problem, and this makes the standard schemes unsuitable for practical calculations.

For the numerical solution of such a problem, Shishkin proposed an approach [4] to the development of a robust difference scheme (a special scheme that converges independently of the parameter $\varepsilon$, i.e. $\varepsilon$-uniformly, in the maximum norm), similar to that used previously for singularly perturbed elliptic and parabolic equations (see, e.g., [5,6]). In this paper, we give a technique for constructing a robust difference scheme using a special piecewise-uniform mesh known in the literature [7] as "Shishkin mesh", with the justification of its $\varepsilon$-uniform convergence on the basis of *a priori* estimates, and first numerical study of solutions of difference schemes is performed for the model problem for singularly perturbed transfer equation. The results of numerical experiments are in agreement with the theoretical ones.

## 2   Problem Formulation. Aim of the Research

On the set
$$\overline{G} = G \cup S, \quad G = D \times (0, T], \quad D = (0, d], \tag{1}$$
where $S = S_0 \cup S^l$, $S_0$ and $S^l$ are the lower and lateral parts of the boundary $S$, $S_0 = \{(x, t) : 0 \le x \le d, \ t = 0\}$, $S^l = \{(x, t) : x = 0, \ 0 < t \le T\}$, and, $d, T \sim \mathcal{O}(1)$, we consider the initial-boundary value problem for a singularly perturbed transport equation [4]

$$L u(x, t) \equiv \left\{ \varepsilon b(x, t) \frac{\partial}{\partial x} + c(x, t) + p(x, t) \frac{\partial}{\partial t} \right\} u(x, t) = f(x, t), \ (x, t) \in G, \tag{2}$$
$$u(x, t) = \varphi(x, t), \ (x, t) \in S.$$

The functions $b(x, t)$, $c(x, t)$, $p(x, t)$, $f(x, t)$ are assumed to be sufficiently smooth on $\overline{G}$, the function $\varphi(x, t)$, $(x, t) \in S$, is sufficiently smooth on the sets $S_0$ and $\overline{S}^l$, and continuous on $S$, and also,[1]

$$b_0 \le b(x, t) \le b^0, \quad 0 \le c(x, t) \le c^0, \quad p_0 \le p(x, t) \le p^0, \tag{3}$$

$$|f(x, t)| \le M, \quad (x, t) \in \overline{G}; \quad |\varphi(x, t)| \le M, \quad (x, t) \in S; \quad b_0, p_0 > 0;$$

---

[1] By $M$ ($m$) we denote sufficiently large (small) positive constants independent of $\varepsilon$. In the case of grid problems, these constants also do not depend on the stencils of the difference schemes.

the parameter $\varepsilon$ takes arbitrary values from the open-closed interval $(0, 1]$. Let the data of problem (2), (1) on the set $S^c = S_0 \cap \overline{S}^l$ which is a corner point, satisfy compatibility conditions ensuring the required smoothness of the solution on $\overline{G}$. In the solution of problem (2), (1), for small values of the parameter $\varepsilon$, a boundary layer appears in a neighborhood of the set $\overline{S}^l$.

**Our aim is** for problem (2), (1) to construct a robust difference scheme, i.e., a special difference scheme convergent $\varepsilon$-uniformly in the maximum norm, to justify its $\varepsilon$-uniform convergence, and to perform numerical testing of the proposed numerical method.

## 3   *A Priori* Estimates of the Solution and Derivatives

Here we give a number of *a priori* estimates for the solution of problem (2), (1) and its derivatives that are used to construct difference schemes and justify their convergence. Note that for the initial-boundary value problem for the singularly perturbed transport equation (2), (1), a maximum principle is valid, that is similar to maximum principle for the initial-boundary value problem for a singularly perturbed parabolic equation [5,6].

We assume that the derivatives $\dfrac{\partial^k}{\partial x^k}\varphi(x, 0)$, $\dfrac{\partial^{k_0}}{\partial t^{k_0}}\varphi(0, t)$ are bounded for $k, k_0 \leq 2$, moreover, in the corner point $(x, t) = (0, 0)$ compatibility conditions are fulfilled: $u \in C^{k,k_0}(\overline{G})$, $k + k_0 \leq 2$, which ensure the required continuity of the solution on $\overline{G}$ and its derivatives with respect to $x$ and $t$ up to the second order. For the solution of problem (2), (1) and its derivatives, we obtain the "standard estimate"

$$\left| \frac{\partial^{k+k_0}}{\partial x^k \partial t^{k_0}} u(x, t) \right| \leq M \varepsilon^{-k}, \quad (x, t) \in \overline{G}, \quad k + k_0 \leq 2. \tag{4}$$

We also give "improved a priori estimates", the derivation of which is performed taking into account the behavior of the solution in a neighborhood of the boundary layer and it is similar to the derivation of estimates for the initial-boundary value problem for a singular perturbed parabolic convection-diffusion equation in [5,6]. The solution of the problem (2), (1) can be represented as the decomposition

$$u(x, t) = U(x, t) + V(x, t), \quad (x, t) \in \overline{G}, \tag{5}$$

where $U(x)$ and $V(x)$ are the regular and singular components of the solution. Here $U(x, t) = U^e(x, t)$, $(x, t) \in \overline{G}$, is the restriction to $\overline{G}$ of the function $U^e(x, t)$, $(x, t) \in \overline{G}^e$, which is a solution of the extended problem

$$L^e U^e(x, t) = f^e(x, t), \quad (x, t) \in G^e, \quad U^e(x, t) = \varphi^e(x, t), \quad (x, t) \in S^e, \tag{6}$$

where $\overline{G}^e = [-d^*, d] \times [0, T]$ is an extension of $\overline{G}$ beyond the lateral boundary $S^l$, the operator $L^e$ and the functions $f^e(x, t)$ and $\varphi^e(x, t)$ are extensions of the operator $L$ to the set $\overline{G}^e$ and the functions $f(x, t)$ and $\varphi(x, t)$ to the sets $\overline{G}^e$

and $S^e$, respectively, with the preservation of properties of the data in problem (2), (1). The singular component $V(x,t)$ is the solution of the problem

$$LV(x,t) = 0, \quad (x,t) \in G, \tag{7}$$

$$V(x,t) = \varphi^l(x,t), \quad \varphi^l(x,t) = \varphi(x,t) - U^e(x,t), \quad (x,t) \in S^l,$$

$$V(x,t) = 0, \quad (x,t) \in S_0.$$

The components $U(x,t)$ and $V(x,t)$, $(x,t) \in \overline{G}$, and its derivatives satisfy the estimates

$$\left| \frac{\partial^{k+k_0}}{\partial x^k \partial t^{k_0}} U(x,t) \right| \le M, \tag{8a}$$

$$\left| \frac{\partial^{k+k_0}}{\partial x^k \partial t^{k_0}} V(x,t) \right| \le M \varepsilon^{-k}, \quad (x,t) \in \overline{G}, \quad k + k_0 \le 2, \tag{8b}$$

moreover,

$$V(x,t) = 0, \quad (x,t) \in \overline{G} \text{ for } x \ge M_1 \varepsilon t, \tag{9}$$

where $M_1 \ge max_{\overline{G}}[b(x,t)\,p^{-1}(x,t)]$. The following theorem holds [4]:

**Theorem 1.** *Let the data of the initial-boundary value problem (2), (1) for the singularly perturbed transport equation satisfy the conditions $b$, $c$, $p \in C^{k,\,k_0}(\overline{G})$, $f \in C^{k,\,k_0}(\overline{G})$, $\varphi(\cdot, 0) \in C^2(S_0)$, $\varphi(0, \cdot) \in C^2(\overline{S}^l)$, $k$, $k_0 \le 2$, $k + k_0 \le 2$. Then the solution $u(x,t)$ of the problem and its components $U(x,t)$ and $V(x,t)$ in the representation (5) satisfy the estimates (8), where $k$, $k_0 \le 2$, $k + k_0 \le 2$; and the relation (9) holds for the singular component $V(x,t)$.*

## 4  Standard Difference Scheme

In this section we construct a standard difference scheme on the basis of a monotone grid approximation of initial-boundary value problem (2), (1) (see, e.g., [2] in the case of a regular equation).

On the set $\overline{G}$ we introduce the rectangular grid

$$\overline{G}_h = \overline{\omega} \times \overline{\omega}_0. \tag{10}$$

Here $\overline{\omega}$ and $\overline{\omega}_0$ are arbitrary, in general, nonuniform grids on the intervals $[0,d]$ and $[0,T]$, respectively. Let $h^i = x^{i+1} - x^i$, $x^i$, $x^{i+1} \in \overline{\omega}$, $h = \max_i h^i$, and $\tau^k = t^{k+1} - t^k$, $t^k$, $t^{k+1} \in \overline{\omega}_0$, $\tau = \max_k \tau^k$. Assume that the condition $h \le M N^{-1}$, $\tau \le M N_0^{-1}$ holds, where $N + 1$ and $N_0 + 1$ are the numbers of nodes in the meshes $\overline{\omega}$ and $\overline{\omega}_0$, respectively.

We approximate problem (2), (1) by the standard difference scheme [1,2]

$$\Lambda z(x,t) \equiv \varepsilon\, b(x,t)\, \delta_{\overline{x}} + c(x,t) + p(x,t)\, \delta_{\overline{t}} = f(x,t), \quad (x,t) \in G_h,$$
$$z(x,t) = \varphi(x,t), \quad (x,t) \in S_h. \tag{11}$$

Here $G_h = G \sqcap \overline{G}_h$, $S_h = S \sqcap \overline{G}_h$, $\delta_{\overline{x}} z(x,t)$ and $\delta_{\overline{t}} z(x,t)$ are the first-order backward difference derivatives in $x$ and $t$.

Difference scheme (11), (10) is monotone $\varepsilon$-uniformly [2]. For this scheme, the grid maximum principle is valid.

In the case of the arbitrary grid[2] $\overline{G}_{h\,(10)}$, taking into account *a priori* estimates (8) and relation (9) for the singular component, we obtain the estimate

$$\|u - z\| \le M \left[(\varepsilon + N^{-1})^{-1} N^{-1} + N_0^{-1}\right], \quad (x,t) \in \overline{G}_{h\,(10)}, M_{(12)} = \mathcal{O}(1). \quad (12)$$

We obtain the similar estimate also on the grid $\overline{G}_h^u$, uniform with respect to $x$ and $t$

$$\overline{G}_h = \overline{G}_h^u \equiv \overline{\omega}^u \times \overline{\omega}_0^u. \quad (13)$$

Schemes (11), (10) and (11), (13) converge under the unimprovable condition $N^{-1} = o(\varepsilon)$, $N_0^{-1} = o(1)$, that is, for their convergence it is required to use grids in $x$ with the number of nodes $N \gg \varepsilon^{-1}$, i.e. unboundedly growing with $\varepsilon \to 0$.

Thus, the standard difference schemes (11), (10) and (11), (13) converge as $N \to \infty$ only for fixed values of the parameter $\varepsilon$ with first-order accuracy, and they do not converge $\varepsilon$-uniformly.

# 5   Special Difference Scheme Convergent $\varepsilon$-Uniformly

Here we construct a robust difference scheme which is a scheme convergent independently of the perturbation parameter $\varepsilon$, i.e., $\varepsilon$-uniformly in the maximum norm. We call it also "special scheme" since this will be the standard scheme on the "special" grid.

On the set $\overline{G}$ we introduce the "special" grid

$$\overline{G}_h = \overline{G}_h^s = \overline{\omega}^s \times \overline{\omega}_0. \quad (14a)$$

Here $\overline{\omega}_0$ is a uniform mesh in $t$, and $\overline{\omega}^s$ is a special piecewise-uniform mesh in $x$ of type "Shishkin mesh", condensing in a neighborhood of the boundary layer region.

When constructing the mesh $\overline{\omega}_{(14)}^s$, we divide the segment $[0, d]$ into two parts $[0, \sigma]$ and $[\sigma, d]$, and, on each of them, we construct uniform meshes with the same number of intervals equal to $2^{-1} N$; $N$ is assumed to be even. The value $\sigma$ (transition point) is defined by the relation dependent on the parameter $\varepsilon$

$$\sigma = \sigma(\varepsilon; d, T) = \min [M_1 \varepsilon T, 2^{-1} d], \quad M_1 = M_{1\,(9)}. \quad (14b)$$

Thus, the mesh-sizes in the mesh $\overline{\omega}_{(14)}^s$ are defined as follows:

$$h_1 = 2\sigma N^{-1}, \ x \in [0, \sigma]; \quad h_2 = 2 (d - \sigma) N^{-1}, \ x \in [\sigma, d]. \quad (14c)$$

Under the condition $\varepsilon \ge 2^{-1} M_1^{-1} T^{-1} d$, the grid $\overline{G}_{h\,(14)}$ becomes uniform.

---

[2] The notation $D_{(i.j)}$ ($L_{(i.j)}$, $m_{(i.j)}$, $M_{(i.j)}$, $D_{h(i.j)}$) means that these sets (operators, constants, grids) were introduced in formula $(i.j)$.

For the solution of the difference scheme (11), (14), taking into account *a priori* estimates (8) and relation (9) for the singular component, and using the grid maximum principle, we obtain the estimate

$$\|u - z\| \leq M\,[\,N^{-1} + N_0^{-1}\,], \quad (x,t) \in \overline{G}_{h\,(14)}. \tag{15}$$

Thus, special difference scheme (11), (14), i.e., the standard difference scheme on the special piecewise-uniform grid, converges $\varepsilon$-uniformly in the maximum norm with first-order accuracy in $x$ and $t$.

## 6   Numerical Study of a Model Problem

In this section we perform a numerical study of a model initial-boundary value problem for a singularly perturbed transport equation.

As a model problem, we consider a simplified initial-boundary value problem (2), (1) on the set $\overline{G}_{(1)}$ for $d = T = 1$ for the following singularly perturbed transport equation:

$$L\,u(x,t) \equiv \left( \varepsilon \frac{\partial}{\partial x} + \frac{\partial}{\partial t} \right) u(x,t) = f(x,t), \quad (x,t) \in G, \tag{16a}$$

$$u(x,t) = \varphi(x,t), \quad (x,t) \in S.$$

Here

$$f(x,t) = \varepsilon\,x - 2^{-1}\,t, \quad (x,t) \in G, \tag{16b}$$

$$\varphi(x,t) = \left\{ \begin{array}{ll} \varphi(0,t) = -2^{-2}\,t^2 + t^3, & x = 0, \quad 0 < t \leq 1 \\ \varphi(x,0) = 2^{-1} x^2, & 0 \leq x \leq 1, \quad t = 0 \end{array} \right\}, \quad (x,t) \in S.$$

The solution of the model problem (16) can be represented as the decomposition

$$u(x,t) = U(x,t) + V(x,t), \quad (x,t) \in G, \tag{17}$$

in which the regular component $U(x,\,t)$ and the singular component $V(x,\,t)$ are defined by the relations

$$U(x,t) = 2^{-1}(x^2 - 2^{-1}t^2), \quad (x,t) \in G; \tag{18}$$

$$V(x,t) = \left\{ \begin{array}{ll} -(\varepsilon^{-1}x - t)^3, & 0 \leq x < \varepsilon t \\ 0, & \varepsilon t \leq x \leq 1 \end{array} \right\}, \quad (x,t) \in \overline{G}. \tag{19}$$

Thus, we have the explicit exact solution $u(x,t)$ of the model problem.

Model problem (16) is approximated by the standard difference scheme

$$(\varepsilon\,\delta_{\overline{x}} + \delta_{\overline{t}})\,z(x,t) = f(x,t), \quad (x,t) \in G_h, \quad z(x,t) = \varphi(x,t), \quad (x,t) \in S_h. \tag{20}$$

We are interested in approximation errors of the solution $u(x,t) = u(x,t;\varepsilon)$ of model problem (16) by the grid solution $z(x,t) = z(x,t;\varepsilon, N = N_0)$ of the standard scheme (20) in two cases for the grid $G_h$, when:

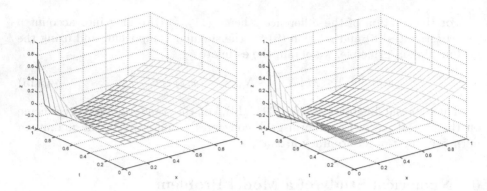

**Fig. 1.** Numerical solutions $z(x,t)$: on the left, for the standard scheme, $(x,t) \in \overline{G}_{h\,(13)}$, on the right, for the special scheme, $(x,t) \in \overline{G}_{h\,(14)}$, for $\varepsilon = 2^{-5}$, $N = N_0 = 16$

**Table 1.** Errors $E(\varepsilon, N = N_0)$ in the solution of the standard scheme on the uniform grid

| $\varepsilon \setminus N$ | 4 | 16 | 64 | 256 | 1024 | 4096 |
|---|---|---|---|---|---|---|
| 1 | 2.832e−1 | 9.050e−2 | 2.467e−2 | 6.310e−3 | 1.586e−3 | 3.972e−4 |
| $2^{-2}$ | **2.666e−1** | 1.661e−1 | 5.226e−2 | 1.407e−2 | 3.580e−3 | 8.990e−4 |
| $2^{-4}$ | 5.463e−2 | **2.183e−1** | 1.450e−1 | 4.517e−2 | 1.208e−2 | 3.070e−3 |
| $2^{-6}$ | 6.055e−2 | 5.326e−2 | **2.098e−1** | 1.403e−1 | 4.355e−2 | 1.162e−2 |
| $2^{-8}$ | 6.201e−2 | 1.550e−2 | 5.781e−2 | **2.079e−1** | 1.391e−1 | 4.315e−2 |
| $2^{-10}$ | 6.238e−2 | 1.559e−2 | 1.213e−2 | 5.907e−2 | **2.074e−1** | 1.388e−1 |
| $2^{-12}$ | 6.247e−2 | 1.562e−2 | 3.904e−3 | 1.460e−2 | 5.939e−2 | **2.073e−1** |
| $2^{-14}$ | 6.249e−2 | 1.562e−2 | 3.906e−3 | 2.956e−3 | 1.522e−2 | 5.947e−2 |
| $\overline{E}(N)$ | **2.832e−1** | **2.183e−1** | **2.098e−1** | **2.079e−1** | **2.074e−1** | **2.073e−1** |

(1) we use the standard uniform grid $\overline{G}_h = \overline{G}^u_{h(13)} \equiv \overline{\omega} \times \overline{\omega}_0$, with the uniform meshes $\overline{\omega}$ in $x$ and $\overline{\omega}_0$ in $t$,
(2) we use the special grid $\overline{G}_h = \overline{G}^s_{h(14)} \equiv \overline{\omega}^s \times \overline{\omega}_0$, where $\overline{\omega}^s$ is the piecewise uniform mesh in $x$ of type "Shishkin mesh", and $\overline{\omega}_0$ is the uniform mesh in $t$.

For this, we apply a technique similar used in [8], and we investigate the behavior of the quantities

$$E(\varepsilon, N = N_0) = \max_{\overline{G}_h} |u(x,t;\varepsilon) - z(x,t;\varepsilon, N = N_0)|,$$

$$\overline{E}(N = N_0) = \overline{E}(N = N_0; u(\cdot)) = \max_{\varepsilon} E(\varepsilon, N = N_0)$$

depending on the value of the parameter $\varepsilon$, $\varepsilon = 2^0 \div 2^{-14}$, for $N = N_0 = 2^2 \div 2^{12}$.

In the Tables 1 and 2, numerical results are given corresponding to the cases (1) and (2). Grid solutions corresponding to the cases (1) and (2) are shown in Fig. 1.

**Table 2.** Errors $E(\varepsilon, N = N_0)$ in the solution of the standard scheme on the piecewise-uniform grid, i.e., the special (robust) scheme

| $\varepsilon \setminus N$ | 4 | 16 | 64 | 256 | 1024 | 4096 |
|---|---|---|---|---|---|---|
| 1 | 2.832e−1 | 9.050e−2 | 2.467e−2 | 6.310e−3 | 1.586e−3 | 3.972e−4 |
| $2^{-2}$ | 2.666e−1 | 1.661e−1 | 5.226e−2 | 1.407e−2 | 3.580e−3 | 8.990e−4 |
| $2^{-4}$ | 2.494e−1 | 1.627e−1 | 5.135e−2 | 1.384e−2 | 3.524e−3 | 8.849e−4 |
| $2^{-6}$ | 2.482e−1 | 1.625e−1 | 5.129e−2 | 1.383e−2 | 3.520e−3 | 8.839e−4 |
| ... | ... | ... | ... | ... | ... | ... |
| $2^{-14}$ | 2.482e−1 | 1.625e−1 | 5.129e−2 | 1.383e−2 | 3.520e−3 | 8.839e−4 |
| $\overline{E}(N)$ | **2.832e−1** | **1.661e−1** | **5.226e−2** | **1.407e−2** | **3.580e−3** | **8.990e−4** |

From Table 1 it follows that, when $N$ grows for $N = N_0$, the grid solution $z(x,t)$ of the standard scheme on the uniform grid converges to the exact solution $u(x,t)$ only for fixed values of the parameter $\varepsilon$ (see $\varepsilon N \gg 1$ above the selected diagonal), and it does not converge $\varepsilon$-uniformly.

From Table 2 it follows that, when $N$ grows for $N = N_0$, the grid solution $z(x,t)$ of the standard scheme on the piecewise-uniform mesh, i.e., the special (robust) scheme, converges to the exact solution $u(x,t)$ $\varepsilon$-uniformly.

Thus, the results of numerical experiments are consistent with the theoretical ones.

# References

1. Marchuk, G.: Methods of Computational Mathematics. Nauka, Moskva (1989). (in Russian)
2. Samarskii, A.: The Theory of Difference schemes. Nauka, Moskva (1989). (in Russian)
3. Kalitkin, N., Koriakin, P.: Numerical Methods. Methods of Mathematical Physics. Publishing Center "Academy", Moskva (2013). (in Russian)
4. Shishkin, G.: Difference scheme for an initial-boundary value problem for a singularly perturbed transport equation. Comput. Math. Math. Phys. **57**(11), 1789–1795 (2017). https://link.springer.com/article/10.1134%2FS0965542517110136
5. Shishkin, G.: Grid Approximations of Singularly Perturbed Elliptic and Parabolic Equations. Russian Academy of Sciences, Ural Branch, Ekaterinburg (1992). (in Russian)
6. Shishkin, G., Shishkina, L.: Difference Methods for Singular Perturbation Problems. Chapman and Hall/CRC Monographs and Surveys in Pure and Applied Mathematics, vol. 140. CRC Press, Boca Raton (2009)
7. Roos, H.-G., Stynes, M., Tobiska, L.: Numerical Methods for Singularly Perturbed Differential Equations. Convection-Diffusion and Flow Problems. Springer, Berlin (1996). https://doi.org/10.1007/978-3-662-03206-0
8. Farrel, P., Hegarty, A., Miller, J., O'Riordan, E., Shishkin, G.: Robust Computational Techniques for Boundary Layers. Chapman and Hall/CRC, Boca Raton (2000)

# A High Order Accurate Overlapping Domain Decomposition Method for Singularly Perturbed Reaction-Diffusion Systems

Joginder Singh[1(✉)], Sunil Kumar[1], and Mukesh Kumar[2]

[1] Department of Mathematical Sciences,
Indian Institute of Technology (BHU) Varanasi, Varanasi, India
virk1516@gmail.com
[2] Department of Mathematics, College of Charleston, Charleston, SC, USA

**Abstract.** In this work, we consider a coupled system of singularly perturbed reaction-diffusion equations (SPRDEs) with distinct small positive parameters, exhibiting overlapping boundary layers at both ends of the domain. In [4], the authors designed an overlapping domain decomposition method that gives almost second order accurate approximations to the solution of coupled systems of SPRDEs. High order methods are of great significance to the numerical community. To this end, for numerically solving the coupled systems of SPRDEs, we designed a high order accurate overlapping domain decomposition method via. defining an appropriate decomposition of the original domain and then considering a hybrid difference scheme on a uniform mesh on each subdomain. More precisely, the method gives almost fourth order accurate approximations to the solution of the problem, as compared to almost second order accurate approximations in [4]. Numerical results are given to demonstrate the effectiveness of the proposed method.

**Keywords:** Singularly perturbed · Coupled system ·
Domain decomposition · Reaction-diffusion

## 1 Introduction

Domain decomposition methods have received a lot of attention in the last two decades. There is a rich literature on effective numerical methods based on domain decomposition for regular boundary value problems, see [9,15] and the references therein, but the same cannot be said for singularly perturbed boundary value problems. To this end, in this paper, we design a high order overlapping domain decomposition method for a coupled system of singularly perturbed reaction-diffusion problems. Such systems are widespread in modeling of various physical problems [6].

© Springer Nature Switzerland AG 2019
I. Dimov et al. (Eds.): FDM 2018, LNCS 11386, pp. 484–492, 2019.
https://doi.org/10.1007/978-3-030-11539-5_56

In this paper, we consider the following model problem

$$\boldsymbol{Lu} := -\mathbf{E}\boldsymbol{u}'' + \boldsymbol{Au} = \boldsymbol{f} \text{ in } \Omega = (0,1), \qquad \boldsymbol{u}(0) = \boldsymbol{a}_1, \quad \boldsymbol{u}(1) = \boldsymbol{a}_2, \qquad (1)$$

where $\mathbf{E} = \text{diag}(\varepsilon_1, \varepsilon_2)$ is a diagonal matrix with small parameters $0 < \varepsilon_i \leq 1, i = 1, 2$, $\boldsymbol{f} = (f_1, f_2)^T$ and $\boldsymbol{u} = (u_1, u_2)^T$. Without loss of generality we consider the ordering $\varepsilon_1 \leq \varepsilon_2$. The coupling matrix $\boldsymbol{A} = (a_{ij}(x))_{2 \times 2}$ is assumed to satisfy the following conditions

$$a_{ij}(x) \leq 0, \quad i \neq j; \quad a_{ii}(x) > 0, \quad \sum_{j=1}^{2} a_{ij}(x) > \alpha > 0, \quad i = 1, 2, x \in \overline{\Omega}. \quad (2)$$

With these assumptions problem (1) exhibits a unique solution having overlapping boundary layers at $x = 0$ and $x = 1$ of $O(\sqrt{\varepsilon_i} \ln(1/\varepsilon_i)), i = 1, 2$; see [8]. Moreover, the solution $\boldsymbol{u}$ can be decomposed as $\boldsymbol{u} = \boldsymbol{v} + \boldsymbol{w}$, with

$$\|v_i^{(s)}\| \leq C(1 + \varepsilon_1^{1-s/2}), \quad s = 0, \ldots, 4, i = 1, 2,$$

$$|w_1^{(s)}(x)| \leq C(\varepsilon_1^{-s/2}\mathcal{B}_{\varepsilon_1}(x) + \varepsilon_2^{-s/2}\mathcal{B}_{\varepsilon_2}(x)), \quad |w_2^{(s)}(x)| \leq C\varepsilon_2^{-s/2}\mathcal{B}_{\varepsilon_2}(x), \quad s = 0, 1, 2,$$

$$|w_i^{(s)}(x)| \leq C\varepsilon_i^{1-s/2}(\varepsilon_1^{-1}\mathcal{B}_{\varepsilon_1}(x) + \varepsilon_2^{-1}\mathcal{B}_{\varepsilon_2}(x)), \quad s = 3, 4,$$

for all $x \in \overline{\Omega}$. Here $\mathcal{B}_{\varepsilon_i}(x) = \exp(-x\sqrt{\alpha/\varepsilon_i}) + \exp(-(1-x)\sqrt{\alpha/\varepsilon_i})$.

Due to overlapping layers special care is required in solving problem (1). More precisely, we need to develop a robust numerical method, where error is independent of $\varepsilon_1$ and $\varepsilon_2$. Some robust domain decomposition methods for singularly perturbed problems have been explored in [1,3,5,7,10–13]. The authors in [4] considered problem (1) and designed an overlapping domain decomposition method that gives almost second order accurate approximations to the solution of the problem. High order methods are of great significance to the numerical community. Therefore, the purpose of this paper is to design a high order accurate overlapping domain decomposition method for problem (1) via. defining an appropriate decomposition of the original domain and then considering a hybrid difference scheme on a uniform mesh on each subdomain. More precisely, the method gives almost fourth order accurate approximations to the solution of the problem, as compared to almost second order accurate approximations in [4]. Numerical results are given to demonstrate the effectiveness of the method.

The paper is arranged as follows. A high order overlapping Schwarz method is designed in Sect. 2. Numerical tests are performed in Sect. 3.

**Notation:** For any functions $g, y_p \in C(\overline{\Omega})$, define $g_j = g(x_j)$, $y_{p;j} = y_p(x_j)$; if $\boldsymbol{g}, \boldsymbol{y}_p \in C(\overline{\Omega})^2$ then $\boldsymbol{g}_j = \boldsymbol{g}(x_j) = (g_{1;j}, g_{2;j})^T$, $\boldsymbol{y}_{p;j} = \boldsymbol{y}_p(x_j) = (y_{p,1;j}, y_{p,2;j})^T$. For any function $g \in C(\overline{\Omega})$, we define $\|g\| = \max_{x \in \overline{\Omega}} |g(x)|$; if $\boldsymbol{g} \in C(\overline{\Omega})^2$, define $\|\boldsymbol{g}\| = \max\{\|g_1\|, \|g_2\|\}$. We denote by $\|.\|_{D^N}$ the analogous discrete maximum norm on the mesh $D^N$.

## 2    Overlapping Domain Decomposition Method

We decompose the domain $\Omega$ into five overlapping subdomains (see Fig. 1)

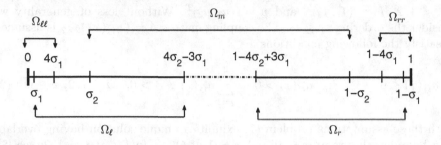

**Fig. 1.** Decomposition of the domain.

$$\Omega_{\ell\ell} = (0, 4\sigma_1), \quad \Omega_\ell = (\sigma_1, 4\sigma_2 - 3\sigma_1), \quad \Omega_m = (\sigma_2, 1 - \sigma_2),$$

$$\Omega_r = (1 - 4\sigma_2 + 3\sigma_1, 1 - \sigma_1), \quad \Omega_{rr} = (1 - 4\sigma_1, 1),$$

where the subdomain parameters $\sigma_1$ and $\sigma_2$ are defined as follows

$$\sigma_2 = \min\left\{\frac{4}{26}, \frac{4\sqrt{\varepsilon_2}}{\sqrt{\alpha}}\ln N\right\} \quad \text{and} \quad \sigma_1 = \min\left\{\frac{\sigma_2}{4}, \frac{4\sqrt{\varepsilon_1}}{\sqrt{\alpha}}\ln N\right\}. \tag{3}$$

Suppose $N = 3 \times 2^n, n \geq 1$. We define a uniform mesh length $h_p = (d - a)/N$ in each subdomain $\Omega_p = (a, d)$, $p = \ell\ell, \ell, m, r, rr$. Let $\overline{\Omega}_p^N : a = x_0 < x_1 < \cdots < x_{N-1} < x_N = d$, with $x_j = a + jh_p$, $j = 0, \ldots, N$. On each subdomain $\Omega_p^N$, $p = \ell\ell, \ell, m, r, rr$, discretization is of the form

$$[\boldsymbol{L}_p^N \boldsymbol{U}_p]_j = [\boldsymbol{\Upsilon}_p \boldsymbol{f}]_j, \tag{4}$$

where

$$[\boldsymbol{L}_p^N \boldsymbol{U}_p]_j := \begin{pmatrix} [L_{p,1}^N \boldsymbol{U}_p]_j \\ [L_{p,2}^N \boldsymbol{U}_p]_j \end{pmatrix} = \begin{pmatrix} [R_p(U_{p,1})]_j + [Q_p(a_{12}U_{p,2})]_j \\ [R_p(U_{p,2})]_j + [Q_p(a_{21}U_{p,1})]_j \end{pmatrix}, \tag{5}$$

$$[\boldsymbol{\Upsilon}_p \boldsymbol{f}]_j := \begin{pmatrix} [\Upsilon_{p,1} \boldsymbol{f}]_j \\ [\Upsilon_{p,2} \boldsymbol{f}]_j \end{pmatrix} = \begin{pmatrix} [Q_p(f_1)]_j \\ [Q_p(f_2)]_j \end{pmatrix} \tag{6}$$

with

$$[R_p(V_i)]_j = r_{p;j}^{i,-} V_{i;j-1} + r_{p;j}^{i,c} V_{i;j} + r_{p;j}^{i,+} V_{i;j+1}; \quad [Q_p(V_i)]_j = q_{p;j}^{i,-} V_{i;j-1} + q_{p;j}^{i,c} V_{i;j} + q_{p;j}^{i,+} V_{i;j+1}. \tag{7}$$

The coefficients $r_{p;j}^{i,\star}$ depend on $q_{p;j}^{i,\star}$, and are defined as follows

$$r_{p;j}^{i,-} = -\frac{\varepsilon_i}{h_p^2} + q_{p;j}^{i,-} a_{ii;j-1}, \quad r_{p;j}^{i,c} = \frac{2\varepsilon_i}{h_p^2} + q_{p;j}^{i,c} a_{ii;j}, \quad r_{p;j}^{i,+} = -\frac{\varepsilon_i}{h_p^2} + q_{p;j}^{i,+} a_{ii;j+1}, (8)$$

The coefficients $q_{p;j}^{i,\star}$ depend on the relation between the mesh size of subdomain and the perturbation parameters. The idea of choosing a scheme based on the relation between the mesh size and perturbation parameter is considered previously in [14] for scalar convection-diffusion problems and in [2] for scalar reaction-diffusion problems. On subdomains $\Omega_{\ell\ell}$ and $\Omega_{rr}$ the coefficients $q_{p;j}^{i,\star}$ are defined by

$$q_{p;j}^{i,-} = \frac{1}{12}, \quad q_{p;j}^{i,c} = \frac{10}{12}, \quad q_{p;j}^{i,+} = \frac{1}{12}. \tag{9}$$

On subdomain $\Omega_m$, when $h_p^2 \|a_{ii}\| \le 12\varepsilon_i$, the coefficients $q_{p;j}^{i,\star}$ are defined by (9). On the other hand, when $h_p^2 \|a_{ii}\| > 12\varepsilon_i$, the coefficients $q_{p;j}^{i,\star}$ are given by

$$q_{p;j}^{i,-} = 0, \quad q_{p;j}^{i,c} = 1, \quad q_{p;j}^{i,+} = 0. \tag{10}$$

On subdomains $\Omega_\ell$ and $\Omega_r$, the coefficients $q_{p;j}^{2,\star}$ are defined by (9). When $h_p^2 \|a_{11}\| \le 12\varepsilon_1$, the coefficients $q_{p;j}^{1,\star}$ are defined by (9), and when $h_p^2 \|a_{11}\| > 12\varepsilon_1$, the coefficients $q_{p;j}^{1,\star}$ are defined by (10).

Then the proposed method is described as follows.

*Step 1.* We start with

$$\boldsymbol{U}^{[0]}(x_j) \equiv \boldsymbol{0}, \quad 0 < x_j < 1, \quad \boldsymbol{U}^{[0]}(0) = \boldsymbol{u}(0), \quad \boldsymbol{U}^{[0]}(1) = \boldsymbol{u}(1).$$

*Step 2.* We get $\boldsymbol{U}_p^{[k]}$, $p = \ell\ell, \ell, m, r, rr$, by solving

$$[\boldsymbol{L}_{\ell\ell}^N \boldsymbol{U}_{\ell\ell}^{[k]}] = [\boldsymbol{\Upsilon}_{\ell\ell} \boldsymbol{f}] \quad \text{in } \Omega_{\ell\ell}^N, \quad \boldsymbol{U}_{\ell\ell}^{[k]}(0) = \boldsymbol{u}(0), \quad \boldsymbol{U}_{\ell\ell}^{[k]}(4\sigma_1) = \mathcal{S}\boldsymbol{U}^{[k-1]}(4\sigma_1),$$

$$[\boldsymbol{L}_{rr}^N \boldsymbol{U}_{rr}^{[k]}] = [\boldsymbol{\Upsilon}_{rr} \boldsymbol{f}] \quad \text{in } \Omega_{rr}^N, \quad \boldsymbol{U}_{rr}^{[k]}(1 - 4\sigma_1) = \mathcal{S}\boldsymbol{U}^{[k-1]}(1 - 4\sigma_1), \quad \boldsymbol{U}_{rr}^{[k]}(1) = \boldsymbol{u}(1),$$

$$[\boldsymbol{L}_\ell^N \boldsymbol{U}_\ell^{[k]}] = [\boldsymbol{\Upsilon}_\ell \boldsymbol{f}] \quad \text{in } \Omega_\ell^N, \quad \boldsymbol{U}_\ell^{[k]}(\sigma_1) = \mathcal{S}\boldsymbol{U}_{\ell\ell}^{[k]}(\sigma_1),$$

$$\boldsymbol{U}_\ell^{[k]}(4\sigma_2 - 3\sigma_1) = \mathcal{S}\boldsymbol{U}^{[k-1]}(4\sigma_2 - 3\sigma_1),$$

$$[\boldsymbol{L}_r^N \boldsymbol{U}_r^{[k]}] = [\boldsymbol{\Upsilon}_r \boldsymbol{f}] \quad \text{in } \Omega_r^N, \quad \boldsymbol{U}_r^{[k]}(1 - 4\sigma_2 + 3\sigma_1) = \mathcal{S}\boldsymbol{U}^{[k-1]}(1 - 4\sigma_2 + 3\sigma_1),$$

$$\boldsymbol{U}_r^{[k]}(1 - \sigma_1) = \mathcal{S}\boldsymbol{U}_{rr}^{[k]}(1 - \sigma_1),$$

$$[\boldsymbol{L}_m^N \boldsymbol{U}_m^{[k]}] = [\boldsymbol{\Upsilon}_m \boldsymbol{f}] \quad \text{in } \Omega_m^N, \quad \boldsymbol{U}_m^{[k]}(\sigma_2) = \mathcal{S}\boldsymbol{U}_\ell^{[k]}(\sigma_2), \quad \boldsymbol{U}_m^{[k]}(1 - \sigma_2) = \mathcal{S}\boldsymbol{U}_r^{[k]}(1 - \sigma_2),$$

where $\mathcal{S}\boldsymbol{U}^{[k]}$ is the cubic $C^0$-spline interpolant of $\boldsymbol{U}^{[k]}$ on $\overline{\Omega}^N := (\overline{\Omega}_{\ell\ell}^N \backslash \Omega_\ell) \cup (\overline{\Omega}_\ell^N \backslash \Omega_m) \cup \overline{\Omega}_m^N \cup (\overline{\Omega}_r^N \backslash \Omega_m) \cup (\overline{\Omega}_{rr}^N \backslash \Omega_r)$. Note that it is computed by considering three adjacent and equidistant intervals $\chi = [x_i, x_{i+3}]$ and fitting a cubic function through the numerical approximation at $x_i, x_{i+1}, x_{i+2}, x_{i+3}$.

**Table 1.** Maximum error $E_{\varepsilon_1,\varepsilon_2}^N$ of the method for Example 1 with fixed $\varepsilon_1 = 10^{-7}$.

| $\varepsilon_2 = 10^{-n}$ | $N = 3 \times 2^4$ | $N = 3 \times 2^5$ | $N = 3 \times 2^6$ | $N = 3 \times 2^7$ | $N = 3 \times 2^8$ | $N = 3 \times 2^9$ |
|---|---|---|---|---|---|---|
| $n=0$ | 2.31E−02 | 7.01E−03 | 1.18E−03 | 1.22E−04 | 1.24E−05 | 1.17E−06 |
| 1 | 2.31E−02 | 7.01E−03 | 1.18E−03 | 1.22E−04 | 1.24E−05 | 1.17E−06 |
| 2 | 2.31E−02 | 7.01E−03 | 1.18E−03 | 1.22E−04 | 1.24E−05 | 1.17E−06 |
| 3 | 2.31E−02 | 7.00E−03 | 1.18E−03 | 1.22E−04 | 1.24E−05 | 1.17E−06 |
| 4 | 2.32E−02 | 7.00E−03 | 1.18E−03 | 1.22E−04 | 1.24E−05 | 1.17E−06 |
| 5 | 2.29E−02 | 6.97E−03 | 1.17E−03 | 1.22E−04 | 1.23E−05 | 1.17E−06 |
| 6 | 1.37E−02 | 3.30E−03 | 4.57E−04 | 4.75E−05 | 4.69E−06 | 4.37E−07 |
| 7 | 1.68E−04 | 2.10E−05 | 2.32E−06 | 2.39E−07 | 2.32E−08 | 2.16E−09 |

*Step 3.* We compute $U^{[k]}$ in the following way

$$U^{[k]}(x_j) = \begin{cases} U_{\ell\ell}^{[k]}(x_j), \ x_j \in \overline{\Omega}_{\ell\ell}^N \setminus \overline{\Omega}_\ell; \quad U_\ell^{[k]}(x_j), \ x_j \in \overline{\Omega}_\ell^N \setminus \overline{\Omega}_m; \\ U_m^{[k]}(x_j), \ x_j \in \overline{\Omega}_m^N; \\ U_r^{[k]}(x_j), \ x_j \in \overline{\Omega}_r^N \setminus \overline{\Omega}_m; \quad U_{rr}^{[k]}(x_j), \ x_j \in \overline{\Omega}_{rr}^N \setminus \overline{\Omega}_r. \end{cases} \tag{11}$$

*Step 4.* If

$$||U^{[k+1]} - U^{[k]}||_{\overline{\Omega}^N} \leq \gamma$$

is satisfied, then we stop; otherwise we go to Step 2. Here $\gamma$ is the prescribed accuracy.

Let $N_0$ be the smallest positive integer such that

$$(16/3\alpha) \max\{||a_{11}||, ||a_{22}||\} \leq N_0^2/\ln^2(N_0).$$

Then, for any $N \geq N_0$, the choice of the coefficients ensures that the matrix associated with $L_p^N$ is an M-matrix and hence it satisfies a discrete maximum principle and it is uniformly stable in the maximum norm.

## 3 Numerical Results

In this section we present some numerical results for overlapping domain decomposition method developed in the previous section. For stopping the algorithm we consider $\gamma = N^{-4} \ln^4 N$. We denote the final iterated solution by $U$.

**Example 1.** Consider the following test problem

$$- \varepsilon_1 u_1'' + 2(x+1)^2 u_1 - (1+x^3)u_2 = 2\exp(x), \qquad u_1(0) = u_1(1) = 0,$$
$$- \varepsilon_2 u_2'' - 2\cos(\pi x/4)u_1 + (1+\sqrt{2})\exp(1-x)u_2 = 10x+1, \quad u_2(0) = u_2(1) = 0.$$

**Table 2.** Iteration counts of the method for Example 1 with fixed $\varepsilon_1 = 10^{-7}$.

| $\varepsilon_2 = 10^{-n}$ | $N = 3 \times 2^4$ | $N = 3 \times 2^5$ | $N = 3 \times 2^6$ | $N = 3 \times 2^7$ | $N = 3 \times 2^8$ | $N = 3 \times 2^9$ |
|---|---|---|---|---|---|---|
| $n = 0$ | 7 | 9 | 11 | 12 | 14 | 17 |
| 1 | 7 | 8 | 10 | 11 | 13 | 14 |
| 2 | 7 | 8 | 9 | 10 | 11 | 13 |
| 3 | 6 | 6 | 7 | 7 | 8 | 8 |
| 4 | 3 | 4 | 4 | 4 | 4 | 4 |
| 5 | 2 | 2 | 2 | 2 | 2 | 2 |
| 6 | 1 | 1 | 1 | 1 | 1 | 1 |
| 7 | 1 | 1 | 1 | 1 | 1 | 1 |

**Table 3.** Uniform error $E^N$ and uniform rate of convergence $\rho^N$ of the method for Example 1.

| | $N = 3 \times 2^4$ | $N = 3 \times 2^5$ | $N = 3 \times 2^6$ | $N = 3 \times 2^7$ | $N = 3 \times 2^8$ | $N = 3 \times 2^9$ |
|---|---|---|---|---|---|---|
| $E^N$ | 2.32E$-$02 | 7.01E$-$03 | 1.18E$-$03 | 1.22E$-$04 | 1.24E$-$05 | 1.17E$-$06 |
| $\rho^N$ | 2.26 | 3.23 | 3.98 | 3.93 | 3.97 | |

We define $SU$ on macro intervals $[x_{3i}, x_{3(i+1)}]$, $i = 0, \ldots, 2N/3 - 1$. As the solution of this test problem is not known, we estimate the error $\|u - U\|_{\overline{\Omega}^N}$ using a variant of the double mesh principle. For this purpose, we compute not only the approximate solution $U$ on $\overline{\Omega}^N$, but also another approximate solution $\widehat{U}$ on $\widehat{\overline{\Omega}}^N$ that is constructed using the same subdomain parameters $\sigma_1$ and $\sigma_2$, but $2N$ intervals in each subdomain. For different values of $\varepsilon_1$, $\varepsilon_2$ and $N$, we take the estimate

$$\|u - U\|_{\overline{\Omega}^N} \approx E^N_{\varepsilon_1, \varepsilon_2} := \|\widehat{U} - U\|_{\overline{\Omega}^N}.$$

Fixing $\varepsilon_1$ by taking $\varepsilon_1 = 10^{-n}$, for some non-negative integer $n$, we compute $E^N_{\varepsilon_1} = \max\{E^N_{\varepsilon_1, 1}, E^N_{\varepsilon_1, 10^{-1}}, \ldots, E^N_{\varepsilon_1, 10^{-n}}\}$. Finally, we compute the uniform error as $E^N = \max_{\varepsilon_1 \in S_{\varepsilon_1}} E^N_{\varepsilon_1}$, where $S_{\varepsilon_1} = \{10^0, 10^{-1}, \ldots, 10^{-9}\}$. Since we have an error bound of the form $(N^{-1} \ln N)^\rho$, we compute the uniform numerical rate of convergence by

$$\rho^N = \frac{\ln E^N - \ln E^{2N}}{\ln(2 \ln N) - \ln(\ln(2N))}.$$

Table 1 represents the errors $E^N_{\varepsilon_1, \varepsilon_2}$ for fixed $\varepsilon_1 = 10^{-7}$ and different values of $\varepsilon_2$ and $N$. The number of iterations required by the proposed algorithm are listed in Table 2 for fixed $\varepsilon_1 = 10^{-7}$ and different values of $\varepsilon_2$ and $N$. One can observe that when perturbation parameters are small (that represents the boundary layer phenomena) and of different magnitude fewer iterations are required to get the desired accuracy, and when perturbation parameters are small and of same magnitude one iteration is needed, as shown in [10]. Table 3 lists the uniform

**Table 4.** Maximum error $E^N_{\varepsilon_1,\varepsilon_2}$ of the method for Example 2 with fixed $\varepsilon_1 = 10^{-7}$.

| $\varepsilon_2 = 10^{-n}$ | $N = 3 \times 2^4$ | $N = 3 \times 2^5$ | $N = 3 \times 2^6$ | $N = 3 \times 2^7$ | $N = 3 \times 2^8$ | $N = 3 \times 2^9$ |
|---|---|---|---|---|---|---|
| $n = 0$ | 4.06E−02 | 5.32E−03 | 6.03E−04 | 6.25E−05 | 6.09E−06 | 5.66E−07 |
| 1 | 4.02E−02 | 5.27E−03 | 5.97E−04 | 6.19E−05 | 6.03E−06 | 5.61E−07 |
| 2 | 3.94E−02 | 5.16E−03 | 5.84E−04 | 6.06E−05 | 5.90E−06 | 5.49E−07 |
| 3 | 3.72E−02 | 4.88E−03 | 5.53E−04 | 5.74E−05 | 5.59E−06 | 5.20E−07 |
| 4 | 3.24E−02 | 4.28E−03 | 4.85E−04 | 5.03E−05 | 4.90E−06 | 4.56E−07 |
| 5 | 2.41E−02 | 3.18E−03 | 3.58E−04 | 3.73E−05 | 3.64E−06 | 3.39E−07 |
| 6 | 5.18E−03 | 6.43E−04 | 7.37E−05 | 7.63E−06 | 7.43E−07 | 6.91E−08 |
| 7 | 9.32E−04 | 1.17E−04 | 1.31E−05 | 1.35E−06 | 1.31E−07 | 1.22E−08 |

**Table 5.** Iteration counts of the method for Example 2 with fixed $\varepsilon_1 = 10^{-7}$.

| $\varepsilon_2 = 10^{-n}$ | $N = 3 \times 2^4$ | $N = 3 \times 2^5$ | $N = 3 \times 2^6$ | $N = 3 \times 2^7$ | $N = 3 \times 2^8$ | $N = 3 \times 2^9$ |
|---|---|---|---|---|---|---|
| $n = 0$ | 8 | 10 | 12 | 14 | 16 | 18 |
| 1 | 7 | 8 | 9 | 11 | 12 | 13 |
| 2 | 6 | 7 | 8 | 8 | 9 | 10 |
| 3 | 4 | 4 | 4 | 5 | 5 | 5 |
| 4 | 2 | 2 | 2 | 2 | 2 | 2 |
| 5 | 1 | 1 | 1 | 1 | 1 | 1 |
| 6 | 1 | 1 | 1 | 1 | 1 | 1 |
| 7 | 1 | 1 | 1 | 1 | 1 | 1 |

**Table 6.** Uniform error $E^N$ and uniform rate of convergence $\rho^N$ of the method for Example 2.

| | $N = 3 \times 2^4$ | $N = 3 \times 2^5$ | $N = 3 \times 2^6$ | $N = 3 \times 2^7$ | $N = 3 \times 2^8$ | $N = 3 \times 2^9$ |
|---|---|---|---|---|---|---|
| $E^N$ | 4.06E−02 | 5.32E−03 | 6.03E−04 | 6.25E−05 | 6.09E−06 | 5.66E−07 |
| $\rho^N$ | 3.85 | 3.95 | 3.98 | 3.99 | 4.00 | |

error $E^N$ and uniform rate $\rho^N$ of the proposed algorithm. It is clear from the table that the proposed algorithm gives almost fourth order accurate results.

**Example 2.** Consider the following test problem

$$\begin{cases} -\varepsilon_1 u_1'' + u_1 - 0.5u_2 = f_1, \\ -\varepsilon_2 u_2'' - 2u_1 + 4u_2 = f_2, \end{cases}$$

with the right hand side and the boundary conditions such that the exact solution $\boldsymbol{u} = (u_1, u_2)^T$ is

$$u_1 = g_1(x) + g_2(x) - x + x^2 + \cos^2(\pi x), \quad u_2 = g_1(x) - g_2(x) + \sin(\pi x),$$

where
$$g_1(x) = \frac{\exp\left(-x/\sqrt{\varepsilon_1}\right) + \exp\left(-(1-x)/\sqrt{\varepsilon_1}\right)}{1 + \exp\left(-1/\sqrt{\varepsilon_1}\right)}$$
and
$$g_2(x) = \frac{\exp\left(-2x/\sqrt{\varepsilon_2}\right) + \exp\left(-2(1-x)/\sqrt{\varepsilon_2}\right)}{1 + \exp\left(-2/\sqrt{\varepsilon_2}\right)}.$$

As the exact solution of this test problem is known, for fixed value of $\varepsilon_1$, $\varepsilon_2$ and $N$, we compute
$$E^N_{\varepsilon_1,\varepsilon_2} := \|\boldsymbol{u} - \boldsymbol{U}\|_{\overline{\Omega}^N}.$$

We then compute uniform error $E^N$ and uniform rate of convergence $\rho^N$ as above. The values of errors $E^N_{\varepsilon_1,\varepsilon_2}$ for fixed $\varepsilon_1 = 10^{-7}$ and different values of $\varepsilon_2$ and $N$ are given in Table 4. The number of iterations required by the proposed algorithm for fixed $\varepsilon_1 = 10^{-7}$ and different values of $\varepsilon_2$ and $N$ are shown in Table 5, and the behaviour is similar to the previous test problem. The values of uniform error $E^N$ and uniform rate $\rho^N$ of the proposed algorithm are given in Table 6. From this table one can observe almost fourth order accuracy of the proposed algorithm. In summary, the numerical results presented in this section show that the proposed algorithm is almost fourth order accurate. It is our intention in future to prove the same theoretically.

# References

1. Boglaev, I.: Domain decomposition for a parabolic convection-diffusion problem. Numer. Methods Partial Differ. Equ. **22**, 1361–1378 (2006)
2. Clavero, C., Gracia, J.: High order methods for elliptic and time dependent reaction diffusion singularly perturbed problems. Appl. Math. Comput. **168**, 1109–1127 (2005)
3. Kopteva, N., Pickett, M., Purtill, H.: A robust overlapping Schwarz method for a singularly perturbed semilinear reaction-diffusion problem with multiple solutions. Int. J. Numer. Anal. Model. **6**, 680–695 (2009)
4. Kumar, S., Kumar, M.: An analysis of overlapping domain decomposition methods for singularly perturbed reaction-diffusion problems. J. Comput. Appl. Math. **281**, 250–262 (2015)
5. Kumar, S., Rao, S.C.S.: A robust overlapping Schwarz domain decomposition algorithm for time-dependent singularly perturbed reaction-diffusion problems. J. Comput. Appl. Math. **261**, 127–138 (2014)
6. Linß, T.: Layer-Adapted Meshes for Reaction-Convection-diffusion Problems. LNM, vol. 1985. Springer, Berlin (2010). https://doi.org/10.1007/978-3-642-05134-0
7. MacMullen, H., Miller, J.J.H., O'Riordan, E., Shishkin, G.I.: A second-order parameter-uniform overlapping Schwarz method for reaction-diffusion problems with boundary layers. J. Comput. Appl. Math. **130**, 231–244 (2001)
8. Madden, N., Stynes, M.: A uniformly convergent numerical method for a coupled system of two singularly perturbed linear reaction-diffusion problems. IMA J. Numer. Anal. **23**, 627–644 (2003)

9. Quarternoni, A., Valli, A.: Domain Decomposition Methods for Partial Differential Equations. Numerical Mathematics and Scientific Computation. Oxford University Press, New York (1999)
10. Rao, S.C.S., Kumar, S.: An almost fourth order uniformly convergent domain decomposition method for a coupled system of singularly perturbed reaction-diffusion equations. J. Comput. Appl. Math. **235**, 3342–3354 (2011)
11. Rao, S.C.S., Kumar, S.: Robust high order convergence of an overlapping Schwarz method for singularly perturbed semilinear reaction-diffusion problems. J. Comput. Math. **31**, 509–521 (2013)
12. Stephens, M., Madden, N.: A parameter-uniform Schwarz method for a coupled system of reaction-diffusion equations. J. Comput. Appl. Math. **230**, 360–370 (2009)
13. Stephens, M., Madden, N.: A Schwarz technique for a system of reaction diffusion equations with differing parameters. In: Hegarty, A., Kopteva, N., O'Riordan, E., Stynes, M. (eds.) BAIL 2008 - Boundary and Interior Layers. LNCSE, vol. 69, pp. 247–255. Springer, Berlin, Heidelberg (2009). https://doi.org/10.1007/978-3-642-00605-0_20
14. Stynes, M., Roos, H.: The midpoint upwind scheme. Appl. Numer. Math. **23**, 361–374 (1997)
15. Toselli, A., Widlund, O.: Domain Decomposition Methods - Algorithms and Theory. SSCM. Springer, Heidelberg (2010)

# Numerical Experiments for Some Markov Models for Solving Boundary Value Problems

Alexander S. Sipin[1]([✉])[iD] and Alexander I. Zeifman[2][iD]

[1] Vologda State University, Vologda, Russia
cac1909@mail.ru
[2] Vologda State University, IPI FRC CSC RAS; VolSC RAS, Vologda, Russia
zai@uni-vologda.ru

**Abstract.** The main purpose of this work is the analysis of some stochastic algorithms to determine values of harmonic functions at points of a bounded domain of Euclidean space. To solve the Dirichlet problem we use a Random Walk on Spheres algorithm. The Neumann problem is solved by means of integral equations of potential theory.

We compare Monte Carlo and quasi-Monte Carlo versions of these algorithms numerically. The desired value of the harmonic function is represented as the sum of a series of integrals on hypercubes whose dimension grows. Therefore, the asymptotic formulas for discrepancy cannot be used for estimation of the error of quasi-Monte Carlo algorithm. New results are obtained about the influence of the smoothness of the domain boundary on the accuracy of calculations.

**Keywords:** Boundary value problem · quasi-Monte Carlo method · Monte Carlo method

## 1 Introduction

Boundary value problems for partial differential equations are mathematical models of many real processes studied in the natural and technical Sciences. Numerical methods for solving boundary value problems are diverse and intensively developed. Along with the finite difference method, variation and projection methods, Monte Carlo methods are successfully applied to solve boundary value problems (see, for example, [1,2]). The numerical algorithms considered in the paper are based on the representation of the solution of the boundary value problem in the form of a mathematical expectation of a random variable determined on the trajectories of a random walk. We use a random walk on spheres to solve the Dirichlet problem and a random walk on the boundary to solve the Neumann problem. To simplify the formulas, we consider only three-dimensional problems for the Laplace equation.

© Springer Nature Switzerland AG 2019
I. Dimov et al. (Eds.): FDM 2018, LNCS 11386, pp. 493–500, 2019.
https://doi.org/10.1007/978-3-030-11539-5_57

## 2   Mean Value Operator Iterations

Let $\mathcal{D}$ be a bounded domain in $R^3$, $\Gamma$ be a boundary of $\mathcal{D}$. Let $u(x)$ be a harmonic function continuous in $\overline{\mathcal{D}}$. The distance from the point $x \in \mathcal{D}$ to the boundary $\Gamma$ we denote by $d(x)$. A random vector uniformly distributed on a sphere of radius 1 centered at zero denote by $\omega$. Let $\omega(1), \omega(2), \ldots$ be a sequence independent random vectors with such a distribution.

Random Walk on Spheres process started in point $x \in \mathcal{D}$ is defined by formulas

$$x(0) = x, \quad x(k+1) = x(k) + d(x(k))\omega(k+1), \quad k = 0, 1, 2, \ldots \quad (1)$$

For any continuous function $v(x)$ we define the Mean Value Operator by the formula

$$Kv(x) = Ev(x + d(x)\omega), \quad x \in \mathcal{D}, \quad (2)$$

where $E$ is a symbol of mathematical expectation of a random variable.

Obviously, the following equalities holds for any harmonic function $u(x)$ and any integer $m$:

$$u(x) = K^m u(x) = Eu(x(m)) \quad (3)$$

Let $\alpha = (\alpha_1, \alpha_2, \ldots, \alpha_{2m})$ be a point uniformly distributed in the $2m$-dimensional unit cube $\overline{I}^{2m} = [0; 1]^{2m}$. Using the next formulas for modeling the vector $\omega(i)$ $(i = 1, 2, \ldots, n)$,

$$\omega_1(i) = 2\alpha_{2i-1} - 1,$$
$$\omega_2(i) = \sqrt{1 - \omega_1^2(i)} \cos(2\pi\alpha_{2i}), \quad (4)$$
$$\omega_3(i) = \sqrt{1 - \omega_1^2(i)} \sin(2\pi\alpha_{2i}),$$

we obtain a representation $u(x)$ in the form of an integral over $2m$-dimensional unit cube $\overline{I}^{2m} = [0; 1]^{2m}$. The integral can be calculated by both Monte Carlo and quasi-Monte Carlo methods. You can use this integral to compare the efficiency of methods. It's a pretty complicated test.

We use multiplicative congruential method to generate pseudorandom points $\alpha(n) \in \overline{I}^{2m}$ and we use Halton points $\alpha^H(n) \in \overline{I}^{2m}$ as quasi-random points. Exactly, for any integer $n \geq 0$

$$\alpha^H(n) = (\phi_{b_1}(n), \phi_{b_2}(n), \ldots \phi_{b_{2m}}(n)), \quad (5)$$

where $\phi_{b_i}(n)$ for $1 \leq i \leq 2m$ is the radical-inverse function [3] in base $b_i$ and $b_i$ is the element number $i$ in a sequence of primes $2, 3, 5, 7, \ldots$ Both methods use the formula

$$u(x) \approx \frac{1}{N} \sum_{n=1}^{N} u(x^{(n)}(m)), \quad (6)$$

where $x^{(n)}(m)$ is the last point for $n$–th trajectory of the Random Walk on Spheres process. It is a function of the point $\alpha(n) \in \overline{I}^{2m}$ or the point $\alpha^H(n) \in \overline{I}^{2m}$.

The results of calculations by formula (6) for two harmonic functions in a cube $[0, 10]^3$ are given in Tables 1 and 2. Table 1 shows the results of the deviation of the found approximate values of the harmonic function $u = 3x_1x_2^2 - x_1^3 + 8x_3$ from its exact value $u(x) = 205.472$ at the point $x = (1.6, 5.4, 8.7)$. Even for $N = 10000$, the Monte Carlo method gives an error of less than 2.2% for all values of $m$. For $N = 10000$, the quasi-Monte Carlo method yields an error of less than 1% for $m \leq 10$ and an error of 4% to 21% for $50 \leq m \leq 250$. It is seen that starting from the dimension of the integral 20 and the sample size $N \leq 10^6$ Monte Carlo method gives better results.

**Table 1.** Deviation of Mean Value Operator iterations for the function $u = 3x_1x_2^2 - x_1^3 + 8x_3$ from the exact value at the point $x = (1.6, 5.4, 8.7)$. Monte Carlo (MC) and quasi-Monte Carlo (QMC) methods.

| m\N | $10^4$ MC | $10^4$ QMC | $10^5$ MC | $10^5$ QMC | $10^6$ MC | $10^6$ QMC | $10^7$ MC | $10^7$ QMC |
|---|---|---|---|---|---|---|---|---|
| 1 | −0.244 | 0.011 | 0.223 | 0.001 | 0.086 | 0 | 0.040 | 0 |
| 10 | −4.559 | −1.661 | −1.000 | 0.076 | 0.028 | 0.118 | −0.003 | 0.010 |
| 50 | −2.463 | 9.781 | −1.679 | 2.375 | −0.179 | 0.406 | 0.071 | 0.027 |
| 100 | −0.179 | 17.631 | −0.847 | 3.367 | 0.047 | 0.524 | 0.118 | 0.055 |
| 150 | −1.710 | 26.884 | −0.183 | 4.298 | 0.069 | 0.611 | 0.045 | 0.064 |
| 200 | −1.592 | 36.482 | 0.349 | 5.275 | 0.241 | 0.708 | 0.176 | 0.074 |
| 250 | −0.403 | 43.434 | −3.184 | 6.016 | 0.001 | 0.782 | 0.078 | 0.082 |

Similar results are obtained in Table 2 for the value $u(x) = 1.715$ of the function $u = 1/\sqrt{(x_1 - 10.1)^2 + x_2^2 + x_3^2}$ at the point $x = (0.2, 0.4, 0.3)$. Both methods yield similar results for $m \geq 50$ and $N \geq 10^6$.

**Table 2.** Deviation of Mean Value Operator iterations for the function $u = 1/\sqrt{(x_1 + 0.1)^2 + x_2^2 + x_3^2}$ from the exact value at the point $x = (0.2, 0.4, 0.3)$. Monte Carlo (MC) and quasi-Monte Carlo (QMC) methods.

| m\N | $10^4$ MC | $10^4$ QMC | $10^5$ MC | $10^5$ QMC | $10^6$ MC | $10^6$ QMC | $10^7$ MC | $10^7$ QMC |
|---|---|---|---|---|---|---|---|---|
| 1 | −0.0030 | 0.0000 | 0.0003 | 0.0000 | −0.0002 | 0.0000 | −0.0001 | 0.0000 |
| 10 | −0.0010 | 0.0004 | −0.0015 | −0.0016 | −0.0005 | −0.0003 | −0.0004 | 0.0000 |
| 50 | 0.0008 | −0.0132 | 0.0018 | −0.0046 | −0.0004 | −0.0002 | −0.0003 | −0.0001 |
| 100 | 0.0005 | −0.0143 | −0.0020 | −0.0054 | −0.0003 | −0.0004 | 0.0001 | −0.0002 |
| 150 | −0.0051 | −0.0150 | −0.0015 | −0.0055 | −0.0003 | −0.0004 | 0.0001 | −0.0002 |
| 200 | −0.0104 | −0.0156 | −0.0039 | −0.0055 | −0.0012 | −0.0004 | −0.0001 | −0.0002 |
| 250 | −0.0042 | −0.0166 | 0.0025 | −0.0056 | −0.0003 | −0.0004 | 0.0001 | 0.0000 |

To study the effect of the smoothness of the domain boundary on the accuracy of quasi-Monte Carlo calculations, a ball of radius 10 centered at zero is considered. The value $u(x) = 0.96674$ at the point $x = (9.2, 0.1, 0.5)$ was calculated for the function $u = 1/\sqrt{(x_1 - 10.1)^2 + x_2^2 + x_3^2}$. The results of calculations are presented in the Table 3. The accuracy of quasi-Monte Carlo algorithm has increased. Already at $N = 10^5$ and $m = 100$ the function $u(x)$ is calculated with the same accuracy by both methods.

**Table 3.** Deviation of Mean Value Operator iterations for the function $u = 1/\sqrt{(x_1 - 10.1)^2 + x_2^2 + x_3^2}$ from the exact value at the point $x = (9.2, 0.1, 0.5)$. Monte Carlo (MC) and quasi-Monte Carlo (QMC) methods.

| m\N | $10^4$ MC | $10^4$ QMC | $10^5$ MC | $10^5$ QMC | $10^6$ MC | $10^6$ QMC | $10^7$ MC | $10^7$ QMC |
|---|---|---|---|---|---|---|---|---|
| 1 | 0.0021 | 0.0002 | −0.0018 | 0.0000 | 0.0002 | 0.0000 | −0.0001 | 0.0000 |
| 10 | −0.0113 | 0.0084 | 0.0030 | 0.0028 | 0.0010 | 0.0000 | 0.0002 | 0.0000 |
| 50 | 0.0087 | −0.0196 | 0.0037 | −0.0005 | −0.0004 | −0.0002 | −0.0007 | 0.0002 |
| 100 | 0.0080 | −0.0852 | −0.0012 | −0.0071 | −0.0009 | −0.0010 | −0.0007 | 0.0001 |
| 150 | −0.0150 | −0.1792 | 0.0029 | −0.0165 | −0.0011 | −0.0020 | 0.0000 | 0.0000 |
| 200 | −0.0129 | −0.3015 | 0.0018 | −0.0286 | 0.0004 | −0.0032 | −0.0001 | −0.0001 |
| 250 | 0.0032 | −0.4198 | 0.0000 | −0.0410 | −0.0012 | −0.0044 | 0.0000 | −0.0003 |

## 2.1   The Dirichlet Problem for Harmonic Function

To determine the harmonic function in the domain $\mathcal{D}$ using its values on the boundary, we apply Random Walk on Spheres algorithm. Let $\varepsilon > 0$ and $\Gamma_\varepsilon$ be an $\varepsilon-$ neighborhood of the boundary $\Gamma$. It is known that Random Walk on Spheres $x(k)$ converges to the boundary $\Gamma$ with probability 1. Let $\tau = \min(k : x(k) \in \Gamma_\varepsilon)$ be the first hitting time of the process $x(k)$ into $\Gamma_\varepsilon$. Then the equality $u(x) = Eu(x(\tau))$ is true. Hence, we can use the formula

$$u(x) \approx \frac{1}{N} \sum_{n=1}^{N} u(x^{(n)}(\tau)) \tag{7}$$

to calculate $u(x)$. Here $x^{(n)}(\tau)$ is the first point lying in $\Gamma_\varepsilon$ for $n$-th trajectory of the Random Walk on Spheres. The results of calculations by formula (7) for two harmonic functions in a cube $[0, 10]^3$ are given in Tables 4 and 5. In both examples the parameter $\varepsilon$ has been chosen equal to 0.001. In the quasi-Monte Carlo method, the value $\tau_m = \min(\tau, m)$  $(m = 1500)$ was used instead of $\tau$. The %–column shows the relative error of the calculated approximate value of the function. Variables $L$ and $QL$ denote the average length of the random walk trajectory. The $Lmax$ and $QLmax$ variables denote the maximum length of the random walk trajectory. Again Monte Carlo results are better than quasi-Monte Carlo results when $N < 10^7$.

**Table 4.** Random Walk on Spheres. Deviation of Mean Value for the function $u = 1/\sqrt{(x_1 + 0.1)^2 + x_2^2 + x_3^2}$ from the exact value $u(x) = 0.41922$ at the point $x = (1.1, 0.5, 2)$. Monte Carlo (MC) and quasi-Monte Carlo (QMC) methods.

| N | MC | % | QMC | % | L | QL | Lmax | QLmax |
|---|---|---|---|---|---|---|---|---|
| $10^4$ | 0.00196 | 0.47 | −0.00376 | 0.90 | 22 | 23 | 108 | 105 |
| $10^5$ | −0.00024 | 0.06 | −0.00120 | 0.29 | 22 | 22 | 140 | 194 |
| $10^6$ | −0.00003 | 0.01 | −0.00021 | 0.05 | 22 | 22 | 201 | 210 |
| $10^7$ | −0.00005 | 0.01 | 0.00001 | 0.002 | 22 | 22 | 201 | 214 |

**Table 5.** Random Walk on Spheres. Deviation of Mean Value for the function $u = 3x_1x_2^2 - x_1^3 + 8x_3$ from the exact value $u(x) = 478$ at the point $x = (3, 7, 8)$. Monte Carlo (MC) and quasi-Monte Carlo (QMC) methods.

| N | MC | % | QMC | % | L | QL | Lmax | QLmax |
|---|---|---|---|---|---|---|---|---|
| $10^4$ | 0.28413 | 0.06 | 13,90295 | 2.91 | 25 | 24 | 122 | 1257 |
| $10^5$ | −0.00024 | 0.29 | 2,75519 | 0.58 | 25 | 25 | 160 | 1257 |
| $10^6$ | −0.17767 | 0.04 | 0,50839 | 0.11 | 25 | 25 | 178 | 1257 |
| $10^7$ | 0.09103 | 0.02 | 0,12939 | 0.03 | 25 | 25 | 262 | 1257 |
| $10^8$ | −0.15897 | 0.03 | 0,02750 | 0.01 | 25 | 25 | 262 | 1257 |

## 3    The Neumann Boundary Value Problem

Now consider the Neumann problem for the Helmholtz equation

$$- \Delta u(x) + c^2 u(x) = f(x), \quad x \in \mathcal{D}; \quad \frac{\partial u(x)}{\partial n} = \varphi(x), \quad x \in \Gamma, \quad (8)$$

where $c > 0$ is a constant, and $n = n_x$ is the external normal to the domain $\mathcal{D}$ at the point $x \in \Gamma$.

We briefly describe an algorithm for the Monte Carlo method for the boundary value problem (8). Details can be found in [2].

### 3.1    Probabilistic Representation of the Solution of the Neumann Problem

For the classical solution of the problem (7) integral representation of the solution is valid:

$$u(x) = \frac{1}{4\pi} \int_\Gamma \frac{(1 + cr)e^{-cr} \cos \varphi_{xy} u(y)}{r^2} d_y S$$

$$+ \frac{1}{4\pi} \int_\mathcal{D} \frac{e^{-cr}}{r} f(y) dy + \frac{1}{4\pi} \int_\Gamma \frac{e^{-cr}}{r} \varphi(y) d_y S, \quad x \in \mathcal{D}, \quad (9)$$

where $\cos \varphi_{xy} = (n_y, y - x)/r$ and $r = |x - y|$ is the distance between points $x$ and $y$. Here $(a, b)$ denotes the scalar product of vectors $a$ and $b$.

For a convex domain, the Gaussian kernel $p(x,y) = \cos\varphi_{xy}/(4\pi r^2)$ is the probability density of the point $Y(x,\omega) \in \Gamma$, chosen in a random direction $\omega$ from the point $x$. Denote $|x - Y|$ by $r(x,\omega)$. It is easy to see that

$$\frac{1}{4\pi}\int_{\mathcal{D}}\frac{e^{-cr}}{r}f(y)dy = E\int_0^{r(x,\omega)}e^{-c\rho}\rho f(x+\rho\omega)d\rho, \tag{10}$$

and

$$\frac{1}{4\pi}\int_{\Gamma}\frac{(1+cr)e^{-cr}\cos\varphi_{xy}u(y)}{r^2}d_yS = E(1+cr(x,\omega))e^{-cr(x,\omega)}u(Y(x,\omega)). \tag{11}$$

Let a random variable $\gamma$ has the gamma probability density $p(t) = te^{-t}$ $(t>0)$, and let $\eta(x)$ be an unbiased estimator to last integral in (9). Then the formula (9) can be written in probabilistic form

$$u(x) = E\left((1+cr(x,\omega))e^{-cr(x,\omega)}u(Y(x,\omega))\chi(x,\omega)\right)$$

$$+E\left((1-\chi(x,\omega))\frac{f(x+\gamma r(x,\omega)\omega)}{c^2}+\eta(x)\right), \tag{12}$$

where $\chi(x,\omega)$ is the indicator function for the interval $(cr(x,\omega);\infty)$. The formula (13) is also valid for the boundary point $x$ if the vector $\omega$ of the random direction in space is replaced by the vector of the random direction in the half-space $((n_x,\omega)\leq 0)$.

## 3.2 Random Walk on the Boundary

Let's define a random walk on the boundary recursively. Let $x(0) = x$, where $x$ be a fixed point $x \in \overline{\mathcal{D}}$. Let $\omega(1)$ be a random vector, if $x \in \mathcal{D}$, and $\omega(1)$ be a random vector in the half-space $((n_x,\omega)\leq 0)$, if $x \in \Gamma$. Then $x(1) = Y(x(0),\omega(1))$. If the trajectory $x(0),x(1),...,x(k)$ of the random walk up to the moment $k$ is constructed, then $x(k+1) = Y(x(k),\omega(k+1))$. Here $\omega(k+1)$ is a random vector in the half-space $((n_{x(k)},\omega)\leq 0)$.

Let $\gamma(1),\gamma(2),...$ be a sequence of independent random variables with probability density $p(t) = te^{-t}$ $(t>0)$, which is also independent of the random walk trajectory. The random walk trajectory is terminated at the random moment $\tau = k$, when the condition $\gamma(k) \leq cr(x(k-1),\omega(k))$ is satisfied for the first time. It is proved in [2] that $E\tau < \infty$. Thus, an unbiased estimator $\xi(x)$ for $u(x)$ is obtained

$$\xi(x) = \sum_{k=0}^{\tau-1}\eta(x(k)) + \frac{f(x(\tau-1)+\gamma(\tau)r(x(\tau-1),\omega(\tau))\omega(\tau))}{c^2}. \tag{13}$$

The random variable $\gamma$ can be obtained as $\gamma = -\ln(\alpha_1) - \ln(\alpha_2)$, where $\alpha_1,\alpha_2$ are uniformly distributed on $[0;1]$ and independent. A random vector in half-space can be obtained from a vector $\omega$ in space by replacing with the $-\omega$, if $(n_x,\omega)\geq 0$.

**The Neumann Problem for the Ball.** Let's discuss the implementation of the algorithm for the ball $(x, x) \leq R^2$. In this case, the unbiased estimator $\eta(x)$ is constructing quite simply.

If $x \in \Gamma$, then $|x - y| = 2R \cos \varphi_{xy}$. Hence,

$$\eta(x) = 2Re^{-cr(x,\omega)}\varphi(Y(x,\omega)),$$

where $\omega$ is random vector in half-space.

If $x$ is close to the center of the ball, for example $|x| < R/2$, you can use estimator

$$\eta(x) = R^2 e^{-cr(x,\omega)}\varphi(Y(x,\omega))/r(x,\omega),$$

where $\omega$ is random vector in space.

Finally, if $R/2 < |x| < R$, then we define $x_0 = (R/|x|)x$. Next we get the random vector $\omega$, which lies in a half-space, which defined by inequality $(n_{x_0}, \omega)$. Now,

$$\eta(x) = Re^{-c\rho}r(x_0,\omega)\varphi(Y(x_0,\omega))/\rho,$$

where $\rho = |x - Y(x_0,\omega)|$.

*Numerical Results.* We use Halton points as a quasi-random points. Each step for quasi-random variant of the Random Walk on the boundary requires six Halton numbers. They built due to six new primes. If the trajectory number of the quasi-random walk is "small" (less than 10), the trajectory does not break. Therefore, we used quasi-random vectors with numbers greater than 100.

Table 6 gives the results of calculations by formula (13) with parameters $c = 2$ and $R = 2$. The %–column shows the relative error of the calculated approximate value of the function. Variables $L$ and $QL$ denote the average length of the random walk trajectory. The $Lmax$ and $QLmax$ variables denote the maximum length of the random walk trajectory. Again Monte Carlo results are better than quasi-Monte Carlo results when $N < 10^7$.

**Table 6.** The Neumann problem for the ball. Deviation of Mean Value for the function $u = x_1^2 + x_2^2 + x_3^2$ from the exact value $u(x) = 3.5825$ at the point $x = (0.15, 1.6, 1)$. Monte Carlo (MC) and quasi-Monte Carlo (QMC) methods.

| N | MC | % | QMC | % | L | QL | Lmax | QLmax |
|---|---|---|---|---|---|---|---|---|
| $10^4$ | −0.02120 | 0.59 | −0.01850 | 0.52 | 2.72 | 2.0 | 8 | 9 |
| $10^5$ | 0.00038 | 0.01 | 0.00760 | 0.21 | 2.7 | 2.0 | 9 | 9 |
| $10^6$ | −0.00161 | 0.05 | 0.00169 | 0.05 | 2.7 | 2.0 | 11 | 12 |
| $10^7$ | −0.00128 | 0.036 | −0.00012 | 0.003 | 2.7 | 2.0 | 13 | 15 |

Table 7 gives the results of calculations by formula (13) with parameters $c = 3$ and $R = 2.5$. The names of the columns are the same as in the Table 6. Monte Carlo results are better than quasi-Monte Carlo results when $10^5 < N < 10^7$.

**Table 7.** The Neumann problem for the ball. Deviation of Mean Value for the function $u = e^{x_1 + 2x_2 + 3x_3}$ from the exact value $u(x) = 403.42879$ at the point $x = (0, 0, 2)$. Monte Carlo (MC) and quasi-Monte Carlo (QMC) methods.

| N | MC | % | QMC | % | L | QL | Lmax | QLmax |
|---|---|---|---|---|---|---|---|---|
| $10^4$ | −56.96896 | 14.12 | −14.61933 | 3.62 | 2.2 | 1.5 | 5 | 5 |
| $10^5$ | 25.16671 | 6.24 | −7.32352 | 1.82 | 2.2 | 1.5 | 7 | 6 |
| $10^6$ | 0.19850 | 0.05 | 0.47395 | 0.12 | 2.7 | 1.5 | 8 | 7 |
| $10^7$ | 0.80401 | 0.20 | 0.89175 | 0.22 | 2.7 | 1.5 | 9 | 9 |
| $10^8$ | −0.34090 | 0.08 | 0.40516 | 0.10 | 2.7 | 1.5 | 9 | 10 |

## 4  Summary

The results of computational experiments allow us to draw the following conclusions:

1. The quasi-Monte Carlo method can be used to compute the sum of the Neumann series for the integral equations of potential theory (Neumann boundary value problem.)
2. The quasi-Monte Carlo method can be used to solve integral equations with a sub-stochastic kernel (Dirichlet boundary value problem).
3. When calculating the sum of the Neumann series, the Monte Carlo method works better than the quasi-Monte Carlo, when $N \leq 10^7$.
4. In practice, it is useful to first perform calculations using the Monte Carlo method for $N = 10^6$ and estimate the error. Subsequent calculations should be made by quasi-Monte Carlo method.

**Acknowledgments.** Supported by the Russian Foundation for Basic Research, projects No. 17-01-00267, 18-47-350002.

## References

1. Sabelfeld, K.K.: Monte Carlo Methods in Boundary Value Problems. Springer, Heidelberg (1991)
2. Ermakov, S.M., Nekrutkin, V.V., Sipin, A.S.: Random Processes for Classical Equations of Mathematical Physics. Kluwer Academic Publishers, Dordrecht (1989)
3. Niederreiter, H.: Random Number Generation and Quasi-Monte Carlo Methods. CBMS-NSF Regional Conference Series in Applied Mathematics, vol. 63. Society for Industrial and Applied Mathematics, Philadelphia (1992)

# Numerical Simulation of Deformations of Basalt Roving

Anastasia V. Sivtseva[1] and Petr V. Sivtsev[2(✉)]

[1] Larionov Institute of the Physical-Technical Problems of the North of the Siberian
Branch of the RAS, 1, Oktyabrskaya str., 677891 Yakutsk, Russia
[2] Ammosov North-Eastern Federal University, 58, Belinskogo,
677000 Yakutsk, Russia
sivkapetr@mail.ru

**Abstract.** In the modern world, there is a growing need for composite
materials that satisfy such rigid, often conflicting requirements as minimizing the weight of structures, maximizing strength, rigidity, reliability,
durability when working under severe loading conditions, low temperatures and in corrosive environments. Compared to traditional materials,
fiber-based composites (glass-, basalt-) have a number of advantages -
corrosion resistance, chemical inertness, low thermal conductivity, high
specific mechanical characteristics, low weight, durability, low installa-
tion costs. Basalt fibers are much cheaper than S-glass fibers.

The Republic of Sakha (Yakutia) is a region with extreme climatic
conditions, rich in mineral resources, but remote from industrial centers
of the Russian Federation. Therefore, special requirements are imposed
on building structural materials used in the Far North conditions, such
as frost resistance, increased toughness, wear resistance, etc. A significant problem in the construction in permafrost conditions and repeated
freeze/thaw cycles is the provision of durability of reinforced foundation
structures and piles, working in conditions that promote the acceleration
of corrosion of steel reinforcement and concrete.

In this work we consider the elasticity equations, which describe stress-
strain basalt roving. For numerical solution we approximate our system
using finite elements method. As the model problem we consider deformations of basalt roving subjected to three point bending test to define
dependence of the strength of the basalt roving on interlacing of fibers.
The results of numerical simulation of the 3D problem and comparison
with experimental data are presented.

## 1 Introduction

In recent decades, interest in the study of composite reinforcement by basalt
plastic for the subsequent replacement of metal rods in load-bearing reinforced
concrete structures in construction has been increased. One of the innovative
trends in the development of the North of Russia are technologies related to the
production of basalt continuous fiber (BCF) and composite materials based on

© Springer Nature Switzerland AG 2019
I. Dimov et al. (Eds.): FDM 2018, LNCS 11386, pp. 501–508, 2019.
https://doi.org/10.1007/978-3-030-11539-5_58

it. The presence of huge stocks of raw materials and the relatively low cost of the fibers produced from it makes it possible to consider their prospects in economic terms.

According to the basic physico-chemical indices, properties of basalt fibers are close to one of glass fibers, and their thermal and chemical resistance, durability, and ecological safety significantly exceed them [1].

The volume of production and application of fiber-reinforced polymer composites around the world is continuously expanding. This is due to their high physical and chemical properties and operational characteristics such as corrosion resistance, frost resistance, low thermal conductivity and others [2,3]. The disadvantages of polymer composite reinforcement are the absence of plastic deformation in tension, the impossibility of bending and welding of reinforcing elements [4].

The main feature of reinforced plastics is a pronounced anisotropy of their mechanical properties, determined by the orientation of the fibers in the matrix in one or several preferential directions [5]. The choice of orientation is determined by the distribution of stresses in the structural elements. Thus, for the effective implementation of elastic-strength properties in designs, a solution to a complex of problems is required, related not only to the choice of materials, but also to the design of the product. In the literature, there is no problem with the study of the properties of braided reinforcement. Thus, the purpose of this paper was to study the flexural strength and the modeling of stresses arising in a unidirectional and braided reinforcement.

The paper is oriented on mathematical modeling of deformations of pipes of different structures created from basalt roving. We describe computational algorithm of solution for linear elasticity problem. This algorithm is based on finite-element approximation of displacement by space [6–8]. In order to calculate deformations of composite structure with thousands of roving, method of numerical homogenization is used [9,10]. Numerical realization of method is performed on collection of free software FEniCS [11]. The results of numerical simulation of the 3D problem and comparison with experimental data are presented.

## 2   Problem Statement

Let us consider mathematical model describing stress-strain state of basalt composite pipe clamped between rails and press

$$\operatorname{div} \boldsymbol{\sigma} + \boldsymbol{b} = 0, \quad \boldsymbol{x} \in \Omega = \Omega_1 \cup \Omega_2, \tag{1}$$

where $\Omega_1$ is subdomain of basalt fibers, $\Omega_2$ is subdomain of binder, $\boldsymbol{\sigma}$ is stress tensor, $\boldsymbol{b} = (b_1, b_2, b_3)$ is vector of volumetric forces in point $\boldsymbol{x} = (x_1, x_2, x_3)$.

Equation (1) is supplemented by relations between displacement vector $\boldsymbol{u}$, stress tendor $\boldsymbol{\sigma}$ and strain tensor $\boldsymbol{\varepsilon}$. For convenience we use Voight notation

$$\boldsymbol{\varepsilon} = \begin{pmatrix} \varepsilon_{11} \\ \varepsilon_{22} \\ \varepsilon_{33} \\ \gamma_{12} \\ \gamma_{13} \\ \gamma_{23} \end{pmatrix} = \begin{pmatrix} \varepsilon_{11} \\ \varepsilon_{22} \\ \varepsilon_{33} \\ 2\varepsilon_{12} \\ 2\varepsilon_{13} \\ 2\varepsilon_{23} \end{pmatrix} = \begin{pmatrix} \frac{\partial u_1}{\partial x_1} \\ \frac{\partial u_2}{\partial x_2} \\ \frac{\partial u_3}{\partial x_3} \\ \frac{\partial u_2}{\partial x_1} + \frac{\partial u_1}{\partial x_2} \\ \frac{\partial u_1}{\partial x_3} + \frac{\partial u_3}{\partial x_1} \\ \frac{\partial u_2}{\partial x_3} + \frac{\partial u_3}{\partial x_2} \end{pmatrix}, \quad \boldsymbol{\sigma} = \begin{pmatrix} \sigma_{11} \\ \sigma_{22} \\ \sigma_{33} \\ \sigma_{12} \\ \sigma_{13} \\ \sigma_{23} \end{pmatrix} = C\boldsymbol{\varepsilon},$$

here $C$ is elastic tensor, which is represented as follows

$$C = \begin{pmatrix} C_{1111} & C_{1122} & C_{1133} & C_{1112} & C_{1113} & C_{1123} \\ C_{2211} & C_{2222} & C_{2233} & C_{2212} & C_{2213} & C_{2223} \\ C_{3311} & C_{3322} & C_{3333} & C_{3312} & C_{3313} & C_{3323} \\ C_{1211} & C_{1222} & C_{1233} & C_{1212} & C_{1213} & C_{1223} \\ C_{1311} & C_{1322} & C_{1333} & C_{1312} & C_{1313} & C_{1323} \\ C_{2311} & C_{2322} & C_{2333} & C_{2312} & C_{2313} & C_{2323} \end{pmatrix}.$$

In the case of isotropic body elastic tensor can be expressed by Lame parameters $\lambda, \mu$, which depend on domain and given as

$$\mu = \mu_i, \ \lambda = \lambda_i, \quad \boldsymbol{x} \in \Omega_i \quad i = 1, 2.$$

Lame parameters can be uniquely determined by Young modulus $E$ and Poisson coefficient $\nu$.

Equation of solid body is supplemented with boundary conditions, related with rail fixation and force of press. Rail fixation is defined by Dricihlet boundary condition:

$$\boldsymbol{u} = (0, 0, 0), \quad \boldsymbol{x} \in \Gamma_D, \tag{2}$$

and press force is defined by Neumann boundary condition

$$\boldsymbol{\sigma_n} = (0, 0, -P), \quad \boldsymbol{x} \in \Gamma_N. \tag{3}$$

To approximate the space, the finite element method is used. The original differential problem reduces to a discrete variational problem. The numerical implementation is performed using the FEniCS computational platform.

## 3    Investigation Object

Our objects of investigation are 10 composite pipes of two different types. The main difference of sample types is distribution of basalt fibers shown in Fig. 1. Fibers in samples of first type are located in parallel, while fibers in samples of second type form three string braid. Samples are made of roving BCF (2400 tex in one bundle, the thickness of one fiber is 13 microns). Basalt concentration for

**Fig. 1.** Sample types before and after addition of binder

**Table 1.** Composite materials elastic parameters

| Material | Young modulus $E, GPa$ | Poisson coefficient $\nu$ |
|---|---|---|
| Basalt fibre | 75 | 0.25 |
| Binder | 3.2 | 0.35 |

different sample types are equal. The thickness of the samples is 6 mm and is impregnated with a binder based on an orthophthalic polyester resin (Yarkopol-110) with Butanox M-50 hardener. (Fig. 1 photo of the finished samples). Elastic properties of basalt fibers and binder are presented in Table 1.

Trials for transverse three-point bending were performed on a universal electromechanical machine "Zwick Z600E" of the "Zwick Roell" firm, Germany, in the presence of the author.

Samples were installed on two supports and loaded with a pressure tip, the samples have the shape of a cylinder. The distance between the supports was set to 64 mm. The tests were carried out according to standart on a universal electromechanical machine "Zwick Z600E", equipped with sensors for measuring load and displacement, combined with a PC. Machine with installed sample is shown in Fig. 2.

## 4    Numerical Averaging of Coefficient

To calculate the effective coefficients, we use the method of numerical averaging over a representative square domain. The basic idea of numerical averaging is

**Fig. 2.** "Zwick Z600E" with installed sample

to find a homogeneous medium equivalent to the original composite material. In other words, it is necessary to find at what coefficients of the homogeneous medium it can approximately show the same behavior as the composite material. This means that for equal average values of the strain tensor we obtain equivalent averaged values of the stress tensor.

The following notation is used for the mean value of the function $\psi$:

$$\langle \psi \rangle = \frac{\int_\omega \psi \, d\boldsymbol{x}}{\int_\omega d\boldsymbol{x}},$$

where $\omega$ is representative domain. Thus, the previously described statement about the averaged coefficient can be written in the following form

$$\langle \boldsymbol{\sigma} \rangle = \langle \boldsymbol{C}\varepsilon \rangle = \boldsymbol{C}^* \langle \varepsilon \rangle,$$

where $\boldsymbol{C}^*$ is effective coefficient. To calculate the components of the effective coefficient, we consider six problems with the following boundary conditions in order to calculate corresponding components of the effective coefficient:

1. $\boldsymbol{u} = (x_1, 0, 0), \boldsymbol{x} \in \partial\omega,$     1. $C^*_{ij11} = \langle \sigma_{ij} \rangle,$
2. $\boldsymbol{u} = (0, x_2, 0), \boldsymbol{x} \in \partial\omega,$     2. $C^*_{ij22} = \langle \sigma_{ij} \rangle,$
3. $\boldsymbol{u} = (0, 0, x_3), \boldsymbol{x} \in \partial\omega,$     3. $C^*_{ij33} = \langle \sigma_{ij} \rangle.$
4. $\boldsymbol{u} = (0.5\, x_2, 0.5\, x_1, 0), \boldsymbol{x} \in \partial\omega,$     4. $C^*_{ij12} = \langle \sigma_{ij} \rangle,$
5. $\boldsymbol{u} = (0.5\, x_3, 0, 0.5\, x_1), \boldsymbol{x} \in \partial\omega,$     5. $C^*_{ij13} = \langle \sigma_{ij} \rangle,$
6. $\boldsymbol{u} = (0, 0.5\, x_3, 0.5\, x_2), \boldsymbol{x} \in \partial\omega.$     6. $C^*_{ij23} = \langle \sigma_{ij} \rangle.$
where $ij = 11, 22, 33, 12, 13, 23.$

Under these boundary conditions, the average values of the stress tensor determine the corresponding components of the effective coefficient.

To calculate effective parameters of braided sample we performed couple procedures of numerical averaging sequentially. First, we calculate effective parameters of parts for one segment of braid. Then we derive elastic parameters of sample itself. These procedures are schematicly represented in Fig. 3.

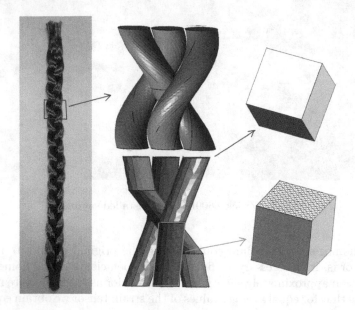

**Fig. 3.** Scheme for calculation of effective elastic parameters of braided sample

## 5    Comparative Analysis

In this section we compare calculated deformations with experimental data. Calculation of three point bending test made for force equal to $100N$. We compare vertical part of deformations since experimental machine provide dependence between applied force and vertical deformations.

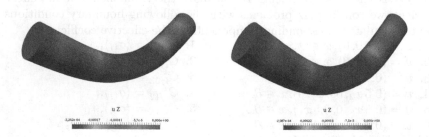

**Fig. 4.** Calculated distribution of vertical displacement for sample with parallel fiber location and two cases of braided sample (move multiplied by 50)

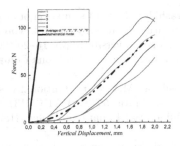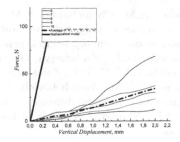

**Fig. 5.** Dependence between applied force and vertical deformations for different types of samples

Calculated distribution of displacement for sample with parallel fiber location and for braided sample are presented in Fig. 4. Experimental and calculated dependence between applied force and vertical deformations for different types of samples are shown in Fig. 5. In these graphs good qualitative agreement between model and experiment can be seen. For both experiment and model braided samples have larger deformations compared to samples with parallel rovings.

## 6    Conclusion

We made numerical solution for linear elasticity problem for composite pipe made of basalt roving. As result of comparison of numerical solutions and experimental data we conclude that there is good qualitative agreement. In the future it is necessary to conduct a more thorough investigation including non-linear models.

**Acknowledgments.** The research was supported by mega-grant of the Russian Federation Government (N 14.Y26.31.0013).

## References

1. Novickij, A.G., Efremov, M.V., Fedotov, G.B.: Issledovanie i sovershenstvovanie processov polucheniya nepreryvnogo bazal'tovogo volokna. Sb. dokl. V Vserossijskoj nauchno-prakticheskoj konferencii ≪Tekhnika i tekhnologiya proizvodstva teploizolyacionnyh materialov iz mineral'nogo syr'ya≫, pp. 12–21. CEHI ≪Himmash≫ (2005)
2. Shaludin, S.A.: Primenenie bazal'toplastikovoy i kompozitnoy armatury kak innovatsionno orientirovannyy instrument obespecheniya sotsial'no-ekonomicheskogo razvitiya stroitel'nogo kompleksa. Vestnik Moskovskogo gosudarstvennogo otkrytogo universiteta. Tekhnika i Tekhnologiya **2**(8), 59–63 (2012)
3. Kustikova, Yu.O., Rimshin, V.I.: Stressed-deformed state of basalt-plastic reinforcement in reinforced concrete structures. Promyshlennoe i grazhdanskoe stroitel'stvo [Ind. Civ. Eng.] **6**, 6–9 (2014)

4. Usachev, A.M., Chorochordin, A.M., Abdurashidov, M.M.: Estimation of mechanical properties of polymeric composition armature. Nauchniy Vestnik Voronezhskogo gosudarstvennogo arhitekturno-stroitelnogo universiteta, pp. 16–20 (2014)
5. Sidorenko, Yu.N., Kashcheeva, O.: Structural and Functional Fibrous Composite Materials, vol. 107. TSU, Tomsk (2006)
6. Afanaseva, N.M., Vabishchevich, P.N., Vasileva, M.V.: Unconditionally stable schemes for convection-diffusion problems. Russ. Math. **57**(3), 1–11 (2013)
7. Lui, S.H.: Numerical Analysis of Partial Differential Equations. Wiley, Hoboken (2012)
8. Kolesov, A.E., et al.: Numerical analysis of reinforced concrete deep beams. In: Dimov, I., Farago, I., Vulkov, L. (eds.) NAA 2016. LNCS, vol. 10187, pp. 414–421. Springer, Cham (2016). https://doi.org/10.1007/978-3-319-57099-0_46
9. Kari, S., Berger, H., Gabbert, U.: Numerical evaluation of effective material properties of randomly distributed short cylindrical fibre composites. Comput. Mater. Sci. **39**(1), 198–204 (2007)
10. Zakharov, P.E., Sivtsev, P.V.: Numerical calculation of the effective coefficient in the problem of linear elasticity of a composite material. Math. Notes NEFU **24**(2), 75–84 (2017)
11. Logg, A., Mardal, K.A., Wells, G.N.: Automated Solution of Differential Equations by the Finite Element Method. LNCSE. Springer, Heidelberg (2012). https://doi.org/10.1007/978-3-642-23099-8

# A Fractional Diffusion Model with Resetting

Ercília Sousa[1(✉)] and Amal K. Das[2]

[1] CMUC, Department of Mathematics, University of Coimbra, Coimbra, Portugal
ecs@mat.uc.pt
[2] Department of Physics, Dalhousie University, Halifax, Canada

**Abstract.** We consider a model that serves as a paradigm for a class of search strategies in which the searcher having explored its environment unsuccessfully for a while, returns to its initial position and begins a new search. The model describes the diffusive motion of a particle, performing a random walk with Lévy distributed jump lengths, which is interrupted at random times when the particle is reset to its initial position. A numerical method is proposed to determine the solutions of this diffusive problem with resetting. The influence of resetting on the solutions is analysed and physical quantities such as the pseudo second moment will be discussed.

## 1 Introduction

Search problems can occur in many different contexts leading to a variety of interesting processes. Recently, search problems with stochastic resetting of random searchers have been investigated [1–4,7]. These processes involve two consecutive steps. Firstly, the searcher goes out to search, and secondly, the searcher is drawn back to the starting point.

We consider a one dimensional resetting model, that describes the movement of the searcher in discrete time on a line, starting from an initial position $x_0$. At time step $t + dt$, the current location $x(t + dt)$ of the searcher is updated via the following stochastic rule: we have $x(t + dt) = x_0$ with probability $rdt$ or $x(t + dt) = x(t) + \eta(t)dt$ with probability $1 - rdt$, where $r$ is the resetting rate. These equations give the probability of a resetting event and the jump lengths described by $\eta(t)$ are independent and identically distributed random variables each drawn from a probability density function that can be a Gaussian white noise with mean zero [4] or a probability density function with a heavy tail [7]. In this work we assume we have a probability density function with a heavy tail. More specifically we consider the class of Lévy stable processes for which the characteristic function is given by $\psi(k) = \mathrm{e}^{-|k|^\alpha}$. The case $\alpha = 2$ corresponds to the Gaussian case described in [4], while the case $0 < \alpha < 2$ describes Lévy flights where the jumps are typically very large.

© Springer Nature Switzerland AG 2019
I. Dimov et al. (Eds.): FDM 2018, LNCS 11386, pp. 509–516, 2019.
https://doi.org/10.1007/978-3-030-11539-5_59

## 2   The Model

The class of Lévy stable processes described by the characteristic function $e^{-D|k|^\alpha}$, for $1 < \alpha \leq 2$, can be represented by the fractional diffusion equation [8,13], that is, the equation for the probability distribution $f(x,t)$, of finding the particle at position $x$ at time $t$, reads

$$\frac{\partial f(x,t)}{\partial t} = \frac{D}{2\Gamma(2-\alpha)|\cos(\pi\alpha/2)|}\frac{\partial^2}{\partial x^2}\int_{-\infty}^{\infty}\frac{f(x',t)}{|x'-x|^{\alpha-1}}dx' =: D\frac{\partial^\alpha f}{\partial|x|^\alpha}(x,t),$$
(1)

where $\Gamma(\cdot)$ is the Gamma function.

If additionally to the diffusive process we add the resetting events, similarly to what has been presented in [2], we arrive at the equation

$$\frac{\partial f(x,t)}{\partial t} = D\frac{\partial^\alpha f}{\partial|x|^\alpha}(x,t) - rf(x,t) + rf_0(x),$$
(2)

where $f_0(x)$ is the initial condition. The second term and third term on the right hand side accounts for the resetting events [2,9,12], denoting the negative probability flux $-rf(x,t)$ from each point $x$ and a corresponding positive probability flux into $x = x_0$.

Consider Eq. (2) and assume $\lim_{x\to\pm\infty} f(x,t) = 0$ and $f_0(x) = \delta(x)$. The resetting rate $r$ should be a positive quantity, on physical grounds. Let $f$ be a sufficiently smooth solution of (2). Then, the characteristic function is given by

$$\hat{f}(k,t) = e^{-(D|k|^\alpha+r)t} + \frac{r}{D|k|^\alpha+r}(1 - e^{-(D|k|^\alpha+r)t}).$$
(3)

From the inverse Fourier transform, we obtain the following solution of (2),

$$f(x,t) = e^{-rt}u_\alpha(x,t) + \int_0^t re^{-r(t-s)}u_\alpha(x,t-s)ds,$$
(4)

where $u_\alpha(x,t)$ is the solution of the fractional diffusion Eq. (1), which does not include the resetting terms.

The solution of the fractional diffusion Eq. (1) can be written in closed form, when the initial condition is $f_0(x) = \delta(x)$. It is given in terms of the Fox functions ([8], p. 27), that is,

$$u_\alpha(x,t) = \frac{1}{\alpha|x|}H_{2,2}^{1,1}\left[\frac{|x|}{(Dt)^{1/\alpha}}\bigg|\begin{matrix}(1,1/\alpha)\ (1,1/2)\\(1,1)\ \ \ (1,1/2)\end{matrix}\right].$$
(5)

It is easy to conclude that for $r > 0$ then $f(x,t) > 0$ for all $x,t$ when $u_\alpha$ is positive and for $r$ positive we have the sum of two positive quantities.

A more general equation can consider a space resetting rate $r(x)$ [2] and can be written as

$$\frac{\partial f}{\partial t}(x,t) = D\frac{\partial^\alpha f}{\partial|x|^\alpha}(x,t) - r(x)f(x,t) + f_0(x)\int_{\mathbb{R}}r(x')f(x',t)dx',$$
(6)

where now $r$ depends on $x$. For $r$ constant we recover (2) when $\int_{\mathbb{R}} f(x',t)dx' = 1$.

In the next section we derive a numerical method to solve the fractional diffusion equation with resetting terms (6) and analyze its efficiency. Then, in the last section, we discuss the influence of resetting in the moments and how it changes the character of the solution.

## 3   The Numerical Method

The equation under consideration can be written in the form

$$\frac{\partial f}{\partial t}(x,t) = \frac{D}{2|\cos(\alpha\pi/2)|}\left(\frac{\partial^\alpha f}{\partial x^\alpha} + \frac{\partial^\alpha f}{\partial(-x)^\alpha}\right)(x,t) - r(x)f(x,t) + s(x,t), \quad (7)$$

where $s(x,t) = f_0(x)\int_{\mathbb{R}} r(x')f(x',t)dx'$ and the fractional Riesz operator has been represented in terms of the left and right Riemann-Liouville derivatives of order $\alpha$, $1 < \alpha < 2$. The left and right Riemann-Liouville fractional derivatives of order $\alpha$, for $x \in [a,b]$, $-\infty \le a < b \le \infty$ are given respectively by

$$\frac{\partial^\alpha f}{\partial x^\alpha}(x,t) = \frac{1}{\Gamma(2-\alpha)}\frac{\partial^2}{\partial x^2}\int_a^x f(\xi,t)(x-\xi)^{1-\alpha}d\xi, \quad (8)$$

$$\frac{\partial^\alpha f}{\partial(-x)^\alpha}(x,t) = \frac{(-1)^2}{\Gamma(2-\alpha)}\frac{\partial^2}{\partial x^2}\int_x^b f(\xi,t)(\xi-x)^{1-\alpha}d\xi. \quad (9)$$

Our problem is defined in the domain $\mathbb{R} \times [0,T]$. We define a uniform discrete domain in space and time, that is, $x_{j+1} = x_j + \Delta x$, $j \in \mathbb{Z}$, $t_n = n\Delta t$, $\Delta t = T/M$, where $\Delta x$ is the space step and $\Delta t$ is the time step.

Let $f_j^n$ denote the approximation of $f(x_j, t_n)$. We approximate the fractional Riesz operator using the approximations discussed in [10,11] which are approximations of the left and right Riemann-Liouville derivatives. These approximations are respectively given by the discrete operators $\delta_l^\alpha f(x_j,t)/\Delta x^\alpha$ and $\delta_r^\alpha f(x_j,t)/\Delta x^\alpha$ where

$$\delta_l^\alpha f(x_j,t) = \sum_{m=-1}^{\infty} \frac{q_m}{\Gamma(4-\alpha)}f(x_{j-m},t), \quad \delta_r^\alpha f(x_j,t) = \sum_{m=-1}^{\infty} \frac{q_m}{\Gamma(4-\alpha)}f(x_{j+m},t).$$
$$(10)$$

The coefficients $q_m$ are defined by

$$q_{-1} = a_0, \quad q_0 = -2a_0 + a_1 \quad q_m = a_{m-1} - 2a_m + a_{m+1}, \ m \ge 1, \quad (11)$$

where $a_0 = 1$ and $a_m = (m+1)^{3-\alpha} - 2m^{3-\alpha} + (m-1)^{3-\alpha}$, $m \ge 1$. Finally, the discrete operator $\delta^\alpha f/\Delta x^\alpha$ approximates the Riesz operator, where

$$\delta^\alpha f(x_j,t) = \frac{1}{2|\cos(\alpha\pi/2)|}\left(\delta_l^\alpha f(x_j,t) + \delta_r^\alpha f(x_j,t)\right). \quad (12)$$

We consider the explicit first order upwind numerical method

$$\frac{f_j^{n+1} - f_j^n}{\Delta t} = D \frac{\delta^\alpha f_j^n}{(\Delta x)^\alpha} - r_j^n f_j^n + s_j^n, \tag{13}$$

where $r_j^n = r(x_j, t_n)$ and $s_j^n$ is the second order approximation of $s(x_j, t_n)$

$$s_j^n = f_0(x_j)\Delta x \left[ \frac{1}{2}(r_0 f_0^n + r_N f_N^n) + \sum_{j=1}^{N-1} r_j f_j^n \right]. \tag{14}$$

We discuss the efficiency of the numerical method in terms of two essential aspects, which are accuracy and stability.

The accuracy of the method comes directly from its formulation and the fact that the approximation we have used for the Riesz operator is known to be second order accurate. This conclusion follows from the next result concerning the approximation of the left fractional derivative.

**Theorem 1** [10,11]: *Let $u$ be a function with sufficiently many continuous spatial derivatives that vanish at infinity in an appropriate manner. Then, we have that*

$$\frac{\partial^\alpha u}{\partial x^\alpha}(x_j) - \frac{\delta_l^\alpha u}{(\Delta x)^\alpha}(x_j) = O((\Delta x)^2).$$

A similar result is valid for the right fractional derivative. From here we can conclude that the Riesz operator is approximated by a second order accurate formula.

In order to derive stability conditions for the finite difference scheme, we apply the von Neumann analysis or Fourier analysis. We have assumed resetting $r$ locally constant. Let

$$s_\alpha = -\sum_{m=-1}^{\infty} q_m \cos(m\theta), \qquad A_\alpha = \frac{\mu_\alpha}{2|\cos(\alpha\pi/2)|}, \qquad \mu_\alpha = \frac{D\Delta t}{(\Delta x)^\alpha}.$$

**Theorem 2.** *The numerical method (13) is von Neumann stable if and only if*

$$s_\alpha A_\alpha \leq 2. \tag{15}$$

**Proof.** Fourier analysis assumes that for a solution defined in the whole real line, the error will be propagated forward in time according to the equation

$$e_j^{n+1} = e_j^n + \mu_\alpha \delta^\alpha e_j^n - r e_j^n. \tag{16}$$

This analysis also assumes the error $e_j^n$ is decomposed into a Fourier series with terms given by $\kappa_p^n e^{i\xi_p(j\Delta x)}$, where $\kappa_p^n$ is the amplitude of the $p$-th harmonic and $\theta$ is the phase angle and covers the domain $[-\pi, \pi]$. Considering a single mode $\kappa^n e^{ij\theta}$, its time evolution is determined by the same numerical scheme as the error $e_j^n$. The stability conditions will be satisfied if the amplification factor $\kappa$ does not grow in time.

We denote by $\kappa(\theta; \mu_\alpha)$ the amplification factor since it will depend on $\mu_\alpha$. If we insert $\kappa^n e^{ij\theta}$ in (16) we obtain the equality for the amplification factor

$$\kappa(\theta; \mu_\alpha) = 1 - r\Delta t + \frac{\mu_\alpha}{2|\cos(\alpha\pi/2)|\Gamma(4-\alpha)} \left[ \sum_{m=-1}^{\infty} q_m e^{-im\theta} + \sum_{m=-1}^{\infty} q_m e^{im\theta} \right]$$

$$= 1 - r\Delta t + \frac{\mu_\alpha}{2|\cos(\alpha\pi/2)|\Gamma(4-\alpha)} \left[ \sum_{m=-1}^{\infty} q_m \cos(m\theta) \right] = 1 - r\Delta t - A_\alpha s_\alpha.$$

Hence, $|\kappa(\theta; \mu_\alpha)| = |1 - A_\alpha s_\alpha - r\Delta t| \leq |1 - A_\alpha s_\alpha| + r\Delta t$. Note that $s_\alpha \geq 0$ (see [11]). Therefore the method is stable if and only if $|1 - A_\alpha s_\alpha| \leq 1$. This inequality is equivalent to $s_\alpha A_\alpha \leq 2$. $\square$

Note that $s_\alpha \leq |q_{-1} + q_1| + |q_0| + |\sum_{m=2}^{\infty} q_m| \leq 8(1 - 2^{1-\alpha}) \leq 8$ and if $A_\alpha \leq 1/4$ the method is stable for all $1 < \alpha \leq 2$. Therefore the numerical method can be implemented efficiently by considering a space step $\Delta x$ and a time step $\Delta t$ that verifies $A_\alpha = 1/4$.

The tests done in the next section for the solution defined in all real line are run in a domain $[-L, L]$. The constant $L$ should be large enough such that the presence of the artificial boundaries at $x = \pm L$ do not affect the accuracy of the approximate solution. In particular, the power law decay verified for anomalous diffusion requires some additional care. Without resetting this would necessarily mean to increase the computational domain as we increase time.

The matricial form of the numerical method is build in the domain $[-L, L]$ and the discrete points in space are $x_j = -L + j\Delta x$, $\Delta x = 2L/N$. We get

$$\mathbf{f}^{n+1} = (\mathbf{I} + \mathbf{A}_\alpha - \mathbf{R} * \mathbf{I})\mathbf{f}^n + \mathbf{s}^n, \tag{17}$$

where $\mathbf{A}_\alpha$ is an $(N-1) \times (N-1)$ matrix related to the anomalous diffusion term, $\mathbf{R}$ is a diagonal matrix with the resetting values $r_j$ as entries and $\mathbf{s}^n$ is the vector that contains the entries $s_j^n$. The matrix $\mathbf{A}_\alpha$ will be of the form $\mathbf{A}_\alpha = \mathbf{L}_\alpha + \mathbf{L}_\alpha^T$, where $\mathbf{L}_\alpha$ is a matrix that contains the coefficients defined in (11).

## 4    The Effect of Resetting

In this section we discuss the effect of resetting in the solutions. We consider the initial condition $\delta_\epsilon(x) = (1/\epsilon\sqrt{\pi})e^{-(x-x_0)^2/\epsilon^2}$ that can be seen as an approximation of the delta Dirac function.

In the absence of resetting, the anomalous or classical diffusion in free space does not have a stationary state (see Fig. 1(a) and (b)). However a nonzero rate of resetting to a fixed position leads to a stationary state when $t \to \infty$ (see Fig. 1(c) and (d)). The steady state solution of (7), when $r(x)$ is constant is $f_{st}(x) = \sqrt{r/D} \exp(-\sqrt{r/D}|x - x_0|)$. The numerical solution, when $\alpha = 2$, plotted in Fig. 1(d) is in agreement to this steady state solution. This stationary distribution is non-gaussian in $x$, contrasting with the fact that if we omit the resetting term, then the stationary solution would be gaussian in $x$. Therefore, the resetting term changes the character of the long time solution.

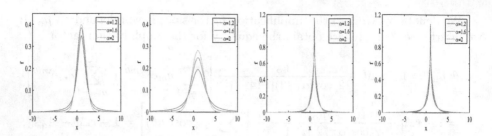

**Fig. 1.** Plots of $f(x,t)$ for $x_0 = 1$, $D = 0.5$ computed with $\Delta x = 0.005$, $\Delta t = 0.01 \Delta x$. Left figures: $r(x) = 0$. (a) $t = 1$; (b) $t = 2$. Right figures: $r(x) = 1$; (c) $t = 1$; (d) $t = 5$. For $r(x) = 1$ and $t = 10$ the solution is similar to (d) indicating we have reached the steady state.

For the sake of clarity in the discussion that follows we assume $x_0 = 0$. It is well known that the first and second moments for the classical diffusion are given respectively by $<x> = 0$ and $<x^2> = 2Dt$. If we consider the diffusion problem with resetting, for $\alpha = 2$, we can find, from the characteristic function given by (3), that the first and second moments are given respectively by

$$M_1(t) := \int_{\mathbb{R}} x f(x,t) dx = 0, \quad M_2(t) := \int_{\mathbb{R}} x^2 f(x,t) dx = \frac{2D}{r} \left( 1 - e^{-rt} \right). \quad (18)$$

At this point we can ask two questions, for the case $\alpha = 2$. Can we see at which time the steady state is reached by looking at the second moment? And if we increase the resetting rate does it mean the solution is pushed quicker to the steady state? In order to answer to the previous questions in Fig. 2 we plot the second moment $M_2(t)$ for the classical diffusion with resetting $r = 1$ and $r = 3$. Theoretically, we have that

$$\lim_{t \to \infty} \frac{2D}{r}(1 - e^{-rt}) = \frac{2D}{r} \quad (19)$$

and this is in agreement to what is shown in Fig. 2. We can observe that for $r = 1$ the second moment starts to look constant around $t = 6$ and according to the value $M_2(t) = 2D/r = 1$. For $r = 3$ is around $t = 2$ where $M_2(t) = 1/3$ which is also according to (19).

Now let us turn to the anomalous diffusion case. For $1 < \alpha < 2$ the second moment for diffusion without resetting is known to be divergent and the first moment is zero. For the case with resetting from the characteristic function, given by (3), we can also conclude that the second moment is divergent and the first moment is zero. The problem of having a diverging second moment encountered in the discussions of Lévy flights can be circumvented by considering a pseudo second moment, that is,

$$M_2^L(t) := <x^2>_L = \int_{L_1(t)}^{L_2(t)} x^2 f(x,t) dx \quad (20)$$

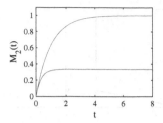

**Fig. 2.** Second moment $M_2(t)$ for the classical diffusion with resetting $r = 1$ (red) and $r = 3$ (blue): $D = 0.5$. (Color figure online)

according to which the walker is considered in an imaginary box with $L = \max\{|L_1|, |L_2|\}$. Without loss of generality, we assume a symmetric box, that is, $L_2(t) = -L_1(t) = L(t)$. Note that the cut-offs of the integral are time dependent and the imaginary box is chosen in the spatial interval. We can say it gives a measure, that a finite portion of the probability is gathered within the given interval $2L(t)$. Pseudo second moments have been considered before in literature. For instance, in [5,6,8], the box considered was $[L_1 t^{1/\alpha}, L_2 t^{1/\alpha}]$. Here, we consider a different type, that is, for a final time T we consider a sufficiently large box $[-L(T), L(T)]$ and then we compute (20) in that box for all $0 < t \leq T$.

Since the density follows the power-law asymptotic behaviour the cut-offs of the integral needs to be chosen such that the asymptotic behaviour of $f(x,t)$ is reached. This can be done, by choosing a box $[-L, L]$ for which the values of $f(x,t)$ are very small in $\pm L$. Note that for smaller values of $\alpha$ we have larger tails, that is, the tail increases with decreasing $\alpha$.

In what follows we compare the estimates of the pseudo second moments, $M_2^L(t)$, for the problems with and without resetting. We plot in Fig. 3(a) and (b) the results for $r = 0$ and in Fig. 3(c) and (d) we plot the results for $r = 1$. From these results we infer that for the problem without resetting we have

$$M_2^L(t) = C_\alpha t \tag{21}$$

with $C_\alpha$ a constant that depends on $\alpha$. Then in the presence of resetting we get the behaviour

$$M_2^L(t) \sim C_\alpha \left( \frac{1 - \exp(-rt)}{r} \right). \tag{22}$$

Another interesting information we can obtain from the pseudo second moments is related to the steady state solutions. The results suggest that in the presence of resetting we reach a steady state for all $\alpha$'s as can be seen in Fig. 3. It also seems that smaller is the value of $\alpha$, it takes longer to reach the steady state. This information was obtained for particular values of $L$. Although the pseudo moment increases with $L$ as expected, it becomes constant at similar instants of time for different values of $L$.

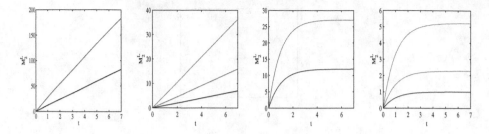

**Fig. 3.** Left figures: Pseudo second moment for anomalous diffusion with resetting $r = 0$ with $L = 100$. (a) $\alpha = 1.2$ (blue), $\alpha = 1.4$ (black). (b) $\alpha = 1.6$ (red), $\alpha = 1.8$ (blue), $\alpha = 2$ (black). Right figures: Pseudo second moment for anomalous diffusion with resetting $r = 1$ with $L = 100$: (c) $\alpha = 1.2$ (blue), $\alpha = 1.4$ (black); (d) $\alpha = 1.6$ (red), $\alpha = 1.8$ (blue), $\alpha = 2$ (black). (Color figure online)

# References

1. Bray, A.J., Majumdar, S.N., Schehr, G.: Persistence and first passage properties in non equilibrium systems. Adv. Phys. **62**, 225–361 (2013)
2. Evans, M.R., Majumbar, S.N.: Diffusion with optimal resetting. J. Phys. A. **44**, 435001 (2011)
3. Evans, M.R., Majumbar, S.N.: Diffusion with stochastic resetting. Phys. Rev. Lett. **106**, 160601 (2011)
4. Evans, M.R., Majumdar, S.N., Mallick, K.: Optimal diffusive search: nonequilibrium resetting versus equilibrium dynamics. J. Phys. A: Math. Theor. **46**, 185001 (2013)
5. Fogedby, H.C.: Lévy flights in quenched random force fields. Phys. Rev. E **58**, 1890 (1998)
6. Jespersen, S., Metzler, R., Fogedby, H.C.: Lévy flights in external force fields: Langevin and fractional Fokker-Planck equations and their solutions. Phys. Rev. E. **59**, 2736 (1999)
7. Kusmierz, L., Majumdar, S.N., Sabhapandit, S., Schehr, G.: First order transition for the optimal search time of Lévy flights with resetting. Phys. Rev. Lett. **113**, 220602 (2014)
8. Metzler, R., Klafter, J.: The random walk's guide to anomalous diffusion: a fractional dynamics approach. Phys. Rep. **339**, 1–77 (2000)
9. Pal, A.: Diffusion in a potential landscape with stochastic resetting. Phys. Rev. E. **91**, 012113 (2015)
10. Sousa, E.: An explicit high order method for fractional advection diffusion equations. J. Comput. Phys. **278**, 257–274 (2014)
11. Sousa, E., Li, C.: A weighted finite difference method for the fractional diffusion equation based on the Riemann-Liouville derivative. Appl. Numer. Math. **90**, 22–37 (2015)
12. Whitehouse, J., Evans, M.R., Majumbar, S.N.: Effect of partial absorption on diffusion with resetting. Phys. Rev. E. **87**, 022118 (2013)
13. Zaslavsky, G.M.: Chaos, fractional kinetics, and anomalous transport. Phys. Rep. **371**, 461–580 (2002)

# Generalized Multiscale Finite Element Method for Unsaturated Filtration Problem in Heterogeneous Medium

D. Spiridonov[1]([✉]) [iD] and M. Vasilyeva[1,2] [iD]

[1] Multiscale Model Reduction Laboratory, North-Eastern Federal University,
Yakutsk, Russia
`d.stalnov@mail.ru`, `vasilyevadotmdotv@gmail.com`
[2] Institute for Scientific Computation, Texas A&M University,
College Station, TX, USA

**Abstract.** We consider a mathematical model for simulation of the unsaturated flow problems in heterogeneous porous medium that describes by the Richards equation. To resolve all heterogeneity, we construct fine grid and construct finite element approximation. For dimension reduction of the discrete system, we construct multiscale solver for coarse grid solution using Generalized Multiscale Finite Element Method (GMsFEM). We generate multiscale basis functions by solution of the local spectral problems. We present numerical result and compare relative error for different number of the multiscale basis functions for 2D and 3D model problems.

**Keywords:** Heterogeneous media · Unsaturated filtration ·
Richards equation · Multiscale method · GMsFEM

## 1 Introduction

Prediction of fluid movement in unsaturated soils is an important problem in many branches of science and engineering. These include soil science, agricultural engineering, environmental engineering, and groundwater hydrology. For simulation of the unsaturated filtration processes in soils, we use Richards equation [1,2,4]. One of the main problems in the construction of a mathematical model is the consideration of highly heterogeneous media, because the grid resolution of heterogeneity leads to a large number of unknowns of the discrete system and requires large computational resources [3,5,8]. To solve such problems, a homogenization techniques or multiscale methods are usually used. In this paper, we construct coarse scale approximation using Generalized Multiscale Finite Element method. Some earlier pioneering works in multiscale finite element methods are presented in [6,7]. The main idea of the GMsFEM is to divide the computation into offline and online stages [9–11]. In the offline stage, we construct a snapshot space and the offline space via spectral decomposition

© Springer Nature Switzerland AG 2019
I. Dimov et al. (Eds.): FDM 2018, LNCS 11386, pp. 517–524, 2019.
https://doi.org/10.1007/978-3-030-11539-5_60

of the snapshot space. The main concept of constructing the snapshot space is that the snapshot vectors preserve some essential properties of the solution and provide a good approximation space. The main idea of the offline space is that it gives a good solution approximation with fewer basis functions. We use GMSH software [12] to construct computational domains and grids. For implementation, we use open-source library FEniCS [13].

We organize the paper as follows. In Sect. 2, we present a model problem. In Sect. 3 we consider fine-scale approximation. In Sect. 4, we describe the construction of the multiscale basis functions and present coarse grid approximation using GMsFEM. Finally, we present numerical results in Sect. 5 for 2D and 3D model problems and conclusion.

## 2    Problem Formulation

Prediction of fluid flow in unsaturated soils is an important problem in science and engineering. The water flow into the porous media is driven by the pressure gradient and describes by the Darcy's Law

$$q = k(x, p)\text{grad}\,(p + z), \tag{1}$$

where $q$ is the velocity vector, $k$ is the unsaturated hydraulic conductivity tensor and $z$ represent the influence of the gravity to the flow.

The fluid flow in domain $\Omega$, we have following equation

$$\frac{\partial \Theta}{\partial t} + \text{div}\,q = 0, \quad x \in \Omega, \tag{2}$$

where $\Theta$ is the water content and represents the fraction of porous medium total volume that is filled with fluid.

Therefore, we have following problem in domain $\Omega$

$$\frac{\partial \Theta}{\partial t} - \text{div}\,(k(x, p)\text{grad}\,p) - \frac{\partial k}{\partial z} = 0, \quad x \in \Omega, \tag{3}$$

with initial condition

$$p(x) = p_0, \quad x \in \Omega, \quad t = 0, \tag{4}$$

and boundary condition

$$p(x) = p_1, \quad x \in \Gamma_D, \quad q \cdot n = 0, \quad x \in \Gamma_N.$$

where $\partial \Omega = \Gamma_D \cup \Gamma_N$.

As constitutive relations, we use Havercamp model [4]

$$\Theta(p) = \frac{\alpha(\Theta_s - \Theta_r)}{\alpha + |p|^\beta} + \Theta_r,$$

$$k(x, p) = k_s(x)\frac{A}{A + |p|^\gamma} + \Theta_r, \tag{5}$$

where $k_s(x)$ is also known as the saturated hydraulic conductivity.

## 3   Fine Grid Approximation

For numerical solution, we construct finite elements approximation by space on the fine grid

$$\int_\Omega \frac{\partial \Theta}{\partial t} v \, dx + \int_\Omega (k \operatorname{grad} p, \operatorname{grad} v) \, dx - \int_\Omega \frac{\partial k}{\partial z} v \, dx = 0, \quad \forall v \in \hat{V}. \qquad (6)$$

where $p \in V$ and

$$V = \{v \in H^1(\Omega) : v(x) = p_1, x \in \Gamma_D\},$$

$$\hat{V} = \{v \in H^1(\Omega) : v(x) = 0, x \in \Gamma_D\}.$$

For approximation by time, we use backward Euler method coupled with a Picard iteration scheme for nonlinear coefficients

$$\int_\Omega \frac{\Theta^{n+1,m+1} - \Theta^n}{\tau} v \, dx$$
$$+ \int_\Omega \left(k^{n+1,m} \nabla p^{n+1,m+1}, \nabla v\right) dx - \int_\Omega \frac{\partial k^{n+1,m}}{\partial z} v \, dx = 0, \qquad (7)$$

where superscripts indicate time $(n)$ and iteration levels $(m)$.

For $\Theta^{n+1,m+1}$, we use Taylor series with respect to $p$

$$\Theta^{n+1,m+1} \approx \Theta^{n+1,m} + \left.\frac{d\Theta}{dp}\right|^{n+1,m} (p^{n+1,m+1} - p^{n+1,m}) \qquad (8)$$
$$= \Theta^{n+1,m} + C^{n+1,m}(p^{n+1,m+1} - p^{n+1,m}),$$

where terms higher than linear are neglected.

Therefore, we obtain following approximation of the nonlinear equation [2]

$$\int_\Omega C^{n+1,m} \frac{p^{n+1,m+1} - p^{n+1,m}}{\tau} v \, dx + \int_\Omega \frac{\Theta^{n+1,m} - \Theta^n}{\tau} v \, dx$$
$$+ \int_\Omega \left(k^{n+1,m} \nabla p^{n+1,m+1}, \nabla v\right) dx - \int_\Omega \frac{\partial k^{n+1,m}}{\partial z} v \, dx = 0. \qquad (9)$$

Let $V_h \subset V$ and $\hat{V}_h \subset \hat{V}$. We have following discrete variational problem: find $p_h \in V_h$ such that

$$\int_\Omega C^{n+1,m} \frac{p_h^{n+1,m+1} - p_h^{n+1,m}}{\tau} v \, dx + \int_\Omega \frac{\Theta^{n+1,m} - \Theta^n}{\tau} v \, dx$$
$$+ \int_\Omega \left(k^{n+1,m} \operatorname{grad} p_h^{n+1,m+1}, \operatorname{grad} v\right) dx - \int_\Omega \frac{\partial k^{n+1,m}}{\partial z} v \, dx = 0, \qquad (10)$$

where $\forall v \in \hat{V}_h$.

We can write approximation, in the matrix form as

$$S^m \frac{p_h^{n+1,m+1} - p_h^{n+1,m}}{\tau} + A^m p_h^{n+1,m+1} = F^m, \tag{11}$$

where

$$S^m = \{s_{ij}^m\}, \quad s_{ij}^m = \int_\Omega C^{n+1,m} \psi_i \, \psi_j \, dx,$$

$$A^m = \{a_{ij}^m\}, \quad a_{ij}^m = \int_\Omega \left(k^{n+1,m} \text{grad}\,\psi_i, \text{grad}\,\psi_j\right) dx,$$

$$F^m = \{f_j^m\}, \quad f_j^m = \int_\Omega \frac{\partial k^{n+1,m}}{\partial z} \psi_j \, dx - \int_\Omega \frac{\Theta^{n+1,m} - \Theta^n}{\tau} \psi_j \, dx,$$

where $\psi_i$ is the linear basis functions.

## 4    Coarse Grid Approximation Using GMsFEM

For construction of the coarse grid solver, we use a Generalized Multiscale Finite Element Method (GMsFEM).

Let $\mathcal{T}_H$ is the coarse grid and $\omega_i$ is the local domains, where $i = 1, ..., N_v$ and $N_v$ is the number of coarse grid nodes. A local domain $\omega_i$ is obtained by the combining all the coarse cells around one vertex of the coarse grid. For the construction of the multiscale basic functions, we solve spectral problems in local domains. Spectral problems identify the most important characteristics of the solution. Using constructed multiscale basis functions, we construct a coarse grid approximation for a given configuration of heterogeneous properties.

In the multiscale basis calculations, we first construct a snapshot space $V_{\text{snap}}^{\omega_i}$. The snapshot space is constructed by the solution of the following local problems

$$-\text{div}\,(k_s(x)\nabla\psi_l) = 0 \quad x \in \omega_i \tag{12}$$

with boundary conditions $\psi_l(x) = \delta_j$ on $\partial\omega_i$ and $\delta_j$ is the function, which takes the value 1 at $x = x_j$ and zero elsewhere.

Next, we solve a local spectral problems on the snapshot space

$$A\varphi^i = \lambda S\varphi^i, \tag{13}$$

where the elements of the matrices $A = \{a_{ij}\}$ and $S = \{s_{ij}\}$ are defined as follow

$$a_{ij} = \int_{\omega_i} (k_s(x)\nabla u, \nabla q) dx, \quad s_{ij} = \int_{\omega_i} k_s(x)\, u\, q\, dx. \tag{14}$$

We solve this spectral problem on the snapshot space

$$\tilde{A}\tilde{\varphi}^i = \lambda \tilde{S}\tilde{\varphi}^i, \quad \tilde{A} = PAP^T, \quad \text{and} \quad \tilde{S} = PSP^T. \tag{15}$$

where $P = \{\psi_0, \psi_1, ..., \psi_{J_i}\}$ and $\varphi_k^i = P^R \tilde{\varphi}_k^i$ for $k = 1, 2, ...$. Then, we choose the smallest $M_i$ eigenvalues from (15) and use them for the construction of multiscale basis functions.

**Table 1.** Relative L2 errors (%) for different number of basis functions

| Number of bases | $t_1$ | $t_{25}$ | $t_{50}$ | $t_{200}$ |
|---|---|---|---|---|
| 1 | 9.94 | 5.49 | 5.51 | 5.51 |
| 2 | 9.84 | 6.25 | 6.28 | 6.28 |
| 4 | 2.55 | 3.05 | 2.95 | 2.95 |
| 8 | 1.63 | 2.35 | 2.17 | 2.27 |
| 12 | 0.58 | 1.32 | 1.34 | 1.29 |
| 16 | 0.48 | 1.17 | 1.19 | 1.15 |

The multiscale space is defined as the span of $\chi_i \varphi_k^i$, where $\chi_i$ is the usual nodal basis function for the node $i$ (linear partition of unity functions). The number of bases can be different, the accuracy of the solution can be improved when we increase the number of bases.

Finally, we create following matrix for each $\omega_i$

$$R^i = \left[ \chi_i \varphi_1^i, \ldots, \chi_i \varphi_{M_i}^i, \chi_i \eta^i \right].$$

and define a transition matrix from a fine grid to a coarse grid to reduce the dimension of the problem

$$R = [R^1, R^2, ..., R^{N_v}],$$

where $N_v$ is the number of local domains $\omega_i$.

Therefore, we have following coarse grid approximation

$$S_c^m \frac{p_H^{n+1,m+1} - p_H^{n+1,m}}{\tau} + A_c^m p_H^{n+1,m+1} = F_c^m, \tag{16}$$

where $S_c^m = R S^m R^T$, $A_c^m = R A^m R^T$ and $F_c^m = R F^m$. Here, using multiscale solution $p_H$, we can reconstruct a fine grid solution $p = R^T p_H$.

## 5    Numerical Results

We consider water infiltration into porous medium in 2D computational domain $\Omega = [0, 10]^2$ meters. As boundary conditions we use $p_1 = -20.7$ on top boundary $\Gamma_1$, $p_2 = -61.5$ on bottom boundary $\Gamma_2$ and $p_0 = -61.5$ for initial conditions.

For soil properties we use $\alpha = 1.511 \times 10^6$, $\beta = 3.96$, $\Theta_s = 0.287$, $\Theta_r = 0.075$, $A = 1.175 \times 10^6$ and $\gamma = 4.74$. For numerical solution we assume that the nonlinearity and heterogeneity are separable, where the latter comes from $k_s(x)$. Coefficient $k_s(x)$ is presented on Fig. 1. We took the maximum time equal to 0.5 s. and 200 time steps.

As hydraulic conductivity, we use $k_s = 0.01$ in blue domain and $k_s = 0.0001$ for red (high-contrast conductivity). Computational domain and fine grid are

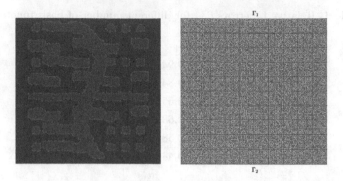

**Fig. 1.** Left: coefficient $k_s(x)$. Right: coarse (blue color) and fine (black color) grids with 31,5 thousands cells and 14,5 thousand vertices (Color figure online)

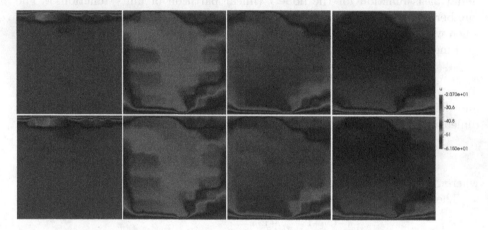

**Fig. 2.** Fine scale solution (top) and multsicale solution using 16 basis functions (bottom) for different time steps $t_1$, $t_{25}$, $t_{50}$ and $t_{200}$

represented on Fig. 1 which contains approximately 31,5 thousands of triangle cells and 14,5 thousands vertices. We use $10 \times 10$ coarse rectangular grid to find multiscale solution.

The fine-scale solution and multiscale solution using 16 basis function are presented on Fig. 2. We consider the solution behavior with different number of basis functions on the 1, 25, 50, 200 time steps (see Table 1). Our comparison showed convergence of the solution, when we increase number of basis functions.

Next, we consider 3D problem. As boundary conditions we use $p_1 = -20.7$ on top boundary surface, $p_2 = -61.5$ on bottom boundary surface and $p_0 = -61.5$ for initial conditions. Coefficient values are same as in previous section. In this case we took maximum time equal to 0.025 s. with 200 time steps. The computational domain and fine grid with 132,5 thousands of tetrahedral cells and 18,6 thousands vertices are presented on Fig. 3. For coarse scale approximation, we use $5 \times 5$ coarse cubic grid.

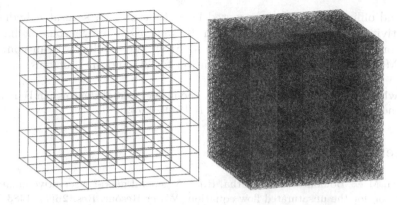

**Fig. 3.** Coarse grid and fine grid with 132,5 thousands cells and 18,6 thousands vertices

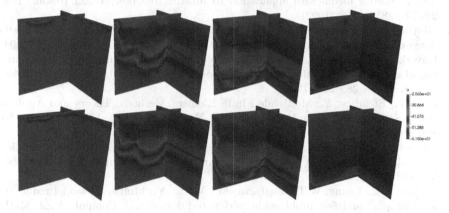

**Fig. 4.** Fine scale solution (top) and multsicale solution using 16 basis functions (bottom) for different time steps $t_1$, $t_{25}$, $t_{50}$ and $t_{200}$

**Table 2.** Relative L2 errors (%) for 16 multiscale basis functions

| Number of bases | $t_1$ | $t_{25}$ | $t_{50}$ | $t_{200}$ |
|---|---|---|---|---|
| 16 | 2.02 | 3.05 | 2.85 | 2.86 |

The fine-scale solution and multiscale solution using 16 basis function are presented on Fig. 4. In Table 2, we present relative L2 errors for different time steps.

## 6   Conclusion

In this paper we describe Generalized Multiscale Finite Element method for Richards equation with Havercamp model. We present construction of snapshot space and offline space. We constructed basis functions for saturated hydraulic conductivity. We solved this problem in 2D and 3D heterogeneous media. We made comparison with fine-scale solution. From the results, we can conclude that GMsFEM can provide accurate solution for 2D and 3D problems.

**Acknowledgments.** Work is supported by the mega-grant of the Russian Federation Government (N 14.Y26.31.0013 and RFBR N 17-01-00732).

# References

1. Celia, M.A., Bouloutas, E.T., Zarba, R.L.: A general mass-conservative numerical solution for the unsaturated flow equation. Water Resour. Res. **26**(7), 1483–1496 (1990)
2. Celia, M.A., Binning, P.: A mass conservative numerical solution for two-phase flow in porous media with application to unsaturated flow. Water Resour. Res. **28**(10), 2819–2828 (1992)
3. Ginting, V.E.: Computational upscaled modeling of heterogeneous porous media flow utilizing finite volume method. Ph.D. thesis, Texas A and M University (2004)
4. Haverkamp, R., Vauclin, M., Touma, J., Wierenga, P.J., Vachaud, G.: A comparison of numerical simulation models for one-dimensional infiltration. Soil Sci. Soc. Am. J. **41**(2), 285–294 (1977)
5. Hou, T., Efendiev, Y.: Multiscale Finite Element Methods: Theory and Applications. STAMS, vol. 4, 2nd edn. Springer, New York (2009). https://doi.org/10.1007/978-0-387-09496-0
6. Akkutlu, I.Y., Efendiev, Y., Vasilyeva, M., Wang, Y.: Multiscale model reduction for shale gas transport in poroelastic fractured media. J. Comput. Phys. **353**, 356–376 (2018)
7. Chung, E.T., Leung, W.T., Vasilyeva, M., Wang, Y.: Multiscale model reduction for transport and flow problems in perforated domains. J. Comput. Appl. Math. **330**, 519–535 (2018)
8. Chung, E.T., Efendiev, Y., Lee, C.S.: Mixed generalized multiscale finite element methods and applications. Multiscale Model. Simul. **13**(1), 338–366 (2015)
9. Chung, E.T., Efendiev, Y., Li, G., Vasilyeva, M.: Generalized multiscale finite element methods for problems in perforated heterogeneous domains. Appl. Anal. **95**(10), 2254–2279 (2016)
10. Efendiev, Y., Hou, T.Y., Ginting, V.: Multiscale finite element methods for nonlinear problems and their applications. Commun. Math. Sci. **2**(4), 553–589 (2004)
11. Efendiev, Y., Galvis, J., Hou, T.Y.: Generalized multiscale finite element methods (GMsFEM). J. Comput. Phys. **251**, 116–135 (2013)
12. Software GMSH. http://geuz.org/gmsh/
13. Logg, A., Mardal, K.-A., Wells, G.: Automated Solution of Differential Equations by the Finite Element Method: The FEniCS Book. LNCSE, vol. 84. Springer, Berlin (2011). https://doi.org/10.1007/978-3-642-23099-8

# A Continuous Model for the Ecological Collapse of Easter Island

Bálint Takács[(✉)] [iD]

Eötvös Loránd University, Budapest, Hungary
takacs.balint.mate@gmail.com

**Abstract.** In this article a mathematical model describing the possible events that could have lead to the ecological catastrophe of Easter Island is extended in a way that instead of the originally spatially discrete model (the domain is split into several regions) now a spatially continuous one is considered (the number of each population is observed at each point of the domain). In other words, the original system of ordinary differential equations is transformed into a system of partial differential equations, and then the effect of the diffusion of the trees is observed, i.e. whether it stabilizes the system like in the original case, or not. It turned out that because the linearized system can be written in a pretty similar form to the matrix of the two dimensional case which was examined in a previous article [10], the same theorems can be said about this system, meaning that the increase of the diffusion of the trees actually stabilizes the system in this case too.

**Keywords:** Differential equations · Stability · Easter Island ·
Modeling · Diffusion

## 1 Introduction

Since its discovery by Europeans in the 18th century, the case of Easter Island has puzzled generations of scientists. There have been numerous theories proposed describing the causes of the fall of a once flourishing civilization, and also the disappearance of the trees from the island. Most of these blamed the irresponsibility of the settlers, who cut down all the trees on the island, and consequently destroyed their own culture.

However, the work of Hunt [3,4] showed that the collapse could have happened much faster than supposed before. Because of this, Hunt proposed that not only the humans were responsible for the ecological collapse, but also the rats of the island. These animals (originally brought to the island by the settlers themselves) ate the seeds of the trees, making the plants unable to cope with the harvest done by the settlers.

The ideas were formulated by Basener et al. [1] as a system of ordinary differential equations as follows. The island is thought of as a disc, meaning that it has a volcano in the middle and only its coast is habitable. This habitable

© Springer Nature Switzerland AG 2019
I. Dimov et al. (Eds.): FDM 2018, LNCS 11386, pp. 525–532, 2019.
https://doi.org/10.1007/978-3-030-11539-5_61

region is split into $N$ equal regions in a way that region 1 and region $N$ are next to each other. In the model $P^s(t)$, $R^s(t)$ and $T^s(t)$ denotes the number of the people, the rats and the trees in region $s$, respectively, $a$, $b$ and $c$ denote the reproduction rate of each group, $f$ is the negative effect of the rats on the trees, $M$ is the maximum number of trees that can live on the island, and $h$ is the constant harvest of the trees done by the settlers. In order to model the movement of the species between regions diffusion terms are also introduced, in which $D_P$ and $D_R$ denotes the diffusion coefficient of people and rats, respectively. Note that in these models $P = 1$ means one person, $T = 1$ means the number of trees used by a person in a year, and $R = 1$ is the number of rats which could be supported by $T = 1$ trees. Thus, the following system of equations are obtained:

$$
\begin{aligned}
\frac{dP^s(t)}{dt} &= aP^s(t)\left(1 - \frac{P^s(t)}{T^s(t)}\right) + D_P(P^{s-1}(t) - 2P^s(t) + P^{s+1}(t)), \\
\frac{dR^s(t)}{dt} &= cR^s(t)\left(1 - \frac{R^s(t)}{T^s(t)}\right) + D_R(R^{s-1}(t) - 2R^s(t) + R^{s+1}(t)), \\
\frac{dT^s(t)}{dt} &= \frac{b}{1 + fNR^s(t)}T^s(t)\left(1 - \frac{T^s(t)}{\frac{M}{N}}\right) - hP^s(t).
\end{aligned}
\tag{1}
$$

It turned out in [2] that if the diffusion of people or the diffusion of the rats was increased, the coexistence equilibrium of the system became unstable. This was a surprising result, since the increase of diffusion in biological or physical systems usually makes it more stable. This strange phenomenon was a bit more understood in [7], in which we added a further term to the equation of the trees, namely the diffusion of the trees, which may sound surprising at first, but because of the constant wind on the island the seeds of the trees can actually travel along the coast, in some cases even on the fur of small animals. In this way we obtained the following system of equations, in which $D_T$ is the aforementioned diffusion coefficient of the trees

$$
\begin{aligned}
\frac{dP^s(t)}{dt} &= aP^s(t)\left(1 - \frac{P^s(t)}{T^s(t)}\right) + D_P(P^{s-1}(t) - 2P^s(t) + P^{s+1}(t)), \\
\frac{dR^s(t)}{dt} &= cR^s(t)\left(1 - \frac{R^s(t)}{T^s(t)}\right) + D_R(R^{s-1}(t) - 2R^s(t) + R^{s+1}(t)), \\
\frac{dT^s(t)}{dt} &= \frac{b}{1 + fNR^s(t)}T^s(t)\left(1 - \frac{T^s(t)}{\frac{M}{N}}\right) - hP^s(t) \\
&\quad + D_T(T^{s-1}(t) - 2T^s(t) + T^{s+1}(t)).
\end{aligned}
\tag{2}
$$

Then we examined the way this extension influences the stability of the coexistence equilibrium, and we found that (for a certain, biologically reasonable range of parameters) the system becomes more stable if we increase $D_T$, i.e. it does not lose its stability (the populations do not die out) upon the increase of the diffusion of the trees, and if it was unstable, it becomes stable at a sufficiently large $D_T$ value.

Another question is how to choose the value of $N$, or in other words, what should be the number of the examined regions. It turned out in [8] that if we are only interested in the behavior of the coexistence equilibrium, it is sufficient to consider only the $N = 2$ case, as the stability region tends to the region of that case as $N \to \infty$. However, usually we are interested not only in the stability of the equilibrium, but also in the numbers of each population of the species. Because of this, it is usually useful to use higher values of $N$. However, in this way we obtain a system of $3N$ equations, which becomes harder to solve if $N$ increases. Because of this, perhaps it is more useful not to consider a discrete problem in space, but to use partial differential equations instead. In this paper we consider the latter, and examine whether the stabilizing property of tree diffusion also holds in this continuous case.

## 2   Constructing the Model

Let us suppose that we have a partition of the coast of the island into $\mathcal{N}$ regions. Because of the assumption that there are no differences in the densities of the species radius-wise (meaning that the number of each of the species is the same along a line drawn from the center of the island towards the coast), we can think of the coast as a one-dimensional line, and model our problem as a one-spatial-dimensional problem with periodic boundary conditions. In this way our domain will be an interval, let this be $\Omega = [0, \mathcal{L})$. Let us denote the density of people, rats and trees at point $x \in \Omega$ at time $t$ as $P(t, x)$, $R(t, x)$ and $T(t, x)$, respectively. Thus, the number of individuals in a region can be calculated in the following way:

$$\int_{\Omega^\alpha} Q(t, x) \, \mathrm{d}x = Q^\alpha(t),$$

where $Q \in \{P, R, T\}$ and $\Omega^\alpha$ denotes the domain corresponding to the region indexed by $\alpha$. Now let us assume that the density of a given species is the same inside a given region. (For a sufficiently large $N$, this is a good approximation.) In this way we get the equation

$$\mathcal{L}^\alpha Q(t, x) = Q^\alpha(t),$$

where $\mathcal{L}^\alpha$ is the length of region $\Omega^\alpha$ and $x \in \Omega^\alpha$ arbitrary. For the sake of simplicity let us suppose that the lengths of the regions are equal. Writing these into equation (1) omitting the diffusion terms, we get the following:

$$\frac{\partial P(t, x)}{\partial t} = aP(t, x) \left( 1 - \frac{P(t, x)}{T(t, x)} \right),$$

$$\frac{\partial R(t, x)}{\partial t} = cR(t, x) \left( 1 - \frac{R(t, x)}{T(t, x)} \right),$$

$$\frac{\partial T(t, x)}{\partial t} = \frac{b}{1 + f\mathcal{L}R(t, x)} T(t, x) \left( 1 - \frac{T(t, x)}{\dfrac{M}{\mathcal{L}}} \right) - hP(t, x),$$

(3)

where we used that $\mathcal{NL}^\alpha = \mathcal{L}$. Let us introduce the notations $\tilde{f} = f\mathcal{L}$ and $\tilde{M} = M/\mathcal{L}$. In order to model the motion of the species we add some diffusion terms to the equations:

$$
\begin{aligned}
\frac{\partial P(t,x)}{\partial t} &= aP(t,x)\left(1 - \frac{P(t,x)}{T(t,x)}\right) + D_P \Delta P(t,x), \\
\frac{\partial R(t,x)}{\partial t} &= cR(t,x)\left(1 - \frac{R(t,x)}{T(t,x)}\right) + D_R \Delta R(t,x), \\
\frac{\partial T(t,x)}{\partial t} &= \frac{b}{1 + \tilde{f}R(t,x)}T(t,x)\left(1 - \frac{T(t,x)}{\tilde{M}}\right) - hP(t,x) + D_T \Delta T(t,x).
\end{aligned}
\tag{4}
$$

As mentioned before, we will assign a periodic boundary condition to the problem, meaning that:

$$
\partial_x^\alpha Q(t,0) = \partial_x^\alpha Q(t,\mathcal{L}), \qquad \alpha \in \{0,1,2\},
$$

where $\partial_x$ denotes partial derivative with respect to the second coordinate, in the sense that at point 0 it is a right derivative, and at $\mathcal{L}$ a left derivative. The previous system can be rewritten as follows:

$$
\begin{cases}
\partial_t Q(t,x) = F(Q(t,x)) + D\Delta Q(t,x), \\
\partial_x^\alpha Q(t,0) = \partial_x^\alpha Q(t,\mathcal{L}), \qquad \alpha \in \{0,1,2\}, \\
Q(0,x) = \varphi(x),
\end{cases}
\tag{5}
$$

where $Q(t,x) = (P(t,x), R(t,x), T(t,x))$, $F$ is the $\mathbb{R}^3 \to \mathbb{R}^3$ function describing the interactions between the species, $\partial_t$ is the time derivative, $D$ denotes the diagonal matrix containing the diffusion coefficients and $\varphi(x)$ gives the initial distribution of the species on the island. In this way we get a reaction-diffusion equation with periodic boundary conditions. We are searching for the classical solution $Q(t,x)$ which is in $C^{1,2}(\Omega)$ i.e. it is continuously differentiable in its time variable once, and in the spatial coordinates twice.

## 3   The Effect of the Diffusion of the Trees

A convenient way to examine the behavior of system (5) is to search for constant stationary solutions of it and check their stability properties, i.e. those solutions which take the same value at every point of the shore. It is easy to see that these are equivalent to the equilibrium points of the system (1). In other words, $Q(t,x) = (P^*, R^*, T^*)$ is a stationary solution of (5) if and only if $Q(t) = (\mathcal{L}P^*, \mathcal{L}R^*, \mathcal{L}T^*)$ is an equilibrium point of (1), where $P^*$, $R^*$ and $T^*$ are constants. Because of this, system (5) has only one coexistence constant stationary solution, which is

$$
P^*(t,x) = R^*(t,x) = T^*(t,x) \equiv \frac{\tilde{M}(b-h)}{b + h\tilde{M}\tilde{f}}, \qquad \forall x \in \Omega, t \in \mathbb{R}^+.
\tag{6}
$$

This is a corollary of the similar property of (1), which can be easily proved.

For the examination of the stability of this solution, we are going to use the following method also described e.g. in [6]. Let us linearize around the aforementioned stationary solution, and search for perturbations in the form

$$w(t, x) = w_0 e^{\lambda t} W(x). \tag{7}$$

Because we have periodic boundary conditions, any solutions can be written as the sum of sinusoidal waves of the form

$$W(x) = c_1 e^{ikx} + c_2 e^{-ikx}.$$

Our steady solution is unstable if any of the growth rates $\lambda$ have positive real parts, and if all of them have negative, then it is stable.

If we substitute (7) into system (5), after a brief calculation we get the following system of equations:

$$\lambda P(t) = -k^2 D_P P(t) - aP(t) + aT(t),$$
$$\lambda R(t) = -k^2 D_R R(t) - cR(t) + cT(t), \tag{8}$$
$$\lambda T(t) = -k^2 D_T T(t) - hP(t) + AR(t) + BT(t),$$

in which we used the notations $A = \dfrac{-\tilde{f}\tilde{M}h(b - h)}{b(1 + \tilde{f}\tilde{M})}$ and $B = \dfrac{\tilde{f}\tilde{M}h - b + 2h}{1 + \tilde{f}\tilde{M}}$.

This system can be rewritten as a matrix in the following way

$$\begin{bmatrix} -\lambda - k^2 D_P - a & 0 & a \\ 0 & -\lambda - k^2 D_R - c & c \\ -h & A & -\lambda - k^2 D_T + B \end{bmatrix} \begin{pmatrix} P(t) \\ R(t) \\ T(t) \end{pmatrix} = \begin{pmatrix} 0 \\ 0 \\ 0 \end{pmatrix}. \tag{9}$$

Our system is stable if $\lambda$ has negative real part, meaning that

$$\mathcal{M} := \begin{bmatrix} -k^2 D_P - a & 0 & a \\ 0 & -k^2 D_R - c & c \\ -h & A & -k^2 D_T + B \end{bmatrix} \tag{10}$$

has negative eigenvalues.

The next key observation is that this matrix $\mathcal{M}$ is very similar to the one which was observed in [10]. The only difference between the two is that in that case the matrix had Neumann eigenvalues instead of the terms $k^2$, but because both are positive, they behave the same way. Consequently, the proofs presented there (using the Routh-Hurwitz criteria) are applicable also for this problem, and thus, we can state similar theorems.

**Theorem 1.** *Let us suppose that the model parameters satisfy the condition*

$$B^2 - Ac - ah < 0, \tag{11}$$

*Moreover, let us suppose that $D_P$ and $D_R$ are fixed positive diffusion values. Then, if system (5) is stable for $D_T = 0$ then it is stable for all positive $D_T$ values.*

*Remark 1.* In [10] the corresponding theorem had two conditions, however, the second one was only needed for the theoretic result used there, which is not necessary here, so only one condition is present here.

*Remark 2.* The values of the parameters originally chosen in [1,2] are the following. ($\mathcal{L}$ is chosen to be 60000, since the coast of the island is approximately 60000 meters long, however, it does not affect the condition.)

$$a = 0.03, \quad b = 1, \quad c = 10, \quad M = 12000, \quad f = 0.001, \quad h = 0.25, \quad \mathcal{L} = 60000.$$

Condition (11) is satisfied if $f > 8.6751 \times 10^{-6}$ (or $\tilde{f} > 0.52051$), which can be assumed since rats have a significant effect on the reproduction rate of the trees.

In this way if the system was stable in the case of no tree diffusion, it will remain so even if $D_T$ is increased, so the region of stability cannot shrink as the diffusion is increased.

**Theorem 2.** *Let $D_P$ and $D_R$ be two fixed positive numbers, and suppose that the following conditions hold:*

$$B < \min\{a + c, A + h\},$$
$$B(Ac + ah + (a + c)^2) < c^2(A + a) + B^2(a + c) + a^2(c + h). \quad (12)$$

*Then there is a positive number $\tilde{D}_T$ such that the system (5) is stable for all $D_T > \tilde{D}_T$ values.*

*Remark 3.* Condition (12) is satisfied with the parameter values described above.

Consequently, the increase of $D_T$ makes our system stable even if it was unstable for small values of the diffusion of the trees.

Now let us examine whether the previous analytic results can be confirmed by some numerical experiments or not. For these, a finite difference scheme is used in the following way. Let us consider a grid $\mathcal{G}$ on our interval $[0, \mathcal{L})$ which consists of $\mathcal{N}$ number of points $\{\omega_i : \frac{i\mathcal{L}}{\mathcal{N}}\}_{i=0}^{\mathcal{N}-1}$, and let us denote the numerical approximation of our function $\mathcal{Q}(t, x)$ at these points by $\mathcal{Q}_i(t)$. Also, let us approximate the diffusion terms $D\Delta\mathcal{Q}(t, x)$ by $D(\mathcal{N}/\mathcal{L})^2(\mathcal{Q}_{i-1}(t) - 2\mathcal{Q}_i(t) + \mathcal{Q}_{i+1}(t))$. Note that this way we got the same system as (2), meaning that the numerical solution of this semi-discretized problem would be the same as the one already calculated in [9]. In that a fourth order Runge-Kutta method is used with initial vector $\mathbf{w} \in \mathbb{R}^{3\mathcal{N}}$ which is defined as

$$w_i := \begin{cases} 0.001 & \text{for } i \in \{3j - 2, \, 3j - 1 \mid j \in \mathbf{Z}^+, \, j \leq 0.1\,\mathcal{N}\}, \\ \tilde{M} & \text{for } 3|i, \\ 0 & \text{otherwise.} \end{cases}$$

This way, we have trees with density of $\tilde{M}$ in every region, and a relatively small density of people and rats in a small part of our coast, which corresponds to the moment when the settlers arrived on the island bringing the rats with them.

With the initial vector **w** we examine the way the previous analytic results occur in the terms of the orbits of solutions. On the left panel of Fig. 1 we can see that for the diffusion parameters $(D_P, D_R, D_T) = (0.02, 0.01, 0.0005)$ the system is unstable, meaning that the populations die out. However, if the diffusion of the trees is increased and the parameters $(0.02, 0.01, 0.001)$ are considered, then the solutions tend to the stationary solution (6), which confirms the analytic results.

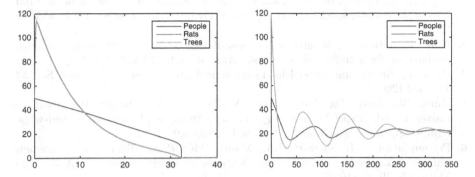

**Fig. 1.** The number of people (blue), rats (red) and trees (yellow) at point 0 with $\mathcal{N} = 100$ spatial mesh points for diffusion parameters $(0.02, 0.01, 0.0005)$ (left) and for parameters $(0.02, 0.01, 0.001)$. As we can see, the increase of the diffusion of the trees stabilizes the stationary solution corresponding to coexistence. Note that the density of rats is very close to the density of the trees, thus it is not visible. (Color figure online)

# 4    Conclusions, Further Work

In this paper the original model from [7] was extended in a way that instead of a discrete spatial domain we considered a continuous one. It turned out that as in the discrete case, the increase of the tree diffusion stabilizes our system. We have also seen that the analytic results can be confirmed by numerical experiments. A further task would be to understand the global behavior of this system better, i.e. decide whether it has periodic solutions, or other, non-constant steady states, and examine their stability properties.

A further extension of this model would be to consider a two dimensional system, since Easter Island does not have one, but two volcanoes, and these are not in the middle, but near the shores of the island. This work was done in [10], in which it turned out that tree diffusion stabilizes the system even in the two dimensional case, meaning that for a sufficiently large value of $D_T$ the populations do not die out, but rather stabilize on a smaller, but positive value. Some authors suggest that the original settlers did not die out before the first Europeans arrived on the island, but had a small, but stable economy (still smaller than before), and the real catastrophe happened when the first slave-traders arrived a few years later, as mentioned in [5].

532    B. Takács

**Acknowledgments.** Supported by the ÚNKP-17-3 New National Excellence Program of the Ministry of Human Capacities.

# References

1. Basener, W., Brooks, B., Radin, M., Wiandt, T.: Rat instigated human population collapse on Easter Island. Nonlinear Dyn. Psychol. Life Sci. **12**(3), 227–240 (2008)
2. Basener, W., Brooks, B., Radin, M., Wiandt, T.: Spatial effects and turing instabilities in the invasive species model. Nonlinear Dyn. Psychol. Life Sci. **15**(4), 455–464 (2011)
3. Hunt, T.: Rethinking the fall of Easter Island: new ecidence points to an alternative explanation for a civilization's collapse. Am. Sci. **94**, 412–419 (2006)
4. Hunt, T.: Rethinking Easter Island's ecological catastrophe. J. Archaeol. Sci. **34**, 485–502 (2007)
5. Hunt, T., Lipo, C.: The Statues Walked - What Really Happened on Easter Island. http://longnow.org/seminars/02013/jan/17/statues-walked-what-really-happened-easter-island/. Accessed 2 Aug 2018
6. Perumpanani, A.J., Sherratt, J.A., Maini, P.K.: Phase differences in reaction-diffusion-advection systems and applications to morphogenesis. IMA J. Appl. Math. **55**, 19–33 (1995)
7. Takács, B., Horváth, R., Faragó, I.: The effect of tree-diffusion in a mathematical model of Easter Island's population. Electron. J. Qual. Theory Differ. Equ. **84**, 1–11 (2016)
8. Takács, B.: Analysis of some characteristic parameters in an invasive species model. Annales. Univ. Sci. Budapest., Sect. Comp. **45**, 119–133 (2016)
9. Takács, B.: Modeling the ecological collapse of Easter Island. Master's thesis (2017). https://web.cs.elte.hu/blobs/diplomamunkak/msc_alkmat/2017/takacs_balint.pdf
10. Takács, B., Horváth, R., Faragó, I.: The effect of tree diffusion in a two dimensional continuous model for Easter Island. Eur. J. Math. (accepted)

# Solving a Singularly Perturbed Elliptic Problem by a Cascadic Multigrid Algorithm with Richardson Extrapolation

Svetlana Tikhovskaya$^{(\boxtimes)}$ [ID]

Sobolev Institute of Mathematics SB RAS,
Omsk Branch, Pevtsova Street, 13, 644099 Omsk, Russia
s.tihovskaya@yandex.ru

**Abstract.** A two-dimensional linear elliptic equation with regular boundary layers is considered in the unit square. It is solved by using an upwind difference scheme on the Shishkin mesh which converges uniformly with respect to a small parameter $\varepsilon$. It is known that the application of multigrid methods leads to essential reduction of arithmetical operations amount. Earlier we investigated the cascadic two-grid method with the application of Richardson extrapolation to increase the $\varepsilon$-uniform accuracy of the difference scheme. In this paper multigrid algorithm of the same structure is studied. We construct an extrapolation of initial guess using numerical solutions on two coarse meshes to reduce the arithmetical operations amount. The application of the Richardson extrapolation method based on numerical solutions on the last three meshes leads to increase the $\varepsilon$-uniform accuracy of the difference scheme by two orders. The different components of a cascadic multigrid method are studied. The results of some numerical experiments are discussed.

**Keywords:** Singularly perturbed elliptic problem ·
Regular boundary layers · Difference scheme · Shishkin mesh ·
$\varepsilon$-uniform accuracy · Cascadic multigrid method ·
Richardson extrapolation

## 1 Introduction

The two-dimensional linear elliptic problem with regular boundary layers in the unit square is considered. It is well known that the application of classical difference schemes for a singularly perturbed problem leads to large errors for small values of the perturbation parameter [1–5]. The uniform convergence of a difference scheme for such problem can be provided by fitting the scheme to a boundary layer component [1] or by using a mesh which is dense in a boundary layer [2,3].

The multigrid methods [6–9] lead to essential reduction of the number of arithmetical operations. These methods can be effective applied for singularly

© Springer Nature Switzerland AG 2019
I. Dimov et al. (Eds.): FDM 2018, LNCS 11386, pp. 533–541, 2019.
https://doi.org/10.1007/978-3-030-11539-5_62

perturbed problems, see [10–17] and the references therein. To increase the accuracy of the difference scheme in multigrid method the Richardson extrapolation [5, 18] can be applied. The two-grid method with the application Richardson extrapolation to increase the $\varepsilon$-uniform accuracy of the difference scheme on the Shishkin mesh is investigated in [15–17]. In [19] the multigrid algorithm of the same structure based on three meshes is considered and to reduce the number of iteration we apply the idea of the extrapolation of numerical solutions on coarse meshes [20, 21].

A cascadic multigrid method for classical elliptic boundary value problems is considered in a number of papers [22–24] and the references therein. In [23] it is shown numerically that the two-level method is computationally effective and has simplicity of programm, may be apply in the cases of the square and L-shape domains for 2D and 3D. In [22] it is shown numerically that the proposed extrapolation cascadic multigrid method is much more efficient comparing to the classical V-cycle and W-cycle multigrid method and it is particularly suitable for solving large scale problems. In [24] it is shown numerically that the proposed EXFMG method improves the solution accuracy and keeps less computational costs, compared to the classical multigrid methods and can also be extended to solve other kinds of partial differential equations. It is shown in [10] that standard multigrid methods are not convergent for singularly perturbed elliptic problem.

The aim of this work is to study a cascadic multigrid method using Richardson extrapolation and its components for a singularly perturbed elliptic problem based on the $\varepsilon$-uniform difference scheme on the Shishkin mesh.

**Notation:** Here $C$, sometimes subscripted, denotes a generic positive constant that is independent of the perturbation parameter $\varepsilon$ and the step size of the mesh.

## 2   Preliminaries

We consider the following boundary value problem:

$$\varepsilon u_{xx} + \varepsilon u_{yy} + a(x)u_x + b(y)u_y - c(x,y)u = f(x,y), \ (x,y) \in \Omega = (0,1)^2, \tag{1}$$
$$u(x,y) = g(x,y), \quad (x,y) \in \Gamma = \overline{\Omega} \setminus \Omega,$$

where the coefficients $a$, $b$, $c$ are bounded and satisfy the condition:

$$a(x) \geqslant \alpha > 0, \quad b(y) \geqslant \beta > 0, \quad c(x,y) \geqslant 0, \tag{2}$$

the perturbation parameter $\varepsilon$ takes arbitrary values in the open-closed interval $(0, 1]$. The coefficients, the right-hand side $f$ and the boundary function $g$ are sufficiently smooth. We also assume that the sufficient compatibility conditions are satisfied.

It is known [3–5] that the solution of problem (1) under condition (2) is uniformly bounded and has two regular boundary layers near $x = 0$ and $y = 0$.

Define the piecewise-uniform mesh [3] in the domain $\overline{\Omega}$:

$$\Omega_N = \{(x_i, y_j),\ i,j = 0,1,\ldots,N,\ h_i = x_i - x_{i-1},\ \tau_j = y_j - y_{j-1}\},$$
$$h_i = 2\sigma_x/N,\quad 1 \leqslant i \leqslant N/2;\qquad h_i = 2(1-\sigma_x)/N,\quad N/2 < i \leqslant N,$$
$$\tau_j = 2\sigma_y/N,\quad 1 \leqslant j \leqslant N/2;\qquad \tau_j = 2(1-\sigma_y)/N,\quad N/2 < j \leqslant N, \qquad (3)$$
$$\sigma_x = \min\left\{\frac{1}{2}, \frac{2\varepsilon}{\alpha}\ln N\right\},\quad \sigma_y = \min\left\{\frac{1}{2}, \frac{2\varepsilon}{\beta}\ln N\right\}.$$

where $\alpha$ and $\beta$ are given in (2).

We consider the upwind difference scheme for the problem (1) on the mesh $\Omega_N$:

$$L^N u^N = \frac{2\varepsilon}{h_i + h_{i+1}}\left(\frac{u_{i+1,j}^N - u_{i,j}^N}{h_{i+1}} - \frac{u_{i,j}^N - u_{i-1,j}^N}{h_i}\right) +$$

$$+\frac{2\varepsilon}{\tau_j + \tau_{j+1}}\left(\frac{u_{i,j+1}^N - u_{i,j}^N}{\tau_{j+1}} - \frac{u_{i,j}^N - u_{i,j-1}^N}{\tau_j}\right) + a(x_i)\frac{u_{i+1,j}^N - u_{i,j}^N}{h_{i+1}} + \qquad (4)$$

$$+b(y_j)\frac{u_{i,j+1}^N - u_{i,j}^N}{\tau_{j+1}} - c(x_i, y_j)u_{i,j}^N = f(x_i, y_j),\quad (x_i, y_j) \in \Omega_N,$$

$$u_{i,j}^N = g(x_i, y_j),\qquad (x_i, y_j) \in \Gamma_N = \Gamma \cap \Omega_N.$$

Let $[u]_\Omega$ be the projection of a function $u(x,y)$ on a mesh $\Omega$. According to [3,4] the difference scheme (4) on the mesh (3) converges $\varepsilon$-uniformly and the following accuracy estimate is satisfied:

$$\max_{0 \leqslant i,j \leqslant N} |u_{i,j}^N - u(x_i, y_j)| = \|u^N - [u]_{\Omega_N}\| \leqslant C\Delta_N,\qquad \Delta_N = \ln N/N, \quad (5)$$

where $C$ is positive constant that is independent of the perturbation parameter $\varepsilon$ and the parameters of the mesh (3).

Notice that the matrix of this system (4) is $M$-matrix and a number of iterative methods for its resolving converge [25,27].

## 3   Cascadic Multigrid Method

We study a cascadic multigrid method with structure like in the papers [15–17,19], and with the number of nodes of the coarsest mesh 4. To improve the accuracy of the difference scheme (4) in the multigrid method we apply Richardson extrapolation [5,18] using three last numerical solutions.

Let $u^N$ be the solution of the difference scheme (4) on the mesh $\Omega_N$. For the application of Richardson extrapolation we use the solutions of the difference scheme (4) on the meshes $\Omega_{N/2}$ and $\Omega_{N/4}$ which ones have the same value of the parameters $\sigma_x$ and $\sigma_y$ as the refined mesh $\Omega_N$. Thus these meshes are nested that is $\Omega_{N/4} = \{(X_l, Y_m)\} \subset \Omega_{N/2} = \{(X_p, Y_q)\} \subset \Omega_N = \{(x_i, y_j)\}$.

Let us define the solution of the difference scheme (4) on the mesh $\Omega_{N/2}$ as $u^{N/2}$ and on the mesh $\Omega_{N/4}$ as $u^{N/4}$. According to the Richardson extrapolation we introduce the function $u^{nN}$ on the mesh $\Omega_N$. At first we define the function $u^{nN}$ at the nodes of the coarse mesh $(X_l, Y_m) \in \Omega_{N/4}$ the formula:

$$u^{nN}(X_l, Y_m) = 8/3\, u^{N/4}(X_l, Y_m) - 2\, u^{N/2}(X_l, Y_m) + 1/3\, u^N(X_l, Y_m).$$

Now let us specify $u^{nN}(x_i, y_j)$ at the nodes of refined mesh $\Omega_N \setminus \Omega_{N/4}$ using the interpolation function. Then we define for each of nodes $(x_i, y_j) \in \Omega_N$ from some cell $S_{l,m} = [X_{l-2}, X_{l+2}] \times [Y_{m-2}, Y_{m+2}]$ as

$$u^{nN}(x_i, y_j) = I([u^{nN}]_{\Omega_{N/4}}, x_i, y_j), \tag{6}$$

where we use $\varepsilon$-uniform interpolation, see [16, 17, 19] and reference therein.

To construct a better initial guess on refined mesh $\Omega_N$ we use the same idea of articles [20, 21] and the proposed formulas in [19]. We can get the values at other nodes on the mesh $\Omega_N \setminus \Omega_{N/2}$, using an appropriate interpolation.

It is stated in the following algorithm.

Step 1.1. Compute the solutions $u_1^{N/4}$, $u_2^{N/2}$ on the coarse meshes and $k = 3$.
Step 1.2. Get an initial guess $u_m^{N/4,(0)} = E_k^N(u_{k-2}^{N/4}, u_{k-1}^{N/2})$.
Step 1.3. Solve till the stop criterion is not fulfill.
Step 1.4. If $k \neq (\log_2(N) - 2)$ then $k := k + 1$ and return Step 1.2 else stop and $u_m^{N,(M)} = u^{nN} = R_k^N(u_{k-2}^{N/4}, u_{k-1}^{N/2}, u_k^N)$.

Note that for $k = 1$ $u_1^{N,(0)} \equiv 0$ at nodes $(X_l, Y_m) \in \Omega_N \setminus \Gamma_N$, $u_1^{N,(0)} \equiv g(X_l, Y_m)$ at nodes $(X_l, Y_m) \in \Gamma_N$.

We proposed to use a residual stop criterion for the coarsest mesh $k = 1$

$$\|L^N u^N - [f]_{\Omega_N}\| \leqslant \Delta_N, \qquad \Delta_N = \ln N/N, \tag{7}$$

and a corrected residual stop criterion for other meshes $k > 1$

$$\|L^N u^N - [f]_{\Omega_N}\| \leqslant \ln N/(\varepsilon N), \tag{8}$$

Note that we can use the stop criterion (8) only since we have sufficiently good guess.

## 4    Results of Numerical Experiments

We consider the following boundary layer problem:

$$\begin{aligned} \varepsilon\, u_{xx} + \varepsilon\, u_{yy} + u_x + u_y &= f(x, y), \qquad (x, y) \in \Omega, \\ u(x, y) &= 0, \qquad (x, y) \in \Gamma, \end{aligned} \tag{9}$$

where $f$ corresponds to the exact solution:

$$u(x, y) = (1 - x)\left(1 - e^{-x/\varepsilon}\right)(1 - y)\left(1 - e^{-y/\varepsilon}\right).$$

**Table 1.** The error norm for a one-grid method (left) and for a CMG (right)

| $\varepsilon$ | $N$ | | | | $\varepsilon$ | $N$ | | | |
|---|---|---|---|---|---|---|---|---|---|
| | 64 | 128 | 256 | 512 | | 64 | 128 | 256 | 512 |
| $10^{-3}$ | 6.16e−2 | 3.72e−2 | 2.17e−2 | 1.24e−2 | $10^{-3}$ | 4.12e−3 | 1.05e−3 | 2.95e−4 | 8.16e−5 |
| $10^{-4}$ | 6.18e−2 | 3.73e−2 | 2.18e−2 | 1.24e−2 | $10^{-4}$ | 4.23e−3 | 1.27e−3 | 3.31e−4 | 8.85e−5 |
| $10^{-5}$ | 6.19e−2 | 3.73e−2 | 2.18e−2 | 1.24e−2 | $10^{-5}$ | 4.24e−3 | 1.27e−3 | 3.28e−4 | 8.80e−5 |
| $10^{-7}$ | 6.19e−2 | 3.73e−2 | 2.18e−2 | 1.24e−2 | $10^{-7}$ | 4.25e−3 | 1.27e−3 | 3.27e−4 | 8.79e−5 |
| $10^{-9}$ | 6.19e−2 | 3.73e−2 | 2.18e−2 | 1.24e−2 | $10^{-9}$ | 4.25e−3 | 1.27e−3 | 3.27e−4 | 8.79e−5 |

**Table 2.** The time of computing for a one-grid method (left) and for CMG without Step 1.2 and with a standard residual stop criterion (right)

| $\varepsilon$ | $N$ | | | | $\varepsilon$ | $N$ | | | |
|---|---|---|---|---|---|---|---|---|---|
| | 32 | 64 | 128 | 256 | | 32 | 64 | 128 | 256 |
| $10^{-2}$ | 0.01 | 0.08 | 0.89 | 12.97 | $10^{-2}$ | 0.02 | 0.05 | 0.44 | 5.59 |
| $10^{-3}$ | 0.01 | 0.08 | 0.99 | 14.77 | $10^{-3}$ | 0.01 | 0.06 | 0.63 | 8.26 |
| $10^{-5}$ | 0.01 | 0.11 | 1.24 | 18.38 | $10^{-5}$ | 0.01 | 0.09 | 0.97 | 12.79 |
| $10^{-7}$ | 0.01 | 0.13 | 1.59 | 22.81 | $10^{-7}$ | 0.01 | 0.12 | 1.29 | 17.08 |
| $10^{-9}$ | 0.02 | 0.17 | 1.92 | 27.29 | $10^{-9}$ | 0.02 | 0.16 | 1.67 | 22.27 |

The solution of the problem (9) is computed based on the difference scheme (4). We define the initial guess as $u^{(0)}(x_i, y_j) = 0$, $(x_i, y_j) \in \Omega_N$.

The difference scheme (4) can be written as the five-point scheme. Notice that the efficiency of Gauss-Seidel method in the case of the problem (1) depends on the ordering of equations and unknowns [25–27]. Therefore we study a line Gauss-Seidel of alternating symmetric type, the lines are processed in $x$- and $y$-directions in forward and backward lexicographical order.

Table 1 contains the error norm for a one-grid method (left table) and for a cascadic multigrid method – CMG (right table) for various values of $\varepsilon$ and $N$.

It follows from Table 1 that the application of Richardson extrapolation (6) in the Step 1.4 increases the accuracy of the difference scheme (4) to $O(\ln^3 N/N^3)$.

Table 2 contains the time of computing for a one-grid method (left table) and for CMG without Step 1.2 and with a standard residual stop criterion (7) (right table) for various values of $N$ and $\varepsilon$.

It follows from Table 2 that the time of computing for one-grid method and CMG without Step 1.2 and with a standard residual stop criterion depends from a small parameter $\varepsilon$ and does not increment linearly and it does not similar to properties multigrid method. But CMG is slightly more computationally efficient than one-grid method.

Table 3 contains the time of computing for CMG with Step 1.2 and a standard residual stop criterion (left table) and for CMG without Step 1.2 and a corrected residual stop criterion (8) (right table) for various values of $N$ and $\varepsilon$.

**Table 3.** The time of computing for CMG with Step 1.2 and a standard residual stop criterion (left) and for CMG without Step 1.2 and a corrected residual stop criterion (right)

| $\varepsilon$ | $N$ | | | | $\varepsilon$ | $N$ | | | |
|---|---|---|---|---|---|---|---|---|---|
| | 64 | 128 | 256 | 512 | | 64 | 128 | 256 | 512 |
| $10^{-2}$ | 0.05 | 0.15 | 1.95 | 28.11 | $10^{-2}$ | 0.05 | 0.42 | 5.20 | 84.22 |
| $10^{-3}$ | 0.04 | 0.37 | 4.06 | 51.26 | $10^{-3}$ | 0.04 | 0.43 | 5.32 | 86.26 |
| $10^{-5}$ | 0.08 | 0.79 | 10.28 | 162.73 | $10^{-5}$ | 0.04 | 0.43 | 5.37 | 85.36 |
| $10^{-7}$ | 0.10 | 1.16 | 14.88 | 244.36 | $10^{-7}$ | 0.04 | 0.42 | 5.34 | 85.59 |
| $10^{-9}$ | 0.14 | 1.48 | 19.60 | 324.44 | $10^{-9}$ | 0.04 | 0.42 | 5.37 | 85.50 |

**Table 4.** The time of computing for CMG with Step 1.2 and a corrected residual stop criterion (left) and for CMG with $N_k = 10$

| $\varepsilon$ | $N$ | | | | $\varepsilon$ | $N$ | | | |
|---|---|---|---|---|---|---|---|---|---|
| | 64 | 128 | 256 | 512 | | 64 | 128 | 256 | 512 |
| $10^{-2}$ | 0.04 | 0.17 | 1.66 | 22.19 | $10^{-2}$ | 0.03 | 0.10 | 0.34 | 1.58 |
| $10^{-3}$ | 0.03 | 0.19 | 1.83 | 21.21 | $10^{-3}$ | 0.03 | 0.09 | 0.36 | 1.68 |
| $10^{-5}$ | 0.03 | 0.19 | 1.82 | 21.04 | $10^{-5}$ | 0.03 | 0.09 | 0.39 | 1.77 |
| $10^{-7}$ | 0.03 | 0.19 | 1.81 | 20.96 | $10^{-7}$ | 0.03 | 0.10 | 0.37 | 1.73 |
| $10^{-9}$ | 0.03 | 0.19 | 1.84 | 20.92 | $10^{-9}$ | 0.03 | 0.09 | 0.38 | 1.71 |

It follows from Tables 2 and 3 that the use Step 1.2 leads to reduce the time of computing. From Table 3 follows that the time of computing for CMG with a corrected residual stop criterion does not depend from a small parameter $\varepsilon$. However, the corrected residual stop criterion (8) does not lead to the time of computing increment linearly.

We note that CMG can have good properties similar to the multigrid method. To achieve it we can use instead of the residual stop criterions (7), (8) in Step 1.3 on each level (the coarse meshes) a fixed number of iterations, for instance, $N_k = 10$ times.

Table 4 contains the time of computing for CMG with Step 1.2 and a corrected residual stop criterion (left table) and the time of computing for CMG with fixed number of iterations $N_k = 10$ (right table) for various values of $N$ and $\varepsilon$. It should be remembered that the number of iteration $N_k = 10$ is a numerical result.

It follows from Table 4 that that the time of computing for CMG with a corrected residual stop criterion does not depend from a small parameter $\varepsilon$ and the use of Step 1.2 leads to essential reduce of the time of computing. Here it should be noted that such a stop criterion (fixed number of iterations) leads to a loss of the accuracy of the numerical solution in the Step 1.4 since the Richardson extrapolation does not work.

**Table 5.** The error norm for CMG with $N_k = 10$ without the Richardson extrapolation (left) and with the Richardson extrapolation (right)

| $\varepsilon$ | $N$ | | | | $\varepsilon$ | $N$ | | | |
|---|---|---|---|---|---|---|---|---|---|
| | 64 | 128 | 256 | 512 | | 64 | 128 | 256 | 512 |
| $10^{-2}$ | 6.06e−2 | 3.89e−2 | 2.54e−2 | 1.74e−2 | $10^{-2}$ | 7.26e−3 | 7.26e−3 | 7.08e−3 | 6.94e−3 |
| $10^{-3}$ | 6.29e−2 | 4.01e−2 | 2.61e−2 | 1.78e−2 | $10^{-3}$ | 7.46e−3 | 7.33e−3 | 7.15e−3 | 7.03e−3 |
| $10^{-5}$ | 6.32e−2 | 4.02e−2 | 2.62e−2 | 1.79e−2 | $10^{-5}$ | 7.49e−3 | 7.38e−3 | 7.22e−3 | 7.09e−3 |
| $10^{-7}$ | 6.32e−2 | 4.02e−2 | 2.62e−2 | 1.79e−2 | $10^{-7}$ | 7.49e−3 | 7.39e−3 | 7.22e−3 | 7.09e−3 |
| $10^{-9}$ | 6.32e−2 | 4.02e−2 | 2.62e−2 | 1.79e−2 | $10^{-9}$ | 7.49e−3 | 7.39e−3 | 7.22e−3 | 7.09e−3 |

Table 5 contains the error norm for CMG with fixed number of iterations $N_k = 10$ without the Richardson extrapolation (left table) and with the Richardson extrapolation (right table) for various values of $N$ and $\varepsilon$.

It follows from Table 5 that the use of extrapolation (6) in the case of a stop criterion with a fixed number of iteration is useless and the accuracy is the same with one-grid method. Thus, we find that the numerical solutions from CMG with Step 1.4 are much more accurate than those from CMG with fixed number of iterations. This may be considered how, to achieve the same accuracy, fewer mesh points are needed. For instance, the numerical solution by CMG with Step 1.4 on the 64 × 64 mesh is comparable to the numerical solution obtained by CMG with fixed number of iterations on the 512 × 512 mesh in Tables 1 and 5.

**Acknowledgments.** Research has been supported by the program of fundamental scientific researches of the SB RAS No I.1.3., project No 0314-2016-0009.

# References

1. Ilyin, A.M.: A difference scheme for a differential equation with a small parameter at the highest derivative. Mat. Zametki. **6**(2), 237–248 (1969). (in Russian)
2. Bakhvalov, N.S.: The optimization of methods of solving boundary value problems with a boundary layer. USSR Comput. Math. Math. Phys. **9**(4), 139–166 (1969). https://doi.org/10.1016/0041-5553(69)90038-X
3. Shishkin, G.I.: Grid Approximations of Singular Perturbation Elliptic and Parabolic Equations. UB RAS, Yekaterinburg (1992). (in Russian)
4. Roos, H.-G., Stynes, M., Tobiska, L.: Robust Numerical Methods for Singularly Perturbed Differential Equations. SSCM. Springer, Berlin (2008). https://doi.org/10.1007/978-3-540-34467-4
5. Shishkin, G.I., Shishkina, L.P.: Difference Methods for Singular Perturbation Problems. Chapman & Hall/CRC, Boca Raton (2009)
6. Fedorenko, R.P.: The speed of convergence of one iterative process. Zh. Vychisl. Mat. Mat. Fiz. **4**(3), 559–564 (1964). (in Russian)
7. Hackbusch, W.: Multigrid convergence for a singular perturbation problem. Linear Algebra Appl. **58**, 125–145 (1984)

8. Shaidurov, V.V.: Multigrid Methods for Finite Elements. MAIA. Springer, Berlin (1995). https://doi.org/10.1007/978-94-015-8527-9
9. Trottenberg, U., Oosterlee, C.W., Schuller, A.: Multigrid. Academic Press Inc., San Diego (2001)
10. Gaspar, F.J., Clavero, C., Lisbona, F.: Some numerical experiments with multigrid methods on Shishkin meshes. J. Comput. Appl. Math. **138**, 21–35 (2002). https://doi.org/10.1016/S0377-0427(01)00365-X
11. Olshanskii, M.A.: Analysis of a multigrid method for convection-diffusion equations with the dirichlet boundary conditions. Comput. Math. Math. Phys. **44**(8), 1374–1403 (2004)
12. Angelova, I.T., Vulkov, L.G.: Comparison of the two-grid method on different meshes for singularly perturbed semilinear problems. In: Applications of Mathematics in Engineering and Economics, pp. 305–312. American Institute of Physics (2008). https://doi.org/10.1063/1.3030800
13. Vulkov, L.G., Zadorin, A.I.: Two-grid algorithms for the solution of 2D semilinear singularly perturbed convection-diffusion equations using an exponential finite difference scheme. In: Application of Mathematics in Technical and Natural Sciences, pp. 371–379. AIP Conference Proceedings (2009). https://doi.org/10.1063/1.3265351
14. MacLachlan, S., Madden, N.: Robust solution of singularly perturbed problems using multigrid methods. SIAM J. Sci. Comput. **35**(5), A2225–A2254 (2013). https://doi.org/10.1016/0024-3795(84)90208-8
15. Tikhovskaya, S.V.: A two-grid method for an elliptic equation with boundary layers on a Shishkin mesh. Lobachevskii J. Math. **35**(4), 391–397 (2014). https://doi.org/10.1134/S1995080214040064
16. Zadorin, A.I., Tikhovskaya, S.V., Zadorin, N.A.: A two-grid method for elliptic problem with boundary layers. Appl. Numer. Math. **93**, 270–278 (2015). https://doi.org/10.1016/j.apnum.2014.06.003
17. Tikhovskaya, S.V.: Investigation of a two-grid method of improved accuracy for elliptic reaction-diffusion equation with boundary layers. Kazan. Gos. Univ. Uchen. Zap. Ser. Fiz.-Mat. Nauki **157**(1), 60–74 (2015). (in Russian)
18. Shishkin, G.I., Shishkina, L.P.: A higher-order richardson method for a quasilinear singularly perturbed elliptic reaction-diffusion equation. Differ. Equ. **41**(7), 1030–1039 (2005). https://doi.org/10.1007/s10625-005-0245-8
19. Tikhovskaya, S.V.: Solving a singularly perturbed elliptic problem by a multigrid algorithm with richardson extrapolation. In: Dimov, I., Faragó, I., Vulkov, L. (eds.) NAA 2016. LNCS, vol. 10187, pp. 674–681. Springer, Cham (2017). https://doi.org/10.1007/978-3-319-57099-0_77
20. Chen, C.M., Hu, H.L., Xie, Z.Q., Li, C.L.: Analysis of extrapolation cascadic multigrid method (EXCMG). Sci. China Ser. A Math. **51**(8), 1349–1360 (2008). https://doi.org/10.1007/s11425-008-0119-7
21. Ming, L., Chen-liang, L., Xiang-zhao, C., Jin-e, Z.: Cascadic multigrid methods combined with sixth order compact scheme for Poisson equation. Numer. Algor. **71**(4), 715–727 (2016). https://doi.org/10.1007/s11075-015-0018-2
22. Pann, K., Hu, D., Hu, H.: An extrapolation cascadic multigrid method combined with a fourth-order compact scheme for 3D poisson equation. J. Sci. Comput. **70**(3), 1180–1203 (2017). https://doi.org/10.1007/s10915-016-0275-9
23. Li, N., Tan, Z., Feng, X.: Novel two-level discretization method for high dimensional semilinear elliptic problems base on RBF-FD scheme. Numer. Heat Transf., Part B: Fundam. **72**(5), 349–360 (2017). https://doi.org/10.1080/10407790.2017.1409511

24. Li, M., Zheng, Z., Pan, K.: An extrapolation full multigrid algorithm combined with fourth-order compact scheme for convection-diffusion equations. Adv. Differ. Equ. **70**, 178-1–178-14 (2018). https://doi.org/10.1186/s13662-018-1631-x
25. Wessiling, P.: An Introduction to Multigrid Methods. Wiley, Chichester (1992)
26. Han, H., Il'in, V.P., Kellogg, R.B.: Flow directed iterations for convection dominated flow. In: Proceeding of the Fifth International Conference on Boundary and Interior Layers, pp. 7–17 (1988)
27. Ilin, V.P.: Finite Difference and Finite Volume Methods for Elliptic Equations. ICMMG Publishers, Novosibirsk (2001). (in Russian)

# Iterative Algorithms for Coupled Physics Electrical Conductivity Imaging

Alex Timonov$^{(\boxtimes)}$

University of South Carolina Upstate, Spartanburg, SC 29303, USA
atimonov@uscupstate.edu

**Abstract.** Coupled Physics Electrical Conductivity Imaging is a group of new imaging modalities in which the electrical conductivity is recovered from the interior data. In this talk the inverse conductivity problem of Current Density Impedance Imaging (CDII) is considered. The variational approach is applied to solve this problem, and the corresponding weighted least gradient problems are analyzed for the Dirichlet and Robin boundary conditions. Three iterative algorithms for constructing the minimizing sequences for the weighted gradient functionals are presented. Their computational effectiveness is demonstrated in numerical experiments. The presentation is based on some results of a long-term project that has been pursued in collaboration with A. Nachman (University of Toronto, Canada) and A. Tamasan (University of Central Florida, USA).

**Keywords:** Current density impedance imaging ·
Weighted least gradient problem · Regularization ·
Iterative algorithms · Numerical experiments

## 1 Introduction

In Current Density Impedance Imaging (CDII) (see [10] and [16]) the inverse conductivity problem is formulated as follows. Let $\Omega$ be a bounded domain in $R^n$, $(n = 2, 3)$ that is filled with a conductive medium. The electrical conductivity $\sigma$ is supposed to be isotropic and positive, and the voltage potential $u$ satisfies the equation

$$\nabla \cdot (\sigma(x)\nabla u) = 0 \text{ in } \Omega \tag{1}$$

subject to a certain boundary condition. Given the function $a = \sigma|\nabla u|$ in $\Omega$ and the boundary condition, find the pair $(\sigma, u)$ in $\Omega$. According to the Ohm's law $J = -\sigma\nabla u$, the function $a$ is interpreted as the magnitude of the current density field inside $\Omega$, and the inverse problem can also be formulated in terms of the weighted 1-Laplacian

$$\nabla \cdot \left( a(x)\frac{\nabla u}{|\nabla u|} \right) = 0 \ \text{ in } \Omega \tag{2}$$

© Springer Nature Switzerland AG 2019
I. Dimov et al. (Eds.): FDM 2018, LNCS 11386, pp. 542–549, 2019.
https://doi.org/10.1007/978-3-030-11539-5_63

subject to a certain boundary condition if there exists its regular solution such that

$$0 < c_1 \leq \frac{a}{|\nabla u|} \leq c_2 < \infty,$$

where $c_1, c_2$ are some constants. The main difficulty is due to the fact that the inverse conductivity problem formulated above is nonlinear and ill-posed in the sense of Hadamard, and the Eq. (2) is singular and elliptic degenerate. In [7] it was first observed the connection with the minimal surfaces in a Riemannian space determined by the interior data and in [8] and [9] the Dirichlet data was also imposed on the boundary. The inverse problem was reduced to a weighted least gradient Dirichlet problem. Both the existence and uniqueness of this problem was studied in [2] and [6]. The structural stability result can be found in [12]. The continuous dependence of $\sigma$ on the interior data $a$ for the given unperturbed Dirichlet data was established in [3] and [4].

However, in case of the Dirichlet boundary data, it is impractical to impose or to measure such data in CDII. This is due to rotation of an object when measuring the interior data. Motivated by this, the Complete Electrode Model (CEM) introduced and developed in [14] and [13] was recently utilized in [11] within the variational approach to CDII. Instead of the boundary voltage potential, the CEM formalizes the physical phenomenon of injecting the direct currents into a conductive object by the surface electrodes. However, it was recently shown in [11] that the corresponding weighted least gradient problem has non-unique solutions. In this presentation, we present a family of the regularizing weighted gradient functionals that utilize the Robin boundary condition. These functionals are used for constructing the minimizing sequences. We outline three iterative procedures for constructing the minimizing sequences. These procedures utilize the so-called simple iteration and Bregman methods. We conclude the presentation with the demonstration of some results of numerical experiments.

## 2    A Weighted Least Gradient Dirichlet Problem

We impose the Dirichlet condition $u = g$ on $\partial\Omega$ and call the pair $(g, a) \in C^{1,\epsilon}(\partial\Omega) \times C^{\epsilon}(\Omega)$ to be admissible if there exists a positive map $\sigma \in L^{\infty}(\Omega)$ bounded away from zero such that if $u \in H^1(\Omega)$ satisfies the Eq. (1) subject to the Dirichlet condition, then $a = \sigma|\nabla u|$. Consider a weighted least gradient problem

$$argmin \left\{ \int_{\Omega} a|\nabla u| dx : u \in W^{1,1}_+(\Omega) \cap C(\overline{\Omega}), u = g \text{ on } \partial\Omega \right\}. \tag{3}$$

The following result establishes existence of a unique solution to this problem

**Theorem 1.** [8] *Let $\Omega$ be a bounded domain in $R^n$ with the connected $C^{1,\epsilon}$-boundary $\partial\Omega$. Let the pair $(g, a)$ be admissible and $a > 0$ a. e. in $\Omega$. Then the variational problem (3) has the unique solution $u^*$ and $\sigma^* = a|\nabla u^*|^{-1}$ is the unique conductivity, for which $a$ is the magnitude of the current density field $J$ and $g$ is the boundary voltage potential.*

Note that at the minimizer $u^*$ the Gateaux derivative vanishes, i.e.,

$$\int_\Omega \frac{|J|}{|\nabla u^*|} \nabla u^* \cdot \nabla v dx = 0, \quad \forall v \in H^1(\Omega).$$

This means that $u^*$ is a weak solution to the Euler-Lagrange Eq. (2).

## 3    A Regularized Weighted Least Gradient Problem

Since imposing or measuring the boundary voltage potential $g$ is impractical, the CEM or a more general Robin condition needs to be utilized when performing CDII. However, in this case the uniqueness result cannot be established for an analogue of the problem (3) with the Robin condition $\sigma(\nabla u \cdot \nu) + \varphi u = f$. Therefore, we regularize the corresponding weighted least gradient problem by introducing a family of the regularizing functionals

$$F_\varepsilon^\alpha(u; a) = \int_\Omega a|\nabla v| dx + \frac{1}{2} \int_{\partial\Omega} \varphi_\varepsilon (u - u_h)^2 ds + \frac{\alpha}{2} \int_\Omega |\nabla(u - u_h)|^2 ds$$

where $\varepsilon > 0, \alpha > 0$ are the small parameters, and $\varphi_\varepsilon$, $f_\varepsilon$ are the $C^2$-smoothers of the given piecewise continuous coefficients $\varphi$ and $f$. Here, $u_h$ is a harmonic function in $\Omega$ with the trace $f_\varepsilon \varphi_\varepsilon^{-1}$ on $\partial\Omega$. Then, for every $\varepsilon > 0$ and $\alpha > 0$ we consider a regularized weighted least gradient problem

$$argmin\{F_\varepsilon^\alpha(u; a) : u \in H^1(\Omega)\}. \tag{4}$$

**Theorem 2.** [15] *Suppose* $a \in L^2(\Omega)$, $a > 0$. *Then for every* $\varepsilon > 0$, $\alpha > 0$ *the variational problem (4) is uniquely solvable.*

## 4    Constructing the Minimizing Sequences

We first describe the process of constructing a minimizing sequence for the problem (3) by the simple iteration method. For a given admissible pair $(a, g)$ let $\sigma_0 = 1$ and $u_0 = h$, where $h$ is a harmonic function in $\Omega$ with the trace $h = g$ on $\partial\Omega$. Assume that we made $(k - 1)$ iterations, and let $u_{k-1} \in H^1(\Omega)$ be its $(k - 1)$st element. Then, we update to $\sigma_k = a|\nabla u_{k-1}|^{-1}$ and determine $u_k$ that satisfies the Eq. (1) with $\sigma_k$ and $u_k = g$ on $\partial\Omega$, etc. The process results in a sequence $\{u_k\}$.

**Theorem 3.** [8] *For* $(k = 1, 2, ....)$ *suppose* $u_{k-1} \in C^{1,\epsilon}(\overline{\Omega})$, $\nabla u_{k-1} \neq 0$. *Then* $u_k \in C^{1,\epsilon}(\overline{\Omega})$, $\nabla u_k \neq 0$, *and* $\sigma_k = a|\nabla u_k|^{-1} \in C^\epsilon(\overline{\Omega})$ *is bounded above and below away from zero. If the sequence* $\{\sigma_k\}$ *is uniformly bounded below and the sequence* $\{G(u_k; a)\}$ *such that*

$$G(u_k; a) = \int_\Omega a|\nabla u_k| dx$$

decreases fast enough, so that $\sum_{k=1}^{\infty} \sqrt{G(u_k; a) - G(u_{k+1}; a)} < \infty$, then there exists $\overline{u} \in H^1(\Omega)$ such that $\overline{u} = g$ on $\partial\Omega$ and

$$\lim_{k \to \infty} \|u_k - \overline{u}\|_{H^1(\Omega)} = 0,$$

and $a|\nabla\overline{u}|^{-1} = \overline{\sigma} \in L_+^{\infty}(\Omega)$.

This procedure is only defined for a boundary voltage potential $g$ that is almost two-to-one, i.e., its set of local maxima consists of either one point or one connected arc. Such a sufficient condition guarantees non-vanishing gradients $\nabla u$ in $\Omega$. If $g$ is not two-to-one, then $\nabla u$ may vanish at some points in $\Omega$. To alleviate this concern, the split Bregman method by Goldstein and Osher [1] was adopted in [5] for any $g \in H^{1/2}(\partial\Omega)$.

Within the concept of the split Bregman method we exploit the convexity property of the energy functional in (3) and Rockafellar-Fenchel duality, which is well suited to a Douglas-Rachford splitting. According to the alternating split Bregman algorithm [5], we represent the voltage potential as $u = v + h$ and initialize $d_0, b_0 \in (L^2)^n$. Beginning with $k = 0$, we start to determine a minimizing sequence $\{u_k\}$ by solving the boundary value problems

$$\triangle v = \nabla \cdot (d_k - b_k) \text{ in } \Omega \quad v = 0 \text{ on } \partial\Omega$$

and updating to $u_{k+1} = v + h$,

$$d_{k+1} = \begin{cases} \max\{|\nabla u_{k+1} + b_k| - a, 0\} \frac{\nabla u_{k+1} + b_k}{|\nabla u_{k+1} + b_k|} & \text{if } |\nabla u_{k+1} + b_k| \neq 0, \\ 0 & \text{if } |\nabla u_{k+1} + b_k| = 0, \end{cases}$$

$$b_{k+1} = b_k + \nabla u_{k+1} - d_{k+1}.$$

**Theorem 4.** [5] *Let $\Omega$ be a bounded domain in $R^n$ with the connected $C^{1,\epsilon}$-boundary $\partial\Omega$, and $a \in L^2(\Omega), a > 0$ a.e. in $\Omega$, $g \in H^{1/2}(\partial\Omega)$. Suppose $\sigma \in C^\epsilon$ and $u^*$ is the unique solution of the weighted least gradient problem (3). Then the sequences $\{u_k\}$, $\{d_k + \nabla h\}$ converge weakly to $u^*$ and $\nabla u^*$, respectively.*

Now we outline constructing a minimizing sequence for the regularized weighted least gradient problem (4). Beginning with $\sigma_0 = 1$ and given $\varphi_\varepsilon, f_\varepsilon, \underline{\sigma}, \overline{\sigma}, \varepsilon > 0, \alpha > 0$, assume that after the $(k-1)$st iteration we obtain $u_{k-1}$. Then we update to

$$\sigma_k = \min\left\{\max\left(\frac{\tilde{a}}{|\nabla u_{k-1}|}, \underline{\sigma}\right), \overline{\sigma}\right\}$$

and solve the problem for $u_k$

$$\nabla \cdot (\sigma_k + \alpha)\nabla u_k = 0 \text{ in } \Omega,$$
$$(\sigma_k + \alpha)(\nabla u_k \cdot \nu) + \varphi_\varepsilon u_k = \alpha(\nabla u_h \cdot \nu) \text{ on } \partial\Omega.$$

The relaxation property for this iterative procedure is established by the following result.

**Theorem 5.** [15] *Suppose the function $v \in H^1(\Omega)$ satisfies the inequalities $\underline{\sigma} \leq \frac{a}{|\nabla v|} \leq \overline{\sigma}$, where $\underline{\sigma}$ and $\overline{\sigma}$ are positive numbers, and the function $u \in H^1(\Omega)$ satisfies the problem*

$$\nabla \cdot \left( \frac{a}{|\nabla v|} + \alpha \right) \nabla u = 0 \ in \ \Omega,$$

$$\left( \frac{a}{|\nabla v|} + \alpha \right) (\nabla u \cdot \nu) + \varphi_\varepsilon u = \alpha(\nabla u_h \cdot \nu) \ on \ \partial\Omega.$$

*Then $F_\varepsilon^\alpha(u; a) \leq F_\varepsilon^\alpha(v; a)$. If the equality holds, then $u = v$.*

It follows from Theorem 2 in [15] that there exists a subsequence $\{u_k\}$ of the sequence of all minimizers of the regularized weighted least gradient problems

$$argmin\{F_\varepsilon^\alpha(u) : u \in H^1(\Omega)\},$$

which converges to the unique minimizer $u^*$ of the following weighted least gradient problem

$$argmin \left\{ \int_\Omega a|\nabla u|dx + \frac{1}{2} \int_{\partial\Omega} \varphi_\varepsilon(u - u_f)ds : u \in H^1(\Omega) \right\}$$

in $L^q$, $0 \leq q \leq n/(n-1)$.

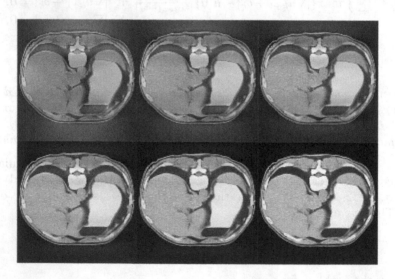

**Fig. 1.** Conductivity reconstruction using the simple iterations algorithm with $N = 1, 5, 10, 30, 50, 100$ iterations (show from the left upper corner to the right lower corner) for the almost two-to-one boundary condition $f = y$.

# 5    Some Results of Numerical Experiments

We present some results of numerical experiments with the iterative algorithms indicated above. Our goal is to demonstrate both the numerical convergence and computational effectiveness of the variational approach to constructing the minimizing sequences for the weighted least gradient problems including its regularized version. It is motivated by the following. As is well known, the Lax-Richtmyer equivalency theorem that consistent numerical approximation, stability and convergence are equivalent does not apply to the nonlinear and, especially, ill-posed problems. Although the convergence and continuous dependence of the approximate solutions on both the interior data and regularization parameter was analytically established, the numerical consistency and stability have to do with the truncation and roundoff errors, respectively, i.e., with perturbations in general. Therefore, the numerical convergence is understood in the sense that a numerical solution gets closer and closer to the original one as the mesh size becomes finer and finer and perturbations of any kind tend to zero. In the numerical experiments we first simulate the interior data, i.e., the magnitude of the current density, by utilizing the CEM. Then, we recover the electrical conductivity $\sigma$ from the perturbed interior data $\tilde{a}$ using both the Dirichlet and Robin boundary conditions and iterative procedures indicated above. The reconstructed images are compared to each other. All computations were performed on the Dell Precision workstation T5400 running under IDL 6.2.

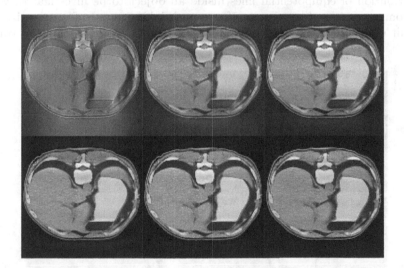

**Fig. 2.** Conductivity reconstruction using the alternating split Bregman algorithm with $N = 1, 5, 10, 30, 50, 100$ iterations (shown from the left upper corner to the right lower corner) for the almost two-to-one boundary condition $f(x, y) = y$.

Since in the numerical experiments the simulated interior data $a$ is stochastically perturbed, we perform reconstructions for samples of 20 realizations, so that a numerical solution is represented by the mean conductivity distribution. In the numerical experiments with the simple iteration and Bregman methods we use the almost two-to-one boundary voltage $f(x,y) = y$. The results obtained by applying the simple iteration method with $N = 1, 5, 10, 30, 50, 100$ iterations are shown in Figs. 1, and 2 shows the mean conductivity reconstructed by the alternating split Bregman algorithm. For two-to-one boundary data, the results of the two algorithms are similar, although the simple iteration method outperforms the alternating Bregman algorithm for a sufficiently large number of iterations. Figure 3 demonstrates performance of the regularizing functionals when recovering the conductivity from the noisy interior data. In the upper row we show the conductivity means for the full electrode apertures and perturbed with the noise levels $10^{-3}$, $10^{-2}$, and $5 \cdot 10^{-2}$ (from left to right). The quasi-optimal parameter $\alpha$ is chosen as $3 \cdot 10^{-2}$, $8 \cdot 10^{-2}$, and $1.2 \cdot 10^{-1}$, respectively. The corresponding relative $l_2$-errors of reconstruction are $4.5 \cdot 10^{-3}$, $7.9 \cdot 10^{-3}$, and $4.3 \cdot 10^{-2}$. In the lower row we show the conductivity means recovered from the interior data simulated for the reduced electrode apertures (in the left corner - a half of the full aperture, in the middle - a quarter of the full aperture, and in the right corner - two step sizes $2h$) with the noise of the level $10^{-2}$ (from left to right). The corresponding relative $l_2$-errors of reconstruction are $1.1 \cdot 10^{-2}$, $1.8 \cdot 10^{-2}$, and $5.3 \cdot 10^{-2}$, respectively. A significant deterioration of accuracy in the lower rows in both figures (from left to right) can be explained as follows. The shrinking of the aperture of electrodes from the full aperture to two step sizes implies a redistribution of equipotential lines inside an object to be investigated, which is becoming more and more nonuniform as shrinking goes to few step sizes. As a result of such a nonuniform spanning, the non two-to-one boundary voltage

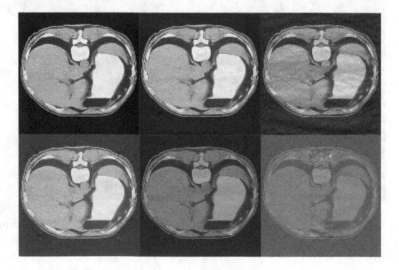

**Fig. 3.** Comparison of the conductivity means recovered from the noisy interior data by the regularized method.

potentials is more typical than exceptional. Apparently, this effect is particularly pronounced for the small electrode apertures and manifests itself in the appearance of some singularity points at the top of the images in the right corner of the lower rowes in the figure.

# References

1. Goldstein, T., Osher, S.: The split Bregman method for L1-regularized problems. SIAM J. Imaging Sci. **2**(2), 323–343 (2009)
2. Jerrard, R.L., Moradifam, A., Nachman, A.: Existence and uniqueness of minimizers of general least gradient problems. J. Reine Angew. Math. **734**, 71–97 (2018)
3. Montalto, C., Stefanov, P.: Stability of coupled physics inverse problems with one internal measurement. Inverse Prob. **29**(12), 125004 (2013)
4. Montalto, C., Tamasan, A.: Stability in conductivity imaging from partial measurements of one interior current. Inverse Prob. Imaging **11**(2), 339–353 (2017)
5. Moradifam, A., Nachman, A., Timonov, A.: A convergent algorithm for the hybrid problem of reconstructing conductivity from minimal interior data. Inverse Prob. **28**, 084003 (2012)
6. Moradifam, A., Nachman, A., Tamasan, A.: Uniqueness of minimizers of weighted least gradient problems arising in conductivity imaging. arXiv:1404.05992 (2017)
7. Nachman, A., Tamasan, A., Timonov, A.: Conductivity imaging with a single measurement of boundary and interior data. Inverse Prob. **23**, 2551–2563 (2007)
8. Nachman, A., Tamasan, A., Timonov, A.: Recovering the conductivity from a single measurement of interior data. Inverse Prob. **25**, 035014 (2009)
9. Nachman, A., Tamasan, A., Timonov, A.: Reconstruction of planar conductivities in subdomains from incomplete data. SIAM J. Appl. Math. **70**(8), 3342–3362 (2010)
10. Nachman, A., Tamasan, A., Timonov, A.: Current density impedance imaging. Contemp. Math. AMS **559**, 1335–1351 (2011)
11. Nachman, A., Tamasan, A., Veras, J.: A weighted minimum gradient problem with complete electrode model boundary conditions for conductivity imaging. SIAM J. Appl. Math. **76**(4), 1321–1343 (2016)
12. Nashed, Z., Tamasan, A.: Structural stability in a minimization problem and applications to conductivity imaging. Inverse Prob. Imaging **5**, 219–236 (2010)
13. Nieminen, J.O., Zevenhoven, K.C.J., Vesanen, P.T., et al.: Current density imaging using ultra low field MRI with adiabatic pulses. Magn. Reson. Imaging **32**, 54–59 (2014)
14. Somersalo, E., Cheney, M., Isaacson, D.: Existence and uniqueness for electrode models for electric current computed tomography. SIAM J. Appl. Math. **54**(1), 1023–1040 (1992)
15. Tamasan, A. and Timonov, A.: A regularized weighted least gradient problem for conductivity imaging. arXiv:1804.00724 (2018)
16. Tamasan, A., Timonov, A.: Coupled physics electrical conductivity imaging. Eurasian J. Math. Comput. Appl. **2**(2), 5–29 (2014)

# Explicit and Conditionally Stable Combined Numerical Method for 1D and 2D Nonlinear Schrödinger Equation

Vyacheslav A. Trofimov$^{(\boxtimes)}$ ⓘ and Evgeny M. Trykin ⓘ

Lomonosov Moscow State University, Leninskie Gory, 119992 Moscow, Russia
vatro@cs.msu.ru, emtrykin@gmail.com

**Abstract.** We propose explicit and conditionally stable combined numerical method based on using of both the conservative finite-difference scheme and non-conservative Rosenbrock method, for solving of 1D and 2D nonlinear Schrödinger equation. Each of these finite-difference schemes has own advantages and disadvantages. The conservative finite-difference scheme is implicit, conservative and possesses the property of asymptotic stability and the second order of approximation. The Rosenbrock method is conditionally conservative, explicit and possesses the same order of approximation in spatial coordinate only. Proposed finite-difference scheme is explicit and more effective for some cases.

The main idea of the combined method consists in using the Rosenbrock's method near the boundaries of the domain. It means, that we introduce certain sub-domains near the boundaries of the domain under consideration. In other part of the domain, the conservative finite-difference scheme is used for computation. The problem solution is provided in several stages. In the first stage the problem solution is computed in the sub-domains at using the finite-difference scheme based on the Rosenbrock method with artificial boundary conditions (ABCs). Note, that for the first time layer we use the initial complex amplitude distribution. The second stage consists in the problem solution at using the conservative finite-difference scheme with boundary conditions (BCs) defined at the previous stage. Then, we repeat these stages by using the results obtained at using the conservative finite-difference scheme as the initial condition and using the results obtained on the base of Rosenbrock method as the BC for the solution computation on the next time layers.

**Keywords:** 1D and 2D nonlinear Schrödinger equation ·
Conservative finite-difference scheme · Rosenbrock method ·
Combined method · Artificial boundary conditions

## 1   Introduction

We focus our attention on the problem of numerical simulation of a laser pulse interaction with inhomogeneous nonlinear medium (optical periodic structure or

© Springer Nature Switzerland AG 2019
I. Dimov et al. (Eds.): FDM 2018, LNCS 11386, pp. 550–557, 2019.
https://doi.org/10.1007/978-3-030-11539-5_64

photonic crystal) which is described in the framework of 1D and 2D nonlinear Schrödinger equation. This work is a continuation of the combined method investigation, which we proposed in [1] for solving this type of problems. Combined method is based on using of the conservative finite-difference scheme, which proposed by Karamzin in 1974 [2], and non-conservative Rosenbrock method, which proposed by Rosenbrock in 1963 [3], for solving 1D or 2D nonlinear Schrödinger equation. Each of these schemes has its own advantages and disadvantages. The conservative finite-difference scheme is implicit, conservative, has the property of symmetry and second order of approximation in any coordinate. Rosenbrock method is conditionally conservative, explicit and have the same order of approximation by spatial coordinate. The proposed combined method in 2D case can significantly accelerate a computer simulation performance due to eliminating of using an iterative process, which is necessary at application of the conservative finite-difference scheme.

## 2  Problem Statement and Invariants for 2D Nonlinear Schrödinger Equation

Let's consider the dimensionless nonlinear Schrödinger equation describing the femtosecond pulse propagation in 2D photonic crystal (PC)

$$\varepsilon(z,x)\frac{\partial A}{\partial t} + \imath D_z\frac{\partial^2 A}{\partial z^2} + \imath D_x\frac{\partial^2 A}{\partial x^2} + \imath\beta(\varepsilon(z,x) + \alpha(z,x)|A|^2)A = 0, \\ t > 0, 0 < z < L_z, 0 < x < L_x \tag{1}$$

with initial complex amplitude distribution

$$A|_{t=0} = e^{-(z-L_{z_c})^2/a_z^2 - (x-L_{x_c})^2/a_x^2 + \imath 2\pi\chi(z-L_{z_c})}, \tag{2}$$

and non-reflecting ABCs for the Eq. (1) are written in the following way

$$\begin{pmatrix} \frac{\partial A}{\partial t} - 2D_z\Omega_{z_l}\frac{\partial A}{\partial z} + \imath D_z\Omega_{z_l}^2 A + \imath\beta(2+\alpha|A|^2)A \end{pmatrix}|_{z=0} = 0, \\ \begin{pmatrix} \frac{\partial A}{\partial t} + 2D_z\Omega_{z_r}\frac{\partial A}{\partial z} + \imath D_z\Omega_{z_r}^2 A + \imath\beta(2+\alpha|A|^2)A \end{pmatrix}|_{z=L_z} = 0, \\ \begin{pmatrix} \frac{\partial A}{\partial t} - 2D_x\Omega_{x_b}\frac{\partial A}{\partial x} + \imath D_x\Omega_{x_b}^2 A + \imath\beta(2+\alpha|A|^2)A \end{pmatrix}|_{x=0} = 0, \\ \begin{pmatrix} \frac{\partial A}{\partial t} + 2D_x\Omega_{x_u}\frac{\partial A}{\partial x} + \imath D_x\Omega_{x_u}^2 A + \imath\beta(2+\alpha|A|^2)A \end{pmatrix}|_{x=L_x} = 0. \tag{3}$$

In (1) $A(t,z,x)$ is a slowly varying amplitude in time only; $z,x$ denote a spatial coordinates; $L_z, L_x$ are their maximal value; $(L_{z_c}, L_{x_c})$ are the coordinates of the laser beam center at initial time moment; $\varepsilon(z,x), \alpha(z,x)$ are the functions of dielectric permittivity and nonlinearity coefficient. Parameters $D_z, D_x, \beta, \chi$ are the real coefficients, which satisfy the following conditions: $D_z = -\frac{1}{4\pi\chi}, \beta = -\pi\chi$. Parameters $\Omega_{z_l}, \Omega_{z_r}, \Omega_{x_b}, \Omega_{x_u}$ are the local wave numbers of laser beam near the low and upper of the artificial boundary along x-coordinate, left and right artificial boundary along z-coordinate, correspondingly. For definiteness we denote $\Omega_{z_l} = \Omega_{z_r} = \Omega_{x_b} = \Omega_{x_u} = \Omega = 2\pi\chi$.

The problem (1)–(3) possesses two following invariants (conservation laws):

$$I_1(t) = \int\limits_0^{L_z}\int\limits_0^{L_x} \varepsilon(z,x)|A|^2 dzdx, \tag{4}$$

$$I_3(t) = \int\limits_0^{L_z} \int\limits_0^{L_x} (-D_z |\frac{\partial A}{\partial z}|^2 - D_x |\frac{\partial A}{\partial x}|^2 + \beta(\varepsilon(z,x) + 0.5\alpha(z,x)|A|^2)|A|^2) dz dx. \quad (5)$$

## 3   Conservative Finite-Difference Scheme

At first, we should introduce a uniform grid $\omega$

$$\omega = \omega_t \times \omega_z \times \omega_x, \omega_t = \{t_m = m\tau, m = 0 \ldots N_t, h = \frac{L_t}{N_t}\}, \omega_z = \{z_j = jh_z,$$
$$j = 0 \ldots N_z, h_z = \frac{L_z}{N_z}\}, \omega_x = \{x_k = kh_x, k = 0 \ldots N_x, h_x = \frac{L_x}{N_x}\}. \quad (6)$$

The complex amplitudes on a grid $\omega$ are defined as

$$A = A_{j,k} = A_{m,j,k} = A(t_m, z_j, x_k), \hat{A} = \hat{A}_{j,k} = \hat{A}_{m+1,j,k}$$
$$= A(t_m + \tau, z_j, x_k), \overset{0.5}{A} = 0.5(\hat{A} + A), |\overset{0.5}{A}|^2 = 0.5(|\hat{A}|^2 + |A|^2). \quad (7)$$

$$(\varepsilon_{j,k}, \alpha_{j,k}) = \begin{cases} (1,0), & j = \overline{0, j_{L_0} - 1}, \\ (\frac{\varepsilon_1}{2}, \frac{\alpha_1}{2}), & j = j_{L_0}, \\ (\varepsilon_1, \alpha_1), & j_{L_0} + (p-1)j_d < j < j_{L_0} + j_{d_1} + (p-1)j_d, \\ (\frac{\varepsilon_1 + \varepsilon_2}{2}, \frac{\alpha_1 + \alpha_2}{2}), & j = j_{L_0} + j_{d_1} + (p-1)j_d, \quad k = \overline{0, Nx} \\ (\varepsilon_2, \alpha_2), & j_{L_0} + j_{d_1} + (k-1)j_d < j < j_{L_0} + pj_d, p = \overline{1, N_{str}} \\ (\frac{\varepsilon_2 + \varepsilon_1}{2}, \frac{\alpha_2 + \alpha_1}{2}), & j = j_{L_0} + pj_d, \\ (\frac{\varepsilon_2 + \varepsilon_3}{2}, \frac{\alpha_2 + \alpha_3}{2}), & j = j_{L_{str}}, \\ (\varepsilon_3, 0), & j = \overline{j_{L_{str}} + 1, N_z}, \end{cases}$$

$$(8)$$

where $d_1, d_2, \varepsilon_1, \varepsilon_2$ and $\alpha_1, \alpha_2$ are thicknesses, dielectric permittivity and nonlinearity coefficients of alternating layers correspondingly, $\varepsilon_3$ is a dielectric permittivity of substrate; $N_{str}$ is a number of layers pairs. $L_0$ is dimensionless coordinate of the PC's face, $L_{str}$ - defines the longitudinal size of PC. $j_{L_0} = [\frac{L_0}{h_z}], j_{d_1} = [\frac{d_1}{h_z}], j_{d_2} = [\frac{d_2}{h_z}], j_d = j_{d_1} + j_{d_2} = [\frac{d_1 + d_2}{h_z}], j_{L_{str}} = [\frac{L_{str}}{h_z}] = j_d N_{str}$. Symbol [.] denotes an integer part of a real number.

For the problem under consideration (1)–(3), we write the following finite-difference scheme:

$$\varepsilon \frac{\hat{A} - A}{\tau} + \imath D_z \Lambda_{\bar{z}z} \overset{0.5}{A} + \imath D_x \Lambda_{\bar{x}x} \overset{0.5}{A} + \imath \beta(\varepsilon + \alpha|\overset{0.5}{A}|^2)\overset{0.5}{A} = 0,$$
$$j = \overline{1, N_z - 1}, k = \overline{1, N_x - 1}, m = \overline{1, N_t}$$
$$\frac{\hat{A}_{0,k} - A_{0,k}}{\tau} - 2D_z \Omega \frac{\overset{0.5}{A}_{1,k} - \overset{0.5}{A}_{0,k}}{h_z} + \imath D_z \Omega^2 \overset{0.5}{A}_{0,k} + \imath \beta(2 + \alpha|\overset{0.5}{A}_{0,k}|^2)\overset{0.5}{A}_{0,k} = 0,$$
$$\frac{\hat{A}_{N_z,k} - A_{N_z,k}}{\tau} + 2D_z \Omega \frac{\overset{0.5}{A}_{N_z,k} - \overset{0.5}{A}_{N_z-1,k}}{h_z} + \imath D_z \Omega^2 \overset{0.5}{A}_{N_z,k}$$
$$+\imath \beta(2 + \alpha|\overset{0.5}{A}_{N_z,k}|^2)\overset{0.5}{A}_{N_z,k} = 0, k = \overline{1, N_x}. \quad (9)$$
$$\frac{\hat{A}_{j,0} - A_{j,0}}{\tau} - 2D_x \Omega \frac{\overset{0.5}{A}_{j,1} - \overset{0.5}{A}_{j,0}}{h_z} + \imath D_x \Omega^2 \overset{0.5}{A}_{j,0} + \imath \beta(2 + \alpha|\overset{0.5}{A}_{j,0}|^2)\overset{0.5}{A}_{j,0} = 0,$$
$$\frac{\hat{A}_{j,N_x} - A_{j,N_x}}{\tau} + 2D_x \Omega \frac{\overset{0.5}{A}_{j,N_x} - \overset{0.5}{A}_{j,N_x-1}}{h_z} + \imath D_x \Omega^2 \overset{0.5}{A}_{j,N_x}$$
$$+\imath \beta(2 + \alpha|\overset{0.5}{A}_{j,N_x}|^2)\overset{0.5}{A}_{j,N_x} = 0, j = \overline{1, N_z}.$$

The 2D problem under consideration is nonlinear one, therefore for its computation we use a two-step iteration process. The first step of the iteration process along z-coordinate has the following form:

$$\varepsilon\frac{\overset{s+1}{\hat{A}}-A}{\tau} + \imath D_z\Lambda_{\bar{z}z}\overset{s+1}{\underset{0.5}{\hat{A}}} + \imath D_x\Lambda_{\bar{x}x}\overset{s}{\underset{0.5}{A}} + \imath\beta(\varepsilon+\alpha|\overset{s}{\underset{0.5}{A}}|^2)\overset{s}{\underset{0.5}{A}} = 0,$$
$$j = \overline{1,N_z-1}, k = \overline{1,N_x-1}, m = \overline{1,N_t}$$

$$\frac{\overset{s+1}{\hat{A}_{0,k}}-A_{0,k}}{\tau} - 2D_z\Omega\frac{\overset{s+1}{\underset{0.5}{A_{1,k}}}-\overset{s+1}{\underset{0.5}{A_{0,k}}}}{h_z} + \imath D_z\Omega^2\overset{s+1}{\underset{0.5}{\hat{A}_{0,k}}} + \imath\beta(2+\alpha|\overset{s}{\underset{0.5}{A_{0,k}}}|^2)\overset{s}{\underset{0.5}{A_{0,k}}} = 0, \quad (10)$$

$$\frac{\overset{s+1}{\hat{A}_{N_z,k}}-A_{N_z,k}}{\tau} + 2D_z\Omega\frac{\overset{s+1}{\underset{0.5}{A_{N_z,k}}}-\overset{s+1}{\underset{0.5}{A_{N_z-1,k}}}}{h_z} + \imath D_z\Omega^2\overset{s+1}{\underset{0.5}{A_{N_z,k}}}$$

$$+\imath\beta(2+\alpha|\overset{s}{\underset{0.5}{A_{N_z,k}}}|^2)\overset{s}{\underset{0.5}{A_{N_z,k}}} = 0, \qquad k = \overline{1,N_x}.$$

The second step of the iteration process along x-coordinate is written in the way:

$$\varepsilon\frac{\overset{s+2}{\hat{A}}-A}{\tau} + \imath D_z\Lambda_{\bar{z}z}\overset{s+1}{\underset{0.5}{\hat{A}}} + \imath D_x\Lambda_{\bar{x}x}\overset{s+2}{\underset{0.5}{\hat{A}}} + \imath\beta(\varepsilon+\alpha|\overset{s}{\underset{0.5}{A}}|^2)\overset{s}{\underset{0.5}{A}} = 0,$$
$$j = \overline{1,N_z-1}, k = \overline{1,N_x-1}, m = \overline{1,N_t}$$

$$\frac{\overset{s+2}{\hat{A}_{j,0}}-A_{j,0}}{\tau} - 2D_x\Omega\frac{\overset{s+2}{\underset{0.5}{A_{j,1}}}-\overset{s+2}{\underset{0.5}{A_{j,0}}}}{h_x} + \imath D_x\Omega^2\overset{s+2}{\underset{0.5}{\hat{A}_{j,0}}} + \imath\beta(2+\alpha|\overset{s}{\underset{0.5}{A_{j,0}}}|^2)\overset{s}{\underset{0.5}{A_{j,0}}} = 0, \quad (11)$$

$$\frac{\overset{s+2}{\hat{A}_{j,N_x}}-A_{j,N_x}}{\tau} + 2D_x\Omega\frac{\overset{s+2}{\underset{0.5}{A_{j,N_x}}}-\overset{s+2}{\underset{0.5}{A_{j,N_x-1}}}}{h_x} + \imath D_x\Omega^2\overset{s+2}{\underset{0.5}{A_{j,N_x}}}$$

$$+\imath\beta(2+\alpha|\overset{s}{\underset{0.5}{A_{j,N_x}}}|^2)\overset{s}{\underset{0.5}{A_{j,N_x}}} = 0, \qquad j = \overline{1,N_z}.$$

The mesh function on the upper layer in time at zero iteration ($s = 0$) is chosen as

$$\overset{s=0}{\hat{A}} = A. \tag{12}$$

The iteration process is stopped if the following condition is valid

$$\max_{z_j,x_k}|\overset{s+2}{\hat{A}} - \overset{s}{\hat{A}}| \leq \tilde{\theta}_1\max_{z_j}|\overset{s}{\hat{A}}| + \tilde{\theta}_2, \tag{13}$$

where $\tilde{\theta}_1, \tilde{\theta}_2 > 0$ are the real constants.

## 4  Combined Method

At first we should note, that the solution of 2D problem at using Rosenbrock method leads to significantly increasing of computation time and complexity because of using the matrix inversion operation. Therefore, we consider the combined method construction.

As mentioned above, the 2D problem (1)–(3) is nonlinear one, therefore for its computation we also use a two-step iteration process. We modify this iteration process in accordance with the definition of a combined method for 1D nonlinear Schrödinger equation (see [1]). Each step of the iteration process (10)–(11) for the combined method consists of two stages: we use, at first, the Rosenbrock method in the domain near the boundaries, then in other parts of the domain the conservative finite-difference scheme is used.

Let's consider the construction of the first step of the method along z-coordinate. With this aim we denote a number of the grid nodes in sub-domains along z-coordinate as $N_{z_R}$ and write the finite-difference scheme, corresponding to the first stage of the first step of the combined method, computed at using Rosenbrock method:

$$
\overset{s+1}{\hat{A}}_{Ros} = \overset{s}{\hat{A}} + \tau_R Re\bar{q}, \qquad j = \overline{0, N_{z_R}}; \overline{N_z - N_{z_R}, N_z}, k = \overline{1, N_x - 1}. \tag{14}
$$

where $\tau_R = \frac{\tau}{M}$ ($M$ is an integer number) is a time step used for computation in sub-domains, $Re\bar{q}$ is a real part of solution for the following linear equation set

$$
\begin{aligned}
(E - \delta\tau_R G_U)\bar{q} &= G(U), & j &= \overline{1, N_R}; \overline{N_z - N_R, N_z - 1}, \\
(E - \delta\tau_R G_{U_b})\bar{q} &= G(U_b), & j &= 0, N_z, k = \overline{1, N_x - 1}.
\end{aligned} \tag{15}
$$

Above a matrix $E$ is the unity matrix, $\delta$ is a parameter of Rosenbrock method, which we take equal to $\delta = 0.5$, $G_U$ is the Jacobian for the right part vector $G(U)$

$$
G(U) = \begin{pmatrix} \frac{D_z}{\varepsilon}\Lambda_{\bar{z}z}\overset{s}{v}_{j,k} + \frac{D_x}{\varepsilon}\Lambda_{\bar{x}x}\overset{s}{v}_{j,k} + \beta(1 + \frac{\alpha}{\varepsilon}(\overset{s}{u}^2_{j,k} + \overset{s}{v}^2_{j,k})\overset{s}{v}_{j,k} \\ -\frac{D_z}{\varepsilon}\Lambda_{\bar{z}z}\overset{s}{u}_{j,k} - \frac{D_x}{\varepsilon}\Lambda_{\bar{x}x}\overset{s}{u}_{j,k} - \beta(1 + \frac{\alpha}{\varepsilon}(\overset{s}{u}^2_{j,k} + \overset{s}{v}^2_{j,k})\overset{s}{u}_{j,k}, \end{pmatrix},
$$
$$
j = \overline{1, N_{z_R}}, \overline{N_z - N_{z_R}, N_z - 1}, \qquad k = \overline{1, N_x - 1} \tag{16}
$$

and $G_{U_b}$ is the Jacobian of the right part vector $G(U_b)$, which corresponds to the ABCs and is written in the following form

$$
G(U_b) = \begin{pmatrix} 2D_z\Omega\frac{\overset{s}{u}_{1,k} - \overset{s}{u}_{0,k}}{h_z} + (D_z\Omega^2 + \beta(2 + \alpha(\overset{s}{u}^2_{0,k} + \overset{s}{v}^2_{0,k})))\overset{s}{v}_{0,k} \\ 2D_z\Omega\frac{\overset{s}{v}_{1,k} - \overset{s}{v}_{0,k}}{h_z} - (D_z\Omega^2 + \beta(2 + \alpha(\overset{s}{u}^2_{0,k} + \overset{s}{v}^2_{0,k})))\overset{s}{u}_{0,k} \\ -2D_z\Omega\frac{\overset{s}{u}_{N_z,k} - \overset{s}{u}_{N_z-1,k}}{h_z} + (D_z\Omega^2 + \beta(2 + \alpha(\overset{s}{u}^2_{N_z,k} + \overset{s}{v}^2_{N_z,k})))\overset{s}{v}_{N_z,k} \\ -2D_z\Omega\frac{\overset{s}{v}_{N_z,k} - \overset{s}{v}_{N_z-1,k}}{h_z} - (D_z\Omega^2 + \beta(2 + \alpha(\overset{s}{u}^2_{N_z,k} + \overset{s}{v}^2_{N_z,k})))\overset{s}{u}_{N_z,k} \end{pmatrix},
$$
$$
k = \overline{1, N_x - 1}. \tag{17}
$$

In formula (16), (17) the functions $\overset{s}{u}(z, x)$ and $\overset{s}{v}(z, x)$ denote the real and imaginary parts of the solution $\overset{s}{\hat{A}}(z, x) = \overset{s}{u}(z, x) + i\overset{s}{v}(z, x)$. The set of equations (14)–(17) is solved by using the Thomas algorithm.

The second stage of the first combined method step along z-coordinate consists in the problem solution by using the conservative finite-difference scheme with BCs defined by the mesh function $\overset{s+1}{\hat{A}}_{Ros}$. Let's denote the problem solution on this stage as $\overset{s+1}{\hat{A}}{}_C$, the solution on the previous step at using the conservative finite-difference scheme as $\overset{s}{\hat{A}}_C$ and the solution on the previous time step is denoted as $A_C$. Thus, the difference problem on the second stage is written as follow:

$$\varepsilon \frac{\overset{s+1}{\hat{A}}_C - A_C}{\tau} + \imath D_z \Lambda_{\bar{z}z} \overset{s+1}{\hat{A}}_C{}^{0.5} + \imath D_x \Lambda_{\bar{z}z} \overset{s}{\hat{A}}_C{}^{0.5} + \imath \beta(\varepsilon + \imath \alpha |\overset{s}{\hat{A}}_C|^2)^{0.5} \overset{s}{\hat{A}}_C{}^{0.5} = 0, \qquad (18)$$
$$j = \overline{1, N_z - 1}, \qquad k = \overline{1, N_x - 1}$$

with following BCs:

$$\overset{s+1}{\hat{A}}_{C0,k} = \overset{s+1}{\hat{A}}_{Ros0,k}, \qquad \overset{s+1}{\hat{A}}_{CN_z,k} = \overset{s+1}{\hat{A}}_{RosN_z,k}, \qquad k = \overline{1, N_x - 1}. \qquad (19)$$

Note, that for the first time layer we use the initial complex amplitude distribution. Thus, we obtain the solution at the first step of the combined method $\overset{s+1}{\hat{A}} = \overset{s+1}{\hat{A}}{}_C$.

The second step of combined method along x-coordinate also has two stages. We denote a number of the grid nodes in sub-domains along x-coordinate as $N_{x_R}$ and write the first stage, corresponding to using Rosenbrock method, in following way:

$$\overset{s+2}{\hat{A}}{}_{Ros} = \overset{s+1}{\hat{A}} + \tau_R Re\bar{q}, \qquad k = \overline{0, N_R}; \overline{N_x - N_R, N_x}, j = \overline{1, N_z - 1}. \qquad (20)$$

where $Re\bar{q}$, is a real part of solution for the linear equation set

$$\begin{aligned}(E - \delta\tau_R G_U)\bar{q} = G(U), \qquad k = \overline{1, N_R}; \overline{N_x - N_R, N_x - 1}, \\ (E - \delta\tau_R G_{U_b})\bar{q} = G(U_b), \qquad k = \overline{0, N_x}, j = \overline{1, N_z - 1}.\end{aligned} \qquad (21)$$

$G_U$ is the Jacobian for the right part vector $G(U)$

$$G(U) = \begin{pmatrix} \frac{D_z}{\varepsilon} \Lambda_{\bar{z}z} \overset{s+1}{v}_{j,k} + \frac{D_x}{\varepsilon} \Lambda_{\bar{x}x} \overset{s+1}{v}_{j,k} + \beta(1 + \frac{\alpha}{\varepsilon}(\overset{s+1}{u^2}_{j,k} + \overset{s+1}{v^2}_{j,k})) \overset{s+1}{v}_{j,k} \\ -\frac{D_z}{\varepsilon} \Lambda_{\bar{z}z} \overset{s+1}{u}_{j,k} - \frac{D_x}{\varepsilon} \Lambda_{\bar{x}x} \overset{s+1}{u}_{j,k} - \beta(1 + \frac{\alpha}{\varepsilon}(\overset{s+1}{u^2}_{j,k} + \overset{s+1}{v^2}_{j,k})) \overset{s+1}{u}_{j,k}, \end{pmatrix}, \qquad (22)$$
$$k = \overline{1, N_{x_R}, N_x - N_{x_R}, N_x - 1}, \qquad j = \overline{1, N_z - 1}$$

and $G_{U_b}$ is the Jacobian of the right part vector $G(U_b)$, which corresponds to the ABCs and is written in the following form

$$G(U_b) = \begin{pmatrix} 2D_x \Omega \frac{\overset{s+1}{u}_{j,1} - \overset{s+1}{u}_{j,0}}{h_x} + (D_x \Omega^2 + \beta(2 + \alpha(\overset{s+1}{u^2}_{j,0} + \overset{s+1}{v^2}_{j,0}))) \overset{s+1}{v}_{j,0} \\ 2D_x \Omega \frac{\overset{s+1}{v}_{j,1} - \overset{s+1}{v}_{j,0}}{h_x} - (D_x \Omega^2 + \beta(2 + \alpha(\overset{s+1}{u^2}_{j,0} + \overset{s+1}{v^2}_{j,0}))) \overset{s+1}{u}_{j,0} \\ -2D_x \Omega \frac{\overset{s+1}{u}_{j,N_x} - \overset{s+1}{u}_{j,N_x-1}}{h_x} + (D_x \Omega^2 + \beta(2 + \alpha(\overset{s+1}{u^2}_{j,N_x} + \overset{s+1}{v^2}_{j,N_x}))) \overset{s+1}{v}_{j,N_x} \\ -2D_x \Omega \frac{\overset{s+1}{v}_{j,N_x} - \overset{s+1}{v}_{j,N_x-1}}{h_x} - (D_x \Omega^2 + \beta(2 + \alpha(\overset{s+1}{u^2}_{j,N_x} + \overset{s+1}{v^2}_{j,N_x}))) \overset{s+1}{u}_{j,N_x} \end{pmatrix},$$
$$j = \overline{1, N_z - 1}.$$

$$(23)$$

In formula (22), (23) the functions $\overset{s+1}{u}(z,x)$ and $\overset{s+1}{v}(z,x)$ denote the real and imaginary parts of the solution $\overset{s+1}{\hat{A}}(z,x) = \overset{s+1}{u}(z,x) + i\overset{s+1}{v}(z,x)$. This set of equations (20)–(23) is solved also by using the Thomas algorithm.

The second stage of second combined method step consists in the problem solution along x-coordinate by using the conservative finite-difference scheme with BCs defined by the mesh function $\overset{s+2}{\hat{A}}_{Ros}$. Thus, the difference problem on the second stage is written as follow:

$$\varepsilon\frac{\overset{s+2}{\hat{A}_C}-A_C}{\tau} + iD_z\Lambda_{\bar{z}z}\overset{s+1}{\underset{0.5}{A_C}} + iD_x\Lambda_{\bar{x}x}\overset{s+2}{\underset{0.5}{A_C}} + i\beta(\varepsilon + i\alpha|\overset{s}{\underset{0.5}{A_C}}|^2)\overset{s}{\underset{0.5}{A_C}} = 0, \qquad (24)$$
$$k = \overline{1, N_x - 1}, \qquad j = \overline{1, N_z - 1}$$

with following BCs:

$$\overset{s+2}{\hat{A}}_{Cj,0} = \overset{s+2}{\hat{A}}_{Rosj,0}, \quad \overset{s+2}{\hat{A}}_{Cj,N_x} = \overset{s+2}{\hat{A}}_{Rosj,N_x}, \qquad j = \overline{1, N_z - 1}. \qquad (25)$$

Thus, we obtain the solution at the upper time layer.

## 5   Conclusion

In this paper we propose a new method for solving 1D and 2D nonlinear Schrödinger equation. This method is explicit, that means the solution on next time layer can be found without using iteration process. The result of computer simulation of this method in comparison with Rosenbrock method and conservative finite-difference scheme has the same order of accuracy (for certain values of parameters $N_{z_R}$, $N_{x_R}$ and $\tau$). The time of computer simulation using combined method is less then time of simulation using Rosenbrock method but more than time of simulation using conservative-finite difference scheme. Using of this method leads to significant advantage in time of simulation compared with Rosenbrock method. So, we can say combined method is more suitable for solving the Schrödinger equation, than the Rosenbrock method. We use two stage iterative process for writing of the set of 1D difference equations at solving 2D problem. This approach leads to decreasing of the computation cost and allow to write the conservative finite-difference scheme for arbitrary boundary conditions.

**Acknowledgments.** Supported by the Russian Science Foundation (Grant 14-21-00081).

## References

1. Trofimov, V.A., Trykin, E.M.: Combined method for solving of 1D nonlinear Schrödinger equation. In: Mastorakis, N., Mladenov, V. (eds.) Computational Problems in Engineering. LNEE, vol. 307, pp. 173–187. Springer, Cham (2014). https://doi.org/10.1007/978-3-319-03967-1_14

2. Karamzin, Y.N.: Difference schemes for computations of three-frequency interactions of electromagnetic waves in a nonlinear medium with quadratic polarization. Zh. Vychisl. Mat. Mat. Fiz. **14**(4), 1058–1062 (1974)
3. Rosenbrock, H.H.: Some general implicit processes for the numerical solution of differential equations. Comput. J. **5**(4), 329–330 (1963). https://doi.org/10.1093/comjnl/5.4.329
4. Tereshin, E.B., Trofimov, V.A., Fedotov, M.V.: Conservative finite difference scheme for the problem of propagation of a femtosecond pulse in a nonlinear photonic crystal with non-reflecting boundary conditions. Comput. Math. Math. Phys. **46**, 154–164 (2006). https://doi.org/10.1134/S0965542506010155
5. Fibich, G.: The Nonlinear Schrödinger Equation: Singular Solutions and Optical Collapse. AMS, vol. 192. Springer, Cham (2015). https://doi.org/10.1007/978-3-319-12748-4
6. Arnold, A., Ehrhardt, M., Sofronov, I.: Discrete transparent boundary conditions for the Schrödinger equation: fast calculation, approximation, and stability. Commun. Math. Sci. **1**(3), 501–556 (2003)
7. Jiang, S., Greengard, L., Sofronov, I.: Efficient representation of nonreflecting boundary conditions for the time-dependent Schrödinger equation in two dimensions. Commun. Pure Appl. Math. **61**(2), 261–288 (2007)
8. Xu, Z., Han, H., Wu, X.: Adaptive absorbing boundary conditions for Schrödinger-type equations: application to nonlinear and multi-dimensional problems. J. Comp. Phys. **225**(2), 1577–1589 (2007). https://doi.org/10.1016/j.jcp.2007.02.004
9. Han, H., Yin, D., Huang, Z.: Numerical solutions of Schrödinger equations in R3. Numer. Methods Part. Differ. Eq. Int. J. **23**(3), 511–533 (2006). https://doi.org/10.1002/num.20193
10. Antoine, X., Besse, C., Mouysset, V.: Numerical schemes for the simulation of the two-dimensional Schrödinger equation using non-reflecting boundary conditions. Math. Comput. **73**(248), 1779–1799 (2004). https://doi.org/10.2307/4100054
11. Zheng, C.: Exact nonreflecting boundary conditions for one-dimensional cubic nonlinear Schrödinger equations. J. Comp. Phys. **215**(2), 552–565 (2006). https://doi.org/10.1016/j.jcp.2005.11.005
12. Soffer, A., Stucchio, C.: Open boundaries for the nonlinear Schrödinger equation. J. Comp. Phys. **225**(2), 1218–1232 (2007). https://doi.org/10.1016/j.jcp.2007.01.020
13. Xu, Z., Han, H.: Absorbing boundary conditions for nonlinear Schrödinger equations. Phys. Rev. E **74**(3), 037704 (2006). https://doi.org/10.1103/PhysRevE.74.037704
14. Fevens, T., Jiang, H.: Absorbing boundary conditions for the Schrödinger equation. SIAM J. Sci. Comput. **21**(1), 255–282 (1999). https://doi.org/10.1137/S1064827594277053
15. Li, H., Guo, Y.: Numerical solution of the general coupled nonlinear Schrödinger equations on unbounded domains. Phys. Rev. E **96**, 063305 (2017). https://doi.org/10.1103/PhysRevE.96.063305
16. Yang, X., Zhang, J.: Computation of the Schrödinger equation in the semiclassical regime on an unbounded domain. SIAM J. Numer. Anal. **52**(2), 808–831 (2014). https://doi.org/10.1137/13090715X
17. Trofimov, V.A., Trykin, E.M.: Construction of adaptive artificial boundary conditions using the invariant ratios for Schrödinger equation. In: 2014 East-West Design Test Symposium (EWDTS), pp. 1–4 . IEEE Conference Publications (2014). https://doi.org/10.1109/EWDTS.2014.7027098

# On a Reliable Numerical Method for a Singularly Perturbed Parabolic Reaction-Diffusion Problem in a Doubly Connected Domain

Irina Tselishcheva$^{(\boxtimes)}$ (iD) and Grigorii Shishkin (iD)

Krasovskii Institute of Mathematics and Mechanics,
Ural Branch of the Russian Academy of Sciences,
S. Kovalevskaya Street, 16, 620990 Yekaterinburg, Russia
{tsi,shishkin}@imm.uran.ru

**Abstract.** In a space-time domain $\overline{G} = \overline{D} \times [0, T]$, where $\overline{D}$ is a doubly connected domain in space—a rectangle $\overline{D}_1$ with a removed circle $D_2$, we consider the Dirichlet initial–boundary value problem for a singularly perturbed parabolic reaction–diffusion equation. As $\varepsilon \to 0$, boundary layers of different types arise in neighborhoods of smooth parts of the lateral boundary and lateral edges. The boundary layers decrease exponentially with distance from the outer and inner lateral boundaries. We discuss an approach for developing a reliable numerical method based on the earlier techniques for simply connected domains. Our aim is to construct an iterative Schwarz method on overlapping subdomains that cover separately the boundary of the parallelepiped or the boundary of the cylinder. It is required that the method converges $\varepsilon$-uniformly in the maximum norm as the number of iterations (and the number of mesh points in the case of a difference scheme) grows. We use the Shishkin meshes that condense in the boundary layers and are piecewise uniform along the normal to the smooth parts of the boundaries. To construct meshes near the outer and inner lateral boundaries, it is proposed to use the Cartesian and cylindrical coordinate systems, respectively.

**Keywords:** Singularly perturbed Dirichlet problem ·
Small parameter · Convection–diffusion equation ·
Biconnected domain · Boundary layers · Difference scheme ·
Piecewise uniform meshes · Domain Decomposition · Schwarz method ·
Parameter-uniform convergence

## 1 Introduction

In recent years, there is a growing interest amongst numerical analysts in the development of reliable numerical methods for solving singularly perturbed problems in the case of domains with complex geometry and rather complicated boundary layers. Reliable methods include those specifically oriented to the accurate resolution of boundary layers in the maximum norm for all possible

© Springer Nature Switzerland AG 2019
I. Dimov et al. (Eds.): FDM 2018, LNCS 11386, pp. 558–565, 2019.
https://doi.org/10.1007/978-3-030-11539-5_65

values of the perturbation parameter. Here, it is attractive to apply domain decomposition methods since this allows us to construct effective numerical approximations that are sufficiently simple on subdomains. The aim of research is to construct iterative schemes based on the alternating Schwartz method, which makes it possible to reduce the solution of the problem to the sequence of problems on simpler subdomains containing singularities of the same type and then to parallelize the solution process in future. To construct finite difference schemes, we use classical monotone approximations and the simplest piecewise uniform meshes condensing in a neighborhood of the boundary layers in each coordinate, i.e., meshes that are piecewise uniform along the normal to the smooth parts of the boundaries of the subdomains.

## 2  Problem Formulation. Aim of the Research

In the biconnected space-time domain (see Fig. 2)

$$\overline{G} = \overline{D} \times [0, T], \tag{1}$$

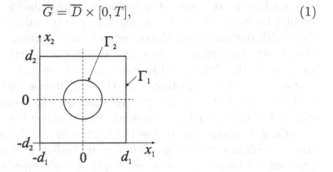

**Fig. 1.** Doubly connected domain $\overline{D}$; $\varGamma_1$ is the boundary of the rectangle, $\varGamma_2$ is the boundary of the circle.

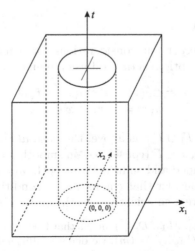

**Fig. 2.** Doubly connected domain $\overline{G}$

where $\overline{D}$ is the rectangle $\overline{D}_1$ with the removed circle $D_2$ (Fig. 1), $D = D_1 \setminus D_2$, $D_1 = (-d_1 < x_1 < d_1) \times (-d_2 < x_2 < d_2)$; $D_2 = \{x : (x_1^2 + x_2^2)^{1/2} < d\}$; $d_1, d_2 > d > 0$, we consider the Dirichlet problem for the singularly perturbed parabolic reaction–diffusion equation[1]

$$L_{(2)}u(x,t) \equiv \varepsilon^2 \Delta u - a(x,t)u - p(x,t)\frac{\partial u}{\partial t} = f(x,t), \quad (x,t) \in G,$$
$$u(x,t) = \varphi(x,t), \quad (x,t) \in S. \tag{2}$$

Here, $S = \overline{G} \setminus G$, the coefficients and the right-hand side $a(x,t)$, $p(x,t)$, $f(x,t)$, $(x,t) \in \overline{G}$, and the boundary function $\varphi(x,t)$, $(x,t) \in S$ (on smooth parts of the boundary $S$) are assumed to be bounded and sufficiently smooth; $a(x,t) \geq 0$, $p(x,t) \geq p_0 > 0$. The parameter $\varepsilon$ takes arbitrrary values in the interval $(0,1]$.

As $\varepsilon \to 0$, boundary layers of different types arise in a neighborhood of the lateral boundary $S^L$. In a neighborhood of the outer boundary, i.e., the lateral faces of the parallelepiped but outside regions near its edges, there appear parabolic boundary layers; in a neighborhood of the edges, the layer is angular. In a neighborhood of the inner boundary (the cylindric surface), there appears a circular boundary layer that is regular. The boundary layers decrease exponentially with distance from the outer and inner parts of $S^L$. The boundary layers of such structure give rise to difficulties in constructing $\varepsilon$-uniform grid approximations to the differential problem (1), (2) and motivate the need for *special* *"connected" grids* consistent with the boundaries and a Schwarz-type method for interfacing between the grid subdomains. To construct grids in the neighborhood of the outer lateral boundary, where the boundary layer is sufficiently smooth, it is reasonable to use the Cartesian coordinate system, while a cylindrical coordinate system is naturally applied in a biconnected region near the inner lateral boundary. Next, we apply the overlapping domain decomposition method with a sufficient width of the overlap so that the subdomains contains either the boundary of the parallelepiped or the boundary of the cylinder.

## 2.1   Model Problem

In what follows, we present our considerations for a model time-independent problem on $\overline{D}$ for the elliptic reaction–diffusion equation

$$L_{(3)}u(x) \equiv \varepsilon^2 \Delta u(x) - a(x)u(x) = f(x). \quad x \in D,$$
$$u(x) = \varphi(x), \quad x \in \Gamma. \tag{3}$$

Here, $\Gamma = \overline{D} \setminus D$, $\Gamma = \Gamma_1 \cup \Gamma_2$; as above, the data $a(x)$, $f(x)$, $x \in \overline{D}$, and the boundary function $\varphi(x)$, $x \in \Gamma$ (considered on smooth parts of the boundary $\Gamma$) are assumed to be bounded and sufficiently smooth; $a(x,t) \geq a_0 > 0$.

For simplicity, we consider that compatibility conditions are fulfilled at the corner points that ensure a sufficiently smooth solution on $\overline{D}$.

---

[1] The notation $L_{(k)}$ $(m_{(k)}, M_{(k)}, D_{h(k)})$ means that this operator (constant, grid) was introduced in formula $(k)$. By $M$ $(m)$, we denote sufficiently large (small) positive constants independent of $\varepsilon$ and the stencils of difference schemes.

## 2.2   A Priori Bounds for the Solution and Derivatives

Let us discuss bounds for the solution of problem (3) and its derivatives used later in the constructions and proof of schemes.

The solution of problem (3) satisfies a rough "standard" estimate

$$\left|\frac{\partial^k}{\partial x_1^{k_1}\partial x_2^{k_2}}u(x)\right| \leq M\varepsilon^{-k}, \quad x \in \overline{D}, \ k \leq K. \tag{4}$$

We represent the solution $u(x)$ as the sum of its regular, $U(x)$, and singular, $V(x)$, components

$$u(x) = U(x) + V(x), \quad x \in \overline{D}, \tag{5}$$

where $U(x)$ and $V(x)$, $x \in \overline{D}$, are solutions of the inhomogeneous and homogeneous equations, respectively. The estimate for $U(x)$ is

$$\left|\frac{\partial^k}{\partial x_1^{k_1}\partial x_2^{k_2}}U(x)\right| \leq M, \quad x \in \overline{D}, \ k \leq K. \tag{6}$$

The function $V(x)$ can be represented as the sum of functions

$$V(x) = V_0(x) + \sum_{j=1}^{4}V_j(x) + \sum_{i,j=1}^{4}V_{ij}(x), \quad x \in \overline{D}. \tag{7}$$

Here $V_0(x)$ is the boundary layer in a neighborhood of the boundary $\Gamma_2$; $V_j(x)$ and $V_{ij}(x)$ are regular and angular elliptic boundary layers in a neighborhood of the boundary $\Gamma_1$.

The singular components from representation (7) satisfy the estimates

$$\left|\frac{\partial^k}{\partial n}V_0(x)\right| \leq M\varepsilon^{-k}\exp(-m\varepsilon^{-k}r(x,\Gamma_2)), \quad x \in \overline{D},$$

$$\left|\frac{\partial^k}{\partial x_1^{k_1}\partial x_2^{k_2}}V_j(x)\right| \leq M\varepsilon^{-k_j}\exp(-m\varepsilon^{-1}r(x,\Gamma_1^j)), \quad x \in \overline{D}, \ 1 \leq j \leq 4, \tag{8}$$

$$\left|\frac{\partial^k}{\partial x_1^{k_1}\partial x_2^{k_2}}V_{ij}(x)\right| \leq M\varepsilon^{-k}\min\left[\exp(-m\varepsilon^{-1}r(x,\Gamma_1^i)), \exp(-m\varepsilon^{-1}[-r(x,\Gamma_1^j))\right],$$

$$x \in \overline{D}, \ k \leq K,$$

where $r(x,\Gamma)$ is the distance from $x$ to the boundary $\Gamma$; $\partial^k/\partial n$ is the derivative along the normal to the boundary $\Gamma_2$; $k_j = 1$ for $j = 1, 3$ and $k_j = 2$ for $j = 2, 4$; $0 < m < m^0$, $m^0 = \min_{\overline{D}}[a^{1/2}(x)]$. By $\Gamma_1^j$, $j = 1, 2, 3, 4$, we denote the sides of the rectangle $D_1$. Assume that $\Gamma_1^1$ and $\Gamma_1^3$ are orthogonal to the $x_1$-axis, while $\Gamma_1^2$ and $\Gamma_1^4$ to the $x_2$-axis; the sides $\Gamma_1^1$ and $\Gamma_1^2$ contain the apex $(-d_1, -d_2)$.

It is easy to see from (8) that the boundary-layer functions decrease exponentially with distance from the corresponding boundaries. In (6) and (8), $K \geq 2$.

# 3   Difference Schemes

**3.1.** Consider the case where $D_2 = \emptyset$.

On $\overline{D}$, we introduce the rectangular grid

$$\overline{D}_h = \overline{\omega}_1 \times \overline{\omega}_2, \tag{9}$$

where $\overline{\omega}_s$ is generally a nonuniform mesh on $[-d_s, d_s]$ on the axis $x_s$, $s = 1, 2$. Define $h_s^i = x_s^{i+1} - x_s^i$, $x_s^i$, $x_s^{i+1} \in \overline{\omega}_s$, $h_s = \max_i h_s^i$, $h = \max_s h_s$, $s = 1, 2$. Let $N_s + 1$ denote the number of mesh points in $\overline{\omega}_s$, $N = \min_s N_s$, and $h \le M N^{-1}$.

On the grid $\overline{D}_h$, we approximate problem (3) by the difference scheme [1]

$$\Lambda_{(10)} z(x) \equiv \left\{ \varepsilon^2 \sum_{s=1,2} \delta_{\overline{x}_s \widehat{x}_s} - a(x) \right\} z(x) = f(x), \quad x \in D_h,$$

$$z(x) = \varphi(x), \quad x \in \Gamma_h. \tag{10}$$

Here $D_h = D \cap \overline{D}_h$, $\Gamma_h = \Gamma \cap \overline{D}_h$, $\delta_{\overline{x}_s \widehat{x}_s} z(x) = z_{\overline{x}_s \widehat{x}_s}(x)$ are the second difference derivatives in $x_s$ on a uniform mesh.

Scheme (10), (9) is monotone. The scheme converges only for fixed values of $\varepsilon$, namely, under the condition $h = o(\varepsilon)$, with an error bound given by

$$| u(x) - z(x) | \le M \left( \varepsilon + N^{-1} \right)^{-1} N^{-1}, \quad x \in \overline{D}_h. \tag{11}$$

Let us introduce the special grid

$$\overline{D}_h^c = \overline{\omega}_1^c \times \overline{\omega}_2^c, \tag{12}$$

where $\overline{\omega}_s^c = \overline{\omega}_s^c(\sigma_s)$ is the piecewise uniform mesh condensing near the endpoints of $[-d_s, d_s]$, $\sigma_s$ is the mesh parameter depending on $\varepsilon$ and $N$, $\sigma_s \le 4^{-1} d_s$. We construct the mesh $\overline{\omega}_s^c(\sigma_s)$ to be uniform on each of the intervals $[0, \sigma_s]$, $[\sigma_s, d_s - \sigma_s]$, and $[d_s - \sigma_s, d_s]$. Its step-size equals $h_s^{(1)} = \sigma_s (N_s/4)^{-1}$ on $[0, \sigma_s]$, $[d_s - \sigma_s, d_s]$, and $h_s^{(2)} = (d_s - 2\sigma_s)(N_s/2)^{-1}$ on $[\sigma_s, d_s - \sigma_s]$. Here, $\sigma_s = \sigma_{s(12)}(\varepsilon, N_s) = \min [d_s/4, M \varepsilon \ln N_s]$ with $M \ge 2(m_{(8)})^{-1}$.

The difference scheme (10), (12) converges $\varepsilon$-uniformly:

$$| u(x) - z(x) | \le M N^{-1} \ln N, \quad x \in \overline{D}_h^c. \tag{13}$$

**Theorem 1.** *Let the solution of problem (3) and its components satisfy estimates (4), (6), (8) for $K = 4$. Then the solution of scheme (10) on grid (12) converges $\varepsilon$-uniformly with bound (13).*

**3.2.** Similarly, we construct an $\varepsilon$-uniformly convergent scheme if $D_2 \ne \emptyset$.

In the case of the biconnected ring

$$\overline{D}_3^* = \overline{D}_3 \setminus D_2, \tag{14}$$

where $\overline{D}_3$ is some $r_0$-neighborhood of $D_2$, passing to the polar coordinates $r$ and $\psi$, $x_1 = r \cos \psi$, $x_2 = r \sin \psi$ (Fig. 3), we consider the boundary value problem

$$L_{(3)}^{r, \psi} u(r, \psi) = f(r, \psi), \quad (r, \psi) \in D_3^*, \quad u(r, \psi) = \varphi(r, \psi), \quad (r, \psi) \in \Gamma_3^*. \tag{15}$$

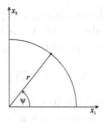

**Fig. 3.** Plane polar coordinates

The problem data in the polar coordinates are assumed to be sufficiently smooth;

$$L_{(3)}^{r,\,\psi} = \varepsilon^2 \left( \frac{\partial^2 u}{\partial r^2} + \frac{1}{r}\frac{\partial u}{\partial r} + \frac{1}{r^2}\frac{\partial^2 u}{\partial \psi^2} \right) - a(r,\psi)u(r,\psi).$$

We approximate problem (15) by the implicit difference scheme

$$\Lambda_{(10)}^{r,\,\psi} z(r,\psi) = f(r,\psi), \quad (r,\psi) \in D_{3h}^*, \quad z(r,\psi) = \varphi(r,\psi), \quad (r,\psi) \in \Gamma_{3h}^*. \quad (16)$$

We obtain an error bound similar to (11) for the solution of scheme (16):

$$| u(r,\psi) - z(r,\psi) | \le M \left( \varepsilon + N_*^{-1} \right)^{-1} N_*^{-1}, \quad (r,\psi) \in \overline{D}_{3h}^*, \quad (17)$$

where $N_* + 1$ is the minimal number of mesh points in $r$ and $\psi$.

On the grid uniform in $\psi$ and piecewise uniform in $r$, we have the bound

$$| u(r,\psi) - z(r,\psi) | \le M N_*^{-1} \ln N_*, \quad (r,\psi) \in \overline{D}_{3h}^*. \quad (18)$$

## 4   Domain Decomposition Schemes

**4.1.** Let us describe the classical alternating Schwarz method on overlapping subdomains that allows us to perform analytic calculations, similar to that considered in [2]. We give conditions to ensure its $\varepsilon$-uniform convergence as the number of iterations grows. The techniques from [3,4] are used in the constructions and proofs.

Let open subdomains

$$D^1, \ D^2, \ldots, D^K \quad (19a)$$

with piecewise smooth boundaries $\Gamma^k$, $\Gamma^k = \overline{D}^k \setminus D^k$, form a covering of $D$: $D = \bigcup_{k=1}^{K} D^k$. The subdomains $D^k$ are assumed convex. By $D^{[k]}$ we denote the union of the subdomains $D^1$, $D^2$ that do not contain the set $D^k$

$$D^{[k]} = \bigcup_{i=1,\ i\neq k}^{K} D^i. \quad (19b)$$

We denote the minimal width of overlapping the pair of sets $D^k$ and $D^{[k]}$ by $\delta^k$. The minimal overlap $\delta$ of the subdomains from (19) is the smallest of $\delta^k$. The value $\delta$, generally speaking, may depend on the parameter $\varepsilon$, $\delta = \delta(\varepsilon)$.

Let

$$u^0(x), \quad x \in \overline{D} \tag{20a}$$

be an arbitrarily given initial function in the iteration process that satisfies the boundary condition from (3). The functions $u^n(x)$, $x \in \overline{D}$, $n = 1, 2, \ldots$, can be found by sequential solving of the boundary value problems

$$L_{(20)}(u^{n+\frac{k}{K}}(x)) \equiv L_{(3)}u(x) - f(x) = 0, \quad x \in D^k, \tag{20b}$$

$$u^{n+\frac{k}{K}}(x) = u^{n+\frac{k-1}{K}}(x), \quad x \in \overline{D} \setminus D^k, \ k = 1, \ldots, K,$$

$$u^{n+1}(x) = u^{n+\frac{K}{K}}(x), \quad x \in \overline{D}, \quad n = 0, 1, 2, \ldots.$$

Each auxiliary function $u^{n+\frac{k}{K}}(x)$, $x \in \overline{D}$, is a solution of the Dirichlet problem on $\overline{D}^k$

$$L_{(20)}(u^{n+\frac{k}{K}}(x)) = 0, \quad x \in D^k, \quad u^{n+\frac{k}{K}}(x) = u^{n+\frac{k-1}{K}}(x), \quad x \in \Gamma^k$$

and coincides with $u^{n+\frac{k-1}{K}}(x)$ on the set $\overline{D} \setminus D^k$. The sequence of the functions $u^n(x)$, $x \in \overline{D}$, $n = 1, 2, \ldots$, is called the solution of iteration process (20), (19), i.e., the Schwarz alternation method.

The condition on the overlap

$$\delta = \delta(\varepsilon) > 0, \quad \inf_{\varepsilon \in (0,1]} \left[ \varepsilon^{-1}\delta(\varepsilon) \right] > 0, \tag{21}$$

which is equivalent to the condition $\delta = \delta(\varepsilon) \geq m_{(21)}\varepsilon$, is sufficient for the $\varepsilon$-uniform convergence of the functions $u^n(x)$ as $n \to \infty$:

$$|u(x) - u^n(x)| \leq Mq^n, \quad x \in \overline{D}, \quad q \leq 1 - m. \tag{22}$$

**Theorem 2.** *Condition (21) is necessary and sufficient for the solution $u^n(x)$ of the iterative Schwarz method (20), (19) to converge $\varepsilon$-uniformly as $n \to \infty$ to the solution $u(x)$ of problem (3). Let the solution of the problem and its components satisfy estimates (4), (6), (8) for $K = 4$. Then, under condition (21), the solution of the Schwarz method (20), (19) satisfies bound (22).*

**4.2.** In a similar way, we construct iterative difference schemes of the Schwarz method on piecewise uniform grids condensing in the boundary layers.

On each of the sets $\overline{D}^k$, we introduce the special grids

$$\overline{D}_h^k \equiv \overline{D}_h^{k\,c} = \overline{D}^k \bigcap \overline{D}_{h(12)}^c. \tag{23}$$

Assume that the sides of the sets $\overline{D}^k$ pass through the nodes of the grid.

Let the function $z^0(x)$, $x \in \overline{D}_h$, be an arbitrary function satisfying the condition

$$z^0(x) = \varphi(x), \quad x \in \Gamma_h. \tag{24a}$$

We find the sequence of auxiliary functions $z^{n+\frac{k}{K}}(x)$, $k = 1, \ldots, K$, $n = 1, 2, \ldots$, by solving such grid problems

$$\Lambda_{(24)}(z^{n+\frac{k}{K}}(x)) \equiv \Lambda_{(8)} z^{n+\frac{k}{K}}(x) - f(x) = 0, \quad x \in D_h^k, \tag{24b}$$

$$z^{n+\frac{k}{K}}(x) = z^{n+\frac{k-1}{K}}(x), \quad x \in \overline{D}_h \setminus D^k, \quad k = 1, \ldots, K,$$

$$z^{n+1}(x) = z^{n+\frac{K}{K}}(x), \quad x \in \overline{D}_h, \quad n = 0, 1, 2, \ldots.$$

Each function $z^{n+\frac{k}{K}}(x)$ is defined on the set $\overline{D}_h$, being a solution of the grid Dirichlet problem, and it coincides with the function $z^{n+\frac{k-1}{K}}(x)$ on the set $\overline{D}_h \setminus D^k$. The function $z^n(x)$, $x \in \overline{D}_h$, $n = 1, 2, \ldots$, is called the solution of iterative grid Schwarz method (24), (23).

Taking bounds (13), (18), and (22) into account, the solution $z^n(x)$ of the grid iterative Schwarz method on overlapping subdomains containing either the boundary of the rectangle or the boundary of the circle converges $\varepsilon$-uniformly:

$$|u(x) - z^n(x)| \le M\left(N^{-1}\ln N + N_*^{-1}\ln N_* + q^n\right), \quad x \in \overline{D}_h, \tag{25}$$

as the numbers of mesh points $N$, $N_*$ and the number $n$ in the Schwarz method grow. In (25), $q \le 1 - m$.

**Theorem 3.** *Let the hypotheses of Theorem 1 hold. Then, under condition (21) the solution of the iterative grid Schwarz method (24) on the grid (23) converges $\varepsilon$-uniformly as $N, N_*, n \to \infty$ to the solution of the boundary value problem (3) with bound (25).*

**Acknowledgments.** The work is supported by the Russian Foundation for Basic Research, grant no. 16-01-00727.

# References

1. Samarskii, A.A.: The Theory of Difference Schemes. Marcel Dekker, New York (2001)
2. Shishkin, G.I., Tselishcheva, I.V.: Parallel methods of solving singularly perturbed boundary value problems for elliptic equations. Matem. Mod. **8**(3), 111–127 (1996). (in Russian)
3. Shishkin, G.I.: Discrete Approximations of Singularly Perturbed Elliptic and Parabolic Equations. UrO RAN, Ekaterinburg (1992). (in Russian)
4. Shishkin, G.I., Shishkina, L.P.: Difference Methods for Singular Perturbation Problems, vol. 140. Chapman & Hall/CRC Monographs and Surveys in Pure and Applied Mathematics. CRC Press, Boca Raton (2009)

# Generalized Multiscale Finite Element Method for Poroelasticity Problems in Heterogeneous Media

A. Tyrylgin[1]($\boxtimes$) ⓘ, M. Vasilyeva[1,2] ⓘ, and D. Brown[3] ⓘ

[1] Multiscale Model Reduction Laboratory, North-Eastern Federal University, Yakutsk, Russia
koc9tk@mail.ru, vasilyevadotmdotv@gmail.com
[2] Institute for Scientific Computation, Texas A&M University, College Station, TX, USA
[3] School of Mathematical Sciences, GeoEnergy Research Center, The University of Nottingham, Nottingham, UK
donaldbrowdr@gmail.com

**Abstract.** In this work, we consider the poroelasticity problems in heterogeneous porous media. Mathematical model contains coupled system of the equations for pressure and displacements. For the numerical solution, we present a Generalized Multiscale Finite Element Method (GMsFEM). This method solves a problem on a coarse grid by construction of the local multiscale basic functions. The procedure begins with construction of multiscale bases for both displacement and pressure in each coarse block. Using a snapshot space and local spectral problems, we construct a basis of reduced dimension. Finally, after multiplying by a multiscale partitions of unity, the multiscale basis is constructed in the online phase and the coarse grid problem then can be solved for arbitrary forcing and boundary conditions. We compare the solutions by choosing different numbers of multiscale basis functions. The results show that GMsFEM can provide good accuracy for two and three dimensional problems in heterogeneous domains.

**Keywords:** Heterogeneous porous media · Poroelasticity ·
Geomechanics · Fluid flow · Coupled system · Multiscale method ·
GMsFEM

## 1 Introduction

The mechanism of fluid flow and deformation of porous media is studied by the theory of poroelasticity [1,10,11]. Basic mathematical models include Lame's equations for displacements of the medium, as well as the law of conservation of mass and Darcy's law for a filtered fluid [5,6]. The most important feature of mathematical models of poroelasticity is that the equations of the system are coupled. The computational algorithms for the approximate solution of the

© Springer Nature Switzerland AG 2019
I. Dimov et al. (Eds.): FDM 2018, LNCS 11386, pp. 566–573, 2019.
https://doi.org/10.1007/978-3-030-11539-5_66

problems of poroelasticity are based on finite-element approximation in space. For coarse grid approximation, we use a Generalized Multiscale Finite Element Method (GMsFEM) [2–4,7,8]. This method solves a problem on a coarse grid by construction of the local multiscale basic functions for pressure and displacements.

The work is organized as follows. In Sect. 2, we provide the mathematical model of the poroelasticity problem in heterogeneous media and construct fine grid approximation using the finite element method. In Sect. 3, we present coarse grid approximation using GMsFEM algorithm, where we describe construction of the multiscale basis functions. Finally, numerical results are presented in Sect. 4.

## 2 Problem Formulation and Fine Grid Approximation

Let $\Omega \subset R^d$ is computational domain. We consider linear poroelasticity problem

$$- \operatorname{div} \sigma(u) + \alpha \operatorname{grad}(p) = 0 \text{ in } \Omega,$$

$$\alpha \frac{\partial \operatorname{div} u}{\partial t} + \frac{1}{M} \frac{\partial p}{\partial t} - \operatorname{div}\left(\frac{k}{\nu} \operatorname{grad} p\right) = f \text{ in } \Omega, \tag{1}$$

where $p$ is the pressure $u$ is displacements, $\sigma$ is the stress tensor, $k$ is the permeability and $f$ is a source term representing injection or production processes. We denote $M$ to be the Biot modulus, $\nu$ is the fluid viscosity, and $\alpha$ is the Biot-Willis fluid-solid coupling coefficient.

In the case of a linear elastic stress-strain constitutive relation we have that the stress tensor and symmetric strain gradient may be expressed as

$$\sigma(u) = 2\mu\varepsilon(u) + \lambda \operatorname{div}(u)\,\mathcal{I}, \quad \varepsilon(u) = \frac{1}{2}\left(\operatorname{grad} u + \operatorname{grad} u^T\right),$$

where $\mu$, $\lambda$ are Lame coefficients, $\mathcal{I}$ is the identity tensor. In the case where the media has heterogeneous material properties the coefficients $\mu$ and $\lambda$ may be highly variable.

For numerical solution of the poroelasticity problem on fine grid we use a standard finite element method

$$a(u,v) + g(p,v) = 0, \quad \text{for all } v \in \hat{V},$$

$$d\left(\frac{du}{dt}, q\right) + c\left(\frac{dp}{dt}, q\right) + b(p,q) = (f,q), \quad \text{for all } q \in \hat{Q}. \tag{2}$$

for $u \in V = [H^1(\Omega)]^d$ and $p \in Q = H^1(\Omega)$. Here for bilinear and linear forms we have define

$$a(u,v) = \int_\Omega \sigma(u)\,v dx, \quad b(p,q) = \int_\Omega \left(\frac{k}{\nu} \operatorname{grad} p, \operatorname{grad} q\right) dx, \quad c(p,q) = \int_\Omega \frac{1}{M} p q\, dx,$$

and

$$g(p,v) = \int_\Omega \alpha(\operatorname{grad} p, v) dx, \quad d(u,q) = \int_\Omega \alpha \operatorname{div} u\, q\, dx, \quad (f,q) = \int_\Omega f q\, dx.$$

## 3  Coarse Grid Approximation Using GMsFEM

In this section, we describe in detail the Generalized Multiscale Finite Element Method (GMsFEM) for solution of the poroelasticity problems in heterogeneous media. GMsFEM contains three steps:

(1) Construction of the coarse and fine meshes and local domains where we construct multiscale basis functions,
(2) Solution of the local spectral problems for multiscale basis functions construction,
(3) Construction and solution of the coarse scale approximation on multiscale space.

Let $\mathcal{T}_H$ and $\mathcal{T}_h$ are the coarse and fine grids. For construction of the multiscale basis functions, we solve a local spectral problems in domain $\omega_i$ for displacement and pressure separately. The local spectral problems:
Displacement

$$a^u(u,v) = \lambda_u, s^u(u,v),$$

$$a^u(u,v) = \int_{\omega_i} \sigma(u), \varepsilon(v)dx, \ s^u = \int_{\omega_i} (\lambda + 2\mu)(u,v)dx,$$

Pressure

$$a^p(p,q) = \lambda_p, s^p(p,q),$$

$$a^p(p,q) = \int_{\omega_i} k\nabla p, \nabla q dx, \ s^p = \int_{\omega_i} k\, p\, q\, dx,$$

where $v, q$ is a trial functions.

The multiscale spaces $V_h, Q_h$, we will form using eigenvectors $\varphi_1, \varphi_2, \ldots,$ $\varphi_L, \psi_1, \psi_2, \ldots, \psi_L$ corresponding to the first smallest $L$ eigenvalues, where $\lambda_1 \leq \lambda_2 \leq \ldots \leq \lambda_L$.

We can write spectral problems in the matrix form:
Displacement

$$A^u\varphi = \lambda_u S^u\varphi, \ x \in \omega_i,$$

Pressure

$$A^p\psi = \lambda_p S^p\psi, \ x \in \omega_i,$$

where $A^u, S^u$ are the stiffness and mass matrices for displacements and $A^p, S^p$ for pressure.

In the fact that the spectral problem must be solved many times. Therefore, it is possible to reduce the dimension of the problem. For solution of the local spectral problem, we use a snapshot space $V_{snap}$ for displacement and $Q_{snap}$ for pressure. Next, we define a transition matrices $R_{snap}^u, R_{snap}^p$: $R_{snap}^u = [\varphi_1^{snap}, \ldots, \varphi_{L_i}^{snap}]$ for displacements and $R_{snap}^p = [\psi_1^{snap}, \ldots, \psi_{L_i}^{snap}]$ for pressure.

Next, we move to spaces $V_{snap}$, $Q_{snap}$ and obtain the following eigenvalue problem:

Displacement

$$\overline{A}_u \overline{\varphi} = \lambda_u \overline{S}_u \overline{\varphi}, \quad x \in V_{snap},$$

where $\overline{A}_u = R_{snap}^u A^u (R_{snap}^u)^T$, $\overline{S}_u = R_{snap}^u S^u (R_{snap}^u)^T$ and $\varphi_j^\omega = (R_{snap}^u)^T \overline{\varphi}_j$.

Pressure

$$\overline{A}_p \overline{\psi} = \lambda_p \overline{S}_p \overline{\psi}, \quad x \in Q_{snap}.$$

where $\overline{A}_p = R_{snap}^p A^p (R_{snap}^p)^T$, $\overline{S}_p = R_{snap}^p S^p (R_{snap}^p)^T$ and $\psi_j^\omega = (R_{snap}^p)^T \overline{\psi}_j$.

For obtaining conforming basis functions we use linear partition of unity functions. We construct transition matrix $R$ from a fine grid to a coarse grid and use it for reducing the dimension of the problem. The transition matrix

$$R = \begin{pmatrix} R_u \\ R_p \end{pmatrix}$$

with

$$R_u = \{\chi^1 \varphi_1^1, \chi^1 \varphi_2^1, \ldots, \chi^1 \varphi_L^1, \ldots, \chi^{N_c} \varphi_1^{N_c}, \chi^{N_c} \varphi_2^{N_c}, \ldots, \chi^{N_c} \varphi_L^{N_c}\}$$

and

$$R_p = \{\chi^1 \psi_1^1, \chi^1 \psi_2^1, \ldots, \chi^1 \psi_L^1, \ldots \chi^{N_c} \psi_1^{N_c}, \chi^{N_c} \psi_2^{N_c}, \ldots, \chi^{N_c} \psi_L^{N_c}, \}.$$

where $\chi^i$ is linear partition of unity functions, $L$ is the number of basis functions and $N_c$ is the number of vertices of a coarse grid.

Then the system of equations can be translated into a coarse grid

$$A_c u_c + G_c p_c = F_c,$$
$$D_c u_c + B_c p_c = Y_c,$$

where $A_c = R_u A_f R_u^T$, $B_c = R_p B_f R_p^T$, $F_c = R_u F_f$, $Y_c = R_p Y_f$, $G_c = R_u G_f R_p^T$, $D_c = R_p D_f R_u^T$ and

$$A_f u = \int_\Omega \sigma(u)\varepsilon(v)dx, \quad B_f p = \frac{1}{M} \int_\Omega \frac{p}{\tau} q dx + \int_\Omega (k\nabla p, \nabla q)dx,$$

$$G_f p = \int_\Omega \alpha \nabla p \cdot v dx, \quad D_f u = \int_\Omega \alpha \frac{\operatorname{div} u}{\tau} q dx,$$

$$F_f = \int_\Omega f_u v dx, \quad Y_f = \int_\Omega f_p q dx.$$

After obtaining of a coarse-scale solution, we can reconstruct fine-scale solution

$$u_{ms} = R_u^T u_c, \quad p_{ms} = R_p^T p_c.$$

As a result, we obtain a solution on a fine grid for the problem of poroelasticity $(u_{ms}, p_{ms})$.

# 4   Numerical Results

## 4.1   Two-Dimensional Problem

In this section, we present numerical results of the poroelasticity problems in heterogeneous media. Calculations will be carried out on a structured computation domain, which is shown in Fig. 1 and we take $\Omega$ as a unit square $[0,1]^2$. Fine grid consists 441 vertices, 800 cells and 1240 facets and coarse grid contains 36 vertices.

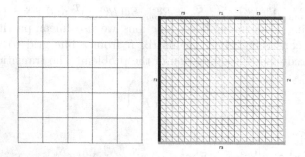

**Fig. 1.** Computation mesh. Left: coarse grid. Right: fine grid

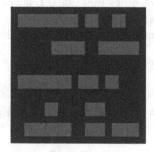

**Fig. 2.** Heterogeneous media for $\lambda, \mu, k$. (Color figure online)

The numerical solution is presented for the following boundary conditions

$$u_x = 0, \ \sigma_y = 0, \ x \in \Gamma_2 \cup \Gamma_4$$

$$u_y = 0, \ \sigma_x = 0, \ x \in \Gamma_3$$

$$\sigma_x = \sigma_y = 0, x \in \Gamma_1$$

for the poroelastic equation we set $p = 0$ on the boundaries $\Gamma_2 \cup \Gamma_3 \cup \Gamma_4 \cup \Gamma_5$. At the boundary $\Gamma_1$, we consider $p_1 = 10$.

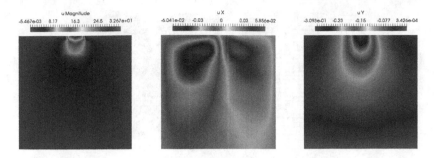

**Fig. 3.** Distribution of pressure, displacement along $X$ and $Y$ directions at the last moment of time.

**Table 1.** Relative errors for displacement and pressure with different numbers of multiscale basis functions in GMsFEM

| $L$ | $L_2^u$ (%) | $H_1^u$ (%) | $L_2^p$ (%) | $H_1^p$ (%) |
|---|---|---|---|---|
| 1 | 95.845 | 96.098 | 97.017 | 99.633 |
| 2 | 40.431 | 56.988 | 61.624 | 66.461 |
| 4 | 13.940 | 23.854 | 22.094 | 39.869 |
| 8 | 1.031 | 5.596 | 2.240 | 12.950 |
| 12 | 0.063 | 0.561 | 0.062 | 0.514 |

For the computations, we set coefficients $\lambda = 5$, $\mu = 4$ and $k = 0.01$ for domain with blue color in Fig. 2 and $\lambda = 10$, $\mu = 8$ and $k = 0.05$ for domain with red color in Fig. 2. We set $\beta = 0.01, \alpha = 1$. The calculation is performed by $T_{max} = 0.1$ with step in time $\tau = 0.01$.

In Fig. 3, we present the distribution of pressure and displacement along the X and Y directions at the last moment of time for the fine grid solution. In Table 1, we present the weighted $L^2$ and $H^1$ errors for multiscale solver.

## 4.2   Three-Dimensional Problem

Let us perform a numerical simulation of the considered problem of poroelasticity in a three-dimensional formulation. Modeling will be carried out in the region $[0, 1] \times [0, 1] \times [0, 1]$ meters at constant boundary conditions for load $p_1 = 10$. The remaining boundary values were chosen similarly to the two-dimensional formulation without $\Gamma_5$.

The calculation is carried out at $T_{max} = 0.001$ with time step $\tau = 0.0001$. The calculated tetrahedral grid contains 9261 vertices, 48000 cells and 102048 facets with the size of the load region $0.2 \times 0.2$ (see Fig. 4).

For the computations, we set coefficients $\lambda = 20$, $\mu = 100$ and $k = 1$ for domain with blue color in Fig. 2 and $\lambda = 100$, $\mu = 200$ and $k = 10$ for domain with red color in Fig. 2. We set $\beta = 1, \alpha = 1$.

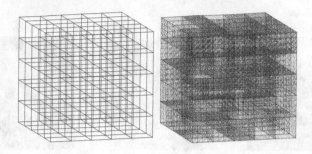

**Fig. 4.** Geometric domain in a tetrahedral computational mesh.

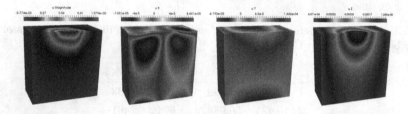

**Fig. 5.** The distribution of pressure and displacement along $X$, $Y$ and $Z$ (from left to right) at the final time.

**Table 2.** Relative errors for displacement and pressure with different numbers of multiscale basis functions in GMsFEM for three-dimensional formulation

| $L$ | $L_2^u$ (%) | $H_1^u$ (%) | $L_2^p$ (%) | $H_1^p$ (%) |
|---|---|---|---|---|
| 1 | 17.423 | 43.768 | 28.158 | 53.484 |
| 2 | 14.137 | 35.863 | 18.698 | 43.129 |
| 4 | 5.437 | 20.492 | 9.294 | 31.934 |
| 8 | 3.124 | 14.134 | 4.306 | 21.463 |
| 12 | 1.643 | 9.635 | 2.369 | 16.109 |

Numerical results of the poroelasticity problem in the three-dimensional formulation are shown in Fig. 5 and illustrate the solution of the problem at the last moment of time on the fine grid. In Table 2 we present the weighted $L^2$ and $H^1$ errors for three-dimensional formulation. This results of the poroelasticity problem show that GMsFEM can provide good accuracy for two and three dimensional problems in heterogeneous domains.

**Acknowledgments.** Work is supported by the grant of the Russian Scientific Found (N 17-71-20055).

# References

1. Meirmanov, A.: Mathematical Models for Poroelastic Flows. Atlantis Press, Paris (2014). https://doi.org/10.2991/978-94-6239-015-7
2. Brown, D.L., Vasilyeva, M.V.: A generalized multiscale finite element method for poroelasticity problems II: nonlinear coupling. J. Comput. Appl. Math. **294**, 372–388 (2016)
3. Brown, D.L., Vasilyeva, M.V.: A generalized multiscale finite element method for poroelasticity problems I: linear problems. J. Comput. Appl. Math. **297**, 132–146 (2016)
4. Akkutlu, I.Y., Efendiev, Y., Vasilyeva, M., Wang, Y.: Multiscale model reduction for shale gas transport in a coupled discrete fracture and dual-continuum porous media. J. Nat. Gas Sci. Eng. **48**, 65–76 (2017)
5. Kolesov, A.E., Vabishchevich, P.N., Vasilyeva, M.V.: Splitting schemes for poroelasticity and thermoelasticity problems. Comput. Math. Appl. **67**, 2185–2198 (2014)
6. Sivtsev, P.V., Vabishchevich, P.N., Vasilyeva, M.V.: Numerical simulation of thermoelasticity problems on high performance computing systems. In: Dimov, I., Faragó, I., Vulkov, L. (eds.) FDM 2014. LNCS, vol. 9045, pp. 364–370. Springer, Cham (2015). https://doi.org/10.1007/978-3-319-20239-6_40
7. Armero, F.: Formulation and finite element implementation of a multiplicative model of coupled poro-plasticity at finite strains under fully saturated conditions. Comput. Methods Appl. Mech. Eng. **171**(3–4), 205–241 (1999)
8. Efendiev, Y., Hou, T.Y.: Multiscale Finite Element Methods: Theory and Applications. STAMS, vol. 4. Springer Science & Business Media, New York (2009)
9. Efendiev, Y., Galvis, J., Hou, T.Y.: Generalized multiscale finite element methods (GMsFEM). J. Comput. Phys. **251**, 116–135 (2013)
10. Terzaghi, K.: Theory of consolidation. In: Theoretical Soil Mechanics, pp. 265–296 (1943)
11. Biot, M.A.: General theory of three-dimensional consolidation. J. Appl. Phys. **12**(2), 155–164 (1941)
12. Lemaitre, J., Chaboche, J.L.: Mechanics of Solid Materials. Cambridge University Press, Cambridge (1994)
13. Salençon, J.: Handbook of Continuum Mechanics: General Concepts thermoelasticity. Springer Science & Business Media, Heidelberg (2001). https://doi.org/10.1007/978-3-642-56542-7

# Past Surface Temperatures Reconstructed by Inversion of the Measured Borehole Temperature-Depth Profiles in Rock

Sergey A. Tyuflin[✉], Oleg V. Nagornov, and Tatiana I. Bukharova

National Research Nuclear University MEPhI,
Kashirskoe Shosse, 31, 115409 Moscow, Russia
satyuflin@mephi.ru

**Abstract.** The measured temperature-depth profiles in boreholes are proxy climate indicators. The climatic temperature signal at the surface penetrates in the rock thickness and disturbs the steady-state temperature of the Earth. The measured temperature-depth profile can be so-called the re-determination condition to solve the inverse problem for the thermal diffusivity equation, and retrieve the past surface temperature history. We study properties of solution for this problem. We derive that the solution is not unique. However, there were done numerous past surface temperature reconstructions. We show that it is needed to take into account some a priori data to determine past temperatures correctly.

**Keywords:** Boreholes · Past surface temperature · Inverse problems

## 1 Introduction

Knowledge of the surface temperature changes is important problem. The development of methods for the past surface temperature reconstructions can give a key in understanding these changes. The systematic instrumental temperature measurements took place no more than two centuries. Thus, indirect estimations of the past temperatures present main information on the past climate. There are different sources of information about paleotemperatures. The most reliable data are kept in the measured borehole temperatures that present response to the surface temperature history.

The underground temperature distribution is mainly determined by two types of processes [1,2]. The first is the surface temperature changes and the second is the heat flux from the Earth that is subjected to the long-time geological processes. The surface temperature changes take place at relatively smaller time scale. Therefore, the measured temperature-depth profiles in the borehole contain information on the climatic changes at the surface. The seasonal temperature variations at the surface are noticeable at depth about 10–15 m while the climatic oscillations reach several hundred meters and more. Thus, the boreholes of several hundred meters can contain information on the past surface temperatures for several hundred years.

© Springer Nature Switzerland AG 2019
I. Dimov et al. (Eds.): FDM 2018, LNCS 11386, pp. 574–579, 2019.
https://doi.org/10.1007/978-3-030-11539-5_67

The heat and mass transfer in rocks is described by the one-dimensional thermal diffusivity equation [1]. The past surface temperature reconstruction is the inverse problem that contains additional redetermination condition. The measured temperature-depth profile presents such condition. We found out that this problem has not the unique and stable solution in general case.

There are several well-known methods of the past surface temperature reconstructions: the Monte-Carlo method [3]; the least-squares inversion method [4], the singular value decomposition method [5] and the control method [6]. We show that these temperature reconstructions are not unique. We consider examples to demonstrate uncertainity in some past surface temperature reconstructions.

## 2   Mathematical Model

The mathematical statement of the inverse problem consists of the one-dimensional thermal conductivity equation, the initial condition, the boundary condition at the bottom of borehole and the measured-temperature-depth profile. The former is used as the redetermination condition, $\chi(z)$, where $z$ is vertical coordinate. Then the inverse problem to find the temperature in the past is the solution of the following problem:

$$
\begin{aligned}
T_t &= a^2 T_{zz}, & 0 < t < t_f,\ 0 < z < H, \\
T(0,t) &= U_s + \mu(t), & 0 < t \le t_f, \\
-k T_z(H,t) &= q, & 0 < t \le t_f, \\
T(z,0) &= U(z), & 0 < z < H, \\
T(z,t_f) &= \chi(z), & 0 \le z \le H,
\end{aligned}
\tag{1}
$$

where $H$ is the borehole depth, $a^2$ is the thermal diffusivity, $k$ is the thermal conductivity, $q$ is the geothermal heat flux at the bottom of borehole, $U(z)$ is the steady-state temperature profile associated with this flux, $U_s$ is the initial temperature, which characterizes the average temperature that was on the surface in the past before the beginning of sharp temperature variations on the surface, $\mu(t)$ is temperature variations on the surface in time with respect to its initial value Us from the moment $t = 0$ to the time of measurements of the borehole temperature profile $t_f$, $\mu(0) = 0$.

Let us represent the borehole temperature profile $T(z,t)$ in the form of the superposition of two temperature profiles: the steady-state temperature profile $U(z)$ associated with the geothermal heat flow from the Earth and the residual temperature profile $V(z,t)$ associated with temperature variations on the surface: $T(z,t) = U(z) + V(z,t)$. Then, the steady-state temperature profile $U(z) = U_s - (q/k)z$.

Let us denote $\theta(z) = \chi(z) - U(z)$ is deviation of the measured temperature profile from the steady-state one. This deviation is associated with the surface temperature changes. Thus, the problem of finding surface temperature history is reduced to the solution of the problem:

$$V_t = a^2 V_{zz}, \quad 0 < t < t_f,\ 0 < z < H,$$
$$V(0,t) = \mu(t), \qquad 0 < t \le t_f,$$
$$V_z(H,t) = 0, \qquad 0 < t \le t_f, \tag{2}$$
$$V(z,0) = 0, \qquad 0 < z < H,$$
$$V(z,t_f) = \theta(z), \qquad 0 \le z \le H.$$

## 3    Properties of Solution

Let us show that the inverse problem (2) in the general case has no the uniqueness solution [7].

**Lemma.** *In addition to the trivial solution* $(V(z,t) \equiv 0; \mu(t) \equiv 0)$, *the inverse problem (2) with* $\theta(z) = 0$ *has a nontrivial solution* $(V(z,t); \mu(t))$.

Let us assume that $\mu(0) = \mu(t_f) = 0$ and $\mu(t) = \sum\limits_{m=1}^{\infty} \alpha_m \sin(\pi m t/t_f)$, where $\alpha_m$ are unknown coefficients. Let $V^{(m)}(z,t)$ be a solution of the direct problem specified as

$$V_t^{(m)} = a^2 V_{zz}^{(m)}, \quad (t,z) \in Q \equiv [0,t_f] \times [0,H],$$
$$V^{(m)}(0,t) = \alpha_m \sin(\pi m t/t_f), \qquad t \in [0,t_f], \tag{3}$$
$$V^{(m)}(z,0) = 0, \qquad z \in [0,H],$$
$$V_z^{(m)}(H,t) = 0, \qquad t \in [0,t_f].$$

The solution of this problem is easily obtained in the form

$$V^{(m)}(z,t) = \alpha_m \sin(\pi m t/t_f) - \alpha_m \pi m t/t_f \cdot$$
$$\sum_{n=1}^{\infty} I_n e_n(z) \int_0^t \exp(-\lambda_n(t-\tau)) \cdot \cos(\pi m \tau/t_f) d\tau \tag{4}$$

where $I_n = \int_0^H e_n(z) dz / \|e_n(z)\|^2 = 2/(H\sqrt{\lambda_n})$, $e_n = \sin(\sqrt{\lambda_n} z)$ and $\lambda_n = a^2 \pi^2 (n-1/2)^2 / H^2$ are the eigenfunctions and eigenvalues of the following Sturm-Liouville problem:

$$a^2 Z''(z) + \lambda Z = 0, \quad z \in (0,H),$$
$$Z(0) = Z'(H) = 0. \tag{5}$$

The asymptotic behavior of the eigenvalues is $|\lambda_n| \sim C \cdot n^2$, $n \to \infty$. Therefore, the series $\sum\limits_{n=1}^{\infty} 1/|\lambda_n|$ converges. Then, the set of the functions $\{\exp(\lambda_n t)\}_{n=1}^{\infty}$ is incomplete in $L_2(0,t_f)$, this is a corollary of Muntz's theorem [7]. Thus, there is a nonzero function $F(t)$ specified at $t \in [0,t_f]$ such that $F(t)$ orthogonal to

$\{\exp(\lambda_n t)\}_{n=1}^{\infty}$ in $L_2(0, t_f)$. Let us expand $F(t)$ into the Fourier series at $t \in [0, t_f]$: $F(t) = \sum_{m=1}^{\infty} \beta_m \sin(\pi m t / t_f)$. Let us prove that $V(z, t) = \sum_{m=1}^{\infty} V^{(m)}(z, t)$ is a solution of the problem specified by Eq. (2) and $\mu(t) = F(t)$, $\alpha_m = \beta_m$. Indeed, $V(z, t)$ satisfies the first of Eq. (2), as well as the initial and boundary conditions. Let us verify the last condition in Eq. (2):

$$V(z, t_f) = - \sum_{m=1}^{\infty} \sum_{n=1}^{\infty} \alpha_m \pi m I_n e_n(z) / t_f \cdot$$

$$\int_0^{t_f} \exp(-\lambda_n (t_f - \tau)) \cdot \cos(\pi m \tau / t_f) d\tau \tag{6}$$

The integration of Eq. (6) by parts yields

$$V(z, t_f) = \sum_{n=1}^{\infty} I_n e_n(z) \lambda_n \exp(-\lambda_n t_f) \cdot$$

$$\int_0^{t_f} \exp(\lambda_n \tau) \cdot \left( \sum_{m=1}^{\infty} \alpha_m \sin(\pi m \tau / t_f) \right) d\tau \tag{7}$$

Since the inner series in Eq. 7 is identically equal to $F(\tau)$ and is orthogonal to $\{\exp(\lambda_n t)\}_{n=1}^{\infty}$, $V(z, t_f) = 0$. Thus, we find the nontrivial solution $V(z, t)$ and the lemma is proved.

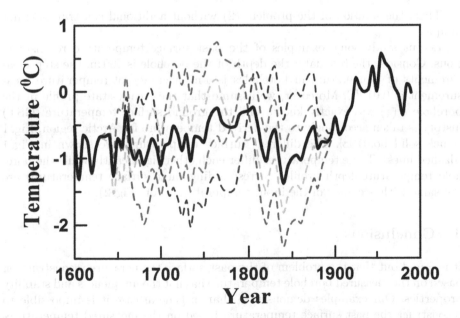

**Fig. 1.** Past surface temperatures (GST - solid line, the other surface temperatures - dashed lines).

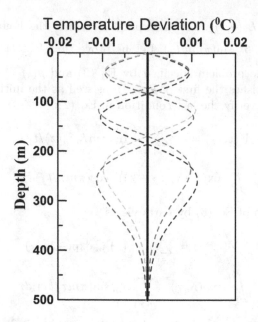

**Fig. 2.** Deviation of the temperature-depth profiles corresponding to GST and the other surface temperatures (Fig. 1).

Thus, the solution of the problem (2) without additional constraints is not unique.

Let us study some examples of the past surface temperature reconstructions. Consider the test data: the depth of the borehole is 500 m, the step measurements of temperature in boreholes is 1 m, accuracy of temperature measurement is 0.01 °C. Moreover, we assume that the steady-state profile in the borehole $U(z)$ was exactly known and the ground surface temperature (GST) history is taken accordingly to the derived temperature for Arctic region, Fig. 1 (black solid bold) [8]. The different surface temperatures are shown in Fig. 1 (dashed lines). These temperatures differ each other significantly while the borehole temperature-depth profiles corresponding these surface temperatures are the same with accuracy of the input temperature error (Fig. 2).

## 4    Conclusions

It is found out that the problem of the past surface temperature reconstruction based on the measured borehole temperature has not the uniqueness and stability properties. Our examples demonstrate that in general case it is impossible to reconstruct the past surface temperature based on the measured temperature-depth profile containing any measurement errors.

# References

1. Pollack, H.N., Huang, S.: Climate reconstruction from subsurface temperatures. Ann. Rev. Earth Planet. Sci. **28**(1), 339–365 (2000)
2. Nagornov, O.V., Tyuflin, S.A.: Inverse problem for paleo-temperature reconstruction based on the tree-ring width and glacier-borehole data. In: Dimov, I., Faragó, I., Vulkov, L. (eds.) Numerical Analysis and Its Applications. LNCS, vol. 10187, pp. 508–516. Springer, Cham (2016). https://doi.org/10.1007/978-3-319-57099-0_57
3. Dahl-Jensen, D., et al.: Past temperatures directly from the Greenland ice sheet. Science **282**, 268–271 (1998)
4. Shen, P.Y., Beck, A.E.: Least squares inversion of borehole temperature measurements in functional space. J. Geophys. Res. **96**, 19965–19979 (1991)
5. Beltrami, H., Cheng, L.Z., Mareschal, J.C.: Simultaneous inversion of borehole temperature data for past climate determination. Geophys. J. Int. **129**, 311–318 (1997)
6. MacAyeal, D.R., Firestone, J., Waddington, E.: Paleothermometry by control method. J. Glaciol. **37**(127), 326–338 (1991)
7. Nagornov, O.V., Tyuflin, S.A., Konovalov, Y.V., Kostin, A.B.: Inverse Problems of Paleothermometry. MEPhI, Moscow (2008)
8. Overpeck, J., et al.: Arctic environmental change of the last four centuries. Science **278**, 1251–1256 (1997)

# Simulation of Cooling Devices and Effect for Thermal Stabilization of Soil in a Cryolithozone with Anthropogenic Impact

N. A. Vaganova[1,2]([envelope]) [iD] and M. Yu. Filimonov[1,2]([envelope]) [iD]

[1] Ural Federal University, Yekaterinburg, Russia
[2] Krasovskii Institute of Mathematics and Mechanics, Yekaterinburg, Russia
{vna,fmy}@imm.uran.ru

**Abstract.** The paper deals with periodic freezing and thawing of soil due to seasonal changes under the climatic conditions that affect the formation of thermal fields in the ground. A model and algorithm for simulation of seasonal cooling systems operating in certain temperature parameters and the effect on thermal stabilization of the soil is proposed. In modeling, the maximum number of climatic and technical parameters of devices are taken into account. The computations make it possible to estimate the efficiency of cooling devices using for keep the properties of the soil.

**Keywords:** Mathematical modelling · Permafrost · Soil stabilization

## 1 Introduction

In Russia, more than 60% of the territory (11 million km$^2$) is occupied by permafrost. Different scenarios of climate changes show, that global warming will intensify in the polar regions of the Arctic, and can lead to significant degradation of permafrost with increasing human impacts. The average thickness of permafrost varies from 10 to 800 m, and the permafrost soils have different physicochemical properties that can vary in all directions. On summer, as a result of positive temperatures and solar radiation effect, there is a seasonal thawing of the upper layer, on winter there is a reverse freezing process. Thawing of ice-saturated rocks due to warming or various technogenic impacts will be accompanied by subsidence of the earth's surface and the development of thermokarst, leading to destruction of various engineering structures [1].

Studying thermal fields on the surface of the soil from underground pipelines in view to determine the damages [2,3], it was found that the solar radiation has a great influence on the formation of thermal fields in the upper layer of the soil. Therefore, in studying non-stationary fields propagation in the soil from various technical systems used in the northern oil and gas fields [4,5], solar radiation was also taken into account.

© Springer Nature Switzerland AG 2019
I. Dimov et al. (Eds.): FDM 2018, LNCS 11386, pp. 580–587, 2019.
https://doi.org/10.1007/978-3-030-11539-5_68

Most of the industrial facilities in the cryolithozone are constructed and operated on the principle of maintaining the frozen state of the foundation soils, in particular, for which piles used for the foundation [6,7]. Therefore, the problem of reducing the intensity of the thermal interaction in a system "heat source–permafrost" is an urgent task, since it has special significance for solving problems of environmental protection, safety, saving costs and improving operational reliability of engineering structures.

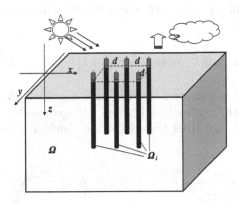

**Fig. 1.** A basic scheme of simulated area

## 2 Heat Impact in Permafrost

The basic human influences on permafrost consists of a combination of disturbance of the structure of the upper layers of the soil and thermal impact of different technical systems.

Simulation of processes of heat distribution in permafrost soil is reduced to solution of three-dimensional diffusivity equation with non-uniform coefficients including localized heat of phase transition—an approach to solve the problem of Stefan type, without the explicit separation of the phase transition in 3D area (Fig. 1) [8–10]. The equation has the form

$$\rho\big(c_\nu(T) + k\delta(T - T^*)\big)\frac{\partial T}{\partial t} = \nabla\left(\lambda(T)\varDelta T\right), \tag{1}$$

with initial condition

$$T(0, x, y, z) = T_0(x, y, z). \tag{2}$$

Here $\rho$ is density [kg/m$^3$], $T^*$ is temperature of phase transition [$K$],

$$c_\nu(T) = \begin{cases} c_1(x, y, z), T < T^*, \\ c_2(x, y, z), T > T^*, \end{cases} \text{is specific heat [J/kg K]},$$

$$\lambda(T) = \begin{cases} \lambda_1(x, y, z), T < T^* \\ \lambda_2(x, y, z), T > T^*, \end{cases} \text{is thermal conductivity coefficient [W/m K]},$$

**Table 1.** Basic annual climatic parameters.

| | Jan | Feb | Mar | Apr | May | Jun | Jul | Aug | Sep | Okt | Nov | Dec |
|---|---|---|---|---|---|---|---|---|---|---|---|---|
| Aver. temperature, °C | −28.1 | −26.3 | −20.0 | −10.8 | −2.0 | 8.5 | 15.1 | 11.5 | 5.1 | −6.5 | −20.8 | −26.2 |
| Solar radiation, W/m²s | 0 | 112 | 282 | 567 | 809 | 865 | 889 | 639 | 355 | 122 | 34 | 0 |

$k = k(x, y, z)$ is specific heat of phase transition, $\delta$ is Dirac delta function.

Balance of heat fluxes at the surface $z = 0$ defines the corresponding nonlinear boundary conditions

$$\gamma q + b(T_{air} - T(x, y, 0, t)) = \varepsilon\sigma(T^4(x, y, 0, t) - T^4_{air}) + \lambda\frac{\partial T(x, y, 0, t)}{\partial z}. \qquad (3)$$

To determine the parameters in boundary condition (3), an iterative algorithm is developed that takes into account the geographic coordinates of the area, lithology of soil and other features of the considered location [7, 8, 11].

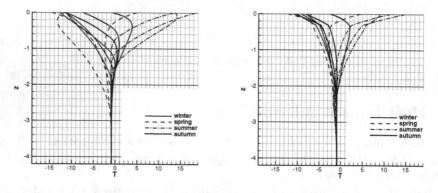

**Fig. 2.** Temperature (°C) in the soil: no riprap (left) and with a 2.5 m riprap (right)

Table 1 shows the data for the considered area. The others parameters in the Eq. (1) and the conditions (2) and (3) are determined as a result of geophysical research. The permafrost temperature lower than the area of influence of seasonal changes (lower than 10 m) is −0.7 °C. In Fig. 2(left) and (right) the temperatures in near-surface layer of soil for different months are shown for the uncoated soil and the soil with a 2.5 m riprap, respectively. The basic thermal parameters of the soil are in the following: thermal conductivity is 1.82 and 1.58 W/(m K), volumetric heat is 2130 and 3140 kJ/(m³ K) for frozen and melted soil, respectively; volumetric heat of phase transition is $1.384 \cdot 10^5$ kJ/(m³ K).

Stabilization of soil is primarily due to the restriction of seasonal effects of changes in air temperature and solar radiation intensity. To do this, as a rule, multi-layer ripraps are used. Figure 2(right) presents temperature profiles for different months with 2.5 m of riprap: 0.3 m of concrete slab, 2 m of sand, 0.2 m

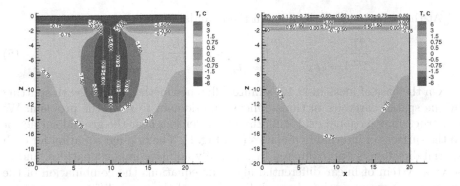

**Fig. 3.** Temperature around SCD's on December (left) and on September (right) in uncoated soil. $n = 3 \times 3$, $d = 1\,\text{m}$.

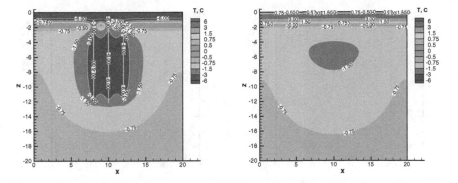

**Fig. 4.** Temperature around SCD's on December (left) and on September (right) in uncoated soil. $n = 3 \times 3$, $d = 2\,\text{m}$.

of foam. The temperature variation range in the near-surface layer of the soil is much less than in uncoated soil, the soil under the riprap is kept longer in a stable frozen state. Also, to preserve the soil in the frozen state seasonal cooling devices (SCDs) are used [12], operating on the basis of physical laws, and creating zones of additional freezing on winter.

Let consider $n$ SCDs which are included in $\Omega$. In Fig. 1 the surfaces of these objects are tubes $\Omega_i(x, y, z)$, $i = 1, \ldots, n$. These surfaces suppose to be inner boundaries with the conditions

$$T\Big|_{\Omega_i} = T_i(t), \quad i = 1, \ldots, n. \tag{4}$$

The computational domain is a three-dimensional box $\Omega$, where $x$ and $y$ axes are parallel to the ground surface and the $z$ axis is directed downward. We assume that the size of the box $\Omega$ is defined by positive numbers $L_x$, $L_y$, $L_z$: $-L_x \leq x \leq L_x$, $-L_y \leq y \leq L_y$, $-L_z \leq z \leq 0$.

At the boundaries of the domain the boundary conditions are given

$$\left.\frac{\partial T}{\partial x}\right|_{x=\pm L_x} = \left.\frac{\partial T}{\partial y}\right|_{y=\pm L_y} = 0, \quad \left.\frac{\partial T}{\partial z}\right|_{z=-L_z} = 0. \tag{5}$$

On the base of ideas in [9,10] a finite difference method is used with splitting by the spatial variables in three-dimensional domain to solve the problem. We construct an orthogonal grid, uniform, or condensing near the ground surface or to the surfaces of internal boundaries ($\Omega_i$, Fig. 1). The original equation for each spatial direction is approximated by an locally additive implicit pattern, and to solve a system of linear differential algebraic equations the combination of the sweep and Newton method is used. Solvability of the same difference problems approximating (1)–(5) is proved in [13].

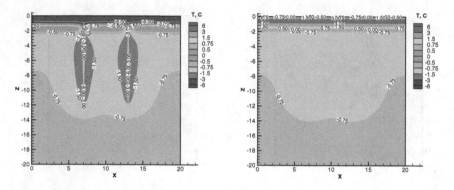

**Fig. 5.** Temperature around SCD's on December (left) and on September (right) in uncoated soil. $n = 2 \times 2$, $d = 4$ m.

# 3   Numerical Results

Let consider a system of several SCDs located in frozen soil (Fig. 1). The SCDs are located at the nodes of a rectangular mesh at a distance of $d$ [m] from each other. A number of SCDs in the mesh will be denoted by $n$, for example, in Fig. 1 $n = 3 \times 2$, where 3 and 2 are numbers of SCDs along the axis $x$ and $y$, respectively. Figures 3, 4, 5, 6, 7 and 8 show the thermal fields around different SCDs systems in the vertical $xz$ section of the soil. In Figs. 3, 4 and 5 the uncoated soil is considered, in Figs. 6, 7 and 8 the riprap consists of 3 layers: a crement slab (0.3 m), a layer of gravelly soil (2.0 m), a layer of foam (0.2 m).

Size of the computational area are $L_x = L_y = L_z = 20$ m, SCD's deep iz 11.5 m, diameter is 0.057 m. The basic soil temperature is $-0.7\,^\circ$C. Numerical grid sizes had varied from $91 \times 91 \times 51$ to $191 \times 191 \times 151$, time step is 24 h. The basic climatic parameters are presented in Table 1.

The thermal fields around SCD's have been stabilized after 3–4 years of simulation. In the following figures there are presented thermal fields on December

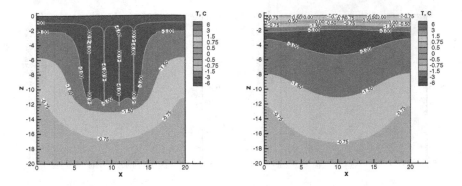

**Fig. 6.** Temperature around SCD's on December (left) and on September (right) in soil with the riprap. $n = 4 \times 4$, $d = 2\,\text{m}$.

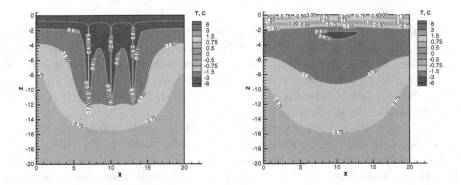

**Fig. 7.** Temperature around SCD's on December (left) and on September (right) in soil with the riprap. $n = 3 \times 3$, $d = 3\,\text{m}$.

(left figures) and September (right figures). December illustrates the soil freezing under SCDs operation and September shows the thermal trace of SCDs effect after the season the soil warmed up.

If the zone covered by the SCDs is sufficiently small (in Fig. 3 the zone is $4\,\text{m}^2$), then the thickness of the layer to be frozen with the help of the SCDs is negligible, and it is not conserved during the warm period, see Fig. 3(right).

For a larger coverage area ($16\,\text{m}^2$) with the same location of the SCDs, with a distance between them of $2\,\text{m}$ (Fig. 4) the freezing zone is observed on September. Temperatures below $-1.5\,°\text{C}$ are located at depths from 4 to $8\,\text{m}$.

With a distance of $4\,\text{m}$ between the SCDs (Fig. 5), the freezing zones do not merge, the SCDs act as a single unit and do not form a conservation area of permafrost.

If on the soil surface a riprap is applied to preserve soil properties, then, as we see in Figs. 6, 7 and 8, the distances between the SCDs in the system can be enlarged without losing the accumulative effect of interaction of SCD's and the merging the freezing zones.

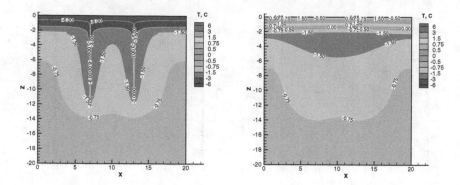

**Fig. 8.** Temperature around SCD's on December (left) and on September (right) in soil with the riprap. $n = 2 \times 2$, $d = 6$ m.

In the last three figures, the area occupied by the SCDs system, is $6 \times 6$ m. In Fig. 6, the distance between the SCDs is 2 m and the depth of the stably frozen layer reaches 10 m. In Fig. 7, the distance between the SCDs is 3 m and the freezing depth is 8 m. And when the riprap is used for a system of four SCDs with a distance of 6 m between each other (Fig. 8), the frozen layer is preserved in September in the depth from 2 to 4.9 m.

In addition, when using a riprap, the top layers of the soil are not affected by the seasonal freezing-thawing regime so much, there are no taliks at the depth corresponding to active layer thickness, that also stabilizes the soil. So, the complex and comprehensive use of ripraps and insulations allow to optimize the SCDs system application for thermal stabilization of permafrost.

**Acknowledgments.** Supported by Russian Foundation for Basic Research 16–01–00401.

# References

1. Nelson, F.E., Anisimov, O.A., Shiklomanov, N.I.: Subsidence risk from thawing permafrost. Nature **410**, 889–890 (2001)
2. Vaganova, N.: Mathematical model of testing of pipeline integrity by thermal fields. In: Venkov, G., Pasheva, V. (eds.) 40th International Conference on Applications of Mathematics in Engineering and Economics, AIP Conference Proceedings, vol. 1631, pp. 37–41. American Institute of Physics, College Park (2014). https://doi.org/10.1063/1.4902455
3. Vaganova, N.: Simulation of thermal fields from an underground pipeline at the ground surface. In: Pasheva, V., Venkov, G., Popivanov, N. (eds.) 43rd International Conference on Applications of Mathematics in Engineering and Economics (AMEE), AIP Conference Proceedings, vol. 1910, Article no. 020005. American Institute of Physics, College Park (2017). https://doi.org/10.1063/1.5013942
4. Filimonov, M.Y., Vaganova, N.A.: Simulation of thermal fields in the permafrost with seasonal cooling devices. In: 2012 Proceedings of the 9th International Pipeline

Conference, vol. 4, pp. 133–141. American Society of Mechanical Engineers, New York (2013). https://doi.org/10.1115/IPC2012-90287

5. Vaganova, N., Filimonov, M.Y.: Computer simulation of nonstationary thermal fields in design and operation of northern oil and gas fields. In: Pasheva, V., Venkov, G., Popivanov, N. (eds.) 41st International Conference on Applications of Mathematics in Engineering and Economics (AMEE), AIP Conference Proceedings, vol. 1690, Article no. 020016. American Institute of Physics, College Park (2015). https://doi.org/10.1063/1.4936694

6. Vaganova, N., Filimonov, M.Y.: Different shapes of constructions and their effects on permafrost. In: Pasheva, V., Venkov, G., Popivanov, N. (eds.) 42nd International Conference on Applications of Mathematics in Engineering and Economics, AMEE 2016, AIP Conference Proceedings, vol. 1789, Article no. 020019. American Institute of Physics, College Park (2016). https://doi.org/10.1063/1.4968440

7. Filimonov, M., Vaganova, N.: Numerical simulation of technogenic and climatic influence on permafrost. In: Advances in Environmental Research, vol. 54. Nova Science Publishers, New York (2017)

8. Vaganova, N., Filimonov, M.: Simulation of freezing and thawing of soil in Arctic regions. In: 2nd International Conference on Sustainable, ICSC 2017, IOP Conference Series: Earth and Environmental Science, vol. 72, Article no. 012005. IOP Publishing, Bristol (2017). https://doi.org/10.1088/1755-1315/72/1/012005

9. Patankar, S.V.: Numerical Heat Transfer and Fluid Flow. Hemisphere, New York (1980)

10. Samarsky, A.A., Vabishchevich, P.N.: Computational Heat Transfer. The Finite Difference Methodology, vol. 2. Wiley, Chichester, New York (1995)

11. Filimonov, M.Y., Vaganova, N.A.: Simulation of technogenic and climatic influences in permafrost for northern oil fields exploitation. In: Dimov, I., Faragó, I., Vulkov, L. (eds.) FDM 2014. LNCS, vol. 9045, pp. 185–192. Springer, Cham (2015). https://doi.org/10.1007/978-3-319-20239-6_18

12. Filimonov, M.Y., Vaganova, N.A.: Simulation of thermal stabilization of soil around various technical systems operating in permafrost. Appl. Math. Sci. 7(144), 7151–7160 (2013)

13. Vaganova, N.A.: Existence of a solution of an initial-boundary value difference problem for a linear heat equation with a nonlinear boundary condition. Proc. Steklov Inst. Math. 261(1), 260–271 (2008)

# The Conjugate Gradient Method for the Dirichlet Problem and Its Modifications

V. I. Vasil'ev$^{(\boxtimes)}$, A. M. Kardashevsky$^{(\boxtimes)}$, and V. V. Popov$^{(\boxtimes)}$

North-Eastern Federal University, Yakutsk, Russia
vasvasil@mail.ru, kardam123@gmail.com, imi.pm.pvvl@mail.ru

**Abstract.** In this paper we consider a numerical solution of the non-classical Dirichlet problem and its modifications for the second-order two-dimensional hyperbolic equations. In order to determine the missing initial condition using an additional condition specified at the final time, an iterative method of conjugate gradient is used. A direct problem is numerically realized at each iteration. The efficiency of the proposed computational algorithm is confirmed by calculations for model two-dimensional problems.

**Keywords:** Hyperbolic equation · Inverse problem ·
Dirichlet problem · Finite difference method · Iterative method ·
Conjugate gradients method · Random errors

## 1 Introduction

The Dirichlet problem for the wave equation is an important problem in mathematical physics. It was for the first time when non-uniqueness of the Dirichlet problem for wave equation was mentioned in the works of Hadamard (1936). A proof of correctness of formulation of the Dirichlet problem for hyperbolic equations of the second order can be found in the books [1–3]. General approach to solving of such ill-posed problems is based on inverse problems with subsequent use of gradient iterative methods. These methods are based on searching for an approximate numerical solution of ill-posed problems from iterative minimization of the corresponding functional. Research works of Kabanikhin [4,5] present a numerical investigation of the Dirichlet problem solution for the wave equation through the Landwebers optimizing iterative methods and the steepest descent method. Besides, a theoretical study of their stability and convergence of an approximate solution to solution of the initial differential problem for one-dimensional and two-dimensional hyperbolic equations of the second order is carried out.

As stated in the monograph [3] iterative algorithms for solving evolutionary inverse problems have great prospects. They take into account a specific character of the considered problems and iteratively specify an initial condition, as on each iteration a correct direct initial-boundary value problem is solved. Thus, for numerical solution of the retrospective inverse heat conduction problem,

© Springer Nature Switzerland AG 2019
I. Dimov et al. (Eds.): FDM 2018, LNCS 11386, pp. 588–595, 2019.
https://doi.org/10.1007/978-3-030-11539-5_69

which is the analog of the Dirichlet problem for parabolic equation, academician Samarskii and his students [6] proposed an iterative method that specifies
an initial condition by iterative methods of variational type. In works [7–9], we
used the iterative method of conjugate gradients for the numerical solution of
the Dirichlet problem.

In this paper, we suggest to apply an approach with iterative improvement
of the missing initial condition using additional information for numerical solution of the Dirichlet problem and its modifications for the second-order two-
dimensional hyperbolic equation. For non-classical problems posed for the given
deviation and velocity distributions at the final time stage using the method of
conjugate gradients we clarify an initial distribution of deviation and velocity.
The results of calculations on model problems with quasi-real solutions demonstrating the possibilities of the method are presented.

## 2   Iterative Solution of Dirichlet Problem

In the parallelepiped $\overline{\Omega} \times (0,T)$ we seek a function $u(x,t)$ – solution of the
Dirichlet problem for the second order hyperbolic equation

$$\frac{\partial^2 u}{\partial t^2} - \sum_{\alpha=1}^{2} \frac{\partial}{\partial x_\alpha}\left(k_\alpha(\mathbf{x})\frac{\partial u}{\partial x_\alpha}\right) = 0, \quad \mathbf{x} \in \Omega, \quad 0 < t < T \tag{1}$$

with homogeneous boundary conditions of the first kind

$$u(\mathbf{x},t) = 0, \qquad \mathbf{x} \in \Gamma, \quad 0 < t \leqslant T \tag{2}$$

and time conditions

$$u(\mathbf{x},0) = \varphi(\mathbf{x}), \ u(\mathbf{x},T) = \nu(\mathbf{x}), \quad \mathbf{x} \in \overline{\Omega}. \tag{3}$$

Its numerical solution is carried out using a symmetric three-layer difference
scheme with a weight factor $\sigma$ [10]:

$$\frac{\hat{y} - 2y + \check{y}}{\tau^2} + A(\sigma\hat{y} + (1 - 2\sigma)y + \sigma\check{y}) = 0, \quad \mathbf{x} \in \omega_{h\tau}, \tag{4}$$

$$y(t_0) = \varphi(\mathbf{x}), \quad y(t_M) = \nu(\mathbf{x}), \quad x \in \overline{\omega}_h, \tag{5}$$

where the weight factor is $0 \leqslant \sigma \leqslant 1$, and the non-alphabetic notation system of
Samarskii was used [10] $\hat{y} = y^{j+1}$, $y = y^j$, $\check{y} = y^{j-1}$, $x \in \overline{\omega}_h$. Here the discrete
analog of the Laplace operator has the form

$$A = \sum_{\alpha=1}^{2} A_\alpha, \ A_1 y = -(a_1(\mathbf{x})y_{\bar{x}_1})_{x_1}, \ A_2 y = -(a_2(\mathbf{x})y_{\bar{x}_2})_{x_2}, \ \mathbf{x} \in \omega_h,$$

putting, for example, $a_\alpha(\mathbf{x}) = k_\alpha(\mathbf{x} - 0.5h_\alpha), \alpha = 1, 2$.

Further, in order to estimate the accuracy of the proposed methods for solving
the Dirichlet problem and its modifications, we solve the direct problem for

oscillation Eq. (1) ($k_\alpha(\mathbf{x}) \equiv 1$, $\alpha = 1, 2$; $0 < t \leqslant T$) with homogeneous boundary conditions (2) and under the initial conditions:

$$u(\mathbf{x}, 0) = \varphi(\mathbf{x}) = e^{-\gamma((x_1 - l/2)^2 + (x_2 - l/2)^2)}, \quad \mathbf{x} \in \overline{\Omega}. \tag{6}$$

$$\frac{\partial u(\mathbf{x}, 0)}{\partial t} = \phi(\mathbf{x}) = e^{-\gamma((x_1 - l/2)^2 + (x_2 - l/2)^2)}, \quad \mathbf{x} \in \overline{\Omega}. \tag{7}$$

In calculations we shall use the following as supplementary conditions at the final time

$$u(\mathbf{x}, 0) = \nu(\mathbf{x}) = y^M(\mathbf{x}), \quad \frac{\partial u(\mathbf{x}, T)}{\partial t} = \eta(\mathbf{x}) = \frac{y^M(\mathbf{x}) - y^{M-1}(\mathbf{x})}{h}, \quad \mathbf{x} \in \overline{\Omega},$$

obtained from the solution of the direct problem (1)–(2), (6)–(7). The Dirichlet problem (1)–(3) belongs to the class of non-classical problems of mathematical physics, and is conditionally correct [1]. For numerical realization of the system of linear algebraic equations (4)–(5), let us use the fastest convergent iterative method of conjugate gradients [11], based on the sequential refinement of the sought initial condition $\nu(\mathbf{x})$ with further solution at each iteration of the discrete analog of the direct problem.

Let us give the corresponding operator formulation to this problem. By successively eliminating the intermediate values of the grid function $y(t)$, $t \in \omega_\tau$ at the given $\phi$ and $\nu$ for a finite time, we obtain

$$y^M = \mathcal{A}\phi + \mathcal{B}\varphi = \nu,$$

where $\mathcal{A}$, $\mathcal{B}$ are respectively operator polynomials of a positive definite, self-adjoint operator. Consequently, the operators $\mathcal{A}$, $\mathcal{B}$ are self-adjoint. Thus, the numerical solution of the Dirichlet problem reduces to the solution of the linear operator equation

$$\mathcal{A}\phi = \nu - \mathcal{B}\varphi. \tag{8}$$

Since we solve the inverse problem, we will subsequently use a random number generator due to the supplementary condition in the Dirichlet problem considered to be given with some error. In computational experiments the given function $\varphi(x)$, $x \in \omega$, was perturbed as follows

$$\varphi_\delta(x) = \varphi + \delta F(x), \quad x \in \omega,$$

where $F(x)$ are random variables uniformly distributed on the interval $(-0.5, 0.5)$. The computational efficiency of the proposed method will be illustrated by numerical solution of the Dirichlet model problem for the equation with constant coefficients $k_\alpha(\mathbf{x}) \equiv 1$, $\alpha = 1, 2$; $0 < t \leqslant t$.

Below are the results of the computational experiment on recovery of the initial velocity with similar values of the initial data set while solving a direct problem. A numerical solution of the inverse Dirichlet problem was performed on the explicit finite difference scheme with weight multiplier $\sigma = 0$, $l_1 = l_2 = l = 1$, $\gamma = 100$ on the space-time grid with parameters $n_1 = n_2 = N =$

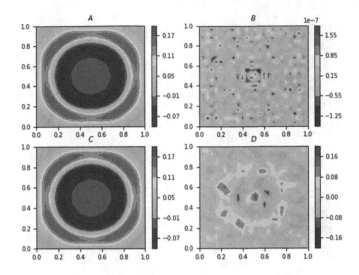

**Fig. 1.** The error of the iterative method for the Dirichlet problem

25, $h_1 = h_2 = h = l/N$, $M = 15$, $\tau = 0.98h/\sqrt{2}$, hence $T = 0.42002$. In Fig. 1A a diagram of the specified value of the solution at the final time stage $\nu(\mathbf{x}) = u(\mathbf{x}, T)$ obtained through solving a direct problem with initial conditions (6)–(7) is presented. In Fig. 1B we see an error of the first initial condition recovered through the considered iterative method. In Fig. 1C a graph of the perturbed solution at the final time stage $\nu_\delta(\mathbf{x})$, is shown, and Fig. 1D-displays an error of the first initial condition calculated through introduction of noisy condition. Here it was required to carry out 75 iterations, and under noise with $\delta = 0.003$ as a supplementary condition $\nu_\delta(x)$, there was carried out 7 iterations.

## 3   Iterative Solution of the First Modification of the Dirichlet Problem

In the parallelepiped $\overline{\Omega} \times (0, T)$ we search for the function $u(x, t)$ – a solution of the Dirichlet problem for the hyperbolic equation of the second order (1)–(2) and time conditions:

$$u(\mathbf{x}, 0) = \varphi(\mathbf{x}), \quad \frac{\partial u(\mathbf{x}, T)}{\partial t} = \eta(\mathbf{x}), \quad \mathbf{x} \in \overline{\Omega}. \tag{9}$$

The discrete analogue of the problem (1), (2), (9) looks like the following:

$$\frac{\hat{y} - 2y + \check{y}}{\tau^2} + Ay = 0, \quad \mathbf{x} \in \omega_{h\tau}, \tag{10}$$

$$y(t_0) = \varphi(\mathbf{x}), \quad y(t_M) = y(t_{M-1}) + \tau\eta(\mathbf{x}), \quad \mathbf{x} \in \overline{\omega}_h. \tag{11}$$

The obtained system of equations was also calculated by conjugate gradients.

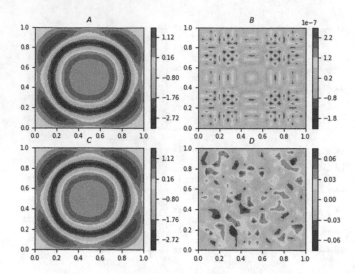

**Fig. 2.** The error of the method for the first modification of the Dirichlet problem

Here we present the results of numerical implementation of the first modification of the Dirichlet problem. In Fig. 2A there is a graph of the given velocity value for solving at the final time stage $\frac{\partial u(\mathbf{x},T)}{\partial t} = \eta(\mathbf{x})$ that is obtained through solving of the direct problem with initial conditions (6)–(7). In Fig. 2B – here is an error of the first initial condition recovered through the iterative method considered.

In Fig. 2C there shown a diagram of the perturbed velocity for the solution at the final time stage $\eta_\delta(\mathbf{x})$, and in Fig. 2D – one can see an error of the first initial condition (6) that has been restored through the considered iterative method (iterations 98), 24 iterations have been carried out under noise with $\delta = 0.003$ as the extra condition $\eta_\delta(x)$.

## 4   Iterative Solution of the Second Modification of the Dirichlet Problem

In the parallelepiped $\overline{\Omega} \times (0, T)$ we search for the function of $u(x, t)$ – a solution of the Dirichlet problem for hyperbolic equation of the second order (1), (2) and with time conditions

$$\frac{\partial u(\mathbf{x}, 0)}{\partial t} = \phi(\mathbf{x}), \ u(\mathbf{x}, T) = \nu(\mathbf{x}), \quad \mathbf{x} \in \overline{\Omega}. \tag{12}$$

Finite-difference analogue of the problem (1), (2), (12) is the following:

$$\frac{\hat{y} - 2y + \check{y}}{\tau^2} + Ay = 0, \quad \mathbf{x} \in \omega_{h\tau}, \tag{13}$$

$$y(t_1) = y(t_0) + \tau\phi(\mathbf{x}), \quad y(t_M) = \nu(\mathbf{x}), \quad \mathbf{x} \in \overline{\omega}_h. \tag{14}$$

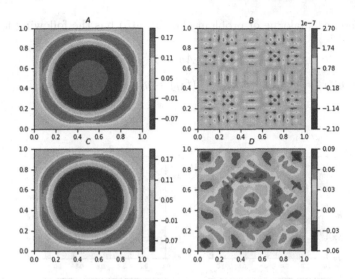

**Fig. 3.** The error of the method for the second modification of the Dirichlet problem

Figure 3 shows the results of the quasi-real computing experiment on recovering of the second initial condition with similar initial data values that have been set while solving a direct problem. In Fig. 3A one can see a diagram of the preset value for solving at the final time stage $\nu(\mathbf{x}) = u(\mathbf{x}, T)$ obtained through solving the direct problem with initial conditions (6)–(7). In Fig. 3B we see an error of the second initial condition $\phi_0(\mathbf{x}) = (y^1(\mathbf{x}) - y^0(\mathbf{x}))/\tau$, $\mathbf{x} \in \overline{\omega}_h$. Recovered through the considered iterative method of the discrete analogue. Figure 3C shows a diagram of the perturbed solution at the final time stage $\nu_\delta(\mathbf{x})$, and in Fig. 3D there is an error of the second initial conditions calculated through the given iterative method [6] under noise condition with $\delta = 0.003$. It was required to carry out 112 iterations and 19 iterations under supplementary noisy condition.

## 5    Iterative Solution of the Third Modification of the Dirichlet Problem

Now let us move on to the third modification. In parallelepiped $\overline{\Omega} \times (0, T)$ we search for function $u(x, t)$ which is a solution of the Dirichlet problem for the hyperbolic equation of the second order (1), (2) and time conditions:

$$\frac{\partial u(\mathbf{x}, 0)}{\partial t} = \phi(\mathbf{x}), \quad \frac{\partial u(\mathbf{x}, T)}{\partial t} = \nu(\mathbf{x}), \quad \mathbf{x} \in \overline{\Omega}. \tag{15}$$

Finite-difference analogue of problem (1), (2), (15) has the following form:

$$\frac{\hat{y} - 2y + \check{y}}{\tau^2} + Ay = 0, \quad \mathbf{x} \in \omega_{h\tau}, \tag{16}$$

$$y(t_1) = y(t_0) + \tau\phi(\mathbf{x}), \quad y(t_M) = y(T) + \tau\nu(\mathbf{x}), \quad \mathbf{x} \in \overline{\omega}_h. \tag{17}$$

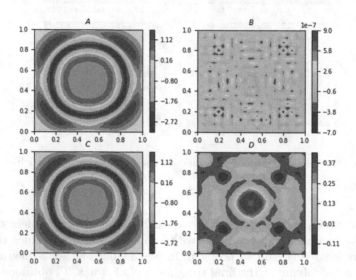

**Fig. 4.** The error of the method for the third modification of the Dirichlet problem

In Fig. 4 we see the results of quasi-real computing experiment for recovering the initial condition having similar initial data values that were set during the direct problem solution. In Fig. 4A there is a graph of the preset velocity value for solving at the final time stage $\frac{\partial u(\mathbf{x},T)}{\partial t} = \eta(\mathbf{x})$, obtained through solving the direct problem with the initial conditions (6)–(7). For Fig. 4B shows an error of the second initial condition $\phi_0(\mathbf{x}) = (y^1(\mathbf{x}) - y^0(\mathbf{x}))/\tau$, $\mathbf{x} \in \overline{\omega}_h$. Recovered through the given iteration of the discrete analogue. In Fig. 4C we present a diagram of the perturbed velocity solution at the final time stage $\nu_\delta(\mathbf{x})$, and in Fig. 4D present an error of the second initial condition calculated through the iterative method considered [6] under noise with $\delta = 0.003$. It was required to conduct 174 iterations and 34 iterations under supplementary noisy condition.

## 6    Conclusion

The article presents an application of the iterative method of conjugate gradients for numerical solution of the finite-difference analogue of the two-dimensional Dirichlet problem for the second-order hyperbolic equation and its three modifications. The results of the computational experiment based on quasi-exact solutions, including noisy conditions, confirm that the recovery of the desired initial condition is performed with any acceptable accuracy: we can achieve a very high accuracy of identification of the desired condition increasing a number of iterations. Moreover, in case of supplementary noisy condition we obtain an acceptable identification in the first and third modifications of the Dirichlet problem. At the same time, we can establish off-iteration cycle condition comparable to the value of "noise". In this case an increase in the number of iterations due to the incorrectness of the Dirichlet problem, can lead to an incorrect numerical result.

**Acknowledgments.** Supported by mega-grant of the Russian Federation Government (14.Y26.31.0013) and Grant of the Russian Foundation for Basic Research (17-01-00732).

# References

1. Kabanikhin, S.I.: Inverse and Ill-Posed Problems. Theory and Applications. De Gruyter, Berlin (2011)
2. Lavrent'ev, M.M., Romanov, V.G., Shishatskii, S.P.: Ill-Posed Problems of Mathematical Physics and Analysis. American Mathematical Society, Providence (1986)
3. Samarskii, A.A., Vabishchevich, P.N.: Numerical Methods for Solving Inverse Problems of Mathematical Physics. De Gruyter, Berlin (2007)
4. Kabanikhin, S.I., Bektemesov, M.A., Nurseitov, D.B., Krivorotko, O.I., Alimova, A.N.: An optimization method in the Dirichlet problems for the wave equation. J. Inverse Ill-Posed Probl. **2**(20), 193–211 (2012)
5. Kabanikhin, S.I., Krivorotko, O.I.: Chislennyj metod reshenia zadachi Dirichle dlia volnovogo uravnenia. Sib. J. Ind. Math. **15**(4), 90–101 (2012)
6. Samarskii, A.A., Vabishchevich, P.N., Vasil'ev, V.I.: Iterative solution of a retrospective inverse problem of heat conduction. Matematicheskoe Modelirovanie **9**(5), 119–127 (1997)
7. Vasil'ev, V.I., Popov, V.V., Eremeeva, M.S., Kardashevsky, A.M.: Iterative solution of a nonclassical problem for the equation of string vibrations. Vest. Mosk. Gos. Univ. im. N. E. Baumana, Estest. Nauki. **3**, 77–87 (2015)
8. Vabishchevich, P.N., Vasil'ev, V.I.: Iteracionnoe reshenie zadachi Dirichle dlia giperbolicheskogo uravnenia, Grid methods for boundary value problems and applications. In: Proceedings of the Tenth International Conference, pp. 162–166. Publishing House of Kazan University, Kazan (2014)
9. Vasil'ev, V.I., Kardashevsky A.M.: Iteracionnoe reshenie nekotoryx obratnyx zadach dlia giperbolicheskix uravnenij vtorogo poriadka. In: Proceedings of the International Conference "Actual Problems of Computational and Applied Mathematics", Novosibirsk, pp. 150–156 (2015)
10. Samarskii, A.A.: The Theory of Difference Schemes. Marcel Dekker, New York (2001)
11. Saad, U.: Iterative Methods for Sparse Linear Systems, 2nd edn. SIAM (2003)

# On a Digital Version of Pseudo-Differential Operators and Its Applications

Vladimir B. Vasilyev[✉][iD]

Belgorod National Research University, Belgorod 308007, Russia
vbv57@inbox.ru

**Abstract.** We develop discrete variant of a theory of pseudo-differential operators and equations. For some canonical domains we obtain solvability results for such equations and use these results to construct approximate solutions.

**Keywords:** Digital pseudo-differential operator ·
Periodic factorization · Approximate solution

## 1 Introduction

Pseudo-differential operators were introduced near a half-century ago, and now these operators play a crucial role in a lot of problems in modern mathematics [1–3]. Such a theory connects different parts of partial differential operators and some kinds of integral operators and corresponding equations. Since the latter equations describes certain natural phenomenons it is very important to find their numerical solution at least. For this purpose there are some computational theories [4,5] to justify existing numerical algorithms. These studies are related as a rule to certain boundary value problem for a partial differential equation or certain boundary integral equation. We will try to develop a similar theory for boundary value problems especially for pseudo-differential equations to apply it to computational problems. Some results in this directions are presented in author's papers [6–12].

## 2 Digitization of Operators

Let $A(\xi)$ be a function defined on $\mathbf{R}^m$ and satisfying the condition

$$c_1(1 + |\xi|)^\alpha \leq |A(\xi)| \leq c_2(1 + |\xi|)^\alpha, \tag{1}$$

where $c_1, c_2$ are positive constants, and $S(\mathbf{R}^m)$ be the Schwartz class of infinitely differentiable rapidly decreasing at infinity functions. Such a function defines the following pseudo-differential operator

$$(Au)(x) = \int\limits_{\mathbf{R}^m} \int\limits_{\mathbf{R}^m} A(\xi)e^{i(x-y)\cdot\xi}u(y)d\xi dy, \quad x \in \mathbf{R}^m, \tag{2}$$

© Springer Nature Switzerland AG 2019
I. Dimov et al. (Eds.): FDM 2018, LNCS 11386, pp. 596–603, 2019.
https://doi.org/10.1007/978-3-030-11539-5_70

it can be introduced first for $u \in S(\mathbf{R}^m)$ and then extended on a more general space. Such function $A(\xi)$ is called a symbol of the pseudo-differential operator $A$.

*Remark 1.* Usually they consider more general pseudo-differential operator (symbol)

$$(Au)(x) = \int\limits_{\mathbf{R}^m} \int\limits_{\mathbf{R}^m} A(x,\xi)e^{i(x-y)\cdot\xi}u(y)d\xi dy, \quad x \in \mathbf{R}^m,$$

generated by the symbol $A(x,\xi)$ defined on $\mathbf{R}^m \times \mathbf{R}^m$. But taking into account the local principle our first goal is studying more simple operator (2) from "discrete" point of view.

Let $A_d(\xi)$ be a periodic function in $\mathbf{R}^m$ such that

$$c_1(1 + |\zeta_h^2|)^{\frac{\alpha}{2}} \le |A_d(\xi)| \le c_2(1 + |\zeta_h^2|)^{\frac{\alpha}{2}}, \tag{3}$$

where $\zeta_h^2 = h^{-2} \sum\limits_{k=1}^{m} (e^{-ih\xi_k} - 1)^2$, and constants $c_1, c_2$ do not depend on $h$.

Let $D \subset \mathbf{R}^m$ be a domain. We consider functions $u_d(\tilde{x})$ defined on $D_d \equiv D \cap h\mathbf{R}^m, h > 0$, and introduce for them the following operator

$$(A_d u_d)(\tilde{x}) = \sum_{\tilde{y} \in h\mathbf{Z}^m} \int\limits_{\hbar T^m} A_d(\xi)u_d(\tilde{y})e^{i(\tilde{x}-\tilde{y})\cdot\xi}h^m d\xi, \quad \tilde{x} \in D_d,$$

where $\hbar \equiv h^{-1}, \mathbf{T}^m \equiv [-\pi,\pi]^m$.

**Definition 1.** *The operator $A_d$ is called a digital pseudo-differential operator or shortly h-operator. The periodic function $A_d(\xi)$ is called its $\hbar$-symbol.*

We remind that a symbol (operator) is called elliptic if

$$ess \inf_{\xi \in \mathbf{R}^m} |A_d(\xi)| > 0,$$

and obviously all our symbols are elliptic.

## 2.1 Discrete Fourier Transform

If $u_d(\tilde{x}), \tilde{x} \in h\mathbf{Z}^m$, is a function of a discrete variable then we call it "discrete function". For such discrete functions one can define the discrete Fourier transform

$$(F_d u_d)(\xi) \equiv \tilde{u}_d(\xi) = \sum_{\tilde{x} \in h\mathbf{Z}^m} e^{-i\tilde{x}\cdot\xi} u_d(\tilde{x})h^m, \quad \xi \in \hbar\mathbf{T}^m,$$

if the latter series converges, and the function $\tilde{u}_d(\xi)$ is a periodic function on $\mathbf{R}^m$ with the basic cube of periods $\hbar\mathbf{T}^m$. This discrete Fourier transform preserves basic properties of the integral Fourier transform, particularly the inverse discrete Fourier transform is given by the formula

$$(F_d^{-1}\tilde{u}_d)(\tilde{x}) = \frac{1}{(2\pi)^m} \int\limits_{\hbar\mathbf{T}^m} e^{i\tilde{x}\cdot\xi}\tilde{u}_d(\xi)d\xi, \quad \tilde{x} \in h\mathbf{Z}^m.$$

The discrete Fourier transform is a one-to-one correspondence between the spaces $L_2(h\mathbf{Z}^m)$ and $L_2(\hbar\mathbf{T}^m)$ with norms

$$||u_d||_2 = \left( \sum_{\tilde{x}\in h\mathbf{Z}^m} |u_d(\tilde{x})|^2 h^m \right)^{1/2} \quad and \quad ||\tilde{u}_d||_2 = \left( \int_{\xi\in\hbar\mathbf{T}^m} |\tilde{u}_d(\xi)|^2 d\xi \right)^{1/2}.$$

## 2.2   Discrete Spaces

Since the definition for Sobolev–Slobodetskii spaces includes partial derivatives we use their discrete analogue, i.e. divided difference of first order

$$(\Delta_k^{(1)} u_d)(\tilde{x}) = h^{-1}(u_d(x_1,\cdots,x_k+h,\cdots,x_m) - u_d(x_1,\cdots,x_k,\cdots,x_m)),$$

for which its discrete Fourier transform looks as follows

$$(\widetilde{\Delta_k^{(1)} u_d})(\xi) = h^{-1}(e^{-ih\cdot\xi_k} - 1)\tilde{u}_d(\xi).$$

Further for the divided difference of second order we have

$$(\Delta_k^{(2)} u_d)(\tilde{x}) = h^{-2}(u_d(x_1,\cdots,x_k+2h,\cdots,x_m)$$

$$-2u_d(x_1,\cdots,x_k+h,\cdots,x_m) + u_d(x_1,\cdots,x_k,\cdots,x_m))$$

and its discrete Fourier transform

$$(\widetilde{\Delta_k^{(2)} u_d})(\xi) = h^{-2}(e^{-ih\cdot\xi_k} - 1)^2\tilde{u}_d(\xi).$$

Thus, for the discrete Laplacian we have

$$(\Delta_d u_d)(\tilde{x}) = \sum_{k=1}^{m}(\Delta_k^{(2)} u_d)(\tilde{x}),$$

so that

$$(\widetilde{\Delta_d u_d})(\xi) = h^{-2} \sum_{k=1}^{m}(e^{-ih\cdot\xi_k} - 1)^2\tilde{u}_d(\xi).$$

Now we will introduce the basic space $S(h\mathbf{Z}^m)$ which consists of discrete functions with finite semi-norms

$$|u_d| = \sup_{\tilde{x}\in h\mathbf{Z}^m} (1 + |\tilde{x}|)^l |\Delta^{(\mathbf{k})} u_d(\tilde{x})|$$

for arbitrary $l \in \mathbf{N}, \mathbf{k} = (k_1,\cdots,k_m), k_r \in \mathbf{N}, r = 1,\cdots,m$, where

$$\Delta^{(\mathbf{k})} u_d(\tilde{x}) = \Delta_1^{k_1} \ldots, \Delta_m^{k_m} u_d(\tilde{x}).$$

In other words the space $S(h\mathbf{Z}^m)$ is a discrete analogue of the Schwartz space $S(\mathbf{R}^m)$ of infinitely differentiable rapidly decreasing at infinity functions. Usually the space of distributions over the basic space $S(\mathbf{R}^m)$ is denoted by $S'(\mathbf{R}^m)$.

**Definition 2.** *The space $H^s(h\mathbf{Z}^m)$ is a closure of the space $S(h\mathbf{Z}^m)$ with respect to the norm*

$$||u_d||_s = \left( \int_{\hbar \mathbf{T}^m} (1 + |\varsigma_h^2|)^s |\tilde{u}_d(\xi)|^2 d\xi \right)^{1/2}. \tag{4}$$

**Definition 3.** *The space $H^s(D_d)$ consists of discrete functions from $H^s(h\mathbf{Z}^m)$ which supports belong to $\overline{D_d}$. A norm in the space $H^s(D_d)$ is induced by a norm of the space $H^s(h\mathbf{Z}^m)$. The space $H_0^s(D_d)$ consists of discrete functions $u_d$ with a support in $D_d$, and these discrete functions should admit a continuation into the whole $H^s(h\mathbf{Z}^m)$. A norm in the $H_0^s(D_d)$ is given by the formula*

$$||u_d||_s^+ = \inf ||\ell u_d||_s,$$

*where infimum is taken over all continuations $\ell$.*

The Fourier image of the space $H^s(D_d)$ will be denoted by $\tilde{H}^s(D_d)$.

Of course all norms (4) are equivalent to the $L_2$-norm but this equivalence depends on $h$. Let us note that all constants below in our considerations do not depend on $h$.

## 3    Solvability of Digital Equations

### 3.1    A Whole Space

Let $D = \mathbf{R}^m$. Let us consider the following equation

$$(A_d u_d)(\tilde{x}) = v_d(\tilde{x}), \quad \tilde{x} \in D_d. \tag{5}$$

Result on a solvability for the Eq. (5) is given by the following theorem.

**Theorem 1.** *If the elliptic symbol $A_d(\xi)$ satisfies the condition (3) then the operator $A_d : H^s(h\mathbf{Z}^m) \to H^{s-\alpha}(h\mathbf{Z}^m)$ is bounded and invertible.*

### 3.2    A Half-Space

Let $D = \mathbf{R}_+^m \equiv \{x \in \mathbf{R}^m : x = (x', x_m), x_m > 0\}$. For studying the discrete half-space case we need a special technique like continue case [1]. It was found for this case [13, 14] the periodic analogue of the Hilbert transform [1] with the parameter $\xi'$

$$(H_{\xi'}^{per} \tilde{u}_d)(\xi', \xi_m) = \frac{h}{2\pi i} v.p. \int_{-\hbar \pi}^{\hbar \pi} \cot \frac{h(\xi_m - \eta_m)}{2} \tilde{u}_d(\xi', \eta_m) d\eta_m,$$

where

$$v.p. \int_{-\hbar \pi}^{\hbar \pi} \cot \frac{h(\xi_m - \eta_m)}{2} \tilde{u}_d(\xi', \eta_m) d\eta_m$$

$$= \lim_{\varepsilon \to 0+} \left( \int_{-\hbar\pi}^{\xi_m - \varepsilon} + \int_{\xi_m + \varepsilon}^{\hbar\pi} \right) \cot \frac{h(\xi_m - \eta_m)}{2} \tilde{u}_d(\xi', \eta_m) d\eta_m$$

This operator generates two projectors [14]

$$P_{\xi'}^{per} = \frac{1}{2}(I + H_{\xi'}^{per}), \quad Q_{\xi'}^{per} = \frac{1}{2}(I - H_{\xi'}^{per}).$$

*Remark 2.* In some previous papers the author has missed the factor $h$ before cot. It does not influent on solvability picture, but it is very important for concrete evaluations (see below).

Let us denote $\Pi_\pm$ half-strips in the complex plane $\mathbf{C}$

$$\Pi_\pm = \{z \in \mathbf{C} : z = s + i\tau, s \in [-\pi, \pi], \pm\tau > 0\}.$$

**Definition 4.** *Periodic factorization of the symbol $A_d(\xi)$ is called its representation in the form*

$$A_d(\xi) = A_{d,+}(\xi) \cdot A_{d,-}(\xi),$$

*where the factors $A_{d,\pm}(\xi)$ admit an analytical continuation into half-strips $\hbar\Pi_\pm$ on the last variable $\xi_m$ for almost all fixed $\xi' \in \hbar\mathbf{T}^{m-1}$ and satisfy the estimates*

$$|A_{d,+}^{\pm 1}(\xi)| \le c_1(1 + |\hat{\zeta}^2|)^{\pm\frac{\alpha}{2}}, \quad |A_{d,-}^{\pm 1}(\xi)| \le c_2(1 + |\hat{\zeta}^2|)^{\pm\frac{\alpha-\alpha}{2}},$$

*with constants $c_1, c_2$ non-depending on $h$,*

$$\hat{\zeta}^2 \equiv \hbar^2 \left( \sum_{k=1}^{m-1} (e^{-ih\xi_k} - 1)^2 + (e^{-ih(\xi_m+i\tau)} - 1)^2 \right), \quad \xi_m + i\tau \in \hbar\Pi_\pm.$$

*The number $æ \in \mathbf{R}$ is called an index of periodic factorization.*

*Remark 3.* For an elliptic symbol $A_d(\xi)$ such periodic factorization exists always (see [1]).

For some simple cases one can use the topological formula [1]

$$æ = \frac{1}{2\pi} \int_{-\hbar\pi}^{\hbar\pi} d \arg A_d(\cdot, \xi_m),$$

where $A_d(\cdot, \xi_m)$ means that $\xi' \in \hbar\mathbf{T}^{m-1}$ is fixed, and the integral is the integral in Stieltjes sense. It means that we need to calculate divided by $2\pi$ variation of the argument of the symbol $A_d(\xi)$ when $\xi_m$ varies from $-\hbar\pi$ to $\hbar\pi$ under fixed $\xi'$.

For a special case we have the following result.

**Theorem 2.** *If the elliptic symbol $\tilde{A}_d(\xi)$ satisfies the condition (3) and admits periodic factorization with index $æ$ so that $|æ - s| < 1/2$ then the Eq. (2) has unique solution in the space $H^s(D_d)$ for arbitrary right-hand side $v_d \in H^{s-\alpha}(D_d)$,*

$$\tilde{u}_d(\xi) = \tilde{A}_{d,+}^{-1}(\xi) P_{\xi'}^{per}(\tilde{A}_{d,-}^{-1}(\xi)\widetilde{\ell v_d}(\xi)). \tag{6}$$

## 4     Computational Aspects

We will consider the corresponding equation

$$(Au)(x) = v(x), \quad x \in D, \tag{7}$$

and will suggest some computational algorithms.

Since we have such solvability pictures for discrete equations in $\mathbf{R}^m$ and $\mathbf{R}^m_+$ we need to choose such digital pseudo-differential operators that they preserve all necessary properties of their initial continuous data.

### 4.1     Standard Approach

Let $P_h$ be an restriction operator on $h\mathbf{Z}^m$, i.e. for $u \in S(\mathbf{R}^m)$

$$(P_h u)(x) = \left\{ \begin{array}{ll} u(\tilde{x}), & x = \tilde{x} \in h\mathbf{Z}^m; \\ 0, & x \notin h\mathbf{Z}^m. \end{array} \right.$$

We have verified this approach for simplest pseudo-differential operators namely Calderon–Zygmund operators [8,9]. These operators are pseudo-differential operators of order zero.

To obtain more preferable approximation we will use another construction.

### 4.2     Spectral Approach

Here we will introduce a new restriction operator $Q_h$ for a function $u \in S(\mathbf{R}^m)$. We take the Fourier transform $\tilde{u}(\xi)$, then we take its restriction on $\hbar\mathbf{T}^m$ and periodically continue it to a whole $\mathbf{R}^m$. Further we use the inverse discrete Fourier transform $F_d^{-1}$, it will be a discrete function and call it $(Q_h u)(\tilde{x}), \tilde{x} \in h\mathbf{Z}^m$. For many reasons such projector $Q_h$ is more convenient than introduced above the projector $P_h$. Both projectors $P_h$ and $Q_h$ are almost the same according to the following result.

**Lemma 1.** *For $u \in S(\mathbf{R}^m), \forall \beta > 0$, we have the following property*

$$|(P_h u)(\tilde{x}) - (Q_h u)(\tilde{x})| \leq Ch^\beta, \quad \forall \tilde{x} \in h\mathbf{Z}^m,$$

*where constant $C$ can depend on the function $u$ only.*

Further, the symbol $A_d(\xi)$ will be constructed in the following way. To construct $A_d(\xi)$ we take a restriction $A(\xi)$ on the cube $\hbar\mathbf{T}^m$ and periodically continue it on a whole $\mathbf{R}^m$. We will consider this $h$-operator as approximating operator for $A$. Therefore, to find a digital approximate solution for the Eq. (5) for the case $D = \mathbf{R}^m$ we can use the following digital equation

$$A_d u_d = Q_h v. \tag{8}$$

Its solution is given by the formula

$$u_d(\tilde{x}) = \frac{1}{(2\pi)^m} \int_{\hbar\mathbf{T}^m} e^{i\tilde{x}\cdot\xi} A^{-1}(\xi)\tilde{v}(\xi)d\xi, \quad \tilde{x} \in h\mathbf{Z}^m,$$

so that we do not need to solve an infinite system of linear algebraic equations [8,9]. It is enough for this case to use a certain cubature formula for the latter integral and another cubature formula for calculating the Fourier transform $\tilde{v}(\xi)$.

Using the Lemma 1 we can give a comparison between continuous and digital solution for "good" right-hand side and symbol.

**Theorem 3.** *If the symbol $A(\xi)$ satisfies the condition (1) and infinitely differentiable in $\mathbf{R}^m$, $u$ is solution of the Eq. (7), $u_d$ is solution of the Eq. (8), then for $v \in S(\mathbf{R}^m)$ we have the following error estimate*

$$|u(\tilde{x}) - u_d(\tilde{x})| \leq Ch^\beta, \quad \forall \tilde{x} \in h\mathbf{Z}^m,$$

*for arbitrary $\beta > 0$.*

### 4.3   Limit Transfer for a Half-Space

It is well known that

$$\cot x = \frac{1}{x} - \sum_{n=1}^{\infty} \frac{2^{2n}|B_{2n}|}{(2n)!}x^{2n-1}, \quad -\pi < x < \pi,$$

where $B_{2n}$ are Bernoulli numbers.

If we will look on the formula (6) then we will see that the kernel of the operator $H_{\xi'}^{per}$, i.e. $h\cot\frac{h\xi_m}{2}$ has the following representation

$$h\cot\frac{h\xi_m}{2} = \frac{2}{\xi_m} - h\sum_{n=1}^{\infty} \frac{2^{2n}|B_{2n}|}{(2n)!}\left(\frac{h\xi_m}{2}\right)^{2n-1},$$

so that we will obtain under $h \to 0$ well-known kernel of the Hilbert transform $\frac{1}{\pi i}\frac{1}{\xi_m}$ with respect to a last variable. It corresponds to a continuous case [1].

Unfortunately for this case obtaining an error estimate for $u$ and $u_d$ is not so easy like Theorem 3. One can assert only that we have convergence $\tilde{u}_d$ to the $u$ under $h \to 0$.

## 5   Conclusion

For a conclusion let us note that we have considered in this paper only two canonical domains $\mathbf{R}^m$ and $\mathbf{R}_+^m$. We have also some results on a solvability of digital equations in the cone $C_+^a = \{x \in \mathbf{R}^m : x = (x', x_m), x_m > a|x'|, a > 0$ [6,7]. We hope to develop these studies in this direction.

**Acknowledgments.** The work is supported by the State contract of the Russian Ministry of Education and Science (contract No 1.7311.2017/8.9).

# References

1. Eskin, G.: Boundary Value Problems for Elliptic Pseudodifferential Equations. AMS, Providence (1981)
2. Taylor, M.E.: Pseudodifferential Operators. Princeton Univ. Press, Princeton (1981)
3. Trevés, F.: Introduction to Pseudodifferential Operators and Fourier Integral Operators. Springer, New York (1980). https://doi.org/10.1007/978-1-4684-8780-0
4. Samarskii, A.A.: The Theory of Difference Schemes. CRC Press, Boca Raton (2001)
5. Ryaben'kii, V.S.: Method of Difference Potentials and its Applications. Springer, Heidelberg (2002). https://doi.org/10.1007/978-3-642-56344-7
6. Vasilyev, V.B.: The periodic Cauchy kernel, the periodic Bochner kernel, and discrete pseudo-differential operators. In: AIP Conference Proceedings, vol. 1863, p. 14014 (2017). https://doi.org/10.1063/1.4992321
7. Vasilyev, V.B.: Discreteness, periodicity, holomorphy, and factorization. In: Constanda, C., Dalla Riva, M., Lamberti, P.D., Musolino, P. (eds.) Integral Methods in Science and Engineering, Theoretical Technique, vol. 1, pp. 315–324. Birkhäuser, Cham (2017). https://doi.org/10.1007/978-3-319-59384-5_28
8. Vasilyev, A.V., Vasilyev, V.B.: Discrete approximations for multidimensional singular integral operators. In: Dimov, I., Faragó, I., Vulkov, L. (eds.) NAA 2016. LNCS, vol. 10187, pp. 706–712. Springer, Heidelberg (2017). https://doi.org/10.1007/978-3-319-57099-0_81
9. Vasilyev, A.V., Vasilyev, V.B.: Two-scale estimates for special finite discrete operators. Math. Model. Anal. **22**(3), 300–310 (2017). https://doi.org/10.3846/13926292.2017.1307790
10. Vasilyev, A.V., Vasilyev, V.B.: On a digital approximation for pseudo-differential operators. Proc. Appl. Math. Mech. **17**(1), 763–764 (2017). https://doi.org/10.1002/pamm.201710349
11. Vasilyev, V.B.: On some equations on non-smooth manifolds: canonical domains and model operators. In: Kalmenov, T., Nursultanov, E., Ruzhansky, M., Sadybekov, M. (eds.) FAIA 2017. PROMS, vol. 216, pp. 363–375. Springer, Heidelberg (2017). https://doi.org/10.1007/978-3-319-67053-9_34
12. Vasilyev, V.B.: Discrete pseudo-differential operators and boundary value problems in a half-space and a cone. Lobachevskii J. Math. **39**(2), 289–296 (2018). https://doi.org/10.1134/S1995080218020270
13. Vasilev, A.V., Vasilev, V.B.: Periodic Riemann problem and discrete convolution equations. Differ. Equ. **51**(5), 652–660 (2015). https://doi.org/10.1134/S0012266115050080
14. Vasilyev, A.V., Vasilyev, V.B.: Discrete singular operators and equations in a half-space. Azerb. J. Math. **3**(1), 84–93 (2013)

# Upscaled Model for Mixed Dimensional Coupled Flow Problem in Fractured Porous Media Using Non-local Multicontinuum (NLMC) Method

Maria Vasilyeva[1,2]($\boxtimes$), Eric T. Chung[3], Yalchin Efendiev[1,4], Wing Tat Leung[5], and Yating Wang[4]

[1] Institute for Scientific Computation, Texas A&M University,
College Station, TX, USA
[2] North-Eastern Federal University, Yakutsk, Russia
vasilyevadotmdotv@gmail.com
[3] Department of Mathematics, The Chinese University of Hong Kong (CUHK),
Shatin, Hong Kong
[4] Department of Mathematics, Texas A&M University, College Station, TX, USA
[5] Institute of Computational Engineering and Sciences,
University of Texas at Austin, Austin, USA

**Abstract.** In this paper, we consider a mixed dimensional discrete fracture model with highly conductive fractures. Mathematically the problem is described by a coupled system of equations consisting a $d$ - dimensional equation for flow in porous matrix and a $(d-1)$ - dimensional equation for fracture networks with a specific exchange term for coupling them. For the numerical solution on the fine grid, we construct unstructured mesh that is conforming with fracture surface and use the finite element approximation. Fine grid approximation typically leads to very large systems of equations since it resolves the fracture networks, and therefore some multiscale methods or upscaling methods should be applied. The main contribution of this paper is that we propose a new upscaled model using Non-local multi-continuum (NLMC) method and construct an effective coarse grid approximation. The upscaled model has only one additional coarse degree of freedom (DOF) for each fracture network. We will present results of the numerical simulations using our proposed upscaling method to illustrate its performance.

**Keywords:** Fractured porous media · Fluid flow · Coupled system · Upscaling · Multiscale method · Non-local multi-continuum method

## 1 Introduction

Mathematical modeling of flow in fractured porous media is an important problem in subsurface simulations. The development of mathematical models and

© Springer Nature Switzerland AG 2019
I. Dimov et al. (Eds.): FDM 2018, LNCS 11386, pp. 604–611, 2019.
https://doi.org/10.1007/978-3-030-11539-5_71

efficient computational algorithms for numerical modeling of such processes is therefore an important research area. The mathematical model should take into account the entire complicated and multiscale processes occurring in fractured porous media. Since the fractures are highly permeable, they have a significant effect on the flow processes. Thus one needs special care in the construction of mathematical models and computational algorithms [10–12,15].

In this paper, we construct a coupled mixed dimensional model for simulation of the flow process in fractured porous media. The system of equations contains a reduced dimensional equation for flow in fracture networks and an equation for flow in the porous matrix. These two mixed dimensional equations are coupled by a specific exchange term which simulates mass exchange between fractures and matrix. Next, we present a fine-scale approximation where we resolve fracture networks by the mesh using discrete fracture network model (DFM). DFM model leads to very expensive fine-scale simulations. In order to reduce size of the system and efficient solution of the presented problem, we propose an upscaled model. Our model is based on the recently developed NLMC and CEM-GMsFEM methods [5,7]. In CEM-GMsFEM, construction of the multiscale space is based on the auxiliary space. The auxiliary space is based on GMsFEM and requires solutions of a local spectral problem [1,2,6,8,9,13]. These basis functions correspond to small (contrast-dependent) eigenvalues and represent the channels (high-contrast features). Upscaled model is directly related to the well-known multi-continuum approaches, which have been commonly used in approximating subgrid effects for flow and transport in fractured media [3,4,14]. For the construction of the upscaling method, we construct multiscale basis functions by the solution of the coupled mixed dimensional problems in local domains subject to specific constraints. We show that our proposed method can achieve a good accuracy with a very few degrees of freedom. We discuss it in our numerical results.

## 2  Mathematical Model and Fine-Grid Approximation

We consider a mixed dimensional mathematical model. This coupled problem describes fluid flow in domain $\Omega \in \mathcal{R}^d$ (d = 2, 3) in the fractured porous medium, where fractures have small width and can be described by a reduced dimensional model on $\gamma \in \mathcal{R}^{d-1}$

$$
\begin{aligned}
c_m \frac{\partial u_m}{\partial t} - \nabla (k_m \nabla u_m) + \sigma(u_m - u_f)\delta_\gamma = f_m, \quad x \in \Omega, \\
c_f \frac{\partial u_f}{\partial t} - \nabla(k_f \nabla u_f) - \sigma(u_m - u_f) = f_f, \quad x \in \gamma,
\end{aligned}
\tag{1}
$$

where $u_m$ is the pressure in the porous matrix, $u_f$ is the pressure in the fractures, $c_l$, $k_l$, $f_l$ and $\sigma$ are positive constants with $l = m, f$. As boundary conditions, we set zero flow conditions $u_l \cdot n = 0$ and $u_l = u_0$ as inital condition $(l = m, f)$.

System of Eq. (1) is coupled by the specific mass transfer functions $\sigma(u_m - u_f)$ that expresses the conservation of the flow rate (the fluid that is lost in the fractures goes into the porous matrix), where $\delta_\gamma$ is the Dirac function on fracture interface $\gamma$ and $\sigma = k^*/b$ ($b$ is the fracture width and $k^*$ is the harmonic average between $k_m$ and $k_f$).

Next, we consider fine-grid approximation of the mixed dimensional coupled problem (1) on the unstructured grid. Let $\mathcal{T}_h$ be the fine grid that is conforming with fractures with triangular or tetrahedral cells for the domain $\Omega$ and $\gamma = \cup \mathcal{E}_\gamma$ is the fracture facets, where $\mathcal{E}_\gamma \subset \mathcal{E}_h$ and $\mathcal{E}_h$ is the all facets of the mesh $\mathcal{T}_h$.

Let $V_m = H^1(\Omega)$ and $V_f = H^1(\gamma)$. Using the finite element method for approximation in space, we have following variational formulation for the coupled system of equations for pressure in fractured porous media: find $(u_m, u_f) \in (V_m, V_f)$ such that

$$\int_\Omega c_m \frac{u_m - \breve{u}_m}{\tau} v_m \, dx + \int_\Omega k_m \nabla u_m \cdot \nabla v_m \, dx$$
$$+ \int_\gamma \sigma(u_m - u_f) v_m \, ds = \int_\Omega f_m v_m \, dx,$$
$$\int_\gamma c_f \frac{u_f - \breve{u}_f}{\tau} v_f \, ds + \int_\gamma k_f \nabla u_f \cdot \nabla v_f \, ds$$
$$- \int_\gamma \sigma(u_m - u_f) v_f \, ds = \int_\gamma f_f v_f \, ds,$$

$$(2)$$

where $(v_m, v_f) \in (V_m, V_f)$. Here we use stable implicit scheme for approximation in time and $(\breve{u}_m, \breve{u}_f)$ is the solution from previous time step.

Let $V_m^h \subset V_m$, $V_f^h \subset V_f$ and

$$u_m = \sum_{i=1}^{N_\Omega} u_{m,i} \phi_i, \quad u_f = \sum_{i=1}^{N_\gamma} u_{f,i} \psi_i,$$

where $N_\Omega$ and $N_\gamma$ are the number of vertices in $\mathcal{T}_h$ and $\mathcal{E}_\gamma$, $\phi_{m,i}$ and $\phi_{f,i}$ are the linear basis functions on fine grid. Therefore in matrix form, we have

$$\begin{pmatrix} \frac{1}{\tau} S_m + A_m + Q & -Q \\ -Q & \frac{1}{\tau} S_f + A_f + Q \end{pmatrix} \begin{pmatrix} u_m \\ u_f \end{pmatrix} = \begin{pmatrix} \frac{1}{\tau} S_m \breve{u}_m \\ \frac{1}{\tau} S_f \breve{u}_f \end{pmatrix} + \begin{pmatrix} F_m \\ F_f \end{pmatrix} \quad (3)$$

where $A_m$ and $A_f$ are the stiffness matrices, $S_m$ and $S_f$ are the mass matrices, $Q$ is the approximation of the transfer term. More precisely,

$$S_m = \{s_{ij}^m = \int_\Omega c_m \phi_i \phi_j\}, \quad S_f = \{s_{ij}^f = \int_\gamma c_f \psi_i \psi_j\},$$

for stiffness matrices

$$A_m = \{a_{ij}^m = \int_\Omega (k_m \nabla \phi_i, \nabla \phi_j)\}, \quad A_f = \{a_{ij}^f = \int_\gamma (k_f \nabla \psi_i, \nabla \psi_j)\},$$

and for transfer term and right hand side vector

$$Q = \{q_{ij} = \int_\gamma \sigma \psi_i \psi_j\}, \quad F_m = \{f_j^m = \int_\Omega f_m \phi_j\}, \quad F_f = \{f_j^f = \int_\gamma f_f \psi_j\},$$

where $\phi_i$ and $\psi_i$ are the linear basis functions for porous matrix and fractures, respectively.

## 3   Coarse-Grid Multi-continuum Upscaled Model

In this work, we construct an accurate approximation of the flow problem in the fractured porous media using Non-local multi-continuum (NLMC) approach. In this method, for the construction of multiscale basis functions, we solve local problems in the oversampled local region subject to the constraint that the local solution vanishes in other continua except the one for which it is formulated for. This technique is similar to the dual porosity (multi-continuum) approaches.

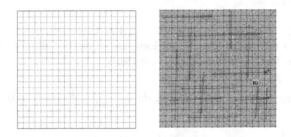

**Fig. 1.** Computational meshes. Left: Coarse mesh $20 \times 20$ with 400 cells. Right: Fine mesh with 13,796 vertices and 27,270 cells (blue color) with illustration of the local cell $K$ (yellow color) and oversampled region $K^+$ (green color) (Color figure online)

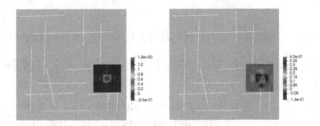

**Fig. 2.** Multiscale basis functions in oversampled region $K^+ = K^2$ (two oversampling coarse cell layers) on coarse mesh $20 \times 20$. Left: Multuscale basis function for porous matrix. Right: Multuscale basis function fracture

We construct simplified basis for perforated domain which has a spatial decay property in local domains and separate background medium and fractures. Let $\mathcal{T}_H$ be the structured coarse mesh and $K_i^+$ be the oversampled region for the coarse cell $K_i$ (see Fig. 1), where oversampled region is constructed by adding several coarse cell layers around $K_i$, $i = \overline{1, N_c}$, $N_c$ is the number of coarse grid cells. Let $\gamma = \cup_{l=1}^{L} \gamma^{(l)}$, where $\gamma^{(l)}$ is a fracture network and $L$ is the total number of fracture networks. Besides, $\gamma_j^{(l)} = K_j \cap \gamma^{(l)}$ is the fracture inside the coarse cell $K_j \in K_i^+$ and $L_j$ is the number of fractures in $K_j$.

For construction of the multiscale basis functions, we find $\phi_l^i = \{\phi_{m,l}^i, \phi_{f,l}^i\}$ by minimizing the energy $a(\phi_l^i, \phi_l^i)$ subject to the following constraint:
(1) background medium ($\phi_{m,0}^i$) :

$$\int_{K_j} \phi_{m,0}^i \, dx = \delta_{i,j}, \quad \int_{\gamma_j^{(r)}} \phi_{f,0}^i \, ds = 0, \quad \forall r = \overline{1, L_j}.$$

(2) $l$-th fracture network in $K_i$ ($\phi_{f,l}^i$):

$$\int_{K_j} \phi_{f,l}^i \, dx = 0, \quad \int_{\gamma_j^{(r)}} \phi_{f,l}^i \, ds = \delta_{i,j} \delta_{r,l}, \quad \forall r = \overline{1, L_j}.$$

where $\phi_{m,l}^i \in V_m^i$, $V_m^i \in H^1(K_i^+)$ and $\phi_{f,l}^i \in V_f^i$, $V_f^i \in H^1(\gamma_i^+)$, $\gamma_i^+ = K_i^+ \cap \gamma$.

Therefore, for multiscale basis functions, we solve the following local problems in $K_i^+$ with zero Dirichlet boundary conditions

$$a(\phi_l^i, v) + \sum_j \lambda_j^m \int_{K_j} v_m \, dx + \sum_j \lambda_j^f \int_{\gamma_j} v_f \, ds = 0,$$

$$\int_{K_j} \phi_{m,l}^i \, dx = \delta_{i,j}, \quad \forall K_j \in K_i^+ \tag{4}$$

$$\int_{\gamma_j} \phi_{f,l}^i \, ds = \delta_{i,j} \delta_{r,l}, \quad \forall \gamma_j = K_j \cap \gamma_l,$$

where

$$a(u, v) = a_m(u_m, v_m) + q(u_m, v_f) + a_f(u_f, v_f) + q(u_f, v_m),$$

with $a_m(u_m, v_m) = \int_{K_i^+} (k_m \nabla u_m, \nabla v_m)$, $a_f(u_f, v_f) = \int_{\gamma_i^+} (k_f \nabla u_f, \nabla v_f)$ and $q(u_m, v_f) = \int_{\gamma_i^+} \sigma u_m v_f\}$.

Therefore, we have the following multiscale space $V_{ms} = \text{span}\{\phi_l^i\}$. In Fig. 2, we depict a multiscale basis function for oversampled region $K_i^+ = K_i^2$ (two oversampling coarse cell layers) on coarse mesh $20 \times 20$.

Based on the property of the multiscale basis, we write our upscaled model that is similar to dual continuum model (embedded fracture model)

$$\frac{1}{\tau} \begin{pmatrix} \bar{S}_m & 0 \\ 0 & \bar{S}_f \end{pmatrix} \begin{pmatrix} \bar{u}_m^{n+1} - \bar{u}_m^n \\ \bar{u}_f^{n+1} - \bar{u}_f^n \end{pmatrix} + \begin{pmatrix} T_{mm} + \bar{Q} & T_{mf} - \bar{Q} \\ T_{fm} - \bar{Q} & T_{ff} + \bar{Q} \end{pmatrix} \begin{pmatrix} \bar{u}_m^{n+1} \\ \bar{u}_f^{n+1} \end{pmatrix} = \begin{pmatrix} \bar{F}_m \\ \bar{F}_f \end{pmatrix} \tag{5}$$

where $\bar{S}_m = \{\bar{s}_{ij}^m\}$, $\bar{S}_f = \{\bar{s}_{ij}^f\}$, $\bar{Q} = \{\bar{q}_{ij}\}$

$$\bar{s}_{ij}^m = \begin{cases} c_m |K_i|, & i = j \\ 0 & else \end{cases}, \quad \bar{s}_{ij}^f = \begin{cases} c_f |\gamma_i|, & i = j \\ 0 & else \end{cases}, \quad \bar{q}_{ij} = \begin{cases} \sigma |\gamma_j|, & K_i \cap \gamma_j \neq 0 \\ 0 & else \end{cases},$$

and right-hand side vector $\bar{q}_i^m = f_m |V_i|$ ($h_0 = 0$ and $h_m = g$ for $m \neq 0$), where $|V_{m=0}| = |K_i|$ and $|V_m| = |\gamma_m|$ for $m \neq 0$. Here $\bar{u} = (\bar{u}_m, \bar{u}_f)$, $\bar{u}_m$ and $\bar{u}_f$ are the average cell solution on coarse grid cell for background medium and for fracture. The stiffness matrix is non-local and provides a good approximation due to the coupled construction, $T = \{\bar{a}_{ij}\}$

$$T = \begin{pmatrix} T_{mm} & T_{mf} \\ T_{fm} & T_{ff} \end{pmatrix}, \quad T = \{\bar{a}_{ij}\}, \quad \bar{a}_{ij} = \begin{cases} -\sum_{j \neq i} \bar{a}_{ij} & i = j \\ \bar{a}_{ij} & else \end{cases},$$

with $\bar{a}_{ij} = a(\psi_i, \psi_j)$.

## 4   Numerical Results

We present numerical results for computational domain $\Omega = [0, 1] \times [0, 1]$ with 14 fractures. In Fig. 1, we present a computational coarse and fine grids, where the fractures are depicted with red color. Fine mesh contains 13,796 vertices and 27,270 cells. These fractures are resolved on the fine grid using DFM for construction of the upscaled model (fine grid depicted by blue color). The coarse grid is uniform and depicted with black color. We consider two coarse grids: (1) coarse grid $20 \times 20$ (400 cells) and (2) coarse grid $40 \times 40$ (1600 cells). The other model parameters used are as follows: $k_m = 10^{-6}$, $k_f = 1$, $\sigma = 10^{-4}$, $c_m = 10^{-5}$ and $c_f = 10^{-6}$. We set a point source on the fractures at the two coarse cells with $f_f = \pm 10^{-3}$. As for the initial pressure, we set $u_0 = 1$. We compute the solution until $t_{max} = 0.1$ with 20 time steps for upscaled and fine-scale solvers.

**Table 1.** Relative errors of the average cell solution on a coarse meshes $20 \times 20$ (left) and $40 \times 40$ (right).

| $K^s$ | $t = 0.02$ | $t = 0.06$ | $t = 0.1$ | $K^s$ | $t = 0.02$ | $t = 0.06$ | $t = 0.1$ |
|---|---|---|---|---|---|---|---|
| $s = 1$ | 0.319 | 1.801 | 3.799 | $s = 1$ | 1.020 | 6.084 | 13.700 |
| $s = 2$ | 0.137 | 0.261 | 0.398 | $s = 2$ | 0.196 | 0.596 | 0.922 |
| $s = 3$ | 0.128 | 0.256 | 0.363 | $s = 3$ | 0.175 | 0.456 | 0.699 |

In Fig. 3, we present the fine scale and upscaled solution for coarse meshes $20 \times 20$ and $40 \times 40$. In the first row, we showed a fine-scale solution with $DOF_f = 14474$. On the second row, we present a multiscale solution on coarse mesh $20 \times 20$ using local domains $K^2$ (oversampled domain $K^+$ with 2 coarse cells layers oversampling). On the third row, we depict a multiscale solution on coarse mesh $40 \times 40$. For both cases, the relative errors of average cell solution for background medium is less than one percent.

In Table 1, we present relative errors for two coarse grids and for different numbers of oversampling layers $K^s$ with $s = 1, 2$ and $3$. From the numerical results, we observe a good convergence, when we take sufficient number of over-sampled layers. For coarse mesh with 400 cells, when we take 2 oversampling

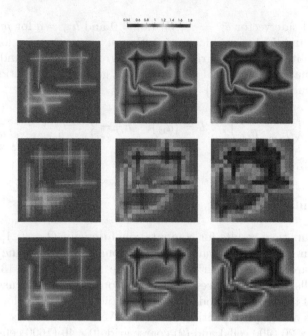

**Fig. 3.** Multiscale solutions for $t = 0.02$, $t = 0.06$ and $t = 0.1$ (from left to right). First row: fine-scale solution, $DOF_f = 14474$. Second row: multiscale solutions on coarse mesh $20 \times 20$ with local domain $K^2$ (two oversampling coarse cells layers), $DOF_c = 533$. Third row: multiscale solutions on coarse mesh $40 \times 40$ with local domain $K^2$ (two oversampling coarse cells layers), $DOF_c = 1853$

layers, we have 0.398% of error at final time. For the coarse mesh with 1600 cells, error is 0.699%. For coarse scale system, we have $DOF_c = 533$ ($20 \times 20$ coarse mesh) and $DOF_c = 1853$ ($40 \times 40$ coarse mesh), where $DOF_f$ and $DOF_c$ are the number of degrees of freedom for fine grid and for upscaled system (coarse grid multicontinuum), respectively. We have huge reduction of the system size with very small errors for unsteady mixed dimensional coupled system.

We note that, the multiscale basis functions are calculated on the offline stage and can be used for the solution of the coarse grid system with different parameters for fixed geometry. Next, we discuss the computational advantages of our approach and consider the online computational cost. Number of degrees of freedom on the fine grid system $DOF_f$ is equal to the sum of vertices for computational meshes for domain $\Omega$ (martrix) and $\gamma$ (fracture), $DOF_f = 13796 + 678$. Number of degrees of freedom for the upscaled system $DOF_c$ is equal to the sum of the number of multiscale basis functions in each local domain, $DOF_c = \sum_i L_i$. For example, on the coarse mesh $20 \times 20$, we have less than one percent of relative error with 3.6% of $DOF_f$ ($DOF_f = 14474$ and $DOF_c = 533$). Note that, for the very complex fracture distribution, we may need very fine grid and $DOF_f$ can be larger. We would like to note that the main computational gain in the online simulations is the low number of multiscale basis functions that calculated adaptively.

**Acknowledgements.** MV's work is supported by the grant of the Russian Scientific Found N17-71-20055. YE's is supported by the mega-grant of the Russian Federation Government (N 14.Y26.31.0013). EC's work is partially supported by Hong Kong RGC General Research Fund (Project 14317516) and CUHK Direct Grant for Research 2016-17

# References

1. Akkutlu, I.Y., Efendiev, Y., Vasilyeva, M., Wang, Y.: Multiscale model reduction for shale gas transport in a coupled discrete fracture and dual-continuum porous media. J. Nat. Gas Sci. Engin. **48**, 65–76 (2017)
2. Akkutlu, I.Y., Efendiev, Y., Vasilyeva, M.: Multiscale model reduction for shale gas transport in fractured media. Computat. Geosci. **20**(5), 1–21 (2015)
3. Barenblatt, G.I., Zheltov, I.P., Kochina, I.N.: Basic concepts in the theory of seepage of homogeneous liquids in fissured rocks [strata]. J. Appl. Math. Mech. **24**(5), 1286–1303 (1960)
4. Chung, E.T., Efendiev, Y., Leung, W., Vasilyeva, M.: Coupling of multiscale and multi-continuum approaches. GEM-Int. J. Geomath. **8**(1), 9–41 (2017)
5. Chung, E.T., Efendiev, Y., Leung, W.: Constraint energy minimizing generalized multiscale finite element method, arXiv preprint arXiv:1704.03193 (2017)
6. Chung, E.T., Efendiev, Y., Leung, W., Vasilyeva, M., Wang, Y.: Online adaptive local multiscale model reduction for heterogeneous problems in perforated domains. Appl. Anal. **96**(12), 2002–2031 (2017)
7. Chung, E.T., Efendiev, Y., Leung, W., Wang, Y., Vasilyeva, M.: Non-local multi-continua upscaling for flows in heterogeneous fractured media, arXiv preprint arXiv:1708.08379 (2017)
8. Chung, E.T., Efendiev, Y., Li, G., Vasilyeva, M.: Generalized multiscale finite element methods for problems in perforated heterogeneous domains. Appl. Anal. **95**(10), 2254–2279 (2016)
9. Efendiev, Y., Galvis, J., Hou, T.: Generalized multiscale finite element methods. J. Comput. Phys. **251**, 116–135 (2013)
10. Efendiev, Y., Hou, T.: Multiscale Finite Element Methods: Theory and Applications. Surveys and Tutorials in the Applied Mathematical Sciences, vol. 4. Springer, New York (2009). https://doi.org/10.1007/978-0-387-09496-0
11. Jenny, P., Lee, S.H., Tchelepi, H.A.: Adaptive multiscale finite-volume method for multiphase flow and transport in porous media. Multiscale Model. Simul. **3**(1), 50–64 (2005)
12. Tene, M., Al Kobaisi, M.S., Hajibeygi, H.: Algebraic multiscale solver for flow in heterogeneous fractured porous media. In: SPE Reservoir Simulation Symposium, Society of Petroleum Engineers (2015)
13. Vasilyeva, M., Stalnov, D.: A generalized multiscale finite element method for thermoelasticity problems. In: Dimov, I., Faragó, I., Vulkov, L. (eds.) NAA 2016. LNCS, pp. 713–720. Springer, Heidelberg (2016). https://doi.org/10.1007/978-3-319-57099-0_82
14. Warren, J.E., Root, P.J., et al.: The behavior of naturally fractured reservoirs. Soc. Pet. Eng. J. **3**(03), 245–255 (1963)
15. D'Angelo, C., Quarteroni, A.: On the coupling of 1D and 3D diffusion-reaction equations: application to tissue perfusion problems. Math. Models Methods Appl. Sci. **18**(08), 1481–1504 (2008)

# Some Features of the Asymptotic-Numerical Method for the Moving Fronts Description in Two-Dimensional Reaction-Diffusion Problems

Vladimir Volkov[✉] and Dmitry Lukyanenko

Department of Mathematics, Faculty of Physics,
Lomonosov Moscow State University, 119991 Moscow, Russia
volkovvt@mail.ru

**Abstract.** This paper develops an analytic-numerical approach for the description of moving fronts in two-dimensional nonlinear singularly perturbed parabolic equations. Asymptotic technique allows to reduce two-dimensional nonlinear reaction-diffusion equation to a series of more simple one-dimensional problems. This decomposition significantly decreases the complexity of numerical calculations and allows the effective use of parallel computing. Some numerical experiments are presented to demonstrate the main features of the proposed method.

## 1 Introduction

Many practical applications in different fields of science (chemical kinetics, biophysics, heat conduction, elasticity and others) can be described by mathematical models which include nonlinear parabolic equations of reaction-diffusion-advection type. If such problems characterized by different physical or geometric scales, small parameters at higher derivatives can arise and solutions of these problems often contain interior layers or moving fronts what do numerical calculations for this class of problems quite difficult or impossible at all. But the presence of small parameters allows to effective use of asymptotic technique and reduce the original nonlinear multi-dimensional problem to a series of more simple problems which: (1) are not singularly perturbed; (2) part of them are linear; (3) the spatial dimension each of them is minimum per one unit less than original problem and in some cases they are not PDE, but ODE or not differential at all.

In this paper we propose the asymptotic-numerical approach for the description of the solutions with moving internal layers (fronts) for some classes of nonlinear two-dimensional reaction-diffusion models. It is important that the layer location and its speed are not known *a priori* and could be determined from the asymptotic procedure by smooth joining of asymptotic expansions ($C^{(1)}$-matching). In quite general cases this procedure can be done explicitly and also explicit asymptotic formulas for the layers location or the front speed can be

© Springer Nature Switzerland AG 2019
I. Dimov et al. (Eds.): FDM 2018, LNCS 11386, pp. 612–620, 2019.
https://doi.org/10.1007/978-3-030-11539-5_72

written [1,2]. But for some classes of reaction-diffusion problems it can not be done explicitly and the asymptotic algorithm needs to be supplemented by the appropriate numerical calculations. It is important, that this $C^{(1)}$–matching procedure must be done only for the normal derivatives at each point $(x;y)$ on some curve $\hat{C}(t)$, so the numerical determination of the layer location or speed can be reduced to a series of one-dimensional problems. This fact allows to optimize computer calculations and effectively use parallel computing technologies.

The basic objective of this paper is, on the one hand, to show the main ideas of the asymptotic approach for the description of moving fronts in two-dimensional case; on the other hand, to outline some problems which require to use numerical calculations on some steps of the asymptotic procedure.

Structure of the paper is the following. Section 2 contains the problem statement and the main conditions. In Sect. 3 the main ideas and steps of the asymptotic algorithm are briefly described and some problems which require the numerical calculations are formulated. In Sect. 4 we explain the method of construction the numerical schemes and perform some numerical experiments.

## 2    Problem Statement

We demonstrate our approach by the following problem:

$$\varepsilon^2 \Delta u - \varepsilon \frac{\partial u}{\partial t} = f\left(u,x,y,\varepsilon\right), \qquad x \in \mathbb{R}^1, \qquad y \in (0,a), \qquad t > 0 \qquad (1)$$

with the boundary and initial conditions

$$\frac{\partial u}{\partial y}\bigg|_{y=0;a} = 0, \quad u(x,y,t,\varepsilon) = u(x+L,y,t,\varepsilon), \quad u(x,y,0,\varepsilon) = u^0(x,y,\varepsilon). \quad (2)$$

In the Eq. (1) functions $u^0(x,y,\varepsilon)$ and $f(u,x,y,\varepsilon)$ are assumed to be sufficiently smooth and $L$-periodic in the variable $x$, $\varepsilon > 0$ is the small parameter. The small parameter before the spatial derivatives is determined by the characteristics of the physical system, while the small parameter before the time derivative determines only the scale of the time, convenient for the further consideration.

It is known [1–3] that under certain conditions the solution of the problem (1) contains internal layer or moving front with the location described by function $y = h(x,t)$. Asymptotic approximation of such type solutions can be built using methods [1,2] and consists of three parts: regular functions that describe the solution far from boundaries $y = 0$ and $y = a$, boundary functions of exponential type near the boundaries $y = 0$ and $y = a$ and some special functions that describe the moving front. For our further considerations we use only the equations which describe the moving front structure and determine its location. While the boundary functions exponentially decrease with the distance from the boundaries $y = 0$ and $y = a$ and do not have a significant effect on the front, below we only briefly outline, how to obtain asymptotic approximation for the moving front and the equation describing its dynamics.

Main ideas for the asymptotic approximation of the moving front type solution are contained in [1–3]. Here we use some general formulas and equations for the description of the front dynamics from [2] and discuss some specific problems of asymptotic procedure.

Suppose the following assumptions are satisfied.

($\mathbf{A_1}$).     (a) *The function* $f(u, x, y, \varepsilon)$ *is such that the reduced equation* $f(u, x, y, 0) = 0$ *has exactly three roots* $u = \varphi^{(\pm)}(x, y)$, $u = \varphi^{(0)}(x, y)$.

(b) *Assume that* $\varphi^{(-)}(x, y) < \varphi^{(0)}(x, y) < \varphi^{(+)}(x, y)$ *for all* $(x, y) \in \bar{D} = (-\infty, +\infty) \times [0, a]$ *and* $f_u(\varphi^{(\pm)}(x, y), x, y, 0) > 0$, $f_u(\varphi^{(0)}(x, y), x, y, 0) < 0$.

It is known from [4] that under condition ($\mathbf{A_1}$) and quite general assumptions for the initial function $u^0(x, y)$ over a time period of order $t_B(\varepsilon) = B\varepsilon |\ln \varepsilon|$ the solution of the problem (1) and (2) quickly generates a thin internal transition layer between the two levels $\varphi^{(-)}(x, y)$ and $\varphi^{(+)}(x, y)$ and located in the neighborhood of some curve $y = h^0(x)$.

($\mathbf{A_2}$). *Assume that the initial function* $u^0(x, y)$ *has the form of a transition layer:* $u^0(x, y, \varepsilon) = \varphi^{(-)}(x, y) + O(\varepsilon)$ *for* $(x, y) \in D_0^{(-)}$, $u^0(x, y, \varepsilon) = \varphi^{(+)}(x, y) + O(\varepsilon)$ *for* $(x, y) \in D_0^{(+)}$ *excluding a small neighborhood of the curve* $y = h^0(x)$.

## 3    Formal Asymptotic Procedure

We will explore a solution to the problem (1) with the moving front: a solution that at every time moment is close to the surface $\varphi^{(-)}(x, y)$ near the line $y = 0$ and to the surface $\varphi^{(+)}(x, y)$ near the line $y = a$, and have a sharp transition from the values on the surface $\varphi^{(-)}(x, y)$ up to the values on the surface $\varphi^{(+)}(x, y)$ in the neighborhood of some moving curve $y = h(x, t)$ sufficiently distant from the boundaries $y = 0$ and $y = a$. We define the location of the front at each time moment $t$ by a curve on which the solution $u(x, y, t, \varepsilon)$ takes the value of $\varphi^{(0)}(x, y)$. The location of this curve is not known a priori and can be found as a series by degrees of the parameter $\varepsilon$

$$h(x, t) = h_0(x, t) + \varepsilon h_1(x, t) + \varepsilon^2 h_2(x, t) + \dots . \tag{3}$$

We denote by $\bar{D}^{(-)} = \{(x, y) : \mathbb{R} \cup [0; h(x, t)]\}$ and $\bar{D}^{(+)} = \{(x, y) : \mathbb{R} \cup [h(x, t); a]\}$ the domains located at two opposite sides of the curve $y = h(x, t)$.

For further description of the internal transition layer location and dynamics we introduce local coordinates $(r, l)$ in a neighborhood of the curve $y = h(x, t)$ by the equations

$$x = l - r \cdot \frac{h_x}{\sqrt{1 + h_x^2}}, \quad y = h(l, t) + r \cdot \frac{1}{\sqrt{1 + h_x^2}}. \tag{4}$$

In (4) $r$ is the distance from $y = h(x, t)$ along the normal to this curve with the sign "+" in the domain $D^{(+)}$ and with "−" in $D^{(-)}$, $l$ is the coordinate of the point on the curve $y = h(x, t)$ from which this normal is going; $\alpha$ – the angle between the axis $y$ and the normal to the curve $y = h(x, t)$ directed to the region $y > h(x, t)$ at each time moment $t$.

Consider two problems (indices $(+)$ and $(-)$ correspond to $D^{(+)}$ and $D^{(-)}$):

$$\varepsilon^2 \Delta u^{(\pm)} - \varepsilon \frac{\partial u^{(\pm)}}{\partial t} - f\left(u^{(\pm)}, x, y, \varepsilon\right) = 0, \qquad (x,y) \in D^{(\pm)}, \quad t > 0,$$

$$u^{(\pm)}(x, y, t, \varepsilon) = u^{(\pm)}(x + L, y, t, \varepsilon), \quad u^{(\pm)}(x, y, 0, \varepsilon) = u^0(x, y, \varepsilon), \quad (5)$$

$$u^{(\pm)}(x, h(x,t), t, \varepsilon) = \varphi^{(0)}(x, h(x,t)), \quad \left.\frac{\partial u^{(-)}}{\partial y}\right|_{y=0} = 0, \quad \left.\frac{\partial u^{(+)}}{\partial y}\right|_{y=a} = 0.$$

Asymptotic approximation $U(x, y, t, \varepsilon)$ for the solutions of (5) can be constructed in domains $\bar{D}^{(-)}$ and $\bar{D}^{(+)}$ separately

$$U = \begin{cases} U^{(-)}(x, y, t, \varepsilon), & (x, y, t) \in \bar{D}^{(-)} \times [0, T], \\ U^{(+)}(x, y, t, \varepsilon), & (x, y, t) \in \bar{D}^{(+)} \times [0, T]. \end{cases}$$

Each of functions $U^{(-)}$ and $U^{(+)}$ we will represent as the sum of three terms

$$U^{(\pm)} = \bar{u}^{(\pm)}(x, y, \varepsilon) + Q^{(\pm)}(\xi, l, h(l, t), \varepsilon) + \Pi^{(\pm)}(x, \eta^{(\pm)}, \varepsilon). \qquad (6)$$

Here $\bar{u}^{(\pm)}(x, y, \varepsilon)$ is the regular part of asymptotic approximation; functions $\Pi^{(\pm)}(x, \eta^{(\pm)}, \varepsilon)$ describe the solution near the boundaries $y = 0$ $y = a$ respectively, $\eta^{(-)} = \frac{y}{\varepsilon}$, $\eta^{(+)} = \frac{y-a}{\varepsilon}$; functions $Q^{(\pm)}(\xi, l, h(l, t), \varepsilon)$ describe the internal layer (moving front), $\xi = \frac{\tilde{x}}{\varepsilon}$ – stretched variable near the curve of the front location; $\xi > 0$ for $D^{(+)}$ and $\xi < 0$ for $D^{(-)}$.

In (6) each of terms are the power series in parameter $\varepsilon$, particularly

$$\bar{u}^{(\pm)}(x, y, \varepsilon) = \bar{u}_0^{(\pm)}(x, y) + \varepsilon \bar{u}_1^{(\pm)}(x, y) + \dots, \qquad (7)$$

$$Q^{(\pm)}(\xi, l, h(l, t), \varepsilon) = Q_0^{(\pm)}(\xi, l, h(l, t)) + \varepsilon Q_1^{(\pm)}(\xi, l, h(l, t)) + \dots. \qquad (8)$$

Functions $U^{(-)}(x, y, t, \varepsilon)$ and $U^{(+)}(x, y, t, \varepsilon)$ and their first normal derivatives are continuous, therefore we have on the curve $y = h(x, t)$ at each time moment $t$:

$$U^{(-)}(x, h(x,t), t, \varepsilon) = U^{(+)}(x, h(x,t), t, \varepsilon) = \varphi^0(x, h(x,t)), \qquad (9)$$

$$\frac{\partial U^{(-)}}{\partial n}(x, h(x,t), t, \varepsilon) = \frac{\partial U^{(+)}}{\partial n}(x, h(x,t), t, \varepsilon). \qquad (10)$$

$C^{(1)}$–matching conditions (9), (10) must be carried out consistently for all degrees of $\varepsilon$ and led to the following equation for the main term of the moving front description $((x, y)$ are parameters located on the curve $y = h(x, t))$

$$\frac{\partial^2 \tilde{u}}{\partial \xi^2} + W \frac{\partial \tilde{u}}{\partial \xi} = f(\tilde{u}, x, y, 0); \quad \tilde{u}|_{\xi=0} = \varphi^{(0)}(x, y), \quad \tilde{u}|_{\xi \to \pm\infty} = \varphi^{(\pm)}(x, y), \quad (11)$$

where $\tilde{u} = \bar{u}_0^{(\pm)}(x, y) + Q_0^{(\pm)}(\xi, l, h(l, t))$ and $W(x, y)$ is the main term of the normal speed of the point $(x, y)$ on the front location curve $y = h(x, t)$.

If $f(\tilde{u}, x, y, \varepsilon)$ satisfies the condition (A$_1$), the existence of a unique pair $(W(x, y), \tilde{u}(\xi; x, y))$ which satisfies problem (11) is proved (for ex., [6]). Also it is known, that $\tilde{u}(\xi; x, y)$ exponentially tends to $\varphi^{(\pm)}(x, y)$ when $\xi \to \pm\infty$.

According to the condition (A$_1$), stationary points $(\varphi^{(\pm)}(x, y), 0)$ of (11) on the phase plane $\left(\tilde{u}; \frac{\partial \tilde{u}}{\partial \xi}\right)$ are the saddle type points, and $(\varphi^{(0)}(x, y), 0)$ is a point of nonstable focus type (or nonstable node type). The solution of (11) exists if there exists the separatrix joining two saddles, which can be obtained for every fixed $(x, y)$ by variation of the parameter $W$.

For the general case of the function $f(\tilde{u}, x, y, \varepsilon)$ in (11) the value of the parameter $W$ which defines the existence of the joining separatrix at phase plane $\left(\tilde{u}; \frac{\partial \tilde{u}}{\partial \xi}\right)$ can not be found explicitly, and asymptotic algorithm must be supplement by some numerical procedure for its determination.

For the moving front location we have the Cauchy problem (see [2])

$$\frac{h_t}{\sqrt{1 + h_x^2}} = W(x, h(x, t)), \quad h(x, 0) = h^0(x), \quad h(x, t) = h(x + L, t), \qquad (12)$$

which defines the main term $h_0(x, t)$ of the asymptotic approximation of the moving front location in (3). $W(x, y)$ is the main term of the normal speed of the point $(x, y)$ on the curve $y = h_0(x, t)$ and it was defined from the condition of solvability of (11).

## 4    Numerical Realization

In the previous Section we have briefly shown how to obtain equations (12) and (11) for the main terms of the asymptotic approximation for the moving front location and its structure. The normal speed of each point of the moving front $W(x, y)$ was defined as the condition of solvability of (11): this speed can be determined as a parameter that realises the smooth joining condition ($C^{(1)}$–matching) for the solutions $\tilde{u}^{(-)}$ and $\tilde{u}^{(+)}$ with the first derivatives of the problem (11) for $\xi < 0$ and $\xi > 0$ at the point $\xi = 0$

$$\left.\frac{d\tilde{u}^{(-)}(\xi, W)}{d\xi}\right|_{\xi=0-0} = \left.\frac{d\tilde{u}^{(+)}(\xi, W)}{d\xi}\right|_{\xi=0+0}. \qquad (13)$$

Parameter $W$ in (11) can be found by the numerical solving of the problem (11) for $\xi < 0$ and $\xi > 0$ and then minimization of absolute value of the difference between the derivatives (13) on the opposite sides of the joining point $\xi = 0$ by the method of golden section:

$$W = \min_v \left| \left.\frac{d\tilde{u}^{(-)}(\xi, v)}{d\xi}\right|_{\xi=0-0} - \left.\frac{d\tilde{u}^{(+)}(\xi, v)}{d\xi}\right|_{\xi=0+0} \right|. \qquad (14)$$

*Calculation of the derivatives.* As we found numerical solution of boundary-values problems (11) we are able to calculate the derivatives in (13) and (14) on the opposite sides of the point $\xi = 0$ for the problems (11) respectively by using following asymmetric approximation with the second order of accuracy:

$$
u'_{n\ left} = \frac{(\tilde{u}_n - \tilde{u}_{n-1})\left(2 + \frac{\xi_{n-1}-\xi_{n-2}}{\xi_n - \xi_{n-1}}\right) - (\tilde{u}_{n-1} - \tilde{u}_{n-2})\left(\frac{\xi_n - \xi_{n-1}}{\xi_{n-1}-\xi_{n-2}}\right)}{\xi_n - \xi_{n-2}},
$$

$$
u'_{n\ right} = \frac{(\tilde{u}_{n+1} - \tilde{u}_n)\left(2 + \frac{\xi_{n+2}-\xi_{n+1}}{\xi_{n+1} - \xi_n}\right) - (\tilde{u}_{n+2} - \tilde{u}_{n+1})\left(\frac{\xi_{n+1}-\xi_n}{\xi_{n+2}-\xi_{n+1}}\right)}{\xi_{n+2} - \xi_n}.
$$

*Solving of the boundary value problems.* For numerical solving of each boundary value problems (11) for $\xi < 0$ and $\xi > 0$ we use the relaxation count method for the problem

$$
\begin{cases}
\dfrac{\partial \tilde{u}^{(\pm)}}{\partial t} = \dfrac{\partial^2 \tilde{u}^{(\pm)}}{\partial \xi^2} + W \dfrac{\partial \tilde{u}^{(\pm)}}{\partial \xi} - f(\tilde{u}^{(\pm)}, x, y, 0), & t > 0, \\
\tilde{u}^{(\pm)}(\pm\infty, t) = \varphi^{(\pm)}(x, y), \quad \tilde{u}^{(\pm)}(0, t) = \varphi^{(0)}(x, y), \\
\tilde{u}^{(\pm)}(\xi, 0) = \tilde{u}_{init}(\xi),
\end{cases}
\tag{15}
$$

For numerical solving of system (15) we apply the stiff method of lines (SMOL) [6] in order to reduce initial PDE to the system of ODEs that can be solved by Rosenbrock scheme with complex coefficient CROS1 [7].

At first, we introduce quasi-uniform mesh [8] only on $\xi$-dimension which has number of nodes $N + 1$ (i.e. $N$ intervals):

$$
\xi_n \equiv \xi(\nu_n) = \frac{c\nu_n}{(1 - \nu_n^2)^m}, \quad \nu_n = a + \frac{b - a}{N} n, \quad n = \overline{0, N},
$$

where $a = -1$, $b = 0$ in the case $\xi < 0$ and $a = 0$, $b = 1$ for $\xi > 0$; $c$ and $m$ – some control parameters, which set density of the quasi-uniform mesh at a neighbourhood of $\xi = 0$ and $\xi = \pm\infty$. We set $c = 1$ $m = 1$ (for more detailed explanations see [8], including explanations about different transformations which are able to generate quasi-uniform meshes).

After finite-difference approximations of the spatial derivatives with second order of accuracy for unbounded domain in (15) we obtain the following system of ODEs from which we should determine $N-1$ unknown functions $\tilde{u}_n \equiv \tilde{u}_n(t) \equiv \tilde{u}(\xi_n, t)$ ($n = \overline{1, N-1}$, $\tilde{u}_0$ and $\tilde{u}_N$ we know from the boundary conditions):

$$
\frac{d\tilde{u}}{dt} = f(\tilde{u}, t), \quad \tilde{u}(0) = \tilde{u}_{init},
\tag{16}
$$

where $\boldsymbol{u} = \begin{pmatrix} \tilde{u}_1 & \tilde{u}_2 & \tilde{u}_3 & \dots & \tilde{u}_{N-1} \end{pmatrix}^T$, $\boldsymbol{f} = \begin{pmatrix} f_1 & f_2 & f_3 & \dots & f_{N-1} \end{pmatrix}^T$ and $\tilde{\boldsymbol{u}}_{init} = \begin{pmatrix} \tilde{u}_{init}(\xi_1) & \tilde{u}_{init}(\xi_2) & \dots & \tilde{u}_{init}(\xi_{N-1}) \end{pmatrix}^T$.

For numerical solving of this system (16) we use Rosenbrock scheme with complex coefficient (CROS1), that has the order of accuracy $(O(\tau^2))$, is monotonic and stable [6]. In order to apply this scheme we introduce uniform mesh (it is also possible to introduce a quasi-uniform mesh for unbounded domain

$[0, +\infty)$ without any change in further algorithm) $T_M$ on $t$–dimension that has number of nodes $M + 1$ (i.e. $M$ intervals): $T_M = \{t_m, \ 0 \leqslant m \leqslant M : \ 0 = t_0 < t_1 < t_2 < \ldots < t_{M-1} < t_M = T^*\}$, where $T^*$ — estimated relaxation time.

Then we are able to apply the CROS1 scheme for solving system (16):

$$\tilde{u}\,(t_{m+1}) = \tilde{u}\,(t_m) + (t_{m+1} - t_m)\,Re\,w, \quad \text{where } w \text{ is a solution of the SLAE}$$

$$\left[ E - \frac{1 + i}{2}\,(t_{m+1} - t_m)\,f_{\tilde{u}}\Big(\tilde{u}\,(t_m), t\Big) \right] w = f\left( \tilde{u}\,(t_m), \frac{t_m + t_{m+1}}{2} \right). \quad (17)$$

Here $E$ is the identity matrix, $f_{\tilde{u}}$ is the Jacobian matrix.

*Parallelization of the Algorithm.* We want to notice that proposed method can be very efficiently parallelized. So, we propose some way of its parallelization (using $N_{procs}$ calculating units indexed as $rank = \overline{0, N_{procs} - 1}$).

1. Process with $rank = 0$ introduces on the segment $[a, b] \equiv [0, 1]$ an uniform mesh $X_N$ that has number of nodes $N + 1$ (i.e. $N$ intervals): $X_N = \{x_n, \ 0 \leqslant n \leqslant N : x_n = a + \frac{b-a}{N}\,n\}$. Set $t := 0$.
2. Process with $rank = 0$ calculates on the mesh $X_N$ coordinates of the points $(x_n, y_n)$ $(N+1$ pairs) which lie on the curve $h_0$ (e.g. $h_0$: $y = 0.5 - 0.2\sin 2\pi x$).
3. Process with $rank = 0$ scatters the pairs $(x_n, y_n)$ among others processor units $(1 \leqslant rank \leqslant N_{procs} - 1)$.
4a Process with $rank = 0$ calculates in each point $(x_n, y_n)$ of the curve slope ratio of the normal with the second order of accuracy, where

$$f'(x_n) = \frac{y_n - y_{n-1}}{x_n - x_{n-1}} + \frac{x_n - x_{n-1}}{x_{n+1} - x_{n-1}} \left( \frac{y_{n+1} - y_n}{x_{n+1} - x_n} - \frac{y_n - y_{n-1}}{x_n - x_{n-1}} \right).$$

We take into account the periodicity conditions: at the point $(x_0, y_0)$ the derivative has to be calculated using nodes with numbers $N{-}1$, 0, (or $N$), 1.
4b. Each process with $rank \neq 0$ for each point $(x_n, y_n)$ from its set of points calculates the normal velocity $W(x_n, y_n)$ of this point (as the solution of (11)). Corresponding numerical scheme was discussed in the previous paragraphs.
5. Process with $rank = 0$ gathers the values of $v_{0n} \equiv v_0(x_n, y_n)$ for each point $(x_n, y_n)$ from others processors units.
6. Process with $rank = 0$ performs transition of each point $(x_n, y_n)$ along normal direction with steep $v_{0n}\tau$ (where $\tau$ — time steep): $x_n := x_n - v_{0n}\cos\alpha_n\,\tau$, $y_n := y_n - v_{0n}\sin\alpha_n\,\tau$ (for calculation of $\cos\alpha_n$ and $\sin\alpha_n$ we use *a priory* information about direction in which each point of the curve is moving, e.g. $0 < \alpha_n < \pi$). After that we check whether this point has left segment $[a, b]$ or not. If $x_n < a$, then using the periodicity conditions: $x_n := x_n + (b - a)$. If $x_n > b$, then: $x_n := x_n - (b - a)$. Set $t := t + \tau$.
7. If $t < T$, go to item 3.

Some examples of calculations with $f(u, x, y, \varepsilon) = (u^2 - 1) \cdot (u - \varphi^0(x, y))$ are represented here. The results are shown below on Figs. 1 and 2.

*Example 1.* The numerical example (see Fig. 1) has been calculated for the following set of parameters: $D = [0, 1] \times [0, 1]$, $\varphi^{(0)}(x, y) = \frac{1}{6}\cos 2\pi x$, the curve $h^0$ is $y = 0.5 - 0.2\sin 2\pi x$, $T = 1$, $M = 100$ ($\tau = T/M$), $N = 100$. The parameters of the relaxation count method: $T^* = 10$, $M = 100$, $N = 100$.

*Example 2.* The numerical example 2 has been calculated for the following set of parameters: $D = [0, 1] \times [0, 1]$, $\varphi^{(0)}(x, y) = (0.9 - y)(y - 0.1)\cos 4\pi x$, the curve $h^0$ is $y = 0.5 - 0.2\sin 2\pi x$, $T = 10$, $M = 1000$ ($\tau = T/M$), $N = 100$. The parameters of the relaxation count method: $T^* = 10$, $M = 100$, $N = 100$.

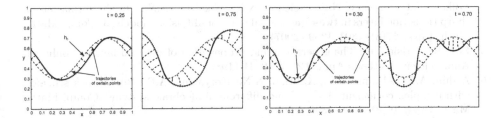

**Fig. 1.** Calculations for Example 1.          **Fig. 2.** Calculations for Example 2.

## 5   Conclusion

Summarizing the main ideas, we can say that asymptotic analysis allows to reduce the spatial dimension of the numerical problem, optimize numerical calculations and save computational resources. Particularly, spatial dimension of the equations for the internal layer location is per one unit less than of original parabolic problem. But asymptotic algorithm includes some joining procedure at each point $(x; y)$ on some curve $y = h(x, t)$ which ultimately determines the normal speed of the front. It can be done explicitly only for some special cases of nonlinearity $f$, for example, $f(u, x, y, \varepsilon) = (u^2 - 1) \cdot (u - \varphi^0(x, y))$ but, in general, requires to implement some numerical calculations. It is important, that this joining procedure must be realized only for the normal derivatives on the curve, so the numerical determination of the layer location or speed can be reduced to a series of one-dimensional problems. This fact allows to optimize computer calculations and effectively use parallel computing technologies.

**Acknowledgements.** This work is supported by RSCF, project No. 18-11-00042.

# References

1. Volkov, V., Nefedov, N., Antipov, E.: Asymptotic-numerical method for moving fronts in two-dimensional R-D-A problems. In: Dimov, I., Faragó, I., Vulkov, L. (eds.) FDM 2014. LNCS, vol. 9045, pp. 408–416. Springer, Cham (2015). https://doi.org/10.1007/978-3-319-20239-6_46
2. Antipov, E.A., Volkov, V.T., Levashova, N.T., Nefedov, N.N.: Moving front solution of the reaction-diffusion problem. Model. Anal. Inf. Syst. **24**(3), 259–279 (2017)
3. Volkov, V., Nefedov, N.: Asymptotic-numerical investigation of generation and motion of fronts in phase transition models. In: Dimov, I., Faragó, I., Vulkov, L. (eds.) NAA 2012. LNCS, vol. 8236, pp. 524–531. Springer, Heidelberg (2013). https://doi.org/10.1007/978-3-642-41515-9_60
4. Volkov, V.T., Grachev, N.E., Nefedov, N.N., Nikolaev, A.N.: On the formation of sharp transition layers in two-dimensional reaction-diffusion models. J. Comp. Math. Math. Phys. **47**(8), 1301–1309 (2007)
5. Fife, P.C., Hsiao, L.: The generation and propagation of internal layers. Nonlinear Anal. Theory Methods Appl. **12**(1), 19–41 (1988)
6. Alshin, A.B., Alshina, E.A., Kalitkin, N.N., Koryagina, A.B.: Rosenbrock schemes with complex coefficients for stiff and differential algebraic systems. Comp. Math. Math. Phys. **46**(8), 1320–1340 (2006)
7. Rosenbrock, H.H.: Some general implicit processes for the numerical solution of differential equations. Comput. J. **5**(4), 329–330 (1963)
8. Kalitkin, N.N., Alshin, A.B., Alshina, E.A., Rogov, B.V.: Computations on Quasi-Uniform Grids, Fizmatlit, Moscow (2005). (in Russian)

# Conservative Semi-Lagrangian Numerical Algorithm with Decomposition of Integration Domain into Tetrahedrons for Three-Dimensional Advection Problem

Alexander Vyatkin[1]([⊠]) and Elena Kuchunova[2]([⊠])

[1] Institute of Computational Modelling of SB RAS,
Akademgorodok, 660036 Krasnoyarsk, Russia
vyatkin@icm.krasn.ru
[2] Siberian Federal University, 79 Svobodny pr., 660041 Krasnoyarsk, Russia
hkuchunova@sfu-kras.ru

**Abstract.** A conservative semi-Lagrangian method is developed in order to solve three-dimensional linear advection equation. It based on balance equation in integral form. Main feature of proposed method consists in way of computation of integral at lower time level. To compute integral, we decompose a domain of integration into several tetrahedrons and approximate integrand by trilinear function.

**Keywords:** Semi-Lagrangian method · Advection equation · Decomposition of integration domain · Local conservation low

## 1 Introduction

Semi-Lagrangian methods was developed from idea of applying a characteristic trajectories in problems for weather prediction [1]. Elaboration of this approach led to a new wide family of semi-Lagrangian methods for solving advection dominated problems [2–7]. Semi-Lagrangian algorithms do not involve the time step restriction [8] which traditionally ensues from the Courant-Friedrichs-Lewy condition for Eulerian schemes. Most of the semi-Lagrangian methods [9–11] are based on the integral equality for unknown function between two neighboring time levels. This integral equality express conservation low. In general case algorithm consists of three main steps: calculation of the integral of an approximate solution at the upper time level; computation of trajectories backward in time to the lower time level; calculation of the integral at the lower time level. There are many ways to implement each step. The improvement of properties of a numerical method usually leads to computational complications. This is especially valid for three-dimensional problems [12,13]. Despite of that there is a wide variety of semi-Lagrangian algorithms, only some of them can be conveniently employed for three-dimensional problem.

© Springer Nature Switzerland AG 2019
I. Dimov et al. (Eds.): FDM 2018, LNCS 11386, pp. 621–629, 2019.
https://doi.org/10.1007/978-3-030-11539-5_73

We present the semi-Lagrangian algorithm for three-dimensional problem. It is based on the conservation low. The focus of described investigation is on constructing conservative algorithm without weight coefficients. Another aim of research is to create method algorithmically accessible for three dimensional problem. For with purpose we make several simplifications. Firstly, we use only uniform cubic space grid. Algorithm can be generalized into space grid with different grid steps only for different space axes. So, we can use only one fixed grid step for one space direction. Therefore, space grid should be uniform. Secondly, we construct method of first order of convergence. There are many semi-Lagrangian methods of convergence order higher than one. All these methods use weight coefficients to make valid the conservation low for numerical solution. Such approach distorts numerical solution. To avoid this weakness, we construct method of first order of convergence. Thirdly, we use time step which is twice smaller than time step corresponding to Courant-Friedrichs-Lewy condition for Eulerian schemes. This restriction reduces the algorithm of integration at lower time level substantially. The algorithm became more convenient for three-dimensional problem. We find numerical solution of problem as a grid function. To make algorithm explicit, we set domain of integration at the upper time level as cube neighborhood of a grid node and approximate this integral by the midpoint cubature rule. Domain of integration at the lower time level is defined by trajectories which are issued out from the domain of integration at the upper time level. To compute integral, we approximate integrand by trilinear function. Due to restriction for time step, domain of integration at the lower time level is a curved cuboid. To approximate it, we compute coordinates of its vertices by one step of Euler method. We decompose this cuboid by planes into several parts. In general case the number of parts is eight. Each part we decompose into several tetrahedrons. To compute integral over a tetrahedron, we use Gauss quadrature rule which is accurate for trilinear function. Different numerical experiments confirm the first-order convergence and validity of conservation low.

## 2    The Formulation of the Problem

We use computational region $D = [0, 1] \times [0, 1] \times [0, 1]$ and its boundary $\partial D$. In the cylinder $[0, T] \times D$ we consider three-dimensional advection equation

$$\frac{\partial \rho}{\partial t} + \frac{\partial(u\rho)}{\partial x} + \frac{\partial(v\rho)}{\partial y} + \frac{\partial(w\rho)}{\partial z} = f. \tag{1}$$

We suppose that components $u(t, \mathbf{x})$, $v(t, \mathbf{x})$, $w(t, \mathbf{x})$, $\mathbf{x} = (x, y, z)$, of the velocity vector $\mathbf{u} = (u, v, w)$ and function $f(t, \mathbf{x})$ are known and sufficiently smooth in $[0, T] \times D$. Function $\rho(t, \mathbf{x})$ is unknown. We denote the inflow boundary by $\Gamma_{\text{in}} = \partial D\big|_{x=0}$, the outflow boundary by $\Gamma_{\text{out}} = \partial D\big|_{x=1}$, and the rigid boundary by $\Gamma_{\text{rig}} = \partial D \setminus (\Gamma_{\text{in}} \cup \Gamma_{\text{out}})$. For the velocity functions we use the conditions

$$\mathbf{u}(t, \mathbf{x})\big|_{\Gamma_{\text{rig}}} = \mathbf{0}, \quad u(t, \mathbf{x})\big|_{\Gamma_{\text{in}}} \geq 0, \quad u(t, \mathbf{x})\big|_{\Gamma_{\text{out}}} \geq 0 \qquad \forall\, t \in [0, T]. \tag{2}$$

Also we set the following boundary and initial conditions for function $\rho(t, \mathbf{x})$:

$$\rho(t, \mathbf{x})\big|_{\Gamma_{\text{in}}} = \rho_{\text{in}}(t, y, z) \quad \forall\, t \in [0, T],\; y \in [0, 1],\; z \in [0, 1], \tag{3}$$

$$\rho(0, \mathbf{x}) = \rho_{\text{init}}(\mathbf{x}) \quad \forall\, \mathbf{x} \in D. \tag{4}$$

Functions $\rho_{\text{in}}(t, y, z)$ and $\rho_{\text{init}}(\mathbf{x})$ are known and sufficiently smooth.

To solve problem (1)–(4), we construct grid $D^h$ with mesh-size $h = 1/N$

$$D^h = \{(x_i, y_j, z_k) : x_i = ih, y_j = jh, z_k = kh;\; i, j, k = 0, 1, \ldots, N\}.$$

We denote $\mathbf{x}_{i,j,k} = (x_i, y_j, z_k)$, $i, j, k = 0, \ldots, N$, and define neighborhood

$$\Omega_{i,j,k} = ([x_i - h/2, x_i + h/2] \times [y_j - h/2, y_j + h/2] \times [z_k - h/2, z_k + h/2]) \cap D$$

and grid cell

$$D_{i,j,k} = [x_i, x_{i+1}] \times [y_j, y_{j+1}] \times [z_k, z_{k+1}].$$

We introduce time grid $T^\tau = \{t_m : t_m = m\tau, m = 0, \ldots, M\}$ with time step $\tau = T/M$. We find solution of problem (1)–(4) as grid function $\rho^h$ defined on $T^h \times D^h$.

## 3   The Numerical Algorithm

We introduce notations $g_{i,j,k}^m = g(t_m, \mathbf{x}_{i,j,k})$ and $g_{i,j,k}^{h,m} = g^h(t_m, \mathbf{x}_{i,j,k})$ for a function $g$ and a grid function $g^h$ defined in $T \times D$ and $T^\tau \times D^h$, respectively. We suppose that $\rho^h$ is already defined at time level $t_{m-1}$ and we need to compute $\rho^h$ at next time level $t_m$. To construct numerical algorithm, we suppose that

$$\tau \le \frac{h}{2\,\max|\mathbf{u}\,(t, \mathbf{x})|}, \tag{5}$$

where $|\mathbf{u}\,(t, \mathbf{x})| = \sqrt{u^2\,(t, \mathbf{x}) + v^2\,(t, \mathbf{x}) + w^2\,(t, \mathbf{x})}$.

### 3.1   The Local Integral Equality

To calculate $\rho_{i,j,k}^{h,m}$, $i = 1, \ldots, N$; $j, k = 0, \ldots, N$, we consider $\Omega_{i,j,k}$ and its boundary $\partial\Omega_{i,j,k}$. For a point $\mathbf{B} = (B_x, B_y, B_z) \in \partial\Omega_{i,j,k}$ at the time level $t_m$ we construct trajectory to the previous time level $t_{m-1}$ backward in time. This trajectory is solution of Cauchy problem for the system of ordinary differential equations

$$\hat{x}'(t) = u\,(t, \hat{x}, \hat{y}, \hat{z})\,,\;\; \hat{y}'(t) = v\,(t, \hat{x}, \hat{y}, \hat{z})\,,\;\; \hat{z}'(t) = w\,(t, \hat{x}, \hat{y}, \hat{z}) \tag{6}$$

with the initial condition

$$\hat{x}(t_m) = B_x, \quad \hat{y}(t_m) = B_y, \quad \hat{z}(t_m) = B_z \tag{7}$$

and $t \in [t_{m-1}, t_m]$. Denote solution of problem (6)–(7) by $\bar{\mathbf{x}}(t; \mathbf{B})$. Set of all points $\mathbf{B} \in \partial\Omega_{i,j,k}$ forms boundary $\partial V_{i,j,k}^m\,(t) = \{\bar{\mathbf{x}}(t; \mathbf{B}) : \mathbf{B} \in \partial\Omega_{i,j,k}\}$ of a region $V_{i,j,k}^m\,(t)$. Thus, $V_{i,j,k}^m(t_m) = \Omega_{i,j,k}$. The following statement is available.

**Statement.** For each $i = 1, ..., N; j, k = 0, ..., N$ is valid

$$\int_{\Omega_{i,j,k}} \rho(t_m, \mathbf{x}) \, d\Omega = \int_{V_{i,j,k}^m(t_{m-1})} \rho(t_{m-1}, \mathbf{x}) \, dV + \int_{t_{m-1}}^{t_m} \int_{V_{i,j,k}^m(t)} f(t, \mathbf{x}) \, dV \, dt. \quad (8)$$

To construct numerical method, we approximate each term of this equality.

## 3.2 The Integration at the Upper Time Level

We compute integral over $\Omega_{i,j,k}$ in (8) approximately in the following way:

$$\int_{\Omega_{i,j,k}} \rho(t_m, \mathbf{x}) \, d\Omega \approx \rho_{i,j,k}^{h,m} \operatorname{meas}(\Omega_{i,j,k}), \quad (9)$$

where $\operatorname{meas}(\Omega_{i,j,k})$ is volume of $\Omega_{i,j,k}$. To approximate the integral for integrand $f(t, \mathbf{x})$ in (8), we employ the rectangle rule in $t$-direction. After that, we get

$$\int_{t_{m-1}}^{t_m} \int_{V_{i,j,k}^m(t)} f(t, \mathbf{x}) \, dV \, dt \approx \tau \int_{V_{i,j,k}^m(t_m)} f(t_m, \mathbf{x}) \, dV \approx \tau f_{i,j,k}^m \operatorname{meas}(\Omega_{i,j,k}). \quad (10)$$

## 3.3 The Integration at the Lower Time Level

To compute integral over $V_{i,j,k}^m(t_{m-1})$ in (8), we consider eight vertices $\mathbf{A}_n = (A_{n,x}, A_{n,y}, A_{n,z})$ of $\Omega_{i,j,k}$. We consistently use $\mathbf{A}_n$ as initial condition for system (6). Thus we get eight Cauchy problems. To find numerical solution of Cauchy problem, we use Euler method. We denote by $\mathbf{W}_n^h = (W_{n,x}^h, W_{n,y}^h, W_{n,z}^h)$ value of numerical solution at the time level $t_{m-1}$. Therefore

$$W_{n,x}^h = A_{n,x}^h - \tau u(t_m, \mathbf{A}_n), \quad W_{n,y}^h = A_{n,y}^h - \tau v(t_m, \mathbf{A}_n),$$
$$W_{n,z}^h = A_{n,z}^h - \tau w(t_m, \mathbf{A}_n).$$

Nodes $W_n^h, n = 1, \ldots, 8$, form an octagon $V_{i,j,k}^{h,m}(t_{m-1})$. We decompose $V_{i,j,k}^{h,m}(t_{m-1})$ by planes $x = x_i$, $y = y_j$, and $z = z_k$ into several octagons $R_n$. If $\mathbf{x}_{i,j,k} \notin \partial D$, then due to (5) the number of octagons is eight If $\mathbf{x}_{i,j,k} \in \partial D$, then the number of octagons is four or two. Here and after we consider case with eight octagons (see Fig. 1). For other cases the reasoning is similar. Due to (5), any $R_n$ lies in one corresponding cell $D_{p,q,r}$. We define trilinear function $\rho_I^h(t_{m-1}, \mathbf{x})$ on $D_{p,q,r}$

$$\rho_I^h(t_{m-1}, \mathbf{x}) = \sum_{a=p}^{p+1} \sum_{b=q}^{q+1} \sum_{c=r}^{r+1} \varphi_a(x) \psi_b(y) \chi_c(z) \rho_{a,b,c}^{h,m-1}, \quad (11)$$

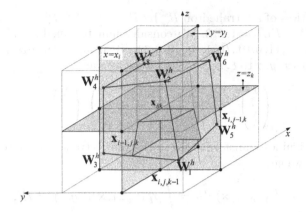

**Fig. 1.** Decomposition of octagon $V_{i,j,k}^{h,m}(t_{m-1})$

where

$$\varphi_a(x) = \begin{cases} \frac{(x-x_{a-1})}{h}, & \text{if } x \in [x_{a-1}, x_a), \\ \frac{(x_{a+1}-x)}{h}, & \text{if } x \in [x_a, x_{a+1}], \\ 0, & \text{otherwise}, \end{cases} \quad \psi_b(y) = \begin{cases} \frac{(y-y_{b-1})}{h}, & \text{if } y \in [y_{b-1}, y_b), \\ \frac{(y_{b+1}-y)}{h}, & \text{if } y \in [y_b, y_{b+1}], \\ 0, & \text{otherwise}, \end{cases}$$

$$\chi_c(z) = \begin{cases} \frac{(z-z_{c-1})}{h}, & \text{if } z \in [z_{c-1}, z_c), \\ \frac{(z_{c+1}-z)}{h}, & \text{if } z \in [z_c, z_{c+1}], \\ 0, & \text{otherwise}. \end{cases}$$

Thus we construct the following approximation:

$$\int_{V_{i,j,k}^m(t_{m-1})} \rho(t_{m-1}, \mathbf{x}) \, dV \approx \sum_{n=1}^{8} \int_{R_n} \rho_I^h(t_{m-1}, \mathbf{x}) \, dR. \qquad (12)$$

To compute integral over $R_n$, we unscramble $R_n$ into two prisms. Each prism we decompose into three tetrahedrons (see Fig. 2). Therefore

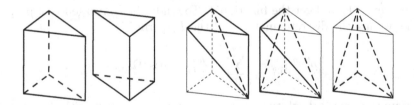

**Fig. 2.** Decomposition of $R_n$ into six tetrahedrons

$$\int_{R_n} \rho_I^h(t_{m-1}, \mathbf{x}) \, dR = \sum_{s=1}^{6} \int_{R_n^s} \rho_I^h(t_{m-1}, \mathbf{x}) \, dR. \qquad (13)$$

We denote vertices of a tetrahedron $R_n^s$ by $P_l = (P_l^x, P_l^y, P_l^z,)$, $l = 0, 1, 2, 3$, and vectors $e_l = P_l - P_0$, $l = 1, 2, 3$. We consider unit tetrahedron $E$ with vertices in points $(0, 0, 0), (1, 0, 0), (0, 1, 0), (0, 0, 1)$. To compute integral over $R_n^s$, we change variables $x, y, z$ to $\xi, \eta, \zeta$ in the following way:

$$\begin{pmatrix} x \\ y \\ z \end{pmatrix} = \begin{pmatrix} P_0^x \\ P_0^y \\ P_0^z \end{pmatrix} + \begin{pmatrix} e_1^x & e_2^x & e_3^x \\ e_1^y & e_2^y & e_3^y \\ e_1^z & e_2^z & e_3^z \end{pmatrix} \begin{pmatrix} \xi \\ \eta \\ \zeta \end{pmatrix}. \tag{14}$$

This formula signifies transformation of the unit tetrahedron $E$ to tetrahedron $R_n^s$. Therefore we get

$$\int_{R_n^s} \rho_I^h (t_{m-1}, \mathbf{x}) \, dR = \int_E \rho_I^h (t_{m-1}, \mathbf{x}(\xi, \eta, \zeta)) \, J \, dE, \tag{15}$$

where $J$ is Jacobian of transformation. To compute the integral over $E$, we use the following Gauss quadrature rule

$$\int_E g(\xi, \eta, \zeta) \, dE \approx \frac{1}{24} g \left( \frac{5 + 3\sqrt{5}}{20}, \frac{5 - \sqrt{5}}{20}, \frac{5 - \sqrt{5}}{20} \right) + \frac{1}{24} g \left( \frac{5 - \sqrt{5}}{20}, \frac{5 + 3\sqrt{5}}{20}, \frac{5 - \sqrt{5}}{20} \right) +$$

$$\frac{1}{24} g \left( \frac{5 - \sqrt{5}}{20}, \frac{5 - \sqrt{5}}{20}, \frac{5 + 3\sqrt{5}}{20} \right) + \frac{1}{24} g \left( \frac{5 - \sqrt{5}}{20}, \frac{5 - \sqrt{5}}{20}, \frac{5 - \sqrt{5}}{20} \right).$$

This quadrature rule is accurate for a trilinear function $g$.

## 4   A Numerical Experiment

To test algorithm, several numerical experiments were carried out. The main purpose was to study order of convergence. We take the components of velocity

$$u = 11y(1 - y)z(1 - z) \left[ \pi/2 - \text{arctg}(x) \right],$$
$$v = w = \text{arctg} \left( x(1 - x)y(1 - y)z(1 - z)(1 + t)/10 \right)$$

and consider the function $\rho = 1.1 + \sin(t\,x\,y\,z)$, $t \in [0, 1]$. We substituted these functions into (1) and get the function $f$. To study the convergence of proposed scheme, we use the discrete analogue of the $L_1$-norm

$$\left\| \rho^{h,m} \right\|_{L_1^h} = \sum_{i,j,k=0}^N \left| \rho_{i,j,k}^{h,m} \right| \text{meas}\left( \Omega_{i,j,k} \right).$$

We compute numerical solution $\rho^h$ on the set of space grids with different grid steps. Denote by $\rho_n^h$ a numerical solution $\rho^{h,M}$ computed on the grid $D^h$ with parameter $N = 10 \cdot 2^n$, $n = 0, \ldots, 5$. Since order of convergence depends on $h$ and $\tau$, we put $\tau = c\,h$. To make inequality (5) valid, we set $c = 0.5$. We define

$$\sigma(n) = \log_2 \left( \frac{\left\| \rho - \rho_{n-1}^h \right\|_{L_1^h}}{\left\| \rho - \rho_n^h \right\|_{L_1^h}} \right)$$

to evaluate the order of convergence. As shown in Fig. 3, the scheme has the first order of convergence.

In next numerical test we check validity of conservation low for numerical solution. For this purpose we consider level set equation for the Enright test [14]

$$\rho_{\text{init}}(\mathbf{x}) = \begin{cases} 1, & \text{if } |\mathbf{x} - \mathbf{x}_c| \leq R^2, \\ 0, & \text{otherwise}, \end{cases} \quad f = 0 \qquad (16)$$

where $R = 0.1, \mathbf{x}_c = (0.35, 0.35, 0.35)$, $t \in [0,1]$. The velocity functions we define in the following way:

$$\begin{aligned} u &= 2\cos(\pi t)\sin^2(\pi x)\sin(2\pi y)\sin(2\pi z), \\ v &= -\cos(\pi t)\sin(2\pi x)\sin^2(\pi y)\sin(2\pi z), \\ w &= -\cos(\pi t)\sin(2\pi x)\sin(2\pi y)\sin^2(\pi z). \end{aligned} \qquad (17)$$

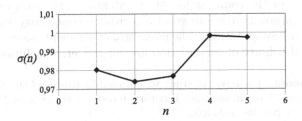

**Fig. 3.** The order of convergence

Computations are performed under condition $\tau = h/5$. To check the validity of conservation low, we consider two integrals

$$I_0 = \int_D \rho_I^h(0, \mathbf{x})\, d\mathbf{x}, \quad I_N = \int_D \rho_I^h(T, \mathbf{x})\, d\mathbf{x}. \qquad (18)$$

Since velocity functions $u, v, w$ are equal to zero on boundary $\partial D$, then values $I_0$ and $I_N$ should be the same for conservative method. It is easy to understand that if $\rho_{i,j,k}^{h,m} \geq 0$, then

$$\int_D \rho_I^h(t_m, \mathbf{x})\, d\mathbf{x} = \sum_{i,j,k=0}^{N} \left| \rho_{i,j,k}^{h,m} \right| \text{meas}(\Omega_{i,j,k}) = \left\| \rho^{h,m} \right\|_{L_1^h}.$$

Therefore to evaluate the validity of conservation low, we consider two items

$$I_0 = \left\| \rho^{h,0} \right\|_{L_1^h}, \quad \delta I = \frac{\left| \left\| \rho^{h,0} \right\|_{L_1^h} - \left\| \rho^{h,0} \right\|_{L_1^h} \right|}{\left\| \rho^{h,0} \right\|_{L_1^h}}.$$

This values presented in the table below. The items of $\delta I$ are equals to zero with precision of computational error (Table 1).

**Table 1.** The numerical result for second test problem

| $n$ | $N$ | $I_0$ | $\delta I$ |
|---|---|---|---|
| 0 | 10 | $8.00E-03$ | $3.89E-09$ |
| 1 | 20 | $3.37E-03$ | $8.50E-10$ |
| 2 | 40 | $3.92E-03$ | $1.88E-10$ |
| 3 | 80 | $4.11E-03$ | $3.02E-11$ |
| 4 | 160 | $4.17E-03$ | $4.26E-12$ |

# 5   Conclusion

We develop new algorithm from the family of semi-Lagrangian methods for the three-dimensional advection equation. The main feature of the scheme consists in the technique of the computation of an integral at the previous time level. The described approach allows to avoid algorithmic complexity and to decrease resource-intensive computations. The numerical experiments show the first-order convergence and validity of conservation low for numerical solution.

**Acknowledgments.** The reported study was funded by Russian Foundation for Basic Research, Government of Krasnoyarsk Territory, Krasnoyarsk Regional Fund of Science to research project No. 18-41-243006.

# References

1. Wiin-Nielson, A.: On the application of trajectory methods in numerical forecasting. Tellus **11**, 180–186 (1959)
2. Robert, A.: A stable numerical integration scheme for the primitive meteorological equations. Atmosphere-Ocean **19**, 35–46 (1981)
3. Morton, K.: Numerical Solution of Convection-Diffusion Problems. Chapman and Hall, London (1996)
4. Russell, T., Celia, M.: An overview of research on Eulerian-Lagrangian localized adjoint methods. Adv. Water Res. **25**, 1215–1231 (2002)
5. Andreeva, E., Vyatkin, A., Shaidurov, V.: The semi-Lagrangian approximation in the finite element method for Navier-Stokes equations for a viscous incompressible fluid. In: AIP Conference Proceedings, vol. 1611 (2014). https://doi.org/10.1063/1.4893794
6. Celledoni, E., Kometa, B., Verdier, O.: High order semi-Lagrangian methods for the incompressible Navier-Stokes equations. J. Sci. Comput. **66**(1), 91–115 (2016)
7. Cameron, A., Raynaud, R., Dormy, E.: Multi-stage high order semi-Lagrangian schemes for incompressible flows in Cartesian geometries. Int. J. Numer. Meth. Fluids **82**, 879–892 (2016). https://doi.org/10.1002/fld.4245
8. Lentine, M., Gretarsson, J., Fedkiw, R.: An unconditionally stable fully conservative semi-Lagrangian method. J. Comput. Phys. **230**, 2857–2879 (2011)
9. Arbogast, T., Wang, W.: Convergence of a fully conservative volume corrected characteristic method for transport problems. SIAM J. Numer. Anal. **48**(3), 797–823 (2010)

10. Efremov, A., Karepova, E., Shaydurov, V., Vyatkin, A.: A computational real-
    ization of a semi-Lagrangian method for solving the advection equation. J. Appl.
    Math. (2014). Article ID 610398. https://doi.org/10.1155/2014/610398
11. Phillips, T., Williams, A.: Conservative semi-Lagrangian finite volume schemes.
    Numer. Meth. Part. Diff. Equ. **17**, 403–425 (2001)
12. Behrens, J., Mentrup, L.: A conservative scheme for 2D and 3D adaptive semi-
    Lagrangian advection. Contemp. Math. **383**, 175–190 (2005)
13. Vyatkin, A.: A semi-Lagrangian algorithm based on the integral transformation
    for the three-dimensional advection problem. In: AIP Conference Proceedings, vol.
    1684, p. 090012 (2015). https://doi.org/10.1063/1.4934337
14. Terekhov, K.M., Nikitin, K.D., Olshanskii, M.A., Vassilevski, Y.V.: A semi-
    Lagrangian method on dynamically adapted octree meshes. Russ. J. Numer. Anal.
    Math. Model. **30**(6), 363–380 (2015)

# Numerical Method for Calculating the Pipe Spatial Vibrations

Natalia Yaparova[✉]

Department of Computational Mathematics and High Performance Computing,
South Ural State University (National Research University),
76, pr.Lenin, Chelyabinsk 454080, Russia
ddjy@math.ac.susu.ru, iaparovanm@susu.ru

**Abstract.** The paper is devoted to the problem of calculation of the spatial vibrations of elementary pipeline sections that appear under shock pulse. The equation of the motion and the boundary conditions describing the pipe deflections are presented as the forth-order partial differential equation with the Dirichlet and Neumann boundary conditions. To solve the problem, the numerical method based on finite-difference equations and a regularization technique is proposed. In order to evaluate the efficiency of the proposed method, the computational experiments were carried out. The results demonstrated sufficient accuracy of numerical solutions and confirm the sensitivity of method to changes in system.

**Keywords:** Finite-difference method · Spatial vibrations of pipe · Regularization · Computational scheme

## 1 Introduction

The problem of calculation of the spatial vibrations in pipes arise in a wide variety of engineering applications such as, acoustic diagnostics, flow measurement, material science and control processes of operating engines. The mathematical foundations of the vibration processes have been formulated by Timoshenko et al. in [1] and Pfeiffer in [2]. These approaches provided the basis for development of the data processing methods. Among the well-known researches devoted to this problem, we should highlight the works of Paidoussis [3], Akulenko [4], Kutin [5]. In regard with the application of theoretical results on actual practice, we also mention the works [6] and [7].

In recent years, the results of mathematical studies have been widely applied for the data processing. The use of mathematical tools allows to solve the engineering problems at a qualitatively higher level. The studies [8–13], provided examples of such research.

In traditional approaches, it is commonly assumed that the general solution of the vibration equation is harmonic in time and it is written in complex form with uncertain coefficients. Next, the required parameters are determined via using the results of mathematical studies. The use of mathematical tools for the

© Springer Nature Switzerland AG 2019
I. Dimov et al. (Eds.): FDM 2018, LNCS 11386, pp. 630–637, 2019.
https://doi.org/10.1007/978-3-030-11539-5_74

development of methods for data processing allows to solve the applied problems at a qualitatively higher level.

The essential feature of the existing approaches is that the application of these methods is only possible under specific assumptions. However, it is difficult to employ them when the initial data are measured with an error or the major characteristics of the process is differ from the generally accepted assumptions.

In this paper, it is proposed the method for calculation of the spatial tube vibrations appearing under the action of the shock pulse. The computational procedure involves numerical solving a forth-order partial differential equation with the Dirichlet and Neumann boundary conditions that characterized the dependence of the vibrations on measured initial data. This method is based on the finite-difference explicit scheme and regularization technique. Using the regularization procedure ensures the computational stability. The proposed approach can be a basis for the data processing of measurements, where the uncertainty of parameters is eliminated. The reliability of the method was verified by the computational experiments. The computational results for the some test data are presented in this paper.

## 2  Statement of the Problem

The mathematical model of tube vibration is based on the Euler-beam theory. The basic assumptions of problem are as follows. The straight tube element of the length $\ell$ has mass per unit length $M_t$ and flexural rigidity $EI$. The element is clamped at both ends and is modelled as the Euler beam. The tube conveys the incompressible, steady fluid flow of mass per unit length $M_f$ with axial velocity $V$. Due to the frictional losses, the tube is loaded with a compressive axial force $P$ with the longitudinal tension $\Gamma$. The force deviation, and the tension deviation per length have the values that can be neglected. The action of an shock pulse is introduced by the initial condition. We use the designations: $t$–time, $\xi$–distance from left end to the tube point, $\xi \in [0, \ell]$. Let introduce the dimensionless variables and parameters using the formulas:

$$x = \frac{\xi}{\ell}, \quad \tau = \frac{t}{\ell^2}\sqrt{\frac{EI}{M_t}}, \quad \beta = \frac{M_f}{M_t},$$

$$\upsilon = V\ell\sqrt{\frac{M_t}{EI}}, \quad \Pi = \frac{\Gamma - P\ell^2}{EI}, \quad \chi = \frac{C\ell^2}{\sqrt{M_t EI}}$$

where $C$ is damping constant. Presume that $\tau \in [0, T]$. Denote $Q_T = (0, 1) \times (0, T)$ and $\overline{Q_T} = [0, 1] \times [0, T]$. According the approaches proposed in [3,5] the small lateral motion equation is written as:

$$\frac{\partial^4 u}{\partial x^4} + \left(\beta \upsilon^2 - \Pi\right)\frac{\partial^2 u}{\partial x^2} + 2\beta\upsilon\frac{\partial^2 u}{\partial x \partial \tau} + (1 + \beta)\frac{\partial^2 u}{\partial \tau^2} + \chi\frac{\partial u}{\partial \tau} = 0, \quad (x, \tau) \in Q_T, \quad (1)$$

Based on the assumptions, we obtain following boundary and initial conditions:

$$u(0,\tau) = u(1,\tau) = 0, \quad \frac{\partial u}{\partial x}\mid_{x=0} = \frac{\partial u}{\partial x}\mid_{x=1} = 0, \quad \tau \in [0,T]. \qquad (2)$$

$$u(x,0) = 0, \quad x \in [0,1]. \qquad (3)$$

We represent the action of an shock pulse as follows:

$$\frac{\partial u}{\partial \tau}\mid_{\tau=0} = g(x), \quad x \in [0,1]. \qquad (4)$$

In this problem, it is required to determine the lateral deflection of the tube. In other words, we must find the function $u(x,\tau)$ satisfying (1)–(4), and then compare the values of function $u(x,\tau)$ at characteristic points $x1, x2, x3$.

The existence of solution $u(x,\tau)$ to the problem (1)–(4) and its uniqueness for $\beta(x,\tau), \upsilon(x,\tau), \Pi(x,\tau) \in H^{\mu,\mu/6}(\overline{Q_T})$, with $\mu \in (0,1)$ and $g(x) \in C^2([0,1])$ have been proved in [8,11]. However, instead of exact values of the function $g(x)$ we are given some approximations $g_\delta$ and an error level $\delta > 0$ such that $\|g_\delta - g\| \leq \delta$.

The physical meaning of the problem allows to assume that $u(x,\tau)$ satisfying to (1)–(4) is smooth function and there exist the constants $\Phi, \gamma, R > 0$ such that the inequality

$$\max_{x \in Q}|u(x,\tau)| \leq \Phi e^{\gamma\tau}, \max\left\{\max_{(x,\tau)\in\overline{Q_T}}\left|\partial_\tau^3 u\right|, \max_{(x,\tau)\in\overline{Q_T}}\left|\partial_x^5 u\right|\right\} \leq R \qquad (5)$$

is valid for any $(x,\tau) \in \overline{Q_T}$. Then we can use the computational scheme for solving the problem (1)–(4) that is based on finite difference equation with regularization effect.

## 3    Numerical Regularization Method

We introduce the finite difference grid $G$ in $\overline{Q_T}$, such that:

$$G = \begin{cases} (x_i, \tau_j) : x = (i-1)h_x, \tau = (j-1)h_\tau, \\ h_x = 1/N_x; h_\tau = T/N_\tau; i = \overline{1, N_x+1}; j = \overline{1, N_\tau+1}; \end{cases}$$

and define the set $V_h = \{v(x_i, \tau_j)\} = \{v_{i,j,k}\}$ of discrete functions. Following Samarskii [14] we approximate the partial derivatives at the each point $(x_i, \tau_j)$ of $G$

$$v_{xxxx}^{i,j} = \frac{v_{i+3,j} - 4v_{i+2,j} - 4v_{i,j} + v_{i-1,j,k}}{h_x^4}, \quad v_{xx}^{i,j} = \frac{v_{i+1,j} - 2v_{i,j} + v_{i-1,j}}{h_x^2},$$

$$v_{\tau\tau}^{i,j} = \frac{v_{i,j+1} - 2v_{i,j} + v_{i,j-1}}{h_\tau^2}, \quad v_{x\tau}^{i,j} = \frac{v_{i+1,j+1} - v_{i,j+1} - v_{i+1,j} + v_{i,j}}{h_x h_\tau}. \qquad (6)$$

Next, using the finite difference analoges (6), we represent the differential equation (1) in equivalent form:

$$v_{xxxx}^{i,j} + \left(\beta v_{i,j}^2 - \Pi\right)v_{xx} + 2\beta v_{i,j}v_{x\tau}^{i,j} + (1+\beta)v_{\tau\tau}^{i,j} = 0. \qquad (7)$$

Setting $Av = v_{xxxx}^{i,j} + \left(\beta v_{i,j}^2 - \Pi\right) v_{xx} + 2\beta v_{i,j} v_{x\tau}^{i,j} + (1+\beta) v_{\tau\tau}^{i,j}$, we represent the Eq. (7) as $Av = 0$.

The basic idea of proposed method is that we calculate the values $v_{i+1,j+1}$ from the Eq. (7) and from the finite difference analogies of initial data. It is known that the computational scheme based on (7) is unsteady. To improve it, we enhance (7) via additional stabilizing functional with the regularization parameter $\alpha$. Thus, the Eq. (7) is represented as

$$Av + \alpha v = 0, \tag{8}$$

Further, we choose the discretization steps and $\alpha$ according to some condition and calculate the values $v_{i+1,j+1}$ from (8). The similar approach has been used for solving the inverse heat conduction problem with unknown initial condition by the discrete regularization method, e.g. [15].

## 4   Computational Results

To exam the principle possibility of determining the spatial vibration function $u(x,t)$ via directly solving the problem (1)–(4) and to evaluate the reliability of the proposed method, the computational experiments were carried out.

In experiments, the numerical solutions $u_\delta(x,\tau)$ to (1)–(4) for various shock pulse functions $g(x)$ were calculated in the domain $\overline{Q_T} = [0,1] \times [0,300]$. The computational results for some numerical examples are presented in this paper.

The validation of mathematical model and the reliability of the computational scheme were verified by comparision of the numerical results $u_\delta(x,t)$ that are calculated at the characteristic points under various shock pulse functions. The characteristic points are $x1 = (0.25,\tau)$, $x1 = (0.5,\tau)$ and $x1 = (0.75,\tau)$, where $\tau \in [0,300]$.

The first stage of experiment involved solving the problem (1)–(4) with $g(x) = 0$ via the proposed computational scheme. The solutions to this problem at the characteristic points are depicted in Fig. 1, the solution in the domain $\overline{Q_T}$ is presented in Fig. 2. The graphs show the sufficient accuracy of the proposed method.

Next, the problem (1)–(4) with various shock pulse functions $g(x)$ were solved. The purpose of this experiment stage was verified the reliability of computation procedure. To test the stability of method, instead the exact function $g(x)$, we emploied the values $g_\delta(x)$ that are given with additional noise. The noised values $g_\delta(x)$ were simulated by the formulas:

$$g_\delta = g(x_i, \tau_j) + erp_{i,j}$$

where the values $erp_{i,j}$, are random variables evenly distributed in $[-\delta, \delta]$ .

Further, the discretization steps are chosen in keeping with the conditions that guarantee the stability of the method. Then, the numerical solution to the problem (1)–(4) was calculated via proposed computational procedure.

The numerical solutions obtained for $g(x) = \frac{\ell}{300} sin(\frac{2\pi x}{\ell})$ are presented in Figs. 3 and 4. The graphs of vibration function at the characteristic points are

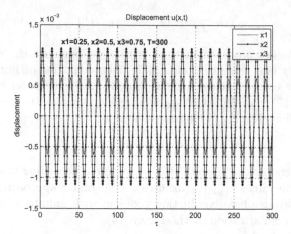

**Fig. 1.** The vibration function at the characteristic points. Results of numerical solving to the problem (1)–(4) with $g(x) = 0$.

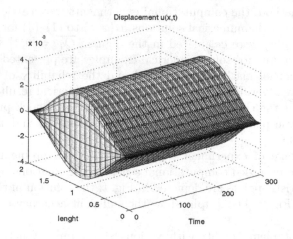

**Fig. 2.** The solution $u(x,t)$ to the problem (1)–(4) for $g(x) = 0$.

depicted in Fig. 3. The surface in Fig. 4 corresponds to the numerical solution $u_\delta(x,t)$ to problem of calculating the spatial tube vibrations, obtained in the domain $\overline{Q_T}$.

Similar design are used for representation of numerical solution to the problem (1)–(4) with the shock pulse function $g(x) = \frac{\ell}{300} sin(\frac{\pi x}{\ell})$. The vibration function at the characteristic points is depicted in Figs. 5 and 6 show the results obtained in the domain $\overline{Q_T}$.

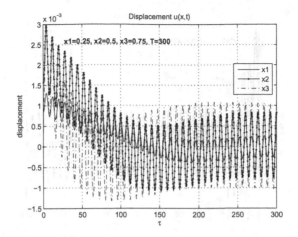

**Fig. 3.** The vibration function at the characteristic points. Results of numerical solving to the problem (1)–(4) with $g(x) = \frac{\ell}{300} sin(\frac{2\pi x}{\ell})$.

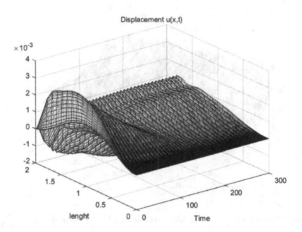

**Fig. 4.** The solution $u(x, t)$ to the problem (1)–(4) for $g(x) = \frac{\ell}{300} sin(\frac{2\pi x}{\ell})$.

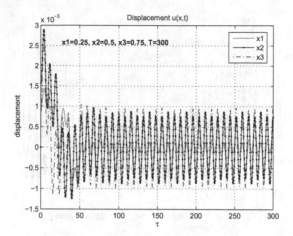

**Fig. 5.** The vibration function at the characteristic points. Results of the numerical solutions to the problem (1)–(4) with $g(x) = \frac{\ell}{300} sin(\frac{\pi x}{\ell})$.

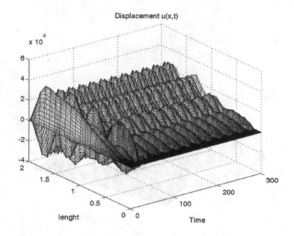

**Fig. 6.** The solution $u(x,t)$ to the problem (1)–(4) for $g(x) = \frac{\ell}{300} sin(\frac{\pi x}{\ell})$.

## 5    Conclusion

In this contribution, the problem of calculating the spatial vibrations of elementary sections of pipe is considered. The method for directly determining the spatial vibrations function from PDE equation is proposed. This method is based on using the regularization technique and the finite-difference equations. The computational scheme involves solving the problem at each time step. The application of explicit scheme allows to simplify the computational procedure. A significant feature of proposed approach is the capability to obtain the numerical solution in explicit form and then use the Fourier transform for determining the required parameters of system. This technique eliminates the uncertainty in the computational process. The computational algorithm is presented and validated by numerical examples. The experimental results confirm the capabilities of proposed computational scheme for solving the spatial pipe vibration problem and indicate the sufficient accuracy of the numerical solutions.

**Acknowledgments.** The work was supported by the Ministry of Education and Science of the Russian Federation within the framework of the basic part of the State task "Development, research and implementation of data processing algorithms for dynamic measurements of spatially distributed objects", Terms of Reference 8.9692.2017/8.9 from 17.02.2017.

# References

1. Timoshenko, S., Young, D.H.: Vibration Problems in Engineering. Van Nostrand, New Jersey (1955)
2. Pfeiffer, P.: Vibrations, of Elastic Bodies. The All-the-Union Scientific and Technical Publishing, Moscow (1934). (Russian; original edition: Handbuch Der Physik, band VI. Mechanik Der Elastischen Korper. Redigiert von R. Grammel. Berlin (1928))
3. Paidoussis, M.P.: Fluid-Structure Interactions: Slender Structures and Axial Flow, vol. 2. Elsevier Academic Press, London (2003)
4. Nesterov, S.V., Akulenko, L.D., Korovina, L.I.: Transverse oscillations of a pipeline with a uniformly moving fluid. Rep. Acad. Sci. Mech. **427**(6), 781–784 (2009)
5. Kutin, J., Bajsic, I.: An analytical estimation of the Coriolis Meter's characteristics based on modal superposition. Flow Meas. Instrum. **12**, 345–351 (2002)
6. Khakimov, A.G., Shakiryanov, M.M.: Spatial oscillations of the pipeline under the influence of variable internal pressure. Bull. UFA State Aviat. Tech. Univ. **14**(37), 30–35 (2014)
7. Mironov, M.A., Pyatakov, P.A., Andreev, A.A.: Forced flexural vibrations of a tube with a liquid flow. Acoust. J. **56**(5), 684–692 (2010)
8. Baderko, E.A.: A method of potential theory in boundary value problems for $2m$-parabolic equations in a semibounded domain with a nonsmooth lateral boundary. Differ. Equ. **24**(1), 1–5 (1988)
9. Prokudina, L.A.: Nonlinear evolution of perturbations in a thin fluid layer during wave formation. J. Exp. Theor. Phys. **118**(3), 480–488 (2014)
10. Samarskii, A.A., Vabishchevich, P.N.: Numerical Methods for Solving Inverse Problems of Mathematical Physics. Walter de Gruyter, Germany (2007)
11. Chernyatin, V.A.: A Substantiation of the Fourier Method in the Mixed Problem for Partial Differential Equations. Moscow State University, Moscow (1991)
12. Kabanikhin, S.I.: Inverse and Ill-Posed Problems: Theory and Applications. Walter de Gruyter, Germany (2011)
13. Tikhonov, A.N., Goncharsky, A.V., Stepanov, V.V., Yagola, A.G.: Numerical Methods for the Solution of Ill-Posed Problems. Kluwer, London (1995)
14. Samarskii, A.A.: The Theory of Difference Schemes. Marcel Dekker Inc., New York (2001)
15. Yaparova, N.: Numerical method for solving an inverse boundary problem with unknown initial conditions for parabolic PDE using discrete regularization. In: Dimov, I., et al. (eds.) NAA 2016. LNCS, vol. 10187, pp. 752–759. Springer, Heidelberg (2017). https://doi.org/10.1007/978-3-319-57099-0_87

# Finite Difference Computation of a Stochastic Aquaculture Problem Under Incomplete Information

Hidekazu Yoshioka[1]([✉]), Kentaro Tsugihashi[2], and Yuta Yaegashi[3,4]

[1] Faculty of Life and Environmental Science, Shimane University,
Nishikawatsu-cho 1060, Matsue 690-8504, Japan
yoshih@life.shimane-u.ac.jp
[2] Graduate School of Life and Environmental Science, Shimane University,
Nishikawatsu-cho 1060, Matsue 690-8504, Japan
a179806@matsu.shimane-u.ac.jp
[3] Graduate School of Agriculture, Kyoto University, Kitashirakawa-oiwake-cho,
Sakyo-ku, Kyoto 606-8502, Japan
yaegashi.yuta.54s@st.kyoto-u.ac.jp
[4] Research Fellow of Japan Society for the Promotion of Science, Tokyo, Japan

**Abstract.** Population dynamics of fishery resource is often uncertain for its manager. In most cases, especially the body growth rate is not easy to know a priori. In this paper, for approaching the issue above, a stochastic aquaculture problem under incomplete information is formulated and its associated Hamilton–Jacobi–Bellman (HJB) equation governing the value function is derived. A finite difference scheme for discretization of the HJB equation with an exponential time-stepping is then presented. The HJB equation is numerically solved with realistic parameter values for aquacultured *Plecoglossus altivelis* (*P. altivelis*, Ayu), a major inland fishery resource in Japan. The scheme naturally handles the boundary conditions and computes numerical solutions that comply with theoretical upper- and lower-bounds of the value function.

## 1  Introduction

There is no doubt that fish have long been an indispensable protein source for human. Aquaculture is one of the most widely-spread ways for farming fish around the world (Tidwell and Allan 2001). Exploring cost-effective management policies for aquacultured fish has been a hot topic in fisheries engineering and science (Pomeroy et al. 2008).

Optimization methods have been effective mathematical tools for tackling the issue above. Finding a cost-effective aquaculture policy can be formulated as a stochastic control problem (Yoshioka and Yaegashi 2017). In this framework, the problem is ultimately reduced to a terminal and boundary value problem of a Hamilton–Jacobi–Bellman (HJB) equation, a nonlinear degenerate-parabolic equation. In real problems, the body growth rate of fish is often uncertain for

© Springer Nature Switzerland AG 2019
I. Dimov et al. (Eds.): FDM 2018, LNCS 11386, pp. 638–645, 2019.
https://doi.org/10.1007/978-3-030-11539-5_75

its manager, and it has to be identified through observations during farming the fish. The problem can then be formulated as a stochastic control problem under incomplete information (Liptser and Shiryaev 2013).

This paper focuses on a simplified stochastic control problem for finding the most cost-effective aquaculture under incomplete information. The corresponding HJB equation has the diffusion and drift coefficients degenerated on the boundaries. A finite difference scheme based on the central differencing as much as possible strategy (Wang and Forsyth 2013) and an exponential time stepping is presented for discretization of the HJB equation. Accuracy and stability of the scheme is then examined against the HJB equation with realistic parameter values for aquacultured *Plecoglossus altivelis* (*P. altivelis*, Ayu), a major inland fishery resource in Japan. The resulting scheme generates non-oscillatory numerical solutions with first-order convergence rate. The computed numerical solutions comply with theoretical upper- and lower-bounds of the value function.

## 2    Mathematical Model

### 2.1    Stochastic Control Problem

The problem considered here is a stochastic control counterpart of the optimization problem of Yoshioka and Yaegashi (2017). A population of fish is aquacultured in an artificial pool, a closed habitat. Let $t \in [0, T]$ be the time and $T > 0$ the fixed terminal time of the aquaculture. The fish is introduced into the pool at $t = 0$. The total number of the population and the representative body weight of the fish at the time $t$ are denoted as $N_t$ and $W_t$, respectively. The dynamics of $N_t$ and $W_t$ are assumed to be decoupled and the density-dependence of the growth and mortality is assumed to be negligible. These assumptions are often justified in real aquaculture systems.

The true (intrinsic) body growth rate of the fish is denoted as $\mu > 0$, which is an unknown for the manager of the fish, the decision-maker. His/her estimate of $\mu$ at the time $t$ is denoted as $\hat{\mu}_t$. This paper assumes the simplest condition where $\mu$ is either high $\mu_H$ or low $\mu_L$ with $\mu_H > \mu_L > 0$. More complicated situations can also be dealt with based on the mathematical framework and the numerical scheme presented in this paper, which will be addressed elsewhere. The decision-maker dynamically updates $\hat{\mu}_t$ based on the observed $N_t$ and $W_t$ up to the time $t$.

The Itô's SDE that governs $N_t$ is set as (Yoshioka and Yaegashi 2017)

$$dN_t = -(R + c_t) N_t dt, \quad t > 0, \quad N_0 \geq 0 \tag{1}$$

and that governs $W_t$ as (Lungu and Øksendal 1997)

$$dW_t = W_t \left(1 - K^{-1} W_t\right) (\mu dt + \sigma dB_t), \quad t > 0, \quad 0 \leq W_0 \leq K. \tag{2}$$

Here, $R > 0$ is the mortality rate, $c_t$ ($0 \leq c_t \leq c_{\max}$) is the harvesting rate serving as a control variable with the maximum value $c_{\max} > 0$, $B_t$ is the 1-D standard Brownian motion, $K > 0$ is the maximum body weight, and $\sigma > 0$ represents the magnitude of stochasticity involved in the dynamics of $W_t$. The SDE (2) is nonlinear and its coefficients are not Lipschitz continuous, but has a unique strong solution such that $0 \leq W_t \leq K$ for $t \geq 0$ a.s.. Its solution properties are analyzed in Lungu and Øksendal (1997) and Lv et al. (2015).

The left-hand side of the SDE (2) is observable for the decision-maker, while its right-hand side is not since it involves the unknown $\mu$. The natural filtration generated by the process $W_t$ is denoted as $\mathscr{G}_t$. The conditional probability $Y_t$ ($t \geq 0$) is defined as $Y_t = \Pr(\hat{\mu}_t = \mu_H | \mathscr{G}_t)$. Clearly, $1 - Y_t = \Pr(\hat{\mu}_t = \mu_L | \mathscr{G}_t)$. Let $\hat{\mu}(Y_t) = \mu_H Y_t + \mu_L (1 - Y_t)$. The SDE (2) is rewritten in the context of partially-observable problems (Theorem 9.1 of Liptser and Shiryaev 2013) as

$$dX_t = g(X_t)\left(\hat{\mu}(Y_t)dt + \sigma d\hat{B}_t\right), \quad X_t = K^{-1}W_t, \quad t > 0, \quad 0 \leq X_0 \leq 1 \quad (3)$$

with $g(x) = x(1-x)$ for $0 \leq x \leq 1$. The dynamics of $Y_t$ is given as

$$dY_t = \sigma^{-1}\Delta\mu g(Y_t)\hat{B}_t, \quad t > 0, \quad 0 \leq Y_0 \leq 1. \quad (4)$$

Here, $\Delta\mu = \mu_H - \mu_L$ and $\hat{B}_t$ is a Brownian motion called innovation process. $\hat{B}_t$ is adapted to $\mathscr{G}_t$. As for the SDE (2), the system (3)–(4) admits a unique strong solution $(X_t, Y_t)$, which can be proven with the help of Lv et al. (2015). Hereafter, $c_t$ is chosen so that it is adapted to $\mathscr{G}_t$ and the system (1), (3), (4) admits a unique strong solution $(N_t, X_t, Y_t)$.

## 2.2 Hamilton-Jacobi-Bellman Equation

The performance index to be maximized by the decision-maker through optimization of $c_t$ is set as (Yoshioka and Yaegashi 2017)

$$J(t, n, x, y; c) = \mathrm{E}\left[\int_t^T \alpha c_s N_s X_s ds - \int_t^T \beta p N_s X_s ds\right]. \quad (5)$$

Here, E represents the expectation conditioned on $\mathscr{G}_t$ and the notations $N_t = n$, $X_t = x$, and $Y_t = y$ have been used. $\alpha > 0$ and $\beta > 0$ are weight constants, and $p > 0$ is the farming cost per unit time. The performance index $J$ simply measures the expectation of the profit by harvesting minus the cost of farming.

The value function $u = u(t, n, x, y)$ is the maximum of $J$:

$$u(t, n, x, y) = \sup_c J(t, n, x, y; c). \quad (6)$$

The optimal $c_t$ that achieves the maximization in (6) is denoted as $c_t^*$. Finding this $c_t^*$ is the goal of the present stochastic control problem.

The linearity of the SDE (1) with respect to $N_t$ allows us to decompose $u$ as $u = n\Psi$ with some function $\Psi = \Psi(t, x, y)$. The dynamic programming principle then leads to the (reduced) HJB equation that governs $\Psi$ as

$$
\begin{aligned}
&\frac{\partial \Psi}{\partial t} - R\Psi + \hat{\mu}(y)g(x)\frac{\partial \Psi}{\partial x} + \frac{\sigma^2}{2}g^2(x)\frac{\partial^2 \Psi}{\partial x^2} + \Delta\mu g(x)g(y)\frac{\partial^2 \Psi}{\partial x \partial y} \\
&+ \frac{(\Delta\mu)^2 g^2(y)}{2\sigma^2}\frac{\partial^2 \Psi}{\partial y^2} - \beta px - \min_{0 \leq c \leq c_{\max}}(c(\Psi - \alpha x)) = 0
\end{aligned}
\quad (7)
$$

for $0 \leq t < T$, $0 \leq x \leq 1$, and $0 \leq y \leq 1$ subject to the terminal condition $\Psi = 0$ at $t = T$. The boundary conditions along $x = 0, 1$ and $y = 0, 1$ are directly specified by considering (7) on these boundaries. Hereafter, the condition $\alpha c_{\max} - \beta p > 0$ is assumed, meaning that there is a possibility to make the profit be larger than the cost by choosing some $c_t$. This is a quite natural assumption. Through the HJB equation (7), the optimal $c_t^*$ is found as the maximizer of its last term, which is expressed with an abuse of notation as $c^*(t, X_t, Y_t)$. Since this $c^*$ does not explicitly depend on the total population $N_t$, the optimal harvesting strategy is based on the growth and its uncertainty.

The HJB equation (7) cannot be solved analytically, but upper- and lower-bounds of $\Psi$ are derived from (5) and (6). Observe that the expectation of $W_t$ with $0 \leq W_0 \leq K$ is increasing with respect to $t$. This is proven at least formally using the Fokker-Planck equation that governs the probability density function of $W_t$. With the help of comparison arguments of the SDE (3) and techniques used in Yoshioka and Yaegashi (2017), we have the following estimates.

**Proposition 1.**

$$\Psi(t, x, y) \geq \frac{\alpha c_{\max} - \beta p}{R + c_{\max}} \left(1 - e^{-(R + c_{\max})(T - t)}\right) x, \ 0 \leq x, y \leq 1, \ 0 \leq t \leq T. \quad (8)$$

**Proposition 2.**

$$\Psi(t, x, y) \leq \frac{\alpha c_{\max} - \beta p}{R + c_{\max}} \left(1 - e^{-(R + c_{\max})(T - t)}\right), \ 0 \leq x, y \leq 1, \ 0 \leq t \leq T. \quad (9)$$

These results are used for accuracy verification of the finite difference scheme.

## 3    Finite Difference Scheme

### 3.1    Discretization

The HJB equation (7) is discretized with a finite difference scheme presented below. The spatio-temporal domain $[0, 1] \times [0, 1] \times [0, T]$ is discretized with vertices $(x_i, y_j, t_l)$ with $x_i = i I^{-1}$, $y_j = j J^{-1}$, $t_l = l L^{-1} T$ for $0 \leq i \leq I$, $0 \leq j \leq J$, $0 \leq l \leq L$, and fixed natural numbers $I, J, L$. The mesh resolution parameters are introduced as $\Delta x = I^{-1}$, $\Delta y = J^{-1}$, $\Delta t = L^{-1} T$. $\Psi$ approximated at $(x_i, y_j, t_l)$ is denoted as $\Psi_{i,j,l}$. The numerical discretization of the HJB equation (7) at each inner vertex $(x_i, y_j, t_l)$ $(0 < i < I, 0 < j < J, 0 \leq l < L)$ is presented below. The following notations are introduced for the sake of brevity:

$$\Delta_{1,0}\Psi = \frac{\Psi_{i+1,j,l} - \Psi_{i,j,l}}{\Delta x}, \ \Delta_{1,1}\Psi = \frac{\Psi_{i+1,j,l} - \Psi_{i-1,j,l}}{2\Delta x}, \ \Delta_{1,2}\Psi = \frac{\Psi_{i,j,l} - \Psi_{i-1,j,l}}{\Delta x},$$

$$(10)$$

$$\Delta_{2,0}\Psi = \frac{\Psi_{i+1,j,l} + \Psi_{i-1,j,l} - 2\Psi_{i,j,l}}{(\Delta x)^2}, \ \Delta_{2,1}\Psi = \frac{\Psi_{i,j+1,l} + \Psi_{i,j-1,l} - 2\Psi_{i,j,l}}{(\Delta y)^2},$$

$$(11)$$

$$\Delta_{2,2}\Psi = \frac{\Psi_{i+1,j+1,l} + \Psi_{i-1,j-1,l} - \Psi_{i-1,j+1,l} - \Psi_{i+1,j-1,l}}{4\Delta x \Delta y}. \tag{12}$$

Then, the finite difference scheme for the HJB equation (7) is presented as

$$\frac{R + c_{i,j,l}}{1 - \bar{e}}\left(\bar{e}\Psi_{i,j,l+1} - \Psi_{i,j,l}\right) - \beta p x_i + A\Delta_{1,3}\Psi \\ + B\Delta_{2,0}\Psi + C\Delta_{2,1}\Psi + D\Delta_{2,2}\Psi = 0 \qquad , \quad \bar{e} = e^{-(R+c_{i,j,l})\Delta t}, \tag{13}$$

where

$$A = \hat{\mu}\left(y_j\right)g\left(x_i\right), \quad c_{i,j,l} = \frac{1 + \operatorname{sgn}\left(\Psi_{i,j,l+1}\right)}{2}c_{\max}, \tag{14}$$

$$B = \frac{\sigma^2}{2}g^2\left(x_i\right), \quad C = \Delta\mu g\left(x_i\right)g\left(y_j\right), \quad D = \frac{\left(\Delta\mu\right)^2 g^2\left(y_j\right)}{2\sigma^2}, \tag{15}$$

$$\Delta_{1,3}\Psi = \begin{cases} \Delta_{1,0}\Psi \ (A\Delta x \geq 2B) \\ \Delta_{1,1}\Psi \ (-2B < A\Delta x < 2B) \ , \\ \Delta_{1,2}\Psi \ (-2B > A\Delta x) \end{cases} \tag{16}$$

the last equation being the discretization based on the central differencing as much as possible strategy (Wang and Forsyth 2012). The boundary conditions are implemented using the same discretization with (13) and $x_i = 0, 1$, $y_j = 0, 1$. The exponential time differencing is a key to generate numerical solutions complying with the theoretical upper- and lower-bounds. The present scheme is consistent for $C^2$-class classical solutions and satisfies the positive coefficient conditions if the term multiplied by $C$ is dropped. The numerical solution at each vertex is computed in the time-backward manner with the terminal condition $\Psi_{i,j,L} = 0$. The optimal control $c_t^*$, which gives the desired harvesting policy, at each vertex is then identified with $c_{i,j,l}$. Therefore, for each $(i, j, l)$, it is optimal to harvest when $c_{i,j,l} = c_{\max}$, while it is not when $c_{i,j,l} = 0$.

### 3.2 Computational Conditions

The model parameters are identified for aquacultured *P. altivelis*, which is a major inland fishery resource in Japan. With the help of Hii River Fisheries Cooperatives in Japan, the authors collected the body growth data of the fish in each year from 2012 to 2016. The details of the data are not presented here, but are partly found in Yoshioka and Yaegashi (2017). The life history of the aquacultured *P. altivelis* is also found in the literature above and the references therein. Based on these data, the parameter values are specified as $T = 120$ (day), $\mu_H = 0.050$ (1/day), $\mu_L = 0.015$ (1/day), $R = 0.005$ (1/day), $\sigma = 0.2$ (1/day$^{1/2}$), $c_{\max} = 0.030$ (1/day), $p = 0.015$ (1/day), $\alpha = 1.0$, and $\beta = 0.2$.

The spatio-temporal domain is uniformly discretized into $t$, $x$, and $y$ directions with $I = J$. The numerical solutions are computed for $I = J = 64 \cdot 2^k$ and $L = 125 \cdot 2^k$ ($k = 0, 1, 2, 3, 4$). The numerical solution with the finest resolution ($k = 4$) is regarded as the reference solution to examine convergence of the numerical solutions with $k = 0, 1, 2, 3$.

## 3.3   Computational Results

Table 1 shows the $l^2$ and $l^\infty$ errors and the convergence rates between the numerical solutions and the reference solution. The results imply that the present scheme has the first-order computational accuracy in both the $l^2$ and $l^\infty$ senses. Figure 1 shows the numerical solution at $T = 0$ and the area where $c^* = c_{\max}$. The numerical solutions do not have spurious oscillations and rigorously comply with the theoretical upper- and lower-bounds of $\Psi$, which is owing to the use of the exponential time stepping. In fact, employing the conventional fully-implicit Euler method for temporal discretization in the present scheme generate numerical solutions that violates the upper-bound near and along $x = 1$. The violation has the magnitude of $O(\Delta t)$.

**Table 1.** $l^2$ and $l^\infty$ errors and the convergence rates between the numerical solutions and the reference solution. "Conv" represents the convergence rate.

| $I = J$ | $L$ | $l^2$ error | $l^\infty$ error | $l^2$ Conv | $l^\infty$ Conv |
|---|---|---|---|---|---|
| 64 | 125 | 8.84E−4 | 1.50E−2 | 1.22E+0 | 7.26E−1 |
| 128 | 250 | 3.78E−4 | 9.08E−3 | 1.30E+0 | 7.90E−1 |
| 256 | 500 | 1.54E−4 | 5.25E−3 | 1.47E+0 | 8.98E−1 |
| 512 | 1000 | 5.56E−5 | 2.82E−3 | | |

On the mixed partial derivative term, employing the biased discretization presented in Chap. 10 of Øksendal and Sulem (2005) instead of the central one in (12) leads to similar computational results. Consequently, for the present HJB equation, no significant difference has been found between the performances of the two discretization methods for the mixed partial derivative term.

Computational results for a variety of parameter values suggest that $c^*$ as a function of $(t, x, y)$ satisfies

$$c^*(t, x, y) = c_{\max} \quad \text{for} \quad x > b(t, y), \quad 0 \le t < T, \quad 0 \le x, y \le 1, \quad (17)$$

where $b = b(t, y)$ is some non-negative, increasing with respect to $t$ and $y$, and continuous function such that $b = 0$ for large $t (\le T)$. Figure 2 shows the computed free boundary $b$ for $0 < t < T$ and $0 < y < 1$. The function $b$ serves as a threshold that theoretically gives whether the fish should be harvested $(X_t > b(t, Y_t))$ or not $(X_t < b(t, Y_t))$. Both harvesting and not harvesting are optimal when $X_t = b(t, Y_t)$. Although not presented here, all the numerical solutions had the free boundaries of the form (17), implying robustness of the present finite difference scheme on discretization of the present HJB equation.

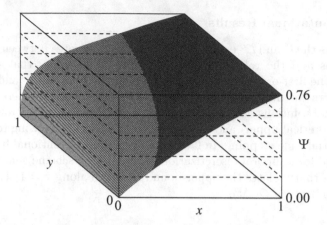

**Fig. 1.** Computed value function $\Psi$ with the resolution $k = 1$ ($I = J = 128$, $L = 250$). $c^* = c_{\max}$ in the colored area. (Color figure online)

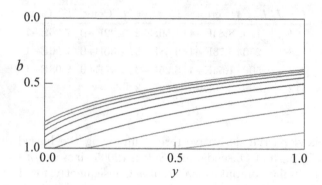

**Fig. 2.** Computed free boundary $b = b(t, y)$ with the resolution $k = 3$ ($I = J = 512$, $L = 1000$). The plotted $b(t, y)$ are for $t = 0.1T, 0.2T, ...0.9T, 1.0T$ from the bottom to the top.

## 4    Conclusions

The HJB equation for finding the stochastic aquaculture policy was formulated and a specialized finite difference scheme for its discretization was presented. The scheme is based on the central differencing as much as possible strategy and an exponential time stepping. Numerical experiments demonstrated that the scheme is first-order accurate and computes numerical solutions complying with the theoretical upper- and lower-bounds of the value function.

There exist a number of theoretical and numerical issues. In reality, the filtering process, namely the observation of the body growth rate, is costly. Incorporating the observation cost into the present model can be achieved through adding the cost of the observation to the performance index as in (Harrison and Sunar 2015). The present modeling framework can also be applied to management problems of the non-renewable released fishes (Yoshioka and Yaegashi 2017). From a numerical viewpoint, it is desirable to employ a discretization that complies with the positive coefficient condition (Forsyth and Vetzal 2012). These topics are currently in progress.

**Acknowledgments.** This work was supported by The River Foundation under grant The River Fund No. 285311020, The Japan Society for the Promotion Science under grant KAKENHI No. 17K15345 and No. 17J09125, and Water Resources Environment Center under grant The WEC Applied Ecology Research Grant No. 2016-02. The authors thank the officers of Hii River Fisheries Cooperatives for providing valuable data and comments for this research.

# References

Tidwell, J.H., Allan, G.L.: Fish as food: aquaculture's contribution: ecological and economic impacts and contributions of fish farming and capture fisheries. EMBO Rep. **2**, 958–963 (2001)

Pomeroy, R., Bravo-Ureta, B.E., Solis, D., Johnston, R.J.: Bioeconomic modelling and salmon aquaculture: an overview of the literature. Int. J. Environ. Poll. **33**, 485–500 (2008)

Yoshioka, H., Yaegashi, Y.: Optimization model to start harvesting in stochastic aquaculture system. Appl. Stoch. Model. Bus. **33**, 476–493 (2017)

Liptser, R.S., Shiryaev, A.N.: Statistics of Random Processes: I. General Theory. Springer, Heidelberg (2013). https://doi.org/10.1007/978-3-662-13043-8

Wang, J., Forsyth, P.A.: Maximal use of central differencing for Hamilton-Jacobi-Bellman PDEs in finance. SIAM J. Numer. Anal. **46**, 1580–1601 (2008)

Lungu, E.M., Øksendal, B.: Optimal harvesting from a population in a stochastic crowded environment. Math. Biosci. **145**, 47–75 (1997)

Lv, J., Wang, K., Jiao, J.: Stability of stochastic Richards growth model. Appl. Math. Model. **39**, 4821–4827 (2015)

Harrison, J.M., Sunar, N.: Investment timing with incomplete information and multiple means of learning. Oper. Res. **63**, 442–457 (2015)

Øksendal, B.K., Sulem, A.: Applied Stochastic Control of Jump Diffusions. Springer, Heidelberg (2005). https://doi.org/10.1007/978-3-540-69826-5

Yoshioka, H., Yaegashi, Y.: Mathematical analysis for management of released fish. Optim. Contr. Appl. Meth. **39**, 1141–1146 (2018)

Forsyth, P.A., Vetzal, K.R.: Numerical methods for nonlinear PDEs in finance. In: Duan, J.C., Härdle, W., Gentle, J. (eds.) Handbook of Computational Finance. Springer Handbooks of Computational Statistics, pp. 503–528. Springer, Heidelberg (2012). https://doi.org/10.1007/978-3-642-17254-0_18

# Asymptotic and Numerical Analysis of Dynamics in a Generalized Free-Interfacial Combustion Model

Jun Yu[1(✉)], Kewang Chen[1], and Laura K. Gross[2]

[1] Department of Mathematics and Statistics, University of Vermont,
Burlington, VT 05401, USA
Jun.Yu@uvm.edu
[2] Departments of Mathematics and Computer Science, Bridgewater State University,
Bridgewater, MA 02325, USA

**Abstract.** Dynamics of temperature distribution and interfacial front propagation in a generalized solid combustion model are studied through both asymptotic and numerical analyses. For asymptotic analysis, we focus on the weakly nonlinear case where a small perturbation of a neutrally stable parameter is taken so that the linearized problem is marginally unstable. Multiple scale expansion method is used to obtain an asymptotic solution for large time by modulating the most linearly unstable mode. On the other hand, we integrate numerically the exact problem by the Crank-Nicolson method. Since the numerical solutions are very sensitive to the derivative interfacial jump condition, we integrate the partial differential equation to obtain an integral-differential equation as an alternative condition. The result system of nonlinear algebraic equations is then solved by the Newton's method, taking advantage of the sparse structure of the Jacobian matrix. Finally, we show that our asymptotic solution captures the marginally unstable behaviors of the solution for a range of model parameters.

## 1 Mathematical Model

In gasless combustion, chemical reactions convert a solid fuel directly into solid products. For example, in self-propagating high-temperature synthesis (SHS), a flame wave advances through powdered ingredients, leaving high-quality ceramic materials or metallic alloys in its wake. (See, for instance, [1–3].) The interplay between heat generation and heat diffusion in the medium controls the development of the flame front. See [4] for a review of models and experiments. Numerical simulations have also captured these dynamics as in [5,7–9].

Here, a generalized free-interfacial combustion model described in [5,6] is considered. We study the temperature distribution $u(x,t)$ in one spatial dimension and the interface position $\Gamma(t) = \{x | x = f(t)\}$ that satisfy the appropriately nondimensionalized free-boundary problem

$$\frac{\partial u}{\partial t} = a_p \frac{\partial^2 u}{\partial x^2} \text{ on } x < f(t), \quad \frac{\partial u}{\partial t} = \frac{\partial^2 u}{\partial x^2} \text{ on } x > f(t), \tag{1}$$

© Springer Nature Switzerland AG 2019
I. Dimov et al. (Eds.): FDM 2018, LNCS 11386, pp. 646–653, 2019.
https://doi.org/10.1007/978-3-030-11539-5_76

$$\lim_{x \to -\infty} |u(x,t)| < \infty, \quad \lim_{x \to \infty} u(x,t) = 0, \tag{2}$$

$$V = \exp\left[\left(\frac{1}{\nu}\right)\frac{u|_\Gamma - 1}{\sigma + (1 - \sigma)\,u|_\Gamma}\right], \tag{3}$$

$$\left.\frac{\partial u}{\partial x}\right|_{\Gamma+} - a_p \left.\frac{\partial u}{\partial x}\right|_{\Gamma-} = -V, \tag{4}$$

where $V = f'(t)$, $a_p$ is the ratio of diffusivities in the product and fresh mixtures

$$a_p = \frac{\kappa_b}{\kappa_u},$$

$\nu$ is the inverse Zel'dovich number, where

$$\frac{1}{\nu} = \frac{T_a}{2T_b}\,(1 - \sigma);$$

and

$$\sigma = \frac{T_0}{T_b},$$

where $T_0$ is the temperature of the fresh mixture, $T_a$ is the activation temperature and $T_b$ is the adiabatic temperature of the combustion products.

The problem (1)–(4) in a front-attached coordinate frame

$$\eta = x - f(t), \quad \tau = t$$

becomes

$$\frac{\partial u}{\partial \tau} = \frac{\partial^2 u}{\partial \eta^2} + \frac{df(\tau)}{d\tau}\frac{\partial u}{\partial \eta}, \quad \eta > 0, \quad \frac{\partial u}{\partial \tau} = a_p\frac{\partial^2 u}{\partial \eta^2} + \frac{df(\tau)}{d\tau}\frac{\partial u}{\partial \eta}, \quad \eta < 0, \tag{5}$$

$$u|_{\eta=0} = 1 + \nu K(V), \tag{6}$$

$$\left.\frac{\partial u}{\partial \eta}\right|_{\eta=0+} - a_p \left.\frac{\partial u}{\partial \eta}\right|_{\eta=0-} = -V, \tag{7}$$

$$\lim_{\eta \to -\infty} |u(\eta,\tau)| < \infty, \quad \lim_{\eta \to \infty} u(\eta,\tau) = 0, \tag{8}$$

where $K$ is the Arrhenius kinetics modified for gasless combustion as in [8]. The problem (5)–(8) has the following basic solution

$$u_{basic}(\eta,\tau) = \begin{cases} e^{-\eta} & \text{if } \eta \geq 0 \\ 1 & \text{if } \eta < 0 \end{cases}, \quad f_{basic}(\tau) = \tau. \tag{9}$$

Linear stability analysis of the problem (5)–(8) around the basic solution has been carried out in [4,6]. It was shown that critical value $\nu_c(a_p)$ of $\nu$ can be solved numerically, at which the traveling-wave solution loses stability for the range of the values of $a_p$.

## 2  Weakly Nonlinear Multiple-Scale Analysis

In this study we consider cases where parameters are only a small deviation from the neutral stability value and moving toward the unstable direction. In particular, we define a small parameter $\epsilon$ such that

$$\epsilon^2 = \nu_c(a_p) - \nu. \tag{10}$$

This choice of the parameter allows for the possibility of a Hopf bifurcation where the magnitude of the solution is on the order of the square root of the bifurcation parameter. Introduce the multiple time scales

$$t_0 = \tau, \quad t_1 = \epsilon\tau, \quad t_2 = \epsilon^2\tau, \tag{11}$$

where $\epsilon$ is given by (10) and $t_1$ and $t_2$ are called slow times.

Let us seek a solution to the problem (5)–(8) as a perturbation about the basic solution as follows:

$$u \sim u_{basic} + \epsilon w_1(\eta, t_0, t_1, t_2) + \epsilon^2 w_2(\eta, t_0, t_1, t_2) + \epsilon^3 w_3(\eta, t_0, t_1, t_2) + \cdots \tag{12}$$

$$f \sim f_{basic} + \epsilon\phi_1(t_0, t_1, t_2) + \epsilon^2\phi_2(t_0, t_1, t_2) + \epsilon^3\phi_3(t_0, t_1, t_2) + \cdots. \tag{13}$$

Proceeding as in [11] we find the following results at $O(\epsilon)$, $O(\epsilon^2)$ and $O(\epsilon^3)$:

For $O(\epsilon)$ problem, we have

$$w_1 = \begin{cases} A(t_1, t_2)e^{i\omega t_0}g_1(\eta) + \text{CC}, & \eta > 0, \\ A(t_1, t_2)e^{i\omega t_0}g_2(\eta) + \text{CC}, & \eta < 0, \end{cases} \tag{14}$$

$$\phi_1 = \left\{A(t_1, t_2)e^{i\omega t_0} + \text{CC}\right\} + B(t_1, t_2), \tag{15}$$

where $g_1(\eta)$, $g_2(\eta)$ are solutions to the related linear problem and CC represents the complex-conjugate terms. The pure imaginary eigenvalue, $\lambda = i\omega$, in the exponential function contributes to a linearly neutral stable part of solution. We look for a marginally unstable solution with the same structure by allowing $A$ and $B$ to vary slowly. In other words, we modulate the amplitudes of the linearly unstable modes by functions that depend only on $t_1$ and $t_2$. From now on we will use the notation $A$ and $B$ to mean $A(t_1, t_2)$ and $B(t_1, t_2)$.

For $O(\epsilon^2)$ problem, we find that the solvability conditions for this problem are

$$\frac{\partial A}{\partial t_1} = 0 \tag{16}$$

and

$$\frac{\partial B}{\partial t_1} = A\bar{A}r_0, \tag{17}$$

where

$$r_0 = -\omega^2\left[\frac{4}{\sqrt{1 - 4a_p i\omega} + \sqrt{1 + 4a_p i\omega}} + K''(1)\right]. \tag{18}$$

For $O(\epsilon^3)$ problem, we obtain the following Landau–Stuart equation for the complex amplitude $A(t_2)$

$$\frac{dA}{dt_2} = \kappa A + \beta A^2 \bar{A},$$  (19)

where

$$\kappa = \chi + \frac{S_1}{S_3},$$

$$\beta = \frac{S_2}{S_3},$$

$$S_1 = -i\omega\,\mathcal{U},$$

$$S_2 = \int_{0^+}^{\infty} P_1^+(\eta)\bar{h}_1(\eta)\,d\eta + \int_{-\infty}^{0^-} P_1^-\bar{h}_2(\eta)\,d\eta + F_1\mathcal{U},$$

$$S_3 = \int_{0^+}^{\infty} \left(g_1(\eta) + e^\eta\right)\bar{h}_1(\eta)\,d\eta + \int_{-\infty}^{0^-} g_2(\eta)\,d\eta - \nu_c\mathcal{U} + \mathcal{V},$$

and

$$\mathcal{U} = \bar{h}_1'(0^+) - a_p\bar{h}_2'(0^-), \quad \mathcal{V} = -\bar{h}_1(0^+).$$

See [11] for more detailed derivations of the above Landau–Stuart equation. Finally, to find $A(t_2)$, we integrate the ordinary differential equation (19) using a fourth-order Runge–Kutta method.

## 3   Numerical Method

In this part, we study numerically the exact problem as given by (5)–(8). We first introduce perturbation variables $u^*$ and $f^*$ defined by

$$u = \begin{cases} e^{-\eta} + \epsilon u^* & \text{if } \eta \geq 0 \\ 1 + \epsilon u^* & \text{if } \eta < 0. \end{cases}, \quad f = t + \epsilon f^*.$$  (20)

As was pointed out in [10], numerical solutions of (5)–(8) are very sensitive to the boundary condition (7). In order to obtain an alternative condition, we integrate (5) with respect to $\eta$ from $-\infty$ to $\infty$ and apply conditions (6)–(8), subsequently. The result equation is

$$\frac{d}{dt}\int_{-\infty}^{\infty} u^*d\eta = u_\eta^*\big|_{\infty} - a_p\,u_\eta^*\big|_{-\infty} - u^*\big|_{-\infty} - \epsilon f_t^*\,u^*\big|_{-\infty}.$$  (21)

We use condition (21) to replace (7). Then we adopt the Crank–Nicolson method for the discretization and approximate the integral on the left-hand side of condition (21) by a composite trapezoidal rule. The computation domain for $\eta$ is $[-10, 10]$ with $dt = d\eta = 0.025$. This produces a nonlinear system of m $(= 800)$ equations.

We solve the nonlinear system of equations using the Newton's method. At each iterating step of the Newton's method, we solve a linear system of equations with Jacobian matrix that has the following sparse structure:

$$
\begin{bmatrix}
\# & \# & \# & \# & \# & \cdots & \# & \# & \# & \# & \cdots & \# & \# \\
\# & \# & \# & 0 & 0 & \cdots & 0 & 0 & 0 & 0 & \cdots & 0 & 0 \\
\# & \# & \# & \# & 0 & \cdots & 0 & 0 & 0 & 0 & \cdots & 0 & 0 \\
\# & 0 & \# & \# & \# & \cdots & 0 & 0 & 0 & 0 & \cdots & 0 & 0 \\
\vdots & \vdots & & \ddots & \ddots & \ddots & \vdots & \vdots & \vdots & \vdots & \cdots & \vdots & \vdots \\
\# & 0 & 0 & \cdots & \# & \# & \# & 0 & 0 & 0 & \cdots & 0 & 0 \\
\# & 0 & 0 & \cdots & \cdots & \# & \# & \# & 0 & 0 & \cdots & 0 & 0 \\
\# & 0 & 0 & \cdots & \cdots & & 0 & \# & 0 & 0 & \cdots & 0 & 0 \\
\# & 0 & 0 & \cdots & \cdots & & 0 & \# & \# & \# & \cdots & 0 & 0 \\
\vdots & \vdots & \vdots & \cdots & \cdots & \cdots & & & \ddots & \ddots & \ddots & \vdots & \vdots \\
\# & 0 & 0 & \cdots & \cdots & \cdots & & \cdots & & \# & \# & \# & 0 \\
\# & 0 & 0 & \cdots & \cdots & \cdots & & \cdots & & 0 & \# & \# & \# \\
\# & 0 & 0 & \cdots & \cdots & \cdots & & \cdots & & 0 & 0 & \# & \#
\end{bmatrix}
\tag{22}
$$

where $\#$ denotes a nonzero element. We take advantage of this sparse structure and apply Gaussian elimination from bottom-up, eliminating the two nonzero elements at $(m-1, m)$ and $(1, m)$ positions first; then the two at $(m-2, m-1)$ and $(1, m-1)$ second; and so on. This reduces the matrix to a lower triangular form. Finally, the solution of the linear system of equations can be obtained by a forward substitution from top-down.

## 4   Comparison Between Asymptotics and Numerics

To start, take $\sigma = 0.46$ in the kinetics function (3). For the remainder of this paper we take the initial condition $A(0) = 0.1$.

For $a_p = 0.5$, Fig. 1 reveals that from $t = 0$ to about $t = 30$, the small front speed perturbation is linearly unstable, and its amplitude grows exponentially in time. As this amplitude becomes large, nonlinearity takes effect. At around $t = 30$, the front speed perturbation has reached steady oscillation. The two solutions are slightly out of phase, and the asymptotic solution oscillates symmetrically about the time axis, while the numerical solution has spiky peaks. The asymptotic solution accurately captures the period in both transient behavior for $t = 0$ to 30 and the long-time behavior after $t = 30$. This is an example in which the weakly nonlinear approach describes rather well the marginally unstable large-time behavior: A single modulated temporal mode captures the period of fluctuations (at frequency $\omega = 1.1406$ for $a_p = 0.5$) in velocity perturbation. Similar result for $a_p = 0.2$ is obtained in Fig. 2.

We then numerically calculated the velocity perturbation data on the time interval $50 < t < 100$, using the parameter values as in Fig. 1. As shown in Fig. 3, the discrete Fourier transform of the data reveals the dominance of one mode. However, the subsequent modes do contribute to the solution as well. The second

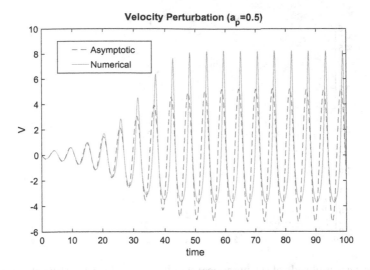

**Fig. 1.** Comparison between numerical (solid line) and asymptotic (dashed line) for Arrhenius kinetics: $a_p = 0.5$, $\sigma = 0.46$, $\epsilon = 0.1$, $A(0) = 0.1$, $\nu_c \approx 0.2703$.

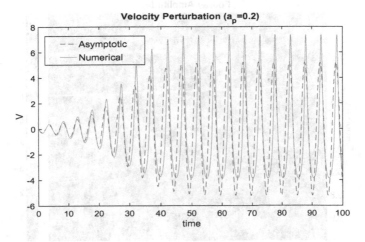

**Fig. 2.** Comparison between numerical (solid line) and asymptotic (dashed line) for Arrhenius kinetics: $a_p = 0.2$, $\sigma = 0.46$, $\epsilon = 0.1$, $A(0) = 0.1$, $\nu_c \approx 0.3041$.

spike in Fig. 3 is about $4/9$ the height of the first, and the third is less than $1/2$ the height of the second. Contributions of higher-order modes may explain some quantitative discrepancies between the numerical and asymptotic solutions in Fig. 1.

For $\epsilon = 0.1$ and $a_p = 0.5$, Fig. 4 summarizes the Fourier transformed velocity data for all physical values of $\sigma \in (0, 1)$. For each $\sigma$ value and each frequency, the color indicates the corresponding amplitude, with the red end of the spectrum standing for larger numbers than the violet end, as the legend to the right of the

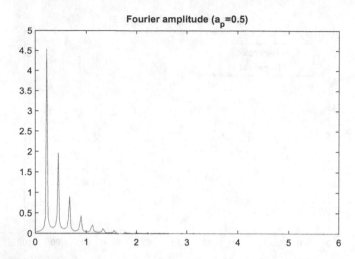

**Fig. 3.** Fourier amplitude of the numerical steady-state velocity perturbation: $a_p = 0.5$, $\sigma = 0.46$, $\epsilon = 0.1$, $A(0) = 0.1$, $50 < t < 100$, $\nu_c \approx 0.2703$.

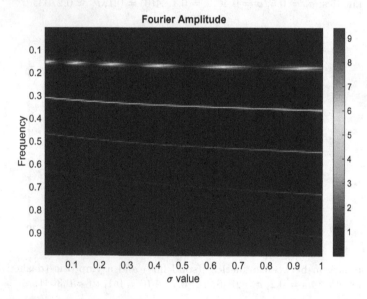

**Fig. 4.** Amplitudes corresponding to each frequency of the Fourier transformed velocity perturbation data for the Arrhenius kinetics parameter: $a_p = 0.5$, $\sigma \in (0, 1)$, $\epsilon = 0.1$, $A(0) = 0.1$, $\nu_c \approx 0.2703$ and $1000 < t < 1500$.

figure illustrates. As predicted by weakly nonlinear analysis, Fig. 4 shows that one mode dominates strongly for all physical values of $\sigma \in (0, 1)$ when $\nu = \nu_c - \epsilon^2$ is sufficiently close to the neutrally stable value.

# 5 Conclusion

We have done both asymptotic and numerical analyses for a generalized solid combustion model. For the chosen parameters (e.g., $\sigma = 4.6, \epsilon = 0.1$ and etc.), our asymptotic solution with a single modulated temporal mode describes well the marginally unstable solution behavior. The Crank-Nicolson finite difference method with the Newton's method, using the sparse structure of the Jacobian matrix, provides an efficient numerical solver for the nonlinear free boundary problem. For future studies, more modes of linear solution may be used in the asymptotic procedure in order to cover more variabilities in the nonlinear solution.

# References

1. Merzhanov, A.G.: SHS processes: combustion theory and practice. Arch. Combust. **1**, 2–48 (1981)
2. Munir, Z.A., Anselmi-Tamburini, U.: Self-propagating exothermic reactions: the synthesis of high-temperature materials by combustion. Mat. Sci. Rep. **3**, 277–365 (1989)
3. Varma, A., Rogachev, A.S., Mukasyan, A.S., Huang, S.: Combustion synthesis of advanced materials: principles and applications. Adv. Chem. Eng. **24**, 7–226 (1998)
4. Yang, Y., Gross, L.K., Yu, J.: Comparison study of dynamics in one-sided and two-sided solid-combustion models. SIAM J. Appl. Math. **70**(8), 3022–3038 (2010)
5. Matkowsky, B.J., Sivashinsky, G.I.: Propagation of a pulsating reaction front in solid fuel combustion. SIAM J. Appl. Math. **35**(93), 465–478 (1978)
6. Chen, K., Gross, L.K., Yu J., Yang, Y.: On a generalized free-interface model of solid combustion. J. Eng. Math. (submitted)
7. Bayliss, A., Matkowsky, B.J.: Two routes to chaos in condensed phase combustion. SIAM J. Appl. Math. **50**(2), 437–459 (1990)
8. Brailovsky, I., Sivashinsky, G.: Chaotic dynamics in solid fuel combustion. Physica D **65**, 191–198 (1993)
9. Frankel, M.L., Roytburd, V., Sivashinsky, G.: Complex dynamics generated by a sharp interface model of self-propagating high-temperature synthesis. Combust. Theory Model. **2**, 1–18 (1998)
10. Frankel, M.L., Roytburd, V.: Dynamical portrait of a model of thermal instability: cascades, chaos, reversed cascades and infinite period bifurcations, Internat. J. Bifur. Chaos Appl. Sci. Eng. **4**(3), 579–593 (1994)
11. Chen, K.: Mathematical analysis of some partial differential equations with applications. Ph.D. thesis, The University of Vermont (in progress)

# Analogue of Cubic Spline for Functions with Large Gradients in a Boundary Layer

Alexander Zadorin[1(✉)] and Igor' Blatov[2]

[1] Sobolev Institute of Mathematics, Siberian Branch RAS,
pr. Akad. Koptyuga 4, Novosibirsk 630090, Russia
zadorin@ofim.oscsbras.ru
[2] Povolzhskiy State University of Telecommunications and Informatics,
ul. Lva Tolstogo 23, Samara 443010, Russia
blatow@mail.ru

**Abstract.** The problem of spline interpolation of functions with large gradients in the boundary layer is studied. It is assumed that the function contains the known up to a factor boundary layer component responsible for the large gradients of this function in the boundary layer. A modification of the cubic spline, based on the fitting to the boundary layer component is proposed. The questions of existence, uniqueness and accuracy of such spline are investigated. Estimates of the interpolation error which are uniform with respect to a small parameter are obtained.

**Keywords:** Function and boundary layer · Uniform grid ·
Generalized spline · Interpolation · Error estimation

## 1 Introduction

Various convective-diffusion processes with predominant convection are modeled on the base of singularly perturbed problems. Solutions to such problems have large gradients in the boundary layer, as a result of which the application of classical difference schemes and polynomial splines can lead to significant errors. To construct difference schemes with the property of uniform convergence in the small parameter, there are two basic approaches: construction of exponential fitted schemes [1] and application classical schemes on meshes condensing in the boundary layer [2,3].

Spline interpolation of functions with large gradients in the boundary layer have been studied much less. In the presence of the exponential boundary layer, the upper and lower bounds of the interpolation error for parabolic and cubic splines on the uniform mesh and on Shishkin's mesh are obtained in [4–6]. It is proved that the spline errors increase without limit when the small parameter $\varepsilon$ is decreasing. Due to the displacement of one interpolation node, the $\varepsilon$-uniform accuracy of these splines is achieved. In [7] a parabolic spline is generalized on

© Springer Nature Switzerland AG 2019
I. Dimov et al. (Eds.): FDM 2018, LNCS 11386, pp. 654–662, 2019.
https://doi.org/10.1007/978-3-030-11539-5_77

the basis that it becomes exact on the boundary layer component. In [8] on a uniform grid the analogue of cubic spline exact on the boundary layer component $\Phi(x)$ is constructed. Here $\Phi(x) = e^{-\alpha x/\varepsilon}$, $x \in [0,1], \alpha > 0, \varepsilon \in (0,1]$. The error of the constructed spline is uniform with respect to the parameter $\varepsilon$.

In this paper we study the analogue of a cubic spline for interpolation of a function having large gradients. We construct the spline that is exact on the known up to multiplier component $\Phi(x)$, responsible for the large gradients of the function in the boundary layer. Through the paper $C$ and $C_j$ denote generic positive constants independent of $\varepsilon$ and mesh step $h$.

## 2    Construction of Spline

The following estimate is known for a cubic spline $S_3(x; u)$ : $|S_3(x; u) - u(x)| \leq C \max |u^{(4)}(x)| h^4$, where $h$ is mesh step. Consequently, the spline has an error of the order of $O(h^4)$, if the derivative $u^{(4)}(x)$ is uniformly bounded. However, if $u(x)$ is a solution of a singularly perturbed problem, then the derivative $u^{(4)}(x)$ is of order $O(\varepsilon^{-4})$ and grows unboundedly with decreasing $\varepsilon$. In this case, there is a need to construct a spline with $\varepsilon$-uniform accuracy. For building such a spline, we will assume that for the interpolated function $u(x)$ with large gradients the following representation holds:

$$u(x) = p(x) + \gamma\Phi(x), \quad x \in [0,1], \tag{1}$$

where the regular component $p(x)$ has bounded derivatives up to some order, the boundary layer component $\gamma\Phi(x)$ is responsible for the large gradients of the function $u(x)$. It is assumed that $\Phi(x)$ is a known function of general form, $\gamma, p(x)$ are not given.

It is known that the representation (1) is valid for the solution of a singular perturbed boundary value problem [9]. The problem is to construct the spline whose error does not depend on $\Phi(x)$ and its gradients. In the case of exponential boundary layer the error must be $\varepsilon$-uniform.

We define a uniform grid $\Omega = \{x_n : x_n = nh, 0 \leq n \leq N, x_0 = 0, x_N = 1\}$, $\Delta_n = [x_{n-1}, x_n]$. We assume that the function $u(x)$ of the form (1) is known at the grid nodes, $u_n = u(x_n)$, $n = 0, 1, 2, \ldots, N$.

For the rest of the paper

$$\Phi^{(3)}(x) \neq 0, x \in (x_{n-1}, x_n), \; n = 1, 2, \ldots, N. \tag{2}$$

When condition (2) is fulfilled, the system of functions $\{1, x, x^2, \Phi(x)\}$ is linearly independent on the interval $[x_{n-1}, x_n]$. Therefore, these functions can be used as a basis when constructing a nonpolynomial spline.

So, we construct the spline $S_\Phi(x; u)$. For an arbitrary grid interval $[x_{n-1}, x_n]$ we define the interpolation of the second spline derivative $S''_\Phi(x; u)$ as follows:

$$S''_\Phi(x; u) = M_{n-1} + \left(M_n - M_{n-1}\right)\frac{\Phi''(x) - \Phi''_{n-1}}{\Phi''_n - \Phi''_{n-1}}, \quad x \in \Delta_n, \tag{3}$$

where $M_n = S''_\Phi(x_n; u)$, $\Phi''_n = \Phi''(x_n)$. By virtue of (2) the relation (3) is correct.

Integrating twice in (3) and taking into account the interpolation conditions $S_\Phi(x_{n-1}; u) = u_{n-1}$, $S_\Phi(x_n; u) = u_n$, we obtain an interpolating spline:

$$S_\Phi(x; u) = \frac{M_n - M_{n-1}}{\Phi''_n - \Phi''_{n-1}}$$

$$\times \left( \Phi(x) - \Phi_{n-1} - \frac{\Phi_n - \Phi_{n-1}}{h}(x - x_{n-1}) - \frac{1}{2}\Phi''_{n-1}(x - x_{n-1})(x - x_n) \right)$$

$$+ \frac{1}{2}M_{n-1}(x - x_{n-1})(x - x_n) + (u_n - u_{n-1})\frac{x - x_{n-1}}{h} + u_{n-1}, \quad x \in \Delta_n. \quad (4)$$

Taking into account the equalities $S'_\Phi(x_n - 0; u) = S'_\Phi(x_n + 0; u)$ and the imposed boundary conditions $S''_\Phi(0, u) = u''(0), S''_\Phi(1; u) = u''(1)$, we get the following system of equations on the coefficients $\{M_n\}$:

$$A_n M_{n-1} + (1 - A_n - B_n)M_n + B_n M_{n+1} = \frac{u_{n+1} - 2u_n + u_{n-1}}{h^2}, \quad 0 < n < N, \quad (5)$$

$$M_0 = u''(0), \quad M_N = u''(1), \quad (6)$$

where

$$A_n = \frac{1}{h^2(\Phi''_n - \Phi''_{n-1})}\left( \Phi_n - \Phi_{n-1} - h\Phi'_n + \frac{h^2}{2}\Phi''_n \right), \quad (7)$$

$$B_n = \frac{1}{h^2(\Phi''_{n+1} - \Phi''_n)}\left( \Phi_{n+1} - \Phi_n - h\Phi'_n - \frac{h^2}{2}\Phi''_n \right). \quad (8)$$

**Lemma 1.** *Let the conditions (2) be satisfied. Then the system of Eqs. (5)–(6) is uniquely solvable.*

*Proof.* The system (5)–(6) can be written in matrix form: $HM = F$.

Let $d_n$ be the diagonal dominance of the $n$-th column of the matrix $H$. Then, for $1 < n < N$ $d_n = 1 - (B_{n-1} + A_n) - (B_n + A_{n+1})$. Let us prove that for each $n$ there holds $d_n > 0$. First, we show that for each $n$

$$1/4 \leq A_{n+1} + B_n < 1/2. \quad (9)$$

We have

$$A_{n+1} + B_n = \frac{1}{2} + \frac{1}{h^2(\Phi''_{n+1} - \Phi''_n)}\left[ 2\Phi_{n+1} - 2\Phi_n - h\Phi'_{n+1} - h\Phi'_n \right].$$

Applying expansions in the Taylor series about node $x_n$ with the remainder term in integral form, we get

$$A_{n+1} + B_n = \frac{1}{2} - \frac{\int\limits_{x_n}^{x_{n+1}} (x_{n+1} - s)(s - x_n)\Phi^{(3)}(s)\,ds}{h^2 \int\limits_{x_n}^{x_{n+1}} \Phi^{(3)}(s)\,ds}. \quad (10)$$

From (2), (10) it follows that $A_{n+1} + B_n < 1/2$. It is obvious that $A_{n+1} + B_n \geq 1/4$. Thus, if condition (2) is satisfied, the double inequality (9) holds. Hence, for any $n$ $d_n > 0$ and the matrix $H$ of the system (5)–(6) has a strict diagonal predominance on the column. Consequently, this matrix is also non-degenerate and the system (5)–(6) is uniquely solvable.    ◇

In accordance with Lemma 1 and (4), the spline $S_{\Phi}(x; u)$ is uniquely determined. Let us prove that the spline $S_{\Phi}(x; u)$ is exact for the boundary layer component $\Phi(x)$. Let us take $u(x) = \Phi(x)$. It is easy to see that $M_j = \Phi''_j$, $j = 0, 1, \ldots, N$ is the unique solution of the system (5)–(6). Then from (4) we get that $S_{\Phi}(x; \Phi) = \Phi(x)$. Thus, $S_{\Phi}(x; u) \in C^2[0, 1]$ and $S_{\Phi}(x; \Phi) = \Phi(x)$.

From (7), (8) we have that

$$A_n = \frac{1}{2} \frac{\int_{x_{n-1}}^{x_n} \Phi^{(3)}(t)(t - x_{n-1})^2 \, dt}{h^2 \int_{x_{n-1}}^{x_n} \Phi^{(3)}(t) \, dt}, \qquad B_n = \frac{1}{2} \frac{\int_{x_n}^{x_{n+1}} \Phi^{(3)}(t)(t - x_{n+1})^2 \, dt}{h^2 \int_{x_n}^{x_{n+1}} \Phi^{(3)}(t) \, dt}. \tag{11}$$

From (11) and (2) we have that $0 < A_n, B_n < 1/2$. We will use the following norm for an $N$-by-$N$ matrix $P$: $||P||_{\infty} = \max_i \sum_{j=0}^{N} |P_{i,j}|$.

**Lemma 2.** *Suppose that the matrix $H$ corresponds to the system (5)–(6). Then for all $x \in [0, 1]$ the following estimates hold*

$$|S''_{\Phi}(x; u) - u''(x)| \leq \gamma$$

$$= ||H^{-1}||_{\infty} \max_{0 < n < N} \int_{x_{n-1}}^{x_{n+1}} \left[ \frac{h}{3} |p^{(4)}(s)| + |p^{(3)}(s)| \right] ds + 2 \max_{0 < n \leq N} \int_{x_{n-1}}^{x_n} |p^{(3)}(s)| \, ds, \tag{12}$$

$$|S'_{\Phi}(x; u) - u'(x)| \leq \gamma h, \qquad |S_{\Phi}(x; u) - u(x)| \leq \gamma \frac{h^2}{4}. \tag{13}$$

*Proof.* Set $Z_n = M_n - u''_n$, $n = 0, 1, \ldots, N$. Then

$$HZ = F - HU, \qquad U_n = u''_n, \quad n = 0, 1, \ldots, N. \tag{14}$$

In accordance with the representation (1) and that in the case $u(x) = \Phi(x)$ the vector $\{\Phi''_n\}, n = 0, 1, \ldots, N$ is the solution of the system (5)–(6), we get

$$A_n Z_{n-1} + (1 - A_n - B_n) Z_n + B_n Z_{n+1} = G_n = \frac{p_{n+1} - 2p_n + p_{n-1}}{h^2}$$

$$-(A_n p''_{n-1} + (1 - A_n - B_n) p''_n + B_n p''_{n+1}), \quad 0 < n < N, Z_0 = 0, Z_N = 0. \tag{15}$$

Consequently,

$$|G_n| \leq \frac{h}{6} \int_{x_{n-1}}^{x_{n+1}} |p^{(4)}(s)| \, ds + \frac{1}{2} \int_{x_{n-1}}^{x_{n+1}} |p^{(3)}(s)| \, ds. \tag{16}$$

Taking into account that $||M||_\infty \le ||H^{-1}||_\infty ||F||_\infty$, and applying the estimate (16), we get from (15)

$$\max_n |M_n - u_n''| \le ||H^{-1}||_\infty \max_n \left[ \frac{h}{6} \int_{x_{n-1}}^{x_{n+1}} |p^{(4)}(s)|\, ds + \frac{1}{2} \int_{x_{n-1}}^{x_{n+1}} |p^{(3)}(s)|\, ds \right]. \quad (17)$$

Suppose that $x \in (x_{n-1}, x_n)$. Using (3), we obtain

$$S_\Phi''(x; u) - u''(x) = (M_{n-1} - u_{n-1}'') + \left( (M_n - u_n'') - (M_{n-1} - u_{n-1}'') \right) \frac{\Phi''(x) - \Phi_{n-1}''}{\Phi_n'' - \Phi_{n-1}''}$$

$$+ \left( u_{n-1}'' + \left( u_n'' - u_{n-1}'' \right) \frac{\Phi''(x) - \Phi_{n-1}''}{\Phi_n'' - \Phi_{n-1}''} - u''(x) \right). \quad (18)$$

Taking into account that the expression in parenthesis is zero if $u(x) = \Phi(x)$, we obtain (12) from (17), (18).

Next we show the first estimate in (13). Suppose that $x \in [x_{n-1}, x_n]$, $r(x) = S_\Phi(x; u) - u(x)$. By virtue of the interpolation conditions $r(x_{n-1}) = 0$ and $r(x_n) = 0$. Then there is a point $s_n \in (x_{n-1}, x_n)$ such that $r'(s_n) = 0$. Then $r'(x) = r'(x) - r'(s_n) = r''(\tau_n)(x - s_n)$. Taking into account the estimate (12), we obtain the required estimate (13). The second estimate in (13) is proved similarly.                                                                                      ◇

## 3    The Case of a Singular Perturbation

Now we consider the case when a function $u(x)$ from (1) can be regarded as a solution of a singularly perturbed problem. In this case we concretize the estimates (12), (13) which depend on the norm of the inverse matrix $H^{-1}$. We consider the boundary layer component of the form:

$$\Phi(x) = g(x, \varepsilon) f(x/\varepsilon), \ \varepsilon \in (0, 1], \quad (19)$$

where $\varepsilon$ can be regarded as the small parameter before the highest derivative [3]. The particular case $\Phi(x) = e^{-\alpha x/\varepsilon}$ was considered in [8].

We define $y = x/\varepsilon$ and make the following assumptions:

(1) the functions $g(x, \varepsilon)$ and $f(y)$ are three times continuously differentiable with respect to $x \in [0, 1]$ and $y \in [0, +\infty)$ respectively and the following inequalities hold

$$0 < C_1 \le |g(x, \varepsilon)| \le C_2, \ f^{(3)}(y) \ne 0, \ |g^{(i)}(x, \varepsilon)| \le C\varepsilon^{1-i}, \ 1 \le i \le 2,$$

$$|f^{(j)}(y)| \le C|f^{(3)}(y)|, \ 0 \le j \le 2, x \in [0, 1], y \in [0, +\infty), \quad (20)$$

(2) $\int_0^{+\infty} y^n |f^{(3)}(y)|\, dy \le C, \ n = 0, 1, 2,$

(3) for k be of non-negative integer and $h/\varepsilon \ge 1$

$$0 < C_1 \le \frac{\int_0^{+\infty} y^n f^{(3)}(y + kh/\varepsilon)\, dy}{\int_0^{+\infty} f^{(3)}(y + kh/\varepsilon)\, dy} \le C_2, \quad n = 1, 2, \quad (21)$$

where the positive constants $C, C_j$ do not depend on $k, \varepsilon, h$.

**Remark.** The representation (19) is valid for a wide class of functions describing the boundary layer. In particular, such a representation with conditions (1)–(3) is valid for a quasipolynomial of the form

$$\Phi(x) = (1 + \sum_{i=1}^{K} \alpha_i x^i)e^{-x/\varepsilon}, \quad \alpha_i \geq 0, K > 0,$$

describing with accuracy $O(\varepsilon^{K+1})$ a boundary layer asymptotics in the case of an exponential boundary layer.

**Lemma 3.** *Let us the function $\Phi(x)$ be of the form (19) and the conditions (1)–(3) be satisfied. Then there is positive constant $C_3$ such that for $h/\varepsilon \geq C_3$ and for some constants $C_4, C_5$ the following estimates hold*

$$C_4 \min\left\{\frac{1}{h}, 1 + \frac{h}{\varepsilon}\right\} \leq \| H^{-1} \|_{\infty} \leq C_5 \min\left\{\frac{1}{h}, 1 + \frac{h}{\varepsilon}\right\}. \tag{22}$$

*Proof.* From (20) we have that

$$\Phi^{(3)}(x) = (g(x, \varepsilon) + O(\varepsilon))f^{(3)}(x/\varepsilon). \tag{23}$$

Further, from (11), (21), (23) for $h/\varepsilon \geq C_3$ we have

$$A_{n+1} = \frac{\varepsilon^2}{2h^2} \frac{\int_0^{h/\varepsilon} y^2(g(\varepsilon y + nh, \varepsilon) + O(\varepsilon))f^{(3)}(y + nh/\varepsilon)dy}{\int_0^{h/\varepsilon}(g(\varepsilon y + nh, \varepsilon) + O(\varepsilon))f^{(3)}(y + nh/\varepsilon)dy} = O\left(\frac{\varepsilon^2}{h^2}\right),$$

$$B_n = \frac{1}{2} - |O^*\left(\frac{\varepsilon}{h}\right)| + O\left(\frac{\varepsilon^2}{h^2}\right).$$

From the last two formulas, it follows that $H = H_1 + H_2$, where $H_1 = M + J$ is upper triangular two diagonal matrix, wherein $M$ is the diagonal part of $H$ and $J$ is the matrix whose non-zero elements are elements of the upper super diagonal of $H_1$. The matrix $H_1$ has a diagonal preponderance $O^*\left(\varepsilon/h\right)$. For the matrix $H_2$ the following estimate holds $\| H_2 \|_{\infty} = O\left(\varepsilon^2/h^2\right)$.

We find the inverse matrix by means of a finite Neumann series

$$H_1^{-1} = (M(I + M^{-1}J))^{-1} = (I + M^{-1}J)^{-1}M^{-1} = \sum_{i=0}^{N-2}(-1)^i(M^{-1}J)^i M^{-1}.$$

Using that $\| M^{-1} \|_{\infty} \leq 2$ and $\| M^{-1}J \|_{\infty} = 1 - O^*(\varepsilon/h)$, we obtain

$$\| H_1^{-1} \|_{\infty} = O^*\left(\| \sum_{i=0}^{N-2}(-1)^i(M^{-1}J)^i \|_{\infty}\right) = O^*\left(\min\left\{\frac{1}{h}, 1 + \frac{h}{\varepsilon}\right\}\right). \tag{24}$$

Using (24) and the decomposition $H = H_1 + H_2$, we obtain (22). ◇

Assume in the addition to (1)–(3) that

(4) Conditions (2)–(3) are also satisfied for $n = 3$,
(5) for $h/\varepsilon \geq 1$ and $n$ of non-negative integers

$$\frac{\int_0^{h/\varepsilon} y^n f^{(3)}(y + kh/\varepsilon)dy}{\int_0^{h/\varepsilon} f^{(3)}(y + kh/\varepsilon)dy} = \tilde{C}_n + O(\varepsilon), \quad n = 1, 2, \tag{25}$$

where $\tilde{C}_1, \tilde{C}_2$ are positive constants independent of $k$.

**Theorem 1.** *Suppose that $u(x)$ corresponds to (1) with $\Phi(x)$ of the form (19), $u(x) \in C^4[0,1]$ and the conditions (1)–(5) hold. Then for some the constant $C$*

$$|S_\Phi^{(i)}(x; u) - u^{(i)}(x)| \leq Ch^{3-i}, \quad x \in [0,1], \ 0 \leq i \leq 2. \tag{26}$$

The conditions (1)–(5) are satisfied for the function $\Phi(x) = e^{-\alpha x/\varepsilon}$.

## 4   Results of Numerical Experiments

We define a following function $u(x)$ of the form (1)

$$u(x) = \cos \frac{\pi x}{2} + (1 + x)e^{-x/\varepsilon}, \ x \in [0,1], \ \varepsilon \in (0,1]. \tag{27}$$

**Table 1.** Error and the convergence rate for the function (27).

| $\varepsilon$ | $N$ | | | | | |
|---|---|---|---|---|---|---|
| | $2^3$ | $2^4$ | $2^5$ | $2^6$ | $2^7$ | $2^8$ |
| $10^{-2}$ | $4.33 \cdot 10^{-4}$ | $4.21 \cdot 10^{-5}$ | $3.01 \cdot 10^{-6}$ | $1.78 \cdot 10^{-7}$ | $1.03 \cdot 10^{-8}$ | $6.12 \cdot 10^{-10}$ |
| | | 3.36 | 3.81 | 4.08 | 4.11 | 4.07 |
| $10^{-3}$ | $4.83 \cdot 10^{-4}$ | $6.02 \cdot 10^{-5}$ | $7.43 \cdot 10^{-6}$ | $8.82 \cdot 10^{-7}$ | $9.21 \cdot 10^{-8}$ | $7.29 \cdot 10^{-9}$ |
| | | 3.00 | 3.09 | 3.08 | 3.26 | 3.65 |
| $10^{-4}$ | $4.84 \cdot 10^{-4}$ | $6.05 \cdot 10^{-5}$ | $7.57 \cdot 10^{-6}$ | $9.45 \cdot 10^{-7}$ | $1.18 \cdot 10^{-7}$ | $1.46 \cdot 10^{-8}$ |
| | | 3.00 | 3.00 | 3.00 | 2.99 | 3.01 |
| $10^{-5}$ | $4.86 \cdot 10^{-4}$ | $6.05 \cdot 10^{-5}$ | $7.57 \cdot 10^{-6}$ | $9.46 \cdot 10^{-7}$ | $1.18 \cdot 10^{-7}$ | $1.47 \cdot 10^{-8}$ |
| | | 3.00 | 3.00 | 3.00 | 2.99 | 3.00 |

Table 1 presents the maximum errors of the spline (4), calculated at the nodes of a condensed grid obtained from the grid $\Omega$ by dividing each interval of the grid into 10 equal parts. The same table contains calculated accuracy orders. This results agree with the estimate (26) for $i = 0$.

Define a function corresponding to the presence of a power boundary layer

$$u(x) = \cos \frac{\pi x}{2} + \sqrt{x + \varepsilon}, \quad x \in [0,1]. \tag{28}$$

Table 2 similarly presents errors in the case of the function (28).

**Table 2.** Error and convergence rate for the function (28).

| $\varepsilon$ | $N$ | | | | | |
|---|---|---|---|---|---|---|
| | $2^2$ | $2^3$ | $2^4$ | $2^5$ | $2^6$ | $2^7$ |
| 1 | $1.62 \cdot 10^{-4}$ | $9.92 \cdot 10^{-6}$ $4.03$ | $6.10 \cdot 10^{-7}$ $4.02$ | $3.79 \cdot 10^{-8}$ $4.01$ | $2.36 \cdot 10^{-9}$ $4.01$ | $1.47 \cdot 10^{-10}$ $4.01$ |
| $10^{-1}$ | $2.74 \cdot 10^{-4}$ | $1.53 \cdot 10^{-5}$ $4.16$ | $9.14 \cdot 10^{-7}$ $4.07$ | $5.57 \cdot 10^{-8}$ $4.04$ | $3.44 \cdot 10^{-9}$ $4.02$ | $2.14 \cdot 10^{-10}$ $4.01$ |
| $10^{-2}$ | $3.03 \cdot 10^{-4}$ | $1.68 \cdot 10^{-5}$ $4.17$ | $9.97 \cdot 10^{-7}$ $4.08$ | $6.08 \cdot 10^{-8}$ $4.04$ | $3.75 \cdot 10^{-9}$ $4.02$ | $2.33 \cdot 10^{-10}$ $4.01$ |
| $10^{-3}$ | $3.06 \cdot 10^{-4}$ | $1.70 \cdot 10^{-5}$ $4.17$ | $1.04 \cdot 10^{-6}$ $4.03$ | $6.40 \cdot 10^{-8}$ $4.02$ | $3.90 \cdot 10^{-9}$ $4.04$ | $2.35 \cdot 10^{-10}$ $4.05$ |
| $10^{-8}$ | $3.07 \cdot 10^{-4}$ | $1.70 \cdot 10^{-5}$ $4.17$ | $1.05 \cdot 10^{-6}$ $4.03$ | $6.53 \cdot 10^{-8}$ $4.01$ | $4.08 \cdot 10^{-9}$ $4.00$ | $2.55 \cdot 10^{-10}$ $4.01$ |

## 5    Conclusion

For the interpolation of functions with large gradients in the boundary layer a generalized spline is constructed. By construction, the spline is exact for the boundary layer component. It is proved that under certain conditions a spline exists and unique. In particular, the case of an exponential boundary layer is considered. The estimates of the error in the approximation of the function and its derivatives are presented. The results of numerical experiments are given.

**Acknowledgements.** Supported by the program of fundamental scientific researches of the SB RAS 1.1.3., project 0314-2019-0009.

## References

1. Il'in, A.M.: Differencing scheme for a differential equation with a small parameter affecting the highest derivative. USSR Math. Notes **6**, 596–602 (1969)
2. Bakhvalov, N.S.: The optimization of methods of solving boundary value problems with a boundary layer. USSR Comput. Math. Math. Phys. **9**, 139–166 (1969)
3. Shishkin, G.I.: Grid Approximations of Singular Perturbation Elliptic and Parabolic Equations. UB RAS, Yekaterinburg (1992). (in Russian)
4. Blatov, I.A., Zadorin, A.I., Kitaeva, E.V.: Parabolic spline interpolation for functions with large gradient in the boundary layer. Siber. Math. J. **58**(4), 578–590 (2017)
5. Blatov, I.A., Zadorin, A.I., Kitaeva, E.V.: On the uniform convergence of parabolic spline interpolation on the class of functions with large gradients in the boundary layer. Numer. Anal. Appl. **10**(2), 108–119 (2017)
6. Blatov, I.A., Zadorin, A.I., Kitaeva, E.V.: Cubic Spline Interpolation of Functions with High Gradients in Boundary Layers. Comput. Math. Math. Phys. **57**(1), 9–28 (2017)

7. Zadorin, A.I.: Spline interpolation of functions with a boundary layer component. Int. J. Numer. Anal. Model Series B **2**(2–3), 262–279 (2011)
8. Blatov, I.A., Zadorin, A.I., Kitaeva, E.V.: On the parameter-uniform convergence of exponential spline interpolation in the presence of a boundary layer. Comput. Math. Math. Phys. **58**(3), 348–363 (2018)
9. Kellogg, R.B., Tsan, A.: Analysis of some difference approximations for a singular perturbation problem without turning points. Math. Comput. **32**, 1025–1039 (1978)

# A Family of Optimal Eighth Order Multiple Root Finders with Multivariate Weight Function

Fiza Zafar[1,2]($\boxtimes$), Alicia Cordero[1], and Juan Ramon Torregrosa[1]

[1] Instituto de Matemática Multidisciplinar,
Universitat Politècnica de València, València 46022, Spain
fizazafar@gmail.com
{acordero,jrtorre}@mat.upv.es
[2] Centre for Advanced Studies in Pure and Applied Mathematics,
Bahauddin Zakariya University, Multan 60800, Pakistan

**Abstract.** Finding repeated zero for a nonlinear equation $f(x) = 0$, $f : I \subseteq R \to R$, has always been of much interest and attention due to it's wide applications in many fields of science and engineering. The modified Newton's method is usually applied to solve this problem. Keeping in view that very few optimal higher order convergent methods exist for multiple roots, we present a new family of optimal eighth order convergent iterative methods for multiple roots with known multiplicity involving multivariate weight function. The numerical performance of the proposed methods is analyzed extensively along with the basins of attractions. Real life models from Life Science, Engineering and Physics are considered for the sake of comparison. The numerical experiments show that our proposed methods are efficient for determining multiple roots of non-linear equations.

**Keywords:** Nonlinear equations · Multiple zeros ·
Optimal iterative methods · Higher order of convergence

## 1 Introduction

It is well-known that Newton's method converges linearly for non-simple roots of a nonlinear function. For obtaining multiple roots of a univariate nonlinear equation with quadratic order of convergence Schröder [1], modified the Newton's method with prior knowledge of the multiplicity $m \geq 1$ of the root as follows:

$$x_{n+1} = x_n - m\frac{f(x_n)}{f'(x_n)}. \tag{1}$$

The scheme (1) can determine the desired multiple root with quadratic convergence and is optimal in the sense of Kung-Traub's conjecture [2], that any multipoint method without memory can reach its convergence order of at most

© Springer Nature Switzerland AG 2019
I. Dimov et al. (Eds.): FDM 2018, LNCS 11386, pp. 663–669, 2019.
https://doi.org/10.1007/978-3-030-11539-5_78

$2^{p-1}$ for $p$ functional evaluations. In the last few decades, many researchers have worked to develop iterative methods for finding multiple roots with greater efficiency and higher order of convergence. In recent years, efforts have been made to obtain an optimal scheme with convergence order greater than four for multiple zeros. Some of them only succeeded to develop iterative schemes of maximum of sixth-order convergence, for example see [3,4]. However, there are only few multi-point iterative schemes with optimal eighth-order convergence for multiple zeros which are proposed most recently (see [5–8]). Motivated by the research going on in this direction and with a need to give more stable optimal higher order methods, we propose a new family of optimal eighth order iterative methods for finding simple as well as multiple zeros of a univariate nonlinear function with multiplicity $m \geq 1$. The derivation of the proposed scheme is based on a univariate and trivariate weight function approach. In addition, our proposed methods not only give the faster convergence but also have smaller residual error. We have demonstrated the efficiency and robustness of the proposed methods by performing several applied science problems for numerical tests and observed that our methods have better numerical results, than the existing methods. The rest of the paper is organized as follows: Sect. 2 provides the construction of the new family of iterative methods, analysis of convergence to prove eighth order of convergence. In Sect. 3, some special cases of the new family are defined. In Sect. 4, numerical performance and comparison of the some special cases of the new family with the existing ones are given. The numerical comparisons has been carried out by using the nonlinear equations that appear in the modelling of predator-prey model, beam designing model, electric circuit modelling, eigenvalue problem.

## 2   Construction of the Family

This section is devoted to the main contribution of this study and convergence analysis of the proposed scheme with main theorem. We consider the following optimal eighth order scheme for multiple zeros with multiplicity $m \geq 1$.

$$y_n = x_n - m\frac{f(x_n)}{f'(x_n)}, n \geq 0,$$

$$z_n = y_n - mu_n G\left(u_n\right)\frac{f(x_n)}{f'(x_n)}$$

$$x_{n+1} = z_n - mu_n H(u_n, t_n, w_n)\frac{f(x_n)}{f'(x_n)}, \tag{2}$$

where $G : \mathbb{C} \to \mathbb{C}$ and $H : \mathbb{C}^3 \to \mathbb{C}$ are an analytic functions in a neighborhood of $(0)$ and $(0,0,0)$ with $u_n = \left(\frac{f(y_n)}{f(x_n)}\right)^{\frac{1}{m}}$, $t_n = \left(\frac{f(z_n)}{f(y_n)}\right)^{\frac{1}{m}}$ and $w_n = \left(\frac{f(z_n)}{f(x_n)}\right)^{\frac{1}{m}}$.

In the next Theorem 1, we demonstrate that the order of convergence of the proposed scheme will reach at optimal order eight without using additional functional evaluations.

**Theorem 1.** *Let us consider $x = \xi$ (say) be a zero with multiplicity $m \geq 1$ of the involved function $f$. In addition, we assume that $f : \mathbb{C} \to \mathbb{C}$ be an analytic function in the region enclosing a multiple zero $\xi$. The proposed scheme defined by (2) has an optimal eighth-order convergence, when it satisfies the following expressions:*

$$G(0) = 1, \; G_1 = G'(0) = 2, \; G_2 = G''(0) = 4 - \frac{G_3}{6}, G_3 = G'''(0),$$

$$H_{000} = 0, \; H_{100} = 0, \; H_{010} = 1, \; H_{101} = 3 - \frac{G_3}{12}, H_{110} = 2 - H_{001},$$

$$H_{011} = 4, \; H_{020} = 1, |G_3| = \left| G''' (0) \right| < \infty, |H_{001}| < \infty, \tag{3}$$

*where* $H_{ijk} = \dfrac{1}{i!j!k!} \dfrac{\partial^{i+j+k}}{\partial u_n^i \partial t_n^j \partial w_n^k} H(u_n, t_n, w_n)|_{(u_n=0, t_n=0, w_n=0)}$ *for* $0 \leq i, k \leq 1, 0 \leq j \leq 2$.

## 3   Some Special Cases of Weight Function

In this section, we will discuss some special cases of our proposed scheme by assigning different kind of weight functions. In this regard, please see following cases, where we have mentioned some different kinds of members of the proposed scheme.

**Case 1:** Let us describe the following polynomial weight functions directly from the Theorem 1:

$$G(u_n) = 1 + 2u_n + \left( 2 - \frac{G_3}{12} \right) u_n^2 + \frac{1}{6} G_3 u_n^3,$$

$$H(u_n, t_n, w_n) = t_n + \left( H_{001} + \left( 3 - \frac{G_3}{12} \right) u_n \right) w_n$$

$$+ ((2 - H_{001}) u_n + 4w_n + t_n) t_n, \tag{4}$$

where $H_{001}$ and $G_3$ are free parameters.

**Case 1A:** We take $H_{001} = 2, G_3 = 0$ in (4).

**Case 2:** Now, we suggest mixture of rational and polynomial weight functions satisfying the conditions (3) as follows:

$$G(u_n) = \frac{1 + a_0 u_n}{1 + (a_0 - 2) u_n + a_2 u_n^2},$$

$$H(u_n, t_n, w_n) = t_n + \left( H_{001} + \left( 3 - \frac{G_3}{12} \right) u_n \right) w_n$$

$$+ ((2 - H_{001}) u_n + 4w_n + t_n) t_n,$$

$$a_2 = -2(a_0 - 1) + \frac{G_3}{12}, \tag{5}$$

where $a_0$, $H_{001}$ and $G_3$ are free parameters.

**Case 2A:** Let us take $a_0 = 2, H_{001} = 2, G_3 = 12$ in (5).

**Case 3**: Now, with another choice of rational weight for $G(u_n)$, we have:

$$G(u_n) = \frac{1 + a_0 u_n}{1 + (a_0 - 2) u_n + a_2 u_n^2 + a_3 u_n^3},$$

$$H(u_n, t_n, w_n) = t_n + \left( H_{001} + \left( 3 - \frac{G_3}{12} \right) u_n \right) w_n$$

$$+ ((2 - H_{001}) u_n + 4 w_n + t_n) t_n,$$

$$a_2 = -2(a_0 - 1) + \frac{G_3}{12}, \tag{6}$$

where $a_0$, $a_3$, $H_{001}$ and $G_3$ are free parameters.

**Case 3A:** We consider the special case of (6) for $a_0 = 4, a_3 = 6, H_{001} = 2, G_3 = 12$.

In the similar fashion, we can develop several new and interesting optimal schemes with eighth-order convergence for multiple zeros by considering new weight functions that satisfy the conditions of Theorem 1.

## 4 Numerical Experiments

This section is devoted to demonstrate the efficiency, effectiveness and convergence behavior of the presented scheme. In this regard, we consider some of the special cases of the proposed scheme namely $NS1$, $NS2$, and $NS3$, given by Case 1A, 2A and 3 A respectively. In addition, we choose a total number of four test problems for comparison: first one is predator-prey model, second one is a beam designing problem, third one is electric circuit modeling for simple zeros, the last is an eigenvalue problem.

Now, we want to compare our methods with other existing robust methods of same order on the basis of difference between two consecutive iterations, residual errors in the function, computational order of convergence $\rho$ and asymptotic error constant $\eta$. We have chosen eighth-order iterative methods for multiple zeros given by Behl et al. [5,8], denoted by $BM1$ and $BM2$.

The Tables 1, 2, 3 and 4 display the number of iteration indices $(n)$, error in the consecutive iterations $|x_{n+1} - x_n|$, computational order of convergence $\rho \approx \dfrac{\log |f(x_{n+1})/f(x_n)|}{\log |(f(x_n)/f(x_{n-1}))|}$, $n \geq 1$, (the formula by Jay [9]), absolute residual error of the corresponding function $(|f(x_n)|)$ and the asymptotic error constant $\eta \approx \left| \dfrac{e_n}{e_{n-1}^8} \right|$. We did our calculations with several number of significant digits (1000 significant digits) to minimize the round off error. We display all the numerical values in Tables 1, 2, 3 and 4 up to 7 significant digits with exponent. Finally, we display the values of approximated zeros up to 30 significant digits in the Examples 1–4, although a precision of 1000 significant digits are available with us.

For the computer programming, all computations have been performed using the programming package *Maple* 16 with multiple precision arithmetic. Further, the meaning of $a(\pm b)$ is $a \times 10^{(\pm b)}$ in the Tables 1, 2, 3 and 4.

We, now, explain the real life models chosen for the sake of comparing the schemes as follows:

*Example 1.* Predator-Prey Model

We consider a predator-prey model with ladybugs as predators and aphids as preys [10] for which we have the following:

$$f_1(x) = 0.7937005260x^3 - 30x^2 + 6349.604208.$$

The desired zero of $f_1$ is 25.198420997897463295344212145564 with m=2. We choose $x_0 = 20$.

*Example 2.* Beam Designing Model

We consider a beam positioning problem (see [11]) where an $r$ meter long beam is leaning against the edge of the cubical box with sides of length 1 m each such that one of its end touches the wall and the other touches the floor which is given as:
$$f_2(x) = x^4 + 4x^3 - 24x^2 + 16x + 16 = 0.$$

The positive solution of the equation is a double root $x = 2$. We consider the initial guess $x_0 = 1.7$.

*Example 3.* The Shockley Diode Equation and Electric Circuit

Let us consider the following equation [12]:

$$-0.5 + 0.1I + 1.4\ln(I+1) = 0.$$

Replacing I with x, we have

$$f_3(x) = -0.5 + 0.1x + 1.4\ln(x+1).$$

The true root of the equation is 0.38997719839007758658645353 2646. We take $x_0 = 0.5$.

*Example 4.* Eigenvalue Problem:

Let us consider the characterstic polynomial for a $9 \times 9$ matrix:

$$f_4(x) = x^9 - 29x^8 + 349x^7 - 2261x^6 + 8455x^5 - 17663x^4$$
$$+ 15927x^3 + 6993x^2 - 24732x + 12960.$$

The above function has one multiple zero at $\xi = 3$ of multiplicity 4 with initial approximation $x_0 = 3.1$.

**Table 1.** Comparison of different multiple root finding methods for $f_1(x)$

|              | BM1             | BM2            | NS1             | NS2            | NS3            |
|--------------|-----------------|----------------|-----------------|----------------|----------------|
| $|x_1 - x_0|$ | 2.064550(1)    | 4.789445       | 1.219414(1)     | 1.214342(1)    | 1.213887(1)    |
| $|f(x_1)|$   | 1.008384(4)     | 4.963523       | 1.739946(3)     | 1.712863(3)    | 1.710446(3)    |
| $|x_2 - x_1|$ | 1.544682(1)    | 4.088744(−1)   | 6.995715        | 6.944984       | 6.940438       |
| $|f(x_2)|$   | 1.967429(−6)    | 3.035927(−7)   | 3.672323(−9)    | 6.792230(−9)   | 4.951247(−9)   |
| $|x_3 - x_2|$ | 2.560869(−4)   | 1.005971(−4)   | 1.106393(−5)    | 1.504684(−5)   | 1.284684(−5)   |
| $|f(x_3)|$   | 5.685107(−81)   | 6.093227(−29)  | 1.223217(−100)  | 5.427728(−98)  | 2.522949(−99)  |
| $\eta$       | 7.900841(−14)   | 0.1287852      | 1.928645(−12)   | 2.780193(−12)  | 2.386168(−12)  |
| $\rho$       | 7.676751        | 3.0078946      | 7.834927        | 7.814388       | 7.825421       |

**Table 2.** Comparison of different multiple root finding methods for $f_2(x)$

|              | BM1            | BM2            | NS1             | NS2            | NS3            |
|--------------|----------------|----------------|-----------------|----------------|----------------|
| $|x_1 - x_0|$ | 1.288477      | 2.734437(−1)   | 7.427026(−1)    | 7.391615(−1)   | 7.388023(−1)   |
| $|f(x_1)|$   | 35.99479       | 1.670143(−2)   | 5.783224        | 5.682280       | 5.672098       |
| $|x_2 - x_1|$ | 9.884394(−1)  | 2.654643(−2)   | 4.427007(−1)    | 4.391589(−1)   | 4.388001(−1)   |
| $|f(x_2)|$   | 3.566062(−8)   | 2.333107(−9)   | 8.652078(−11)   | 1.664205(−10)  | 1.1624462(−10) |
| $|x_3 - x_2|$ | 3.854647(−5)  | 9.859679(−6)   | 1.898691(−6)    | 2.633282(−7)   | 2.200800(−6)   |
| $|f(x_3)|$   | 7.225712(−77)  | 5.512446(−30)  | 2.306147(−95)   | 1.620443(−92)  | 4.8729521(−94) |
| $\eta$       | 4.230427(−5)   | 3.997726(7)    | 1.286982(−3)    | 1.903372(−3)   | 1.601202(−3)   |
| $\rho$       | 7.629155       | 3.0090640      | 7.812826        | 7.7859217      | 7.800775       |

**Table 3.** Comparison of different multiple root finding methods for $f_3(x)$

|              | BM1             | BM2            | NS1             | NS2             | NS3            |
|--------------|-----------------|----------------|-----------------|-----------------|----------------|
| $|x_1 - x_0|$ | 1.100228(−1)   | 1.100228(−1)   | 1.100228(−1)    | 1.100228(−1)    | 1.100228(−1)   |
| $|f(x_1)|$   | 3.213611(−12)   | 1.902432(−10)  | 7.591378(−11)   | 4.728795(−10)   | 1.626799(−10)  |
| $|x_2 - x_1|$ | 2.902439(−12)  | 1.718220(−10)  | 6.856308(−11)   | 4.270907(−10)   | 1.469276(−10)  |
| $|f(x_2)|$   | 9.512092(−97)   | 6.797214(−81)  | 2.215753(−84)   | 2.393956(−77)   | 1.758525(−81)  |
| $|x_3 - x_2|$ | 8.591040(−97)  | 6.139043(−81)  | 2.001202(−84)   | 2.162151(−77)   | 1.588247(−81)  |
| $|f(x_3)|$   | 5.604505(−773)  | 1.805114(−644) | 1.1671510(−672) | 1.032863(−615)  | 3.278426(−649) |
| $\eta$       | 1.705849(−4)    | 8.081072(−3)   | 4.097965(−3)    | 1.953099(−2)    | 7.312887(−3)   |
| $\rho$       | 7.999999        | 7.999999       | 7.999999        | 7.999999        | 7.999999       |

**Table 4.** Comparison of different multiple root finding methods for $f_4(x)$

|              | BM1             | BM2             | NS1             | NS2            | NS3            |
|--------------|-----------------|-----------------|-----------------|----------------|----------------|
| $|x_1 - x_0|$ | 1.577283(−1)   | 9.9275251(−2)   | 1283418(−1)     | 1.283182(−1)   | 1.283180(−1)   |
| $|f(x_1)|$   | 9.361198(−4)    | 2.205656(−11)   | 5.299339(−5)    | 5.281568(−5)   | 5.281425(−5)   |
| $|x_2 - x_1|$ | 5.772837(−2)   | 7.247474(−4)    | 2.834188(−2)    | 2.831824(−2)   | 2.831805(−2)   |
| $|f(x_2)|$   | 9.059481(−49)   | 7.0590148(−38)  | 2.755794(−55)   | 8.779457(−55)  | 5.772523(−55)  |
| $|x_3 - x_2|$ | 3.262145(−13)  | 2.278878(−10)   | 7.661066(−15)   | 1.023515(−14)  | 9.216561(−15)  |
| $|f(x_3)|$   | 4.543117(−408)  | 2.278878(−117)  | 4.807225(−457)  | 1.869778(−452) | 4.077620(−454) |
| $\eta$       | 2.644775(−3)    | 2.264227(15)    | 1.840177(−2)    | 2.474935(−2)   | 2.228752(−2)   |
| $\rho$       | 7.981915        | 3.000250        | 7.989789        | 7.988696       | 7.989189       |

In the Tables 1, 2, 3 and 4, we show the numerical results obtained by applying the different methods for approximating the multiple roots of $f_1 - f_4$. The obtained values confirm the theoretical results. From the tables it can be observed that our proposed scheme $NS1, NS2, NS3$ have better performance to approximate the multiple root of $f_1$, $f_2$ and $f_4$ among other similar methods. Only in the case of example for simple zeros, Behl's scheme $BM1$ is performing slightly better than the other schemes.

**Acknowledgments.** This work is supported by Schlumberger Foundation-Faculty for Future Program, by Ministerio de Economía y Competitividad under grants MTM 2014-52016-C2-2-P and Generalitat Valenciana PROMETEO/2016/089.

# References

1. Schroder, E.: Uber unendlich viele algorithmen zur auflosung der gleichungen. Math. Ann. **2**, 317–365 (1870)
2. Kung, H.T., Traub, J.F.: Optimal order of one-point and multipoint iteration. J. Assoc. Comput. Mach. **21**, 643–651 (1974)
3. Geum, Y.H., Kim, Y.I., Neta, B.: A class of two-point sixth-order multiple-zero finders of modified double-Newton type and their dynamics. Appl. Math. Comput. **270**, 387–400 (2015)
4. Geum, Y.H., Kim, Y.I., Neta, B.: A sixth-order family of three-point modified Newton-like multiple-root finders and the dynamics behind their extraneous fixed points. Appl. Math. Comput. **283**, 120–140 (2016)
5. Behl, R., Cordero, A., Motsa, S.S., Torregrosa, J.R.: Aneighth-order family of optimal multiple root finders and its dynamics. Numer. Algor. **77**(4), 1249–1272 (2017). https://doi.org/10.1007/s11075-017-0361-6
6. Zafar, F., Cordero, A., Rana, Q., Torregrosa, J.R.: Optimal iterative methods for finding multiple roots of nonlinear equations using free parameters. J. Math. Chem. **56**(7), 1884–1901 (2018). https://doi.org/10.1007/s10910-017-0813-1
7. Behl, R., Zafar, F., Alshomrani, A.S., Junjua, M., Yasmin, N.: An optimal eighth-order scheme for multiple zeros of univariate function. Int. J. Comput. Math. **15**(3), 14 (2018). https://doi.org/10.1142/S0219876218430028
8. Behl, R., Alshomrani, A.S., Motsa, S.S.: An optimal schemefor multiple roots of nonlinear equations with eighth-order convergence. J. Math. Chem. **56**(7), 1–6 (2018). https://doi.org/10.1007/s10910-018-0857-x
9. Jay, L.O.: A note on Q-order of convergence. BIT Numer. Math. **41**, 422–429 (2001)
10. Edelstein-Keshet, L.: Differential Calculus for the Life Sciences. Univeristy of British Columbia, Vancouver (2017)
11. Zachary, J.L.: Introduction to Scientific Programming: Computational Problem Solving Using Maple and C. Springer, New York (2012)
12. Khoury, R., Harder, D.H.: Numerical Methods and Modelling for Engineering. Springer, Switzerland (2017)

# Multi-step Iterative Algorithm for Mathematical Modeling of Light Bullets in Anisotropic Media

Irina G. Zakharova[1(✉)], Aleksey A. Kalinovich[1], Maria V. Komissarova[1], and Sergey V. Sazonov[1,2]

[1] Faculty of Physics, M.V. Lomonosov Moscow State University,
Leninskie Gory, Moscow 119991, Russia
{zaharova,komissarova}@physics.msu.ru, kalinovich@gmail.com,
sazonov.sergey@gmail.com
[2] National Research Centre "Kurchatov Istitute",
1 Kurchatova Square, Moscow 123182, Russia

**Abstract.** To perform an analytical and numerical investigation of optical bullets in a focusing bulk waveguide with quadratic nonlinearity we use the well-known quasi-optical approach. We give an approximate soliton solution representing a two-component light bullet. To investigate numerically the regimes of the formation and propagation of two-component optical bullets we construct a conservative difference scheme. To realize the multi-dimensional nonlinear difference scheme we propose a multi-step effective iterative solver. This method allows us to carry out an accurate and efficient modeling of the considered processes.

**Keywords:** Multi-step algorithm · Light bullets

## 1 Introduction

Multi-component multi-dimensional solitons at quadratic nonlinearity are attractive for a higher stability and a lower excitation threshold in comparison with solitons at cubic nonlinearity [1]. The investigation of such solitons started in early 1974 when the possibility of their existence was demonstrated for $(1 + 1)D$ case [2]. In 1981 it was proved that the self-focusing wave collapse under conditions of multi-dimensional second-harmonic generation could be precluded [3] and the history of "light bullets" in media with quadratic nonlinearity began. Further spatial-temporal solitons in homogeneous media with quadratic nonlinearity were investigated intensively [4,5].

Nowadays light bullets promise to be successfully applied in different spheres, for instance in ultrafast optics and highly precise interferometry [6,7]. Recently we developed the detailed theory of "breathing" light bullets propagating in the medium with anomalous dispersion [8,9]. Provided that the phase- and group-matching conditions hold and the second harmonic group velocity dispersion

© Springer Nature Switzerland AG 2019
I. Dimov et al. (Eds.): FDM 2018, LNCS 11386, pp. 670–677, 2019.
https://doi.org/10.1007/978-3-030-11539-5_79

(GVD) coefficient is twice as large as the GVD coefficient of the fundamental wave, the averaged Lagrangian method allowed us to obtain an analytical solution in the form of a two-component spatiotemporal soliton.

In order to enlarge the frequency range where light bullets can be observed one should test media with normal dispersion. The problem is how compensate both linear effects of diffraction and dispersion and nonlinear effect of decompression. Waveguide geometry may play a focusing role.

In this paper we analyze the possibility of the formation and stable propagation of light bullets in waveguides with quadratic nonlinearity. We describe the propagation of light in a waveguide with quadratic nonlinearity with the help of the quasi-optical approach. This method yields a system of multi-dimensional nonlinear parabolic equations with coordinate-dependent coefficients. We derive the integrals of motion inherent for the system.

To perform direct numerical simulations we must construct conservative numerical methods preserving difference analogues of the motion integrals. Such approach was developed in [10] and adapted for systems of coupled Schrodinger equations in [11]. Besides that, in multi-dimensional case one should use an effective method saving computational time. Usually the splitting technique is applied to this purpose (see [10, 12, 13]). The mentioned class of methods allows a significant reduction in the calculation time. But it fails when it is necessary to construct a method preserving several motion integrals [14]. The latter is especially important when modeling a soliton formation and propagation. In addition to the splitting method two other general approaches to the construction of numerical methods for the Schrodinger equations are discussed in [13]. They are called Fast Fourier Transform (FFT) (see also [15]) and multigrid (MG) techniques. In [13] the efficiencies of three named strategies are compared with each other and the conclusion is done that the MG method is comparable with the FFT approach in efficiency. FFT methods were successfully applied to the problems of nonlinear optics [4, 16, 17]. The investigated models in this field are the systems of evolutionary coupled nonlinear $(2 + 1)D$ or $(3 + 1)D$ equations. Thus, even FFT solvers become less efficient. In [18, 19] the two-step iteration algorithm is proposed for the process of femtosecond optical pulse propagation in semiconductors. This approach combines the advantages of splitting technique and nonlinear conservative Crank-Nicolson method.

In the present paper we develop a conservative nonlinear difference scheme for the problem under consideration and propose a relevant multi-step iteration process. This numerical algorithm, which can be generalized to $(3 + 1)D$ case, allows us to carry out an accurate and efficient in time modeling of the formation and propagation of two-component light bullets in quadratically nonlinear waveguides.

## 2   Problem Statement

Propagation of spatial-temporal pulses of fundamental and second harmonics in a focusing bulk waveguide is governed by the following dimensionless system of nonlinear parametrically coupled equations:

$$i\left[\frac{\partial A_1}{\partial z} - D_{a1}(x^2 + dy^2)\frac{\partial A_1}{\partial \tau}\right] + D_\tau\frac{\partial^2 A_1}{\partial \tau^2} - A_1^* A_2 = \\ -D_{q1}(x^2 + dy^2)A_1 + \tfrac{1}{2}D_x \Delta_\perp A_1, \tag{1}$$

$$i\left[\frac{\partial A_2}{\partial z} - D_{a2}(x^2 + dy^2)\frac{\partial A_2}{\partial \tau}\right] + 2D_\tau\frac{\partial^2 A_2}{\partial \tau^2} - \tfrac{1}{2}A_1^2 = \\ -2D_{q2}(x^2 + dy^2)A_2 + \tfrac{1}{4}D_x \Delta_\perp A_2, \tag{2}$$

where: $A_1$ is the slowly-varying complex amplitude of the fundamental harmonic, $A_2$ is the slowly-varying complex amplitude of the second harmonic, $\Delta_\perp = \partial^2/\partial x^2 + d\partial^2/\partial y^2$, $z$ is the spatial longitudinal coordinate (physical variable was normalized for nonlinear length $L_{nl}$) determining the distance of complete energy exchange between harmonics. $x, y$ are the spatial transverse coordinate (physical variables were normalized for the input soliton radii $R_0$ which are equal each other for both directions), $\tau$ is the temporal coordinate (physical variable was normalized for the input soliton duration $\tau_0$), $D_\tau$ is the coefficient determining by the ratio of nonlinear $L_{nl}$ and dispersion $L_{dis}$ lengths. The latter is inversely proportional to $\beta_2$ which is the coefficient of the GVD at the second harmonic depending on the refractive index, $D_x = L_{nl}/L_D$ is the ratio of nonlinear $L_{nl}$ and diffraction $L_D$ lengths. $L_D$ is proportional to $R_0^2$. $D_{aj}$, $D_{qj}$ are the coefficients of waveguide strength and they are both proportional to the ratio $L_{nl}/L_{wg}$. $L_{wg}$ characterizes waveguide transverse inhomogeneity. System (1–2) is written in the both cases of a focusing bulk ($d = 1$) and planar ($d = 0$) waveguide with parabolic profile of the refractive index provided phase and group-velocity synchronism. Refractive index decreases from center to periphery in such waveguide. Firstly we consider the case $d = 0$. The system (1–2) is valid in the domain:

$$0 < z < L_z, (x, \tau) \in \Gamma, \Gamma = \{0 < x < L_x\} \times \{0 < \tau < L_\tau\}. \tag{3}$$

Boundary and initial conditions are as follows.

$$A_{1,2}(z, 0, \tau) = A_{1,2}(z, L_x, \tau) = A_{1,2}(z, x, 0) = A_{1,2}(z, x, L_\tau) = 0, \\ A_1(0, x, \tau) = A_{10}(x, \tau), A_2(0, x, \tau) = A_{20}(x, \tau). \tag{4}$$

In (1–4) $L_z$ is the length of nonlinear medium, $L_\tau$ is the time interval during which laser pulse interaction with a medium is analyzed. $L_x$ is the length of the transversal domain. All these lengths are dimensionless. Since we deal with a finite pulse, it is naturally to take a sufficiently long time interval whose boundaries are not influenced by the pulse. Besides, the transversal size of the studied bullet is also finite due to narrow laser radiation. Thus, we can choose a wide transversal domain with boundaries which are not affected by radiation. Obviously, this choice assumes the need to monitor the fulfillment of zero boundary conditions, and, if necessary, a widening of the computational domain. It may cause specific computational difficulties, which nevertheless, can be eliminated by imposing artificial boundary conditions (see [20]).

To get an approximate analytical solution of the system (1–4), we apply the averaged Lagrangian method [8]. The developed analytical solution is as follows

$$
\begin{aligned}
A_1 &= D_\tau R^{-2/3} sech^2\left(x/2R\right) sech^2\left(sech\left(x/2R\right)\tau/2R^{1/3}\right) exp\left(iQ\right), \\
A_2 &= 0.5 D_\tau R^{-2/3} sech^2\left(x/2R\right) sech^2\left(sech\left(x/2R\right)\tau/2R^{1/3}\right) exp\left(2iQ\right).
\end{aligned}
\tag{5}
$$

In (5) $Q$ is proportional to $z$, $R(z)$ is the transverse pulse radius depending on the propagation distance $z$.

## 3    Numerical Approach in Planar Case

To confirm analytical results presented in the previous section we performed a direct numerical simulation of the system (1–4). Before constructing a numerical algorithm we derive integrals of motion of the system (1–2). First of them is the law of energy conservation.

$$
I_1 = \int_0^{L_\tau}\int_0^{L_x}\left(|A_1|^2 + 2|A_2|^2\right)dxd\tau.
\tag{6}
$$

The second one concerns the evolution of the phases of harmonics:

$$
\begin{aligned}
I_3 = \int_0^{L_\tau}\int_0^{L_x}&[Re\left(A_1^2 A_2^*\right) + D_{a1}|A_1|^2\tfrac{\partial\Phi_1}{\partial\tau} + D_{a2}|A_1|^2|A_2|^2\tfrac{\partial\Phi_2}{\partial\tau} \\
&-0.25 D_x\left(2\left|\tfrac{\partial A_1}{\partial x}\right|^2 + \left|\tfrac{\partial A_2}{\partial x}\right|^2\right) + D_\tau\left(\left|\tfrac{\partial A_1}{\partial\tau}\right|^2 + 2\left|\tfrac{\partial A_2}{\partial\tau}\right|^2\right) \\
&+D_{q1}x^2|A_1|^2 + 2D_{q2}x^2|A_2|^2]dxd\tau.
\end{aligned}
\tag{7}
$$

In (7) $\Phi_{1,2}$ are the wave phases. Developing a finite difference scheme for the problem (1–4) we will check through the fulfillment of difference analogs of (6–7).

We introduce uniform grids in the domain $\Gamma$ and in the $z$ domain:

$$
\begin{aligned}
\omega_\Gamma = \omega_x \times \omega_\tau = &\left\{(x_j, \tau_k) = (jh_x, kh_\tau); j = 1, 2, ..., N_x - 1; \right. \\
&\left. k = 1, 2, ..., N_\tau - 1; h_x = L_x/N_x; h_\tau = L_\tau/N_\tau\right\}, \\
&\omega_z = \{z_l = lh_z; l = 1, 2, ..., N_z - 1, h_z = L_z/N_z\}.
\end{aligned}
\tag{8}
$$

On the grid $\omega_z \times \omega_\Gamma$ we consider a numerical approximation to the exact solution of the problem (1–4) $A_{1,2}^{l,j,k} = A_{1,2}(z_l, x_j, \tau_k)$ which we denote by $\Psi_{1,2}^{l,j,k} = \Psi_{1,2}(z_l, x_j, \tau_k)$. To approximate first- and second-order derivatives with respect to $x$ and $\tau$ we use the standard expressions [10]. We also introduce:
$\Psi_{1,2}^{l,j,k} \overset{0.5}{=} \left(\Psi_{1,2}^{l+1,j,k} + \Psi_{1,2}^{l,j,k}\right)/2$. Then we write down the following nonlinear symmetric finite difference scheme:

$$\frac{\Psi_1^{l+1,j,k}-\Psi_1^{l,j,k}}{h_z} + D_{a1}x_j^2\,\overset{0.5}{\Psi_{1_0^{l,j,k}}} -iD_\tau\,\overset{0.5}{\Psi_{1_{\mp\tau}}^{l,j,k}} +i\left(\overset{0.5}{\Psi_1^{l,j,k}}\right)^* \overset{0.5}{\Psi_2^{l,j,k}}=$$

$$+iD_{q1}x_j^2\,\overset{0.5}{\Psi_1^{l,j,k}} -0.5iD_x\,\overset{0.5}{\Psi_{1_{\bar{x}x}}^{l,j,k}},$$

$$\frac{\Psi_2^{l+1,j,k}-\Psi_2^{l,j,k}}{h_z} + D_{a2}x_j^2\,\overset{0.5}{\Psi_{2_0^{l,j,k}}} -2iD_\tau\,\overset{0.5}{\Psi_{2_{\mp\tau}}^{l,j,k}} +0.5i\left(\overset{0.5}{\Psi_1^{l,j,k}}\right)^2 =$$

$$+2iD_{q2}x_j^2\,\overset{0.5}{\Psi_2^{l,j,k}} -0.25iD_x\,\overset{0.5}{\Psi_{2_{\bar{x}x}}^{l,j,k}},$$

$$(9)$$

Initial and boundary conditions are approximated exactly.

$$\Psi_{1,2}^{l+1,0,k} = \Psi_{1,2}^{l+1,N_x,k} = \Psi_{1,2}^{l+1,j,0} = \Psi_{1,2}^{l+1,j,N_\tau} = 0,$$
$$\Psi_1(0,x_j,\tau_k) = A_{10}(x_j,\tau_k), \Psi_2(0,x_j,\tau_k) = A_{20}(x_j,\tau_k).$$

$$(10)$$

This scheme is known to be of the second order of approximation with respect to all coordinates. The scheme (9–10) is conservative: it preserves difference analogues of the integrals (6–7). To resolve the nonlinear scheme (9–10) it is necessary to linearize it. It is possible to do this with the help of an iteration process [10]. But the direct matrix inversion after linearization is rather bulky because of computational complexity [9,13]. We could use the FFT technique [13,17] but it is not straightforward in the case under consideration due to the x-dependent coefficients of the first order finite difference derivatives with respect to $\tau$ in (9–10). Thus, we choose the approach proposed in [18,19] and develop it applying to (9–10). According to this approach we write down the following two-step iteration process. Firstly we seek for $\overset{s+1}{\Psi_{1,2}^{l+1,j,k}}$:

$$\frac{\overset{s+1}{\Psi_1^{l+1,j,k}}-\Psi_1^{l,j,k}}{h_z} + D_{a1}x_j^2\,\overset{s+1}{\overset{0.5}{\Psi_{1_0^{l,j,k}}}} -iD_\tau\,\overset{s+1}{\overset{0.5}{\Psi_{1_{\mp\tau}}^{l,j,k}}} +i\left(\overset{s}{\overset{0.5}{\Psi_1^{l,j,k}}}\right)^* \overset{s}{\overset{0.5}{\Psi_2^{l,j,k}}}=$$

$$+iD_{q1}x_j^2\,\overset{s}{\overset{0.5}{\Psi_1^{l,j,k}}} -0.5iD_x\,\overset{s}{\overset{0.5}{\Psi_{1_{\bar{x}x}}^{l,j,k}}},$$

$$\frac{\overset{s+1}{\Psi_2^{l+1,j,k}}-\Psi_2^{l,j,k}}{h_z} + D_{a2}x_j^2\,\overset{s+1}{\overset{0.5}{\Psi_{2_0^{l,j,k}}}} -2iD_\tau\,\overset{s+1}{\overset{0.5}{\Psi_{2_{\mp\tau}}^{l,j,k}}} +0.5i\left(\overset{s}{\overset{0.5}{\Psi_1^{l,j,k}}}\right)^2 =$$

$$+2iD_{q2}x_j^2\,\overset{0.5}{\Psi_2^{l,j,k}} -0.25iD_x\,\overset{0.5}{\Psi_{2_{\bar{x}x}}^{l,j,k}},$$

$$(11)$$

$$\overset{s+1}{\Psi_{1,2}^{l+1,0,k}}=\overset{s+1}{\Psi_{1,2}^{l+1,N_x,k}}=\overset{s+1}{\Psi_{1,2}^{l+1,j,0}}=\overset{s+1}{\Psi_{1,2}^{l+1,j,N_\tau}}=0,$$

$$\overset{s=0}{\Psi_{1,2}^{l+1,j,k}}=\Psi_{1,2}^{l,j,k}.$$

$$(12)$$

At the second step we complete the procedure finding $\overset{s+2}{\Psi_{1,2}^{l+1,j,k}}$:

$$\frac{\overset{s+2}{\Psi_1^{l+1,j,k}}-\Psi_1^{l,j,k}}{h_z} + D_{a1}x_j^2\,\overset{s+1}{\underset{0.5}{\Psi_{1\underset{0}{\tau}}^{l,j,k}}} - iD_\tau\,\overset{s+1}{\underset{0.5}{\Psi_{1\bar{\tau}\tau}^{l,j,k}}} + i\left(\overset{s+1}{\underset{0.5}{\Psi_1^{l,j,k}}}\right)^*\overset{s+1}{\underset{0.5}{\Psi_2^{l,j,k}}} =$$

$$+iD_{q1}x_j^2\,\overset{s+1}{\underset{0.5}{\Psi_1^{l,j,k}}} - 0.5iD_x\,\overset{s+2}{\underset{0.5}{\Psi_{1\bar{x}x}^{l,j,k}}},$$

$$\frac{\overset{s+2}{\Psi_2^{l+1,j,k}}-\Psi_2^{l,j,k}}{h_z} + D_{a2}x_j^2\,\overset{s+1}{\underset{0.5}{\Psi_{2\underset{\tau}{0}}^{l,j,k}}} - 2iD_\tau\,\overset{s+1}{\underset{0.5}{\Psi_{2\bar{\tau}\tau}^{l,j,k}}} + 0.5i\left(\overset{s+1}{\underset{0.5}{\Psi_1^{l,j,k}}}\right)^2 =$$

$$+2iD_{q2}x_j^2\,\overset{s+1}{\underset{0.5}{\Psi_2^{l,j,k}}} - 0.25iD_x\,\overset{s+2}{\underset{0.5}{\Psi_{2\bar{x}x}^{l,j,k}}},$$

(13)

$$\overset{s+2}{\Psi_{1,2}^{l+1,0,k}}=\overset{s+2}{\Psi_{1,2}^{l+1,N_x,k}}=\overset{s+2}{\Psi_{1,2}^{l+1,j,0}}=\overset{s+2}{\Psi_{1,2}^{l+1,j,N_\tau}}=0. \tag{14}$$

We see that at each step of the iteration process we deal with one-dimensional problem. Thus, matrix inversion in (11–14) can be made with the help of the tridiagonal matrix algorithm, and in general, the proposed method is time-saving. The iteration procedure stops when the criterion $max_{0\leq j\leq N_x,0\leq l\leq N_\tau}\left|\overset{s+2}{\Psi_{1,2}^{l+1,j,k}} - \overset{s}{\Psi_{1,2}^{l+1,j,k}}\right| \leq \epsilon$ where $\epsilon$ is a constant determining computation accuracy. The constructed conservative difference scheme may be generalized and applied to the investigation of $(3+1)D$ spatiotemporal solitons in inhomogenious medium described by the system (1–2) with $d=1$.

## 4    Numerical Simulation

Modeling the system when $d = 0$ (1–4) we set the initial pulse in the form $A_1 = sech^2(x)sech^2(\tau)$, $A_2 = 0$. Numerical solution presented in Fig. 1 is close to a two-component soliton predicted theoretically for defocusing nonlinearity and focusing waveguide. Beam-pulse propagates in the oscillation regime up to $z = 500L_{nl} = 250L_D = 5L_{dis}$ preserving the central part in both time and space coordinates. When waveguide is absent($D_{a1,2} = 0$) the pulse-beam spreads out at the distance $z < 1$. Method (9–14) was used as a base for modeling the general system (1–2). Incident beam was set as follows: $A_{10} = E_1(x + iy)\exp\left(-x^2 - y^2 - \tau^2\right)$, $A_{20} = 0$. Typical results are given in Fig. 2. Propagating in the waveguide the pulse-beam preserves its vortex structure in x and y coordinates. At that, we distinguish profile distortions but due to waveguide influence the pulse-beam is at the same position with respect to the transversal coordinates. The less is waveguide influence, the greater is pulse spreading. Due to waveguide geometry on the periphery of the plane, there is no threshold value of the waveguide coefficients at which the pulse-beam infinitely spreads. Each value of waveguide coefficient determines a finite radius of pulse spreading. It is worth noticing a quite visible pulse spreading in time. This results in the intensity decrease at the cross-section $\tau = 0$.

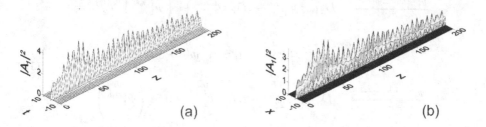

**Fig. 1.** Light bullets at defocusing nonlinearity and focusing waveguide (fundamental harmonic). Intensity distribution at the cross-sections $tz$ (a) and $xz$ (b). $D_\tau = 0.01$, $D_{a1,2} = -0.01$, $D_{q1} = -25$, $D_{q2} = -12.5$, $D_x = 1$.

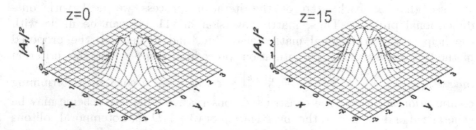

**Fig. 2.** Dynamics of the intensity space distribution of the wave at the fundamental frequency. Parameters: $E_1 = 10$, $D_x = 1$, $D_\tau = -0.05$, $D_{q1} = 10$, $D_{q2} = 5$.

## 5    Conclusion

Using the averaged Lagrangian method, we obtain approximate analytical spatiotemporal solutions to $(2+1)D$ equations for second-harmonic-generation in a planar waveguide. In contrast to homogeneous medium the waveguide geometry allows us to observe stable two-component light bullets in media with normal dispersion which corresponds to the defocusing nonlinearity. A conservative nonlinear finite difference scheme is developed and a multi-step iteration algorithm is proposed to realize it. Results of computer simulation are discussed and a stable propagation of light bullets at normal dispersion is demonstrated numerically. We investigate also a volumetric waveguide where waveguide geometry provides novel interesting regimes of pulse propagation and interaction.

Work was supported by Russian Science Foundation (Project No. 17-11-01157).

# References

1. Kivshar, Y.S., Agrawal, G.: Optical Solitons: From Fibers to Photonic Crystals. Academic press, Cambridge (2003)
2. Karamzin, Y.N., Sukhorukov, A.P.: Nonlinear interaction of diffracting light beams in a medium with quadratic nonlinearity; focusing of beams and limiting the efficiency of optical frequency converters. Eksp Zh. Teor. Fiz. **20**, 734 (1974)
3. Kanashov, A.A., Rubenchik, A.M.: On diffraction and dispersion effect on three wave interaction. Phys. D **4**, 122 (1981)
4. Malomed, B.A., Drummond, P., He, H., et al.: Spatiotemporal solitons in multi-dimensional optical media with a quadratic nonlinearity. Phys. Rev. E **56**, 4725 (1997)
5. Skryabin, D.V., Firth, W.J.: Generation and stability of optical bullets in quadratic nonlinear media. Opt. Commun. **148**, 79 (1998)
6. McLeod, R., Wagner, K., Blair, S.: (3+1)-dimensional optical soliton dragging logic. Phys. Rev. A **52**, 3254 (1995)
7. McDonald, G.D., et al.: Bright solitonic matter-wave interferometer. Phys. Rev. Lett. **113**, 013002 (2014)
8. Sazonov, S.V., Mamaikin, M.S., Komissarova, M.V., Zakharova, I.G.: Planar light bullets under conditions of second-harmonic generation. Phys. Rev. E **96**, 022208 (2017)
9. Sazonov, S.V., Mamaikin, M.S., Zakharova, I.G., Komissarova, M.V.: Planar spatiotemporal solitons in a quadratic nonlinear medium. Phys. Wave Phenom. **25**, 83 (2017)
10. Samarskii, A.A.: The Theory of Difference Schemes Marcel. Dekker Inc., New York (2001)
11. Karamzin, Y.N.: Difference schemes for calculating the three-frequency interactions of electromagnetic waves in a non-linear medium with quadratic polarization. USSR Comput. Math. Math. Phys. **14**(4), 236–241 (1974)
12. Ciegis, R., Mirinavicius, A., Radziunas, M.: Comparison of split step solvers for multidimensional Schrödinger problems. Comput. Methods Appl. Math. **13**(1), 237–250 (2013)
13. Gaspar, F.J., Rodrigo, C., Ciegis, R., Mirinavicius, A.: Comparison of solvers for 2D Schrodinger problems. Int. J. Numer. Anal. Model. **11**(1), 131–147 (2014)
14. Trofimov, V.A., Loginova, M.M.: Difference scheme for the problem of femtosecond pulse interaction with semiconductor in the case of nonlinear electron mobility. J. Comput. Math. Math. Phys. **45**(12), 2185–2196 (2005)
15. Shizgal, B.: Spectral Methods in Chemistry and Physics. SC. Springer, Dordrecht (2015). https://doi.org/10.1007/978-94-017-9454-1
16. Drummond, P.D.: Central partial difference propagation algorithms. Comput. Phys. Commun. **29**, 211 (1983)
17. Agrawal, G.: Nonlinear Fiber Optics, 5th edn. Academic press, Cambridge (2012)
18. Trofimov, V.A., Loginova, M.M., Egorenkov, V.A.: Influence of external electric field on laser- induced wave process occurring in semiconductor under the femtosecond pulse acting. In: Proceedings of SPIE, vol. 9127, p. 912709 (2014)
19. Trofimov, V.A., Loginova, M.M., Egorenkov, V.A.: New two-step iteration process for solving the semiconductor plasma generation problem with arbitrary BC in 2D case. WIT Trans. Model. Simul. **59** (2015). https://doi.org/10.2495/CMEM150081
20. Antoine, X., Besse, C.: Unconditionally stable discretization schemes of nonreflecting boundary conditions for the one-dimensional Schrodinger equation. J. Comput. Phys. **188**, 157–175 (2003)

# Absolute Stability and Implementation of the Two-Times Repeated Richardson Extrapolation Together with Explicit Runge-Kutta Methods

Zahari Zlatev[1] , Ivan Dimov[2], István Faragó[3,4] , Krassimir Georgiev[2] , and Ágnes Havasi[3,4(✉)]

[1] Department of Environmental Science, Aarhus University, Roskilde, Denmark
zz@envs.au.dk
[2] Institute of Information and Communication Technologies,
Bulgarian Academy of Sciences, Sofia, Bulgaria
ivdimov@bas.bg, georgiev@parallel.bas.bg
[3] MTA-ELTE Numerical Analysis and Large Networks Research Group,
Budapest, Hungary
[4] Department of Applied Analysis and Computational Mathematics,
Eötvös Loránd University, Budapest, Hungary
{faragois,havasia}@cs.elte.hu

**Abstract.** Efficient implementation of the Two-times Repeated Richardson Extrapolation is studied in this paper under the assumption that systems of ordinary differential equations (ODEs) are solved numerically by Explicit Runge-Kutta Methods (ERKMs). The combinations of the Two-times Repeated Richardson Extrapolation with the ERKMs are new numerical methods. The computational cost per step of these new numerical methods is higher than the computational cost per step of the underlying ERKMs. However, the order of accuracy of the combined methods becomes very high: if the order of accuracy of the underlying ERKM is $p$, then the order of accuracy of its combination with the Two-times Repeated Richardson Extrapolation is at least $p+3$ when the right-hand-side function of the system of ODEs is sufficiently many times continuously differentiable. Moreover, the stability properties of the new methods are always better than those of the underlying numerical methods when $p = m$ and $m = 1, 2, 3, 4$ (where $m$ is the number of stage vectors in the chosen ERKM). These two useful properties, higher accuracy and better stability, are often giving a very reasonable compensation for the increased computational cost per step, because the same degree of accuracy can be achieved by applying a large stepsize which leads to a considerable reduction of the number of steps when the Two-times Repeated Richardson Extrapolation is used. This fact is verified by several numerical experiments.

**Keywords:** Systems of ordinary differential equations (ODEs) ·
Explicit Runge-Kutta Methods ·
Two-times Repeated Richardson Extrapolation ·
Absolute stability properties

© Springer Nature Switzerland AG 2019
I. Dimov et al. (Eds.): FDM 2018, LNCS 11386, pp. 678–686, 2019.
https://doi.org/10.1007/978-3-030-11539-5_80

# 1    Definition of the Two-Times Repeated Richardson Extrapolation (2TRRE)

Assume that systems of first-order ordinary differential equations (ODEs):

$$\frac{dy}{dt} = f(t, y), \ t \in [a, b], y \in \mathbb{R}^s, \ s \geq 1, \ f \in D \subset \mathbb{R} \times \mathbb{R}^s, \ y(a) = \eta \qquad (1)$$

are to be solved on the following set of equidistant grid-points:

$$t_0 = a, \ t_n = t_{n-1} + h \ (n = 1, 2, \ldots, N), \ t_N = b, \ h = (b - a)/N \qquad (2)$$

by applying an arbitrary one-step numerical method of order $p$. If the Two-times Repeated Richardson Extrapolation (2TRRE) is to be used, then the approximation $y_n$ can be computed by using the previous approximation $y_{n-1}$ and by performing successively the following five calculation processes:

**Calculation Process 1:** Compute an approximation $z_n^{[1]}$ of the solution of (1) at the point $t = t_n$ by using the selected one-step numerical method with a stepsize $h$ (i.e., one step is to be performed during this process).

**Calculation Process 2:** Compute an approximation $z_n^{[2]}$ of the solution of (1) at the point $t = t_n$ by using the selected one-step numerical method with a stepsize $h/2$ (i.e., two steps are to be performed during this process).

**Calculation Process 3:** Compute an approximation $z_n^{[3]}$ of the solution of (1) at the point $t = t_n$ by using the selected one-step numerical method with a stepsize $h/4$ (i.e., four steps are to be performed during this process).

**Calculation Process 4:** Compute an approximation $z_n^{[4]}$ of the solution of (1) at the point $t = t_n$ by using the selected one-step numerical method with a stepsize $h/8$ (i.e., eight steps are to be performed during this process).

**Calculation Process 5:** Compute an approximation $y_n$ of the solution of (1) at the point $t = t_n$ by calculating a linear combination of the approximations $z_n^{[1]}, z_n^{[2]}, z_n^{[3]}$ and $z_n^{[4]}$ obtained in the previous four calculation processes.

The weights in the linear combination use in step 5 are to be specified so as to get a method of order $p + 3$. If the right-hand side function $f$ in (1) is three times continuously differentiable, then the following relations hold:

$$y(t_n) - z_n^{[1]} = h^p K_1 + h^{p+1} K_2 + h^{p+2} K_3 + \mathcal{O}(h^{p+3}) \qquad (3)$$

$$y(t_n) - z_n^{[2]} = (0.5h)^p K_1 + (0.5h)^{p+1} K_2 + (0.5h)^{p+2} K_3 + \mathcal{O}(h^{p+3}) \qquad (4)$$

$$y(t_n) - z_n^{[3]} = (0.25h)^p K_1 + (0.25h)^{p+1} K_2 + (0.25h)^{p+2} K_3 + \mathcal{O}(h^{p+3}) \qquad (5)$$

$$y(t_n) - z_n^{[4]} = (0.125h)^p K_1 + (0.125h)^{p+1} K_2 + (0.125h)^{p+2} K_3 + \mathcal{O}(h^{p+3}) \qquad (6)$$

where $y(t_n)$ is the exact value of the solution of (1) at the point $t = t_n$, while $K_1$, $K_2$ and $K_3$ are constants, which do not depend on the stepsize $h$. Now one has to eliminate the terms in the right-hand-sides of (3)–(6) that contain these three constants. Long but straight-forward transformations lead to the following relationship:

$$y(t_n) = \frac{2^{3p+3}z_n^{[4]} - 7 \cdot 2^{2p+1}z_n^{[3]} + 7 \cdot 2^p z_n^{[2]} - z_n^{[1]}}{2^{3p+3} - 7 \cdot 2^{2p+1} + 7 \cdot 2^p - 1} + \mathcal{O}(h^{p+3}), \qquad (7)$$

which show that

$$y(t_n) = y_n + \mathcal{O}(h^{p+3}), \qquad (8)$$

where

$$y_n := \frac{2^{3p+3}z_n^{[4]} - 7 \cdot 2^{2p+1}z_n^{[3]} + 7 \cdot 2^p z_n^{[2]} - z_n^{[1]}}{2^{3p+3} - 7 \cdot 2^{2p+1} + 7 \cdot 2^p - 1} \qquad (9)$$

According to (8), the order of accuracy of the approximation $y_n$ is greater than or equal to $p + 3$, while (9) shows how this approximation can be calculated. It should be mentioned here that the Classical RE (based essentially on the first two calculation processes) was first introduced in [3].

We shall demonstrate, by using two numerical examples, that the high accuracy of the 2TRRE is sometimes allowing us to increase the stepsize and, by solving the problem (1) with a sufficiently large stepsize, to achieve both the required accuracy and a very good compensation for the need to use much more computations in the performance of the five calculation processes. However, it is necessary first to study the stability of the Two-times RRE.

## 1.1   Absolute Stability of ERKMs Combined with the 2TRRE

The absolute stability of a one-step numerical method on a fixed mesh can be investigated by using Dahlquist's test problem [1]

$$\frac{dy}{dt} = \lambda y, \ t \in [0, \infty), y \in \mathbb{C}, \ \lambda = \alpha + \beta i \in \mathbb{C}^-, \ \alpha \le 0, \ y(0) = \eta \in \mathbb{C} \qquad (10)$$

on the grid

$$t_0 = 0, \ t_n = t_{n-1} + h = t_0 + nh \ (n = 1, 2, \ldots) \qquad (11)$$

The exact solution of (10) is $y(t) = \eta e^{\lambda t}$, and it is bounded when the condition $\alpha \le 0$ is satisfied. Therefore, it is necessary to require that the selected numerical method produces arbitrarily long sequences of approximations $\{y_1, y_2, \ldots, y_N\}$ which are also bounded when $\alpha \le 0$. The application of any one-step numerical method in the treatment of (10) leads to the following recursive relation [2]:

$$y_n = R(\nu)y_{n-1} = [R(\nu)]^n y_0, \ \ \nu = \lambda h, \ n = 1, 2, \ldots \qquad (12)$$

The function $R(\nu)$ from (12) is called stability function. Clearly, if $|R(\nu)| \le 1$ for some value of $\nu = h\lambda \in \mathbb{C}^-$, then the selected one-step numerical method will produce a bounded sequence $\{y_1, y_2, \ldots, y_N\}$ of approximations to the solution of (10) for the applied value $h$ of the time-stepsize. The set of all values $\nu \in \mathbb{C}^-$ for which the relationship $|R(\nu)| \le 1$ holds is forming the absolute stability region of the chosen one-step numerical method [2].

Now the following theorem can be proved:

**Theorem 1.** *The stability function $\tilde{R}(\nu)$ of the 2TRRE, when it is applied to solve (10) in combination with an arbitrary explicit one-step numerical method with a stability function $R(\nu)$, is given by the following expression:*

$$\tilde{R}(\nu) = \frac{2^{3p+3}[R(\frac{\nu}{8})]^8 - 7 \cdot 2^{2p+1}[R(\frac{\nu}{4})]^4 + 7 \cdot 2^p[R(\frac{\nu}{2})]^2 - R(\nu)}{2^{3p+3} - 7 \cdot 2^{2p+1} + 7 \cdot 2^p - 1} \tag{13}$$

*Proof.* It can be concluded from (12) that the following four relationships hold:

$$z_n^{[1]} = R(\nu)y_{n-1}, \quad z_n^{[2]} = [R(\frac{\nu}{2})]^2 y_{n-1}, \quad z_n^{[3]} = [R(\frac{\nu}{4})]^4 y_{n-1}, \quad z_n^{[4]} = [R(\frac{\nu}{8})]^8 y_{n-1}.$$

The following relationship can be obtained by using the above equalities and (9):

$$y_n = \frac{2^{3p+3}[R(\frac{\nu}{8})]^8 - 7 \cdot 2^{2p+1}[R(\frac{\nu}{4})]^4 + 7 \cdot 2^p[R(\frac{\nu}{2})]^2 - R(\nu)}{2^{3p+3} - 7 \cdot 2^{2p+1} + 7 \cdot 2^p - 1} y_{n-1}. \tag{14}$$

It is clear that the coefficient of $y_{n-1}$ in (14) is the stability polynomial $\tilde{R}(\nu)$ of the 2TRRE. In the following we shall introduce: (a) the Explicit Runge-Kutta Methods (the ERKMs), (b) their stability polynomials together with the stability polynomials of the 2TRRE and (c) their absolute stability regions again together with the absolute stability regions of the Two-times RRE (for comparison the absolute stability regions of the of the Classical RE and the RRE will also be given).

The $m$-stage Explicit Runge-Kutta Methods are one-step numerical methods for solving systems of ODEs (1) that are based on the following formula:

$$y_n = y_{n-1} + h \sum_{i=1}^{m} c_i k_i^n. \tag{15}$$

The coefficients $c_i$ are constants, while the stage vectors $k_i^n$ are defined by

$$k_1^n = f(t_{n-1}, y_{n-1}), \quad k_i^n = f(t_{n-1} + ha_i, y_{n-1} + h \sum_{j=1}^{i-1} b_{ij} k_j^n), a_i = \sum_{j=1}^{i-1} b_{ij}, \tag{16}$$

where $i = 2, 3, \ldots, m$ and $b_{ij}$ are constants depending on the particular numerical method.

We shall assume that the order of accuracy $p$ of the chosen ERKM is equal to the number $m$ of the stage vectors $k_i^n$, $i = 1, 2, \ldots, m$, which is possible only if $m \leq 4$ (the relationship $p < m$ is always satisfied when $m > 4$). If $p = m$, then the stability polynomial $R(\nu)$ associated with the selected ERKM is given (see [2], p. 202) by:

$$R(\nu) = 1 + \nu + \frac{\nu^2}{2!} + \frac{\nu^3}{3!} + \ldots + \frac{\nu^p}{p!}, \quad p = m, \quad m = 1, 2, 3, 4. \tag{17}$$

The stability polynomials $\tilde{R}(\nu)$ related to the 2TRRE can easily be obtained by substituting the expression of $R(\nu)$ from (17) in the right-hand-side of (14).

The absolute stability regions of the ERKMs when $p = m$ and $m = 1, 2, 3, 4$ will be compared with the absolute stability regions of the corresponding new numerical one-step methods obtained when these ERKMs are combined with the 2TRRE. The absolute stability regions were obtained numerically by the procedure described in [4] and [5]. The results are given in Fig. 1 (a)–(d), and can be considered as a graphical proof of the following theorem.

**Theorem 2.** *The absolute stability regions of numerical methods that are combinations of ERKMs with the 2TRRE are always larger than the absolute stability regions of the underlying ERKMs when the conditions $p = m$ and $m = 1, 2, 3, 4$ are satisfied.*

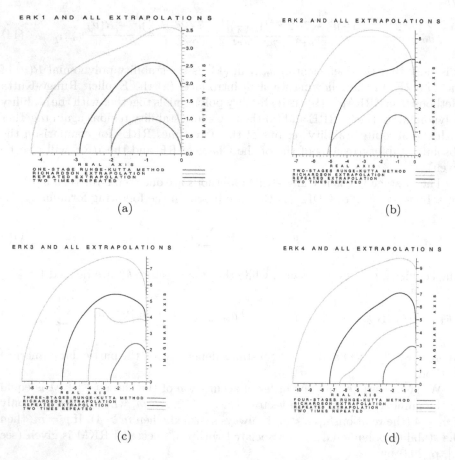

**Fig. 1.** Absolute stability regions of any particular method from the class of (a) one-stage first-order, (b) two-stage second-order, (c) three-stage third-order, (d) four-stage fourth-order ERKMs and their combinations with the classical RE, the RRE and the 2TRRE.

## 2   Numerical Examples

From the ERKMs satisfying the conditions $p = m$ and $m = 1, 2, 3, 4$ we chose the following methods for our numerical experiments (see, for example, in [2]):

- for $p = m = 1$ the Forward Euler Method,
- for $p = m = 2$ the Improved Euler Formula,
- for $p = m = 3$ Heun's three-stage third-order formula, and
- for $p = m = 4$ the most popular and commonly used RK method (see [2]), the well-known four-stage fourth-order ERKM.

We shall present results obtained in the treatment of two numerical examples, each of which is a system of three ODEs. The absolute value of the real eigenvalue of the Jacobian matrix of the first example is much greater than the absolute value of the two complex eigenvalues. The opposite situation occurs in the second example: the absolute value of each of the two complex eigenvalues is dominant. Thus, the length of the absolute stability interval on the negative real axis is important for the first example, while the size of the absolute stability region (and more precisely its height over the negative part of the real axis in the neighbourhood of the imaginary axis) is essential for the second one. Since both examples have oscillatory components, they must be handled by accurate methods. Moreover, the selected examples are mildly stiff.

### 2.1   Example 1: The Real Eigenvalue Is Dominant

Consider the system of three linear ODEs defined in the following way:

$$\frac{dy}{dt} = Ay, \ t \in [0, 13.1072], \ y(0) = (1, 0, 2)^T, \tag{18}$$

with the coefficient matrix $A \in \mathbb{R}^{3 \times 3}$

$$A = \begin{bmatrix} 741.4 & 749.7 & -741.7 \\ -765.7 & -758.0 & 757.7 \\ 725.7 & 741.7 & -734.0 \end{bmatrix}. \tag{19}$$

The eigenvalues of matrix $A$ are $\mu_1 = -750$, $\mu_2 = -0.3 + 8i$, $\mu_3 = -0.3 - 8i$. The absolute value of the real eigenvalue $\mu_1$ is much larger than the absolute value of the two complex eigenvalues of matrix $A$. This means that the computations used in the solution of (18) will be stable when $|\nu_1| = h|\mu_1|$ is smaller than the length of the stability interval on the negative part of the real axis (from the plots given in Fig. 1 (a)–(d) it is clearly seen that this length is smaller than 3 for all four ERKMs studied in this paper). In fact, all three points $h\mu_1$, $h\mu_2$ and $h\mu_3$ should lie in the absolute stability region of the method, but the last two points are not very important (these two points will be inside the absolute stability regions when this is true for the first one).

Eleven runs with different stepsizes were successively completed with each of the above numerical methods, both directly and in combination with the Classial

**Table 1.** Global errors obtained in the treatment of the first example when the first-order one-stage explicit Runge-Kutta method (the Forward Euler Formula) and its combinations with (a) the Classical RE, (b) the RRE and (c) the 2TRRE are run with eleven different stepsizes. "N.S." means that the method is not stable for the stepsize used. In parentheses the ratios between two successive errors are given where applicable (the perfect convergent rates are 2, 4, 8 and 16). "V.L." means very large (due presumably to the fact that the stepsize used in the previous run was close to the boundary of the stability region).

| Steps | Direct | RE | RRE | 2TRRE |
|---|---|---|---|---|
| 1280 | N.S. | N.S. | N.S. | N.S. |
| 2560 | N.S. | N.S. | N.S. | 1.29E−03 |
| 5120 | 2.01E−01 | 4.22E−02 | 4.48E−06 | 9.23E−09 (V.L.) |
| 10240 | 9.21E−02 (2.18) | 2.91E−04 (145.02) | 5.60E−07 (8.00) | 5.77E−10 (16.00) |
| 20480 | 4.41E−02 (2.09) | 7.27E−05 (3.84) | 7.00E−08 (8.00) | 3.61E−11 (15.98) |
| 40960 | 2.16E−02 (2.04) | 1.82E−05 (3.99) | 8.75E−09 (8.00) | 2.26E−12 (15.97) |
| 81920 | 1.07E−02 (2.02) | 4.54E−06 (4.01) | 1.09E−09 (8.03) | 1.41E−13 (16.02) |
| 163240 | 5.32E−03 (2.01) | 1.14E−06 (3.91) | 1.37E−10 (7.96) | 8.81E−15 (16.00) |
| 327680 | 2.65E−03 (2.01) | 2.84E−07 (4.01) | 1.71E−11 (8.01) | 5.51E−16 (15.99) |
| 655360 | 1.33E−03 (1.99) | 7.10E−08 (4.00) | 2.14E−12 (7.99) | 3.44E−17 (16.02) |
| 1310720 | 6.62E−04 (2.01) | 1.78E−08 (3.99) | 2.67E−13 (8.01) | 2.15E−18 (16.00) |

RE, the RRE and the 2TRRE. The maximum norm errors for the Forward Euler Method are given in Table 1. (The errors were evaluated for the three other methods as well, but because of the page limit restrictions we are able to include only a few results.) Four important conclusions can be drawn:

(a) The expected rates of convergence are nearly always achieved.
(b) The 2TRRE is always giving most accurate results.
(c) Sometimes the 2TRRE is stable in runs with large stepsizes for which the underlying ERKMs are unstable.
(d) One would expect that the order of accuracy obtained by the 2TRRE applied in combination with the two-stage second-order ERKM would be five. However, the actually achieved order is six. This means probably that the coefficient of the $\mathcal{O}(h^5)$ term is equal to zero for this example.

## 2.2   Example 2: The Complex Eigenvalues Are Dominant

Consider the system of ODEs defined in the following way:

$$\frac{dy}{dt} = Ay + b, \ t \in [0, 13.1072], \ y(0) = (1, 3, 0)^T, \tag{20}$$

with $b = (-4e^{-0.3t} \sin 4t, -8e^{-0.3t} \sin 4t, 4e^{-0.3t} \sin 4t)^T$ and

$$A = \begin{bmatrix} -937.575 & 562.425 & 187.575 \\ -187.650 & -187.650 & -562.350 \\ -1124.925 & 375.075 & -375.075 \end{bmatrix}. \tag{21}$$

The eigenvalues of matrix $A$ are: $\mu_1 = -750+750i$, $\mu_2 = -750-750i$, $\mu_3 = -0.3$. The absolute value of the two complex eigenvalues $\mu_1$ and $\mu_2$ of $A$ is much larger than the absolute value of the real eigenvalue $\mu_3$. This means that the computations will be stable when $\nu_1 = h\mu_1$ is inside the absolute stability region.

Similar conclusions as those drawn in the previous subsection, can also be drawn by studying the results for Example 2. However, the results obtained in the runs with Example 2 are more accurate than the corresponding results obtained in runs with Example 1 (the second example is not so oscillatory as the first one). Moreover, the accuracy of the 2TRRE combined with the two-stage second-order ERKM is now five (not six as in the previous example).

# 3    General Conclusions

The Two-times Repeated Richardson Extrapolation (2TRRE) is giving very accurate results. Its order of accuracy is at least $p+3$ when the order of accuracy of the underlying method is only $p$. This property often allows us to use large stepsizes and to reduce considerably the computational cost, achieving at the same time much higher accuracy than that obtained by using the EKRMs (and also than that achieved when the Classical RE and the RRE are used).

We found that (a) the 2TRRE have better stability properties than those of the ERKMs when $p = m$ and $m = 1, 2, 3, 4$ and (b) it can sometimes produce stable results in cases where the underlying ERKM is unstable. The fact that high accuracy can be obtained by specifying larger stepsizes during the computations should additionally be exploited in the efforts to achieve greater efficiency.

The use of the Classical RE and the RRE together with Explicit Runge-Kutta Methods (ERKMs) is only briefly mentioned in this paper, but we have presented the absolute stability regions of the methods based on the use of the Classical RE and the RRE as well as numerical results obtained by applying these two methods. Much more details about the application of the Classical RE and the RRE together with ERKMs can be found in [5], Chap. 2, see also [6].

**Acknowledgement.** The work is partially supported by the Bulgarian National Science Fund under grant DFNI 12/5 and by the Hungarian Research Fund OTKA under grant no. K-112157 and SNN-125119.

# References

1. Dahlquist, G.: A special stability problem for linear multistep methods. BIT **3**, 27–43 (1963)
2. Lambert, J.D.: Numerical Methods for Ordinary Differential Equations: The Initial Values Problem. Wiley, New York (1991)

3. Richardson, L.F.: The deferred approach to the limit, i-single lattice. Philos. Trans. R. Soc. Lond., Ser. A **226**, 299–349 (1927)
4. Zlatev, Z., Georgiev, K., Dimov, I.: Studying absolute stability properties of the Richardson extrapolation combined with explicit Runge-Kutta methods. Computers and Mathematics with Applications **67**(12), 2294–2307 (2014)
5. Zlatev, Z., Dimov, I., Faragó, I., Havasi, Á.: Richardson Extrapolation: Practical Aspects and Applications. De Gruyter, Berlin (2017)
6. Zlatev, Z., Dimov, I., Faragó, I., Georgiev, K., Havasi, Á.: Stability properties of the repeated Richardson extrapolation combined with some explicit Runge-Kutta methods. In: Talk Presented at the SIAM Conference in Sofia, 22 December 2017

# Author Index